Ironmaking and Steelmaking

Ironmaking and Steelmaking

Special Issue Editors

Zushu Li
Claire Davis

MDPI • Basel • Beijing • Wuhan • Barcelona • Belgrade

MDPI

Special Issue Editors

Zushu Li
University of Warwick
UK

Claire Davis
University of Warwick
UK

Editorial Office
MDPI
St. Alban-Anlage 66
4052 Basel, Switzerland

This is a reprint of articles from the Special Issue published online in the open access journal *Metals* (ISSN 2075-4701) from 2018 to 2019 (available at: https://www.mdpi.com/journal/metals/special_issues/ironmaking_steelmaking).

For citation purposes, cite each article independently as indicated on the article page online and as indicated below:

LastName, A.A.; LastName, B.B.; LastName, C.C. Article Title. *Journal Name* **Year**, *Article Number, Page Range*.

ISBN 978-3-03921-329-0 (Pbk)
ISBN 978-3-03921-330-6 (PDF)

Contents

About the Special Issue Editors

Zushu Li, Dr., is a Reader and EPSRC Fellow in Manufacturing at the Advanced Steel Research Centre, WMG, of the University of Warwick, United Kingdom. He joined the University of Warwick in April 2016 as a Principal Research Fellow after being awarded the prestigious Engineering and Physical Sciences Research Council (EPSRC) Manufacturing Fellowship (Sustainable Steel Manufacturing). He was previously a Principal Scientist of Steelmaking in Tata Steel R&D (based in the UK) and has worked in Tata Steel for over 9 years. Prior to working for Tata Steel, Dr. Li worked as a member of academic staff (Associate Professor) at Chongqing University, China, and as a researcher in Japan (Kyushu Institute of Technology) and the UK (Imperial College London). He received his Ph.D. degree in Metallurgy from Chongqing University, China. Dr. Li specializes in studying the high temperature reactions/phenomena of multicomponent systems in metal manufacturing processes. He leads the research area in the fields of low carbon steel manufacturing, clean steel, energy and materials recovery in steelmaking, and materials recycling.

Claire Davis, Professor, holds a Royal Academy of Engineering/Tata Steel Chair in Low Energy Steel Processing and also leads the Advanced Steel Research Centre at WMG, the University of Warwick, United Kingdom. Before joining WMG, she was the Professor of Ferrous Metallurgy at the University of Birmingham, UK. She obtained her undergraduate degree and Ph.D. at the University of Cambridge, UK. Her area of research is on the relationships between composition–processing–microstructure properties in steels, with a particular focus on solidification and thermomechanical processing. She also leads the research group on electromagnetic sensors for steel characterization.

metals

MDPI

Editorial

Ironmaking and Steelmaking

Zushu Li and Claire Davis

WMG, University of Warwick, Coventry CV4 7AL, UK; z.li.19@warwick.ac.uk (Z.L.);
Claire.davis@warwick.ac.uk (C.D.); Tel.: +44-247-652-4706 (Z.L.); +44-247-657-3517 (C.D.)

Received: 29 April 2019; Accepted: 7 May 2019; Published: 7 May 2019

1. Introduction and Scope

Steel is a critical material in our society and will remain an important one for a long time into the future. In the last two decades, the world steel industry has gone through drastic changes and this is predicted to continue in the future. The Asian countries (e.g., China) have been dominant in the production of steel, creating global over-capacity, while the steel industry in the developed countries have made tremendous efforts to reinforce its global leadership in process technology and product development, and remain sustainable and competitive. The global steel industry is also facing various grand challenges in strict environmental regulation, new energy and materials sources, and ever-increasing customer requirements for high quality steel products, which has been addressed accordingly by the global iron and steel community.

This Special Issue, Ironmaking and Steelmaking, released by the Journal Metals is solely dedicated to articles from the international iron and steel community to cover the state-of-the-art in ironmaking and steelmaking processes. The Guest Editors will not go into each individual paper in detail, however they will briefly overview some interesting points in the special issue.

2. Contributions

This Special Issue published 33 high quality articles from 10 countries (according to the country of the corresponding author) with the number of contributions in brackets: China (22) [1–22], Japan (1) [23], Korea (1) [24], Canada (1) [25], Sweden (2) [26,27], Italy (1) [28], UK (1) [29], France (1) [30], Austria (1) [31], and Slovakia (2) [32,33]. This clearly reflects the enormous investment in R&D in China and the resultant outstanding outcomes to support its steel industry with ~50% of global steel production. On the other hand, it also demonstrates that the western countries are continuing to invest in R&D for sustainable steel manufacturing.

International collaboration is another feature reflected by the authorship of the published papers. A UK–Netherland–USA contribution (Slater et al. [29]) studied the solidification phenomena of a molten steel surface by using infrared thermography and a Sweden–Russia contribution (Sidorova et al. [27]) reported the modification of non-metallic inclusions in oil-pipeline steels by Ca-treatment. A paper co-authored by scientists from Sweden and Egypt (El-Tawil et al. [26]) investigated the thermal devolatilisation of different bio-coals for the purpose of reducing fossil CO_2 emission in the steel industry. Other international collaborations include France–Canada–Albania (Kanari et al. [30]), Slovakia–Czech Republic (Neslušan et al. [33]), China–Ukraine (Cao et al. [1]), and China–UK (Ge et al. [2]).

The international iron and steel community is conducting intensive research and development to reduce CO_2 emissions from steel manufacturing, which is clearly highlighted by the papers in this special issue. One paper from China (Song et al. [3]) compared the energy consumption and CO_2 emissions between integrated steel plant (ISP) with conventional blast furnace (ISP + BF), ISP with top gas recycling oxygen blast furnace (ISP + TGR–OBF), and ISP with COREX (ISP + COREX). They found that the ISP + TGR–OBF has the lowest net CO_2 emissions compared with the other two process routes.

Another excellent contribution from China (Dong and Wang [4]) analysed the utilisation of CO_2 gas in various steel manufacturing steps from sintering, through blast furnace, steelmaking, ladle furnace, continuous casting to the smelting process of stainless steel. The paper concluded that the quantity of CO_2 utilization is expected to be more than 100 kg per ton of steel. Further, 10 papers covered various aspects of gas-based or carbothermal reduction of various iron ores, in particular the iron ores that are difficult to be treated in conventional processes. The production of iron using hydrogen as a reducing agent is an alternative to the conventional ironmaking process with potential benefit of substantial decrease in CO_2 emissions. Naseri Seftejani and Schenk [31] analysed the thermodynamics of the hydrogen plasma smelting reduction of iron ore in the process of using hydrogen in a plasma state to reduce iron oxides. The other nine papers covered experimental studies (El-Tawil et al. [26], Fukushima, and Takizawa [23], Wu et al. [5], Zhou et al. [6], Zhang et al. [7]), modelling predictions (Gao et al. [8], Tang et al. [9]), and kinetic analysis (Wang et al. [10], Chen et al. [11]).

An outstanding paper from Canada (Kadrolkar and Dogan [25]) developed a model for refining rates in oxygen steelmaking, with a focus on the impact and slag-metal bulk zones. This is of particular interest to the control of oxygen steelmaking, which is producing over 70% of global crude steel.

Five papers investigated continuous casting process-related topics, covering aspects from solidification mechanism (Slater et al. [29]), low fluorine mould flux (Li et al. [12], Zeng et al. [13]), transient fluid flow in the mould (Zhang et al. [14]) and high Al steels (Cui et al. [15]).

An interesting contribution from Italy (Marcias et al. [28]) thoroughly analysed the occupational exposure to fine particles and ultrafine particles in a steelmaking foundry. This is a critical aspect of steel industry considering the environment of the steel manufacturing lines.

3. Conclusions and Outlook

The objective of this Special Issue is to provide a scientific platform for the recent progress in ironmaking and steelmaking and these 33 articles excellently highlight the diversity of the recent research and development in the field. The steel industry is facing significant challenges and opportunities as well, from strict environmental legislations to new energy and raw material sources and rapid development of data science, and it becomes obvious that there are still plenty of exciting topics and research outcomes to publish. It is hoped that the creation of this special issue as a scientific platform will help drive the iron and steel community to build a sustainable steel industry.

Conflicts of Interest: The authors declare no conflict of interest.

References

1. Cao, Y.; Jiang, Z.; Dong, Y.; Deng, X.; Medovar, L.; Stovpchenko, G. Research on the bonding interface of high speed steel/ductile cast iron composite roll manufactured by an improved electroslag cladding method. *Metals* **2018**, *8*, 390. [CrossRef]
2. Ge, Y.; Zhao, S.; Ma, L.; Yan, T.; Li, Z.; Yang, B. Inclusions control and refining slag optimization for fork flat steel. *Metals* **2019**, *9*, 253. [CrossRef]
3. Song, J.; Jiang, Z.; Bao, C.; Xu, A. Comparison of energy consumption and CO_2 emission for three steel production routes—Integrated steel plant equipped with blast furnace, oxygen blast furnace or COREX. *Metals* **2019**, *9*, 364. [CrossRef]
4. Dong, K.; Wang, X. CO_2 utilization in the ironmaking and steelmaking process. *Metals* **2019**, *9*, 273. [CrossRef]
5. Wu, T.; Zhang, Y.; Zhao, Z.; Yuan, F. Effects of Fe_2O_3 on reduction process of Cr-containing solid waste self-reduction briquette and relevant mechanism. *Metals* **2019**, *9*, 51. [CrossRef]
6. Zhou, X.; Luo, Y.; Chen, T.; Zhu, D. Enhancing the reduction of high-aluminum iron ore by synergistic reducing with high-manganese iron ore. *Metals* **2019**, *9*, 15. [CrossRef]
7. Zhang, Y.; Xue, Q.; Wang, G.; Wang, J. Phosphorus-containing mineral evolution and thermodynamics of phosphorus vaporization during carbothermal reduction of high-phosphorus iron ore. *Metals* **2018**, *8*, 451. [CrossRef]

8. Gao, Q.; Zhang, Y.; Jiang, X.; Zheng, H.; Shen, F. Prediction model of iron ore pellet ambient strength and sensitivity analysis on the influence factors. *Metals* **2018**, *8*, 593. [CrossRef]

9. Tang, H.; Yun, Z.; Fu, X.; Du, S. Modeling and experimental study of ore-carbon briquette reduction under CO–CO_2 atmosphere. *Metals* **2018**, *8*, 205. [CrossRef]

10. Wang, G.; Wang, J.; Xue, Q. Kinetics of the volume shrinkage of a magnetite/carbon composite pellet during solid-state carbothermic reduction. *Metals* **2018**, *8*, 1050. [CrossRef]

11. Chen, J.; Chen, W.; Mi, L.; Jiao, Y.; Wang, X. Kinetic studies on gas-based reduction of vanadium titano-magnetite pellet. *Metals* **2019**, *9*, 95. [CrossRef]

12. Li, Z.; You, X.; Li, M.; Wang, Q.; He, S.; Wang, Q. Effect of substituting CaO with BaO and CaO/Al_2O_3 ratio on the viscosity of CaO–BaO–Al_2O_3–CaF_2–Li_2O mold flux system. *Metals* **2019**, *9*, 142. [CrossRef]

13. Zeng, J.; Long, X.; You, X.; Li, M.; Wang, Q.; He, S. Structure of solidified films of CaO-SiO_2-Na_2O based low-fluorine mold flux. *Metals* **2019**, *9*, 93. [CrossRef]

14. Zhang, T.; Yang, J.; Jiang, P. Measurement of molten steel velocity near the surface and modeling for transient fluid flow in the continuous casting mold. *Metals* **2019**, *9*, 36. [CrossRef]

15. Cui, H.; Zhang, K.; Wang, Z.; Chen, B.; Liu, B.; Qing, J.; Li, Z. Formation of surface depression during continuous casting of high-Al TRIP steel. *Metals* **2019**, *9*, 204. [CrossRef]

16. Fan, H.; Chen, D.; Liu, T.; Duan, H.; Huang, Y.; Long, M.; He, W. Crystallization behaviors of anosovite and silicate crystals in high CaO and MgO titanium slag. *Metals* **2018**, *8*, 754. [CrossRef]

17. Zuo, H.; Wang, Y.; Wang, X. Damage mechanism of copper staves in a 3200 m^3 blast furnace. *Metals* **2018**, *8*, 943. [CrossRef]

18. Wang, R.; Yang, J.; Xu, L. Improvement of heat-affected zone toughness of steel plates for high heat input welding by inclusion control with Ca deoxidation. *Metals* **2018**, *8*, 946. [CrossRef]

19. Xu, L.; Yang, J.; Wang, R. Influence of Al Content on the inclusion-microstructure relationship in the heat-affected zone of a steel plate with Mg deoxidation after high-heat-input welding. *Metals* **2018**, *8*, 1027. [CrossRef]

20. Zhang, K.; Zhang, Y.; Wu, T. Distribution ratio of sulfur between CaO-SiO_2-Al_2O_3-Na_2O-TiO_2 slag and carbon-saturated iron. *Metals* **2018**, *8*, 1068. [CrossRef]

21. Yang, X.; Li, J.; Zhang, M.; Yan, F.; Duan, D.; Zhang, J. A Further evaluation of the coupling relationship between dephosphorization and desulfurization abilities or potentials for CaO-BASED slags: Influence of slag chemical composition. *Metals* **2018**, *8*, 1083. [CrossRef]

22. Lu, Y.; Jiang, Z.; Zhang, X.; Wang, J.; Zhang, X. Vertical section observation of the solid flow in a blast furnace with a cutting method. *Metals* **2019**, *9*, 127. [CrossRef]

23. Fukushima, J.; Takizawa, H. In situ spectroscopic analysis of the carbothermal reduction process of iron oxides during microwave irradiation. *Metals* **2018**, *8*, 49. [CrossRef]

24. Kang, Y. Desiliconisation and dephosphorisation behaviours of various oxygen sources in hot metal pre-treatment. *Metals* **2019**, *9*, 251. [CrossRef]

25. Kadrolkar, A.; Dogan, N. Model development for refining rates in oxygen steelmaking: Impact and slag-metal bulk zones. *Metals* **2019**, *9*, 309. [CrossRef]

26. El-Tawil, A.A.; Ahmed, H.M.; Ökvist, L.S.; Björkman, B. Devolatilization kinetics of different types of bio-coals using thermogravimetric analysis. *Metals* **2019**, *9*, 168. [CrossRef]

27. Sidorova, E.; Karasev, A.; Kuznetsov, D.; Jönsson, P. Modification of non-metallic inclusions in oil-pipeline steels by Ca-treatment. *Metals* **2019**, *9*, 391. [CrossRef]

28. Marcias, G.; Fostinelli, J.; Sanna, A.; Uras, M.; Catalani, S.; Pili, S.; Fabbri, D.; Pilia, I.; Meloni, F.; Lecca, L.; et al. Occupational exposure to fine particles and ultrafine particles in a steelmaking foundry. *Metals* **2019**, *9*, 163. [CrossRef]

29. Slater, C.; Hechu, K.; Davis, C.; Sridhar, S. Characterisation of the solidification of a molten steel surface using infrared thermography. *Metals* **2019**, *9*, 126. [CrossRef]

30. Kanari, N.; Menad, N.; Ostrosi, E.; Shallari, S.; Diot, F.; Allain, E.; Yvon, J. Thermal behavior of hydrated iron sulfate in various atmospheres. *Metals* **2018**, *8*, 1084. [CrossRef]

31. Naseri Seftejani, M.; Schenk, J. Thermodynamic of liquid iron ore reduction by hydrogen thermal plasma. *Metals* **2018**, *8*, 1051. [CrossRef]

32. Fröhlichová, M.; Ivanišin, D.; Findorák, R.; Džupková, M.; Legemza, J. The effect of concentrate/iron ore ratio change on agglomerate phase composition. *Metals* **2018**, *8*, 973. [CrossRef]

33. Neslušan, M.; Trško, L.; Minárik, P.; Čapek, J.; Bronček, J.; Pastorek, F.; Čížek, J.; Moravec, J. Non-destructive evaluation of steel surfaces after severe plastic deformation via the Barkhausen noise technique. *Metals* **2018**, *8*, 1029. [CrossRef]

metals **MDPI**

Article

Thermodynamic of Liquid Iron Ore Reduction by Hydrogen Thermal Plasma

Masab Naseri Seftejani * and Johannes Schenk

Department of Metallurgy, Montanuniversitaet Leoben, 8700 Leoben, Austria; Johannes.Schenk@unileoben.ac.at
* Correspondence: naseri.masab@outlook.com; Tel.: +43-677-61812528

Received: 9 November 2018; Accepted: 4 December 2018; Published: 11 December 2018

Abstract: The production of iron using hydrogen as a reducing agent is an alternative to conventional iron- and steel-making processes, with an associated decrease in CO_2 emissions. Hydrogen plasma smelting reduction (HPSR) of iron ore is the process of using hydrogen in a plasma state to reduce iron oxides. A hydrogen plasma arc is generated between a hollow graphite electrode and liquid iron oxide. In the present study, the thermodynamics of hydrogen thermal plasma and the reduction of iron oxide using hydrogen at plasma temperatures were studied. Thermodynamics calculations show that hydrogen at high temperatures is atomized, ionized, or excited. The Gibbs free energy changes of iron oxide reductions indicate that activated hydrogen particles are stronger reducing agents than molecular hydrogen. Temperature is the main influencing parameter on the atomization and ionization degree of hydrogen particles. Therefore, to increase the hydrogen ionization degree and, consequently, increase of the reduction rate of iron ore particles, the reduction reactions should take place in the plasma arc zone due to the high temperature of the plasma arc in HPSR. Moreover, the solubility of hydrogen in slag and molten metal are studied and the sequence of hematite reduction reactions is presented.

Keywords: hydrogen plasma; smelting reduction; HPSR; iron oxide; plasma arc; ionization degree

---

1. Introduction

The average CO_2 emissions from iron and steel industry is 1900 kg/ton liquid steel (tLS) [1]. The integrated blast furnace–basic oxygen furnace steelmaking route produces approximately 2120 kg CO_2/tLS, whereas the integrated HYL3—Electric arc furnace rout produces 1125 kg CO_2/tLS which is the minimum amount among the different steelmaking integrated routes [2]. The reduction of iron ores with hydrogen has been considered a future alternative process for CO_2-free steelmaking [3–8]. However, existing studies have focused mainly on the reduction of iron ore in a solid state, and there are not many studies in the field of liquid iron ore reduction using hydrogen [9–12].

Laboratory facilities of hydrogen plasma smelting reduction (HPSR) are available at the laboratory of the Chair of Ferrous Metallurgy of Montanuniversitaet Leoben. Figure 1 shows the basic process flow sheet of the HPSR laboratory set up and the reactor layout. During this process, a mixture of iron ore with additives, mainly lime, is fed to the reactor through a hollow graphite electrode by a screw feeder. The gas used in this process can be pure hydrogen or a mixture of hydrogen and argon or hydrogen and nitrogen. Therefore, a mixture of hydrogen, argon or nitrogen, and iron ore are injected into the reactor. Hydrogen as a reducing agent plays the main role in the reduction process. Therefore, the hydrogen utilization degree defines process efficiency. According to the results of the previous studies [13–15] at the Chair of Ferrous metallurgy of Montanuniversitaet Leoben, the concentration of hydrogen in the gas mixture should be lowered to increase the hydrogen utilization. Hence, the flow rate of the gas mixture and the ratio of hydrogen to argon or nitrogen are the main influencing parameters on the process efficiency. In fact, there are two possible methods of iron ore reduction using hydrogen. The first is inflight reduction, which occurs from the tip of the electrode and the slag surface where iron ore and gas particles are at high temperatures. The second is the reduction of liquid iron oxide on the slag surface. Despite the iron oxide reduction by hydrogen, a small amount of iron oxide is reduced by carbon. Carbon can be entered into the melt from the graphite electrode and reduce iron oxide due to the high temperature of the electrode. The graphite electrode is eroded and the eroded particles are introduced into the melt.

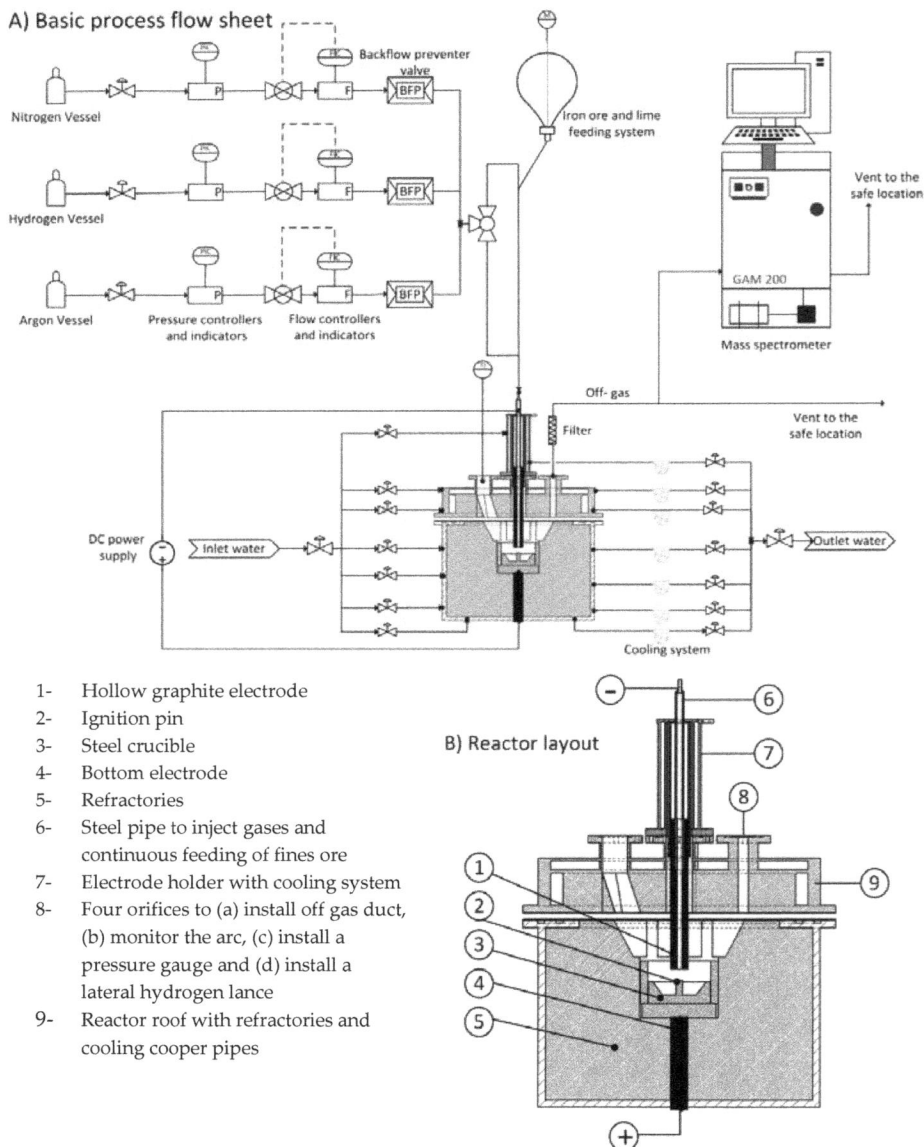

1- Hollow graphite electrode
2- Ignition pin
3- Steel crucible
4- Bottom electrode
5- Refractories
6- Steel pipe to inject gases and continuous feeding of fines ore
7- Electrode holder with cooling system
8- Four orifices to (a) install off gas duct, (b) monitor the arc, (c) install a pressure gauge and (d) install a lateral hydrogen lance
9- Reactor roof with refractories and cooling cooper pipes

Figure 1. (**A**) A basic process flow sheet of the laboratory-scale plasma facility at Montanuniversitaet Leoben and (**B**) rector layout with the main components.

The off gas contains Ar or N_2, H_2O, H_2, CO, and CO_2, which leaves the reactor from the off gas duct. In order to analyze the chemical composition of the off gas and, accordingly, calculate the hydrogen utilization degree, reduction rate, and reduction degree of iron oxide, a mass spectrometer was installed in the laboratory. Electricity power was supplied by a DC power supply with a power maximum of 8 kW. All sections of the plasma reactor were cooled by a water-cooling system. To monitor the arc, an optical spectrometer with a fiber was used to monitor the arc.

2. Thermodynamic Properties of Thermal Plasma

In HPSR, the gas particles are ionized by the generation of the plasma arc at the tip of the graphite electrode inside the HPSR plasma reactor [3,13,16]. The plasma arc can activate molecular hydrogen. Therefore, molecular H_2; atomic H; ionic hydrogen H^+, H_2^+, and H_3^+; and excited state H* are present in the plasma arc zone [17]. Hence, the reduction reaction of hematite is represented by

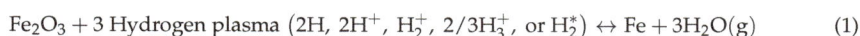

$$Fe_2O_3 + 3 \text{ Hydrogen plasma } \left(2H, 2H^+, H_2^+, 2/3H_3^+, \text{ or } H_2^*\right) \leftrightarrow Fe + 3H_2O(g) \tag{1}$$

Metal oxide and H_2O–H_2, H_2O–H, and H_2O–H^+ lines over the temperature are presented by the Ellingham diagram, which provides an estimation of the possibility of metal oxide reduction by hydrogen in terms of thermodynamic characteristics. In this diagram, the H_2O–H^+ line lies below the other lines. Consequently, hydrogen in the ionized state can reduce not only the iron oxides but also all other metal oxides [18–20].

If the temperature of the particles in plasma (molecules, atoms, ions and electrons) are the same and each process is balanced with its revers process, the plasma is in complete thermodynamic equilibrium (CTE). Plasma can be divided into two different categories: thermal or equilibrium plasmas and cold or nonequilibrium plasmas. In thermal plasmas, the temperature of electrons and ions are equal. However, not only laboratory scale plasmas but also some of the natural plasmas cannot meet all conditions of CTE. In the center of an electric arc, the deviations from equilibrium occur, and then it is more probable to be in a local thermodynamic equilibrium (LTE) state. In HPSR, the particles that diffuse into the plasma arc zone have enough time to equilibrate or to be at the same temperature. Therefore, hydrogen arc plasma is a thermal plasma and it is under LTE conditions [21–23].

Robino et al. [16] represented the standard Gibbs free energy changes for different mole fractions of monoatomic hydrogen in a mixture of H and H_2. The results show that by increasing the mole fraction of monoatomic hydrogen, the standard free energy markedly declines. Despite the low mole fraction of ionic hydrogen, its reduction ability is significantly high. In other words, monoatomic hydrogen (H) is able to reduce metal oxides more readily.

Zhang et al. [24] compared the Gibbs free energies changes for forming water by different hydrogen species as a function of temperature. Based on this, the reduction potentials are ordered as follows

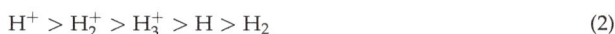

$$H^+ > H_2^+ > H_3^+ > H > H_2 \tag{2}$$

Figure 2 shows the Gibbs free energy changes for reduction of Fe_2O_3, Fe_3O_4, and FeO by various hydrogen species over temperature, which were calculated using FactSage™ 7.1 (Database: FactPS 2017). It confirms the order of the reduction ability of hydrogen plasma species, which is in a good agreement with the Zhang et al. [24] diagram. This diagram also shows that when using hydrogen as a reducing agent, FeO is more stable than the other forms of iron oxides.

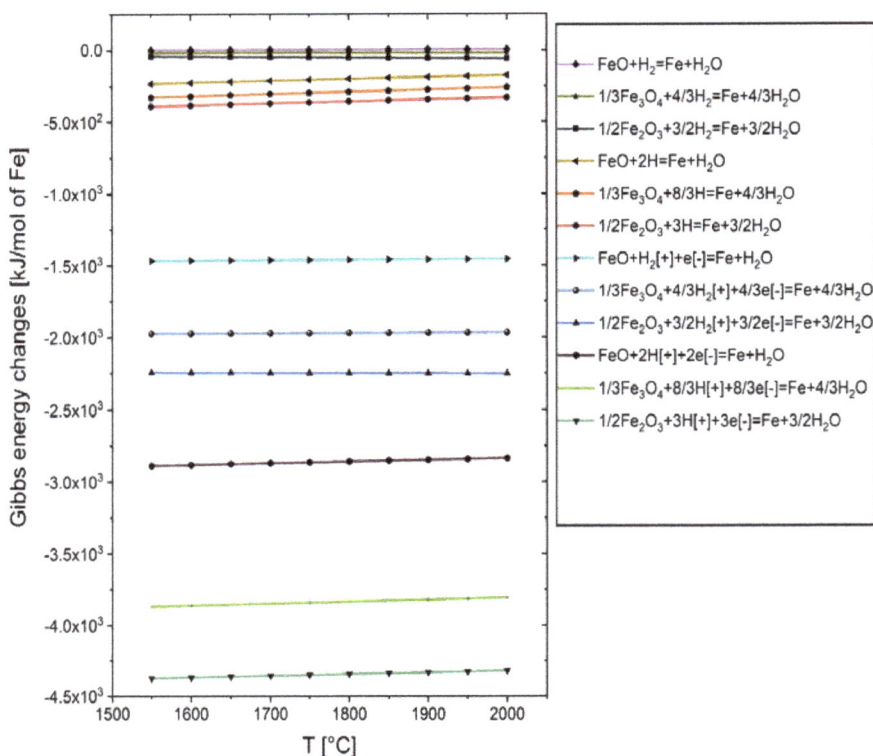

Figure 2. ΔG°–T curve for the reduction of iron oxides with different chemically active hydrogen species calculated using FactSage™ 7.1 (Database: FactPS 2017).

Consequently, to have a high reduction rate, the reduction reaction of iron oxides should occur with hydrogen-activated particles (i.e., atomized, ionized, and excited state of hydrogen particles). HPSR has been investigated extensively at Montanuniversitaet Leoben [3,4,13–15]. Badr [13] studied the characteristics of the HPSR process in terms of thermodynamics, kinetics, and the possibility of industrialization. His results have confirmed the observations of previous researchers [20,25,26].

According to the Saha equation, hydrogen molecules begin to dissociate when the temperature rises above 3000 K. The dissociation and ionization of 0.5 mol of hydrogen and 0.5 mol of argon at equilibrium were calculated by FactSage™ (Toronto, ON, Canada) 7.1 thermochemical software, and the results are shown in Figure 3. The results are in good agreement with Kahne et al. [27] and Lisal et al. [28] works. This figure indicates that dissociation and ionization are two separate processes. Above 5000 °C, hydrogen is completely dissociated and, above 15,000 °C, the ionization process is the dominant process.

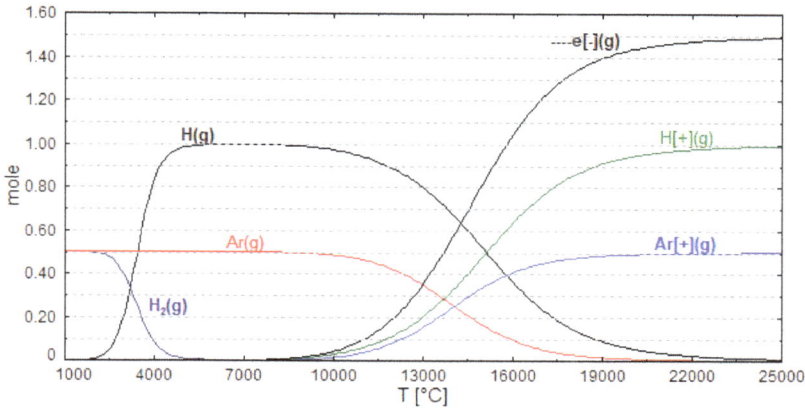

Figure 3. Gas composition of a H_2-Ar mixture over the temperature at 100 kpa (FactSage™ 7.1, Database; FactPS 2017).

In the HPSR process, hydrogen in the plasma zone at high temperatures is partially ionized which leads to create two different gases, light electrons, and heavy ions. n_e and n_i are the electron and heavy ion individual densities, respectively. Those densities can be used to define the ionization degree. The ionization degree is defined by the rates of ionization and recombination. Charge carriers in plasma are lost through the different processes. In HPSR, the main recombination processes are drift to the anode (liquid iron oxide), diffusion to the reactor refractories, and volume recombination. Some volume ionization and recombination processes of hydrogen are shown in Table 1 [29–31].

Table 1. Ionization and recombination of hydrogen atom [24,29–31].

$e[-] + H \rightarrow H^+ + 2e[-]$.	Collisional ionization
$e[-] + H_2 \rightarrow H_2^+ + 2e[-]$	Collisional ionization
$e[-] + H_2 \rightarrow H_2^* + e[-]$	Collisional excitation
$hv + H \rightarrow H^+ + e[-]$	Photoionization
$H^+ + 2e[-] \rightarrow H + e[-]$	Three-body recombination
$H^+ + e[-] \rightarrow H$	Two-body recombination
$H^+ + wall \rightarrow 1/2H_2 + e[-]$	Wall recombination

HPSR involves an equilibrium or thermal plasma (hot plasma) for which $T_e = T_h$ and a chemical equilibrium exists. Due to the collision frequency at high temperatures, the energy distribution is uniform among all particles. The mean kinetic energy of the ions can define the temperature of the ion particles. The kinetic energy or velocity of the individual particles is defined by the collisional processes. Therefore, the total mean kinetic energy is obtained by the summation of the energies of all particles [32].

Several species are present in the plasma zone of the HPSR process, namely photons, free electrons, hydrogen atoms, hydrogen ions, and molecules [17,33,34]. In the plasma arc, not only iron and iron oxide can be released from the iron ore and liquid bath but also carbon is released from graphite electrode. The amount of iron, iron oxide and carbon vapor depends on the process parameters [35]. Bohr's model is used to describe the structure of hydrogen energy levels [32]. The collisional process, which is the dominant ionization process in the HPSR, gives rise to the atomization and ionization of the hydrogen and argon molecules [36]. Excitation occurs when a ground state electron of an atom or a molecule absorbs sufficient energy to transition to a higher energy level. Atoms or molecules in these states are known as excited state X*. The excited state lifetimes of hydrogen particles are between 10^{-8} and 10^{-6} s. Ionization is the process by which an atom or a molecule acquires sufficient energy to gain or lose an electron to form ions. The hydrogen ionization energy of an already excited particle (electron in the second orbit) is less than 13.6 eV, as given by Bohr's theory,

$$\delta\varepsilon_i = \Delta\varepsilon_i \left(1 - \frac{1}{n^{*2}}\right) = 13.6\left(1 - \frac{1}{2^2}\right) = 10.2 \text{ eV} \rightarrow \Delta\varepsilon_i - \delta\varepsilon_i = 13.6 - 10.2 = 3.4 \text{ eV} \tag{3}$$

where $\delta\varepsilon_i$ is the refinement of the ionization energy, $\Delta\varepsilon_i - \delta\varepsilon_i$ is the ionization energy of the already excited particles, and n^* is the quantum number in the excited state [32,37]. The collision cross-section is the effective area in which two particles must meet to scatter from each other. Since atoms do not have a well-defined size, Bohr derived a radius for atoms to calculate the cross-section, defined as

$$r = \frac{n^2 r_1}{z} \tag{4}$$

where n is the principal quantum number, r_1 is the radius of the first Bohr orbit, and z is the atomic number. Therefore, the cross-section for atomic hydrogen is $3.53 \times 10^{-20} \text{ m}^2$ [32].

There are two types of collisions: elastic and inelastic. In an elastic collision, the atom absorbs a fraction of the electron initial momentum without any changes in its energy state. The probability of elastic collision depends on the equivalent cross-section, σ. For an inelastic collision, the degree of ionization is defined by the cross-section area, σ. According to the Maxwell–Boltzmann distribution function, the mean velocity of the particles depends on the square root of the temperature [23,38].

$$\overline{w} = \sqrt{\frac{8 \cdot K \cdot T}{\pi \cdot m}} \tag{5}$$

where m is the mass of the particles, K is the Boltzmann constant ($1.38054 \times 10^{-23} \text{ JK}^{-1}$), and T is the temperature of the particles. The density of the active particles, ψ_i, in an electrically insulated surface and in the absence of an externally applied electric field is given by

$$\psi_i = \frac{n_i \omega_i}{4} \tag{6}$$

where n_i is the density of particles i. Combining Equations (5) and (6) gives the following formula, which is used for the calculation of the activated particles.

$$\psi_i = 1.48 \times 10^{-12} n_i \left(\frac{T_i}{m_i}\right)^{1/2} \left[\text{m}^{-2}\text{s}^{-1}\right] \tag{7}$$

Therefore, with the increase of the temperature and the number density, the density of the activated particles will be increased.

3. Effect of Charge Polarity on the Iron Ore Reduction Reactions

A plasma-confining surface can be positively charged, negatively charged, or neutral. The density of the ions and electrons change while reaching a plasma-confining surface. In a typical thermal plasma, the thermal boundary layer is located near the surface. At the bottom of this layer, the plasma sheath is located. It was found that the plasma sheath is a narrow layer in which particles of opposite polarity are attracted and those with the same polarity are repelled [3,18,38].

The density gradients in the vicinity of the surface depend on the order of the Debye length. The layer near the surface with the charge imbalances is called the Debye sheath [23]. Plasma sheaths in the HPSR reactor are on refractory surfaces, on the liquid iron oxide surface, and on the graphite electrode surface. The thickness of the plasma sheath is defined by

$$\lambda_D = \left(\frac{\varepsilon_0 k T_e}{e^2 n_e}\right)^{1/2} == 69.1\left(\frac{T_e}{n_e}\right) \text{ [m]} \tag{8}$$

where λ_D is the Debye length in m, ε_0 is dielectric constant and equals 8.86×10^{-12} A$\frac{s}{Vm}$, e is electron charge and equals 1.6×10^{-19} As, and k is Boltzmann constant. The Debye length in thermal plasma is between 10^{-8} to 10^{-7} m [23]. The temperature of 25,000 K is the average temperature of the electrons in close vicinity to the liquid slag surface, and the electron density is 10^{23} m^{-3}. Thus, the Debye length on the slag surface in the plasma reactor is 3.5×10^{-8} m. The thickness of the plasma sheath is approximately equal to the Debye length. However, the sheath edge, which is a collision-less transition layer between plasma and sheath, is between 1 and 10 Debye length [23,39]. The thickness of thermal boundary layer is several orders of magnitude larger than that of the plasma sheath. In the thermal boundary layer, recombination occurs, and the concentration of the excited particles decreases.

In HPSR, the reduction reactions of iron ore can take place during two different times: (1) inflight reduction of iron ore fines within the distance of the arc length and (2) the reduction on the surface of the liquid slag. Solid fine particles do not have any applied electrical field. However, a positive polarity is applied to the liquid slag. Therefore, the polarity of the slag surface is one of the main influencing parameters on the efficiency of the reduction process. Dembovsky [38] has described the effect of surface polarity on the thermodynamic variables in metallurgical reactions.

A surface where there is no applied electrical field (i.e., no net current flow) repels the electrons and absorbs positive ions. Because the electrons can first touch the surface due to their higher velocity, the surface is charged negatively. Consequently, positive ions are attracted to the surface, and the electrons are repelled by the negative surface in the plasma sheath. Therefore, the density of positive ions increases.

Figures 4 and 5 show a schematic of active particles in a plasma state reaching positively and negatively charged reaction surfaces, respectively. When the surface is positively charged, the density of the electrons is higher than the density of the ions in the plasma sheath, and vice versa.

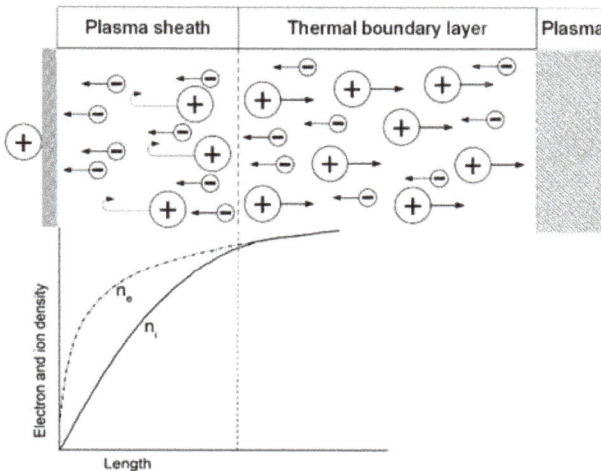

Figure 4. Motion of the activated particles near a positive reaction surface [23,38].

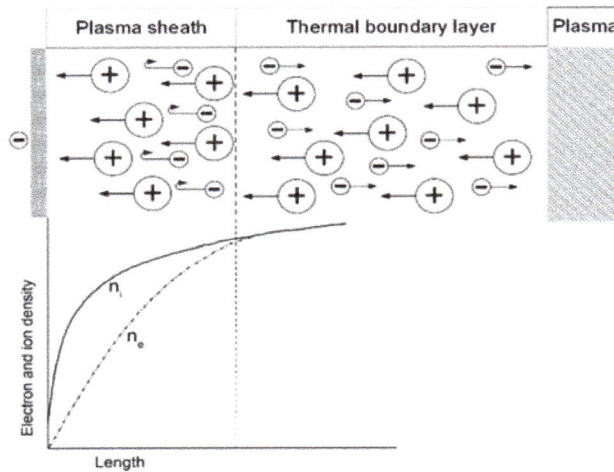

Figure 5. Motion of the activated particles near a negative reaction surface [23,38].

Considering the direct current straight polarity of the HPSR arc, ionized hydrogen atoms are repelled by the positive molten slag, leading to a decrease in the reduction rate of iron oxides. The effect of surface polarity on the Gibbs free energy changes for the reduction reaction of iron oxide using hydrogen at 10,000 K was represented by Dembovsky [38]. The pertinent reaction is given by

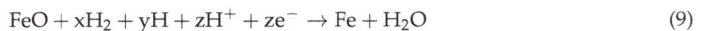

$$FeO + xH_2 + yH + zH^+ + ze^- \rightarrow Fe + H_2O \tag{9}$$

where x, y, and z are the molar fractions of molecular, atomic, and ionized hydrogen. Hydrogen is atomized and ionized at high temperatures. Therefore, the atomization and ionization degree of hydrogen and the polarity define the molar fractions of the particles reaching the reaction surface. Consequently, the reduction reaction rate depends on the molar fraction of particles reaching the surface [3,18,38]. Dembovsky [38] compared changes in the Gibbs free energy for the reduction of iron oxide at various polarities over temperature. He showed that when the surface polarity was negative, Gibbs free energy was more negative than the other cases. Therefore, the reduction reaction proceeds at a higher rate.

To assess the effect of the polarity on the reduction reaction of iron oxide, the Gibbs free energy changes of three reduction reactions were calculated by FactSage™ 7.1 (Database: FactPS 2017), and the results are shown in Figure 6.

$$FeO + H_2 \rightarrow Fe + H_2O \tag{10}$$

$$FeO + 2H \rightarrow Fe + H_2O \tag{11}$$

$$FeO + 2H^+ + 2e^- \rightarrow Fe + H_2O \tag{12}$$

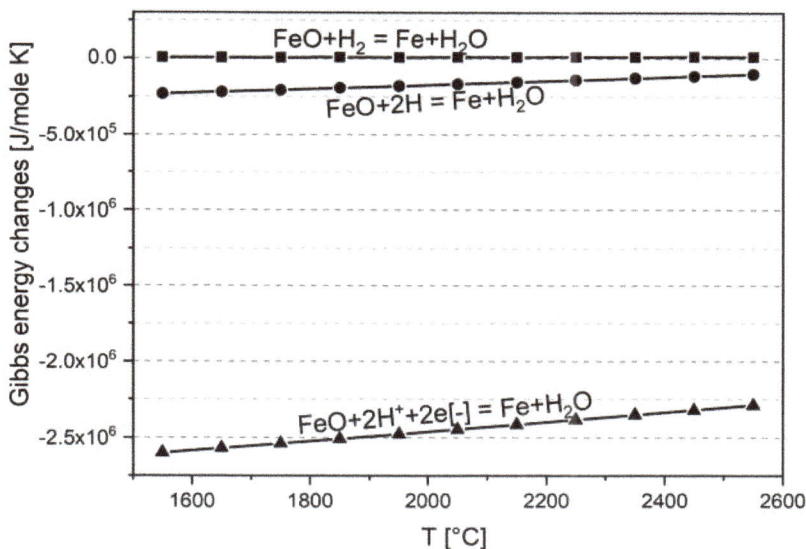

Figure 6. ΔG°–T curve for the reduction of FeO with different hydrogen species calculated by FactSage™ 7.1 (Database: FactPS 2017).

If the positive polarity is used in the process, all ionized hydrogen particles cannot reach the reaction surface. Therefore, the Gibbs free energy is derived to increase and be more positive. In the case of the negative polarity, hydrogen positive ions can reach the reaction surface more readily. Therefore, the Gibbs free energy changes are more negative.

4. Ionization Degree of Hydrogen

The collision of electrons with atoms leads to their ionization or excitation. To calculate the rate of production of the new electrons, the ionization collision frequency was multiplied by the electron density (n_e) obtaining the so-called source rate (S_e).

$$S_e = n_e n_n \langle \sigma_{ion} v_e \rangle \tag{13}$$

where n_e and n_n are the electron and neutral atom densities, respectively. σ_{ion} is the cross-section for the electron-impact ionization and v_e is the electron velocity [40,41]. The ionization rate $\langle \sigma_{ion} v_e \rangle$ of hydrogen atoms for the Maxwell distribution of electron temperature, T_e, is shown in Figure 7. The maximum ionization rate $\langle \sigma_{ion} v_e \rangle$ is reached with energies above hydrogen ionization energy, which is 13.6 eV.

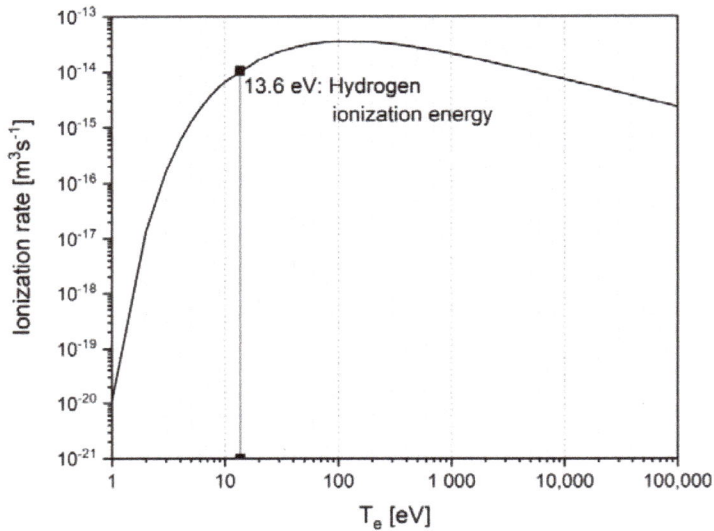

Figure 7. Ionization rate of the hydrogen atoms versus electron temperature T_e [40].

5. Solubility of Hydrogen

In the HPSR process, iron ore in the plasma reactor is melted and then continuously reduced by hydrogen. The density of slag is lower than the density of liquid iron. Therefore, there is expected to be a slag layer on the surface of the liquid metal during the reduction process. However, in the interface of arc and molten metal, where the reduction reactions take place, hydrogen species can reach the liquid iron surface and be absorbed. Hence, the study of the solubility of hydrogen in liquid iron is important.

Several researchers [40,42–45] have studied hydrogen solubility and the mechanism of hydrogen absorption in liquid iron. This study is important for steel makers as it explores the negative effects of hydrogen on the mechanical properties of steel. However, in HPSR, it is more important due to the existence of an enormous amount of hydrogen reaching liquid phases. In HPSR, hydrogen exists in molecular, atomic, and ionized forms. To be dissolved, hydrogen in a liquid iron should first be dissociated. The solubility of atomized hydrogen in liquid iron is defined by [46]

$$[\%H] = K_{H_2}.P_{H_2}^{1/2} \tag{14}$$

where K_{H_2} is the equilibrium constant and $P_{H_2}^{1/2}$ is the hydrogen partial pressure. This equation is valid when hydrogen exists only in a molecular state. The solubility of hydrogen in a plasma state in liquid iron was presented by Dembovsky [42]. Badr [13] then revised the equation and represented by

$$[\%H] = K_{H_2}.P_{H_2}^{1/2} + K_H P_H + K_{H^+}.(P_{H^+} + P_e) \tag{15}$$

where P_H, P_{H^+}, and P_e are the partial pressures of atomized hydrogen, ionized hydrogen, and the electrons, respectively. K_H and K_{H^+} are the equilibrium constants for the dissolution of atomized and ionized hydrogen particles, respectively. Dembovsky [42] experimentally and theoretically showed that the solubility of gases in the plasma state in liquid iron are higher than the solubility of gases in the molecular state due to the lower activation energy for the dissolution of ionized and atomized particles.

During the operation of HPSR, a layer of slag covers the molten metal. Therefore, the slag can pick up hydrogen and water vapor, then transfer it to the liquid metal. Many studies [47–50] have been done to investigate the solubility of hydrogen in slags. Walsh et al. [43] studied the solubility of hydrogen in steelmaking slags. They reported that hydrogen is not dissolved in slag significantly when hydrogen gas is applied. Slags can dissolve small amount of hydrogen via the reaction between slag components and water vapor. They showed that water vapor in the molecular state is not dissolved in slags. Russell [51] studied the dissolution of the water vapor molecularly in molten glasses. He showed that water vapor is dissolved in molten glasses, however, he could not define the form of the dissolved hydrogen. Walsh et al. [29] reported that hydrogen solubility in acidic open hearth slags is less than that of basic slags, which agrees with Wahlster's [52] work.

In HPSR, partial pressure of molecular hydrogen, atomic hydrogen, and ionized hydrogen define the solubility of hydrogen in liquid metal. Therefore, dissociation and ionization degree of hydrogen particles are important to take into account.

6. Mechanism of the Hematite Reduction Reaction in HPSR

To study the reduction of hematite using hydrogen at high temperatures, the equilibrium of Fe_2O_3 and H_2 has been assessed by FactSage™ 7.1. For the assessment of the equilibrium, the range of the equilibrium temperature should first be determined. In HPSR, hematite is reduced by hydrogen. Hydrogen in the plasma arc zone is partially atomized and ionized. As what has already been discussed, the activated hydrogen species are stronger reducing agents than molecular hydrogen in reducing iron ore. The temperatures at the center of arc, at the vicinity of the arc, and on the liquid metal surface mainly depend on the amperage, voltage, arc length, and the gas composition. Murphy et al. [53] simulated the temperatures, velocities, and the vaporization of iron ore in the arc zone for a 150 A tungsten inert gas (TIG) welding arc. They showed that, with the use of helium, Fe is vaporized and the concentration of Fe can reach 7 mol % due to the high temperature of the weld pool, which is approximately 2773 °C. With the use of argon as a plasma gas, the temperature of the liquid metal at the interface and the Fe concentration are 2273 °C and 0.2%, respectively. The reason for this is that helium conducts heat better than argon. The temperature of the gas–liquid metal interface of HPSR is not yet defined. However, it seems to be much higher than the helium welding arc plasma not only because of the higher power of the electric supply but also the use of a high percentage of hydrogen in the gas mixture. Therefore, for the calculations of equilibrium, the temperature range between 1550 and 3000 °C was considered. The lower part of the range (i.e., 1550 °C) was considered in order to be above the melting temperature of pure iron, which is 1537 °C.

The reduction reactions of hematite using hydrogen occurs in two steps which are given by

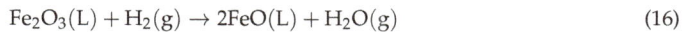

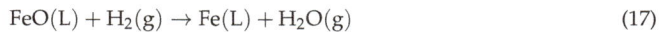

$$Fe_2O_3(L) + H_2(g) \rightarrow 2FeO(L) + H_2O(g) \tag{16}$$

$$FeO(L) + H_2(g) \rightarrow Fe(L) + H_2O(g) \tag{17}$$

This means that, at the first step of the reduction process, FeO is formed. Then, the reduction of wustite to form Fe by hydrogen takes place continuously during operation. To prove the reduction sequences, the Gibbs energy changes were calculated by FactSage™ 7.1 and the results are shown in Figure 8. The figure shows that the Gibbs energy changes of Equation (16) is more negative than that of Equation (17).

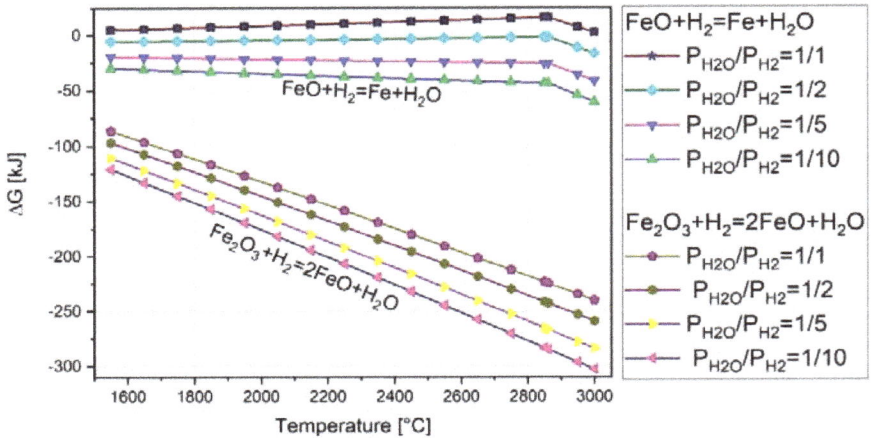

Figure 8. $\Delta G°$–T curve for the reduction of Fe_2O_3 using hydrogen in four different pressure ratios of P_{H_2O}/P_{H_2} calculated by FactSage™ 7.1 (Database: FToxid 2017).

Figure 8 shows the Gibbs energy changes of the two reactions in four different pressure ratios of P_{H_2O}/P_{H_2}. The ratio of P_{H_2O} to P_{H_2} was considered to be 1/1, 1/2, 1/5, and 1/10. The graph shows that, with the increase of water vapor partial pressure, the Gibbs free energy decreased due to the lack of hydrogen to reduce iron oxides. Consequently, at the first step, hematite is reduced to wustite, and then continuously reduced to Fe.

To reduce hematite using hydrogen 3 mol of hydrogen is required for each mole of hematite. Therefore, the pertinent equilibrium has been studied. Figure 9 shows the equilibrium of 1 mol of hematite and 3 mol molecular hydrogen. The results show that hydrogen utilization at equilibrium is 43% at the temperature of 1600 °C. The pertinent reduction reaction is given by

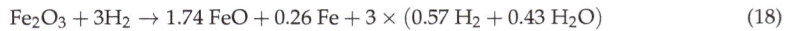

$$Fe_2O_3 + 3H_2 \rightarrow 1.74\,FeO + 0.26\,Fe + 3 \times (0.57\,H_2 + 0.43\,H_2O) \tag{18}$$

Therefore, the maximum hydrogen utilization degree using molecular hydrogen is 43%. However, it is expected to be higher when using hydrogen in a plasma state. In order to completely reduce FeO, further hydrogen should be injected into the reactore. Theoreticaly, regarding 43% hydrogen utilization degree, 2.34 mol of hydrogen is required to reach 100% of iron oxide reduction degree.

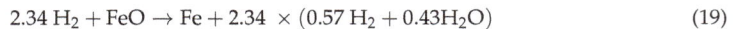

$$2.34\,H_2 + FeO \rightarrow Fe + 2.34 \times (0.57\,H_2 + 0.43H_2O) \tag{19}$$

The complete reduction degree of hematite can be reached by 6.98 mol of hydrogen

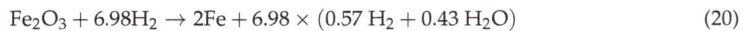

$$Fe_2O_3 + 6.98H_2 \rightarrow 2Fe + 6.98 \times (0.57\,H_2 + 0.43\,H_2O) \tag{20}$$

Similar to the conventional steelmaking processes, FeO concentration in slag and its influence on the reduction rate should be taken into accont. With the decrease of FeO concentration in slag, the reduction rate and accordingly hydrogen utilization degree is decreased.

Kamiya et al. [26] studied the reduction of molten iron oxide using H_2-Ar plasma. They reported that the hydrogen utilization degree can be 60% at low concentration of hydrogen in the gas mixture. Nagasaka et al. [54,55] studied the kinetics of molten iron oxide reduction using hydrogen. They compared the reduction rate of iron oxide with different reducing agent. They reported that the reduction rate of iron oxides using hydrogen is one or two orders of magnitude higher than those by other reductants.

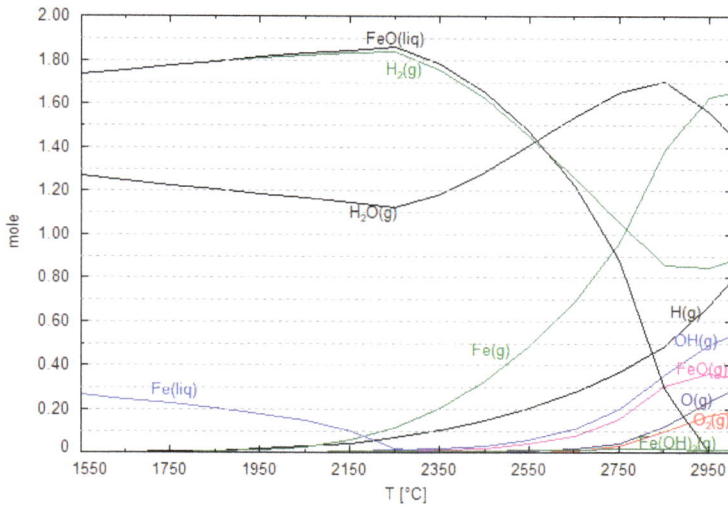

Figure 9. Equilibrium of 3 mol of hydrogen and 1 mol of Fe_2O_3 with a total pressure of 1 atm assessed by FactSage™ 7.1 (Database FactPS 2017).

The figure shows that, with the increase of the temperature, liquid iron begins to be vaporized and molecular hydrogen begins to be dissociated. H_2O and Fe (liquid) decrease gradually with the increase of the temperature until 2268 °C, and FeO and H_2 increase in the same rate. This means that the reduction rate is decreasing. The reason is that water vapor is dissociated, and H_2, O_2, and OH are formed. Consequently, Fe is oxidized by the produced O_2. To prove this assumption, the equilibrium of 1 mol of H_2O was calculated, and the results are presented in Figure 10.

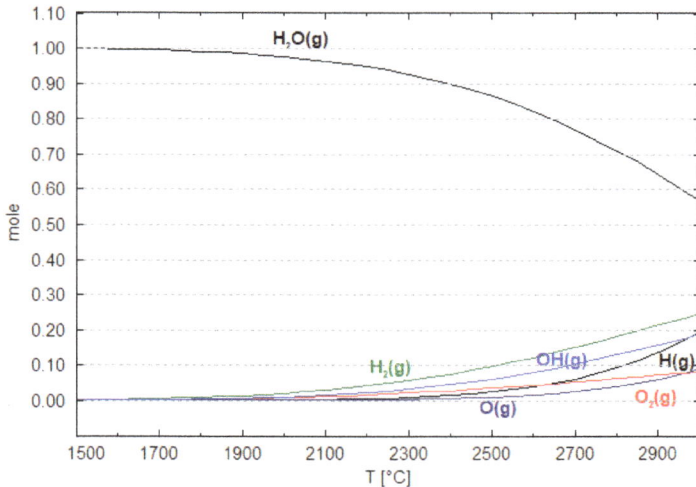

Figure 10. Equilibrium of 1 mol of H_2O at high temperatures (FactSage™ 7.1, Database FactPS 2017).

Baykara et al. [56] produced hydrogen with water thermolysis process at the temperature of 2227 °C and a pressure of 1 atm. Mass and energy balance calculations were done to define the chemical composition of material of each stream. The chemical composition of dissociated water is shown in Table 2. Their results are in a good agreement with the present work which have been calculated theoretically.

Table 2. Chemical composition of H_2O at 2227 °C at the pressure of 1 atm.

Result of	Unit	H_2O	H_2	O_2	H	O	OH
Baykara et al. [56]	mol %	91.14	4.27	1.55	0.53	0.19	2.33
Present work	mol %	92.0	4.3	1.6	0.51	0.18	2.33

The assessment of the water vapor equilibrium shows that it is dissociated and produces molecular oxygen and hydrogen. Therefore, the following reactions take place, respectively.

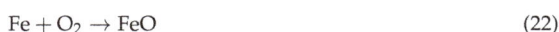

$$2H_2O \rightarrow 2H_2 + O_2 \tag{21}$$

$$Fe + O_2 \rightarrow FeO \tag{22}$$

Consequently, the dissociation of water vapor at high temperatures causes a slight drop in the reduction rate of iron oxide.

To assess the reduction reaction of hematite with 3 mol of hydrogen, the hydrogen utilization degree was calculated at the temperature of 2750 °C, and the results are given by

$$Fe_2O_3 + 3H_2 \rightarrow 0.88\,FeO\,(L) + 4.45 \times (0.24\,H_2 + 0.37H_2O + 0.21\,Fe(g) + 0.08\,H + 0.04\,OH + 0.03\,FeO(g)) \tag{23}$$

The utilization degree of hydrogen can be calculated by the sum of the H_2O and OH in the reaction. At this temperature, the hydrogen utilization degree is 58%.

In Figure 9, at the temperatures between 2268 and 2850 °C, Fe is vaporized. FeO is reduced to form Fe (g) and H_2O. The amount of water vapor produced by the reduction reaction is more than that of released by the dissociation process. Hence, water vapor in this temperature range is increased. There is a peak for the H_2O at the temperature around 2850 °C in Figure 9. Above this temperature, FeO (liquid) is going vanish, and the rate of the reduction decreases. Therefore, amount of molecular hydrogen is kept approximately constant and water vapor is decreased due to the dissociation process.

7. Summary and Conclusions

In summary, excited particles, such as electrons and ions, exist in a plasma arc. The reduction rate of iron oxide depends on the kind of hydrogen species present in the plasma state. It was shown that the reduction ability of each species is different and that the ionized hydrogen H^+ is the strongest reducing agent. The plasma arc temperature and particle density are the main parameters influencing the hydrogen ionization rate and, consequently, iron oxide reduction rate. At temperatures above 15,000 °C, most of the hydrogen and argon particles are in the ionized state, making the reduction of iron oxide more feasible. Regarding thermodynamic aspects, the Gibbs free energy changes were calculated for different iron oxide reduction reactions with different hydrogen species. It was found that the reduction ability of the hydrogen species is in the following order: $H^+ > H_2^+ > H_3^+ > H > H_2$. Moreover, the study of hematite and hydrogen at equilibrium shows that, at 1600 °C, hydrogen utilization is 43%. With the increase of the temperature above 1550 °C, water vapor dissociates. The highest reduction rate can be reached when the reduction reactions take place at the plasma arc zone, which is at high temperatures. Moreover, hydrogen positive ions can reach the liquid iron oxide easier with a negative polarity; therefore, the reduction rate of iron oxide could be increased.

Author Contributions: Conceptualization, M.N.S. and J.S.; methodology, M.N.S. and J.S.; software, M.N.S.; validation, M.N.S. and J.S.; formal analysis, M.N.S. and J.S.; investigation, M.N.S.; resources, M.N.S. and J.S.; writing (original draft preparation), M.N.S.; writing (review and editing), visualization, M.N.S. and J.S.; supervision, M.N.S. and J.S.

Funding: The SuSteel Project supported this research. The SuSteel project was funded by the Austrian Research Promotion Agency (FFG). Montanuniversitaet Leoben, K1–Met GmbH, voestalpine Stahl Donawitz GmbH, and voestalpine Stahl Linz GmbH are the partners of the SuSteel project.

Conflicts of Interest: The authors declare no conflict of interest.

References

1. Kundak, M.; Lazic, L.; Crnko, J. *CO₂ Emissions in the Steel Industry*; Croatian Metallurgical Society (CMS): Zagreb, Croatia, 2009.
2. Lisienko, V.G.; Chesnokov, Y.N.; Lapteva, A.V.; Noskov, V.Y. Types of greenhouse gas emissions in the production of cast iron and steel. *IOP Conf. Ser. Mater. Sci. Eng.* **2016**, *150*, 12023. [CrossRef]
3. Badr, K.; Bäck, E.; Krieger, W. Reduction of iron ore by a mixture of Ar-H_2 with CO and CO_2 under plasma application. In Proceedings of the 18th International Symposium on Plasma Chemistry, Kyoto, Japan, 26–31 August 2007.
4. Kirschen, M.; Badr, K.; Pfeifer, H. Influence of direct reduced iron on the energy balance of the electric arc furnace in steel industry. *Energy* **2011**, *36*, 6146–6155. [CrossRef]
5. Lin, H.-Y.; Chen, Y.-W.; Li, C. The mechanism of reduction of iron oxide by hydrogen. *Thermochim. Acta* **2003**, *400*, 61–67. [CrossRef]
6. Turkdogan, E.T.; Vinters, J.V. Gaseous reduction of iron oxides: Part I. Reduction of hematite in hydrogen. *Metall. Mater. Trans. B* **1971**, *2*, 3175–3188. [CrossRef]
7. Hou, B.; Zhang, H.; Li, H.; Zhu, Q. Study on Kinetics of Iron Oxide Reduction by Hydrogen. *Chin. J. Chem. Eng.* **2012**, *20*, 10–17. [CrossRef]
8. Ranzani da Costa, A.; Wagner, D.; Patisson, F. Modelling a new, low CO_2 emissions, hydrogen steelmaking process. *J. Clean. Prod.* **2013**, *46*, 27–35. [CrossRef]
9. Ban-Ya, S.; Ighuchi, Y.; Nagasaka, T. Rate of reduction of liquid wustite with hydrogen. *Tetsu-to-Hagane* **1984**, *70*, 1689–1696. [CrossRef]
10. Gilles, H.L.; Clump, C.W. Reduction of iron ore with hydrogen in a direct current plasma jet. *Ind. Eng. Chem. Proc. Des. Dev.* **1970**, *9*, 194–207. [CrossRef]
11. Gold, R.G.; Sandall, W.R.; Cheplick, P.G.; MacRae, D.R. Plasma reduction of iron oxide with hydrogen and natural gas at 100 kW and 1 MW. *Ironmak. Steelmak.* **1977**, *4*, 10–14.
12. Hayashi, S.; Iguchi, Y. Hydrogen reduction of liquid iron oxide fines in gas-conveyed systems. *ISIJ Int.* **1994**, *34*, 555–561. [CrossRef]
13. Badr, K. Smelting of Iron Oxides Using Hydrogen Based Plasmas. Ph.D. Thesis, University of Leoben, Leoben, Austria, 2007.
14. Bäck, E. Schmelzreduktion von Eisenoxiden Mit Argon-Wasserstoff-Plasma. Ph.D. Thesis, University of Leoben, Leoben, Austria, 1998.
15. Plaul, J.F. Schmelzreduktion von hämatitischen Feinerzen im Wasserstoff-Argon-Plasma. Ph.D. Thesis, University of Leoben, Leoben, Austria, 2005.
16. Robino, C.V. Representation of mixed reactive gases on free energy (Ellingharn-Richardson) diagrams. *Metall. Mater. Trans. B* **1996**, *27*, 65–69. [CrossRef]
17. Boulos, M.I. Thermal plasma processing. *IEEE Trans. Plasma Sci.* **1991**, *19*, 1078–1089. [CrossRef]
18. Sabat, K.C.; Rajput, P.; Paramguru, R.K.; Bhoi, B.; Mishra, B.K. Reduction of oxide minerals by hydrogen plasma: An Overview. *Plasma Chem. Plasma Process.* **2014**, *34*, 1–23. [CrossRef]
19. Hiebler, H.; Plaul, J.F. Hydrogen plasma- smelting reduction- an option for steel making in the future. *METABK* **2004**, *43*, 155–162.
20. Kitamura, T.; Shibata, K.; Takeda, K. In-flight reduction of Fe_2O_3, Cr_2O_3, TiO_2 and Al_2O_3 by Ar-H_2 and Ar-CH_4 plasma. *ISIJ Int.* **1993**, *33*, 1150–1158. [CrossRef]
21. Trelles, J.P.; Heberlein, J.V.R.; Pfender, E. Non equilibrium Modeling of Arc Plasma Torches. *J. Phys. D Appl. Phys.* **2007**, *40*, 5937. [CrossRef]
22. Bentley, R.E. A departure from local thermodynamic equilibrium within a freely burning arc and asymmetrical Thomson electron features. *J. Phys. D Appl. Phys.* **1997**, *30*, 2880–2886. [CrossRef]
23. Boulos, M.I.; Fauchais, P.; Pfender, E. *Thermal Plasmas. Fundamentals and Applications/Maher I. Boulos, Pierre Fauchais, and Emil Pfender. Vol.1*; Plenum Press: New York, NY, USA; London, UK, 1994.

24. Zhang, Y.; Ding, W.; Guo, S.; Xu, K. Reduction of metal oxide in nonequilibrium hydrogen plasma. *China Nonferrous Metal* **2004**, *14*, 317–321.

25. Nakamura, Y.; Ito, M.; Ishikawa, H. Reduction and dephosphorization of molten iron oxide with hydrogen-argon plasma. *Plasma Chem. Plasma Process.* **1981**, *1*, 149–160. [CrossRef]

26. Kamiya, K.; Kitahara, N.; Morinaka, I.; Sakuraya, K.; Ozawa, M.; Tanaka, M. Reduction of molten iron oxide and FeO bearing slags by H_2-Ar plasma. *ISIJ Int.* **1984**, *24*, 7–16. [CrossRef]

27. Kanhe, N.S.; Tak, A.K.; Bhoraskar, S.V.; Mathe, V.L.; Das, A.K. Transport properties of Ar-Al plasma at 1 atmosphere. In *SOLID STATE PHYSICS: Proceedings of the 56th DAE Solid State Physics Symposium 2011, SRM University, Kattankulathur, Tamilnadu, India, 19–23 December 2011*; AIP: Melville, NY, USA, 2012; pp. 1025–1026.

28. Lisal, M.; Smith, W.; Bures, M.; Vacek, V.; Navratil, J. REMC computer simulations of the thermodynamic properties of argon and air plasmas. *Mol. Phys.* **2002**, *100*, 2487–2497. [CrossRef]

29. Piel, A. *Plasma Physics. An Introduction to Laboratory, Space, and Fusion Plasmas/Alexander Piel*; Springer: Heidelberg, Germany, 2010.

30. Keidar, M.; Beilis, I. *Plasma Engineering. Applications from Aerospace to Bio and Nanotechnology*; Elsevier Science: San Diego, CA, USA, 2013.

31. Janev, R.K.; Reiter, D.; Samm, U. *Collision Processes in Low-Temperature Hydrogen Plasmas*; Institut für Plasmaphysik, Forschungszentrum Jülich GmbH, EURATOM Association: Jülich, Germany, 2003.

32. Feinman, J. *Plasma Technology in Metallurgical Processing*; Iron and Steel Society: Warrendale, PA, USA, 1987.

33. Pfender, E. Thermal plasma technology: Where do we stand and where are we going? *Plasma Chem Plasma Process* **1999**, *19*, 1–31. [CrossRef]

34. Sabat, K.C.; Murphy, A.B. Hydrogen Plasma Processing of Iron Ore. *Metall. Mater. Trans. B* **2017**, *48*, 1561–1594. [CrossRef]

35. Baeva, M.; Uhrlandt, D.; Murphy, A.B. A collisional-radiative model of iron vapour in a thermal arc plasma. *J. Phys. D Appl. Phys.* **2017**, *50*, 22LT02. [CrossRef]

36. Lichtenberg, A.J.; Lieberman, M.A. *Principles of Plasma Discharges and Materials Processing*, 2nd ed.; Wiley-Interscience: Hoboken, NJ, USA; Chichester, UK, 2005.

37. Mills, R. Spectral emission of fractional quantum energy levels of atomic hydrogen from a helium–hydrogen plasma and the implications for dark matter. *Int. J. Hydrogen Energy* **2002**, *27*, 301–322. [CrossRef]

38. Dembovsky, V. How the polarity of a surface reacting with a low temperature plasma affects the thermodynamic variables in metallurgical reactions. *Achta Phys. Slov.* **1984**, *34*, 11–18.

39. Allen, J.E. The plasma–sheath boundary: Its history and Langmuir's definition of the sheath edge. *Plasma Sources Sci. Technol.* **2009**, *18*, 14004. [CrossRef]

40. Goldston, R.J.; Rutherford, P.H. *Introduction to Plasma Physics*; Institute of Physics Pub.: Bristol, UK, 1995.

41. Naseri Seftejani, M.; Schenk, J. Kinetics of molten iron oxides reduction using hydrogen. *La Metall. Ital.* **2018**, *7/8*, 5–14.

42. Dembovský, V. Thermodynamics of dissolution and liberation of gases in the atomization of molten metals by plasma-induced expansion. *J. Mater. Process. Technol.* **1997**, *64*, 65–74. [CrossRef]

43. Walsh, J.H.; Chipman, J.; King, T.B.; Grant, N.J. Hydrogen in Steelmaking Slags. *JOM* **1956**, *8*, 1568–1576. [CrossRef]

44. Jo, S.-K.; Kim, S.-H. The solubility of water vapour in CaO-SiO_2-Al_2O_3-MgO slag system. *Steel Res.* **2000**, *71*, 15–21. [CrossRef]

45. Gedeon, S.A.; Eagar, T.W. Thermochemical Analysis of Hydrogen Absorption in Welding: A new model that addresses the shortcomings of Sievert's A new model that addresses the shortcoming of Sievert's law for predicting hydrogen absorption is proposed. *Weld. J.* **1990**, 264–271.

46. Fruehan, R.J. *The Making, Shaping, and Treating of Steel. [Vol. 2], Steelmaking and Refining Volume*, 11th ed.; AISE Steel Foundation: Pittsburgh, PA, USA, 1998.

47. Jung, I.-H. Thermodynamic Modeling of Gas Solubility in Molten Slags (II)—Water. *ISIJ Int.* **2006**, *46*, 1587–1593. [CrossRef]

48. Brandberg, J.; Sichen, D. Water vapor solubility in ladle-refining slags. *Metall. Mater. Trans. B* **2006**, *37*, 389–393. [CrossRef]

49. Stoephasius, J.-C.; Reitz, J.; Friedrich, B. ESR Refining Potential for Titanium Alloys using a CaF2-based Active Slag. *Adv. Eng. Mater.* **2007**, *9*, 246–252. [CrossRef]

50. Park, J.-Y.; Park, J.G.; Lee, C.-H.; Sohn, I. Hydrogen Dissolution in the TiO_2–SiO_2–FeO and TiO_2–SiO_2–MnO Based Welding-Type Fluxes. *ISIJ Int.* **2011**, *51*, 889–894. [CrossRef]

51. Russell, L.E. *Solubility of Water in Molten Glass*; Massachusetts Institute of Technology: Cambridge, MA, USA, 1955.

52. Wahlster, M.; Reichel, H.-H. Die Wasserstofflöslichkeit von Schlacken des Systems CaO-FeO-SiO$_2$. *Arch. Eisenhüttenwes.* **1969**, *40*, 19–25. [CrossRef]

53. Murphy, A.B.; Tanaka, M.; Yamamoto, K.; Tashiro, S.; Lowke, J.J. CFD modeling of arc welding: The importance of the arc plasma. In Proceedings of the 7th International Conference on CFD in the Mineral and Process Industries, Melbourne, Australia, 9–11 December 2009.

54. Nagasaka, T.; Ban-ya, S. Rate of reduction of liquid iron oxide. *Tetsu-to-Hagane* **1992**, *78*, 1753–1767. [CrossRef]

55. Nagasaka, T.; Hino, M.; Ban-ya, S. Interfacial kinetics of hydrogen with liquid slag containing iron oxide. *Metall. Mater. Trans. B* **2000**, *31*, 945–955. [CrossRef]

56. Baykara, S.; Bilgen, E. An overall assessment of hydrogen production by solar water thermolysis. *Int. J. Hydrogen Energy* **1989**, *14*, 881–891. [CrossRef]

metals

MDPI

Article

In Situ Spectroscopic Analysis of the Carbothermal Reduction Process of Iron Oxides during Microwave Irradiation

Jun Fukushima * and Hirotsugu Takizawa

Department of Applied Chemistry, Tohoku University, Aoba Aramaki, Sendai, Miyagi 980-8579, Japan; takizawa@aim.che.tohoku.ac.jp
* Correspondence: fukushima@aim.che.tohoku.ac.jp; Tel.: +81-22-795-7226

Received: 25 December 2017; Accepted: 9 January 2018; Published: 11 January 2018

Abstract: The effects of microwave plasma induction and reduction on the promotion of the carbothermal reduction of iron oxides (α-Fe_2O_3, γ-Fe_2O_3, and Fe_3O_4) are investigated using in situ emission spectroscopy measurements during 2.45 GHz microwave processing, and the plasma discharge (such as CN and N_2) is measured during microwave E-field irradiation. It is shown that CN gas or excited CN molecules contribute to the iron oxide reduction reactions, as well as to the thermal reduction. On the other hand, no plasma is generated during microwave H-field irradiation, resulting in thermal reduction. Magnetite strongly interacts with the microwave H-field, and the reduction reaction is clearly promoted by microwave H-field irradiation, as well as thermal reduction reaction.

Keywords: ironmaking; microwaves; carbothermal reduction; iron oxides; emission spectrum

---

1. Introduction

Electromagnetic waves such as microwaves can supply energy to materials effectively, and are thus, expected to be applicable to novel ironmaking processes [1–6]. In the conventional blast-furnace method, the combustion of coke and high-temperature CO gas requires the delivery of energy to the sites of the reduction reaction. In this method, sintered iron ore and coke are essential to ensure access to CO gas. This iron ore is walnut-sized, and the carbothermal reduction reaction of iron ore is slow. In contrast, in the microwave steelmaking method, microwaves can replace this extra carbon combustion, because they can supply energy to the iron ore or powder directly [4,7,8]. Using the microwave process, the reduction of CO_2 emissions is estimated to be 55%. Furthermore, by using microwaves as a heat source, it is possible to rapidly produce a pig iron containing about 13% carbon at 1380 °C, which is 200–300 °C lower than the temperature of the current blast furnace manufacturing method. In addition, this low-temperature reaction suppresses the reduction of phosphorus oxides, and the amount of generated slag is reduced. A recent in situ study using time-resolved X-ray powder diffraction synchrotron radiation revealed that the reduction of wustite to iron can be achieved at 770 °C in a microwave H-field [9], compared with over 1100 °C in an E-field [10]. In summary, the ironmaking method using carbothermal microwave reduction can reduce the CO_2 emissions, reaction temperature, and reaction time.

The effects of microwave irradiation on the ironmaking reaction are not restricted to heating only. However, the non-thermal effects on the reduction process of microwave ironmaking are still unknown. Matsubara et al. [11] confirmed the luminescence of Fe at temperatures greater than 1260 °C using in situ luminescence spectroscopy during the microwave ironmaking process, suggesting that reduction occurs in this low-temperature range. In addition, they reported in another work [12] that a CO peak was observed at 1400 °C, indicating that even the Boudouard reaction proceeds under microwave irradiation. However, these reports showed no clear evidence of microwave specific effects. Meanwhile, another effect of microwaves on the reduction reaction has been pointed out in previous studies that have used transition-metal oxides. In the reduction of TiO_2, oxygen plasma was detected by spectroscopy during microwave irradiation at 900 W for 10 min at the maximum intensity of the microwave electric field in a 10^{-1} Pa vacuum, and the weak reduction of TiO_2 was confirmed [13]. Other papers have reported the reduction of various transition-metal oxides by plasma [14–16]. Sabat et al. [15] mentioned that, typically, the activation energy of plasma-assisted reactions is lower than that of the corresponding standalone reactions. Moreover, the reduction of copper oxide (CuO) occurred during microwave irradiation, and the reduction behavior was different between irradiation by the microwave electric field and the magnetic field [17]. These studies suggest that the effects of plasma and microwave electric and magnetic fields itself can contribute to ironmaking reactions, and these effects result in low-temperature reactions and a shortening of the period of the carbothermal reduction reaction.

In this investigation, to clarify the effects—other than heating—of microwave irradiation, the behavior of the carbothermal reduction reaction was investigated. In particular, in order to verify the reduction effect of plasma arising from an electric field concentration, we used an experimental device capable of in situ visible image acquisition and emission spectroscopy during microwave irradiation.

2. Materials and Methods

In this study, we observed the plasma spectrum arising from a mixture of iron oxides and carbon during microwave irradiation. Hematite (99.9% purity, 0.3 μm, Kojundo Chemical Laboratory Co., Ltd., Saitama, Japan), maghemite (99% purity, 20–40 nm, Kojundo Chemical Laboratory Co., Ltd., Saitama, Japan), and magnetite (99.9% purity, <1 μm, Kojundo Chemical Laboratory Co., Ltd., Saitama, Japan) were used as iron oxides. Each powder was mixed with pure carbon (99.9% purity, 50 μm, Kojundo Chemical Laboratory Co., Ltd., Saitama, Japan), and 0.15 g of the mixture was compacted to a 6-mm diameter pellet. The molar retio between iron oxides and carbon was 2:3 when hematite and maghemite was used, and 1:2 when magnetite was used. Figure 1 shows a schematic view of the experimental setup.

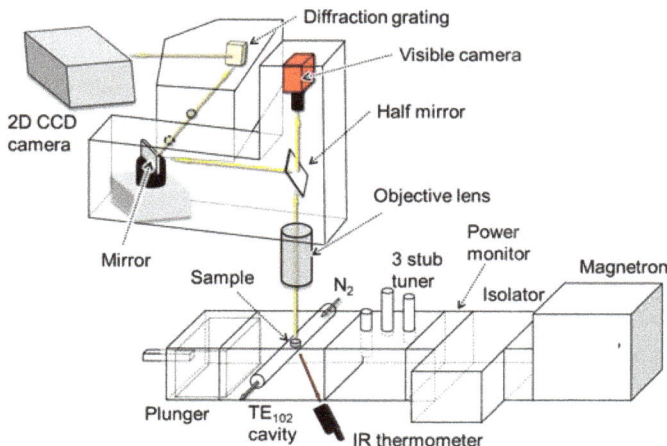

Figure 1. Schematic view of the experimental setup.

The pellet was placed in a quartz test tube, and a flow of N_2 gas at 0.1 L/min was used. Each pellet was placed at the maximum point of the E-field or H-field microwave intensity in a TE102 single-mode cavity. The sample was irradiated with microwaves by using a magnetron (IMH-20A259, IDX Co., Ltd., Tochigi, Japan, 2455 ± 15 MHz). An IR thermometer (FTK9-P300R, Japan Sensor Co., Tokyo, Japan) was used to measure the sample temperature from the side of the cavity ($\varepsilon = 0.95$). During microwave irradiation, visible images and emission spectra were obtained by using an integrated microscopic imaging spectrometer (IMIS, Bunkoukeiki Co., Ltd., Tokyo, Japan) [12]. These images and spectra were measured on the upside of the pellet. The spectrometric wavelength ranged from 308 nm to 492 nm, where the CN line spectra can be observed [18,19]. The measurement conditions of the emission spectra and photograph are listed in Table 1. Line spectra were assigned based on the literature [20–23]. After microwave irradiation, the pellet was ground, and the phases were analyzed by XRD.

Table 1. Measurement conditions of the emission spectra and photograph.

	Spectrometer	Visible Camera
Exposure time	0.5 s	0.5 s
Spatial resolution	4.9 μm^2	25 μm^2
Wavelength resolution	0.5 nm	-

3. Results and Discussions

3.1. Spectroscopic Measurements during Microwave Irradiation

Figure 2 shows the temperature and microwave power profiles of the hematite and carbon mixture irradiated with an E-field (a) and an H-field (b).

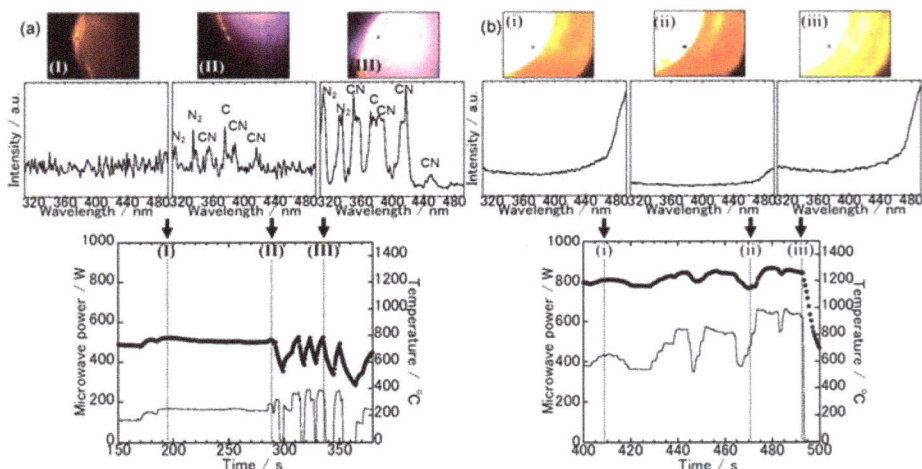

Figure 2. Temperature and microwave power profiles of the hematite and carbon mixture irradiated with an E-field (**a**) and an H-field (**b**).

In addition, the visible images and spectra at each time interval, which are indicated by broken lines in each temperature profile, are shown. The spectroscopic measurement points are indicated by cross marks in each visible image. During E-field irradiation, only part of the pellet was heated to a high temperature, and the maximum temperature was about 800 °C in the measurement area. On the other hand, high-temperature spots can be seen in photo (I). The spectroscopic spectrum of (I) was almost flat between 308–492 nm. However, the temperature of the bright spot was not high, because a continuous spectrum of Planck radiation was not observed in the higher wavelength range. From the spectroscopic result at point (II), weak line spectra corresponding to CN, N_2, and C were observed. However, as shown in the photograph of point (II), purple plasma was generated outside of the acquisition area of the visible image; thus, it is considered that the plasma did not affect the reduction reaction of the entire pellet. From the visible image and spectra at point (III), white-purple plasma was generated, which consisted of CN, N_2, and C. This plasma was concentrated only in the upper left part of the pellet. From the above results, although the area of plasma was limited to the upper left part of the pellet, it suggests that the excited molecules of CN and N_2 had some influence on the reduction reaction, in addition to the carbothermal reduction in the E-field processing of hematite.

To examine the relationship between the microwave power, sample temperature, and plasma discharge, the microwave power was increased and decreased several times between 300–350 s. This plasma was discharged when the microwave power reached about 250 W, and the temperature was over 800 °C. In the hematite and carbon system, the microwave E-fields only coupled with carbon, which absorbs the energy. In addition, carbon is a conductive material, and electric fields concentrated between the particles of carbon powder, leading to field electron emission from the carbon. Thus, sufficient electrons for plasma discharge were emitted by thermionic emission at around 800 °C, and by field electron emission over the electric fields of the 250 W microwaves.

In the H-field processing of the hematite–carbon mixture, as shown in the visible image of Figure 2b, the temperature of the whole pellet was increased, and became higher than that in the E-field. From the spectroscopic spectrum at (I), a continuous Planck radiation spectrum appeared in the high-wavelength range, confirming that the sample temperature was higher than that in the E-field. The temperature profiles and spectra at (II) and (III) indicated that the pellet temperature was related to the intensity of the continuous spectra. On the other hand, no line spectra were observed, indicating that excited molecules, such as CN*, were not generated. Therefore, in the H-field processing of the hematite–carbon system, the reduction reaction was dominant in the carbothermal reduction process.

The results for the maghemite–carbon case are shown in Figure 3.

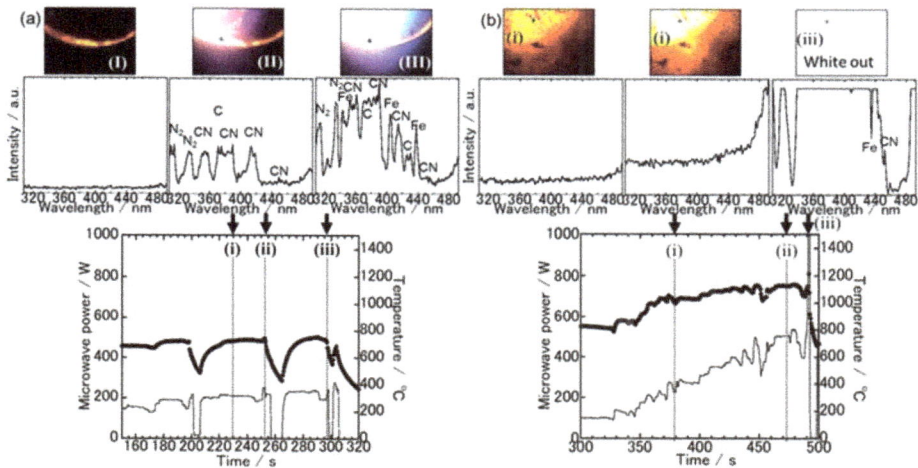

Figure 3. Temperature and microwave power profiles of maghemite and carbon mixture irradiated with an E-field (**a**) and an H-field (**b**). The visible images and spectroscopic spectra at each time, which are indicated by broken lines in each temperature profile, are shown.

Just as in the hematite case, in an E-field irradiation, spot heating occurred in the peripheral edge of the pellet, as shown in Figure 3a (I). The spectra at the hot spot were almost flat in the range from 308 nm to 492 nm, indicating that the spot temperature was not high. In the visible image at point (II), it can be seen that white-purple plasma appeared in the lower left part of the pellet. In addition, relatively strong line spectra were observed at the same time for CN, N_2, and C. At point (III), white blue plasma was generated in the same area as (II). The intensity of the line spectra was stronger than that at (II) as a whole, and the line spectra of Fe appeared in addition to CN, N_2, and C. This plasma was discharged when the microwave power exceeded 250 W, and the sample temperature measured by an IR thermometer exceeded 700 °C. Therefore, as with the hematite case, excited molecules such as CN* had some influence on the reduction reaction, as well as on the carbon thermal reduction reaction. Furthermore, because the line spectra of Fe were observed, the reduction of the surface of the pellet was estimated to have proceeded to a greater degree than in the hematite case.

The sample temperature increased upon irradiation in a magnetic field compared with irradiation in the electric field. From the visible image at point (I), it was found that the temperature of the pellet increased overall. A continuous spectrum was observed on the higher wavelength side, and it was confirmed that the sample temperature was higher than that when heating in an electric field. From the temperature and spectrum of point (II), it was found that the temperature was higher than that of point (III), and the pellet temperature and spectrum shape were related. In addition, from the respective spectral shapes, no emission line spectra were observed; thus, excited molecules of CN, N_2, etc., were not present. However, when the microwave power was increased to 600 W, the visible image was a whiteout, and the spectra were partially saturated. We found that when high-intensity microwaves were used, plasma was generated, even in a magnetic field. The spectral saturation was not confirmed between 460–480 nm, but it is inferred that the continuous spectrum is superimposed, and the sample temperature was high. Thus, in irradiation in a magnetic field, the reduction reaction is dominated by the carbothermal reduction process, but the plasma can be excited by increasing the microwave power intensity.

The results of the magnetite case are shown in Figure 4.

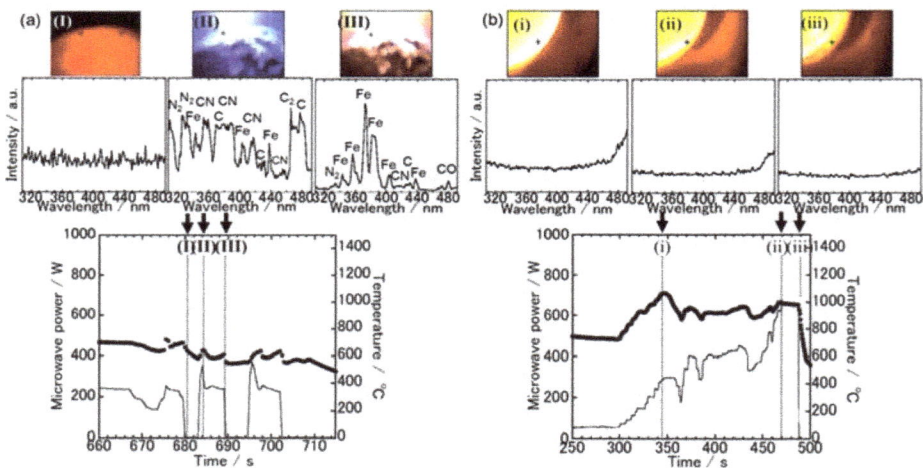

Figure 4. Temperature and microwave power profiles of the magnetite and carbon mixture irradiated with an E-field (**a**) and an H-field (**b**). The visible images and spectroscopic spectra at each time, which are indicated by broken lines in each temperature profile, are shown.

In an E-field irradiation, unlike the other case, the temperature of the sample increased relatively uniformly, as shown in the visible image of point (I). The spectrum at that point was almost linear in the range of 308 nm to 492 nm. Accordingly, the temperature at the cross point was under 1100 °C. At point (II), white-blue plasma was generated in the upper part of the pellet, and spectra corresponding to CN, N_2, C, and Fe were observed in the spectroscopic results at that time. On the other hand, at point (III), white-orange plasma was generated. This plasma consisted of mainly strong line spectra of Fe and CO, N_2, CN, and C plasma. This transition of the plasma color occurred from 685 s to 690 s, and the intensity of the Fe line spectra increased with increasing time. This result suggests that the reduction reaction proceeded. This plasma was ignited at microwave powers greater than 250 W and temperatures greater than 600 °C in the area of IR measurement, and the area was confined to the upper part of the pellet. Therefore, this result suggests that the plasma induced the reduction reaction, resulting in an increase in the intensity of the Fe line spectra with increasing plasma exposure time.

In H-field irradiation with the magnetite–carbon mixture, the sample temperature became higher than that irradiated by the electric field, but the temperature was lower than that in the cases of hematite and maghemite. From the visible image at point (I), it was observed that the temperature in the peripheral edge of the pellet was higher than that in the center of the pellet. In the spectrum, continuous spectra were observed on the high wavelength side, and the sample temperature was confirmed to be higher than that in an E-field. The intensity of the continuous spectrum was weaker than that of hematite and maghemite, which is consistent with the temperature profiles. However, after 350 s, the heating behaviour of the sample changed, and significant power was required to increase the sample temperature. From the temperature and spectra of points (II) and (III), the change of the sample temperature corresponded to the intensity of the spectra. In any case, line spectra were not observed, suggesting no generation of plasma.

3.2. The Sample Reduction State after Microwave Irradiation

From the spectroscopy results and the temperature profiles, it is considered that the reduction behavior differed with the microwave irradiation mode. Therefore, XRD results are shown in each irradiation mode, and not for each iron oxide.

Figure 5 shows the XRD results and photographs of the sample after irradiation with the E-field.

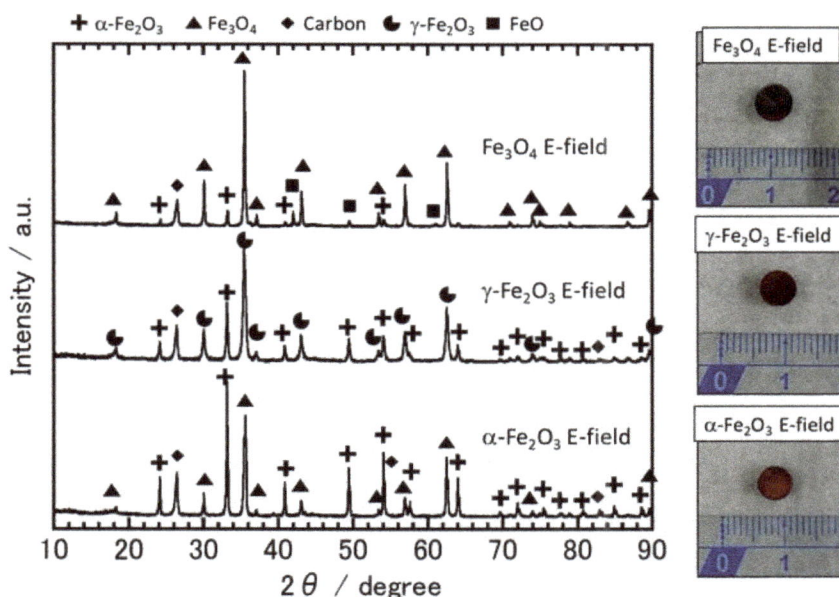

Figure 5. XRD results and photographs of the sample after irradiation with E-fields.

The hematite sample was reddish brown as a whole, and the pristine hematite and carbon remained present, as illustrated by the XRD result. This is because most of the pellet area was as low as

800 °C. On the other hand, as can be inferred from the visible image and the spectroscopic results shown in Figure 2a, the color of the upper left part of the pellet was partially discolored and turned black. A small amount of magnetite was observed in the XRD pattern, suggesting that the reduction reaction was driven by the thermal energy of the plasma or by excited C and CN molecules. The reduction of hematite with carbon or CN gas to produce magnetite can be represented by Equations (1) and (2).

$$6Fe_2O_3 + C \rightarrow 4Fe_3O_4 + CO_2, \tag{1}$$

$$6Fe_2O_3 + CN \rightarrow 4Fe_3O_4 + CO_2 + N_2, \tag{2}$$

At 927 °C, the $\Delta G \circ$ of Equations (1) and (2) are -144.24 and -152.66 kJ/mol, respectively [24]. In addition, it is considered that excited CN molecules can transfer their energy to other molecules or atoms, and the energy can be used for the reduction reaction. Therefore, the reduction reaction of hematite with CN gas would proceed easily.

The XRD results of maghemite indicated that the pristine sample remained and was partly oxidized to hematite. However, from the spectroscopic results of maghemite, as shown in Figure 3a at point (III), it is inferred that the sample was reduced. This is because it was hard to distinguish between trace amounts of magnetite and maghemite from the XRD results. In addition, because the area of plasma was limited to the lower left part of the pellet, maghemite was oxidized to hematite in the other areas.

The XRD results for magnetite illustrate that the sample remained unchanged: partially oxidized to hematite and partly reduced to wustite. From the photograph, the upper part of the pellet is discolored and gray. The lower part is reddish brown. The upper part corresponds to the area of the plasma discharge, having been reduced and converted to wustite. Conversely, the lower part became reddish brown, and it was thought to be oxidized. When plasma is discharged, microwaves tend to be absorbed or reflected by the plasma, resulting in a temperature decrease at the outside of the discharged area of the plasma. Therefore, in the lower part of the pellet, where plasma was not discharged, the sample temperature became relatively low, and oxidation proceeded.

Figure 6 shows the XRD results and photographs of the sample after irradiation with H-fields.

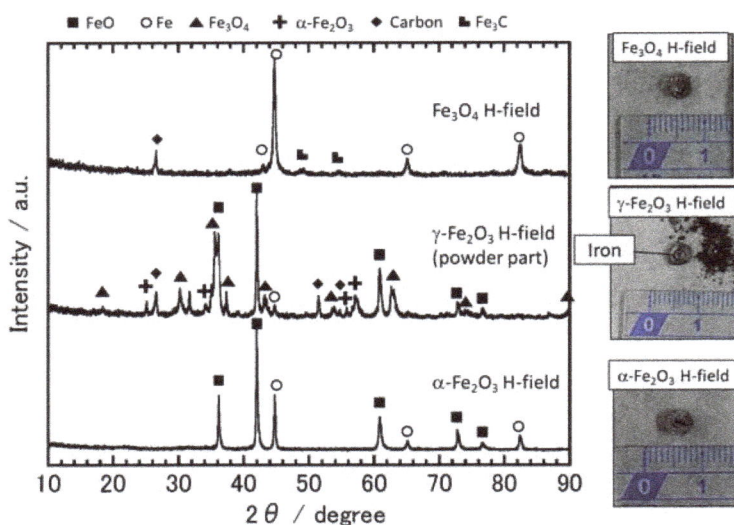

Figure 6. XRD results and photographs of the sample after irradiation with H-fields.

Hematite was reduced by irradiation with H-fields, leading to the formation of wustite and an iron phase. Residual carbon was not observed, despite the incomplete reduction reaction. On the other hand, from the XRD results for magnetite, it was observed that carbon remained, even though the magnetite had been completely reduced to iron. The reason is that when the hematite was irradiated with an H-field, the reaction temperature was relatively high: over 1200 °C. Ishizaki et al. reported that when the temperature is about 1000 °C, CO gas is produced from CO_2 by the Boudouard reaction [25]. Thus, in the H-field processing of hematite, the reaction between solid carbon and CO_2 gas proceeded well by the Boudouard reaction, and all of the solid carbon was consumed.

As shown in the photograph, the maghemite pellet was partially crushed during the heating process, and the crushed part was analyzed by XRD. The crushed powder contained hematite, magnetite, carbon, wustite, and iron. On the other hand, the pellet part was iron, suggesting that the reduction reaction proceeded sufficiently.

From the XRD results for magnetite, almost all of the sample was reduced to iron, although some residual carbon and a trace amount of Fe_3C was observed. In contrast to the hematite case, residual carbon was observed in the sample after irradiation, even though all of the magnetite had been completely reduced to iron. The results suggest that magnetite was reduced by the special effect of the microwave H-fields, as well as by the carbothermal reduction. Previous reports have shown that the reduction reactions of some transition-metal oxides, such as CuO and TiO_2, are promoted by microwave irradiation [17,26,27]. In particular, the reduction reaction of $NiMn_2O_4$, which has a similar crystal structure to magnetite, is promoted by microwave H-field irradiation, and the author emphasized that the promotion effect was derived from the interaction between the spins of the sample and the microwave H-field. Other papers have shown that the interaction between the spins and H-fields remain above the Curie temperature [28]. Therefore, it is suggested that microwave H-fields promoted the reduction of magnetite at temperatures greater than 1000 °C.

4. Conclusions

In situ emission spectroscopy during microwave processing was conducted using 2.45 GHz single-mode microwave irradiation furnace. Plasma discharge was observed during microwave E-field irradiation using the in situ spectroscopic measurement device. We believe that CN gas or excited CN molecules contributed to the reduction reaction of iron oxides, in addition to thermal reduction. On the other hand, no excited molecules were generated during microwave H-field irradiation. Thus, thermal reduction mainly occurred. Magnetite strongly interacts with the microwave H-field, suggesting that the reduction reaction of magnetite was promoted by microwave H-field irradiation, as well as by the thermal reduction reaction. We believe that the energy consumption for the formation of iron by microwave processing would be decreased by controlling the plasma discharge by changing the powder particle size, mixed state, and microwave irradiation condition.

Acknowledgments: This work was supported by a JSPS Grant-in-Aid for Scientific Research (S) No. JP17H06156.

Author Contributions: J.F. and H.T. conceived and designed the experiments; J.F. performed the experiments; J.F. analyzed the data; H.T. contributed reagents/materials/analysis tools; J.F. wrote the paper.

Conflicts of Interest: The authors declare no conflict of interest.

References

1. Ishizaki, K.; Nagata, K.; Hayashi, T. Localized Heating and Reduction of Magnetite Ore with Coal in Composite Pellets Using Microwave Irradiation. *ISIJ Int.* **2007**, *47*, 817–822. [CrossRef]
2. Kashimura, K.; Sato, M.; Hotta, M.; Agrawal, D.K.; Nagata, K.; Hayashi, M.; Mitani, T.; Shinohara, N. Iron production from Fe_3O_4 and graphite by applying 915 MHz microwaves. *Mater. Sci. Eng. A* **2012**, *556*, 977–979. [CrossRef]
3. Castro, E.R.D.; Moura, M.B.; Jermolovicius, Ã.L.A.; Takano, C. Carbothermal reduction of iron ore applying microwave energy. *Steel Res. Int.* **2012**, *83*, 131–138. [CrossRef]

4. Hara, K.; Hayashi, M.; Sato, M.; Nagata, K. Continuous Pig Iron Making by Microwave. *J. Microw. Power Electromagn. Energy* **2011**, *45*, 137–147. [CrossRef] [PubMed]
5. Hara, K.; Hayashi, M.; Sato, M.; Nagata, K. Pig Iron Making by Focused Microwave Beams with 20 kW at 2.45 GHz. *ISIJ Int.* **2012**, *52*, 2149–2157. [CrossRef]
6. Sabelström, N.; Hayashi, M.; Yokoyama, Y.; Watanabe, T.; Nagata, K. XRD In Situ Observation of Carbothermic Reduction of Magnetite Powder in Microwave Electric and Magnetic Fields. *Steel Res. Int.* **2013**, *84*, 975–981. [CrossRef]
7. Nagata, K.; Sato, M.; Hara, K.; Hotta, T.; Kitamura, Y.; Hayashi, M.; Kashimura, K.; Mitani, T.; Fukushima, J. Microwave Blast Furnace and Its Refractories. *J. Tech. Assoc. Refract. Jpn.* **2014**, *34*, 66–73.
8. Kashimura, K.; Nagata, K.; Sato, M. Concept of Furnace for Metal Refining by Microwave Heating—A Design of Microwave Smelting Furnace with Low CO_2 Emission. *Mater. Trans.* **2010**, *51*, 1847–1853. [CrossRef]
9. Ishizaki, K.; Stir, M.; Gozzo, F.; Catalá-Civera, J.M.; Vaucher, S.; Nicula, R. Magnetic microwave heating of magnetite-carbon black mixtures. *Mater. Chem. Phys.* **2012**, *134*, 1007–1012. [CrossRef]
10. Stir, M.; Ishizaki, K.; Vaucher, S.; Nicula, R. Mechanism and kinetics of the reduction of magnetite to iron during heating in a microwave E-field maximum. *J. Appl. Phys.* **2009**, *105*, 124901. [CrossRef]
11. Matsubara, A.; Takayama, S.; Okajima, S.; Sato, M. Evolution of the Near-UV Emission Spectrum Associated with the Reduction Process in Microwave Iron Making. *J. Microw. Power Electromagn. Energy* **2008**, *42*, 4–8. [CrossRef] [PubMed]
12. Matsubara, A.; Nakayama, K.; Okajima, S.; Sato, M. Microscopic and Spectroscopic Observations of Plasma Generation in the Microwave Heating of Powder Material. *Plasma Fusion Res.* **2010**, *5*, 041. [CrossRef]
13. Sonobe, T.; Mitani, T.; Shinohara, N.; Hachiya, K.; Yoshikawa, S. Plasma emission and surface reduction of titanium dioxides by microwave irradiation. *Jpn. J. Appl. Phys.* **2009**, *48*, 116003. [CrossRef]
14. Sabat, K.C.; Paramguru, R.K.; Mishra, B.K. Reduction of Oxide Mixtures of (Fe_2O_3 + CuO) and (Fe_2O_3 + Co_3O_4) by Low-Temperature Hydrogen Plasma. *Plasma Chem. Plasma Process.* **2017**, *37*, 979–995. [CrossRef]
15. Sabat, K.C.; Rajput, P.; Paramguru, R.K.; Bhoi, B.; Mishra, B.K. Reduction of Oxide Minerals by Hydrogen Plasma: An Overview. *Plasma Chem. Plasma Process.* **2014**, *34*, 1–23. [CrossRef]
16. Kitamura, T.; Shibata, K.; Takeda, K. In-fright Reduction of Fe_2O_3, Cr_2O_3, TiO_2 and Al_2O_3 by Ar-H_2 and Ar-CH_4 Plasma. *ISIJ Int.* **1993**, *33*, 1150–1158. [CrossRef]
17. Fukushima, J.; Kashimura, K.; Takayama, S.; Sato, M.; Sano, S.; Hayashi, Y.; Takizawa, H. In-situ kinetic study on non-thermal reduction reaction of CuO during microwave heating. *Mater. Lett.* **2013**, *91*, 252–254. [CrossRef]
18. Dong, M.; Lu, J.; Yao, S.; Zhong, Z.; Li, J.; Li, J.; Lu, W. Experimental study on the characteristics of molecular emission spectroscopy for the analysis of solid materials containing C and N. *Opt. Express* **2011**, *19*, 17021–17029. [CrossRef] [PubMed]
19. Wasowicz, T.J.; Kivimäki, A.; Coreno, M.; Zubek, M. Superexcited states in the vacuum-ultraviolet photofragmentation of isoxazole molecules. *J. Phys. B Atom. Mol. Opt. Phys.* **2012**, *45*, 205103. [CrossRef]
20. Pearse, R.W.B.; Gaydon, A.G. *The Identification of Molecular Spectra*, 3rd ed.; Chapman and Hall: London, UK, 1965.
21. Voevodin, A.A.; Jones, J.G.; Zabinski, J.S.; Hultman, L. Plasma characterization during laser ablation of graphite in nitrogen for the growth of fullerene-like CN_x films. *J. Appl. Phys.* **2002**, *92*, 724–735. [CrossRef]
22. Bourquard, F.; Maddi, C.; Donnet, C.; Loir, A.-S.; Barnier, V.; Wolski, K.; Garrelie, F. Effect of nitrogen surrounding gas and plasma assistance on nitrogen incorporation in a-C:N films by femtosecond pulsed laser deposition. *Appl. Surf. Sci.* **2016**, *374*, 104–111. [CrossRef]
23. Kraus, M.; Egli, W.; Haffner, K.; Eliasson, B.; Kogelschatz, U.; Wokaun, A. Investigation of mechanistic aspects of the catalytic CO_2 reforming of methane in a dielectric-barrier discharge using optical emission. *Phys. Chem. Chem. Phys.* **2002**, *4*, 668–675. [CrossRef]
24. Chase, M.W.J. (Ed.) *NIST-JANAF Thermochemical Tables*, 4th ed.; American Institute of Physics: College Park, MD, USA, 1998.
25. Ishizaki, K.; Nagata, K. Microwave Induced Solid—Solid Reactions between Fe_3O_4 and Carbon Black Powders. *ISIJ Int.* **2008**, *48*, 1159–1164. [CrossRef]
26. Fukushima, J.; Kashimura, K.; Sato, M. Chemical bond cleavage induced by electron heating Gas emission behavior of titanium-metalloid compounds (titanium nitride and oxide) in a microwave field. *Mater. Chem. Phys.* **2011**, *131*, 178–183. [CrossRef]

27. Fukushima, J.; Kashimura, K.; Takayama, S.; Sato, M. Microwave-energy Distribution for Reduction and Decrystallization of Titanium Oxides. *Chem. Lett.* **2012**, *41*, 39–41. [CrossRef]

28. Tanaka, M.; Kono, H.; Maruyama, K. Selective heating mechanism of magnetic metal oxides by a microwave magnetic field. *Phys. Rev. B* **2009**, *79*, 104420. [CrossRef]

Article

Kinetic Studies on Gas-Based Reduction of Vanadium Titano-Magnetite Pellet

Junwei Chen, Weibin Chen, Liang Mi, Yang Jiao and Xidong Wang *

Department of Energy and Resources Engineering, College of Engineering, Peking University, Beijing 100871, China; 1401111582@pku.edu.cn (J.C.); chenwb@pku.edu.cn (W.C.); liangmicoming@163.com (L.M.); jiaoyang518@foxmail.com (Y.J.)
* Correspondence: xidong@pku.edu.cn; Tel.: +86-10-8252-9083

Received: 5 January 2019; Accepted: 14 January 2019; Published: 16 January 2019

Abstract: Vanadium titano-magnetite (VTM) is a significant resource in China—analysis shows that China possesses approximately 10 billion tons of VTM. In this study, we characterize VTM's isothermal reduction mechanisms in the mixture of H_2, CO, and N_2 where the variables considered include reduction time, reduction temperature, gas composition, and pellet size. The kinetics of the reduction process were studied following a shrinking core model. The results indicate that the reduction degree of oxidized VTM pellets increases with increases of reduction time and reduction temperature but decreases with increasing pellet size. Moreover, we found that an increase of $H_2/(H_2 + CO)$ ratio induced an increase of the reduction degree. We discuss the transformation of main Ti-bearing mineral phases, and we consider the most probable reaction mechanism. For the entire reduction process, the kinetic results confirm the existence of an early and later stages that are controlled by interface chemical reaction and diffusion, respectively. Furthermore, the results show that the diffusion-control step can be observably shortened via decreased pellet size because a thinner product layer is formed during the reduction process. Our study thus provides a valuable technical basis for industrial applications of VTM.

Keywords: vanadium titano-magnetite; gas-based reduction; carbon monoxide; hydrogen; and nitrogen; kinetics; pellet size

1. Introduction

Vanadium titano-magnetite (VTM) is a type of multi-elements-coexistent mineral that contains iron (Fe), titanium (Ti), vanadium (V), and various rare metals [1–3]. VTM is becoming increasingly important because of its significant value in the high-tech industries [4–6]. According to statistics, more than 10 billion tons VTM resources are located in China. The abundant reserves ranks China third in the world following Russia and South Africa [7,8]. The VTM with TiO_2 and V_2O_5 is abundant in Panxi Area of China, where the amount of TiO_2 and V_2O_5 present account for more than 90% and 80%, respectively, of the total quantity of China [9–11]. Hence, it is important to utilize the VTM in the Panxi Area of China as the main supply of Ti and V. Especially, the extraction of valuable elements such as vanadium in VTM has attracted the attention of many scholars [12,13]. In particular, vanadium plays a major role in many fields such as biological and medicine fields [14]. The crystalline structure of VTM is complicated because the Ti and Fe are closely associated with each other in the ore, and V is as isomorphism in the lattice of VTM [15]. The key to utilizing the resource lies in how to separate the Fe, V and Ti efficiently. The processes involved in utilizing VTM fall into two categories: blast furnace (BF) processes and non-BF processes [16,17]. The BF technology is widely recognized because it has been employed for a long time in China and Russia. However, this process does have several drawbacks. First, the continual shortage of coke resource limits development of BF technology. In addition, the introduction of limestone as a solvent in blast furnaces decreases the grade of TiO_2, and the reaction of TiO$_2$ with limestone to produce perovskite leads to almost no recycling of titanium available in VTM. Currently, much research has been conducted on the comprehensive utilization of VTM by non-BF processes, and several technological processes for the utilization of VTM have already been developed. The pre-reduction electric furnace smelting process is the most promising processes due to the high recovery rate of valuable elements and low cost [15].

Reduction is an essential procedure in the pre-reduction electric furnace smelting process. Hence, it is extremely important to investigate the reduction behavior and kinetics of VTM. In recent decades, a large number of studies have been carried out to investigate reduction behavior and kinetics of VTM. Most previous research on the reduction behavior and kinetics of VTM was mainly focused on the process that used coal as the reducing agent [12,13,18–22]. However, the rate of the reduction reaction is relatively slow in coal-based reduction process. In addition, with political pressure focused on environmental protection, gas-based reduction of VTM is becoming increasingly prevalent. Compared with coal-based reduction, gas-based reduction has a number of advantages, such as higher reduction degree, better processing capacity as well as less pollution [12,13,23–26]. Although a few papers have dealt with the gas-based reduction of VTM, there are two important problems in the previous research on the gas-based reduction behavior and kinetics of VTM. Firstly, many reduction experiments were based on the use of pure hydrogen (H$_2$) or pure carbon monoxide (CO). Nevertheless, most of the reaction is between the VTM and a gas mixture of CO, H$_2$ and N$_2$ in the actual industrial process. Second, the previous studies did not consider the particle size of oxidized VTM pellets; this can play an important role in the reduction kinetics. Therefore, previous experiments were not carried out by simulating the actual reduction gas, and thus they are problematic for guiding the industrial application of VTM.

In this paper, the gas-based reduction kinetics of oxidized VTM pellets was examined systematically. The effects of reduction time, reduction temperature, gas composition of different H$_2$/(H$_2$ + CO) ratios with 25% N$_2$, and pellet size on the VTM reduction kinetics were studied using unreacted nuclear shrinkage model. Our study is expected to provide a more valuable technical basis for industrial application of VTM.

2. Materials and Methods

2.1. Preparation of Oxidized VTM Pellets

The VTM sample used in this study was obtained from the Panzhihua and Xichang areas of Sichuan Province in China. Sample crystal phase of sample was characterized by X-ray powder diffraction (XRD; Bruker D8 ADVANCE, Germany, CuKα radiation, 40 kV, 40 mA). The XRD patterns of the VTM are shown in Figure 1a. As can be seen from Figure 1a, the main mineral phases of the VTM are magnetite (Fe$_3$O$_4$) and Ilmenite (FeTiO$_3$). Elements Ti and Fe are closely associated to each other in the mineral phase.

The oxidized VTM pellets were prepared as follows. Then the VTM was fully mixed with 1 wt % binder and 8.5 wt % water and then pelletized to the diameters of 6–8 mm, 8–10 mm, 10–12 mm, 12–14 mm, and 14–16 mm in a disc pelletizer (Hong Xing Company, Shaoxing, China). Then, the VTM pellets were loaded into a quartz reactor after being dried at 110 °C for 4 h. Finally, the VTM pellets were calcined at 1350 °C for 20 min under the air conditions. The main chemical compositions are displayed in Table 1, while the XRD patterns of the oxidized VTM pellets are shown in Figure 1b. It can be seen that the main mineral phases of oxidized VTM pellets were hematite (Fe$_2$O$_3$) and pseudobrookite (Fe$_2$TiO$_5$). It can be concluded that Fe$_3$O$_4$ and FeTiO$_3$ transformed to hematite Fe$_2$O$_3$ and Fe$_2$TiO$_5$ during the pre-oxidation process.

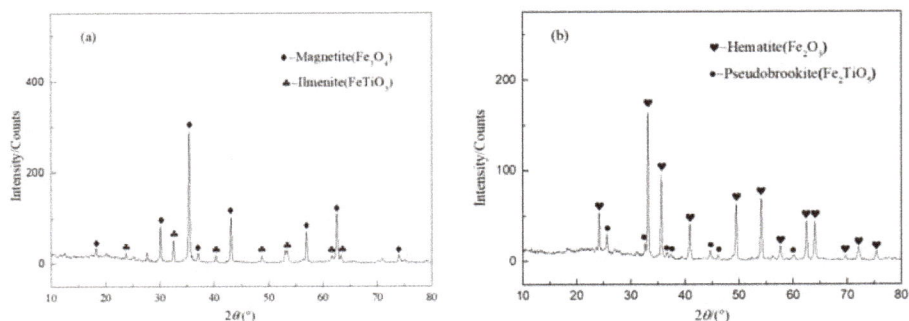

Figure 1. XRD patterns of the sample: (**a**) the VTM; (**b**) the oxidized VTM pellets.

Table 1. Main chemical compositions of oxidized VTM pellets (wt %).

TFe	FeO	TiO$_2$	SiO$_2$	Al$_2$O$_3$	MgO	CaO	V$_2$O$_5$	MnO	S
45.50	0.59	13.40	8.42	6.54	3.28	1.40	0.54	0.32	0.037

2.2. Experimental Measurements

In this study, 200 g of heated oxidized VTM pellets directly react with reducing gas in the furnace. The experimental setup is shown in Figure 2. The equipment (Beijing Hengjiu Experimental Equipment Co., Ltd., Beijing, China) consisted of a reducing gas flow control cabinet, a tube furnace, a high temperature resistant reactor made of steel alloy, a coal gas analyzer (3012H Automatic Smoke and Gas analyzer" produced in China), an electronic balance and a computer.

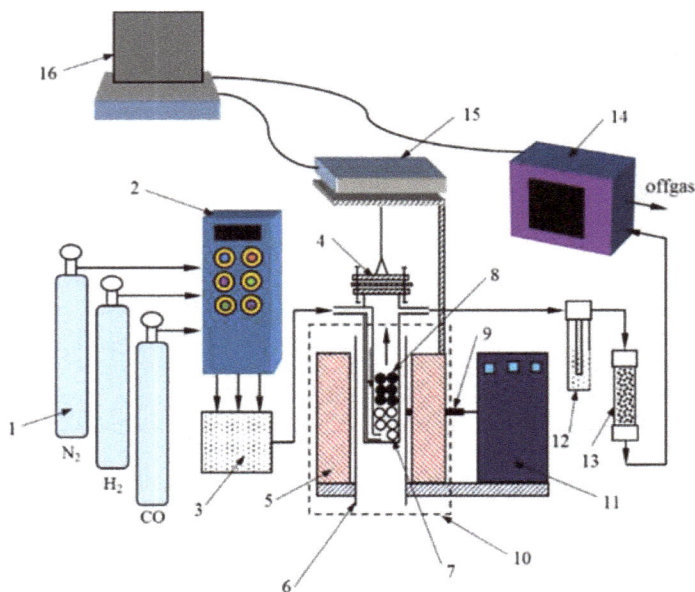

Figure 2. Schematic of the experimental apparatus: 1—gas cylinder, 2—flow control cabinet, 3—gas mixing chamber, 4—reactor, 5—fever zone, 6—alundum tube, 7—corundum ball, 8—oxidized VTM pellets, 9—thermocouple, 10—tube furnace, 11—temperature controlling cabinet, 12—wash bottle, 13—drying bottle, 14—coal gas analyzer, 15—electronic balance, 16—computer.

Firstly, 400 g corundum balls with average diameter of 1 cm were placed at the bottom of the reactor to disperse the reducing gases, and the oxidized VTM pellets were laid evenly above the corundum balls. A gasket made of silica gel was placed between the flanges to improve the air tightness of the reactor. Then the blind flange of the reactor was then closed. The inlet pipe and the exit pipe were connected with the reactor, and the reactor was filled with N_2 to check the air tightness. Upon satisfactory air tightness, the reactor was suspended on the bottom of the electronic balance placed above the tube furnace. The tube furnace was turned on, and the temperature program was as the heat was applied. During the entire heating process, the reactor was filled with N_2. When the temperature reached a specified level, the electronic balance and the coal gas analyzer were turned on and the data were recorded. The N_2 was replaced by the reducing gas to carry out the reduction. Afterward, the reducing gas was switched for N_2. Finally, the sample was taken from furnace the after cooling.

H_2, CO and N_2 controlled by flow-controlled cabinet were fully mixed in the mixing chamber in the reduction reaction, and the reaction pressure was 1 atm. The gas mixture entered the bottom of the reactor through the inlet pipe along the inner wall of reactor. Afterward, the reducing gas mixture met with the oxidized VTM pellets. The mass change was continuously recorded by the electronic balance and stored in the computer. It was critical to completely remove the dust and vapor in the offgas to avoid damaging the coal-gas analyzer. Thus, the offgas firstly entered into the water-wash bottle followed by a drying bottle. Later, we measured the chemical components of offgas online using the coal-gas analyzer. Finally, the off-gas was emptied. After the samples being reduced for 240 min, the VTM pellets were cooled to room temperature under an N_2 atmosphere for later testing.

The reduction degree of oxidized VTM pellets can be calculated as follows [15]:

$$R = \frac{m(O_L)}{m_0(O_L)} = \left(\frac{0.11w(FeO)}{0.43w(TFe)} + \frac{m_1 - m_2}{m_1 \times 0.43w(TFe)} \right) \times 100\% \qquad (1)$$

In which R is reduction degree of iron, $m_0(O_L)$ is the total weight of O which bonds with Fe in oxidized VTM pellets, $m(O_L)$ is the weight loss of O that bonds with Fe during reduction, $w(TFe)$ and $w(FeO)$ are the mass fraction of TFe and FeO respectively, in oxidized VTM pellets, m_1 is the mass of the VTM pellet before reduction, and m_2 is the mass of the VTM pellet in reduction process. The parameters 0.11 and 0.43 represent the oxygen demand conversion coefficients when converting FeO and Fe to Fe_2O_3, respectively.

2.3. Reduction Kinetics Analysis

Based on previous research [15,16], it can be concluded that the reduction of the oxidized VTM pellets proceed topochemically. Therefore, the kinetic model used to describe the iron ore reduction was the un-reacted shrinking core model. This model includes the processes of external diffusion of gaseous species, intrinsic chemical reactions, and the diffusion of gas species. Under the condition of the mixture of H_2, CO and N_2, the total reduction time can be calculated according to the Equation (2), which is based on the un-reacted shrinking core model. Therefore, the rate equations can be obtained as shown in Equations (3) and (4). According to Equations (3) and (4), multiple graphs can be drawn to investigate the actual rate-controlling step of the reduction process. If the process is controlled by the intrinsic chemical reaction, the plot of $1 - (1 - R)^{\frac{1}{3}}$ vs. time should be a straight line. For the diffusion of gas species controlling step, the plot of $1 - \frac{2}{3}R - (1 - R)^{\frac{2}{3}}$ vs. time should be a straight line. In Equation (2), Equation (3) and Equation (4), t is the total reduction time (min), k is the reduction rate constant (cm/min), r_0 is characteristic initial radius of the pellet (cm), ρ_0 is initial oxygen concentration in the pellet (mol/cm^3), k_0 is the parameter constant, c_0 and c_q are reduction gas concentration at granule surface and in equilibrium respectively (mol/cm^3), and D_e is the effective diffusion coefficient (cm^2/min).

$$t = \frac{r_0 \rho_0}{k_0 (c_0 - c_q)} \left[1 - (1 - R)^{\frac{1}{3}} \right] + \frac{r_0^2 \rho_0}{D_e (c_0 - c_q)} \left[1 - \frac{2}{3} R - (1 - R)^{\frac{2}{3}} \right] \tag{2}$$

$$kt = 1 - (1 - R)^{\frac{1}{3}} \tag{3}$$

$$kt = 1 - \frac{2}{3} R - (1 - R)^{\frac{2}{3}} \tag{4}$$

2.4. Characterization

Sample crystal phase of sample was characterized by X-ray powder diffraction (XRD; Bruker D8, CuKα radiation, 40 kV, 40 mA).

3. Results and Discussion

3.1. Reduction Temperature

The reduction of VTM pellets was carried out in the temperature range of 973 K–1373 K with the intervals of 100 K. Other specific experimental conditions were controlled: The total gas flow was 5 L·min^{-1}, the proportion of N_2 was 25%, the proportion of H_2 and CO was 75% (the $H_2/(H_2 + CO)$ = 1/2), and the diameter of oxidized VTM pellets was 10–12 mm. The experimental results are shown in Figure 3.

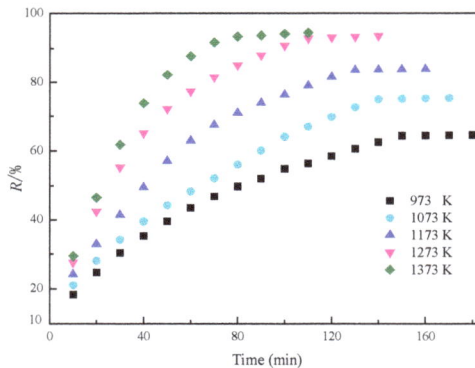

Figure 3. The reduction degree curves of oxidized VTM pellets at 973 K–1373 K in $H_2/(H_2 + CO)$ = 1/2 atmosphere.

It can be seen that the reduction degree of oxidized VTM pellets gradually increased with the reduction time. The reduction percentages were 49.8%, 56.0%, 71.0%, 85.0% and 93.3% at 973 K, 1073 K, 1173 K, 1273 K and 1373 K, respectively, for 80 min in $H_2/(H_2 + CO)$ = 1/2 atmosphere, indicating that the reaction temperature can also strongly influence the reduction rate of oxidized VTM pellets. The reduction degree can be elevated with the increase of temperature because crystal lattice has a higher energy in higher temperature. In addition, H_2 and CO have higher reduction thermodynamic potential energy as the temperature increases. Thus, the reduction temperature of oxidized VTM pellets is recommended to exceed 1273 K in actual industrial production.

Figure 4 shows the XRD patterns of the VTM pellets after the reduction process. The results indicate that Fe can be produced at 973 K–1373 K. In addition, the Ti-bearing main mineral phase changes with the increase in temperature. The phase of Fe_2TiO_5 disappears, and the phase of the reduction product is Fe_2TiO_4 at 973 K. As the temperature increases, Fe_2TiO_4 is reduced to $FeTiO_3$.

With the temperature further increasing to 1173 K, $FeTiO_3$ is reduced to $FeTi_2O_5$. Therefore, the most probable reaction mechanism of the reduction process can be described in Equations (5)–(16).

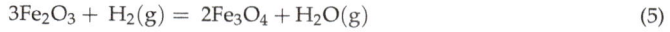

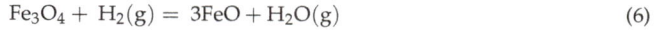

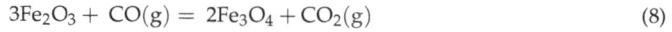

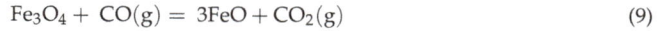

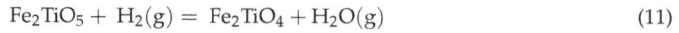

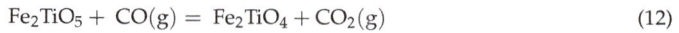

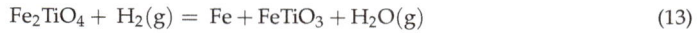

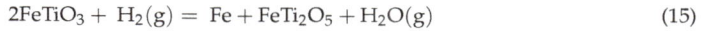

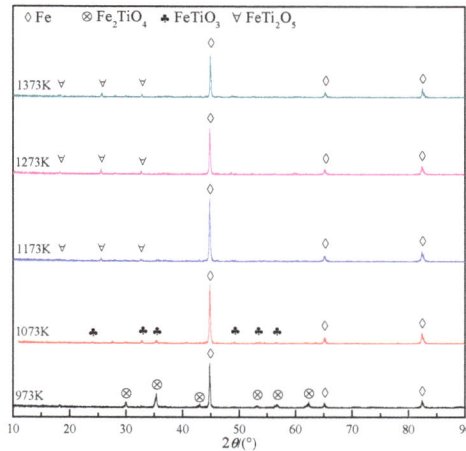

$$3Fe_2O_3 + H_2(g) = 2Fe_3O_4 + H_2O(g) \tag{5}$$

$$Fe_3O_4 + H_2(g) = 3FeO + H_2O(g) \tag{6}$$

$$FeO + H_2(g) = Fe + H_2O(g) \tag{7}$$

$$3Fe_2O_3 + CO(g) = 2Fe_3O_4 + CO_2(g) \tag{8}$$

$$Fe_3O_4 + CO(g) = 3FeO + CO_2(g) \tag{9}$$

$$FeO + CO(g) = Fe + CO_2(g) \tag{10}$$

$$Fe_2TiO_5 + H_2(g) = Fe_2TiO_4 + H_2O(g) \tag{11}$$

$$Fe_2TiO_5 + CO(g) = Fe_2TiO_4 + CO_2(g) \tag{12}$$

$$Fe_2TiO_4 + H_2(g) = Fe + FeTiO_3 + H_2O(g) \tag{13}$$

$$Fe_2TiO_4 + CO(g) = Fe + FeTiO_3 + CO_2(g) \tag{14}$$

$$2FeTiO_3 + H_2(g) = Fe + FeTi_2O_5 + H_2O(g) \tag{15}$$

$$2FeTiO_3 + CO(g) = Fe + FeTi_2O_5 + CO_2(g) \tag{16}$$

Figure 4. XRD patterns of the reduction products at 973 K–1373 K in $H_2/(H_2 + CO) = 1/2$ atmosphere.

On the basis of Equations (3) and (4), kinetic results of the VTM pellets are shown in Figure 5a,b. The figure clearly shows that there are two stages in the whole reduction process: an early stage and a later stage. The correlation coefficients (R^2) of the straight lines are 0.982, 0.998, 0.983, 0.985 and 0.986 at the early stage. The correlation coefficient (R^2) of the straight lines are 0.999, 0.998, 0.985, 0.993 and 0.997 at the later stage. Thus, the strong linear relationships indicate that the reduction processes of the early stage and the later stage are controlled by interface chemical reaction and diffusion, respectively. The values of the reduction rate constant of two stages are presented in Table 2. The reduction rate constants of both the early stage and the later stage increase with increasing temperature. This indicates that an increase in temperature can effectively improve the reduction process.

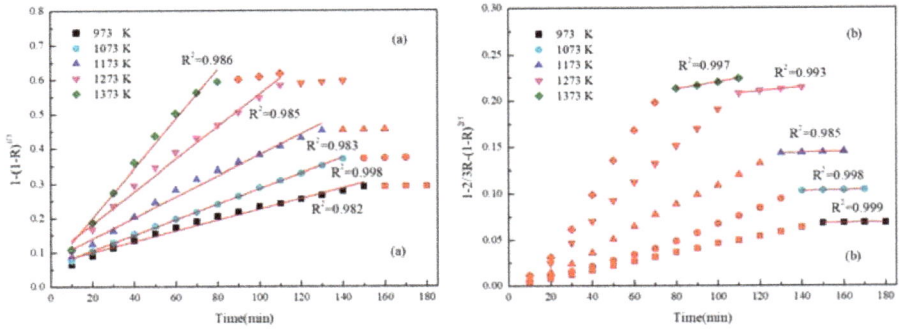

Figure 5. Plot of $1 - (1 - R)^{\frac{1}{3}}$ vs. time (**a**) and plot of $1 - \frac{2}{3}R - (1 - R)^{\frac{2}{3}}$ vs. time (**b**) at 973 K–1373 K in $H_2/(H_2 + CO) = 1/2$ atmosphere.

Table 2. The values of the reduction rate constant at 973 K–1373 K in $H_2/(H_2 + CO) = 1/2$ atmosphere.

Temperature	973 K	1073 K	1173 K	1273 K	1373 K
Intrinsic Chemical Reaction Control	0.00156	0.00224	0.00302	0.00469	0.00717
Diffusion Control	1.34×10^{-5}	3.88×10^{-5}	6.15×10^{-5}	2.13×10^{-4}	3.70×10^{-4}

According to the Arrhenius equation, the relationship between the reaction temperature (T) and the reduction rate constant (k) can be obtained as shown in Equation (17). In Equation (17), E is the activation energy (kJ·mol^{-1}), k_0 is the frequency factor.

$$k = k_0 \exp\left(-\frac{E}{RT}\right) \tag{17}$$

Taking the natural logarithm of both sides in Equation (17):

$$\ln k = -\frac{E}{RT} + \ln k_0 \tag{18}$$

The plot of ln$k$ vs. $1/T$ in $H_2/(H_2 + CO) = 1/2$ atmosphere is shown in Figure 6a,b. The activation energies of intrinsic chemical reaction control stage and diffusion control stage can be evaluated based on the data of Figure 6 and Equation (18), and the calculated values are 41.65 kJ/mol and 92.45 kJ/mol, respectively. Therefore, the value of activation energy in intrinsic chemical reaction control stage is lower than that in diffusion control stage for gas-based reduction of oxidized VTM pellets.

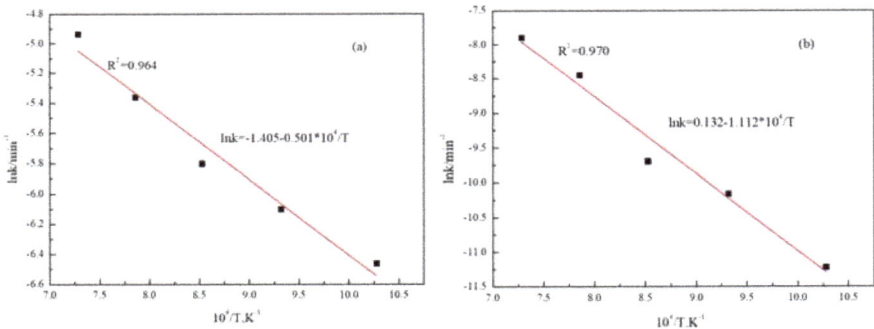

Figure 6. The Arrhenius plot of reaction degree constant k vs. temperature (**a**) intrinsic chemical reaction control step and (**b**) diffusion control step at 973 K–1373 K in $H_2/(H_2 + CO) = 1/2$ atmosphere.

3.2. Gas Composition

To investigate the effect of gas composition on the reduction of oxidized VTM pellets, the experiment was carried out under the $H_2/(H_2 + CO)$ ratios of 0, 1/4, 1/3, 1/2, 2/3, 3/4, and 1. Other specific experimental conditions were controlled as follows: The reduction temperature was 1273 K, the total gas flow was 3 L·min^{-1}, the proportion of N_2 was 25%, the proportion of H_2 and CO was 75%, the oxidized VTM pellets diameter was 10–12 mm. The experimental results are shown in Figure 7. For reducing under the condition of $H_2/(H_2 + CO) = 0$ for 150 min, the reduction degree of VTM pellets was only 72.6%. As the ratio of $H_2/(H_2 + CO)$ increased to 1, the reduction degree of VTM pellets increases to 92.1%. This shows that the reduction degree of VTM pellets increases with the increasing $H_2/(H_2 + CO)$ ratio. Previous research has confirmed that H_2 has higher reduction capacity and utilization than CO at high temperature ($T > 1084$ K) [27,28].

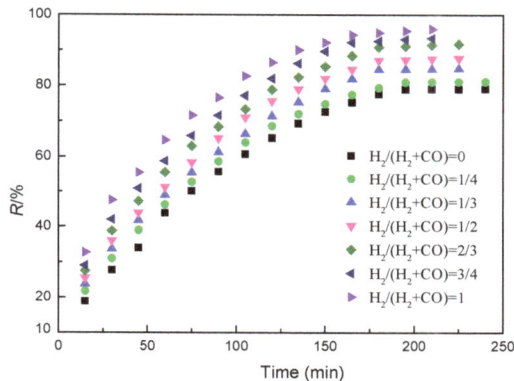

Figure 7. The reduction degree curves of oxidized VTM pellets at 1273 K with $H_2/(H_2 + CO)$ ratios.

Figure 8 presents the XDR patterns of the reduction VTM pellets products that were obtained after being reduced by the reduction gas with different $H_2/(H_2 + CO)$ ratios at 1273 K for 240 min. The pattern indicates that when the proportion of $H_2/(H_2 + CO)$ is small, the phase of Fe_2TiO_5 disappears, and the main phase of the reduction product is $FeTiO_3$. The most probable reaction mechanism is given by Equations (11–14). As the ratio of $H_2/(H_2 + CO)$ increases to 2, $FeTiO_3$ is reduced to $FeTi_2O_5$. The most probable reactions are Equation (15), and Equation (16).

Figure 9a,b show the plot of $1 - (1 - R)^{\frac{1}{3}}$ vs. time and plot of $1 - \frac{2}{3}R - (1 - R)^{\frac{2}{3}}$ vs. time, respectively, at 1273 K in different $H_2/(H_2 + CO)$ ratios. There are also both the early stage and the later stage, these are controlled by interface chemical reaction and by diffusion, respectively. The plots show that the diffusion-control step was significantly shortened as the ratio of $H_2/(CO + H_2)$ increased. Therefore, the kinetic conditions of reduction can be improved at 1273 K by increasing the $H_2/(H_2 + CO)$ ratio. The hydrogen-rich reduction gas is definitely beneficial to the reduction of oxidized VTM pellets in the actual industrial production.

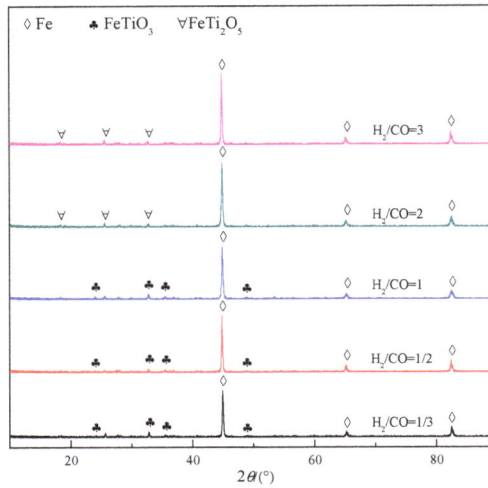

Figure 8. XRD patterns of reduction products at 1273 K with $H_2/(H_2 + CO)$ ratios.

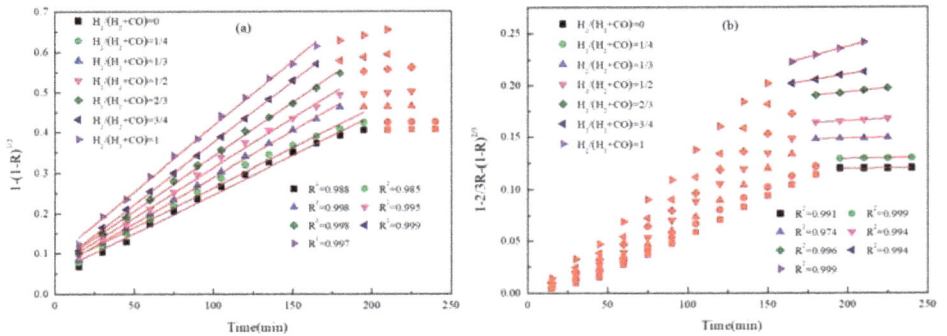

Figure 9. The reduction degree curves of oxidized VTM pellets at 1273 K with $H_2/(H_2 + CO)$ ratios. (**a**) the plot of $1 - (1 - R)^{\frac{1}{3}}$ vs. time and (**b**) plot of $1 - \frac{2}{3}R - (1 - R)^{\frac{2}{3}}$ vs. time.

3.3. Pellet Size

To investigate the effect of pellet size on the reduction of oxidized VTM pellets, the reduction experiments were carried out with the oxidized VTM pellets having diameters of 6–8 mm, 8–10 mm, 10–12 mm, 12–14 mm, and 14–16 mm. Other specific experimental conditions were as follows: The reduction temperature was 1273 K, the total gas flow was 5 L·min^{-1}, the proportion of N_2 was 25%, the proportion of H_2 and CO was 75% (the $H_2/(H_2 + CO) = 1/2$).

As the pellet size changed from 6–8 mm to 14–16 mm, the resulting reduction degree of the oxidized VTM pellets is shown in Figure 10. It can be seen that the pellet size of oxidized VTM pellets has a strong influence on the reduction degree. The reduction degree increased with the decreasing pellet size.

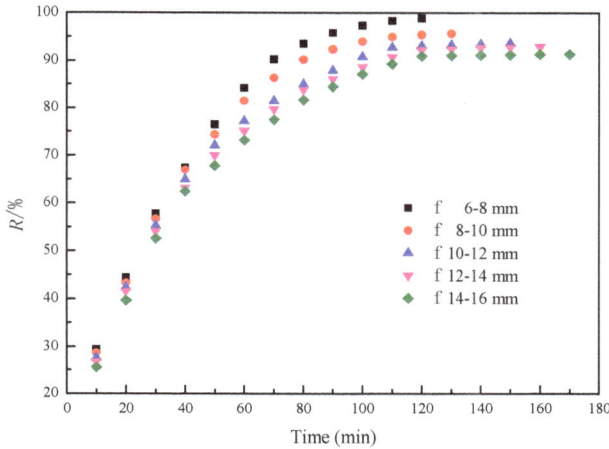

Figure 10. The reduction degree curves of oxidized VTM pellets at 1273 K with different oxidized VTM pellets diameter in $H_2/(H_2 + CO) = 1/2$ atmosphere.

Figure 11 shows the XRD patterns of the reduction VTM pellets products that were obtained at 1273 K for 240 min in $H_2/(H_2 + CO) = 1/2$ atmosphere. It can be seen that the phase of $FeTiO_3$ can be transferred into that of Fe_2TiO_5 with the decrease in pellet diameter. The most probable reaction mechanism is described in Equations (11–14). Until the pellet size decreases to 10–12 mm, $FeTiO_3$ is reduced to $FeTi_2O_5$. The most probable reactions are Equation (15), and Equation (16). With the continuous decrease of pellet size, the diffractive peaks of $FeTi_2O_5$ become weaker and the peak metallic iron getting stronger, resulting in the enhancement of reduction degree.

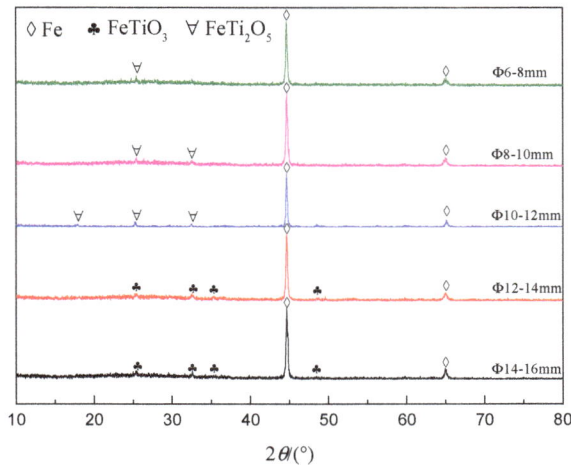

Figure 11. The reduction degree curves of oxidized VTM pellets at 1273 K with different oxidized VTM pellets diameter in $H_2/(H_2 + CO) = 1/2$ atmosphere.

Figure 12a,b show the Plots of $1 - (1 - R)^{\frac{1}{3}}$ and $1 - \frac{2}{3}R - (1 - R)^{\frac{2}{3}}$ vs. time with different diameters of oxidized VTM pellets. It is obvious that early stage and the later stage also exists during the reduction process, these are controlled by interface chemical reaction and diffusion, respectively. It is can be seen that the pellet size has strong influence on the later stage. The thickness of the product can be decreased with decreasing pellet size during the reduction process. The diffusion degree of gas decreased quickly with the increasing thickness of product layer. This result shows that the diffusion-control step can be shortened via decreasing of the pellet size. Therefore, it is important to select the appropriate pellet size in order to optimize the reduction process. From Figure 12, the values of reaction rate constant can be obtained, and these are listed in Table 3. The results indicate that the kinetic conditions of reduction can be improved at 1273 K in $H_2/(H_2 + CO) = 1/2$ atmosphere via a decrease in the pellet size.

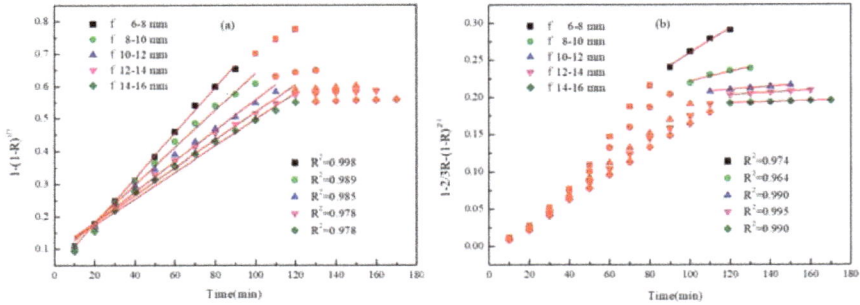

Figure 12. Plot of $1 - (1 - R)^{\frac{1}{3}}$ vs. time (**a**) and plot of $1 - \frac{2}{3}R - (1 - R)^{\frac{2}{3}}$ vs. time (**b**) at 1273 K with different oxidized VTM pellets diameter in $H_2/(H_2 + CO) = 1/2$ atmosphere.

Table 3. The values of the reduction rate constant at 1273 K with different oxidized VTM pellets diameter.

Pellets Diameter	6–8 mm	8–10 mm	10–12 mm	12–14 mm	14–16 mm
Intrinsic Chemical Reaction Control	0.00694	0.0057	0.00469	0.00423	0.00405
Diffusion Control	1.66×10^{-3}	6.35×10^{-4}	2.12×10^{-4}	1.58×10^{-4}	7.64×10^{-5}

4. Conclusions

Gas-based reduction of oxidized VTM pellets in the mixture of H_2, CO, and N_2 was systematically investigated at 973 K–1373 K. The reduction degree increased with increasing reduction time, reduction temperature, and $H_2/(H_2+CO)$ ratios, and decreasing pellet size. The Ti-bearing main mineral phase transformation of oxidized VTM pellets in $H_2/(H_2 + CO) = 1/2$ atmosphere is described as $Fe_2TiO_5 \rightarrow Fe_2TiO_4 \rightarrow FeTiO_3 \rightarrow FeTi_2O_5$. The Ti-bearing main mineral phase transformation of the reduction VTM pellets products at 1273 K is described as $Fe_2TiO_5 \rightarrow FeTiO_3 \rightarrow FeTi_2O_5$ with the increase of $H_2/(H_2 + CO)$ ratios and the decrease of pellet size. The most probable reaction mechanism is provided based on the reduction process. The kinetics of oxidized VTM pellet in the reduction process is successfully modeled as a shrinking unreacted-core. For the entire reduction process, the kinetics study indicated that there are an early stage and a later stage that are controlled by interface chemical reaction and diffusion, respectively. In addition, the diffusion-control step can be observably shortened via decreasing the pellet size because the thickness of the product layer becomes thinner in the reduction process.

Author Contributions: Conceptualization, X.W.; data curation, L.M.; formal analysis, J.C.; funding acquisition, X.W.; investigation, J.C. and W.C.; methodology, J.C. and Y.J.; resources, X.W.; software, X.W.; supervision, X.W.; visualization, J.C.; writing—original draft, J.C.; writing—review & editing, W.C.

Funding: This work was supported by the ministry of land and resources public welfare industry research project (201511062-02).

Conflicts of Interest: The authors declare no conflict of interest.

References

1. Alfantazi, A.M.; Moskalyk, R.R. Processing of indium: A review. *Miner. Eng.* **2003**, *16*, 687–694. [CrossRef]
2. Zhou, L.H.; Zeng, F.H. Reduction mechanisms of vanadium–titanomagnetite–non-coking coal mixed pellet. *Ironmak. Steelmak.* **2014**, *38*, 59–64. [CrossRef]
3. Zhang, J.L.; Xing, X.D.; Cao, M.M.; Jiao, K.X.; Wang, C.L.; Ren, S. Reduction kinetics of vanadium titano-magnetite carbon composite pellets adding catalysts under high temperature. *J. Iron. Steel Res. Int.* **2013**, *20*, 1–7. [CrossRef]
4. Guo, L.; Yu, J.T.; Tang, J.K.; Lin, Y.H.; Guo, Z.C.; Tang, H.Q. Influence of coating MgO on sticking and functional mechanism during fluidized bed reduction of vanadium titano-magnetite. *J. Iron. Steel Res. Int.* **2015**, *22*, 464–472. [CrossRef]
5. Bai, Y.Q.; Cheng, S.S.; Bai, Y.M. Analysis of Vanadium-bearing titanomagnetite sintering process by dissection of sintering bed. *J. Iron Steel Res. Int.* **2011**, *18*, 8–15. [CrossRef]
6. Lv, X.; Lun, Z.; Yin, J.; Bai, C. Carbothermic reduction of vanadium titanomagnetite by microwave irradiation and smelting behavior. *ISIJ Int.* **2013**, *53*, 1115–1119. [CrossRef]
7. Zheng, F.; Chen, F.; Guo, Y.; Jiang, T.; Travyanov, A.Y.; Qiu, G. Kinetics of hydrochloric acid leaching of titanium from titanium-bearing electric furnace slag. *JOM* **2016**, *68*, 1–9. [CrossRef]
8. Sui, Y.L.; Guo, Y.F.; Jiang, T.; Qiu, G.Z. Sticking behaviour of vanadium titano-magnetite oxidised pellets during gas-based reduction and its prevention. *Ironmak. Steelmak.* **2016**, *44*, 185–192. [CrossRef]
9. Chen, D.S.; Song, B.; Wang, L.N.; Qi, T.; Wang, Y.; Wang, W.J. Solid state reduction of panzhihua titanomagnetite concentrates with pulverized coal. *Miner. Eng.* **2011**, *24*, 864–869. [CrossRef]
10. Guo, Y.F.; Gao, Y.; Tao, J.; Qiu, G.Z. Solid-state reduction behavior of panzhihua ilmenite. *J. Cent. South Univ.* **2010**, *41*, 1639–1644.
11. Cao, M.M.; Zhang, J.L.; Xing, X.D.; Wang, C.L.; Bai, Y.N.; Wen, Y.C. Reduction mechanism of vanadium titano-magnetite carbon composite pellets. *Iron Steel* **2012**, *47*, 5–12. [CrossRef]
12. Chen, D.S.; Zhao, H.; Hu, G.P.; Qi, T.; Yu, H.D.; Zhang, F.Z.; Wang, L.N.; Wang, W.J. An extraction process to recover vanadium from low-grade vanadium-bearing titanomagnetite. *J. Hazard. Mater.* **2015**, *294*, 35–40. [CrossRef]
13. Tang, J.; Chu, M.S.; Ying, Z.W.; Li, F.; Feng, C.; Liu, Z.G. Non-isothermal gas-based direct reduction behavior of high chromium vanadium-titanium magnetite pellets and the melting separation of metallized pellets. *Metals* **2017**, *7*, 153. [CrossRef]
14. Tripathi, D.; Mani, V.; Pal, R.P. Vanadium in biosphere and its role in biological processes. *Biol. Trace Elem. Res.* **2018**, *186*, 52–67. [CrossRef] [PubMed]
15. Sui, Y.L.; Guo, Y.F.; Jiang, T.; Qiu, G.Z. Reduction kinetics of oxidized vanadium titano-magnetite pellets using carbon monoxide and hydrogen. *J. Alloy. Compd.* **2017**, *706*, 546–553. [CrossRef]
16. Sui, Y.L.; Guo, Y.F.; Jiang, T.; Xie, X.L.; Wang, S.; Zheng, F.Q. Gas-based reduction of vanadium titano-magnetite concentrate: Behavior and mechanisms. *Int. J. Miner. Metall. Mater.* **2017**, *24*, 10–17. [CrossRef]
17. Zhu, Z.; Zhang, W.; Cheng, C.Y. A synergistic solvent extraction system for separating copper from iron in high chloride concentration solutions. *Hydrometallurgy* **2012**, *113–114*, 155–159. [CrossRef]
18. Liu, S.S.; Guo, Y.F.; Qiu, G.Z.; Jiang, T.; Chen, F. Preparation of Ti-rich material from titanium slag by activation roasting followed by acid leaching. *Trans. Nonferrous Met. Soc. China* **2013**, *23*, 1174–1178. [CrossRef]
19. Tang, J.; Chu, M.S.; Feng, C.; Tang, Y.T.; Liu, Z.G. Melting separation behavior and mechanism of high-chromium vanadium-bearing titanomagnetite metallized pellet got from gas-based direct reduction. *ISIJ Int.* **2016**, *56*, 210–219. [CrossRef]
20. Mehdizadeh, A.M.; Klausner, J.F.; Barde, A.; Mei, R. Enhancement of thermochemical hydrogen production using an iron–silica magnetically stabilized porous structure. *Int. J. Hydrogen Energy* **2012**, *37*, 8954–8963. [CrossRef]
21. Piotrowski, K.; Mondal, K.; Lorethova, H.; Stonawski, L.; Szymański, T.; Wiltowski, T. Effect of gas composition on the kinetics of iron oxide reduction in a hydrogen production process. *Int. J. Hydrogen Energy* **2005**, *30*, 1543–1554. [CrossRef]

22. Zhao, W.; Chu, M.; Wang, H.; Liu, Z.; Tang, J.; Ying, Z. Volumetric shrinkage characteristics and kinetics analysis of vanadium titanomagnetite carbon composite hot briquette during isothermal reduction. *ISIJ Int.* **2018**, *58*, 823–832. [CrossRef]

23. Guo, D.; Hu, M.; Pu, C.; Xiao, B.; Hu, Z.; Liu, S. Kinetics and mechanisms of direct reduction of iron ore-biomass composite pellets with hydrogen gas. *Int. J. Hydrogen Energy* **2015**, *40*, 4733–4740. [CrossRef]

24. Huitu, K.; Helle, M.; Helle, H.; Kekkonen, M.; Saxén, H. Optimization of midrex direct reduced iron use in ore-based steelmaking. *Steel Res. Int.* **2015**, *86*, 456–465. [CrossRef]

25. Kromhout, J.A.; Ludlow, V.; Mckay, S.; Normanton, A.S.; Thalhammer, M.; Ors, F.; Cimarelli, T. Physical properties of mould powders for slab casting. *Ironmak. Steelmak.* **2002**, *29*, 191–193. [CrossRef]

26. Li, W.; Fu, G.Q.; Chu, M.S.; Zhu, M.Y. Oxidation induration process and kinetics of hongge vanadium titanium-bearing magnetite pellets. *Ironmak. Steelmak.* **2016**, *44*, 294–303. [CrossRef]

27. Sun, H.Y.; Dong, X.J.; She, X.F.; Xue, Q.G.; Wang, J.S. Reduction mechanism of titanomagnetite concentrate by carbon monoxide. *J. Min. Metall.* **2013**, *49*, 263–270. [CrossRef]

28. Huang, Z.C.; Ling-Yun, Y.I.; Hu, P.; Jiang, T. Effects of roast temperature on properties of oxide pellets and its gas-based direct reduction. *J. Cent. South Univ.* **2012**, *43*, 2889–2895.

metals

MDPI

Article

Enhancing the Reduction of High-Aluminum Iron Ore by Synergistic Reducing with High-Manganese Iron Ore

Xianlin Zhou [1,2,3]🄳, Yanhong Luo [1,3,]*, Tiejun Chen [1,3] and Deqing Zhu [2,]*

[1]   Hubei Key Laboratory for Efficient Utilization and Agglomeration of Metallurgic Mineral Resources, Wuhan University of Science and Technology, Wuhan 430081, China; xlzhou@wust.edu.cn (X.Z.); chentiejun@wust.edu.cn (T.C.)

[2]   School of Mineral Processing and Bioengineering, Central South University, Changsha 410083, China

[3]   School of Resource and Environmental Engineering, Wuhan University of Science and Technology, Wuhan 430081, China

*   Correspondence: yhluo@wust.edu.cn (Y.L.); dqzhu@csu.edu.cn (D.Z.); Tel.: +86-027-6886-2204 (Y.L.); +86-731-8883-6942 (D.Z.)

Received: 29 November 2018; Accepted: 19 December 2018; Published: 22 December 2018

Abstract: How to utilize low grade complex iron resources is an issue that has attracted much attention due to the continuous and huge consumption of iron ores in China. High-aluminum iron ore is a refractory resource and is difficult to upgrade by separating iron and alumina. An innovative technology involving synergistic reducing and synergistic smelting a high-aluminum iron ore containing 41.92% Fe_{total}, 13.74% Al_2O_3, and 13.96% SiO_2 with a high-manganese iron ore assaying 9.24% Mn_{total} is proposed. The synergistic reduction process is presented and its enhancing mechanism is discussed. The results show that the generation of hercynite ($FeAl_2O_4$) and fayalite (Fe_2SiO_4) leads to a low metallization degree of 66.49% of the high-aluminum iron ore. Over 90% of the metallization degree is obtained by synergistic reducing with 60% of the high-manganese iron ore. The mechanism of synergistic reduction can be described as follows: MnO from the high-manganese ore chemically combines with Fe_2SiO_4 and $FeAl_2O_4$ to generate Mn_2SiO_4, $MnAl_2O_4$ and FeO, resulting in higher activity of FeO, which can be reduced to Fe in a CO atmosphere. The main products of the synergistic reduction process consist of Fe, Mn_2SiO_4, and $MnAl_2O_4$.

Keywords: high-aluminum iron ore; synergistic reduction; high-manganese iron ore; hercynite; fayalite

---

1. Introduction

In 2017, 87.47% of 1229 million tons of iron ore consumed by the Chinese iron steel industry were imported [1]. With the increasing consumption of good-quality iron ores, the poor, fine, and complex domestic iron ore resources cannot meet the huge demand of the iron and steel industry in China. Thus, it is important to utilize low grade iron resources efficiently [2], such as high-aluminum iron resources including high-aluminum limonite and red mud, which is a residue generated after the clarification of bauxite [3].

High-aluminum iron ore is a typical refractory iron resource, which is difficult to upgrade by physical processes due to the superfine size and close dissemination of iron minerals with gangue minerals [4]. Some were directly used as a sintering raw material at a low ratio, which affected the sintering process adversely [5]. Therefore, many separating approaches have been published for high-aluminum iron resources, which can be classified as: (1) physical processes like gravity concentration and magnetic separation [6,7] and flotation [8], (2) pyrometallurgical processes containing solid-state reduction [4,9–12], and smelting [12–15]. So far, iron and aluminum cannot be separated completely by physical processes and flotation processes, and more than 10% of sodium additive is required by the solid-state sodium roasting process, leading to a high risk of the degradation of refractory materials in furnaces [16] and higher costs. In the smelting process, iron ore was mixed with coal and binder to make briquettes and then smelted in a melter to separate the iron and slag over 1550 °C [17], where iron and aluminum can be isolated entirely. Nevertheless, Al_2O_3 affects the viscosity and desulfurizing capacity of blast furnace slag a lot [18], causing a limit usage of the high-aluminum resources in a blast furnace. In order to elevate the ratio of high-aluminum iron resources in the smelting process, a novel technology of synergistic reducing and synergistic smelting the high-aluminum iron ore with a high-manganese iron ore is proposed, and the synergistic reduction process is presented in this paper to discuss its enhancing mechanism.

2. Materials and Methods

2.1. Raw Materials

The chemical analysis of a high-aluminum iron ore (HA ore) and a high-manganese iron ore (HM ore) are given in Table 1. The two iron ores are both low iron grade, and the HA ore contains high contents of alumina and silica, while high manganese content was observed in the HM ore, assaying 9.24%. XRD results in Figure 1 illustrate that iron minerals in the two ores consist of hematite and goethite, the aluminum and silica minerals are kaolinite and gibbsite, and the manganese mineral in the HM ore is pyrolusite. Former results showed that hematite in the HA ore is closely included with kaolinite [19]. Figure 2 indicates that iron mineral is surrounded by kaolinite, and that pyrolusite and kaolinite are closely associated with each other. Complex mineral compositions and microstructures in the two iron ores led to significant difficulties in the separation of iron and gangue.

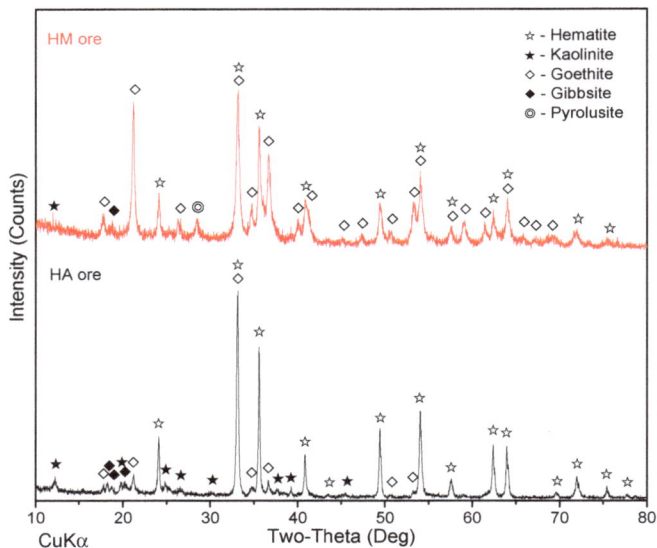

Figure 1. X-ray diffraction (XRD) pattern of high-aluminum iron ore (HA ore) and high-maganese iron ore (HM ore).

A soft coal was used as the reducer during the tests, which contained 52.12% fixed carbon on an air dry basis (FCad), 30.41% volatile matter on a dry ash free (Vdaf) basis, 4.49% ash on an air dry basis (Aad), 0.58% S and a melting temperature of 1376 °C. The size distribution of the soft coal is 100%, passing at 5 mm.

Table 1. Chemical compositions of raw materials (wt. %).

Ores	Fe_{total}	Mn_{total}	Al_2O_3	SiO_2	CaO	MgO	Pb	Zn	P	S	LOI
High-aluminum iron ore (HA ore)	41.92	1.24	13.74	13.96	0.13	0.88	0.64	0.21	0.130	0.014	7.20
High-manganese iron ore (HM ore)	42.32	9.24	6.60	4.22	0.20	0.20	1.86	0.98	0.065	0.018	11.05

Note: LOI, loss on ignition.

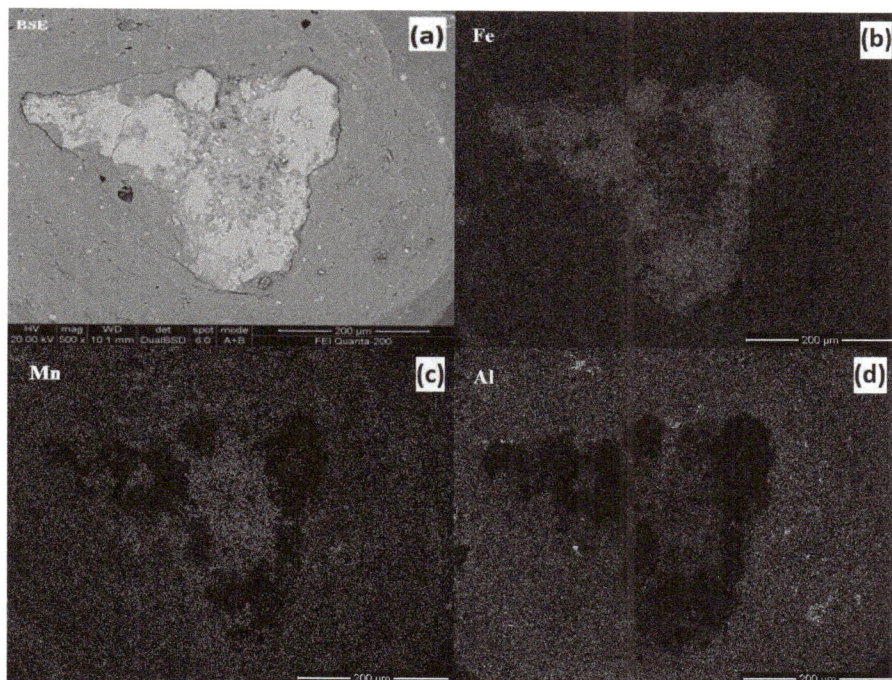

Figure 2. Representative SEM-BSE micrographs of HM ore. (BSE in (**a**), Back Scattered Electron Imaging; Fe element in (**b**); Mn element in (**c**); Al element in (**d**)).

2.2. Experimental Procedures

The reduction process includes procedures as follows: mixing the two iron ores at a given ratio, pelletizing of the mixture, and reduction roasting of dried pellets. Different experimental conditions of the reduction tests are summarized in Table 2.

Mixtures were prepared by mixing the two iron ores under different ratios, where the fraction of added HM ore is referred to the mixture of two ores. Then, green balls were made by balling the mixtures in a disc pelletizer of 0.8 m in diameter and a 0.2 m rim depth, rotating at 38 rpm and being inclined at 47° to the horizontal. The screened green balls of 8–16 mm were loaded into the drying oven to dry at 105 °C for 2 h until the weight was unchanged.

The dried pellets were put into a stainless steel crucible and covered by some soft coal, where the mass of soft coal was determined by the C/Fe mass ratio, which was calculated on the whole available iron content of the pellets. The crucible was loaded into a vertical furnace diagramed in Figure 3 (model: SK-8-13, The Great Wall Furnace, Changsha, China) and roasted for a given reduction time while the reducing temperature was elevated to the target value. After that, the reduced pellets were unloaded and cooled down by covering with pulverized coal.

Table 2. Different experimental conditions of the reduction tests.

Series No.	High-Aluminum Iron Ore: High-Maganese Iron Ore (HA:HM) Ratio	C/Fe Mass Ratio	Temperature	Reduction Duration
1	100:0	1.5	800 °C, 850 °C, 900 °C, 950 °C, 1000 °C, 1050 °C, 1100 °C	60 min
2	100:0	1.5	1050 °C	15 min, 30 min, 45 min, 60 min, 75 min, 90 min, 120 min
3	100:0	0.5, 1.0, 1.5, 2.0	1050 °C	90 min
4	100:0, 80:20, 60:40, 40:60, 20:80, 0:100	1.5	1050 °C	90 min

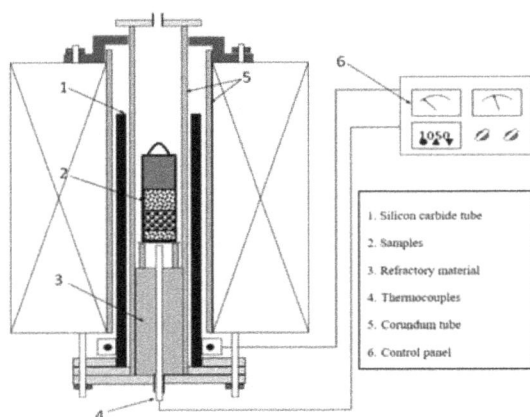

Figure 3. Schematic of vertical furnace for reduction.

X-ray fluorescence spectroscopy (XRF, PANalytical Axios mAX, PANalytical B.V., Almelo, The Netherlands) and chemical analysis were applied for the chemical compositions of raw materials and the reduced pellets. The crystalline phase compositions of the materials were detected by an X-ray diffractometer (XRD, D/Max-2500, Rigaku Co., Tokyo, Japan). Proximate analysis of coal was conducted by the Chinese standards GB/T212-2008. Microstructures of raw materials were observed by a scanning electron microscope (SEM, FEI Quanta-200, FEI Company, GG Eindhoven, The Netherlands) and an optical microscope (DMI4500P, Leica, Wetzlar, Germany), respectively. The compositional analyses were carried out using an energy dispersion system (EDAX-TSL, Ametek Inc., Paoli, CO, USA) within the SEM. Microstructures of reduced pellets were observed by an optical microscope. Metallization degree was applied to evaluate the reduction results, where the metallization degree was determined by the ratio between metallic iron on total iron of the reduced pellets according to ISO 11258:2015.

3. Results

3.1. Effect of Reduction Temperature

The single HA ore pellets were reduced by optimizing the reduction parameters, to reveal the reducibility of HA ore. The effect of reduction temperature on the metallization degree of HA reduced pellets is shown in Figure 4 under reducing for 60 min with C/Fe mass ratio of 1.5. It was discovered that the metallization degree rose at first, then decreased, and the peak appeared at 1050 °C. Only 53.68% of the best value of the metallization degree was obtained, declaring a weak reducibility of the HA ore.

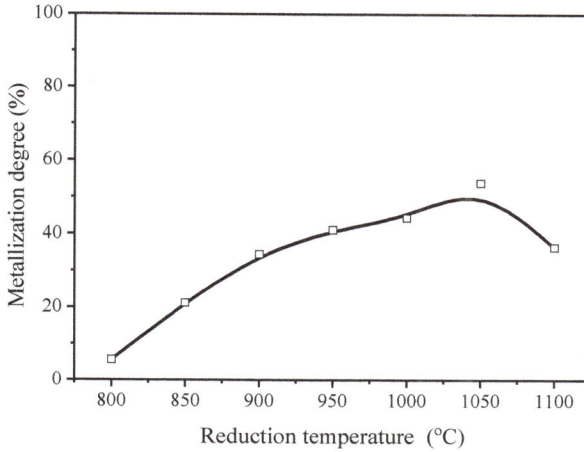

Figure 4. Effect of reduction temperature on metallization degree of reduced pellets of high-aluminum iron ore (HA ore) (reducing for 60 min with C/Fe mass ratio of 1.5).

3.2. Effect of Reduction Duration

Figure 5 presents the reduction duration effect result on the metallization of reduced pellets. By prolong the reduction from 15 min to 30 min, an obvious improvement from 41.86% to 55.29% is achieved. Generally, a subtle enhanced trend is found. However, still low metallization degrees are shown in Figure 5. Only 66.49% of the metallization degree is obtained by reducing the HA pellets for 120 min, which is far lower than that of high grade iron concentrate pellets [20].

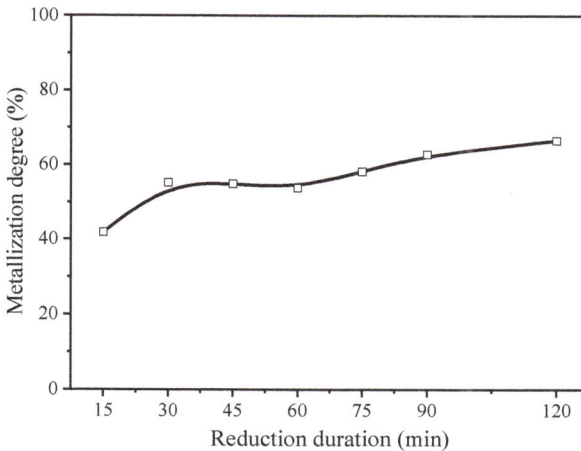

Figure 5. Effect of reduction duration on metallization of reduced pellets of high-aluminum iron ore (HA ore) (reducing at 1050 °C with C/Fe mass ratio of 1.5).

3.3. Effect of Reductant Ratio

As shown in Figure 6, the metallization degree of reduced pellets of HA ore increases slightly and then remains steady when the C/Fe mass ratio is elevated from 0.5 to 2.0. However, only part of the iron was reduced to metallic iron, where the best value only 62.72% is obtained.

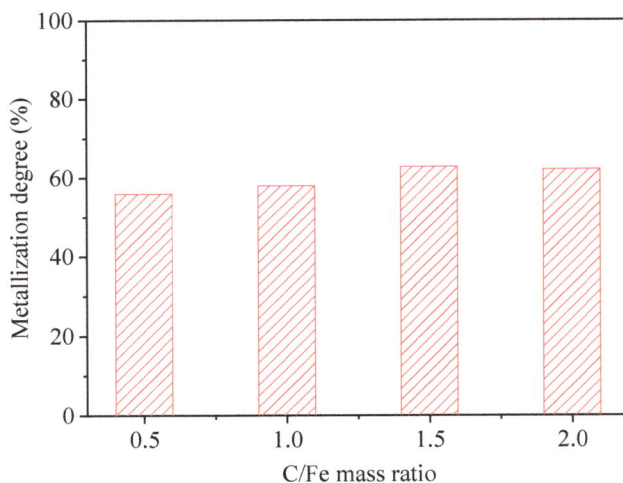

Figure 6. Effect of C/Fe mass ratio on metallization of reduced pellets of high-aluminum iron ore (HA ore) (reducing at 1050 °C for 90 min).

3.4. Effect of HM Ore Ratio

From the above optimization results of the reduction process, it can be concluded that the HA ore has poor reducibility. Thus, the HM ore was blended with the HA ore to investigate the effect of HM ore on the reduction behavior of HA ore. It can be observed from Figure 7 that the metallization degree increases continuously by raising the HM ore ratio. Meanwhile, the reduction of pellets is promoted by prolonging the reduction duration. The distributions of metallic iron grains in Figure 8 signify that more obvious iron grains form by blending with more HM ore.

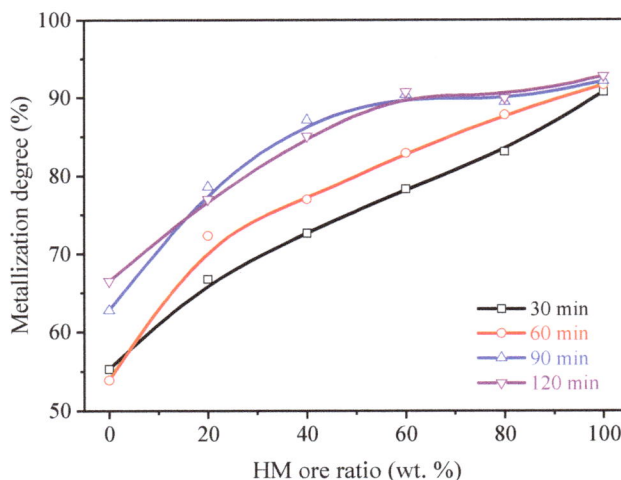

Figure 7. Effect of ratio of high-manganese iron ore (HM ore) in blend on metallization degree of reduced pellets (reducing at 1050 °C with C/Fe mass ratio of 1.5).

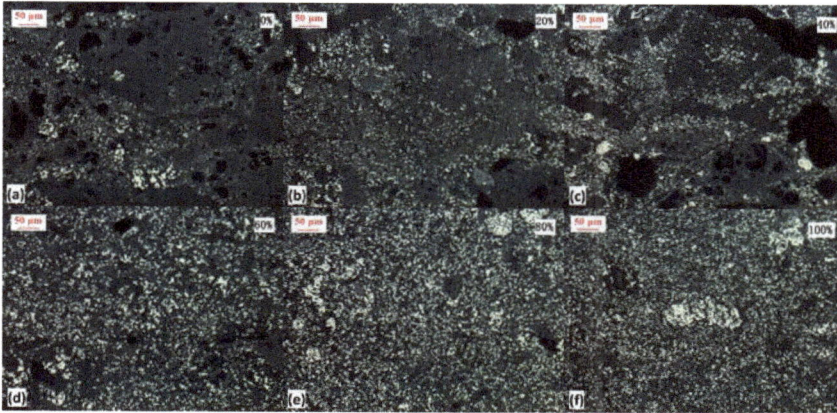

Figure 8. Effect of high-manganese iron ore (HM ore) ratio from 0% to 100% (**a–f**) on distribution of metallic iron grains (white color) in reduced pellets (reducing at 1050 °C for 90 min with C/Fe mass ratio of 1.5).

4. Discussion

XRD results of HA reduced pellets under different temperatures are illustrated in Figure 9. Note that fayalite can easily form during the reduction of HA ore. Meanwhile, the diffraction peak intensities of hercynite increase sharply with ascending temperatures, while those of fayalite weaken. According to the thermodynamic criterion, both fayalite and hercynite could not be reduced by CO when the temperature ranged from 800 °C to 1100 °C, causing low metallization of HA reduced pellets. Moreover, weaker peak intensity of iron and stronger of fayalite were detected at 1100 °C. Dynamically, the reduction from FeO to Fe is the restrictive step of reduction of iron ore [21]. Thus, there is more probability of forming fayalite by the chemical combination of FeO and SiO_2 during the reduction process, leading to a decrease of the metallization degree of the reduced HA pellets at 1100 °C.

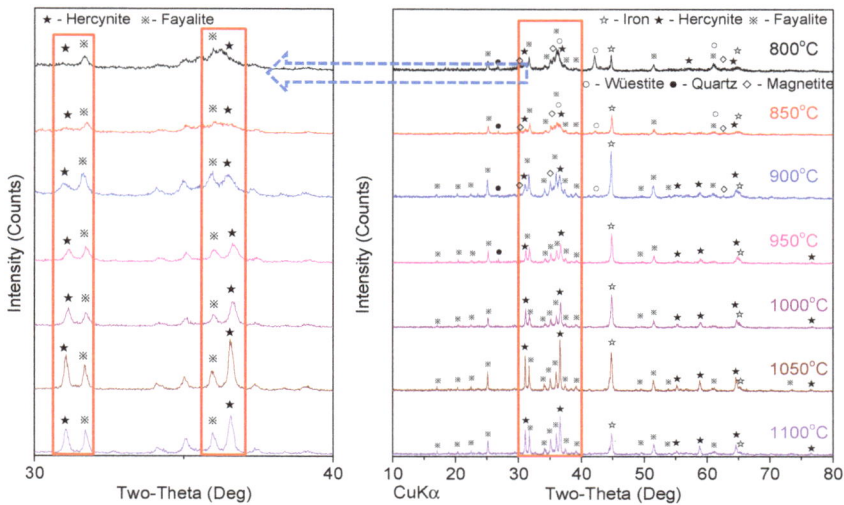

Figure 9. Effect of reduction temperature on mineral compositions of high-aluminum iron ore (HA) reduced pellets (reducing for 60 min with C/Fe mass ratio of 1.5).

A thermodynamic simulation of a system based on the compositions of the HA ore was processed by FactSage7.1 (Thermfact/CRCT, Montreal, QC, Canada; GTT-Technologies, Herzogenrath, Germany), and the result is depicted in Figure 10 as phase fraction as a function of temperature. The simulation is based on reaching equilibrium at every temperature, and only solid phases and gaseous (not shown) are considered. The Fe_2O_3 is already reduced to Fe at 800 °C when the simulation starts. However, Fe_2SiO_4 and $FeAl_2O_4$ also appear. $Fe_2Al_4Si_5O_{18}$ appears at 850 °C and the content increases. The largest variation in concentration is at 975 °C, when $Mn_2Al_4Si_5O_{18}$ totally disappear to the benefit of $Fe_2Al_4Si_5O_{18}$ and $Mn_3Al_2Si_3O_{12}$. According to the simulation result, less than 50% of Fe_2O_3 was reduced to Fe by CO between 800 °C to 1100 °C, while other iron was composed of Fe_2SiO_4, $FeAl_2O_4$ and $Fe_2Al_4Si_5O_{18}$. This indicates that it is difficult to further achieve a better metallization degree only by an optimization of a combination of time, temperature, and C/Fe ratio.

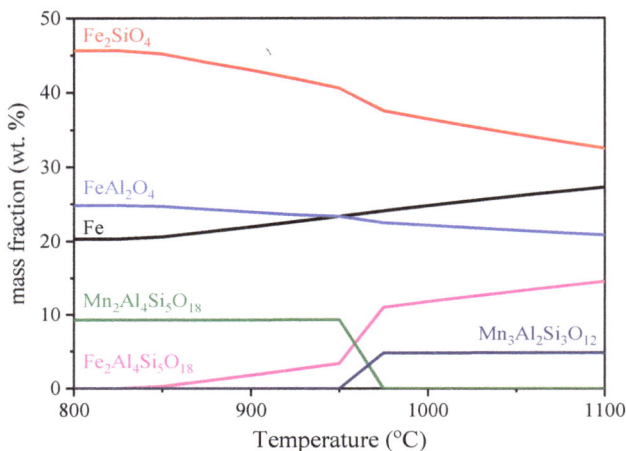

Figure 10. Thermodynamic simulation of a system consisting of Fe_2O_3, CO, SiO_2, Al_2O_3 and MnO carried out between 800 °C and 1100 °C in CO atmosphere at ambient pressure.

Combining the experimental results and simulation results, the mechanism of reducing HA ore shown in Figure 11 can be described as follows: (a), the main iron oxide hematite is deoxidized by CO and the intermediate product FeO forms; (b), FeO is reduced to Fe in CO atmosphere; (c), the reduction rate of (b) is slow dynamically [21], and the intermediate product FeO reacts with the gangue minerals SiO_2 and Al_2O_3 to produce Fe_2SiO_4 and $FeAl_2O_4$. The main products consist of Fe, Fe_2SiO_4 and $FeAl_2O_4$.

As shown in Figure 12, by synergistic reducing with HM ore, hercynite and fayalite are replaced by galaxite and tephroite, respectively. No fayalite but fayalite manganoan $((Fe, Mn)_2SiO_4)$ is detected when the HM ore ratio reaches 40%. Meanwhile, the intensity of hercynite peak gets lower. Thermodynamically, galaxite and tephroite can be generated more easily than hercynite and fayalite, as represented in Figure 13. Thus, FeO can be dissociated from hercynite and fayalite by adding with MnO, where the activity of FeO is improved and the reducibility of HA ore is enhanced.

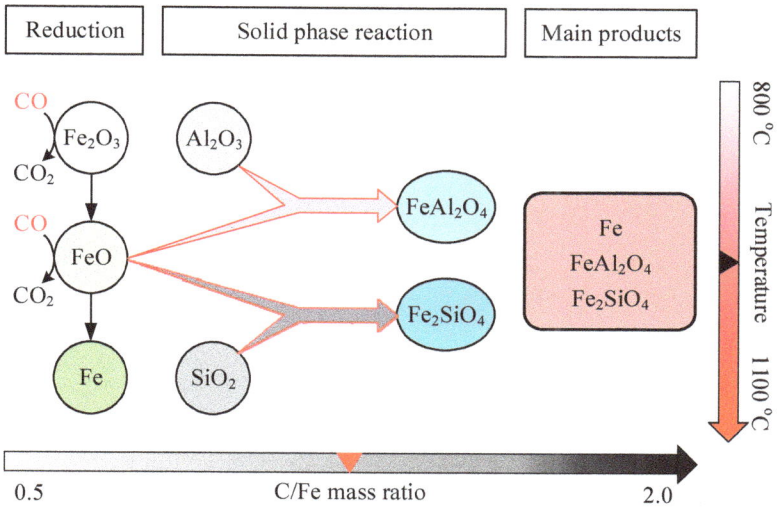

Figure 11. The mechanism of reducing high-aluminum iron ore (HA ore) by CO.

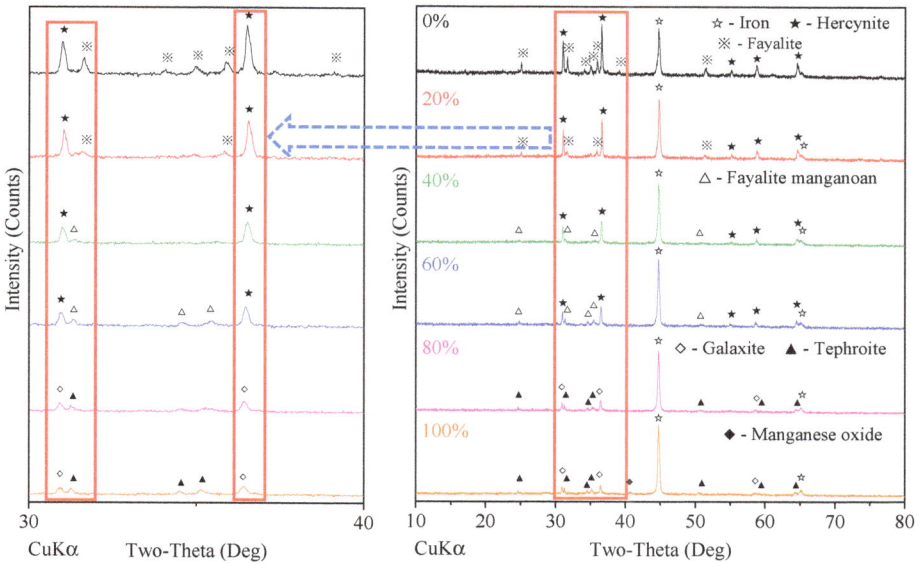

Figure 12. XRD patterns of reduced pellets with different ratio of high-maganese iron ore (HM ore) (reducing at 1050 °C for 90 min with C/Fe mass ratio of 1.5).

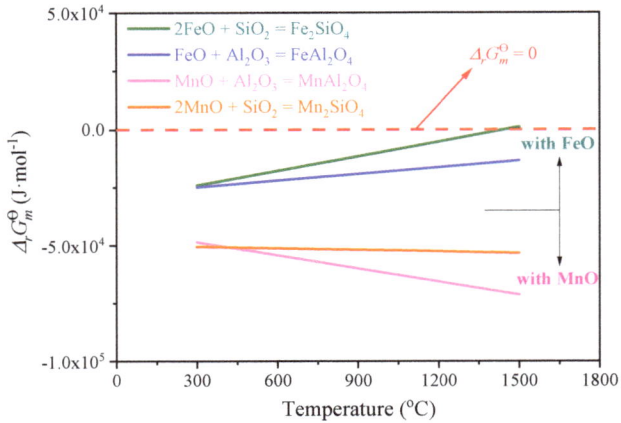

Figure 13. Comparison of standard Gibbs free energy change of reactions between Al_2O_3 and SiO_2 with MnO and FeO.

Another simulation based on the compositions of different HM ore ratio was processed by FactSage7.1, and the results are plotted in Figure 14. Less Fe_2SiO_4 and $FeAl_2O_4$ are produced by increasing the HM ore ratio, while by contrast, more Fe and Mn-Al-Si compounds are generated.

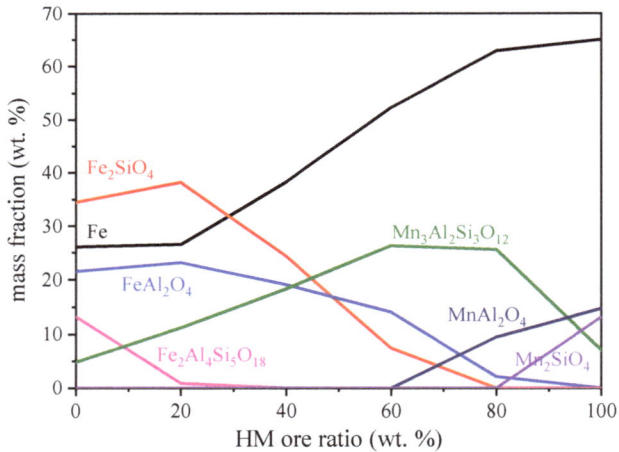

Figure 14. Thermodynamic simulation of a system consisting of Fe_2O_3, CO, SiO_2, Al_2O_3 and MnO (with different high-manganese iron ore (HM ore) ratio) carried out at 1050 °C in CO atmosphere at ambient pressure.

Accordingly, the mechanism of synergistic reducing of HA ore and HM ore shown in Figure 15 can be described as follows: (a) The main iron oxide hematite is deoxidized by CO and the intermediate product FeO forms; (b) FeO reacts with the gangue minerals SiO_2 and Al_2O_3 to produce Fe_2SiO_4 and $FeAl_2O_4$; (c) MnO from HM ore combines with Fe_2SiO_4 and $FeAl_2O_4$ to form Mn_2SiO_4, $MnAl_2O_4$ and FeO; (d) FeO is reduced to Fe in CO atmosphere. The main products consist of Fe, Mn_2SiO_4 and $MnAl_2O_4$.

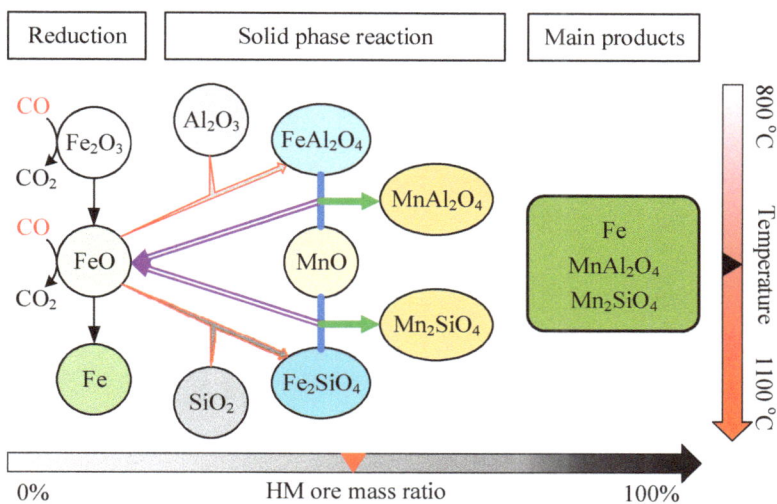

Figure 15. The mechanism of synergistic reducing of high-aluminum iron ore (HA ore) and high-manganese iron ore (HM ore) by CO.

5. Conclusions

In this study, a mechanism for enhancing the reduction of high-aluminum iron ore by synergistic reducing with high-manganese iron ore was investigated. The conclusions can be summarized as follows:

(1) Because of the generation of hercynite and fayalite, only 66.49% of metallization degree is obtained by reducing the high-aluminum iron ore containing 41.92% Fe_{total}, and 13.74% Al_2O_3 under 1050 °C for 120 min with C/Fe mass ratio of 1.5. By synergistic reducing with a high-manganese iron ore assaying 9.24% Mn_{total}, a higher metallization degree is achieved. Over 90% of the metallization degree is obtained by adding with 60% of the high-manganese iron ore.

(2) The mechanism of reducing HA ore can be described as follows: (a) the main iron oxide hematite is deoxidized by CO and the intermediate product FeO forms; (b) FeO reacts with the gangue minerals SiO_2 and Al_2O_3 to produce Fe_2SiO_4 and $FeAl_2O_4$; (c) FeO is reduced to Fe in CO atmosphere. The main products consist of Fe, Fe_2SiO_4 and $FeAl_2O_4$.

(3) The mechanism of synergistic reducing of HA ore and HM ore can be described as: (a) the main iron oxide hematite is deoxidized by CO and the intermediate product FeO forms; (b) FeO reacts with the gangue minerals SiO_2 and Al_2O_3 to produce Fe_2SiO_4 and $FeAl_2O_4$; (c) MnO from HM ore combines with Fe_2SiO_4 and $FeAl_2O_4$ to form Mn_2SiO_4, $MnAl_2O_4$ and FeO; (d) FeO is reduced to Fe in CO atmosphere. The main products consist of Fe, Mn_2SiO_4 and $MnAl_2O_4$.

Author Contributions: X.Z. and DZ. conceived of and designed the experiments. X.Z. and Y.L. performed the experiments and analyzed the data. D.Z. contributed materials. X.Z. and Y.L. wrote the paper. D.Z. and T.C. modified the paper.

Funding: This research was funded by Hubei Key Laboratory for Efficient Utilization and Agglomeration of Metallurgic Mineral Resources (No. 2017zy002, No. 2017zy012), Wuhan University of Science and Technology (No. 2017xz013) and National Natural Science Foundation of China (No. 51574281).

Acknowledgments: This work was financially supported by Hubei Key Laboratory for Efficient Utilization and Agglomeration of Metallurgic Mineral Resources, Wuhan University of Science and Technology and National Natural Science Foundation of China.

Conflicts of Interest: The authors declare no conflict of interest.

References

1. World Steel Association Economics Committee. *Steel Statistical Yearbook 2018*; World Steel Association: Brussels, Belgium, 2018; p. 104.
2. Zhang, K.; Kleit, A.N. Mining rate optimization considering the stockpiling: A theoretical economics and real option model. *Resour. Policy* **2016**, *47*, 87–94. [CrossRef]
3. Samouhos, M.; Taxiarchou, M.; Pilatos, G.; Tsakiridis, P.E.; Devlin, E.; Pissas, M. Controlled reduction of red mud by H$_2$ followed by magnetic separation. *Miner. Eng.* **2017**, *105*, 36–43. [CrossRef]
4. Chun, T.; Long, H.; Li, J. Alumina-Iron Separation of High Alumina Iron Ore by Carbothermic Reduction and Magnetic Separation. *Sep. Sci. Technol.* **2015**, *50*, 760–766. [CrossRef]
5. Lu, L.; Holmes, R.J.; Manuel, J.R. Effects of alumina on sintering performance of hematite iron ores. *ISIJ Int.* **2007**, *47*, 349–358. [CrossRef]
6. Raghukumar, C.; Tripathy, S.K.; Mohanan, S. Beneficiation of Indian High Alumina Iron Ore Fines—A Case Study. *Int. J. Min. Eng. Miner. Proc.* **2012**, *1*, 94–100. [CrossRef]
7. Liu, Z.; Li, H. Metallurgical process for valuable elements recovery from red mud-A review. *Hydrometallurgy* **2015**, *155*, 29–43. [CrossRef]
8. Thella, J.S.; Mukherjee, A.K.; Srikakulapu, N.G. Processing of high alumina iron ore slimes using classification and flotation. *Powder Technol.* **2012**, *217*, 418–426. [CrossRef]
9. Chun, T.; Li, D.; Di, Z.; Long, H.; Tang, L.; Li, F.; Li, Y. Recovery of iron from red mud by high-temperature reduction of carbon-bearing briquettes. *J. S. Afr. Inst. Min. Metall.* **2017**, *117*, 361–364. [CrossRef]
10. Li, G.; Jiang, T.; Liu, M.; Zhou, T.; Fan, X.; Qiu, G. Beneficiation of High-Aluminium-Content Hematite Ore by Soda Ash Roasting. *Min. Proc. Extr. Met. Rev.* **2010**, *31*, 150–164. [CrossRef]
11. Chun, T.J.; Zhu, D.Q.; Pan, J.; He, Z. Preparation of metallic iron powder from red mud by sodium salt roasting and magnetic separation. *Can. Metall. Quart.* **2014**, *53*, 183–189. [CrossRef]
12. Sutar, H.; Mishra, S.C.; Sahoo, S.K.; Chakraverty, A.P.; Maharana, H.S. Progress of red mud utilization: An overview. *Am. Chem. Sci. J.* **2014**, *4*, 255–279. [CrossRef]
13. Wang, H.; She, X.; Zhao, Q.; Xue, Q.; Wang, J. Production of iron nuggets using iron-rich red mud by direct reduction. *Chin. J. Proc. Eng.* **2012**, *12*, 816–821.
14. Borra, C.R.; Blanpain, B.; Pontikes, Y.; Binnemans, K.; Van Gerven, T. Smelting of Bauxite Residue (Red Mud) in View of Iron and Selective Rare Earths Recovery. *J. Sustain. Met.* **2016**, *2*, 28–37. [CrossRef]
15. He, P.; Ju, D.; Shen, P.; Jin, H. Experimental research on comprehensive utilization of red mud based on direct reduction and melting by RHF iron bead technology. *Energ. Metall. Ind.* **2017**, *36*, 57–60.
16. Stjernberg, J.; Olivas-Ogaz, M.A.; Antti, M.L.; Ion, J.C.; Lindblom, B. Laboratory scale study of the degradation of mullite/corundum refractories by reaction with alkali-doped deposit materials. *Ceram. Int.* **2013**, *39*, 791–800. [CrossRef]
17. Gan, L.; Xu, J.; Zhang, Z.; Liang, Y.; Wei, M.; Wen, C. Red Mud Pellets and Its Preparation Method. CN103602805A, 26 February 2014.
18. Shankar, A. Studies on High Alumina Blast Furnace Slags. Ph.D. Thesis, Royal Institute of Technolog, Stockholm, Sweden, 15 June 2007.
19. Zhou, X.; Zhu, D.; Pan, J.; Luo, Y.; Liu, X. Upgrading of High-Aluminum Hematite-Limonite Ore by High Temperature Reduction-Wet Magnetic Separation Process. *Metals* **2016**, *6*, 57. [CrossRef]
20. Donskoi, E.; Olivares, R.I.; McElwain, D.L.S.; Wibberley, L.J. Experimental study of coal based direct reduction in iron ore/coal composite pellets in a one layer bed under nonisothermal, asymmetric heating. *Ironmak. Steelmak.* **2006**, *33*, 24–28. [CrossRef]
21. Zhu, D.; Luo, Y.; Pan, J.; Zhou, X. Reaction Mechanism of Siderite Lump in Coal-Based Direct Reduction. *High Temp. Mater. Procees.* **2016**, *35*, 185–194. [CrossRef]

metals

MDPI

Article

Devolatilization Kinetics of Different Types of Bio-Coals Using Thermogravimetric Analysis

Asmaa A. El-Tawil [1,*], Hesham M. Ahmed [1,2] [ID], Lena Sundqvist Ökvist [1,3] and Bo Björkman [1]

[1] MiMeR, Luleå University of Technology, 97187 Luleå, Sweden; Hesham.ahmed@ltu.se (H.M.A.); lena.sundqvist-oqvist@ltu.se (L.S.Ö.); Bo.Bjorkman@ltu.se (B.B.)
[2] Central Metallurgical Research and Development Institute, P.O Box 87, Helwan, Cairo 11421, Egypt
[3] Swerim AB, 97125 Luleå, Sweden
* Correspondence: asmaa.el-tawil@ltu.se; Tel.: +46-920-493131

Received: 17 January 2019; Accepted: 29 January 2019; Published: 1 February 2019

Abstract: The interest of the steel industry in utilizing bio-coal (pre-treated biomass) as CO_2-neutral carbon in iron-making is increasing due to the need to reduce fossil CO_2 emission. In order to select a suitable bio-coal to be contained in agglomerates with iron oxide, the current study aims at investigating the thermal devolatilization of different bio-coals. A thermogravimetric analyzer (TGA) equipped with a quadrupole mass spectrometer (QMS) was used to monitor the weight loss and off-gases during non-isothermal tests with bio-coals having different contents of volatile matter. The samples were heated in an inert atmosphere to 1200 °C at three different heating rates: 5, 10, and 15 °C/min. H_2, CO, and hydrocarbons that may contribute to the reduction of iron oxide if contained in the self-reducing composite were detected by QMS. To explore the devolatilization behavior for different materials, the thermogravimetric data were evaluated by using the Kissinger–Akahira–Sonuse (KAS) iso-conversional model. The activation energy was determined as a function of the conversion degree. Bio-coals with both low and high volatile content could produce reducing gases that can contribute to the reduction of iron oxide in bio-agglomerates and hot metal quality in the sustained blast furnace process. However, bio-coals containing significant amounts of CaO and K_2O enhanced the devolatilization and released the volatiles at lower temperature.

Keywords: devolatilization; torrefied biomass; bio-coal; volatile matter; iso-conversional method

---

1. Introduction

The blast furnace (BF) is the most widely used technology for producing hot metal for steelmaking. In the BF, iron oxide is reduced to metallic iron by fossil carbon resources (coke, coal, oil, natural gas, etc.). The reducing conditions in the furnace are created by top-charged (coke) and tuyere-injected (pulverized coal, oil, etc.) reducing agents [1]. The total consumption of coke is about 300 kg/t hot metal [1,2] depending on the amount of auxiliary reducing agents [3]. The steel industry aims to reduce coke consumption and minimize CO_2 emissions by improving the energy efficiency of the process and by investigating the use of carbon-neutral materials such as bio-coal (pre-treated biomass) to substitute part of the fossil sources. CO_2 emitted for every ton of steel produced was on average 1.83 tons in 2017. According to the World Steel Association, the iron and steel industry accounts for approximately 7–9% of total world CO_2 emission [4]. The European Union (EU) has set a target to cut 20% of the CO_2 emissions, to achieve a 20% improvement in energy efficiency, and to increase the renewable energy by 20% by 2020; by 2050 the aim is to cut 80% of the CO_2 emissions [5]. Several studies report the decrease in fossil CO_2 emissions by using biomass [6,7]. Use of biomass resources is a possible alternative in Sweden, as forestland amounts to about 28.1 million hectares [8].

The use of raw biomass as a reducing agent in a BF is not possible because of the high moisture content, low content of fixed carbon (C_{fix}), as well as high contents of volatile matter (VM) and oxygen [9,10]. Different technologies can be applied to pre-treat biomass into products with properties suitable for the metallurgical industry. Pre-treatment methods like pyrolysis [11], torrefaction [12], etc., result in higher content of C_{fix}, lower VM and oxygen, and improved grindability, properties which, overall, correspond to a product whose quality resembles that of coal [12].

Top-charging of self-reducing composites constituted of iron ore and bio-coal into the BF is one way to introduce bio-coal and partially replace fossil carbon. Furthermore, this approach has the potential to lower the thermal reserve zone temperature of the BF as a result of the direct reduction of iron oxide in the agglomerates. The indirect reduction of FeO with CO at a lower temperature shifts towards the formation of more metallic iron at a specific partial pressure of CO. This results in higher gas efficiency of the BF, and the CO_2 emission is decreased. However, to achieve the maximum positive effect, the VM present in bio-coals contained in the composites should contribute to the reduction and not be released and lost with the BF top gas.

One of the features of biomass is the presence of considerable amounts of metal oxides of potassium, calcium, and magnesium [13] in the ash. The ash content is often less than 1% in woody bio-mass but may vary up to 15% in some herbaceous biomass and forestry residues [13,14]. The effect of inorganics on the thermal degradation of biomass has been extensively studied [15–17]. In general, inorganics retained in char during pyrolysis act as catalysts for char forming and fragmentation reactions [13,16,18].

There are several available methods for analyzing the kinetic data of devolatilization reactions [19]. These methods may be classified according to the experimental conditions selected and the mathematical analysis performed. Experimentally, either isothermal or non-isothermal methods are employed. The main mathematical approaches employed can be divided into model-fitting and iso-conversional methods [19]. The devolatilization of solid materials, such as biomass, is classified as a heterogeneous chemical reaction. The reaction kinetics of a heterogeneous reaction can be affected by, for example, the breakage and redistribution of chemical bonds, changing reaction geometry, and the interfacial diffusion of reactants and products [20]. Iso-conversional methods are considered an appropriate means of estimating the apparent activation energy, E_a, of heterogeneous reactions.

There are numerous studies on biomass pyrolysis using thermal analysis [21–29]. The effects of important parameters such as particle size and heating rate have been studied by Mani et al. [26] and Biagini et al. [27]. The devolatilization kinetics of main components of raw biomass and its dependence on lignin, cellulose, and hemicellulose under different conditions (inert or oxidizing) [21–25] have been investigated using the iso-conversional method. Tharaka et al. [28] applied the iso-conversional model on the devolatilization of raw and torrefied eucalyptus at different heating rates (5 °C/min to 20 °C/min) with temperatures ranging from 150 to 700 °C. The kinetic analysis showed that torrefied · eucalyptus has higher E_a values than raw eucalyptus. Tran et al. [29] studied the pyrolysis of torrefied stump materials using the Distributed Activation Energy Model (DAEM) as well as a model involving the three components lignin, cellulose, and hemicellulose. It was observed that the torrefied stump has a higher activation energy than the raw stump. Up to now, there is limited information in the literature about the relation between devolatilization kinetics of bio-coals and their properties for use in self-reducing bio-agglomerates.

The purpose of this work was to investigate the devolatilization behavior and the related kinetics of different types of bio-coals when used in self-reducing composites. In the BF, the temperature of the top-charged material will increase rather quickly during the descent in the upper part. Ideally, all the carbon and hydrogen contained in the top-charged bio-coal should contribute to the reduction. The characterization of the devolatilization of the bio-coal is thus important to understand the behavior of bio-coal included in iron ore composites charged into a BF.

2. Kinetic Theory

The devolatilization of lignocellulosic biomass is complex, as several reactions occur simultaneously during its thermal decomposition. Predicting the exact reaction mechanism is difficult, but the kinetic parameters of biomass devolatilization can be calculated by assuming a simplified single-step reaction mechanism. This reaction can be described by the rate equation, as presented in Equation (1) [30]:

$$\frac{d\alpha}{dt} = k(T)f(\alpha) \tag{1}$$

The dependency of the rate constant on temperature can be expressed as:

$$k(T) = Ae^{\left(-\frac{E_\alpha}{RT}\right)} \tag{2}$$

By combining Equation (1) and Equation (2), Equation (3) will be:

$$\frac{d\alpha}{dt} = Ae^{(-E_\alpha/RT)}f(\alpha) \tag{3}$$

where α, R, A, E_α, and $f(\alpha)$ denote the extent of conversion, gas constant, pre-exponential factor, activation energy, and reaction function, respectively. The reaction model can be expressed by using different mathematical forms that are tabulated elsewhere [30] and known as model fitting.

The conversion degree (α) is defined as follows:

$$\alpha = \frac{m_i - m_a}{m_i - m_f} \tag{4}$$

where m_i is the initial mass of the sample, m_a is the actual mass at a specific time, and m_f is the mass after finalized devolatilization.

For devolatilization experiments performed non-isothermally at a constant heating rate (β_I), the temperature (T) changes linearly with time (t) as in Equation (5):

$$\beta_I = \frac{dT}{dt} \tag{5}$$

where the index I denotes different heating rates.

For a constant heating rate, Equation (3) can be modified to Equation (6):

$$\frac{d\alpha}{dT} = \frac{A}{\beta}e^{(-E_\alpha/RT)}f(\alpha) \tag{6}$$

Equation (6) can be written in an integral form to give Equation (7):

$$g(\alpha) = \int_0^\alpha \frac{d\alpha}{f(\alpha)} = \frac{A}{\beta}\int_0^T e^{(-E/RT)}dT \tag{7}$$

where $g(\alpha) = \int_0^\alpha [f(\alpha)]^{-1}d\alpha$ is the integral form of the reaction model [30]. The drawbacks of model-fitting to determine a reliable kinetic triplet of A, E, and $f(\alpha)$ or $g(\alpha)$ can be resolved by using iso-conversional methods.

The iso-conversional principle states that the reaction rate $[\left(\frac{d\alpha}{dt}\right)_{\alpha,I}]$ at a specific extent of conversion is a function of temperature only, and $f(\alpha)$ does not depend on T [31]. Assuming that $f(\alpha)$ at a given conversion degree is independent of temperature, the kinetic parameters can be obtained from a set of runs from the relation between reaction rate versus temperature (differential method) or from that between weight loss versus temperature (integral method). The drawback of differential methods compared to integral methods is that the former are sensitive to inaccuracies in the experimentally determined reaction rate; therefore, integral methods are often preferred [32].

Among the integral methods, the Kissinger–Akahira–Sunose (KAS) iso-conversional model is common because of its relatively higher accuracy in estimated kinetic [33]. The KAS model is represented by a linear Equation (8) [22], in which the apparent E_α can be obtained by plotting $ln \frac{\beta_I}{T_{\alpha I}^2}$ versus $\frac{1000}{T_{\alpha I}}$ for a given value of conversion, α, where the slope is equal to $\frac{-E_\alpha}{R}$.

$$ln\frac{\beta_I}{T_{\alpha I}^2} = ln\left(\frac{A_\alpha R}{E_\alpha g_\alpha}\right) - \frac{E_\alpha}{RT_{\alpha I}} \tag{8}$$

3. Materials and Methods

3.1. Materials and Characterization

The seven different types of bio-coals selected for this study are presented in Table 1. The bio-coals were pulverized and, by sieving, the fraction of 75–150 μm to be used in TGA was obtained. After drying, the pulverized materials were stored in a desiccator. The proximate and ultimate analyses of each bio-coal analyzed by ALS Scandinavia AB using standard methods are given in Table 2. Bio-coals (TFR, TW, and TSD) prepared at low temperature had high content of VM % and less C_{fix} %. The opposite was seen for bio-coals (HTT, PA, PB, and CC) prepared at high temperature.

Table 1. Selected bio-coals with preparation temperatures and times.

Bio-Coal Type	Bio-Coal	Origin	Temperature, °C	Time, min	Abbreviation
Highly volatile bio-coals	Torrefied forest residue	Top and branches pine /spruce	286	6	TFR
	Torrefied saw dust	Spruce	297	6	TSD
	Torrefied willow	Willow	330	6	TW
Lowly volatile bio-coals	High-temperature torrefied	50 % Pine/50% spruce	350	8	HTT
	Pine A	Pine	350	14	PA
	Pine B	Pine	400	14	PB
	Charcoal	Mixture of pine, birch, alder, aspen	550	-	CC

Table 2. Proximate and ultimate values of the used bio-coals materials (dry basis).

Bio-Coals	Proximate Analysis (wt %)			Ultimate Analysis (wt %)				
	C_{fix}	VM	Ash	C_{tot}	H	N	S	O
TFR	23.6	73.2	3.2	52.0	5.9	0.57	0.035	35.2
TSD	24.0	75.6	0.45	57.1	5.9	0.12	0.004	36.4
TW	24.7	73.3	2.0	52.7	5.8	0.30	0.021	39.2
HTT	60.8	38.2	1.0	75.3	4.9	0.10	0.008	18.8
PA	70.3	28.7	1.0	78.6	4.4	0.23	<0.01	15.8
PB	79.1	19.7	1.2	85.0	3.8	0.30	<0.01	9.70
CC	80.7	18.6	0.70	87.0	3.4	0.25	<0.004	8.30

C_{Fix}: Fixed carbon; VM: volatile matter; C_{tot}: Total carbon; H: hydrogen; N: nitrogen; S: sulphur; O: oxygen.

The contents of metal oxides in bio-coals are presented in Table 3. Higher content of the metal oxides K_2O, CaO, and MgO were present in TFR and TW than in TSD.

Table 3. Metal oxides in bio-coals (Wt %, dry basis).

Bio-coals	Al_2O_3	CaO	SiO_2	Fe_2O_3	K_2O	MgO	MnO	Na_2O	P_2O_5	TiO_2
TFR	0.047	0.872	0.618	0.044	0.238	0.124	0.062	0.014	0.151	0.004
TSD	0.003	0.122	0.033	0.011	0.051	0.046	0.013	0.003	0.008	0.000
TW	0.004	0.494	0.019	0.003	0.230	0.076	0.004	0.005	0.142	0.000
HTT	0.024	0.310	0.305	0.079	0.145	0.061	0.040	0.023	0.032	0.001
PA	0.032	0.270	0.235	0.007	0.148	0.074	0.035	0.010	0.021	0.001
PB	0.013	0.339	0.067	0.006	0.167	0.096	0.043	0.004	0.024	0.000
CC	0.006	0.317	0.028	0.009	<0.002	0.112	0.044	<0.009	0.006	0.001

3.2. Experimental Method

The experiments were performed using a TGA, Netzsch STA 409 instrument (sensitivity ±1 µg) (Netzsch, Selb, Germany) attached to a Quadrupole Mass Spectroscopy (QMS, Netzsch, Selb, Germany) to monitor the mass loss and off-gases, respectively. The TGA used in this study is described in detail in reference [34]. In the mass spectrometer, compounds are ionized and separated on the basis of the mass/charge (m/z) ratio number. During the devolatilization experiments, 50–53 mg of sample was placed in an alumina crucible and heated from room temperature up to 1200 °C at the pre-specified heating rates of 5, 10, and 15 °C/min under argon gas (99.999%), with a flow rate of 200 mL/min. The samples were then cooled down at a rate of 20 °C/min. To eliminate errors from the buoyancy effect, correction measurements were carried out. Repeated devolatilization tests in the TGA showed consistent results without any significant variation.

4. Results and Discussion

4.1. Thermogravimetric Analysis

Figure 1 shows the TG curves for bio-coals at the heating rate of 5 °C/min. As seen from the plots, the weight loss for TFR, TW, and TSD began at ~200 °C and proceeded rapidly up to 800 °C, while the weight loss of HTT and PA started at ~300 °C, and that of PB and CC started at ~450 °C. As expected, the pre-treatment temperature affected the starting temperature of devolatilization.

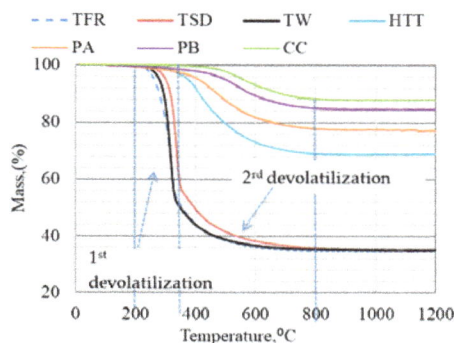

Figure 1. TGA of bio-coals conducted in Ar gas at a heating rate of 5 °C/min up to 1200 °C.

The devolatilization of bio-coal is linked to the presence of three main biomass components: hemicellulose, cellulose, and lignin [35]. The main DTG (derivative thermogravimetry) peak is, according to Marion et al. [36], attributed to cellulose decomposition, accompanied by a shoulder at the lower temperature, which is related to hemicellulose decomposition, and a flat tail at high temperature corresponding to lignin decomposition. The DTG curves deduced for TFR, TW, and TSD indicated two devolatilization steps, as can be seen in Figure 2. The first step of devolatilization started at ~200 °C and continued until 350 °C, and the second step started at 350 °C and continued until 800 °C. These two steps of devolatilization are in agreement with findings by other researchers [35,37]. On the basis of findings by Marion et al. [36], the higher peak in the first devolatilization step for TFR, TW, and TSD can be likely attributed to the presence of cellulose residues. On the other hand, the devolatilization of HTT, PA, and PB occurred in one step, starting from 300 °C and continuing until 800 °C. Similarly, CC only had one step of devolatilization, which started at ~450 °C. The main peak and flat tailing section for HTT, PA, PB, and CC shifted to a higher temperature, compared to highly volatile bio-coals. This was likely due to the decomposition of lignin which, according to the literature, occurs slowly over a broad temperature range (160–900 °C) [35].

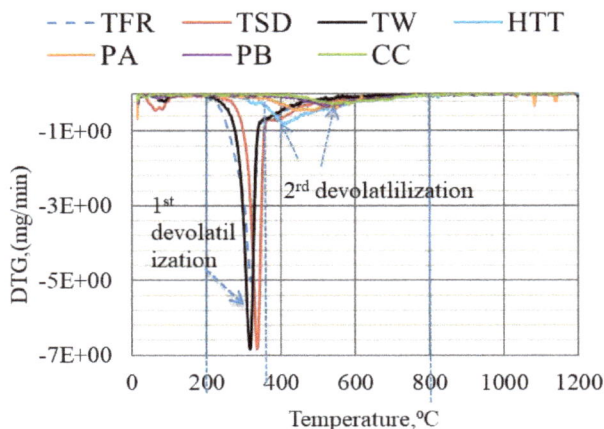

Figure 2. DTG of bio-coals conducted in Ar gas at a heating rate of 5 °C/min up to 1200 °C.

4.2. Off-Gas Analysis during the Thermal Decomposition of Bio-Coals

The thermal decomposition of bio-coals is caused by breaks of chemical bonds and release of VM as temperature increases. When and how this occurs influences the possible contribution of volatiles to reduction. It is important to note at which temperature interval different bio-coals will release the volatile matter. Gases (CO, CO_2, and H_2) and ionized hydrocarbons with one to four carbon atoms per molecule (C_1–C_4) were detected during devolatilization, as shown in Figure 3. However, the lengths of the carbon chains in the released hydrocarbons were probably initially longer before thermal decomposition and excitation in the QMS. The release of hydrocarbon chains in the low-temperature region was especially pronounced for TW but also evident for TFR. The release of H_2 and CO was not pronounced in the low-temperature region. Off-gas analysis of HTT, PA, PB, and CC showed less intensity of ionized hydrocarbons like CH_4^+, $C_2H_5^+$, $C_4H_9^+$ (see Figure 3d–g). H_2 and CO were detected at 500–800 °C for high-temperature pre-treated bio-coals. According to Yang [35,38], this is likely caused by cracking of residual lignin, which contains aromatic rings.

It has earlier been found that the reduction rate of iron oxide was higher when charcoal containing 18% volatiles was used in agglomerates compared to when only coke was used [39]. It has also been reported that H_2 gas released up to 500 °C can at least partly be utilized in the reduction of iron ore [40,41]. H_2 gas released at higher temperatures is known to improve the reduction efficiency and thus can reduce carbon consumption [42].

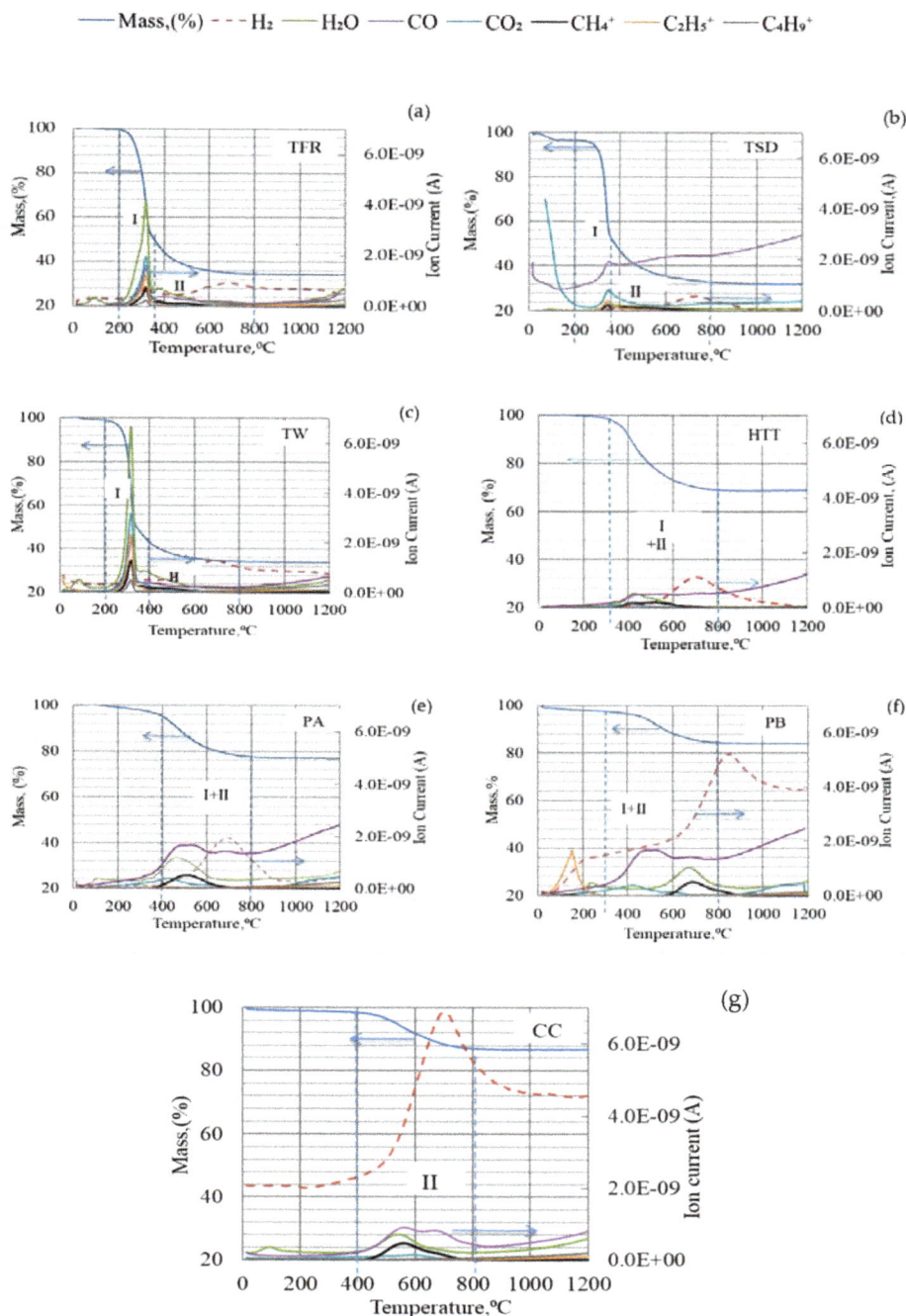

Figure 3. TGA–QMS analysis of different types of bio-coals: (**a**) TFR; (**b**) TSD; (**c**) TW; (**d**) HTT; (**e**) PA; (**f**) PB; (**g**) CC in Argon at a heating rate of 5 °C /min up to 1200 °C.

5. Kinetic Analysis

The mass loss at three heating rates (5, 10, and 15 °C/min) was recorded for each of the bio-coal materials; data recorded for TFR are shown in Figure 4. The extent of conversion degree, α, was calculated according to Equation (4) for all bio-coals tested.

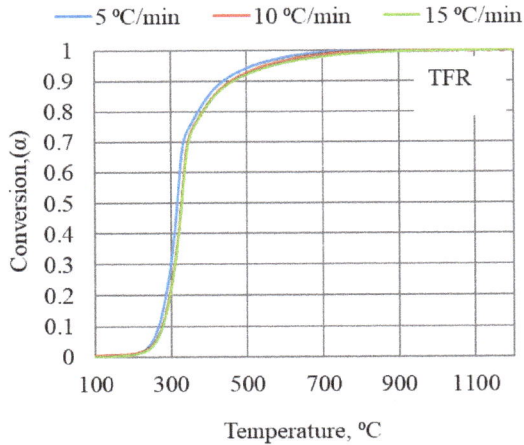

Figure 4. Conversion degrees for torrefied forest residue as a function of temperature in TGA at three heating rates up to 1200 °C.

The results obtained from TGA were elaborated according to the iso-conversional KAS model to calculate the kinetic parameters according to Equation (8) for a given value of conversion, α, in the range from 0.1 to 0.99 for all different heating rates. The KAS plot of $ln \frac{\beta_I}{T_{\alpha I}^2}$ versus $\frac{1000}{T_{\alpha I}}$ K^{-1} for different values of conversion is shown in Figure 5. The apparent activation energies were obtained from the slope, and the correlation coefficients, R^2, are given in Table 4. The correlation coefficients corresponding to linear fittings were in the range from 0.996 to 1.000.

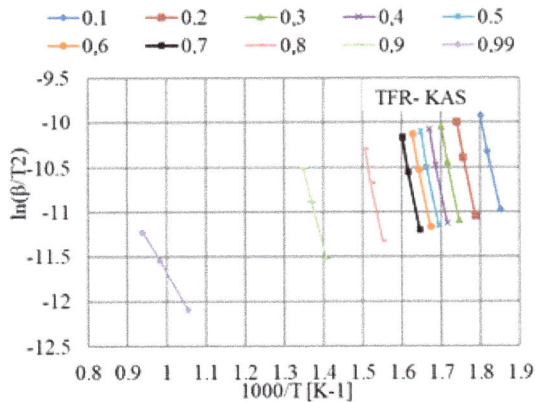

Figure 5. Plot deduced by applying the KAS model for the calculation of the activation energy for TFR at different heating rates.

Table 4. Kinetic parameters obtained from the KAS model for different bio-coals, E_a, activiation energy in kJ/mol. R^2, correlation coefficients.

Conversion, α	TFR		TSD		TW		HTT		PA		PB		CC	
	E_a	R^2	E_a	R^2	E_a	R^2	E_a	R^2	E_a	R^2	E_a	R^2	E_a	R^2
0.1	169	0.999	149	0.998	111	1.000	361	0.981	145	1.000	449	0.996	297	0.999
0.2	179	1.000	158	0.997	120	0.999	278	1.000	189	1.000	411	0.996	271	1.000
0.3	185	0.998	162	0.997	123	0.999	246	0.999	222	0.997	272	1.000	271	1.000
0.4	191	0.997	158	0.999	124	1.00	238	0.997	232	1.000	225	0.999	249	1.000
0.5	185	0.998	159	0.997	124	0.999	274	1.000	269	1.000	215	1.000	260	0.997
0.6	186	0.999	191	1.000	123	0.999	263	1.000	235	0.999	204	1.000	215	0.998
0.7	192	0.998	269	1.000	124	0.999	161	1.000	250	0.999	202	0.999	195	0.998
0.8	194	0.997	385	0.999	124	0.999	135	1.000	243	1.000	158	1.000	194	0.997
0.9	143	1.000	252	1.000	90	0.999	99	0.999	221	0.997	139	1.000	139	0.999
0,99	61	0.999	285	0.997	78	0.999	58	1.000	122	0.996	77	1.000	38	1.000
Average value	169	-	217	-	114	-	211	-	213	-	235	-	213	-

As can be seen in Table 4, the highly volatile bio-coals TFR and TW had quite stable values of E_a up to the conversion degree of 0.8, while TSD showed an increasing trend in E_a from 0.5, reaching a maximum at the conversion of 0.8. Both TFR and TW had low E_a at a high conversion degree. HTT, PB, and CC showed a similar behavior, with a high E_a value at a low conversion degree, decreasing with the increase of the conversion degree, whereas PA behaved differently. However, the general levels of E_a at an intermediate conversion degree were quite similar for all four lowly volatile bio-coals. In general, these also had comparably high E_a at the initial conversion degrees due to volatiles with weak bonds already being removed during torrefaction or pyrolysis. For highly volatile bio-coals, TW differed from TFR and TSD by having, in general, lower E_a values. Variations in E_a with the conversion degree for a bio-coal are likely due to different reactions occurring in parallel.

The KAS model assumes the same reaction occurring at a specific conversion degree. For complex materials like bio-coals, it is difficult to make detailed comparisons between materials, as different reactions may occur at the same conversion degree at different temperatures. Furthermore, there are overlapping reactions occurring at a given conversion degree and temperature. Differences in the conversion degree for different heating rates are in general not very large, as can be seen in Figure 4. Thus, measurement errors can have a significant influence on the calculated E_a values. Further parameters influencing the reaction rate parameter are the origin of wood used in the actual processing of the bio-coal as well as the presence of catalyzing elements in the ash.

Among the tested materials, the content of catalyzing components such as CaO and K_2O varied as did the content of acid components like SiO_2. This was especially significant for TW, which contains substantial amounts of CaO and K_2O and very low amounts of SiO_2. TSD has comparably low contents of oxides, including CaO and K_2O. This might be one reason for the lowest E_a estimate for TW in comparison to other bio-coals with higher volatile content, which is in agreement with reference [18].

One important property when selecting a bio-coal for bio-agglomerates is the release of volatiles at temperatures at which they can be used for reduction. TW is likely not the best selection, because of the ease of volatilization at low temperature; with high contents of K_2O and P_2O_5, the use of both TW and TFR will be restricted from a process and product quality point of view, respectively. Both TSD and CC are suitable for BF applications because of their low contents of K_2O and P_2O_5. In addition, CC is devolatilized at a higher temperature, forming H_2 and CO that will contribute to the reduction reactions. On the basis of their devolatilization behavior, bio-coals pretreated at high temperature are considered suitable for use in agglomerates, provided the ash chemistry is suitable for the BF. Although TSD is a highly volatile bio-coal, its higher E_a indicates that the release of VM might be slightly delayed and may contribute to the reduction at higher temperature. An additional advantage of TSD is a comparably high yield when produced from raw biomass; in addition, TSD is available in sufficient amounts for upscaling studies.

The QMS analysis, summarized in Figure 6, indicated that CH_4 formed at lower temperatures reacted with CO_2, forming CO and H_2, while hydrocarbons released at higher temperatures reacted with H_2O, forming H_2 and some CO. CO formed also at low temperatures may contribute to the reduction of hematite already in the shaft, whereas H_2 formed at higher temperatures will contribute to the reduction of lower iron oxides.

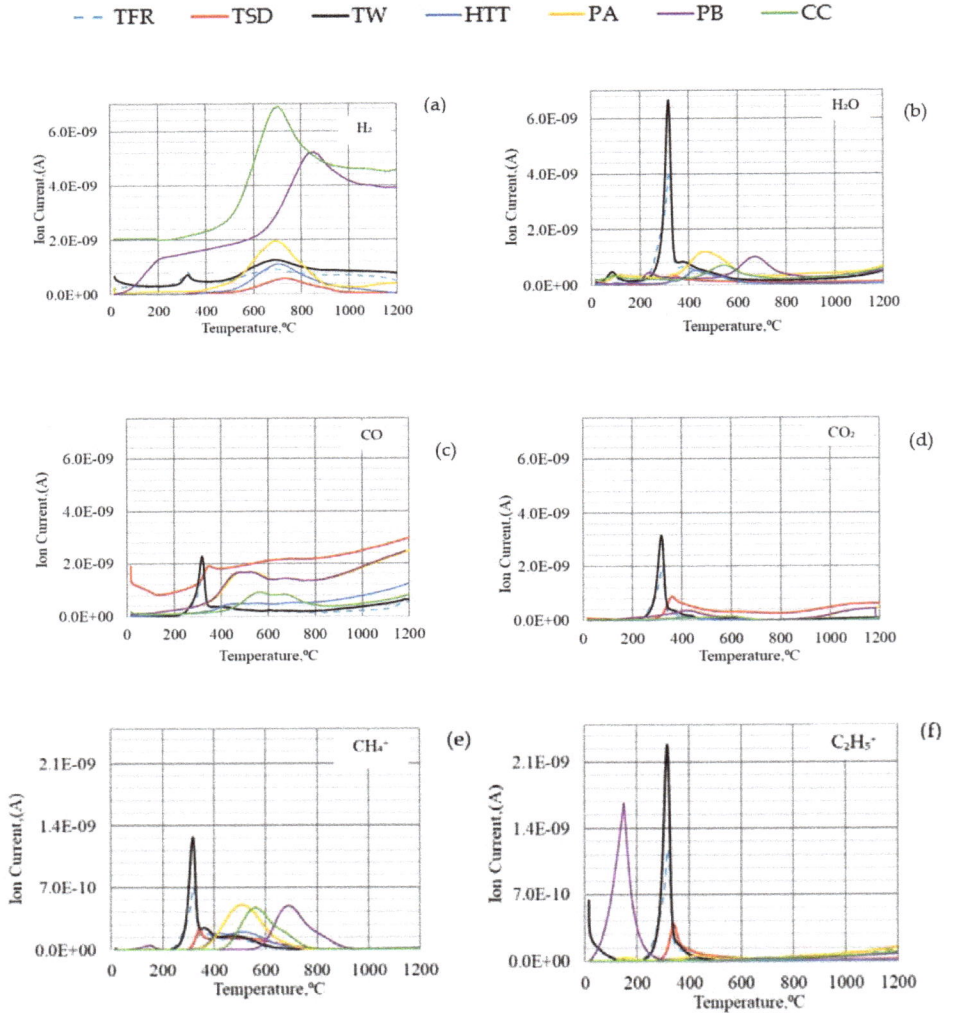

Figure 6. Off-gas analysis of (**a**) H_2; (**b**) H_2O; (**c**) CO; (**d**) CO_2; (**e**) CH_4^+; (**f**) $C_2H_5^+$ for different types of bio-coals.

6. Conclusions

It was found that devolatilization is affected by the bio-coal properties that are linked to volatile contents and ash components.

Bio-coal with a low volatile content will release the volatiles at a comparatively high temperature, ensuring that the generated reducing gases can contribute to the reduction in the agglomerates with iron oxide. However, bio-coals with high K_2O and P_2O_5 should be avoided because of their effect on process and product quality.

Bio-coals with a high volatile content and, at the same time, a high content of the ash component catalyzing the devolatilization, should be avoided, as the volatile will be released at low temperature when the contribution to the reduction is limited.

Bio-coasl with a high volatile content and a low content of catalyzing components will release their volatiles later, and it is possible that this will occur at the temperature at which the formed CO can reduce higher iron oxides. When used in the BF, this material has the advantage of high yield from raw biomass.

High volatile content and high content of K_2O and CaO in bio-coal result in low E_a for devolatilization. Bio-coals with low content of volatiles have in general high E_a for devolatilization, as volatiles with weak bonds are already released during torrefaction or pyrolysis.

Among the tested materials, TSD appears to be the most suitable material to be utilized in bio-agglomerates in the BF, in addition to CC. Both of these materials have low contents of the catalyzing ash components and P_2O_5.

Author Contributions: A.A.E.-T. conceived and design the experiments, performed the experiments, analyzed the data, wrote the first draft of paper, reviewed and wrote the paper; L.S.O., supervision, reviewed and wrote the paper; B.B., supervision, reviewed and wrote the paper; H.M.A., supervision.

Funding: The research was funded by Bio-agglomerate project, grant number 156334.

Acknowledgments: Financial support from the Swedish Energy Agency (Energimyndigheten) for research within the bio-agglomerate project is gratefully acknowledged. For additional support, the following companies and institutions are acknowledged: BioEndev, Swerim AB, SSAB Merox, LKAB and SSAB special steel. The research has been partly financed by CAMM, Center of Advanced Mining and Metallurgy at LTU.

Conflicts of Interest: The authors declare no conflict of interest.

References

1. Geerdes, M.; Chaigneau, R.; Kurunov, I. *Modern Blast Furnace Ironmaking: An Introduction*, 3rd ed.; IOS Press: Amsterdam, The Netherlands, 2015; pp. 1–215.
2. Xu, C.; Cang, D. A Brief Overview of Low CO_2 Emission Technologies for Iron and Steel Making. *J. Iron Steel Res. Int.* **2010**, *17*, 1–7. [CrossRef]
3. Hannu, S.; Timo, F. Towards More Sustainable Ironmaking—An Analysis of Energy Wood Availability in Finland and the Economics of Charcoal Production. *Sustainability* **2013**, *5*, 1188–1207.
4. Steel's Contribution to a Low Carbon Future and Climate Resilient Societies. Available online: https://www.worldsteel.org/publications/bookshop/product-details~{}Steel-s-Contribution-to-a-Low-Carbon-Future--2018-update-~{}PRODUCT~{}Steel-s-Contribution-to-a-Low-Carbon-Future~{}.html (accessed on 14 April 2018).
5. Wang, C.; Mellin, P.; Lövgren, J.; Nilsson, L.; Yang, W.; Salman, H.; Hultgren, A.; Larsson, M. Biomass as blast furnace injectant—Considering availability, pretreatment and deployment in the Swedish steel industry. *Energy Convers. Manag.* **2015**, *102*, 217–226. [CrossRef]
6. Feliciano-Bruzual, C. Charcoal injection in blast furnaces (Bio-PCI): CO_2 reduction potential and economic prospects. *J. Mater. Res. Technol.* **2014**, *3*, 233–243. [CrossRef]
7. Alexander, B.; Dieter, S.; Miguel, F. Charcoal Behaviour by Its Injection into the Modern Blast Furnace. *ISIJ Int.* **2010**, *50*, 81–88.
8. Skogsdata 2018. Available online: https://www.slu.se/en/Collaborative-Centres-and-Projects/the-swedish-national-forest-inventory/forest-statistics/skogsdata/ (accessed on 12 December 2018).
9. Fick, G.; Mirgaux, O.; Neau, P.; Patisson, F. Using biomass for pig iron production: A technical, environmental and economical assessment. *Waste Biomass Valoris* **2014**, *5*, 43–55. [CrossRef]
10. Chen, W.; Du, S.; Tsai, C.; Wang, Z. Torrefied biomasses in a drop tube furnace to evaluate their utility in blast furnaces. *Bioresour. Technol.* **2012**, *111*, 433–438. [CrossRef] [PubMed]
11. Bridgwater, A. The production of biofuels and renewable chemicals by fast pyrolysis of biomass. *Int. J. Glob. Energy Issues* **2007**, *27*, 160–203. [CrossRef]
12. Van der Stelt, M.J.C.; Gerhauser, H.; Kiel, J.H.A.; Ptasinski, K.J. Biomass upgrading by torrefaction for the production of biofuels: A review. *Biomass Bioenergy* **2011**, *35*, 3748–3762. [CrossRef]

13. Keown, D.M.; Favas, G.; Hayashi, J.I.; Li, C.Z. Volatilisation of alkali and alkaline earth metallic species during the pyrolysis of biomass: Differences between sugar cane bagasse and cane trash. *Bioresour. Technol.* **2005**, *96*, 1570–1577. [CrossRef]

14. Agblevor, F.A.; Besler, S. Inorganic compounds in biomass feedstocks. 1. Effect on the quality of fast pyrolysis oils. *Energy Fuels* **1996**, *10*, 293–298. [CrossRef]

15. Jensen, P.A.; Frandsen, F.J.; Dam-Johansen, K.; Sander, B. Experimental Investigation of the Transformation and Release to Gas Phase of Potassium and Chlorine during Straw Pyrolysis. *Energy Fuels* **2000**, *14*, 1280–1285. [CrossRef]

16. Nowakowski, D.J.; Jones, J.M.; Brydson, R.M.D.; Ross, A.B. Potassium catalysis in the pyrolysis behaviour of short rotation willow coppice. *Fuel* **2007**, *86*, 2389–2402. [CrossRef]

17. Raveendran, K.; Ganesh, A.; Khilar, K.C. Influence of mineral matter on biomass pyrolysis characteristics. *Fuel* **1995**, *74*, 1812–1822. [CrossRef]

18. Fahmi, R.; Bridgwater, A.V.; Darvell, L.I.; Jones, J.M.; Yates, N.; Thain, S.; Donnison, I.S. The effect of alkali metals on combustion and pyrolysis of Lolium and Festuca grasses, switchgrass and willow. *Fuel* **2007**, *86*, 1560–1569. [CrossRef]

19. Vyazovkin, S.; Wight, C.A. Kinetics in solids. *Annu. Rev. Phys. Chem.* **1997**, *48*, 125–149. [CrossRef] [PubMed]

20. Galwey, A.K.; Brown, M.E. Kinetic Background to Thermal Analysis and Calorimetry. *Princ. Pract.* **1998**, *1*, 147–223.

21. Cai, J.M.; Bi, L.S. Kinetic analysis of wheat straw pyrolysis using isoconversional methods. *J. Therm. Anal. Calorim.* **2009**, *98*, 325–330. [CrossRef]

22. Slopiecka, K.; Bartocci, P.; Fantozzi, F. Thermogravimetric analysis and kinetic study of poplar wood pyrolysis. *Appl. Energy* **2012**, *97*, 491–497. [CrossRef]

23. Brachi, P.; Miccio, F.; Miccio, M.; Ruoppolo, G. Isoconversional kinetic analysis of olive pomace decomposition under torrefaction operating conditions. *Fuel Process. Technol.* **2015**, *130*, 147–154. [CrossRef]

24. Aboyade, A.O.; Hugo, T.J.; Carrier, M.; Meyer, E.L.; Stahl, R.; Knoetze, J.H.; Görgens, J.F. Non-isothermal kinetic analysis of the devolatilization of corn cobs and sugar cane bagasse in an inert atmosphere. *Thermochim. Acta* **2011**, *517*, 81–89. [CrossRef]

25. Kongkaew, N.; Pruksakit, W.; Patumsawad, S. Thermogravimetric Kinetic Analysis of the Pyrolysis of Rice Straw. *Energy Procedia* **2015**, *79*, 663–670. [CrossRef]

26. Mani, T.; Murugan, P.; Abedi, J.; Mahinpey, N. Pyrolysis of wheat straw in a thermogravimetric analyzer: Effect of particle size and heating rate on devolatilization and estimation of global kinetics. *Chem. Eng. Res. Des.* **2010**, *88*, 952–958. [CrossRef]

27. Biagini, E.; Fantei, A.; Tognotti, L. Effect of the heating rate on the devolatilization of biomass residues. *Thermochim. Acta* **2008**, *472*, 55–63. [CrossRef]

28. Doddapaneni, T.R.K.C.; Konttinen, J.; Hukka, T.I.; Moilanen, A. Influence of torrefaction pretreatment on the pyrolysis of Eucalyptus clone: A study on kinetics, reaction mechanism and heat flow. *Ind. Crop. Prod.* **2016**, *92*, 244–254. [CrossRef]

29. Tran, K.Q.; Bach, Q.V.; Trinh, T.T.; Seisenbaeva, G. Non-isothermal pyrolysis of torrefied stump—A comparative kinetic evaluation. *Appl. Energy* **2014**, *136*, 759–766. [CrossRef]

30. Vyazovkin, S.; Wight, C.A. Isothermal and non-isothermal kinetics of thermally stimulated reactions of solids. *Int. Rev. Phys. Chem.* **1998**, *17*, 407–433. [CrossRef]

31. Sbirrazzuoli, N. Determination of pre-exponential factors and of the mathematical functions f (a) or G (a) that describe the reaction mechanism in a model-free way. *Thermochim. Acta* **2013**, *564*, 59–69. [CrossRef]

32. Vyazovkin, S.; Burnham, A.K.; Criado, J.M.; Pérez-Maqueda, L.A.; Popescu, C.; Sbirrazzuoli, N. ICTAC Kinetics Committee recommendations for performing kinetic computations on thermal analysis data. *Thermochim. Acta* **2011**, *520*, 1–19. [CrossRef]

33. Starink, M.J. The determination of activation energy from linear heating rate experiments: A comparison of the accuracy of isoconversion methods. *Thermochim. Acta* **2003**, *404*, 163–176. [CrossRef]

34. Ahmed, H.M.; Persson, A.; Okvist, L.S.; Bjorkman, B. Reduction Behaviour of Self-reducing Blends of In-plant Fines in Inert Atmosphere. *ISIJ Int.* **2015**, *55*, 2082–2089. [CrossRef]

35. Yang, H.; Yan, R.; Chen, H.; Lee, D.H.; Zheng, C. Characteristics of hemicellulose, cellulose and lignin pyrolysis. *Fuel* **2007**, *86*, 1781–1788. [CrossRef]

36. Carrier, M.; Auret, L.; Bridgwater, A.; Knoetze, J.H. Using Apparent Activation Energy as a Reactivity Criterion for Biomass Pyrolysis. *Energy Fuels* **2016**, *30*, 7834–7841. [CrossRef]
37. Ren, S.; Lei, H.; Wang, L.; Bu, Q.; Chen, S.; Wu, J. Thermal behaviour and kinetic study for woody biomass torrefaction and torrefied biomass pyrolysis by TGA. *Biosyst. Eng.* **2013**, *116*, 420–426. [CrossRef]
38. Yang, H.; Yan, R.; Chen, H.; Lee, D.H.; Liang, D.T.; Zheng, C. Mechanism of Palm Oil Waste Pyrolysis in a Packed Bed. *Energy Fuels* **2006**, *20*, 1321–1328. [CrossRef]
39. Konishi, H.; Ichikawa, K.; Usui, T. Effect of residual volatile matter on reduction of iron oxide in semi-charcoal composite pellets. *ISIJ Int.* **2010**, *50*, 386–389. [CrossRef]
40. Lin, H.Y.; Chen, Y.W.; Li, C. The mechanism of reduction of iron oxide by hydrogen. *Thermochim. Acta* **2003**, *400*, 61–67. [CrossRef]
41. Pineau, A.; Kanari, N.; Gaballah, I. Kinetics of reduction of iron oxides by H2: Part I: Low temperature reduction of hematite. *Thermochim. Acta* **2006**, *447*, 89–100. [CrossRef]
42. Biswas, A.K. *Principles of Blast Furnace Ironmaking*; Cootha Publishing House: Brisbane, Australia, 1981; pp. 1–512.

metals

MDPI

Article

Vertical Section Observation of the Solid Flow in a Blast Furnace with a Cutting Method

Yuanxiang Lu [1], Zeyi Jiang [1,2,*], Xinru Zhang [1,3], Jingsong Wang [4] and Xinxin Zhang [1,2]

[1] School of Energy and Environmental Engineering, University of Science and Technology Beijing, Beijing 100083, China; luyuanxiang2008@163.com (Y.L.); xinruzhang@ustb.edu.cn (X.Z.); xxzhang@ustb.edu.cn (X.Z.)

[2] Beijing Key Laboratory of Energy Saving and Emission Reduction for Metallurgical Industry, Beijing 100083, China

[3] Beijing Engineering Research Center of Energy Saving and Environmental Protection, Beijing 100083, China

[4] State Key Laboratory of Advanced Metallurgy, Metallurgical and Ecological Engineering, University of Science and Technology Beijing, Beijing 100086, China; wangjingsong@ustb.edu.cn

* Correspondence: zyjiang@ustb.edu.cn; Tel.: +86-010-6233-2741

Received: 6 December 2018; Accepted: 21 January 2019; Published: 25 January 2019

Abstract: The solid flow plays an important role in blast furnace (BF) ironmaking. In the paper, the descending behavior of solid flow in BFs was investigated by a cold experimental BF model and numerical simulation via the discrete element method (DEM). To eliminate the flat wall effect on the structure of solid flow in lab observations, a cutting method was developed to observe the vertical section of the solid flow by inserting a transparent plate into the experimental BF model. Both the experimental and numerical results indicated that plug flow is the main solid flow pattern in the upper and middle zones of BFs during burden descending. Meanwhile, a slight convergence flow and a deadman zone form at the lower part of the bosh. In addition, the boundary between the plug flow and convergence flow in BFs was determined by analyzing the velocity of the burden in vertical directions and the Wilcox–Swailes coefficient (U_{ws}). The results indicated that the U_{ws} can be defined as a critical value to determine the solid flow patterns. When $U_{ws} \geq 0.65$, the plug flow is dominant. When $U_{ws} < 0.65$, the convergence flow is dominant. The findings may have important implications to understand the structure of the solid flow in BFs.

Keywords: blast furnace; solid flow; cold experiment; direct element method; Wilcox–Swailes coefficient

1. Introduction

Blast furnaces (BFs) are complex metallurgical reactors that produce pig iron. During the ironmaking process for BFs, the layered coke and ore particles are charged into the top of the BF. The hot gases are injected into the raceway from tuyeres at the bottom of the BF. Then, the coke particles descend to the raceway and the hearth, and are gasified and combusted mainly in the lower part of the BF. Meanwhile, the ore pellets are reduced during the process of burden descending and gradually become small and soft until they melt to liquid iron. In such a complicated multiphase chemical reaction system, the descending behaviors of coke and ore in BFs directly affects the gas flow distribution, heat–mass transfer, and gas–solid reactions in the BF, which all play a significant role in achieving a smooth operation of the BF. Therefore, it is necessary and important to understand the descending behaviors of coke and ore particles in BFs.

In fundamental aspects, the descending of coke and ore particles in BFs is a typical solid flow. Due to the difficulty of experiments, the solid flow in BFs has been extensively studied by various mathematical models in the past decades, as reviewed by Yagi et al. [1], Dong et al. [2],

Ariyama et al. [3], and Kuang et al. [4]. In recent years, with the development in computer technology, numerical approaches (i.e., computational fluid dynamics (CFD) and the discrete element method (DEM)) have been increasingly adopted as important research tools to investigate the solid flow in BFs. However, it is difficult to track the detailed properties of single particles from the flow field at the grain scale by CFD. The DEM was first proposed by Cundall and Strack [5] in 1979, then rapidly developed by many scholars as a result of its significant advantages in micromechanics of granular materials. To date, many researchers have studied the influences of various factors on the structure of solid flow in BFs by the DEM method, including the discharging velocity and particle shapes [6], burden layer structure [7–9], molten slag trickle flow [10,11], segregation behavior [12,13], shaft-injected gas distribution [14], the softening and melting behaviors of ferrous burdens [15], the air pressure drop [16], and the flow and wall stress [17]. The DEM has been recognized as an effective method to study the fundamental behavior of solid flow in BFs, and most of the studies have adopted the corresponding simplified conditions and models to reduce the calculation time.

Some scholars have also studied more comprehensive models. Adema et al. [18] compared three types of BF geometry (slot models and a pie-slice) with different particle shapes, and concluded that the geometry used should be carefully chosen as it has a very large influence on the solid flow. Ping [19] et al. evaluated different burden descent models under four charging patterns. These models could predict the positions and shapes of different timelines in BFs, and the results showed that the C/O charging pattern can influence the shape of the cohesive zone and the deadman. Fu et al. [20] proposed two models, i.e., the geometric profile (GP) model and the potential flow (PF) model, to consider the non-uniform descending speed of the burden. These models can obtain the descending speed with different C/O ratios and can be applied for online prediction of operating blast furnaces. Using a numerical fluid–solid coupled method, Xu et al. [21] and Yang et al. [11] simulated the fluid and solid phases to study their interaction, and Hou et al. [22] established a quasi-steady virtual experimental BF model that considers the operation and energy efficiency of a BF. These numerical models reveal the main profile of burden structure combined with the experimental results [23–26], and the cold experimental results in the laboratory can provide the basis for the numerical simulation of the hot state. To date, many findings have indicated that the solid flow in BFs can be divided into four characteristic flow regions, i.e., plug flow, stagnant, funnel flow, and quasi-stagnant zones. In addition, Wright et al. [25] analyzed the solid flow regions by comparing the 3D model with slot models. They found that the stagnant zone in the slot model is smaller than that in the 3D model due to the wall effect, and the slot model may not fully describe the solid flow behavior. On the basis of these studies, Yang et al. [27] compared three types of models, i.e., full 3D models, slot models, and sector models, and proved that slot models cannot describe the anisotropic solid flow in the tangential direction as a result of the wall effect. The sector model, which was a more reliable simplified model, should be used in the future studies. Up to now, many researchers have found that it is necessary to eliminate the wall effect during the numerical study of the solid flow in BFs. However, few experiments have been developed to evaluate and verify the influence of the wall effect on the solid flow in BFs.

To address this gap, herein, a cold experimental BF model was established to analyze the descending behavior of solid flow in BFs, i.e., a 3D half-circle BF (180°) model, which was made of transparent acrylic material. To eliminate the wall effect, a novel cutting method for the burden was developed to observe internal structure of the solid flow in BFs. Furthermore, two numerical DEM sector models were used to verify the experimental models and examine the characteristics of solid flow under different burden layer conditions, such as full coke, layered coke/ore pellet, and mixed coke/ore pellet. In addition, based on the cutting method, the solid flow patterns in BFs were explored. The findings may have important implications to understand the structure of the solid flow in BFs.

2. Materials and Methods

2.1. Experimental Design

The experiments of solid flow in blast furnaces are often conducted in the laboratories, where the experimental results of the burden structure can be recorded directly by high-speed cameras. Generally, the experimental furnace body is a semicircular structure. There are two walls on the periphery of such a furnace body. One is a semicircular peripheral wall and the other one is a flat wall. The observation surface is the flat wall, and the results are recorded from this surface.

However, due to the observation of the burden structure from the flat wall, there is a physical impact on the solid flow. Compared to the full furnace model without the influence of the flat wall, the half furnace model may have some potential errors on the structure of the solid flow. But it is difficult to observe the internal structure of the burden in the full furnace model. In order to eliminate the impact of the flat wall on solid flow as much as possible, an experimental method is developed in this paper. On the basis of a half BF model, a further exploration of the internal burden structure can be studied. In this way, it can not only observe the descent process of the burden, but also ensure the reliability of the experimental results. This part is arranged in the Appendix A in order to describe the feasibility of this method.

2.2. Experiment Setup

A 3D experimental platform of solid flow in BFs was designed and setup through the above experimental idea. The schematic of the cold experimental BF model is illustrated in Figure 1. The BF model was composed of a BF body, a storage bin, a charging device (inlet), and a discharging device which could control the flow rate of solids in the BF using a spiral discharger. The geometry parameters of the BF body model are listed in Table 1. In order to observe the structure of solid flow, an experimental BF model (i.e., a 3D half-circle BF model made of transparent acrylic material) was established. It should be noted that, the experimental BF models were designed with the scale of 1:15 based on the geometric dimensions and operating conditions of a commercial BF (inner volume 125 m^3 with 8 tuyeres). Generally, during the ironmaking process of BFs, the coke combustion in the raceway is considered as the main driving force of burden descending. Here, in the cold experimental BF model, the discharge process of particles below the tuyere is regarded as the consumption process of coke combustion. The melting of iron ore in the cohesive zone is ignored.

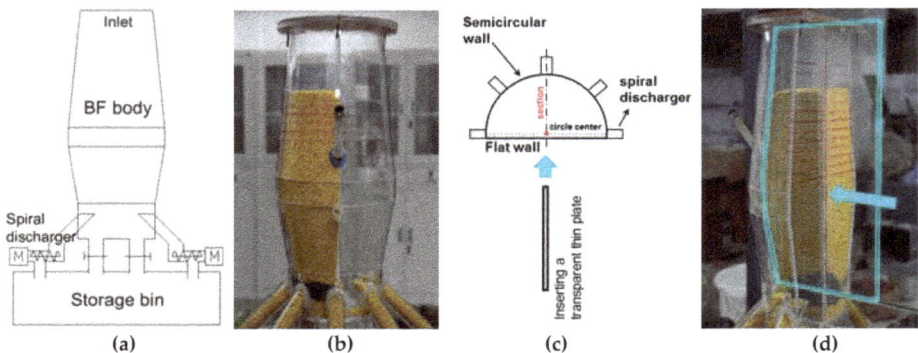

Figure 1. The 3D half-circle blast furnace (BF) model. (**a**) Schematic of the 3D experimental BF model. (**b**) The image of the 3D half-circle experiment BF model. (**c**) Schematic of inserting a transparent plate into the 3D half-circle experiment BF model. (**d**) The image of the vertical section of solid flow obtained by the cutting method in the 3D half-circle experiment BF model.

Table 1. Dimensions of the experimental model.

Parameters	Value
Diameter of hearth, m	0.213
Diameter of belly, m	0.267
Diameter of throat, m	0.200
Height of hearth, m	0.213
Height of bosh, m	0.160
Height of belly, m	0.054
Height of shaft, m	0.333
Height of throat, m	0.100

2.3. Particle Properties

In the experiment, two different cylindrical polyethylene particles were used as the burden particles and tracer particles, which could be used to display the position of burden motion and the distribution of descending velocity. Specifically, in the 3D half-circle BF, the yellow particles and the small dark blue particles were used as the burden particles and tracer particles, respectively. During the experiment, the timelines were formed by these tracer particles, which could be used to analyze the structure of solid flow. It should be noted that, all of these cylindrical polyethylene particles had a repose angle of 40°, because the repose angle of coke generally ranged from 35° to 45°. In addition, these particles had a diameter of 3–5 mm, a real density of 910 kg/m³, a burden porosity of 0.35, and an elastic modulus of 1.07 GPa. The diameter of the tracer particles was 2.5mm. The static friction coefficient between the polyethylene particles and the outer wall of acrylic (i.e., $f_{p\text{-}w}$) is 0.156. The static friction coefficient between polyethylene particles (i.e., $f_{p\text{-}p}$) is 0.21.

2.4. Experimental Procedure

Prior to the experiment, the particle descending speed was determined. Generally, considering the particle Froude similarity, the velocities of burden can be calculated by the Froude number [25]:

$$Fr_s = \frac{\rho_g}{\rho_p - \rho_g} \cdot \frac{u_s^2}{g d_p} \tag{1}$$

where ρ_g and ρ_p are the densities of the gas and the solid, respectively. d_p is the equivalent diameter of the particle. u_s is the particle descending velocity at the furnace throat. The modified Fr_s relates the inertial forces acting on the solid and gas phases. According to the operational data of the BF, the particle descending speed in the model was controlled at 3.45×10^{-4} m/s.

Before the experiment, a verification experiment was carried out to test the reliability of this cutting method as shown in the Appendix A. In the experiment, firstly, the 3D half-circle BF was filled up with the yellow polyethylene particles. Then, the continuous charging and discharging system was initiated. The first layer of the tracer particles were evenly charged from the top of the experimental BF model. Subsequently, the burden particles (yellow) and tracer particles (dark blue and black) were fed into the BF model alternately. The tracer particles were charged into the BF model at intervals of 3 min, thus forming several thin tracer layers acting as timelines. The charging and discharging system was shut off when the first layer of the tracer particles reached the outlets. By this time, a frozen burden body was formed, which could be used to study the structure of the solid flow.

Furthermore, to eliminate the flat wall effect, a cutting method was developed to observe the internal structure of the solid flow in the frozen burden body. In the cutting method, as shown in Figure 1c, a transparent plate was vertically inserted into the flat wall of the 3D half-circle BF. Then, the solid particles on the other side of the transparent plate were removed, as shown in Figure 1d. After that, the structure of the solid flow on the section in the BF model could be photographed. It should be noted that, as shown in Figure 1c, the shape of the half-circle model was not a true semicircle, but a lengthened semicircle. Specifically, a half furnace plus a lengthened part just met the span of the five tuyeres, as shown in Figure 1b,c, and the edge of the flat wall just matched the two tuyeres on each end. This design ensured that the burden lines above the tuyeres could be observed from the flat wall, and the internal structure above the tuyere on the section could be observed after the cutting method. The transparent plate inserted into the BF model was along the direction of the reserved smooth grooves, and at the top of the model there was a top cover with a smooth groove, as shown in Figure 1d. Under the action of instantaneous thrust, the transparent plate could cut the particles in the half furnace into "two parts" through two smooth grooves on the top and bottom, and the burden lines were not affected too much by this cutting method.

2.5. Discrete Element Method

The DEM method by EDEM commercial software (EDEM™, version 2018, DEM solutions, Edinburgh, England) was adopted to simulate the solid flow in the experimental BF model. Since the BF possesses an axisymmetric structure with eight tuyeres on the lower part of the wall, two 3D sector models of a one-eighth circle (45° as shown in Figure 2) were constructed by considering the efficiency and accuracy of calculation. As shown in Figure 2, in the sector model 1, the wall effect was eliminated (i.e., each f_{p-w} is set as zero). Meanwhile, in the sector model 2, the wal -effect was considered (i.e., one of f_{p-w} was set as 0.156). Additionally, in the DEM simulation, the internal angle of the sector was adjusted to a round arc of 5 mm diameter as shown by the cross-section wall in Figure 2, which can weaken the friction effect of the acute angle.

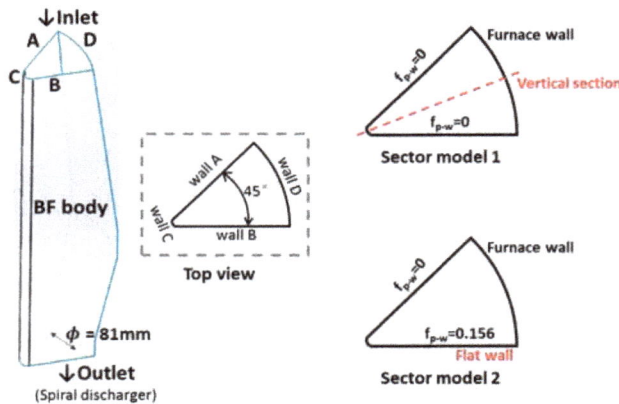

Figure 2. Direct element method (DEM) geometry of the two BF sector models.

Generally, the motions of a particle mainly include the translational and rotational motions, which satisfy Newton's second law and the laws of rotation, respectively. The motion of a particle which is pursued by a multi-body interaction based on the soft sphere approximation can be decided by the contact forces from the geometry and particles. The Hertz–Mindlin [28] model including a spring and dashpot was used in DEM to calculate the contact forces. The trajectory of a particle is obtained by considering the translational and rotational motions of a particle [14]. In detail, the governing equations for a particle *i* can be expressed as follows:

$$m_i \frac{dv_i}{dt} = \sum_j \left(F_{cn,ij} + F_{dn,ij} + F_{ct,ij} + F_{dt,ij} \right) + m_i g \tag{2}$$

$$I_i \frac{d\omega_i}{dt} = \sum_j \left(M_{t,ij} + M_{r,ij} \right) \tag{3}$$

where v_i and ω_i denote the translational velocity (m/s) and rotational velocity (rad/s) of the particle *i*, respectively. m_i is the particle mass in kg/m³. I_i is the moment of inertia of the particle in kg·m², which is given by $I_i = 0.4m_i R^2$. The forces include the gravitational force, $m_i g$, and the contact forces between the particles and particle walls. The contact forces and the damping forces in the normal and tangential directions involved are: $F_{cn,ij}$, $F_{ct,ij}$, $F_{dn,ij}$, and $F_{dt,ij}$ (in units of N), respectively. The torque acting on particle i are: $M_{t,ij}$, which causes particle i to rotate by the tangential force, (in units of N·m) and $M_{r,ij}$, so called the "rolling friction torque", which slows down the relative rotation between particles by the normal force (in units of N·m). The forces and torques used in the model are listed in Table 2.

Table 2. Forces and torques acting on particles i.

Forces and Torques	Symbols	Equations
Normal contact force	$F_{cn,ij}$	$-2/3 S_n |\delta_n| n$
Normal damping force	$F_{dn,ij}$	$-2\sqrt{5/6}\beta\sqrt{S_n m^*} v_{n,ij}$
Tangential contact force	$F_{ct,ij}$	$-S_t |\delta_t| t$
Tangential damping force	$F_{dt,ij}$	$-2\sqrt{5/6}\beta\sqrt{S_t m^*} v_{t,ij}$
Coulomb friction force	$F_{t,ij}$	$-\mu_s |F_{cn,ij} + F_{dn,ij}| t$
Torque by tangential forces	$M_{t,ij}$	$R^* n \times (F_{ct,ij} + F_{dt,ij})$
Rolling friction torque	$M_{r,ij}$	$-\mu_r |F_{cn,ij} + F_{dn,ij}| R \hat{\omega}_i$

Notes: $S_n = 2E^* \sqrt{R^* |\delta_n|}$, $n = \frac{\delta_n}{|\delta_n|}$, $\frac{1}{m^*} = \frac{1}{m_i} + \frac{1}{m_j}$, $\beta = \frac{\ln e}{\sqrt{\ln^2 e + \pi^2}}$, $S_t = 8G^* \sqrt{R^* \delta_n}$, $t = \frac{\delta_n}{|\delta_n|}$, $\frac{1}{E^*} =$ $\frac{1-v_i^2}{E_i} + \frac{1-v_j^2}{E_j}$, $\frac{1}{R^*} = \frac{1}{|R_i|} + \frac{1}{|R_j|}$, $\frac{1}{G^*} = \frac{2(1+v_i)(1-v_i^2)}{E_i} + \frac{2(1+v_j)(1-v_j^2)}{E_j}$, $v_{n,ij} = (v_{ij} \cdot n) \cdot n$, $v_{t,ij} = (v_{ij} \cdot t) \cdot t$, and $\hat{\omega}_i = \frac{\omega_i}{|\omega_i|}$, $v_{ij} = v_j - v_i + \omega_j \times R_j - \omega_i \times R_i$. E^*, δ_n, m^*, R^*, e, G^*, and S_t mean the equivalent Young's modulus, normal amount of overlap, equivalent mass, equivalent radius of the particles, coefficient of restitution, equivalent shear modulus, and tangential stiffness of particles, respectively.

In the DEM simulation, the motion of each particle was traced, and the collision force between the particles was calculated. Equations (2) and (3) were used to record and calculate the velocity of the particles, and thus the motion characteristics of the solid flow were simulated. Specifically, the parameters of the particles in the simulation were shown in Table 3.

Table 3. Parameters of the particles in the DEM simulation.

Parameters	Sector Model 1	Sector Model 2
Particle shape	Spherical	Spherical
Particle motion state	Moving bed	Moving bed
Particle diameter, mm	2.5 (c), 1.25 (o)	2.5 (polyethylene)
Particle density, kg/m^3	1100 (c), 4000 (o)	910 (polyethylene)
Wall density, kg/m^3	7600 (furnace wall)	1200 (acrylic)
Time step, s	1×10^{-4}	1×10^{-4}
Total number	Variable	35,000
Poisson's ratio	0.21 (c), 0.24 (o)	0.49 (polyethylene)
Shear modulus, Pa	1e + 07	1e + 07
Coefficient of restitution	0.3	0.3
Coefficient of interparticle static friction	0.63 (c-c), 0.4 (c-o), 0.32 (o-o)	0.21
Coefficient of interparticle rolling friction	0.05	0.05
Coefficient of static friction (p - wall A)	0	0
Coefficient of static friction (p - wall B)	0	0
Coefficient of static friction (p - wall C)	0	0.156
Coefficient of static friction (p - wall D)	0.56 (c-w), 0.31 (o-w)	0.156
Coefficient of rolling friction (p - w)	0.05	0.05

(c: coke particle; o: ore pellet; p: particle; w: wall)

3. Results

3.1. Vertical Section of the Solid Flow in the Experimental BF Model

To eliminate the flat wall effect, the vertical section of the solid flow was observed by inserting a transparent plate into the experimental BF model using the cutting method. Figure 3 shows the distributions of burden particles and tracer particles in the 3D half-circle BF model observed through the outer wall (flat wall) and the section, in which nine timelines represent the structure of the solid flow.

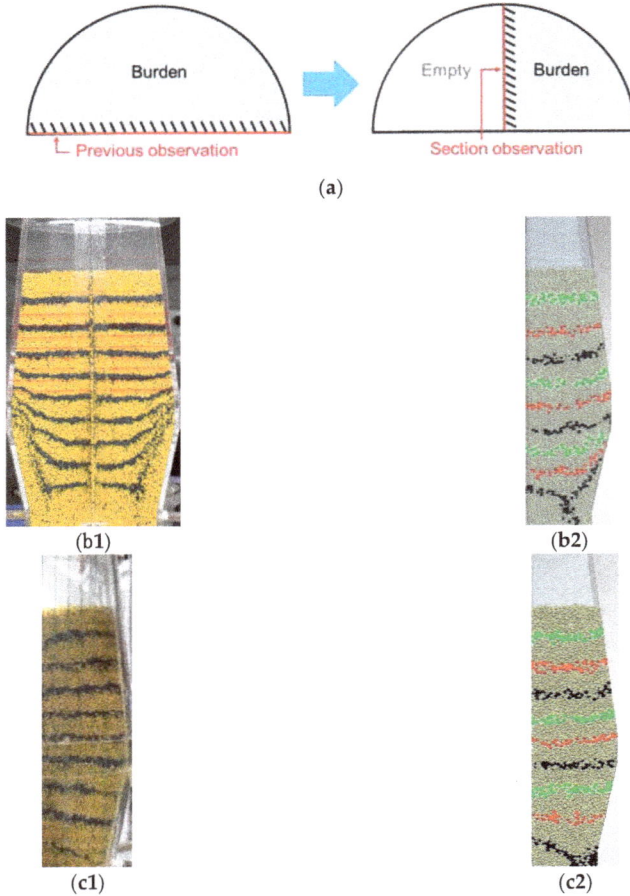

(a)

(b1)

(b2)

(c1)

(c2)

Figure 3. (**a**) The section observation by the cutting method in the 3D half-circle BF model. The distributions of burden particles and tracer particles observed through the flat wall (**b1**) and the internal surface after the cutting method (**c1**) in the experiment. Similarly, the flat wall (**b2**) and the internal surface (**c2**) simulated by sector model 2 and sector model 1, respectively.

Figure 3a shows the section observation by the cutting method. The distributions of burden particles and tracer particles in the 3D half-circle BF model were observed through the vertical section after cutting. As shown in Figure 3b, the solid flow in the bosh region, which was observed from the flat wall, appears as an obvious "W"-shaped convergence flow. Furthermore, in order to further study the solid flow patterns in the BF, the vertical section of the solid flow in the 3D half-circle BF model was observed. Figure 3c1 shows the distributions of burden particles and tracer particles in the 3D half-circle BF model observed through the vertical section after cutting. Clearly, there is a significant difference in the timelines of tracer particles with the results shown in Figure 3b1,c1. Meanwhile, as shown in Figure 3c, the timelines which were observed from the vertical section of solid flow after cutting show approximate straight lines in the shaft area and oblique lines in the bosh area. Compared with the literature [25], there is a longer bosh in this work. The result of the flat wall shows a larger quasi-stagnation zone in the bosh, which illustrates that the shape and length of the bosh can affect the shape of this quasi-stagnation zone. In the hearth region, the experimental result in Figure 3c1 shows that more inclined lines exist in this region, which is different from the results in the literature (the curved lines) [25]. Evidently, the plug flow zone can be extended to the middle of the bosh area. As for the phenomena shown in Figure 3b1, it is likely that, due to the 90° angle between the flat and semicircular walls, the increase in particle velocity at the center and the decrease in particle velocity near the wall led to a long "W" shape in the gradually narrowed bosh region. In addition, it can be found that a vertebral-body deadman exists at the center of the hearth region, and thus, the W shape was obvious at the bottom of the BF model. However, as for the phenomena shown in Figure 3c1, it can be found that when the wall effect was eliminated using the cutting method, the plug flow zone covers most of the area in the BF, which agrees well with the observation shown in the 3D full-circle BF model [27].

Similarly, Figure 3c2 shows an image obtained through the vertical section of the solid flow in the 3D half-circle BF model simulated by sector model 1 via DEM. Evidently, by comparing with the results shown in Figure 3b2, it can be found that, when the wall effect was eliminated, the plug flow is dominant in the BF. Accordingly, the results shown in Figure 3 indicate that only the vertical section of the solid flow can reflect the actual situation of the solid flow due to the elimination of the wall effect. Therefore, in future experiments and simulations, the vertical section of the solid flow should be used to analyze the characteristics of the plug and convergence flows, which is consistent with the literature [25,26]. The experimental cutting method can solve the problem of the wall effect on the burden descending in shaft furnaces, which might have important implications for various industrial applications [24–27].

3.2. Vertical Section of Solid Flow Under Different Burden Layer Conditions

To further understand the solid flow behavior under different burden layer conditions, the solid flows for three types of burden layers were simulated by the sector model 1 via DEM, including the full coke, layered coke/ore pellet, and mixed coke/ore pellet. The physical properties of the coke and ore particles are shown in Table 3. Figure 4a–c shows the images of the vertical section of the solid flow, when the burdens are the full coke, layered coke/ore pellet, and mixed coke/ore pellet, respectively. Evidently, the solid flow behaviors under these three types of burden layers are almost the same, i.e., in the middle and upper part of the BF, the plug flow is dominant, whereas, the convergence flow appears below the middle of the bosh.

It is worth noting that the vertical section of solid flow shown in Figure 4a is similar to that shown in Figure 3c2, indicating that the formation of the timelines in the BF was less affected by the properties of the particles [23,24]. However, it can be found that, after adding ore pellets into the BF models, the timelines in the bosh became straighter. Meanwhile, comparing with the vertical section of solid flow shown in Figure 4b,c, we found that a larger quasi-stagnant zone appeared in Figure 4a. It is likely that ore pellets have smaller scale and smaller roughness than coke particles, and therefore the effect of mixing the coke and ore pellets is becoming more and more obvious in the bosh under gravity. In addition, the relatively poor rolling characteristics of the pure coke pellets may be balanced by the mixing process of the coke and ore pellets, which makes the quasi-stagnant zone further smaller. Nevertheless, the vertical sections of solid flow under different burden layer conditions, as shown in Figure 4, all indicate that the plug flow is the main solid flow pattern in the upper and middle zones during burden descending, whereas a slight convergence flow and a deadman zone form at the lower part of the bosh. Therefore, when analyzing the heat–mass transfer and metallurgical reactions in the BF, the plug flow should be used to model the solid phase, which agrees well with the literature [29]. The results can provide important information for building the one-dimensional model and the two-dimensional model used to simulate the movement and reaction of coke/ore in BFs.

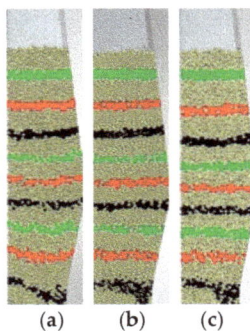

Figure 4. Internal wall of solid flow under three burden layer conditions simulated by the sector model 1 via DEM, i.e., full coke (**a**), layered coke and ore pellet (**b**) and mixed coke and ore pellet (**c**).

3.3. The Boundary of the Plug Flow and Convergence Flow in BF

Due to the significant influence of the solid flow patterns on the smooth operation of BFs, the boundary between the plug flow and convergence flow in BF was determined. As shown in Figure 5a, ten equidistant annular regions (1–10) at different heights along the same section were selected to calculate the regional mean axial velocity of burden. It should be noted that these ten regions are selected from the center of the furnace to the furnace wall, and the data shown in Figure 5a was determined by averaging the values obtained from the three types of the solid flow in Figure 4. Moreover, it can be found that the error bars became large at the lower part of the BF model, i.e., the velocity fluctuations occurred in regions 7–10 at heights of 20–70 mm.

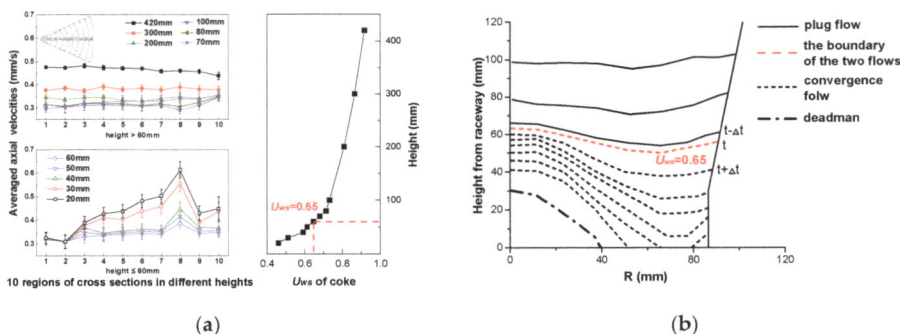

Figure 5. (**a**) Averaged axial velocity in 10 regions at different heights of the BF, and the (U_{ws}) distributions along the height of the BF. (**b**) The boundary between the plug flow and convergence flow in the BF, which was determined by U_{ws}.

Undoubtedly, in an actual production of a BF, the burden timeline in bosh is affected by various factors, such as the cohesive zone and tuyere position. Flow uniformity is an important evaluation index for solid flow. A favorable explanation of the relationship between the appearance of the convergence flow and the differentiation of the particle velocity is necessary. In the paper, based on standard deviation, the Wilcox–Swailes coefficient (i.e., U_{ws}) [30–32] was introduced to evaluate the characteristics of the solid flow, which can be expressed as,

$$U_{ws} = 1 - \frac{\sqrt{\frac{1}{n}\sum_{i=1}^{n}(v_i - \bar{v})^2}}{\bar{v}} \tag{4}$$

where v_i is the local velocity of the labeled particle in m/s, $\bar{v}$ is the average velocity of all particles on the measuring cross section in m/s, and n is the number of particles.

Generally, the velocity stability is favorable when U_{ws} is large, and the maximum value of U_{ws} is 1. As shown in Figure 5a, due to the large U_{ws}, it can be found that the solid flow at the top region of the BF model may be uniform. Meanwhile, the results indicate that, with the burden descending, the U_{ws} decreases drastically in the lower part of the bosh. By combining the regional velocity standard deviation with the U_{ws} of the different working conditions, it can be found that the velocity in the vertical direction changes suddenly at a height of approximately 40–80 mm, which corresponds to $U_{ws} \le 0.7$, as shown in Figure 5a. In addition, Figure 5b shows the timelines in the lower part of the bosh. Evidently, with the burden descending, the plug flow changes to the convergence flow. In detail, the timeline with a U_{ws} of 0.65 is expressed by a red dotted line. The previous time interval $(t - \Delta t)$ and the next time interval $(t + \Delta t)$ are above and below this red dotted line, respectively. It can be found that a roughly equal interval occurred between the timeline $t - \Delta t$ and the timeline t. However, there is a significant difference between the timeline $t + \Delta t$ and the timeline t. Therefore, the U_{ws} with the value of 0.65 can be defined as a critical value to determine the solid flow patterns, i.e., when U_{ws} is higher than 0.65, a plug flow is dominant, whereas, when U_{ws} is lower than 0.65, a convergence flow is dominant. According to the calculation results, the position of the demarcation line for the plug flow and the convergence flow shown in Figure 5b is in agreement with the experimental and simulation results shown in Figures 3 and 4. This method may be used to analyze the motion behavior of large-scale particles in shaft furnaces.

4. Conclusions

To eliminate the wall effect on the structure of solid flow in BFs, a cutting burden method was developed to observe the vertical section of the solid flow by inserting a transparent plate into the experimental BF models. By combining the observations for vertical section of solid flow in experimental BF models via cutting with the simulation results by DEM, we found that the plug flow is the main solid flow pattern in the upper and middle zones during burden descending, whereas a slight convergence flow and a deadman zone form at the lower part of the bosh. The quasi-stagnant zone of the mixed charging method is smaller than that of single particle method. The boundary between the plug flow and convergence flow in the BF can be determined by the velocity distribution of burden and Wilcox–Swailes coefficient, i.e., U_{ws}. When U_{ws} is higher than 0.65, a plug flow is dominant, whereas when U_{ws} is lower than 0.65, a convergence flow is dominant. The findings may have important implications to understand the structure of the solid flow in BFs.

Author Contributions: Data curation, Y.L.; funding acquisition, X.Z.; supervision, J.W.; riting—original draft, Y.L.; writing—review & editing, Z.J. and X.Z. All authors read and approved the manuscript.

Funding: This work was supported by the National Key Research and Development Program of China (2018YFB0605903, 2016YFB0601301).

Conflicts of Interest: The authors declare no conflict of interest.

Appendix A

Before the experiment, a verification experiment was carried out to test the reliability of this cutting method. The transparent thin plate in the experiment is made of organic glass, and this material has a certain bending resistance and strength. The thickness of the plate is 4 mm, and the thickness of the cutting part is less than 1.5 mm.

Firstly, a certain height of yellow particles was placed in the BF model, and then the tracer particles were placed on the top of yellow particle layer. The thickness of the trace particles is 1cm, as shown in Figure A1.

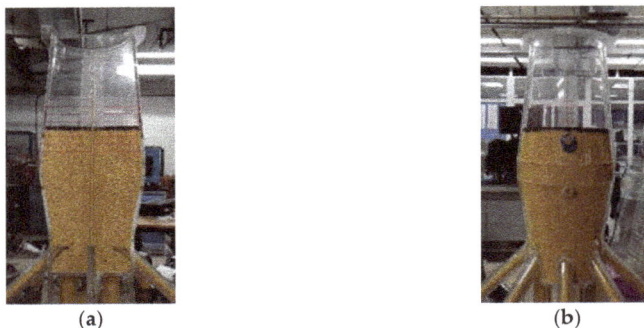

(**a**)	(**b**)

Figure A1. Placing the tracer layer. The view of (**a**) the flat wall and (**b**) the semicircular wall.

In the second step, the yellow particles were filled up in the BF body, as shown in Figure A2.

(**a**)	(**b**)

Figure A2. Filling up the yellow particles. The view of (**a**) the flat wall and (**b**) the semicircular wall.

For the last step, in the process of the cutting method, the plate was instantly inserted into the burden with a short time. As shown in Figure A3, an experimenter held the BF model, and the other experimenter pushed the transparent thin plate into the burden. The plate slides into the burden along the preset chutes on the upper and lower sides of the BF body. At last, the solid particles on the one side of the transparent plate were removed, and the internal surface was exposed, as shown in Figure A3b.

The average thickness of the tracer particles was 1.133 cm after cutting, as shown in Figure 4b, which is increased by 13% compared with the thickness before cutting, as shown in Figure A4a. Though the shape of the tracing line appears to have a little subsidence phenomenon in the position of the entrance, the shape of the half-circle BF model is a little more than the semicircle. Therefore, the selection of the tracing line on the internal surface needs to remove that part of the entrance, like in Figure A4b. From Figure A4, after the cutting method, no significant changes have taken place to the tracer line, with only slight changes in thickness, so it is considered that this method will not destroy the burden structure.

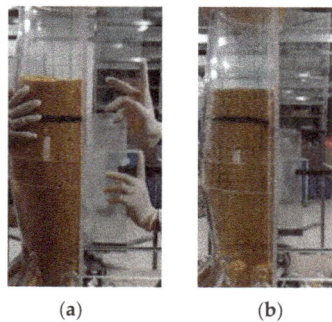

(a) (b)

Figure A3. The cutting method in experiment. (**a**) The process of the cutting method; (**b**) the internal surface after the cutting method.

(a) (b)

Figure A4. Features of the tracer line before and after the cutting method. (**a**) Tracer line on the flat wall; (**b**) tracer line on the section after the cutting method.

References

1. Yagi, J.-I. Mathematical modeling of the flow of four fluids in a packed bed. *ISIJ Int.* **1993**, *33*, 619–639. [CrossRef]
2. Dong, X.; Yu, A.; Yagi, J.-I.; Zulli, P. Modelling of multiphase flow in a blast furnace: Recent developments and future work. *ISIJ Int.* **2007**, *47*, 1553–1570. [CrossRef]
3. Ariyama, T.; Natsui, S.; Kon, T.; Ueda, S.; Kikuchi, S.; Nogami, H. Recent progress on advanced blast furnace mathematical models based on discrete method. *ISIJ Int.* **2014**, *54*, 1457–1471. [CrossRef]
4. Kuang, S.B.; Li, Z.Y.; Yu, A.B. Review on modelling and simulation of blast furnace. *Steel Res. Int.* **2018**, *89*, 1700071. [CrossRef]
5. Cundall, P.A.; Strack, O.D. A discrete numerical model for granular assemblies. *Geotechnique* **1979**, *29*, 47–65. [CrossRef]
6. Ichida, M.; Nishihara, K.; Tamura, K.; Sugata, M.; Ono, H. Influence of ore/coke distribution on descending and melting behavior of burden in blast furnace. *ISIJ Int.* **1991**, *31*, 505–514. [CrossRef]
7. An, X.-W.; Wang, J.-S.; Lan, R.-Z.; Han, Y.-H.; Xue, Q.-G. Softening and melting behavior of mixed burden for oxygen blast furnace. *J. Iron Steel Res. Int.* **2013**, *20*, 11–16. [CrossRef]
8. Liu, J.; Xue, Q.; She, X.; Wang, J. Investigation on interface resistance between alternating layers in the upper of blast furnace. *Powder Technol.* **2013**, *246*, 73–81. [CrossRef]
9. Fu, D.; Chen, Y.; Zhao, Y.; D'Alessio, J.; Ferron, K.J.; Zhou, C.Q. CFD modeling of multiphase reacting flow in blast furnace shaft with layered burden. *Appl. Therm. Eng.* **2014**, *66*, 298–308. [CrossRef]
10. Natsui, S.; Ueda, S.; Fan, Z.; Andersson, N.; Kano, J.; Inoue, R.; Ariyama, T. Characteristics of solid flow and stress distribution including asymmetric phenomena in blast furnace analyzed by discrete element method. *ISIJ Int.* **2010**, *50*, 207–214. [CrossRef]

11. Yang, W.; Zhou, Z.; Yu, A.; Pinson, D. Particle scale simulation of softening–melting behaviour of multiple layers of particles in a blast furnace cohesive zone. *Powder Technol.* **2015**, *279*, 134–145. [CrossRef]
12. Yu, Y.; Saxén, H. Experimental and DEM study of segregation of ternary size particles in a blast furnace top bunker model. *Chem. Eng. Sci.* **2010**, *65*, 5237–5250. [CrossRef]
13. Wu, S.; Kou, M.; Xu, J.; Guo, X.; Du, K.; Shen, W.; Sun, J. DEM simulation of particle size segregation behavior during charging into and discharging from a Paul-Wurth type hopper. *Chem. Eng. Sci.* **2013**, *99*, 314–323. [CrossRef]
14. Dong, Z.; Wang, J.; Zuo, H.; She, X.; Xue, Q. Analysis of gas–solid flow and shaft-injected gas distribution in an oxygen blast furnace using a discrete element method and computational fluid dynamics coupled model. *Particuology* **2017**, *32*, 63–72. [CrossRef]
15. Zhang, H.-J.; She, X.-F.; Han, Y.-H.; Wang, J.-S.; Zeng, F.-B.; Xue, Q.-G. Softening and Melting Behavior of Ferrous Burden under Simulated Oxygen Blast Furnace Condition. *J. Iron Steel Res. Int.* **2015**, *22*, 297–303. [CrossRef]
16. Fan, Z.; Igarashi, S.; Natsui, S.; Ueda, S.; Yang, T.; Inoue, R.; Ariyama, T. Influence of blast furnace inner volume on solid flow and stress distribution by three dimensional discrete element method. *ISIJ Int.* **2010**, *50*, 1406–1412. [CrossRef]
17. Samsu, J.; Zhou, Z.; Pinson, D. Flow and wall stress analysis of granular materials around blocks attached to a wall. *Powder Technol.* **2018**, *330*, 431–444. [CrossRef]
18. Adema, A.T.; Yang, Y.; Boom, R. Discrete element method-computational fluid dynamic simulation of the materials flow in an iron-making blast furnace. *ISIJ Int.* **2010**, *50*, 954–961. [CrossRef]
19. Ping, Z.; Shi, P.-Y.; Song, Y.-P.; Tang, K.-L.; Dong, F. Evaluation of burden descent model for burden distribution in blast furnace. *J. Iron Steel Res. Int.* **2016**, *23*, 765–771.
20. Fu, D.; Chen, Y.; Zhou, C.Q. Mathematical modeling of blast furnace burden distribution with non-uniform descending speed. *Appl. Math. Model.* **2015**, *39*, 7554–7567. [CrossRef]
21. Xu, B.; Yu, A.; Chew, S.; Zulli, P. Numerical simulation of the gas–solid flow in a bed with lateral gas blasting. *Powder Technol.* **2000**, *109*, 13–26. [CrossRef]
22. Hou, Q.; Dianyu, E.; Kuang, S.; Li, Z.; Yu, A. DEM-based virtual experimental blast furnace: A quasi-steady state model. *Powder Technol.* **2017**, *314*, 557–566. [CrossRef]
23. Khodak, L.Z.; Borisov, Y.I. Velocity and pressure distributions of moving granular materials in a model of a shalt kiln. *Powder Technol.* **1971**, *4*, 187–194. [CrossRef]
24. Ho, C.-K.; Chen, Y.-M.; Lin, C.-I.; Jeng, J.-R. On the flow of granular material in a model blast furnace. *Powder Technol.* **1990**, *63*, 13–21. [CrossRef]
25. Wright, B.; Zulli, P.; Zhou, Z.; Yu, A. Gas–solid flow in an ironmaking blast furnace—I: Physical modelling. *Powder Technol.* **2011**, *208*, 86–97. [CrossRef]
26. Zhou, Z.; Zhu, H.; Wright, B.; Yu, A.; Zulli, P. Gas–solid flow in an ironmaking blast furnace-II: Discrete particle simulation. *Powder Technol.* **2011**, *208*, 72–85. [CrossRef]
27. Yang, W.; Zhou, Z.; Yu, A. Discrete particle simulation of solid flow in a three-dimensional blast furnace sector model. *Chem. Eng. J.* **2015**, *278*, 339–352. [CrossRef]
28. Johnson, K.J.C. *Contact Mechanics*; Cambridge University Press: Cambridge, UK, 1985.
29. Jin, P.; Jiang, Z.; Bao, C.; Lu, Y.; Zhang, J.; Zhang, X. Mathematical modeling of the energy consumption and carbon emission for the oxygen blast furnace with top gas recycling. *Steel Res. Int.* **2016**, *87*, 320–329. [CrossRef]
30. Mateos, L. Assessing whole-field uniformity of stationary sprinkler irrigation systems. *Irrig. Sci.* **1998**, *18*, 73–81. [CrossRef]
31. Mateos, L. A simulation study of comparison of the evaluation procedures for three irrigation methods. *Irrig. Sci.* **2006**, *25*, 75–83. [CrossRef]
32. Allaire-Leung, S.; Wu, L.; Mitchell, J.; Sanden, B. Nitrate leaching and soil nitrate content as affected by irrigation uniformity in a carrot field. *Agric. Water Manag.* **2001**, *48*, 37–50. [CrossRef]

metals

MDPI

Article

Distribution Ratio of Sulfur between CaO-SiO$_2$-Al$_2$O$_3$-Na$_2$O-TiO$_2$ Slag and Carbon-Saturated Iron

Kanghui Zhang, Yanling Zhang * and Tuo Wu

State Key Laboratory of Advanced Metallurgy, University of Science and Technology Beijing, Beijing 100083, China; kanghuiz@126.com (K.Z.); wutuo90@163.com (T.W.)
* Correspondence: zhangyanling@metall.ustb.edu.cn; Tel.: +86-10-8237-5191

Received: 28 November 2018; Accepted: 12 December 2018; Published: 15 December 2018

Abstract: To explore the feasibility of hot metal desulfurization using red mud, the sulfur distribution ratio (L_S) between CaO-SiO$_2$-Al$_2$O$_3$-Na$_2$O-TiO$_2$ slag and carbon-saturated iron is evaluated in this paper. First, the theoretical liquid areas of the CaO-SiO$_2$-Al$_2$O$_3$ (-Na$_2$O-TiO$_2$) slag are discussed and the fluxing effects of Al$_2$O$_3$, Na$_2$O, and TiO$_2$ are confirmed. Then, L_S is measured via slag-metal equilibrium experiments. The experimental results show that L_S significantly increases with the increase of temperature, basicity, and Na$_2$O content, whereas it decreases with the increase of Al$_2$O$_3$ and TiO$_2$ content. Na$_2$O in the slag will volatilize with high temperatures and reducing conditions. Furthermore, based on experimental data for the sulfur distribution ratio between CaO-SiO$_2$-Al$_2$O$_3$-Na$_2$O-TiO$_2$ slag and the carbon-saturated iron, the following fitting formula is obtained: $\log L_S = 45.584\Lambda + \frac{10568.406 - 17184.041\Lambda}{T} - 8.529$.

Keywords: sulfur distribution ratio; liquid area; carbon-saturated iron

1. Introduction

Sulfur often deteriorates metal properties [1,2], especially metals' toughness. To realize the deep desulfurization of steel [3,4], the process of hot metal desulfurization has become an economical and efficient method [2]. During this process, traditional lime-based slag has a high melting point. Therefore, CaF$_2$ is widely added as an additive to decrease the melting point and improve the solubility of lime [5]. However, because CaF$_2$ is toxic to the environment and human health, its use has been strictly restricted [6]. Hence, a desulfurizing slag with a much lower melting point, especially under conditions of high basicity, is required. Previous research [5–12] showed that other additives such as Al$_2$O$_3$, Na$_2$O, and TiO$_2$, could improve the desulfurization efficiency of lime-based slag. Niekerk and Dippenaar [5] determined that the Na$_2$O equivalent of CaO was 0.30, and adding Na$_2$O could significantly increase the sulfide capacity of silicate and lime-based slag. Pak and Fruehan [6] reported that the addition of Na$_2$O lowered the melting point and improved the fluidity of lime-based slag. Zhang's experiments [7] obtained good slag fluidity and a much better separation between the slag and the melt phases, attributed to the fact that Al$_2$O$_3$ and Na$_2$O could act as a flux and decrease the melting point of the slag. Yajima et al. [8] found that, with the addition of Al$_2$O$_3$ to the CaO-SiO$_2$-FeO$_x$ slag system at an oxygen partial pressure of 1.8×10^{-3} Pa, the liquid areas were enlarged. Park et al. and Sohn et al. [10,11] confirmed that TiO$_2$ decreased the viscosity of blast furnace slag by depolymerizing the slag structure.

Red mud is the residue discharged by the aluminum industry after the extraction of alumina. In addition to CaO, there is an abundance of Al$_2$O$_3$, Na$_2$O, and TiO$_2$ in red mud. However, it is piled up, pollutes the environment, and increases the burden on enterprises. The sulfur distribution ratio between CaO-SiO$_2$-Al$_2$O$_3$-Na$_2$O-TiO$_2$ slag and carbon-saturated iron is examined in this study to explore the feasibility of hot metal desulfurization using red mud. First, the liquid areas of the $CaO\text{-}SiO_2\text{-}Al_2O_3$ slag system, along with the effects of Al_2O_3, Na_2O, and TiO_2, are investigated thermodynamically using FactSage7.0 software. Then, the sulfur distribution ratio between these slag systems and carbon-saturated iron is examined using an equilibrium experiment in the laboratory.

2. Liquid Areas of the $CaO\text{-}SiO_2\text{-}Al_2O_3$ ($\text{-}Na_2O\text{-}TiO_2$) Slag System

The liquid areas of $CaO\text{-}SiO_2\text{-}Al_2O_3$ slag, simulated by the FactSage7.0 software (developed by CRCT, Montreal, QC, Canada and GTT-Technologies, Herzogenrath, Germany), are shown in Figure 1. Figure 1a shows that the liquid areas at 1400 °C, 1500 °C, and 1600 °C account for about 1/11, 1/4, and 1/2 of the whole diagram area, respectively. This indicates that high temperature is beneficial to the melting of slag. In Figure 1a, there are three lines that correspond to the CaO/SiO_2 of 0.25, 1, and 4, respectively. When CaO/SiO_2 is constant, the melting point of the slag decreases at first and then increases with an increase of Al_2O_3 content, indicating that adding Al_2O_3 can promote the melting of slag at low concentrations, whereas melting deteriorates above a certain content of Al_2O_3. When CaO/SiO_2 is 0.25, 1, and 4, the critical value of Al_2O_3 content is 13.7–19.3%, 15.3–19.1%, and 36.1–43.4%, respectively. Therefore, Al_2O_3 has the effect of a flux, but its concentration should be controlled. Figure 1b,c show the effects of Na_2O and TiO_2 on the liquid areas of $CaO\text{-}SiO_2\text{-}Al_2O_3$ slag at 1500 °C, respectively. Figure 1d shows the effect of a Na_2O and TiO_2 mixture (denoted as "NT") at a mass ratio of $Na_2O/TiO_2 = 2:1$ on the liquid areas of $CaO\text{-}SiO_2\text{-}Al_2O_3$ slag at 1500 °C. The liquid areas become enlarged with the increase of Na_2O, TiO_2, and NT content, showing that these additives could all promote the melting of the slag. However, Na_2O and NT could promote the melting of the slag more effectively than TiO_2.

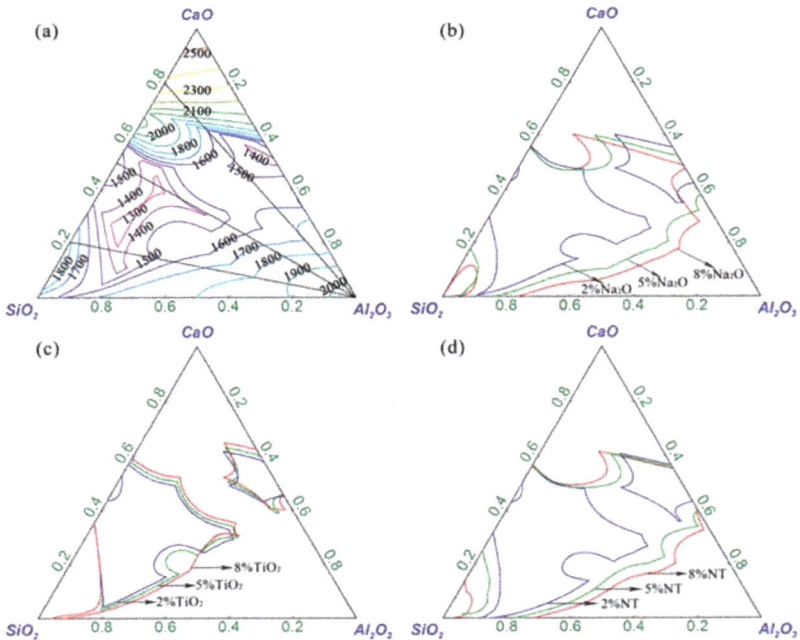

Figure 1. (**a**) Liquid areas of the $CaO\text{-}SiO_2\text{-}Al_2O_3$ slag system between 1200 °C and 2600 °C, (**b**) effect of Na_2O on the $CaO\text{-}SiO_2\text{-}Al_2O_3$ slag system at 1500 °C, (**c**) effect of TiO_2 on the $CaO\text{-}SiO_2\text{-}Al_2O_3$ slag system at 1500 °C, (**d**) effect of NT on the $CaO\text{-}SiO_2\text{-}Al_2O_3$ slag system at 1500 °C.

Therefore, when the temperature is relatively low, it is possible to obtain a good melting effect on slag with the addition of Al_2O_3, Na_2O, and TiO_2. For the actual desulfurization capacity of slag systems and the feasibility of hot metal desulfurization using red mud, slag–metal equilibrium experiments were carried out.

3. Experimental

3.1. Experimental Materials

Iron was prepared by melting electrolytic iron (Fe > 99.6%), high-purity FeS, and graphite particles in an induction furnace. The final iron contained 4.0% carbon and 0.3% sulfur. The base slag was obtained by mixing analytical-grade reagents (i.e., CaO, SiO_2, Al_2O_3, Na_2SiO_3, and TiO_2), and its composition varied around the composition of red mud. Na_2SiO_3 was added as a source of Na_2O [6]. With the help of a muffle furnace, CaO, SiO_2, Al_2O_3, and TiO_2 were calcined at 1000 °C for 2 h, and Na_2SiO_3 was roasted at 300 °C for 2 h to remove carbonates and hydroxides prior to use. The base slag was uniformly mixed in an agate mortar and then formed into cylinders with a diameter of 18 mm, at a pressure of 30 MPa for 2 min. The iron samples were shaped into cylinders with a diameter of 18 mm by wire cutting. The weight of the base slag was about 12 g, and that of the iron sample was about 15 g. The mass ratio of slag to metal was 0.8:1. A graphite crucible (OD = 25 mm, ID = 20 mm, H = 30 mm) was employed in the experiment.

3.2. Experimental Scheme

The experiment mainly consisted of two parts. For experiments T1–Ti3, the effect of temperature, basicity (i.e., CaO/SiO_2), and Al_2O_3, Na_2O, and TiO_2 content on the sulfur distribution ratio were measured by changing a single factor. For experiments 1–21, we referred to the method of uniform design [12]. Multiple factors were changed simultaneously, including temperature, basicity, Al_2O_3, Na_2O, and TiO_2 content. The test points were distributed as evenly as possible within the test range, so that each test point could be representative. The slag composition after equilibration is shown in Table 1.

3.3. Experimental Equipment

The experimental equipment included a horizontal resistance furnace, a gas-purification system, and a water-cooling device, whose schematic diagram is shown in Figure 2. A proportional-integral-derivative (PID) controller controlled the furnace with $MoSi_2$ heating elements. After being calibrated, a Pt-30%Rh/Pt-6%Rh thermocouple was used to measure the temperature. The temperature control range of the furnace was 25–1700 °C, and the temperature accuracy of the heating zone was ± 2 °C. The water-cooling device was circulated with cooling water to control the temperature at the end of the furnace tube. The gas-purification system consisted of allochroic silica gel, a molecular sieve for dehydration, and copper and magnesium pieces (heated to 500 °C) for deoxidation. Through the gas-purification system, high-purity argon (Ar > 99.99%) was introduced into the horizontal furnace tube to protect the samples and graphite crucibles from being oxidized until the experiment's conclusion.

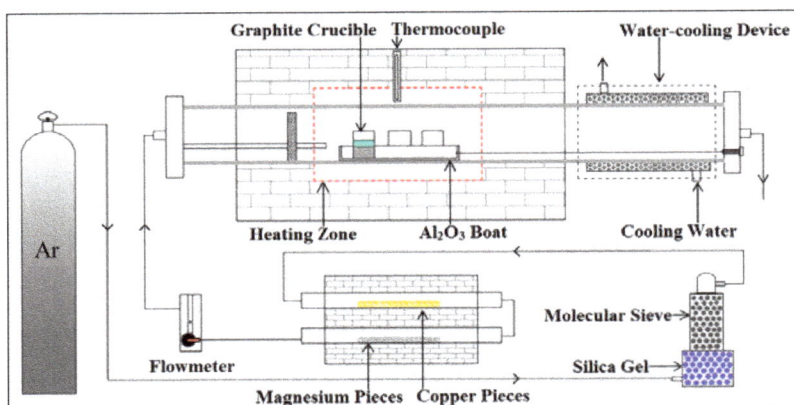

Figure 2. Schematic diagram of the experimental device.

Table 1. Experimental data after equilibration.

| No. | Temp./°C | Slag and Metal-Phase Composition (mass%) | | | | | | | Basicity | Λ | L_S | $\log L_S$ |
		CaO	SiO$_2$	Al$_2$O$_3$	Na$_2$O	TiO$_2$	[%S]	(%S)				
T1	1400	48.29	27.87	17.84	1.80	4.20	0.01345	0.3788	1.73	0.7033	28.16	1.4496
T2	1450	48.99	27.99	17.92	0.89	4.21	0.01049	0.3784	1.75	0.7017	36.07	1.5571
T3	1500	49.47	27.70	18.13	0.40	4.30	0.00914	0.3817	1.79	0.7018	41.76	1.6208
R1		39.14	37.17	17.47	2.13	4.09	0.01376	0.3769	1.05	0.6599	27.39	1.4376
R2		48.99	27.99	17.92	0.89	4.21	0.01049	0.3784	1.75	0.7017	36.07	1.5571
R3		54.04	22.29	18.78	0.55	4.34	0.00475	0.3815	2.42	0.7285	80.32	1.9048
A1		51.51	30.22	13.11	0.96	4.20	0.00783	0.3778	1.70	0.7063	48.25	1.6835
A2		48.99	27.99	17.92	0.89	4.21	0.01049	0.3784	1.75	0.7017	36.07	1.5571
A3	1450	46.20	25.97	22.11	1.34	4.38	0.01300	0.3820	1.78	0.6980	29.38	1.4681
N1		49.74	28.38	17.38	0.33	4.17	0.01126	0.3727	1.75	0.7011	33.10	1.5198
N2		48.99	27.99	17.92	0.89	4.21	0.01049	0.3784	1.75	0.7017	36.07	1.5571
N3		48.01	27.44	18.79	1.46	4.30	0.00947	0.3695	1.75	0.7018	39.02	1.5913
Ti1		50.06	28.57	18.11	1.19	2.07	0.00973	0.3718	1.75	0.7049	38.21	1.5822
Ti2		48.99	27.99	17.92	0.89	4.21	0.01049	0.3784	1.75	0.7017	36.07	1.5571
Ti3		48.19	27.39	17.27	0.62	6.53	0.01062	0.3763	1.76	0.6995	35.43	1.5494
1		45.36	27.11	18.66	2.85	6.02	0.01205	0.3762	1.67	0.6997	31.22	1.4946
2		53.86	23.43	13.88	3.41	5.42	0.00671	0.3854	2.30	0.7386	57.44	1.7591
3		42.81	31.01	19.46	2.03	4.69	0.01461	0.3797	1.38	0.6809	25.99	1.4147
4	1400	53.22	26.69	15.78	0.74	3.57	0.01066	0.3733	1.99	0.7177	35.02	1.5444
5		36.00	34.27	22.51	3.50	3.72	0.01864	0.3801	1.05	0.6598	20.39	1.3095
6		47.94	27.35	19.13	2.51	3.07	0.01180	0.3831	1.75	0.7058	32.47	1.5114
7		56.74	24.49	14.77	1.78	2.22	0.00704	0.3778	2.32	0.7388	53.66	1.7294
8		43.07	30.69	19.17	1.14	5.93	0.01328	0.3785	1.40	0.6790	28.50	1.4549
9		50.80	25.66	16.76	0.92	5.86	0.01083	0.3774	1.98	0.7127	34.85	1.5420
10		36.43	33.92	22.06	2.32	5.27	0.02115	0.3793	1.07	0.6576	17.93	1.2536
11	1450	48.99	27.99	17.92	0.89	4.21	0.01049	0.3784	1.75	0.7015	36.07	1.5571
12		58.63	23.87	13.15	0.74	3.61	0.00671	0.3873	2.46	0.7427	57.72	1.7613
13		46.35	31.45	18.91	0.68	2.61	0.01175	0.3827	1.47	0.6856	32.57	1.5128
14		54.27	25.09	15.83	2.29	2.52	0.00746	0.3865	2.16	0.7310	51.81	1.7142
15		39.55	32.00	21.95	0.74	5.76	0.01783	0.3789	1.24	0.6644	21.25	1.3274
16		49.71	27.19	16.98	0.68	5.44	0.00814	0.3731	1.83	0.7050	45.84	1.6611
17		57.58	24.03	13.39	0.26	4.74	0.00672	0.3696	2.40	0.7367	55.00	1.7401
18	1500	44.20	30.36	19.98	0.46	5.00	0.01606	0.3719	1.46	0.6803	23.16	1.3645
19		54.97	25.67	15.06	0.21	4.09	0.00827	0.3744	2.14	0.7237	45.27	1.6560
20		44.44	31.18	21.14	0.39	2.85	0.01289	0.3811	1.43	0.6790	29.57	1.4708
21		54.60	27.13	15.77	0.12	2.38	0.00708	0.3662	2.01	0.7187	51.72	1.7138

3.4. Experimental Procedure

Pre-prepared slag and iron were placed in the graphite crucible. At the same temperature, an Al_2O_3 boat can hold five graphite crucibles simultaneously. They were positioned in the heating zone of the furnace with the help of a molybdenum bar. Ar gas was introduced into the furnace tube at the flow rate of 500 mL/min. The furnace was then switched on. Before this, an experiment had been carried out to determine the equilibrium time. Figure 3 shows the change of sulfur content with time. (i.e., in experiment T1, the mass ratio of slag to metal was 0.3:1.) It was found that the sulfur content was almost unchanged after 4 h. Therefore, an equilibrium time of 4.5 h was chosen to ensure a complete reaction [13,14]. When equilibrium was reached, the Al_2O_3 boat was pulled out of the furnace tube immediately for quenching. After being cooled, the slag and iron were separated. The slag was dried and then crushed into 200 mesh particles. The iron was washed with a steel brush and ultrasonically cleaned to remove surface residues. The sulfur content in slag and iron was detected by a carbon-sulfur analyzer (EMIA-920V2, HORIBA, Kyoto, Japan), and the composition of the slag was determined by an X-Ray fluorescence spectrometer (XRF-1800, Shimadzu, Kyoto, Japan). All experimental results are listed in Table 1.

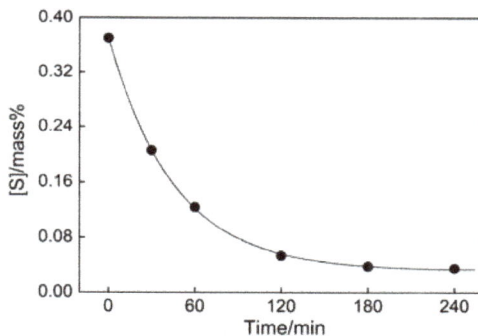

Figure 3. The change of sulfur content in the metal with time.

4. Results and Discussion

4.1. Effect of Temperature

The temperature dependence of $logL_S$ is shown in Figure 4. The data in Figure 4a are the results of experiments T1–T3, shown in Table 1. For these experiments, the slag compositions were similar. Figure 4b shows the dependence of $logL_S$ on the temperature for all five components of the slag system (CaO-SiO_2-Al_2O_3-Na_2O-TiO_2), whose compositions varied significantly (i.e., experiments 1–21 in Table 1). As it can be seen from Figure 4a, $logL_S$ increased from 1.45 to 1.62 when the temperature increased from 1400 °C to 1500 °C, for similar slag compositions. The desulfurization reaction is endothermic. Thus, the high temperature promotes the migration of sulfur from the metal into the slag phase. Simeonov et al. [13] investigated the sulfur distribution ratio between CaO-SiO_2-Al_2O_3-CaF_2 slag and carbon-saturated iron from 1450 °C to 1600 °C and learned that $logL_S$ increased with the increase in temperature. The value of $logL_S$ in Simeonov's experiment was much larger than that in this paper, mainly because of the higher basicity in their study. (The basicity in their study was 4.4, whereas, here, it was only about 1.75). Lin et al. [15] measured the sulfur distribution ratio between CaO-SiO_2-Al_2O_3-MgO-TiO_2 slag and carbon-saturated iron from 1450 °C to 1550 °C. The value of $logL_S$ in Lin's experiment was much larger than that in this paper, because of the low content of weak acid oxides ($Al_2O_3 = 10\%$, $TiO_2 = 0$–8%).

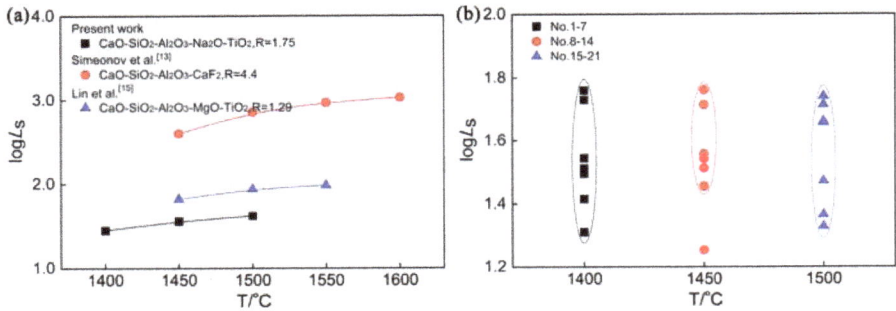

Figure 4. Effect of temperature on $\log L_S$.

However, this trend did not hold for the data shown in Figure 4b, which suggests that the temperature has no significant effect on $\log L_S$ when the slag compositions vary significantly. Despite the different requirements of temperature, a similar tendency was observed in the research [16] on the phosphorus distribution ratio between the $CaO-FeO-SiO_2-Al_2O_3-Na_2O-TiO_2$ slag and the carbon-saturated iron, which suggests that the phosphorus distribution ratio showed much greater dependency on the slag composition, such as on basicity rather than on temperature. Together, these results indicate that, for hot metal pretreatment, slag compositions tend to have significant effects on the desulfurization/dephosphorization efficiency.

4.2. Effect of Basicity

Figure 5a shows the effect of basicity (i.e., CaO/SiO_2) on $\log L_S$. The data are the results of experiments R1–R3, for which the other factors of the slag are similar, but the basicity is different. The value of $\log L_S$ increased from 1.44 to 1.90 with an increase in basicity from 1.05 to 2.42. This trend is in accordance with previous research [17–20]. The value of $\log L_S$ in Huang's experiments [17] (shown in Figure 5a) was relatively low, mainly because of the high content of weak acid oxides ($Al_2O_3 = 15\%$, $TiO_2 = 25\%$). The increase of basicity implies the disintegration of the silicate network structure [18] and the increase of the free O^{2-} concentration in the slag. A high concentration of free O^{2-} thermodynamically promotes the desulfurization reaction [11].

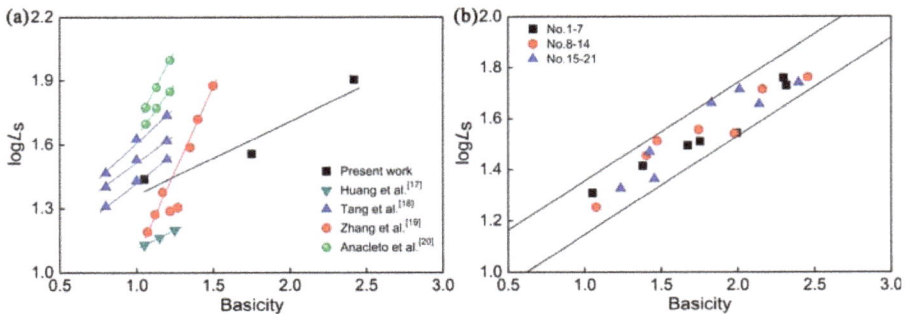

Figure 5. Effect of basicity on $\log L_S$.

The data in Figure 5b are the results of experiments 1–21. The temperature had three gradients, and the composition of slag was changed simultaneously at each temperature. Irrespective of Al_2O_3, Na_2O, and TiO_2 content and temperature, $\log L_S$ showed a strong dependence on the basicity of the slag, which linearly increased with the increase in basicity. This suggests that basicity has a much stronger effect on $\log L_S$ than other influencing factors, such as temperature, under this experimental condition.

4.3. Effect of Al₂O₃

Figure 6a shows the effect of Al_2O_3 content on $logL_S$. The data are the results of experiments A1–A3, for which all slag factors were similar, with the exception of Al_2O_3 content. As can be seen, the value of $logL_S$ decreased from 1.68 to 1.47 with an increase in Al_2O_3 content from 13.11% to 22.11%. This was due to the fact that Al_2O_3 acted as an acidic oxide in the basic slag. Al_2O_3 consumed the free O^{2-} to form $[AlO]_4^{5-}$-tetrahedron, which decreased free O^{2-} concentration and weakened the desulfurization capacity of the slag [21]. The same experimental trend was observed by Zhang et al. [19]. They obtained a much lower sulfur distribution ratio, even at a higher temperature (1500 °C), than the present work (1450 °C), mainly because of the lower basicity (1.17) of the slag in their study.

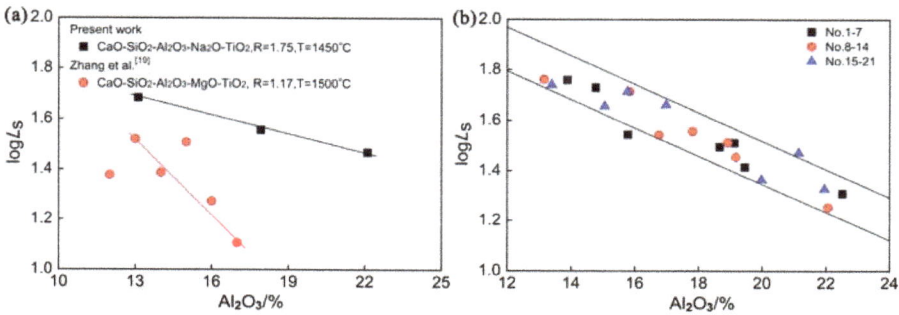

Figure 6. Effect of Al_2O_3 on $logL_S$.

The data in Figure 6b are the results of experiments 1–21. The temperature had three gradients, and the composition of slag was changed simultaneously at each temperature. Irrespective of basicity, Na_2O and TiO_2 content, and temperature, $logL_S$ also showed a strong dependence on Al_2O_3 content. This suggests that Al_2O_3 is also an important factor affecting the sulfur distribution ratio under this experimental condition.

4.4. Effect of Na₂O

Figure 7a shows the effect of Na_2O content on $logL_S$. The data are the results of experiments N1–N3, for which the other factors of the slag are similar. As can be seen, $logL_S$ linearly increased from 1.52 to 1.59, with an increase in Na_2O content from 0.33% to 1.46%. This was because Na_2O is a strong basic oxide, and Na^+ tends to have a strong affinity for S^{2-}. Niekerk et al. [5] pointed out that the addition of Na_2O significantly increased the sulfide capacity of silicate and lime-based slag. Additionally, the Na_2O equivalent of CaO was determined to be 0.30. Pak and Fruehan [6,22] suggested that Na_2O could lower the melting point of a lime-based slag, reduce the consumption of acid oxides to CaO, and significantly increase the activity of CaO. Subsequently, the desulfurization capacity of the slag was enhanced.

The data in Figure 7b are the results of experiments 1–21. The temperature had three gradients, and the composition of the slag was changed simultaneously at each temperature. When other factors (e.g., temperature) changed, Na_2O content in the slag had no significant effect on $logL_S$. This could be due to the experimental conditions. Because of the strong reduction potential of hot metal, Na_2O can be reduced to metal Na, which can easily evaporate into the gas phase, especially at a high temperature. Therefore, the higher the temperature, the more Na vaporizes into the gas phase, and less residual Na_2O content remains in the slag. As shown in Figure 7b, the Na_2O content was 0.74–3.50% (square symbols in Figure 7b) at the lower temperature (1400 °C), while at a high temperature (1500 °C), the Na_2O content was only 0.12–0.74% (triangular symbols in Figure 7b). The promoting effect on desulfurization of higher Na_2O content was eventually counteracted by lower temperatures, as in the case where the negative influence of lower Na_2O content tended to be neutralized by higher temperatures. Comprehensively, Na_2O content showed no significant effect on $logL_S$ under this experimental condition.

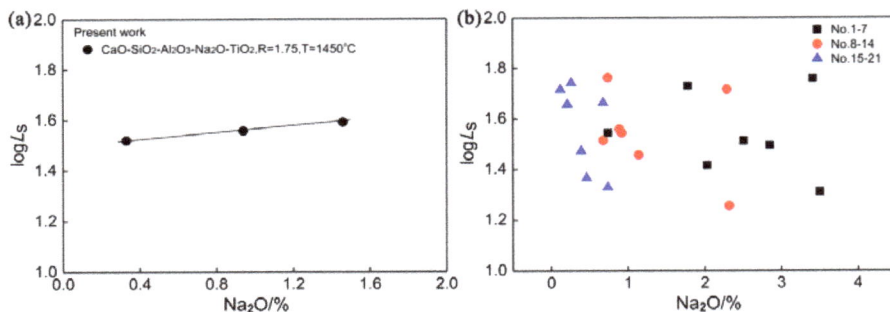

Figure 7. Effect of Na_2O on $logL_S$.

However, a different tendency was observed in the case of hot metal dephosphorization. The results of hot metal dephosphorization by Li [16] showed that, when changing various factors, such as temperature and slag compositions, the distribution ratio of phosphorus between the slag and iron-saturated iron (L_P) increased significantly with the increase of Na_2O content. This was mainly because hot metal dephosphorization was carried out in an oxidizing atmosphere, and Na_2O in the slag was hard to be reduced to metal Na. The volatilization of Na_2O is much weaker than that of metal Na. Finally, more Na_2O is maintained, and it plays the role of a strong basic oxide in the slag phase. Consequently, hot metal dephosphorization is promoted by high Na_2O content.

4.5. Effect of TiO_2

Figure 8a shows the effect of TiO_2 content on $logL_S$. The data are the results of experiments Ti1–Ti3, for which all factors of the slag were similar except for TiO_2 content. As can be seen, the value of $logL_S$ decreased slightly from 1.58 to 1.55 with an increase in TiO_2 content from 2.07% to 6.53%. This is in accord with the results of Lin et al. and Tang et al. [15,18], who investigated the sulfur distribution ratio between CaO-SiO_2-Al_2O_3-MgO-TiO_2 slag and carbon-saturated iron at 1500 °C. The electrostatic potential of Ti^{4+} (5.88) is lower than that of Si^{4+} (9.76). Therefore, the Ti-O bond is weaker than the Si-O bond, which causes TiO_2 to exist as $[TiO]_6^{8-}$-octahedron in a basic slag. Sommerville et al. [23] also suggested that TiO_2 acted as an acidic oxide in a basic slag and existed in the form of $[TiO]_6^{8-}$-octahedron, which decreased the amount of free O^{2-} and weakened the desulfurization capacity of the slag.

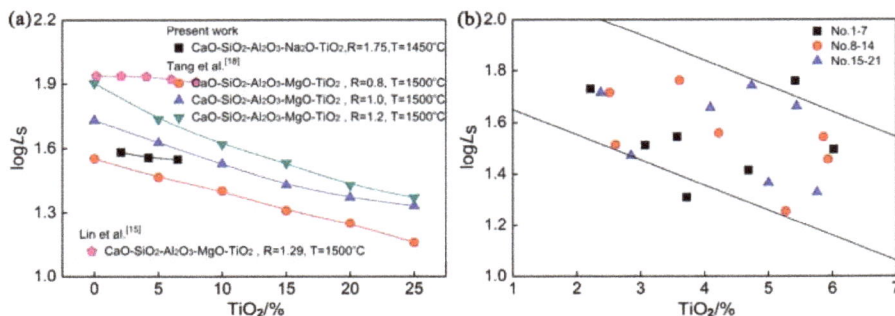

Figure 8. Effect of TiO_2 on $logL_S$.

The data in Figure 8b are the results of experiments 1–21. The temperature had three gradients, and the composition of the slag was changed simultaneously at each temperature. Although a relatively large fluctuation was observed, the total logL_S decreased with an increase in the TiO$_2$ content. This indicates that TiO$_2$ in the slag has also a thermodynamically negative effect on the desulfurization capacity of the slag.

4.6. Regression Analysis of logL$_S$

To further study the relationship between logL_S, Λ, and T, based on the data of experiments 1–21 in Table 1, the following regression equation was obtained by using the SPSS (Statistical Product and Service Solutions) software (SPSS10.91, SPSS Inc., Chicago, IL, USA) [24]. A comparison between the experimental values and the calculated values of logL_S is shown in Figure 9, which shows good agreement.

$$\log L_S = 15.584\Lambda + \frac{10568.406 - 17184.041\Lambda}{T} - 8.529, \ (r = 0.963) \tag{1}$$

where L_S is the sulfur distribution ratio; T is the temperature, °C; Λ is the optical basicity of the slag [25]; r denotes the correlation coefficient.

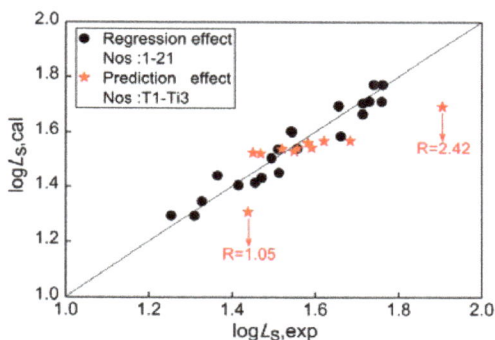

Figure 9. Comparison of experimental values and calculated values.

When the temperature was 1400–1500 °C, basicity was 1.0–2.5, Al$_2$O$_3$ content was 12–22%, Na$_2$O content was 0–3%, and TiO$_2$ content was 2–6%. The equation could be helpful in predicting the sulfur distribution ratio between CaO-SiO$_2$-Al$_2$O$_3$-Na$_2$O-TiO$_2$ slag and carbon-saturated iron.

5. Conclusion

The sulfur distribution ratio between CaO-SiO$_2$-Al$_2$O$_3$-Na$_2$O-TiO$_2$ slag and carbon-saturated iron was re-evaluated in this study. Based on the theoretical calculation and equilibrium experiments, the conclusions are summarized as follows.

(1) The thermodynamic calculation shows that high temperature helps the melting of the slag. Al$_2$O$_3$ is beneficial as a flux. However, the content should be controlled. Na$_2$O can promote the melting effect on the slag more effectively than TiO$_2$.

(2) The experimental data suggest that the distribution ratio of sulfur between the slag and the carbon-saturated iron strongly increases with the increase of temperature, basicity, and Na$_2$O content, whereas it decreases with the increase of Al$_2$O$_3$ and TiO$_2$ content. Na$_2$O in the slag will volatilize because of the high temperature and the reducing conditions.

(3)   Based on the experimental data, for the distribution ratio of sulfur between CaO-SiO$_2$-Al$_2$O$_3$-Na$_2$O-TiO$_2$ slag and carbon-saturated iron, the following fitting formula is obtained:

$$\log L_S = 15.584\Lambda + \frac{10568.406 - 17184.041\Lambda}{T} - 8.529$$

Author Contributions: K.Z. and Y.Z. designed the experiments; K.Z. performed the experiments, analyzed the data, and compiled the text; T.W. reviewed the manuscript.

Funding: The authors would like to acknowledge the financial support of National Key R&D Program of China (No. 2017YFC0210301) and the National Natural Science Foundation of China (No. 51474021) for this work.

Conflicts of Interest: The authors declare no conflict of interest.

References

1.   Mohassab-Ahmed, M.Y.; Sohn, H.Y.; Kim, H.G. Sulfur distribution between liquid iron and magnesia-saturated slag in H$_2$/H$_2$O atmosphere relevant to a novel green ironmaking technology. *Ind. Eng. Chem. Res.* **2012**, *51*, 3639–3645. [CrossRef]

2.   Iwamasa, P.K.; Fruehan, R.J. Effect of FeO in the slag and silicon in the metal on the desulfurization of hot Metal. *Metall. Mater. Trans. B* **1997**, *28*, 47–57. [CrossRef]

3.   Gandarias, A.; Lopez de Lacalle, L.N.; Aizpitarte, X.; Lamikiz, A. Study of the performance of the turning and drilling of austenitic stainless steels using two coolant techniques. *Int. J. Mach. Mach. Mater.* **2008**, *3*, 1–17. [CrossRef]

4.   Rodríguez, A.; Lopez de Lacalle, L.N.; Calleja, A.; Fernández, A.; Lamikiz, A. Maximal reduction of steps for iron casting one-of-a-kind parts. *J. Clean. Prod.* **2012**, *24*, 48–55. [CrossRef]

5.   Van Niekerk, W.H.; Dippenaar, R.J. Thermodynamic aspects of Na$_2$O and CaF$_2$ containing lime-based slags used for the desulphurization of hot-metal. *ISIJ Int.* **1993**, *33*, 59–65. [CrossRef]

6.   Pak, J.J.; Fruehan, R.J. The effect of Na$_2$O on dephosphorization by CaO-based steelmaking slags. *Metall. Mater. Trans. B* **1991**, *22*, 39–46. [CrossRef]

7.   Zhang, Y.; Li, F.; Wang, R.; Tian, D. Application of bayer red mud-based flux in the steelmaking process. *Steel Res. Int.* **2017**, *88*, 304–313. [CrossRef]

8.   Yajima, K.; Matsuura, H.; Tsukihashi, F. Effect of simultaneous addition of Al$_2$O$_3$ and MgO on the liquidus of the CaO-SiO$_2$-FeO$_x$ system with various oxygen partial pressures at 1573 K. *ISIJ Int.* **2010**, *50*, 191–194. [CrossRef]

9.   Chan, A.H.; Fruehan, R.J. The sulfur partition ratio with Fe-C$_{SAT}$ melts and the sulfide capacity of CaO-SiO$_2$-Na$_2$O-(Al$_2$O$_3$) slags. *Metall. Mater. Trans. B* **1989**, *20*, 71–76. [CrossRef]

10. Park, H.; Park, J.Y.; Kim, G.H.; Sohn, I. Effect of TiO$_2$ on the viscosity and slag structure in blast furnace type slags. *Steel Res. Int.* **2012**, *83*, 150–156. [CrossRef]

11. Sohn, I.; Wang, W.; Matsuura, H.; Tsukihashi, F.; Min, D.J. Influence of TiO$_2$ on the viscous behavior of calcium silicate melts containing 17 mass% Al$_2$O$_3$ and 10 mass% MgO. *ISIJ Int.* **2012**, *52*, 158–160. [CrossRef]

12. Fang, K.T.; Wang, Y. *Number-Theoretic Methods in Statistics*; CRC Press: Boca Raton, FL, USA, 1993.

13. Simeonov, S.R.; Ivanchev, I.N.; Hainadjiev, A.V. Sulphur equilibrium distribution between CaO-CaF$_2$-SiO$_2$-Al$_2$O$_3$ slags and carbon-saturated Iron. *ISIJ Int.* **1991**, *31*, 1396–1399. [CrossRef]

14. Tsao, T.; Katayama, H.G. Sulphur distribution between liquid iron and CaO-MgO-Al$_2$O$_3$-SiO$_2$ slags used for ladle refining. *ISIJ Int.* **1986**, *26*, 717–723. [CrossRef]

15. Lin, Y.; Yin, W. Desulfurization ability of the blast furnace slag with low titanium oxide content. *Iron Steel.* **1987**, *22*, 4–9. [CrossRef]

16. Li, F.; Li, X.; Yang, S.; Zhang, Y. Distribution ratios of phosphorus between CaO-FeO-SiO$_2$-Al$_2$O$_3$/Na$_2$O/TiO$_2$ slags and carbon-saturated iron. *Metall. Mater. Trans. B* **2017**, *48*, 2367–2378. [CrossRef]

17. Huang, Z.; Yang, Z.; Guo, T. Effect of MgO in blast furnance type slags containing TiO$_2$ on slag desulpurization. *J. Northeast Univ. Nat. Sci.* **1987**, *3*, 335–339.

18. Tang, X.; Xu, C. Sulphur distribution between CaO-SiO$_2$-TiO$_2$-Al$_2$O$_3$-MgO slag and carbon-saturated iron at 1773 K. *ISIJ Int.* **1995**, *35*, 367–371. [CrossRef]

19. Zhang, J.; Lv, X.; Yan, Z.; Oin, Y.; Bai, C. Desulphurisation ability of blast furnace slag containing high Al_2O_3 and 5 mass% TiO_2 at 1773 K. *Ironmak. Steelmak.* **2016**, *43*, 378–384. [CrossRef]
20. Anacleto, N.M.; Lee, H.-G.; Hayes, P.C. Sulphur partition between CaO-SiO_2-Ce_2O_3 slags and carbon-saturated iron. *ISIJ Int.* **1993**, *33*, 549–555. [CrossRef]
21. Park, J.H.; Min, D.J.; Song, H.S. Amphoteric behavior of alumina in viscous flow and structure of CaO-SiO_2(-MgO)-Al_2O_3 slags. *Metall. Mater. Trans. B* **2004**, *35*, 269–275. [CrossRef]
22. Pak, J.J.; Ito, K.; Fruehan, F.J. Activities of Na_2O in CaO-based slags used for dephosphorization of steel. *ISIJ Int.* **1989**, *29*, 318–323. [CrossRef]
23. Sommerville, I.D.; Bell, H.B. The behaviour of titania in metallurgical slags. *Can. Metall. Q.* **1982**, *21*, 145–155. [CrossRef]
24. Yockey, R.D. *SPSS Demystified: A Step by Step Approach*; Prentice Hall: Upper Saddle River, NJ, USA, 2010.
25. Yang, Y.D.; Sommerville, I.D.; McLean, A. Some fundamental considerations pertaining to oxide melt interactions and their influence on steel quality. *Trans. Indian Inst. Met.* **2006**, *59*, 655–669.

metals

MDPI

Article

Kinetics of the Volume Shrinkage of a Magnetite/Carbon Composite Pellet during Solid-State Carbothermic Reduction

Guang Wang *, Jingsong Wang and Qingguo Xue

State Key Laboratory of Advanced Metallurgy, University of Science and Technology Beijing, Beijing 100083, China; wangjingsong@ustb.edu.cn (J.W.); xueqingguo@ustb.edu.cn (Q.X.)
* Correspondence: wangguang@ustb.edu.cn; Tel.: +86-010-82376018

Received: 21 October 2018; Accepted: 1 December 2018; Published: 11 December 2018

Abstract: The volume shrinkage evolution of a magnetite iron ore/carbon composite pellet during solid-state isothermal reduction was investigated. For the shrinkage, the apparent activation energy and mechanism were obtained based on the experimental results. It was found that the volume shrinkage highly depended on the reduction temperature and on dwell time. The volume shrinkage of the pellet increased with the increasing reduction temperature, and the rate of increment was fast during the first 20 min of reduction. The shrinkage of the composite pellet was mainly due to the weight loss of carbon and oxygen, the sintering growth of gangue oxides and metallic iron particles, and the partial melting of the gangue phase at high temperature. The shrinkage apparent activation energy was different depending on the time range. During the first 20 min, the shrinkage apparent activation energy was 51,313 J/mol. After the first 20 min, the apparent activation energy for the volume shrinkage was only 19,697 J/mol. The change of the reduction rate-controlling step and the automatic sintering and reconstruction of the metallic iron particles and gangue oxides in the later reduction stage were the main reasons for the aforementioned time-dependent phenomena. The present work could provide a unique scientific index for the illustration of iron ore/carbon composite pellet behavior during solid-state carbothermic reduction.

Keywords: carbon composite pellet; direct reduction; shrinkage; kinetics; rotary hearth furnace

1. Introduction

Carbon composite pellets are made from a mixture of pulverized iron ore and carbonaceous material by briquetting or pelletizing at room temperature [1]. The iron oxide and the reducing agent contact closely with the large contact area, and the reduction rate is very fast during heating at temperatures from 1100 to 1300 °C. It can be used for processing metallurgical dust [2], for producing direct reduced iron (DRI) [3], and for the utilization of complex iron ores [4–6]. If the reduction temperature is further increased above 1350 °C, the composite pellet will achieve melting separation and the slag-free pig iron nugget may be obtained, which is a high-grade burden of electric arc furnace (EAF) steelmaking [7,8]. Generally, the carbon composite pellet is charged on the smooth refractory hearth of the rotary hearth furnace (RHF), in no more than two pellet layers, and the pellet is heated by the thermal radiation of the furnace's gas and wall.

Most of the researchers in the field focus on the reduction rate and mechanism of the carbon composite pellet. However, one of the critical shortcomings of the RHF for iron ore reduction is its poor heat diffusion efficiency, when compared with the gas–solid countercurrent blast furnace. The heat from the furnace's gas and wall cannot efficiently be transported to the lower layers of a multilayer bed [9], which results in lower productivity. How to resolve the problem is also very meaningful for the further development of the technology. Fortunately, a few studies on the shrinkage of carbon composite pellets have been reported in the literature. Halder [9] found that if the volume of the top-layer pellets of a pellet bed shrinks by 30%, then the external heat transfer to the second layer would increase by about six times. Therefore, the selection of easily shrinking iron ore and the development of an efficient method to strengthen the volume shrinkage of the pellet is of great significance for RHF reduction technology. Generally, the composite pellet shrinks considerably during the reduction due to the loss of carbon and oxygen from the system, sintering of the iron oxide, and the formation of a molten slag phase inside the pellet. Some researchers used empirical correlation to denote the pellet shrinkage when building the mathematical model of carbon composite pellet reduction [10,11]. However, each iron ore has its own unique shrinkage characteristics and it should be described in a more scientific method.

The shrinkage activation energy is a unique index for the illustration of the volume shrinkage of a carbon composite pellet during carbothermic reduction from the scientific viewpoint. However, few researchers have discussed and calculated the so-called volume shrinkage activation energy in their research work [9–11]. In the present paper, the reduction experiment of a carbon composite pellet with ordinary magnetite has been conducted at the laboratory scale, and then, for the shrinkage, the apparent activation energy was obtained by kinetics analysis. The present work can also provide a reference for studies on the reduction of other kinds of iron ores, as well as to help improve the development of highly efficient RHF reduction technology.

2. Materials and Methods

2.1. Raw Materials

The iron concentrate used in the present study was ordinary magnetite. The chemical composition of the ore sample and reducing agent are shown in Table 1. The mineralogical analysis of the concentrate was investigated by X-ray diffraction (XRD) and showed that the main crystalline phases were magnetite (Fe_3O_4) and a small amount of quartz (SiO_2), which is shown in Figure 1. The particle size of the iron ore concentrate was tested by a particle size analyzer and the results are shown in Figure 2, which illustrates that about 93% of the ore particles are smaller than 0.074 mm. The reducing agent is a kind of anthracite with high fixed carbon content and low sulfur content. The fineness of the reducing agent is 100% passing through a 0.18-mm mesh. The morphology of the magnetite concentrate powder is shown in Figure 3.

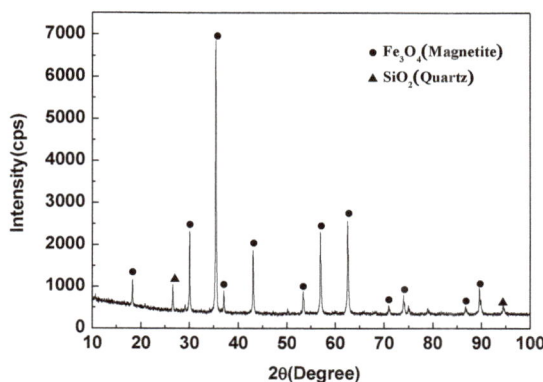

Figure 1. Results of the X-ray diffraction analysis of the iron concentrate.

Table 1. Chemical composition of the raw materials (wt %).

Iron Concentrate	TFe	FeO	SiO$_2$	Al$_2$O$_3$	CaO	MgO	P	S
	64.6	29.3	7.21	0.18	0.17	0.20	0.01	0.21

Reducing Agent	FC$_d$	V$_d$	A$_d$	S				
	81.40	6.40	11.10	0.34				

Note: TFe is the total iron content, FC$_d$ stands for the fixed carbon (dry basis), V$_d$ is the volatile matter (dry basis), A$_d$ denotes ash (dry basis), and S is the total sulfur.

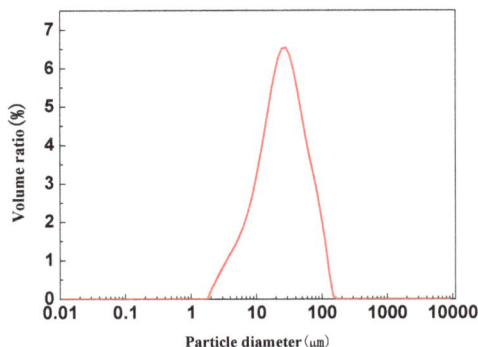

Figure 2. Particle size distribution of the iron concentrate.

Figure 3. Particle morphology of the iron concentrate.

2.2. Experimental Procedure

In the preparation of the raw materials, the molar ratio between the fixed carbon in the reducing agent and the oxygen in the iron oxide in the iron concentrate was set as 1.0. Then, the iron concentrate and the reducing agent were fully mixed together to make the composition of the raw mixture uniform. The moisture of the raw material was set as 7%. The pelletizing process was performed through a manual ball press under the pressure of 15 MPa. The pellet presented as column-like in shape and the size was 20 mm (diameter) × 10 mm (height). The green pellet was dried at 105 °C for 12 h before the reduction test. The reduction experiment was performed in a thermogravimetric system with a shaft MoSi$_2$ resistance furnace, under the protection of high-purity N$_2$ with a flow of 3 L/min. The dry composite pellet was put into a corundum crucible, and the crucible was suspended by Fe–Cr–Al–Mo wire attached to an electronic balance and was heated at different temperatures. The heating temperatures were set at 1000 °C, 1050 °C, 1100 °C, 1150 °C, and 1200 °C in the reduction experiment. The full reduction time was 30 min. Once the reduction finished, the reduced pellet was taken out of the furnace and was quickly cooled to room temperature under the protection of high-purity N_2. The course of reaction can be expressed in terms of the "reaction fraction" (f), defined as in Equation (1):

$$f = W_t/W_0 \tag{1}$$

where W_t is the weight loss measured at a given time t (g) and W_0 is the maximum possible weight loss of the pellet (g).

The content of total iron (TFe) and metallic iron (MFe) of the reduced pellet were obtained by chemical analysis, and the index of metallization degree (η) was calculated by the following formula (2):

$$\eta = MFe/TFe \times 100\% \tag{2}$$

The volume of the reduced pellet was measured by the wax immersion method [12], which can be calculated as follows:

$$V = (P - S)/\rho_0 - (P - W)/\rho \tag{3}$$

where V is the volume of the sample pellet (cm³), P is the weight of the wax-coated specimen (g), W is the initial weight of the specimen using a balance (g), S is the weight of the wax-coated specimen when suspended in water (g), ρ is the density of wax (g/cm³), and ρ_0 is the density of water (g/cm³).

The mineral phases and microstructure of the reduced pellets were characterized by XRD and scanning electron microscope with energy-dispersive spectroscopy (SEM-EDS).

3. Results

3.1. Variation of the Reaction Fraction and Mineral Phase Composition of the Pellet during Reduction

The reaction fraction of the reduced composite pellet with time at different temperatures is shown in Figure 4. It can be clearly observed that the reaction fraction highly depends on the heating temperature and the reaction fraction increases with the increasing temperature. The line A–A′ is the theoretical metallic iron emerging time (i.e., when the iron oxide in the pellet has been reduced into FeO). The reaction fraction is 26.87% on this line. At the temperatures of 1000 °C and 1050 °C, the reduction rate is low and the reaction does not reach a stable state at the time of 30 min. When the temperature is greater than 1100 °C, the reduction reaction can reach a stable state at the time of 30 min. It can also be seen that the reduction rate increases remarkably with the increasing of temperature when the temperature is lower than 1150 °C. The metallic iron has appeared in all of the final reduced samples. The metallic iron appears earlier if the reduction temperature is higher. The fast reduction of the iron ore/carbon composite pellet will result in the formation of gaseous product and the loss of mass, which mainly leads to the consequent volume shrinkage.

The mineral phase compositions of the composite pellet reduced at different temperatures for 30 min were analyzed by XRD (Rigaku, Tokyo, Japan); the results are shown in Figure 5. The mineral phase components in the reduced pellets are relatively simple. For the 1000 °C-reduced pellet, the main phases are metallic iron (Fe), wustite (FeO), and a small amount of silicon oxide (SiO_2); magnetite (Fe_3O_4) has disappeared; and unreacted SiO_2 does not exist in the form of quartz. In the XRD pattern of the 1050 °C-reduced pellet, the peaks associated with the silicon oxide phase disappear, and the intensities of the wustite peaks decrease. In the pattern of the 1100 °C-reduced pellet, the wustite peaks disappear, new peaks associated with fayalite (Fe_2SiO_4) appear, and the intensities of the metallic iron peaks increase. In the pattern of the 1150 °C-reduced pellet, the intensities of peaks associated with the fayalite and metallic iron phases begin to decrease a little. In the pattern of the 1200 °C-reduced pellet, the intensities of the peaks associated with the fayalite become very weak and almost disappear, and metallic iron becomes the only major existing phase.

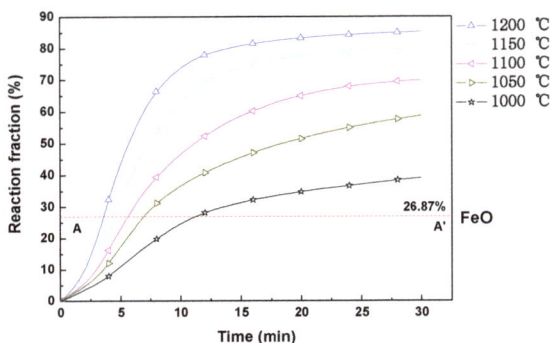

Figure 4. Results of the isothermal reduction experiment.

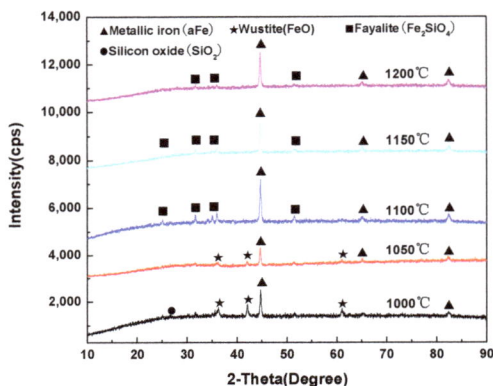

Figure 5. X-ray diffraction analysis of pellets reduced at different temperatures.

3.2. Volume Shrinkage of Pellets during Reduction

Figure 6 shows the variation in volume shrinkage of the composite pellets as a function of time at different temperatures. It can be seen that the volume shrinkage of the reduced pellet highly depends on the reduction temperature. Figure 7 shows the cross-sectional SEM images of composite pellets reduced at different temperatures for 30 min. At the temperature of 1000 °C, the shrinkage value is small because the reduction degree of the pellet is very low (39.2%). The inner structure of the reduced pellet is still in the loose powdery state. The fine metallic iron particles have appeared around the edges of iron ore particles. It also can be seen that the edges of the quartz particles have partially reacted with FeO, which forms from the Fe_3O_4 reduction, to form a new mineral phase. For the 1050 °C-reduced pellet, the structure of the pellet becomes more denser and the particle size of metallic iron becomes obviously larger. More quartz particles have been reacted. Therefore, the pellet volume has shrunk a lot, from 12.2% to 25.6%. For the 1100 °C-reduced pellet, the structure of the pellet begins to change obviously, and the ore particles have sintered together and formed a lot of fayalite. The reaction fraction is approximately 69.8%, and the metallic iron particles become larger. Thus, the volume shrinkage of the reduced pellet further increases to 35.8%; a relatively large ratio. For the 1150 °C- and 1200 °C-reduced pellets, the internal structures of the pellets become much denser. Although the peaks of fayalite are very weak in the XRD patterns of the 1150 °C- and 1200 °C-reduced pellets, the fayalite still exists according to the SEM-EDS analysis. The SEM image of 1200 °C-reduced pellet shows that the gangue phase is in a uniform state and might have melted to some extent, because the melting point of Fe_2SiO_4 is about 1205 °C. The size of the metallic iron particles becomes much larger compared with that of the 1100 °C-reduced pellet. The metallic iron particles begin to sinter together and form large-sized metallic crystal structures. Therefore, the reduced pellets continue to shrink.

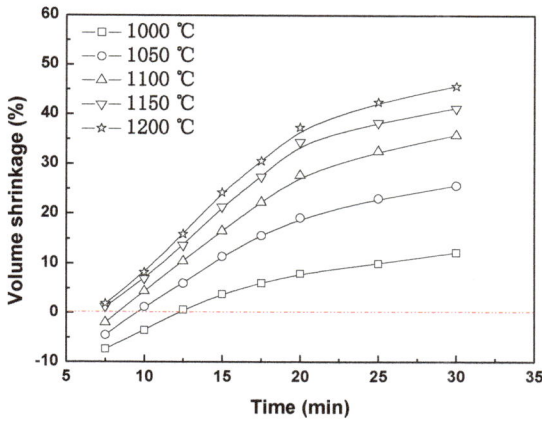

Figure 6. Volume shrinkage of the carbon composite pellets during reduction.

A summary of the variation of some properties, such as metallic iron particle size, fayalite fraction, and density, of the final reduced pellets at different temperatures is listed in Table 2, which can provide additional explanation for the reduction and shrinking process. The grain size of metallic iron gradually increases from 3.5 µm to 16.5 µm with the increasing temperature. Based on the SEM and XRD analysis, it can be concluded that the amount of fayalite is only small and the formation reaction of fayalite is just at the beginning stage owing to the relatively lower temperatures (i.e., 1000 °C and 1050 °C). The volume content of fayalite gradually decreases at the temperatures ranging from 1100 °C to 1200 °C due to the direct reduction of fayalite by solid carbon. The density of the final reduced pellets gradually increases from 2.36 g/cm^3 to 3.02 g/cm^3.

The volume shrinkage firstly increases rapidly before the 20-min timepoint, and then, the rate of change of shrinking is lower. At higher temperatures, the volume shrinkage rate increases because of a higher rate of carbon and oxygen loss from the pellet and a higher rate of sintering of the metallic iron particles and gangue oxides. It is also found that the shrinkage rate is negative at the beginning stage of the reduction process. This is because the iron oxide in the iron concentrate is magnetite (Fe$_3$O$_4$) and the reduction of iron oxides follows in the sequence of magnetite (Fe$_3$O$_4$) → wustite (FeO) → metallic iron (Fe). It is reported that there will be a 7 to 13% increase in volume during the transformation of Fe$_3$O$_4$ to FeO [13]. The pellets begin to shrink once the metallic iron particles appear in relatively large quantities. The highest volume shrinkage is about 45.6%, when reduced at the temperature of 1200 °C under the present experimental conditions.

Figure 8 shows the variation of the volume shrinking rate of the composite pellets with time at different temperatures. The general trend is that the shrinking rate increases with the increasing reduction temperature. The obtained maximum shrinking rates are 1.60%/min (1000 °C), 2.30%/min (1050 °C), 2.55%/min (1100 °C), 2.87%/min (1150 °C), and 3.20%/min (1200 °C), respectively. Thus, the higher the reduction temperature the pellets experience, the larger the volume shrinking rate becomes. The shapes of the shrinking rate curves indicate that the shrinking mechanism changes when the reduction temperature is greater than 1100 °C. It can be concluded that the changing of the shrinking mechanism may be closely related to the reduction rate of iron oxide, the presence or absence of independent iron-containing minerals, and formation of the molten slag phase, among other factors.

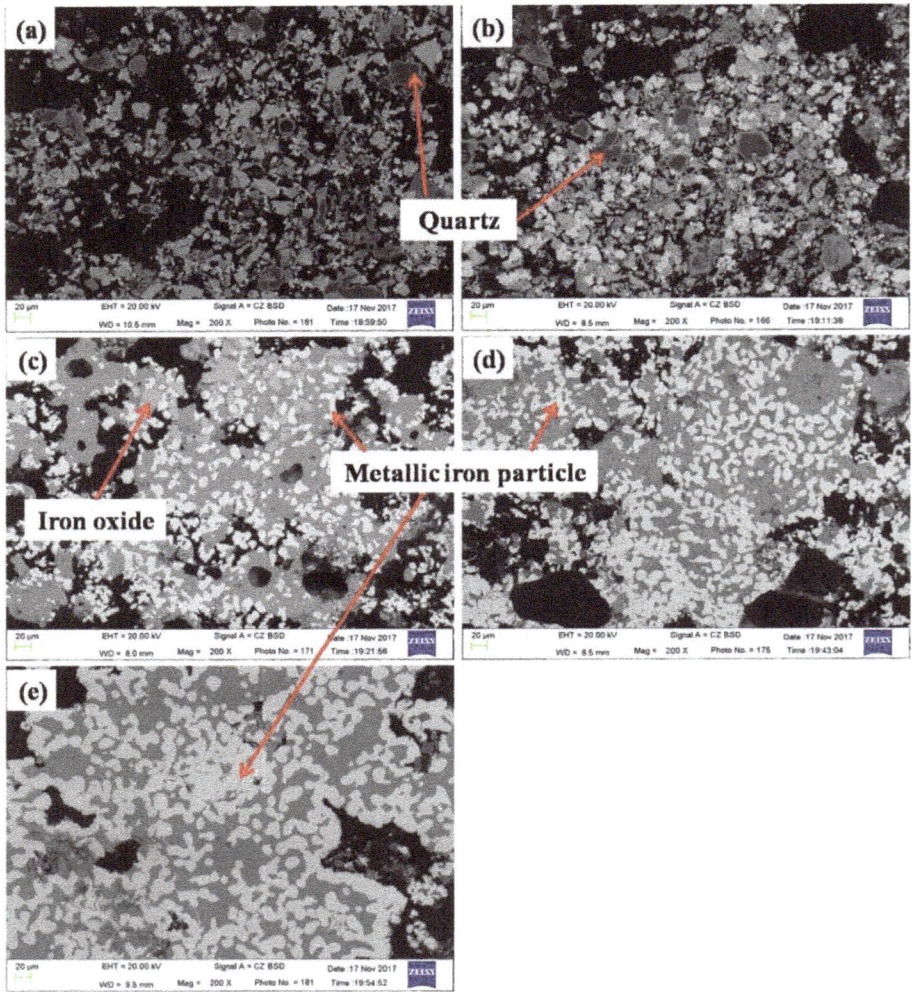

Figure 7. SEM images of the reduced pellets (at 30 min): (**a**) 1000 °C, (**b**) 1050 °C, (**c**) 1100 °C, (**d**) 1150 °C, and (**e**) 1200 °C.

Table 2. Summary of some properties of the final reduced pellets.

Temperature (°C)	Metallic Iron Particle Size (μm)	Fayalite Fraction (vol %)	Density (g/cm³)
1000	3.5	not detected/no data	2.36
1050	6.1	not detected/no data	2.58
1100	7.3	36.9	2.81
1150	9.8	34.2	2.91
1200	16.5	29.9	3.02

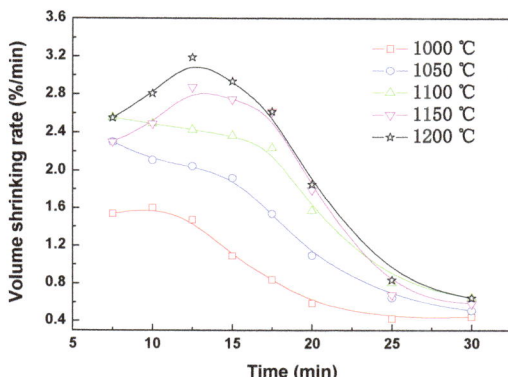

Figure 8. Volume shrinking rate of the carbon composite pellets during reduction.

3.3. Shrinkage Kinetics Analysis

The following Equation (4), which can be expressed as a function of time and temperature based on the work of McAdam [14], has been used to depict the volume shrinking behavior of iron ore/coal composite pellets in the present research.

$$Sh = k_0 t^{2/5} \exp\left(-\frac{E}{RT}\right) = kt^{2/5} \tag{4}$$

where *Sh* is the value of volume shrinkage (%), r_0 is the initial pellet radius (m), *k* is the reaction rate constant, k_0 is the frequency factor, *t* is the time (min), *R* is the ideal gas constant (8.314 J/(mol·K)), *E* is the shrinkage apparent activation energy (J/mol), and *T* is the temperature (K).

Equations (5) and (6) can be obtained from Equation (4) and the Arrhenius equation. The value ln(*k*) can be calculated through the linear fitting between ln(*Sh*) and ln(*t*) at certain temperatures. In the end, the value of *E* can be obtained based on Equation (6).

$$Sh = a(T) + kt^{2/5} \tag{5}$$

$$\ln(k) = \ln(k_0) - \frac{E}{RT} \tag{6}$$

where *a*(*T*) is the coefficient factor depending on the temperature and time range.

Equation (4) assumes that the pellet has a spherical shape; however, the pellet used in the present research has a columnar shape. It can be concluded that the volume shrinkage of the composite pellets during reduction is uniform and has little relation with the shape of the green pellet, because the chemical composition of the pelletizing raw material is uniform. On the other hand, the column-shaped pellet can be converted into a spherical pellet of the same volume, and then the equivalent radius value of the pellet can be obtained and implemented in the equation. Therefore, Equation (4) can also be utilized in the study of the shrinkage of nonspherical pellets.

The data of the 1000 °C-reduced sample has not been taken into consideration due to the small shrinkage value. The linear fitting lines between *Sh* and $t^{2/5}$ in the range from 1050 °C to 1200 °C are listed in Figure 9. The reduction time is selected from 10 min to 30 min. It can be seen that the linearity of the curves obviously changes at the time of 20 min. Therefore, the curves are cut off at the time of 20 min and linear analyses are conducted separately. The linear correlation is perfect for each fitting. The calculated values of *k* of the pellet volume shrinkage during reduction at the different temperatures are also listed in Figure 9. The obtained Arrhenius plot is shown in Figure 10. It can be seen that it is in better linearity when the pellet is reduced within 20 min. The value of the apparent activation energy for the volume shrinkage during reduction is 51,313 J/mol when the reduction time ranges from 10 min to 20 min, and the apparent activation energy is 19,697 J/mol when the pellet is reduced for more than 20 min. The equations describing the fractional shrinkage of the composite pellets at different reduction stages can be obtained according to the calculated data and are listed as follows. The $a(T)_1$ and $a(T)_2$ are coefficient factors depending on the specific temperature and time range, whose values are listed in Table 3.

$$Sh_1 = a(T)_1 + 2505.62 \times t^{2/5} \exp\left(-\frac{51313}{RT}\right) \qquad (10-20 \text{ min}, \ 1050\,°C \text{ to } 1200\,°C) \qquad (7)$$

$$Sh_2 = a(T)_2 + 68.98 \times t^{2/5} \exp\left(-\frac{19697}{RT}\right) \qquad (>20 \text{ min}, \ 1050\,°C \text{ to } 1200\,°C) \qquad (8)$$

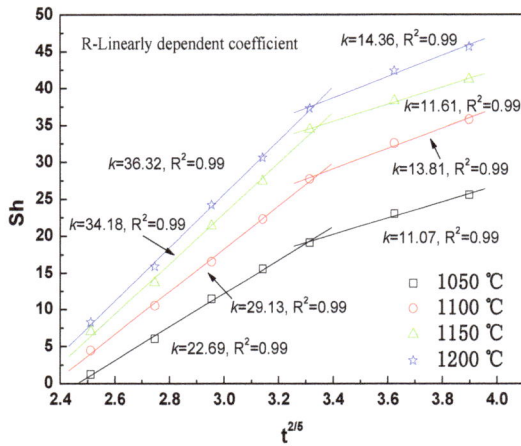

Figure 9. ln(*Sh*) vs ln(*t*) curves. *Sh*: the value of volume shrinkage (%), *t*: time (min), *k*: reaction rate constant.

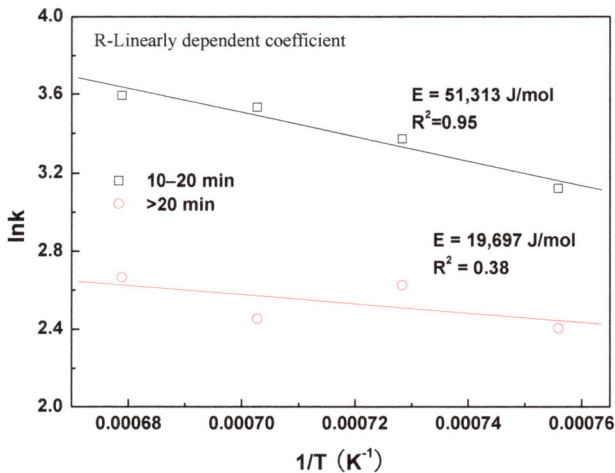

Figure 10. Arrhenius plot of the rate constant. *E*: the shrinkage apparent activation energy.

Table 3. Coefficient factors ($a(T)$) for different temperatures and time ranges.

Temperature(°C)	10–20 min ($a(T)_1$)	>20 min ($a(T)_2$)
1050	−55.89	−17.41
1100	−69.17	−17.88
1150	−79.54	−4.00
1200	−83.33	−10.11

4. Discussion

4.1. Shrinkage Driving Forces

Generally speaking, there are two main driving forces for the automatic growth and sintering together of metallic iron particles [15]. Firstly, the smaller particles contain relatively higher concentration of metallic iron atoms based on the Ostwald equation. Therefore, a concentration difference should exist for metallic iron between differently sized particles. Secondly, particles of relatively larger grain size tend to reduce the surface free energy because of the principle of minimum free energy. As a result of the two main theoretical reasons, Fe atoms will diffuse from tiny particles to bigger ones, and bigger particles will gradually sinter together and ultimately form large crystals. Therefore, small iron particles diminish and disappear, while larger particles reunite and grow as reduction progresses. The gangue oxide phase may also abide by the aforementioned two rules.

For the gangue oxide phase, at the initial reduction stage, the gangue minerals will sinter together and form new minerals simultaneously. As the magnetite contains quartz, the gangue minerals will partially melt, especially when the heating temperature approaches 1200 °C. The formation of the molten slag phase will reduce the porosity of the pellet and make the pellet structure much denser after most of the iron oxide has been reduced into metallic iron.

4.2. Time-Dependent Phenomena of Shrinkage Kinetics

The above result shows that the apparent activation energy of shrinkage of the composite pellets changes obviously at the timepoint of 20 min during reduction. It can be concluded that the phenomena must be in close relation with the reduction rate and the evolution of the microstructure of the reduced pellets. The reduction and shrinkage behaviors of the composite pellets at 1200 °C are used to make the discussion in this part.

Variation of the metallization degree of the reduced pellets with time at 1200 °C is shown in Figure 11. The metallization degree obviously increases during the time before 20 min, and then increases slowly with the further increasing reduction time. Figure 12 illustrates the relationship between the volume shrinkage value and metallization degree of the pellets reduced at 1200 °C. The correlation degree is perfect when using the quadratic equation of one unknown, which demonstrates that the volume shrinkage is highly dependent on the pellets' metallization degree. The reduction rate and rate-determining step must have changed since the 20-min timepoint. Therefore, it can be concluded that the shrinkage mechanism should have also changed simultaneously with the change of the reduction rate-determining step, considering the close relationship between the metallization degree and volume shrinkage value.

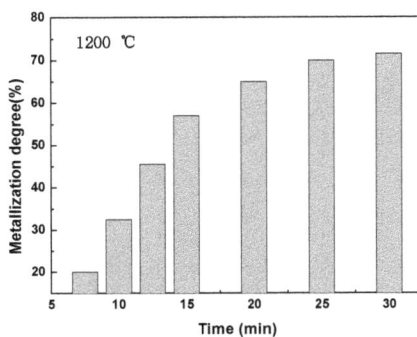

Figure 11. Metallization degree of the pellets reduced at 1200 °C.

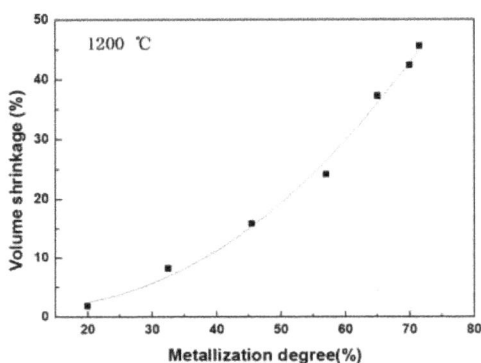

Figure 12. Correlation of volume shrinkage vs metallization degree of the pellets reduced at 1200 °C.

Variation of the microstructure of the reduced pellets with time at 1200 °C is shown in Figure 13. With the reduction progressing, the internal microstructure also changes significantly from a loose powdery mixture to an increasingly dense sintered metallic iron–gangue phase. Once the composite pellet is heated, the reactions, such as iron oxide reduction, carbon gasification, metallic iron nucleation, iron grain growth, and new gangue mineral (i.e., fayalite) formation, will occur simultaneously and the volume of the pellet will shrink quickly during the time before 20 min. At the timepoint of 20 min, the internal microstructure of the reduced pellet has obviously changed compared with the 15 min-reduced pellet. The gangue mineral has developed into a lath shape from a granular shape, and the grain size of the metallic iron has become much larger. During the stage from 10 min to 20 min in the reduction process, the reactions are so complicated and difficult to occur; therefore, the shrinkage apparent activation energy is as high as 51,313 J/mol. With the reduction progressing to the later stage, the reduction of iron oxide has been almost completed, and the shrinkage of the reduced pellet mainly results from spontaneous sintering and reconstruction of the metallic iron particles and gangue oxides, and even partial melting of the gangue phase if the reduction temperature approaches the meting point of fayalite. The above reactions occur relatively easily and some occur spontaneously; therefore, the shrinkage apparent activation energy in the later reduction stage is only 19,697 J/mol.

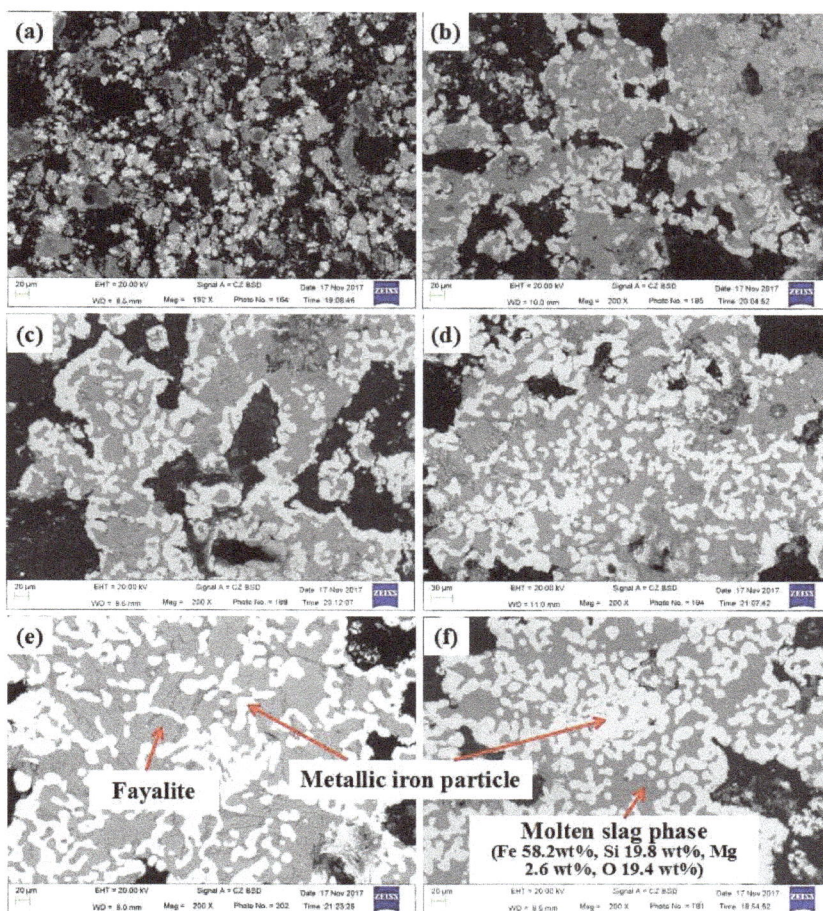

Figure 13. Microstructure of the pellets reduced at 1200 °C: at (**a**) 7.5 min, (**b**) 10 min, (**c**) 12.5 min, (**d**) 15 min, (**e**) 20 min, and (**f**) 25 min.

4.3. Sensitivity Analysis for Shrinkage Kinetics Results

In order to predict the response of the obtained kinetics equation, the calculation of the volume shrinkage with the variation of temperature and time is performed based on Equations (7) and (8), and the result is shown in Figure 14. It can be seen that the variation trend of the predicted values is similar to that of the measured ones. The root mean square error (RMSE) is simply computed and used to evaluate the accuracy of the kinetics equation. The RMSE calculation formula is given as follows [16]:

$$RMSE = \sqrt{\frac{\sum_i^N \left(Sh_i - Sh_i'\right)^2}{N}} \tag{9}$$

where Sh_i is the measured value of volume shrinkage (%) at different times under a given temperature, Sh_i' is the predicted value of volume shrinkage (%) at different times under a given temperature, and N is the number of pellets selected for volume measurement.

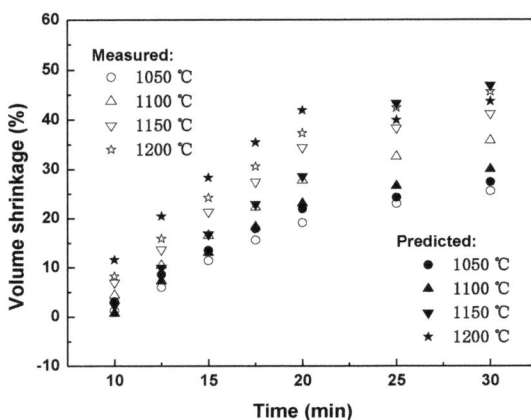

Figure 14. Comparison of measured and predicted volume shrinkage values.

The calculated RMSE values are listed in Table 4. The RMSE value at 1050 °C is the smallest, which indicates that the predicted volume shrinkage is comparatively more exact under the present condition. The RMSE values at 1100 °C and 1150 °C are relatively larger, which means they have worse accuracy.

Table 4. Root mean square error (RMSE) values.

1050 °C	1100 °C	1150 °C	1200 °C
2.15	4.50	4.88	3.85

5. Conclusions

(1) Volume shrinkage of the composite pellet increases with the increasing of reduction temperature, and the shrinkage rate is faster within the initial 20 min. The values of shrinkage apparent activation energy are different at different time ranges. During pellet reduction before the 20-min timepoint, the shrinkage apparent activation energy is 51,313 J/mol due to the complicated reactions. After the 20-min timepoint, the apparent activation energy of volume shrinkage is only 19,697 J/mol.

(2) The volume shrinkage of the composite pellet during reduction mainly results from the mass loss of carbon and oxygen and the sintering of the metallic iron particles and gangue oxides. With the increasing of reduction temperature and time, the formation of large metallic iron crystals and molten slag phase plays an important role in the further shrinkage of the reduced pellet in the final reduction stage.

Author Contributions: G.W. designed and performed the experiments and results analysis and acquired the funding. J.W. and Q.X. administrated the project, supervised the experiments, and reviewed the manuscript.

Funding: This research was funded by the National Natural Science Foundation of China, grant number 51804024, and the State Key Laboratory of Advanced Metallurgy of the University of Science and Technology Beijing, grant number 41618022.

Acknowledgments: The authors wish to acknowledge the contributions of associates and colleagues at the University of Science and Technology Beijing, China, and the financial support of the National Natural Science Foundation of China and the State Key Laboratory of Advanced Metallurgy of the University of Science and Technology Beijing.

Conflicts of Interest: The authors declare no conflict of interest.

References

1. Yang, J.; Mori, T.; Kuwabara, M. Mechanism of carbothermic reduction of hematite in hematite-carbon composite pellets. *ISIJ Int.* **2007**, *47*, 1394–1400. [CrossRef]
2. Bauer, K.; Huette, D.; Lehmkuehler, H.; Schmauch, H. Recycling of iron and steelworks wastes using the Inmetco direct reduction process. *Metall. Plant. Technol.* **1990**, *13*, 74–87.
3. McClelland, J.M.; Metius, G.E. Recycling ferrous and nonferrous waste streams with FASTMET. *JOM* **2003**, *55*, 30–34. [CrossRef]
4. Wang, G.; Wang, J.S.; Ding, Y.G.; Ma, S.; Xue, Q.G. New separation method of boron and iron from ludwigite based on carbon bearing pellet reduction and melting technology. *ISIJ Int.* **2012**, *52*, 45–51. [CrossRef]
5. Ding, Y.G.; Wang, J.S.; Wang, G.; Xue, Q.G. Innovative methodology for separating of rare earth and iron from Bayan Obo complex iron ore. *ISIJ Int.* **2012**, *52*, 1772–1777. [CrossRef]
6. Chen, Y.; Hwang, T.; Marsh, M.; Williams, J.S. Mechanically activated carbothermic reduction of ilmenite. *Metall. Mater. Trans. A* **1997**, *28*, 1115–1121. [CrossRef]
7. Anameric, B.; Kawatra, S.K. Laboratory study related to the production and properties of pig iron nuggets. *Miner. Metall. Proc.* **2006**, *23*, 52–56. [CrossRef]
8. Kikuchi, S.; Ito, S.; Kobayashi, I.; Tsuge, O.; Tokuda, K. ITmk3 process. *Kobelco. Techno. Rev.* **2010**, *29*, 77–84.
9. Halder, S.; Fruehan, R.J. Reduction of iron-oxide-carbon composites: part III. shrinkage of composite pellets during reduction. *Metall. Mater. Trans. B* **2008**, *39*, 809–817. [CrossRef]
10. Donskoi, E.; McElwain, D.L.S. Mathematical modelling of non-isothermal reduction in highly swelling iron ore-coal char composite pellet. *Ironmak. Steelmak.* **2001**, *28*, 384. [CrossRef]
11. Donskoi, E.; McElwain, D.L.S. Estimation and modeling of parameters for direct reduction in iron ore/coal composites: Part I. Physical parameters. *Metall. Mater. Trans. B* **2003**, *34*, 93. [CrossRef]
12. Wang, G.; Xue, Q.G.; Wang, J.S. Volume shrinkage of ludwigite/coal composite pellet during isothermal and non-isothermal reduction. *Thermochim. Acta* **2015**, *621*, 90–98. [CrossRef]
13. Prakash, S. Reduction and sintering of fluxed iron ore pellets-a comprehensive review. *J. South. Afr. Inst. Min. Metall.* **1996**, *96*, 3–16.
14. McAdam, G.; O'Brien, D.; Marshall, T. Rapid reduction of New Zealand ironsand. *Ironmak. Steelmak.* **1977**, *4*, 1–9.
15. Sun, Y.S.; Gao, P.; Han, Y.X.; Ren, D.Z. Reaction behavior of iron minerals and metallic iron particles growth in coal-based reduction of an oolitic iron ore. *Ind. Eng. Chem. Res.* **2013**, *52*, 2323–2329. [CrossRef]
16. Vu-Bac, N.; Lahmer, T.; Zhuang, X.; Nguyen-Thoi, T.; Rabczuk, T. A software framework for probabilistic sensitivity analysis for computationally expensive models. *Adv. Eng. Softw.* **2016**, *100*, 19–31. [CrossRef]

metals

MDPI

Article

The Effect of Concentrate/Iron Ore Ratio Change on Agglomerate Phase Composition

Mária Fröhlichová [1,*], Dušan Ivanišin [2], Róbert Findorák [1], Martina Džupková [1] and Jaroslav Legemza [1]

[1] Faculty of Materials, Metallurgy and Recycling, Technical University of Košice, 04200 Košice, Slovakia; robert.findorak@tuke.sk (R.F.); martina.dzupkova@tuke.sk (M.D.); jaroslav.legemza@tuke.sk (J.L.)

[2] U.S. Steel Košice s.r.o, Vstupný areál, 04454 Košice, Slovakia; DIvanisin@sk.uss.com

* Correspondence: maria.frohlichova@tuke.sk; Tel.: +421-55-602-3152

Received: 17 October 2018; Accepted: 19 November 2018; Published: 21 November 2018

Abstract: The work is focused on studying the influence of the ratio of concentrate to iron ore on the phase composition of the iron ore agglomerate. The concentrate has significantly higher iron content than used iron ore, and is a determining factor, which influences the richness of the batch and consequently, the richness of the agglomerate. The increased iron content in the agglomerate can be achieved by adjusting the raw material ratio in which iron ore materials are added to the agglomeration mixture. If the ratio is in favor of iron ore this reflects in lower iron content in the resulting agglomeration mixture. If the ratio is in favor of a concentrate, which is finer, the fraction share of less than 0.5 mm will be increased, the permeability of the batch will be reduced, the performance of the sintering belt will decrease and the presence of solid pollutants will increase. The possibility of concentrate replacement by iron-rich iron ore with granulometry similar to that of concentrate was experimentally verified. The effect of the concentrate replacement by the finer iron-rich ore was tested in a laboratory sintering pan. There were performed six sinterings, with gradually changing ratio concentrate/iron ore (C/O). The change in the ratio of concentrate to iron ore, does not cause the occurrence of new phases, only the change in their prevalence, which does not bring a significant change of the qualitative indicators of the compared agglomerates. Concentrate replacement by iron ore up to 50% was optimal from technological, quality, and environmental aspects.

Keywords: concentrate; iron ore; agglomerate; structure; phase analysis

---

1. Introduction

Improving the quality of the iron ore agglomerate and reducing the energy intensity of blast-furnace metallurgical processes are an important prerequisite for improving the technical and economic indicators of blast furnace work. The requirements for the quality of agglomerate, the basic metal-bearing part of the blast furnace batch, are constantly increasing. With the current raw material base, when the amount of fine-grained iron ore materials is increasing, this can only be achieved by intensifying the agglomeration process. A good quality agglomerate is characterized by a suitable iron content, high reducibility, good strength and low fine grain shares content prior to charging into blast furnace and high strength after reduction in the blast furnace shaft [1].

In general, a wide range of materials is required to produce the agglomerate. For this reason, their preparation requires a great deal of attention to ensure the correct quality of the agglomerate and its suitability for the blast furnace process.

The mixture required to produce the agglomerate is composed of different types of ores, concentrates and secondary raw materials. It is characterized by fluctuations in the physico-chemical properties and chemical composition and therefore it requires a thorough homogenization.

Homogenization of batch materials is a demanding process due to the use of large volumes and variety of raw materials. Therefore, great emphasis is placed on the knowledge of the qualitative impacts of the batch, as well as on managing the whole technology of agglomeration batch preparation and processing.

For all components of raw materials suitable for agglomerate production, one of the most important aspects of assessment is granulometry. The maximum grain size for the agglomeration process must not exceed 10 mm [2,3]. The most suitable grain size for the sintering process is below 5 mm, or up to 3 mm for some raw materials [4].

The demand for increased iron content in agglomerates has opened up a possibility for the applying higher amounts of concentrates, which are generally richer in total iron than normal iron ores. Their increased proportion in agglomeration batch, on the other hand, brings problems associated with the pre-treatment of the mixture. Such increased requirements on the packing of a mixture with a higher share of concentrates, regarding the pre-pelletizing quality and packing time, present some challenges for the optimization of the sintering process [5].

The batch composition affects not only the quality of the agglomerate but also the ecology of the production. In addition to agglomerate quality monitoring, attention must be paid to the environmental aspect of production. It is important to look for links between the various production parameters. The emissions load on the environment can be reduced by ensuring the appropriate quality of batch raw materials as well as regulating the selected parameters of sintering, especially the height of the sintered layer, quantity and speed of the sucked air, ratio of the concentrate to iron ore [6–8].

Complex physico-chemical processes take place during agglomeration production and no generally-valid rules have been derived which might predict the composition of the finished product. However, there are certain known relationships and dependencies, according to which the desired agglomerate can be produced. The final properties of the agglomerate are determined not only by its chemical composition but also by its mineralogical composition, i.e., phases [9]. Mineralogical phases may have different structural forms depending on the conditions of agglomeration formation. Basically, there are crystalline and glass phases. A relatively large proportion of the mass is in the form of solid solutions of varying composition, making it difficult to quantitatively and qualitatively assess the contribution of the individual phases to the quality of the agglomerate. In view of this, only general conclusions can be made. It follows that when making agglomerates, we must take into account not only the chemical composition of the input components of the batch, but also the physico-mechanical properties and technological parameters. In general, all the technological parameters affect the phase composition and structure of the agglomerate and, in turn, its mechanical properties as well.

The theoretical prediction of the parameters of sintering and the properties of the finished product, as well as the environmental impact, are influenced by a number of factors which influence each other. This study focuses on the one of these factors, i.e., the effect of agglomerate properties of the replacement of concentrate with iron ore which approximating concentrate with its composition and granulometry. For this purpose, experiments were designed and performed on a laboratory sintering pan.

2. Materials and Methods

A total of six iron ore raw materials were used in the experiments: commonly-used iron ores AR-1, AR-2, AR-3, AR-4, concentrate KC1, and iron ore AR-5 with fine granulometry. AR-5 is just produced raw material (as delivered). All iron ores come from Ukrainian mines. The ores used in the experiment were not specially selected; they make part of the batch for the production of agglomerate used in the operating conditions. The number of iron ores used in the experiments is comparable to those utilized in commercial practice.

The water content of these ores, as delivered, is up to 5%, which is the most efficient for transport. The chemical composition of iron ores used for the experiment is shown in Table 1.

Table 1. The Chemical composition of iron ores used for experiment.

Iron Ores	Humidity	Fe	FeO	Fe₂O₃	SiO₂	CaO	MgO	Al₂O₃	Mn	P
					(%)					
AR-1	3.63	57.58	0.48	82.66	15.13	0.08	0.21	0.97	0.020	0.035
AR-2	4.16	60.89	0.37	86.57	8.86	0.22	0.37	1.41	0.026	0.093
AR-3	4.20	60.39	0.52	85.38	11.07	0.07	0.21	0.90	0.027	0.028
AR-4	4.82	61.12	1.09	86.59	8.69	0.77	0.37	1.07	0.054	0.034

Iron Ores	Basicity	S	Na₂O	K₂O	TiO₂	Pb	Zn	As	C	Cl
					(%)					
AR-1	0.018	0.016	0.117	0.047	0.037	0.002	0.005	0.001	0.047	0.20
AR-2	0.058	0.016	0.243	0.120	0.057	0.001	0.003	0.001	0.217	0.07
AR-3	0.024	0.015	0.154	0.053	0.038	0.001	0.001	0.001	0.070	0.21
AR-4	0.117	0.021	0.055	0.054	0.034	0.001	0.001	0.001	0.078	0.32

The most important aspect of the chemical composition of these iron ores is their iron content. In the experiment were used ores, which iron content is standard for Ukrainian or Russian ores. High content of SiO_2 occurs mainly in AR-1 (15.13%) and AR-3 (11.07%). On the other hand, the content of basic components is up to three times lower than for other ores. The content of adverse elements is at the standard level and does not exceed the quality requirements. The disadvantage of the iron ores used iron ores is their very low alkalinity, which directly affects the richness of the resulting agglomerate.

Another batch component is iron ore concentrate. Only one sort was used in the experiment to replace some part of the concentrate with tested iron ore (AR-5). Normally, two to three sorts of concentrate are used in a single homogenized pile at the plant. The chemical composition of the concentrate used in the experiment is shown in Table 2.

Table 2. Chemical composition of concentrate used in experiment.

Concentrate	Humidity	Fe	FeO	Fe₂O₃	SiO₂	CaO	MgO	Al₂O₃	Mn	P
					(%)					
KC1	9.79	67.95	27.80	66.16	4.89	0.12	0.24	0.16	0.03	0.011

Concentrate	Basicity	S	Na₂O	K₂O	TiO₂	Pb	Zn	As	C	Cl
					(%)					
KC1	0.070	0.123	0.029	0.060	0.015	0.001	0.001	0.002	0.152	0.050

The concentrate used has a significantly higher iron content compared to iron ores. The humidity is almost 10%, which is less efficient in terms of transport than in the case of iron ores. An advantage is that making homogenized heaps requires no additional damping. Another advantage of the concentrate used is the low SiO_2 content. In terms of adverse content, this concentrate is considered standard.

The last used iron-ore component was tested iron ore (AR-5), Table 3. From the chemical point of view, this iron ore has high iron content. Its main advantage is the low SiO_2 content, which significantly influences the alkalinity of the mixture. The basicity of this ore far exceeds the concentrates and significantly exceeds the iron ores used. Moreover, the MgO and CaO contents are the highest in comparison with the other iron ore raw materials. The negative feature is the increased sulfur content relative to standard iron ores, but even so the sulfur content is half that of other concentrates. The very low alkali content is positive.

Table 3. Chemical composition of the tested iron ore used in the experiment.

Tested Iron Ore	Humidity	Fe	FeO	Fe$_2$O$_3$	SiO$_2$	CaO	MgO	Al$_2$O$_3$	Mn	P
					(%)					
AR-5	4.78	62.22	4.45	84.04	5.54	0.78	0.57	1.60	0.06	0.30

Tested Iron Ore	Basicity	S	Na$_2$O	K$_2$O	TiO$_2$	Pb	Zn	As	C	Cl
					(%)					
AR-5	0.190	0.059	0.040	0.010	0.10	0.001	0.008	0.001	0.480	-

From the granulometric point of view, all the raw materials were sieved on 8 mm, 5.6 mm, 4 mm, 2 mm, 1 mm, 500 µm, 250 µm, 125 µm, 63 µm sieves. Although the production and technological regulations just require certain amounts of raw materials above and below 10 mm for iron ores, for the needs of these experiments it was necessary to screen the raw materials into ten fractions in order to obtain a complete view of the granulometry. A detailed analysis of the granulometric composition of the used raw materials is given in Table 4, Figure 1.

Table 4. Granulometric composition of used raw materials.

Raw Materials	Above 8 mm	5.6–8 mm	4–5.6 mm	2–4 mm	1–2 mm	500 µm^{-1} mm	250–500 µm	125–250 µm	63–125 µm	Under 63µm
						(%)				
					Standard Iron Ores					
AR-1	4.5	6.7	6.4	12.4	11.5	11.4	11.2	16.8	12.30	6.80
AR-2	10.3	7.8	6.8	5.9	12.2	6.9	13.4	11.8	13.09	11.81
AR-3	16.5	9.3	6.4	9.8	8.6	9.6	12.0	12.2	10.90	4.60
AR-4	8.4	6.9	6.7	17.4	18.7	15.7	15.6	7.4	2.30	0.90
					Concentrate					
KC1	0	0	0	0.1	0.1	0	9.5	8.5	52.4	29.4
					Tested Iron Ore					
AR-5	3.6	4.6	5.4	10.9	9.9	8.1	9.9	22.2	13.2	12.2

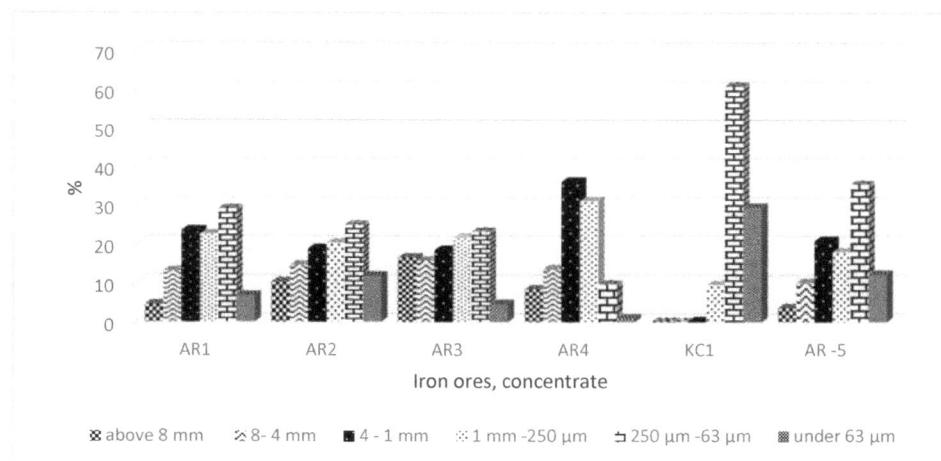

Figure 1. Granulometric composition of used raw materials.

From the granulometric analysis of the iron ore raw materials used in these experiments, it is evident that iron ores contain from 26% to 47% of the fraction below 500 μm. The fraction below 500 μm makes up almost 100% weight of the concentrates. In the case of tested iron ore, which formed the concentrate part in the experiments, the fraction below 500 μm made up 57.5%. Thus, it is a raw material which by the granulometric composition of its finer part is belongs among the conventional agglomerates and concentrates. Conversely, the fraction above 500 μm constitutes, in this tested iron ore, considerably smaller part than in other iron ores. Iron ores contain up to 40% of fractions above 2 mm. In the tested iron ore this fraction is less than 25%. Thus, based on screen analyses, we can say that this raw material is granulometrically on the boundary between standardly-used iron ores and standardly-used concentrates.

To study the effect of changing the ratio of concentrate to iron ore on the agglomerate phase composition, six agglomeration batches were prepared with a set percentage replacement of the concentrate by the tested iron ore AR-5. The compositions of the individual batches used in the experiments are presented in Table 5.

Table 5. The composition of individual experimental batches.

Batch		A	B	C	D	E	F
Concentrate Replacement by Tested Iron Ore AR-5 (%)		0	25	50	66	75	100
				(%)			
Iron Ores							
AR-1		5.88	5.84	5.80	0	5.76	5.73
AR-2		2.35	2.34	2.32	0	2.31	2.29
AR-3		16.06	15.96	15.86	21.31	15.75	15.65
AR-4		11.75	11.68	11.61	13.70	11.53	11.45
Total		36.04	35.82	35.59	35.01	35.35	35.12
Concentrate							
KC1		36.04	26.86	17.79	11.91	8.84	0
Tested Iron Ore							
AR-5		0	8.95	17.79	23.1	26.51	35.12
Total		36.04	35.81	35.58	35.01	35.35	35.12
Other Materials							
Micropellets		0.78	0.78	0.77	0.76	0.77	0.76
Slag-C		1.57	1.56	1.55	1.52	1.54	1.53
Manganese Ore		1.18	1.17	1.16	1.43	1.17	1.16
Dolomite		8.35	8.23	8.01	7.86	7.97	7.85
Limestone		12.64	13.26	13.94	15.02	14.48	15.07
Coke		3.38	3.38	3.38	3.38	3.38	3.38

The ores used in the experiments were not specially selected, they formed part of the batch for the production of agglomerate standardly used in the plant. The number of iron ores used in the experiments is comparable to commercial practice.

From the granulometric (composition) point of view, all the raw materials were sieved (screened) into several fractions. Although the production–technological regulations for iron ores state the sole condition of a certain amount of raw material above and below 10 mm, for the needs of these experiments it was necessary to screen the raw materials into ten fractions in order to obtain a complete overview of granulometry.

Based on screen analyses, therefore, we can conclude that iron ore AR-5 is granulometrically at the border between standard iron ores and commonly-used concentrates. From the point of view of the experiments needed to meet the goals of this work, it is a suitable raw material. A detailed analysis of the granulometric composition of the raw materials used is given in Table 4, and the conditions of the experimental sintering are presented in Table 6.

The weight of one batch in the sintering pan is about 250 kg. The sucked area is 0.25 m^2. The height of the layer is 400 mm. The measured vacuum in individual tests during the entire experiment was from 7.0 kPa to 7.4 kPa. A constant-rpm fan was used for suction, representing an effort to maintain the same conditions in all tests.

The permeability of the mixtures was a determining factor for mixing. This was set at the 0.92 or 0.93 m·s^{-1} for the reference mixture. In all other tests, the same permeability was achieved. Thus, the main parameter for packing raw materials into individual batches was permeability. This is also the reason for the different moisture content of individual batches. Humidity was determined by calculation. After adding the calculated amount of water, a control sample was taken from the packed mixture, and from this, the permeability of the mixture was determined. In the case of a difference in mixture permeability greater than 0.05 m·s^{-1}, additional water was added to the prepared mixture. The humidity of the mixtures ranged from 7.0% to 7.6%.

The temperature was measured along the height of the agglomeration pan at three levels, T1 max, T2 max, T3 max, while T4 max was the exhaust gas temperature, Figure 2. The thermocouples were positioned at levels 100, 200 and 300 mm from the top of the pan.

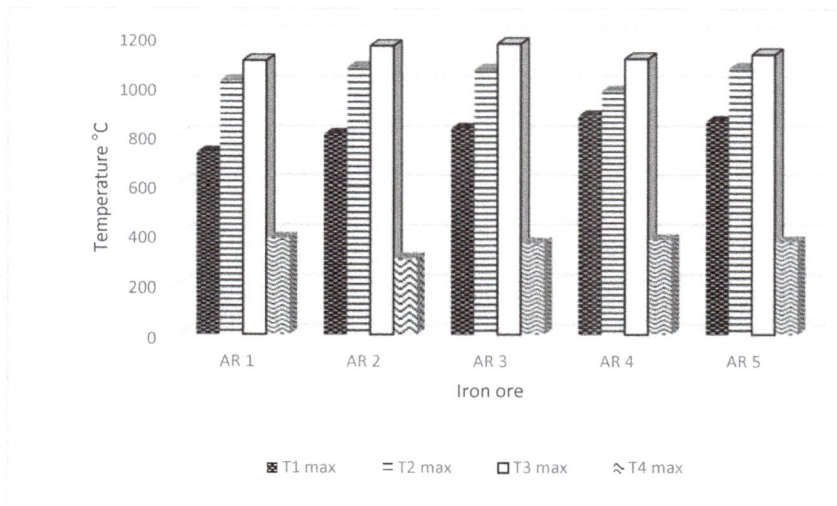

Figure 2. Temperature along the height of the agglomeration pan.

The experiments were carried out on a 250 kg agglomeration pan. Total dosage ranged from 235 kg to 258 kg. The differences in weights were due to the different specific weights of the concentrate portions of the mixture when the raw materials ratio was changed.

Table 6. Conditions of experimental sintering.

Amount of AR-5 (%)	Moisture of Packed Mixture (%)	Batch Weight (kg)	Permeability (m·s^{-1})	Sintering Time (min)
0	7.40	253.20	0.92	62
25	7.40	253.00	0.91	52
50	7.20	248.10	0.92	44
66	7.00	237.90	0.92	37
75	7.60	249.30	0.90	42
100	7.60	235.90	0.93	31

Each test was repeated twice to rule out any possible measurement error during sintering. The pair-sintering measurements were averaged and the pair of sinterings was evaluated as a single sintering, whereby the variance of the values of two sinterings is minimal. Altogether, twelve sinterings, i.e., six pair tests, took place. The basicity was between 1.4 and 1.6 for all tests.

X-ray diffraction phase analysis was used to quantify the phase composition of agglomerates prepared under laboratory conditions. The samples were analysed by Seifert XRD 3003 PTS. Diffraction records were evaluated by DIFFRAC.EVA (Search-Match, KARLSRUHE & VIERSEN, Germany, company Bruker) with PDF2 and TOPAS software (version 4) using the Rietveld method. The results of the phase composition analysis are summarised in Tables 7–9.

Table 7. The phase composition of produced agglomerates—iron oxides.

AR-5 (%)	Chemical Formula	$(Fe,Mg,Mn)_3O_4$	Fe_2O_3	FeO
	Mineralogical Name	Magnetite	Hematite	Wüstite
0	A1	20.20	43.00	0.85
25	B1	15.80	42.40	1.05
50	C1	10.70	37.75	1.15
66	D1	12.60	33.40	0.75
75	E1	9.60	42.45	0.90
100	F1	11.20	37.15	0.90

Table 8. Phase composition of agglomerates—calcium ferrites.

AR-5 (%)	Chemical Formula	$Ca_2Fe_2O_5$	$Ca_4Fe_9O_{17}$	$CaFe_5O_7$	$Ca_2Fe_{1.89}AlO.11O_5$	$Ca_2Fe_{22}O_{33}$	
	Mineralogical Name	Srebrodolskit	Calcium Ferrite	Calcium Ferrite	Brownmillerit	Calcium Ferrite	
0	A1	1.65	2.85	3.30	0.80	4.90	13.50
25	B1	1.30	3.50	3.20	1.55	6.10	15.65
50	C1	2.75	4.55	2.65	2.00	18.70	30.65
66	D1	3.05	2.40	6.50	1.55	8.40	21.90
75	E1	2.60	4.00	3.25	2.15	15.75	27.25
100	F1	3.45	3.85	2.65	3.05	17.60	30.60

Table 9. Phase composition of agglomerates—silicates.

AR-5 (%)	Chemical Formula	Ca_2SiO_4	SiO_2	$CaFeSi_2O_6$
	Mineralogical Name	Larnit	Quartz	Hedenbergit
0	A1	8.80	3.35	5.10
25	B1	9.25	4.75	4.55
50	C1	7.90	3.80	1.40
66	D1	8.10	3.50	9.25
75	E1	7.45	4.75	2.15
100	F1	7.75	3.75	2.45

The technological conditions of agglomeration batch sintering as well as qualitative indicators of the obtained agglomerate have been published in the literature [10].

3. Results and Discussions

The phase composition and final structure of the agglomerate results from the course of the agglomeration process. This process takes place in the heterogeneous system of gaseous–liquid–solid phases, where the gas phase assures the course of critical processes (fuel burning, heat transfer, reduction–oxidation processes). The liquid phase, resulting from the melting of the fine-grained ore and non-ore particles and the partial melting of the coarse ore particles, enables the crystallization of magnetite, hematite, calcium ferrites and partially also of silicates. After solidification it then ensures the connection of these phases into the porous sinter-agglomerate [11,12].

Due to the chemical inhomogeneity in the microvolumes of the produced melt and the high rate of its solidification, relatively large variability occurs in the composition of the microvolumes of the agglomerates and the diversity of their microtextures. The result of this process is the iron ore agglomerate, characterized by the following phases: magnetite, wüstite, hematite, calcium ferrites, ferrocalcium olivines, pyroxenes, dicalcium silicate, tricalcium silicate, residual quartz, periclase, and free CaO [13].

The assessed agglomerate in our experiments also corresponds with this phase composition. Based on the qualitative assessment of the reference agglomerate (without addition of AR-5) and the agglomerates with increasing concentrate replacement by AR-5 iron ore, it was found that any ratio of concentrate substitution with granulometrically finer iron ore led to an increase in the quality of the final product. The agglomerate quality was assessed based on the amount of the return agglomerate, the agglomerate strength TI (+6.3 mm) and the FeO content in the agglomerate [10].

The effect of changing ratio concentrate/iron ore (C/O) ratio on the phase composition of the agglomerate is less pronounced. From the qualitative point of view, the phases in all of the studied agglomerates were identical with those in the reference agglomerate, but their quantity changed, (see Tables 7–9), which ultimately affected the mechanical and metallurgical properties of the agglomerates.

Magnetite Fe_3O_4 forms by crystallisation from melt or hematite reduction in solid phase. It contains isomorphic admixtures of CaO and MgO. The magnetite content had a decreasing trend as the tested iron ratio was increased. The reason for this was the change in magnetite concentration as the magnetite concentrate was replaced with hematite iron ore AR-5.

The lowest magnetite content was obtained in samples C1 and E1, from which we can assume that these agglomerates should have the best reducibility. Conversely, the worst reducibility from the examined agglomerates could be expected in the reference sample, where the magnetite content is the highest, (see Table 7).

The agglomerate matrix consists of hematite, which has a decreasing trend as the iron ore/concentrate ratio increases. The decrease of hematite content in the agglomerate structure is caused by the reaction of fine Fe_2O_3 grains with CaO forming a ferritic bonding phase. Hematite Fe_2O_3 occurs either as remnant of unreacted fragments of hematite ores, in which it forms typical granular, sparsely porous to compact aggregates or individual grains or grains clusters, whereby its crystals grow at the expense of the silicate matrix and also supress magnetite crystals, or alternatively it forms thin slats within the individual magnetite grains and arises only after the basic silicate matter has solidified by secondary oxidation of the previously—crystallized magnetite.

Wüstite is individually represented only at a low level of about 1%. It is one of the most employed indicators of agglomerate quality, while the higher proportion of FeO is associated with higher strength and worse reducibility. It follows from the above that most of the FeO in the studied agglomerates was bound up in magnetite. If we consider, based on these facts, that the FeO content in the agglomerate is directly related to the magnetite content, we can assign the same significance to Fe_3O_4 as to FeO.

Besides magnetite and hematite, minerals based on calcium ferrites and silicates were identified in the structure of the agglomerate (see Tables 8 and 9). Calcium ferrites already begin to form at temperatures 500 to 600 °C. They occur in places with a hematite base, i.e., in cracks and pores, through which the gas phase flows. The sintering reactions usually begin in the solid state between small particles of limestone and ore grains in close contact, which then fuse together. Subsequently there occurs congruent melting of the products accompanied with the assimilation of the nucleus parts. The period during which the temperature does not drop below 1100 °C is the time required for melt formation during sintering [14].

With increasing temperature, the primary melt dissolves the gangue component and hematite to form complex compounds, silicoferrites of calcium and aluminum (SFCA). The basic component is hemicalciumferrite CaO_2 Fe_2O_3, in which SiO_2 and Al_2O_3 are isomorphically admixed. In the literature this phase is referred to as the SFCA ferrite [15].

As the percentage of concentrate replacement with iron ore AR-5 increased in our experiments, the content of the ferritic phase changes. The lowest value was reached in the standard agglomerate. In agglomerates with 50%, and 100% substitution, it achieved almost identical values of 30.65%, and 30.60%. The agglomerate with 25% substitute was the closest to the standard agglomerate, and the agglomerates with 66% and 75% substitution were in the range from 21% to 27%. However, it can be stated that the replacement of the concentrate by AR-5 in all percentages increases the number of ferritic phases in the agglomerate (see Table 8)

Formation of calcium ferrites is impeded by the presence of SiO_2, with which CaO reacts more readily, forming slag silicate melts. SiO_2 already begins to react with magnetite at 650 °C in solid state, producing fayalite. The intense reaction occurs above 1100 °C. Yet a small amount of fayalite produced by the solid-state reaction causes melting of the batch, because above the liquidus line there is an unlimited solubility between fayalite and magnetite. Fayalite may also form in hematite batches, but only at higher temperatures when hematite decays into magnetite. The products of iron oxide reaction with SiO_2 in the Ca-Si-Fe-O system are therefore olivines (see Table 8). The amount of calcium ferrites and dicalcium silicate phases depends on the basicity of the agglomeration mixture; with lower basicity glassy silicates and secondary hematite prevail, as well as magnetite, and calcium ferrites and dicalcium silicates phases at the higher end [16]. The basicity at which our sinterings were performed was set in the range from 1.4 to 1.6. This range was chosen because of the formation of dicalcium silicate in the basicity range from 1.6 to 1.8, causing the agglomerate to become brittle, which could result in distortion of the results.

The silicate components in the agglomerates are ferro-calcium olivines, pyroxins, calcium silicates, and solid glassy silicates. Ferro-calcium olivines $(Ca_XFe_{1-X})_2 \cdot SiO_4$ form solid solutions between fayalite $2FeO \cdot SiO_2$ and dicalcium silicate $2CaO.SiO_2$. The phase composition of silicate phases in studied agglomerates is presented in Table 9.

As can be seen from Table 9, the most represented phase of silicate phases is Ca_2SiO_4 (larnite).

The amount of larnite in the agglomerate depends on the agglomerate basicity. With increasing basicity, its amount increases. In the area of basicity of about 1.8, which corresponds to the Ca_2SiO_4 molecule, the agglomerate loses its strength (decays). The disintegration of the agglomerate is attributed to the polymorphic conversion of Ca_2SiO_4, which is associated with an increase in volume of about 12%. For this reason, a lower level of basicity was chosen in the experiment, i.e., 1.4 to 1.6, where although larnite is present, its amount ranging from 7.45% to 9.25% is so small that it does not cause the agglomerate to break down and thus does not affect its strength [17,18].

The results of the experimental study show that in all agglomerates, in qualitative sense, present phases were identical to the reference agglomerate, only their intensity was changed. The change in phase intensity did not reduce the qualitative parameters of the compared agglomerates.

4. Conclusions

Based on the theoretical knowledge from the study of the structure and phase composition of the agglomerate, the latter can be defined as a multiphase complex whose properties depend on the features and volume concentration of the individual phases and their relative distribution. Furthermore, based on the knowledge of the properties of the individual phases, the nature of their impact on the properties of the final agglomerate can be generally assessed.

On the basis of the results obtained from this study of the phase composition of a conventional agglomerate and agglomerate with changed concentrate/iron ore ratio it can be stated that no new phases occur, and only the intensity of the individual phases changes.

In the standard agglomerate and agglomerate with 25%, 50%, 66%, 75% and 100% replacement of concentrate by the iron ore AR-5, the following phases were identified by X-ray analysis: magnetite, wüstite, hematite, calcium ferrites, ferro-calcium olivines, pyroxenes, dicalcium silicate, residual quartz.

Replacement of magnetite concentrate with hematite ore significantly reduces the content of magnetite. While in standard agglomerate, its quantity is 20.2%, its share in the agglomerate with the

AR-5 iron ore substitution ranges from 9.6% to 15.8%. In the agglomerate without concentrate, the amount of magnetite dropped to 11.2%, which is almost a 50% decrease.

The change in the amount of hematite is not so pronounced. In a standard agglomerate the amount of hematite corresponds to 43%, while in other compared agglomerates, it is ranging from about 33% to 42%. A slight decrease in hematite is probably due to the reaction of fine grains of Fe_2O_3 with CaO forming a ferritic binding phase.

The amount of ferritic phases changes significantly. While the standard agglomerate reaches 13.5%, the agglomerates with hematite ore AR-5 reach values in the range of about 15% to 30% of the ferritic phases.

Wüstite content is at the level of about 1%, and it is assumed that most of it is bound in magnetite in the agglomerates.

The silicate phases are represented by ferro-calcium olivines, larnite and residual quartz. The amount of silicate phases stays within a narrow range. Of these silicate phases, larnite deserves the most attention, as it undergoes polymorphic transformation during cooling which is accompanied by an increase in volume by about 12%, and as a result, the agglomerate breaks down, leading to the loss of strength. The amount of larnite in all compared agglomerates ranged between 7% and 9%. The amount of larnite is small and it is therefore assumed that it becomes is physically stabilized in the ferritic binding phase and does not affect the agglomerate's strength.

Based on the results of the phase composition of the compared agglomerates, it can be stated that the matrix of agglomerates with the changed ratio of concentrate/ore AR-5 is formed by hematite, and the binding phase is mainly composed of ferritic phases and, to a lesser extent, silicate phases. Such a phase composition, in terms of quality, ensures good metallurgical properties. These agglomerates are easily reducible, which also corresponds to lower consumption of metallurgical coke in the process of pig iron production.

High quality agglomerate for blast-furnace processing must therefore consist of mineral phases guaranteeing a good strength of and its easy reducibility, both of which were achieved in this study. Based on the qualitative assessment of the reference agglomerate (without the addition of AR-5) and agglomerates with increasing concentrate replacement by iron ore AR-5, it can be stated that any ratio of replacement of the concentrate by granulometrically finer iron ore leads to an increase in the quality of the final product.

This research has shown that also in the case of replacement of the concentrate with ore approaching the properties of the concentrate in its composition and granulometry it is possible to ensure, in the sintered layer, conditions for the formation of mineral phases in such ratio that there is no great degradation of the properties of the finished agglomerate. In terms of the main objective, which is the quality of agglomerate and reduction of the environmental burden from agglomeration plants, change in the C/O ratio or concentrate replacement with the above-defined ore may be recommended for the iron ore agglomerate production process as the basic blast furnace batch component.

Author Contributions: M.F. performed the methodology, investigation, conceptualization, writing original draft preparation, writing review and editing; D.I. investigation, formal analysis, resources; R.F. investigation, validation; M.D. investigation, visualization, project administration; J.L. investigation, writing review and editing.

Funding: This research was funded by [APVV] Slovak Research and Development Agency, Slovak Republic number APVV–16-0513 and [VEGA MŠ SR a SAV] grant number 1/0847/16.

Conflicts of Interest: The authors declare no conflicts of interest.

References

1. Pustějovská, P.; Brožová, S.; Jursová, S. Environmental benefits of coke consumption decrease. Proceedings of Metal 2010: 19th Anniversary International Conference on Metallurgy and Materials, Brno, Czech Republic, 18–20 May 2010; pp. 79–83.
2. Fernández-González, D.; Ruiz-Bustinza, I.; Mochón, J.; González-Gasca, C.; Verdeja, L.F. Iron Ore Sintering: Raw Materials and Granulation. *Miner. Process. Extr. Met. Rev.* **2017**, *38*, 36–46. [CrossRef]

3. Cores, A.; Muñiz, M.; Ferreira, S.; Robla, J.I.; Mochón, J. Relationship between sinter properties and iron 353 ore granulation index. *Ironmak. Steelmak.* **2012**, *39*, 85–94. [CrossRef]
4. Cohen, R.G. The strength of granules and agglomerates. In Proceedings of the International Symposium Philadelphia, Philadelphia, PA, USA, 12–14 April 1961.
5. Findorák, R.; Fröhlichová, M.; Legemza, J.; Bakaj, F. Effect of high content of concentrate in sinter mixture on chosen emission parameters. *Prace IMZ* **2009**, *61*, 5.
6. Nakano, M.; Okazaki, J. Influence of operational conditions on dust emission from sintering bed. *ISIJ Int.* **2007**, *2*, 240–244. [CrossRef]
7. Kasama, S.; Kitaguchi, H.; Yamamura, Y.; Watanabe, K. Analysis of exhaust gas visibility in iron ore sintering plant. *ISIJ Int.* **2006**, *7*, 1027–1032. [CrossRef]
8. Roubicek, V.; Pustejovska, P.; Bilík, J.; Janík, I. Decreasing CO_2 emissions in metallurgy. *Metalurgija* **2007**, *46*, 53–59.
9. Malysheva, T.Ya.; Detkova, T.V.; Loginov, I.V.; Gorshkolepova, A.V. Phase compositiona and structural features of severstal cherepovets metallurgical combine industrial agglomerate. *Metallurgist* **2010**, *54*, 278–284. [CrossRef]
10. Fröhlichová, M.; Ivanišin, D.; Mašlejová, A.; Findorák, R.; Legemza, J. Iron-ore sintering process optimization. *Arch. Metall. Mater.* **2015**, *60*, 2895–2899. [CrossRef]
11. Button, R.A.; Lundh, P.A. Mineralogy and mineral formation in iron ore sinter with addition of magnetite fines. *Ironmak. Steelmak.* **1989**, *16*, 151–164.
12. Ahsan, S.N. Structure of fluxed sinter. *Ironmak. Steelmak.* **1983**, *10*, 54–64.
13. Kret, J.; Mojžíšek, J. *Mikrostruktura hutních surovín*; VŠB-Technická Univerzita Ostrava: Ostrava, Czech Republic, 2003; pp. 34–47.
14. Loo, Ch.; Tame, N.; Penny, G.C. Effect of iron and sintering conditions on flame front properties. *ISIJ Int.* **2012**, *52*, 967–976. [CrossRef]
15. Mežibrický, R.; Fröhlichová, M. Silico-ferrite of calcium and aluminum characterization by crystal morphology in iron ore sinter microstructure. *ISIJ Int.* **2016**, *56*, 1111–1113. [CrossRef]
16. Fröhlich, L.; Fröhlichová, M. Analysis of stress in Ca_2SiO_4 crystal grid following the crystal and chemical stabilization. *Metalurgija* **1994**, *33*, 99–103.
17. Frohlichova, M.; Majerčák, Š.; Mihok, L. Dicalcium silicate and desintegration capacity of sinter. *Hutnické Listy* **1992**, *45*, 144–148.
18. Fröhlichová, M. Divápenatý kremičitan v štruktúre aglomerátu bohatého na obsah železa/-1991. *Hutnické Listy* **1991**, *46*, 11–12.

metals

MDPI

Article

Prediction Model of Iron Ore Pellet Ambient Strength and Sensitivity Analysis on the Influence Factors

Qiangjian Gao [1,*], Yingyi Zhang [2,*], Xin Jiang [1], Haiyan Zheng [1] and Fengman Shen [1]

[1] School of Metallurgy, Northeastern University, Shenyang 110819, China; Jiangx@smm.neu.edu.cn (X.J.); Zhenghy@smm.neu.edu.cn (H.Z.); Shenfm@mail.neu.edu.cn (F.S.)

[2] School of Metallurgical Engineering, Anhui University of Technology, Ma'anshan 243002, China

* Correspondence: gaoqj@smm.neu.edu.cn (Q.G.); zhangyingyi@cqu.edu.cn (Y.Z.);
Tel.: +86-24-83681506 (Q.G. & Y.Z.)

Received: 2 July 2018; Accepted: 25 July 2018; Published: 30 July 2018

Abstract: The Ambient Compressive Strength (CS) of pellets, influenced by several factors, is regarded as a criterion to assess pellets during metallurgical processes. A prediction model based on Artificial Neural Network (ANN) was proposed in order to provide a reliable and economic control strategy for CS in pellet production and to forecast and control pellet CS. The dimensionality of 19 influence factors of CS was considered and reduced by Principal Component Analysis (PCA). The PCA variables were then used as the input variables for the Back Propagation (BP) neural network, which was upgraded by Genetic Algorithm (GA), with CS as the output variable. After training and testing with production data, the PCA-GA-BP neural network was established. Additionally, the sensitivity analysis of input variables was calculated to obtain a detailed influence on pellet CS. It has been found that prediction accuracy of the PCA-GA-BP network mentioned here is 96.4%, indicating that the ANN network is effective to predict CS in the pelletizing process.

Keywords: iron ore pellets; compressive strength (CS); prediction model; artificial neural network; principal component analysis

---

1. Introduction

Iron-bearing materials should enable the reliable production of hot metal (HM) from a blast furnace (BF) or directly reduced iron (DRI) from a shaft furnace [1], particularly at minimum cost and at a large scale. Pellets are roughly spherical, thermally and/or chemically bonded agglomerates with 10 to 16 mm in diameter. They have several preferred metallurgical properties [2]. For instance, the particle size of a pellet is relatively uniform, which is favorable for burden permeability in the BF process; the pellet iron-grade is higher, favorable for effective iron content input in burden column and enhancing production efficiency; the reduction of pellet is easy, favorable for increasing gas utilization rate of a BF/shaft furnace, etc. Therefore, the consumption of pellets in BF is gradually increasing in many ironmaking plants—20% in China and even 100% in some BFs in Europe and North America.

At present, the pellet induration processes can be carried out by a straight grate indurating machine or grate-rotary kiln [3]. However, no matter which induration facility is used, the production process of iron ore pellets can be summarized as the following four stages: agglomeration of green pellets, drying of green pellets, firing of green pellets obtaining fired pellets, and cooling of fired pellets. Generally, in the BF or shaft furnace process, the pellets undertake double forces containing the extrusion force from the top layer and the friction force by high-speed gas flow beneath the furnace [4]. Consequently, the compressive strength (CS) of fired pellets is often referred to as a principal metallurgical property that helps withstand these two forces. A lower pellet CS may lead to pellet disintegration in the BF process, negatively affecting the permeability of the burden column. Obviously, pellets with higher strength can decrease fines and dust generation, and this in turn helps increase the productivity of the ironmaking unit. The CS standard [5] in China rules that CS should be higher than 2200 N/pellet for the BF process and higher than 2500 N/pellet for a gas-based shaft furnace process.

Previous papers have identified several factors in the pelletizing process that affect pellet CS [6]; they include particle size of ore, bentonite dosage, moisture of green pellet, basicity (CaO/SiO_2), and firing temperature, among others. It is a challenge to comprehensively control CS in pelletizing production. Furthermore, the pelletizing process is a complex physical and chemical process [7] including mass transfer, heat transfer, gas-solid reaction, and crystalline grain change. It is therefore difficult to simulate the whole process of pellet induration. Batterham [8] attempted to predict the strength using shrinking models and thereby deriving the shrinkage-to-strength relationship, but the universality of the model is limited due to various operation conditions in different ironmaking plants. No single cogent metallurgical theory is available to numerically describe the development of CS, and few models can directly work out the effect of critical operational parameters on CS. ANN [9], as an attractive technique, is available to capture complex and non-linear relationships between the input and output in the complex industry system. The ANN model, via enough learning and training, are computationally efficient and no prior domain knowledge is required for the process to be modeled [9]. The BP network is one of the most widely used ANN. It is a multilayer feed forward network, trained by error inverse propagation. It includes an input layer, a hidden layer, and an output layer [10]. Several successful examples have been reported in metallurgical modeling processes using the BP network. For example, Stanford University Neural Network Corporation and North Star Steel Corporation cooperatively developed an intelligent electric arc furnace [11]. The BP network was adopted to predict the temperature distribution of BF top gas at Kobe steel, Japan [12]. Similarly, the BF heat mode and hot air volume mode were carried out based on BP neural network at Kawasaki Steel, Japan [13]. However, with progress of ANN technology, two obvious drawbacks of the BP model were found [14]. The first one is that the BP network appears to converge slowly and with difficultly, as the input variables are too many or some of them are actually correlative. The second is that the weights and thresholds of the BP network may stay at local optima in the training process, which may also limit the convergence of the network. Thus, in order to make the BP model accurate and stable, some necessary measures should be carried out to achieve dimensionality reduction of correlative variables and avoid local optima in the training process.

PCA [15,16] is a method that can be used to reduce the dimensionality of input samples. The new PCA samples after dimensionality reduction are uncorrelated. There is no double that the PCA is available to solve the challenge of the first drawback of the BP network mentioned above. Furthermore, Genetic algorithm (GA) [17] belongs to a class of heuristic methods known collectively as evolutionary computation. It depends on the concept of evolving a more optimal solution to a problem through the repeated application of selection, crossover, and mutation operation. GA shares the ability to escape from local optima by moving through valleys of lower fitness in the search space. Obviously, GA can be utilized to optimize the initial weights in the BP network and escape the local optima in the training process.

In the present work, a prediction model based on the ANN was developed to predict pellets' CS. Firstly, the PCA was applied to fulfill dimensionality reduction of correlative variables. Secondly, the new PCA variables were used as the input variables for the BP network, which was upgraded by the GA; after training and testing with production data in a typical ironmaking plants of China, the PCA-GA-BP neural network was built. Finally, the prediction accuracy of the PCA-GA-BP network mentioned here was investigated. Moreover, sensitivity analysis was carried out to identify the critical factors that affect CS. The target of this work is to provide a reliable CS control strategy in the pelletizing process.

2. Materials and Methods

2.1. Pelletizing Process

The pelletizing process of iron ore is a consolidation method to obtain high-strength pellet; this includes granulation of the mixed materials (such as iron ore, bentonite and flux), drying of green pellet, firing of green pellet with combustion of fuel in burden bed, and cooling. The schematic diagram of the pelletizing process is shown in Figure 1.

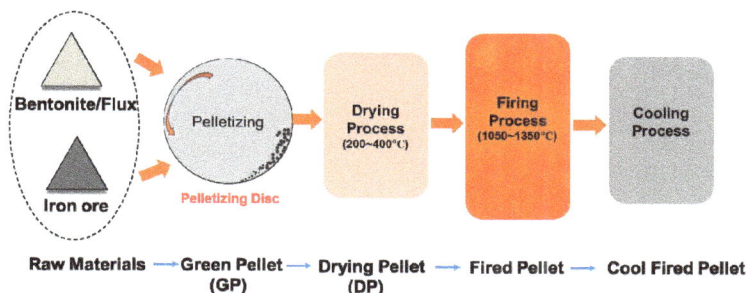

Figure 1. Schematic diagram of the pelletizing process.

The pellet is indurated with high CS, which is detected according to the standard ISO4700 [5]. The general method is this: 64 selected pellets with diameter of 10–12.5 mm are tested in a compressive tester, and the two maximum values and two minimum values are deleted; the average of the remaining 60 values is regarded as the final pellet CS. Several factors in a pelletizing unit can substantially affect CS. Therefore, the prediction of CS should highly consider influencing factors. Based on previous practical experience at the typical ironmaking plants of China, 19 factors were selected as the initial input variables, including particle size of ore, bentonite dosage, moisture of green pellet, CS of green pellet, dropping strength, pellet diameter, porosity of green pellet, FeO%, MgO%, CaO%, Al$_2$O$_3$%, basicity (CaO/SiO$_2$), burst temperature of green pellet, CS of drying pellet, bed depth of burden, firing temperature, firing time, gas consumption, and charging rate. The annual averages (the typical ironmaking plants) of the input variables are listed in Table 1.

Table 1. Initial input variables of the pelletizing process.

Input Variables		Average/Year	Standard Deviation
Raw Materials	Particle size of ore, −74 μm/%	76.88	5.3
	Bentonite, %	1.22	0.1
Green Pellet	Moisture of green pellet, %	8.0	0.36
	CS of green pellet, N/pellet	15	2.0
	Dropping strength, Times/pellet	6	0.6
	Pellet diameter, mm	13	1.0
	Porosity of green pellet, %	30	1.9
Chemical Composition	FeO, %	0.7	0.02
	MgO, %	0.6	0.03
	CaO, %	0.5	0.01
	Al$_2$O$_3$, %	1.5	0.04
	Basicity (CaO/SiO$_2$), Dimensionless	0.09	0.001

Table 1. *Cont.*

Input Variables		Average/Year	Standard Deviation
Drying Process	Burst temperature, C	450	20.0
	CS of drying pellet, N/pellet	80	5.0
	Bed depth, mm	540	10.0
Firing Process	Firing temperature, C	1250	15.0
	Firing time, min	30	2.0
	Gas consumption, m³/t	50	3.5
	Charging rate, ton/h	600	15.0

2.2. Principal Component Analysis

The PCA is a method [16] used to achieve dimensionality reduction of initial variables (19 variables mentioned in Table 1). The new non-linear PCA variables can be brought out to substitute the initial variables. The PCA operation creates linear combinations of initial input variables, so all the input information is retained in the new variables. In order to achieve the dimensionality reduction, only a subset of new variables is chosen as the PCA variables, so there is a loss of initial variables. It is worthwhile mentioning that the lost information is very limited and thus will not affect the accuracy of the following results due to the fact that the cumulative contribution rate of the PCA variables is required to be more than 90%. In summary, the merit of the PCA is that the information, expressed by initial variables, will be retained in the new PCA variables as much as possible. It is acknowledged that copious factors could affect CS, and part of them are correlative, such as firing temperature match with fuel gas consumption, and basicity match with CaO%. Overabundant initial variables without PCA operation will make the model more complex and decrease prediction accuracy. In the present work, the new independent PCA variables based on the PCA operation were finalized as the input variables in the BP model. The flow chart of PCA operation is shown in Figure 2.

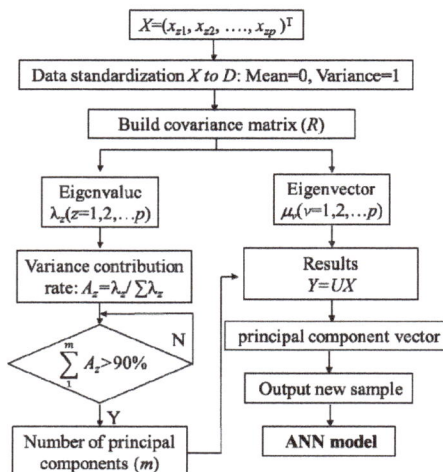

Figure 2. Flow chart of PCA.

(1) The sample matrix is collected as shown in Equation (1):

$$X = (x_{z1}, x_{z2}, \ldots, x_{zp})^T, \text{ or } X = \begin{matrix} x_{11} & x_{12} & \cdots & x_{1p} \\ x_{21} & x_{22} & \cdots & x_{2p} \\ \cdot & \cdot & \cdot & \cdot \\ x_{z1} & x_{z2} & \cdots & x_{zp} \end{matrix} \tag{1}$$

where X is the sample matrix, n is number of samples, $z = 1, 2, \ldots, n$, 300 days production date were used in present work, $n = 300$, p is the number of initial variables, $p = 19$.

(2) Date standardization of initial variable matrix (X) to normalization matrix (D) as shown in Equation (2):

$$D_{zv} = \frac{x_{zv} - \overline{x_v}}{s_v}, \ z = 1, 2, \ldots, n; \ v = 1, 2, \ldots, p \tag{2}$$

where $\overline{x_v} = \frac{\sum\limits_{z=1}^{n} x_{zv}}{n}, s_v = \sqrt{\frac{\sum\limits_{z=1}^{n} (x_{zv} - \overline{x_v})^2}{n-1}}$.

(3) Obtain the correlation coefficient matrix (R) based normalization matrix using Equation (3):

$$R = \frac{D^T D}{n-1} \tag{3}$$

(4) Obtain the eigenvalue (λ_z) and the eigenvector (μ_v) based on the characteristic equation of R in Equation (4):

$$|R - \lambda_z \mu_v| = 0 \tag{4}$$

(5) Obtain the contribution rate $A_z = \frac{\lambda_z}{\sum\limits_{v=1}^{p} \lambda_z}$, and cumulative contribution rate $\sum\limits_{v=1}^{m} A_z$. As the cumulative contribution rate:

$$\sum_{v=1}^{m} A_z = \frac{\sum\limits_{v=1}^{m} \lambda_z}{\sum\limits_{v=1}^{p} \lambda_z} \geq 90\% \tag{5}$$

The m is obtained based on the cumulative contribution rate. In addition, m is the new variable number after dimensionality reduction.

(6) Output the new sample matrix after PCA in Equation (6):

$$Y = UX \tag{6}$$

where U is the eigenvector matrix comprised m eigenvectors.

2.3. Artificial Neural Network

The new m variables obtained from the PCA were utilized as the input variables in the BP model. The BP network has a merit that the information feeds forward and the error is inverse propagation [18]. In order to make the model accurate and stable, the GA was adopted to optimize the initial weights and thresholds in the BP network. The flow chart of the ANN model is shown in Figure 3. Three operations were adopted consisting of PCA, GA, and BP.

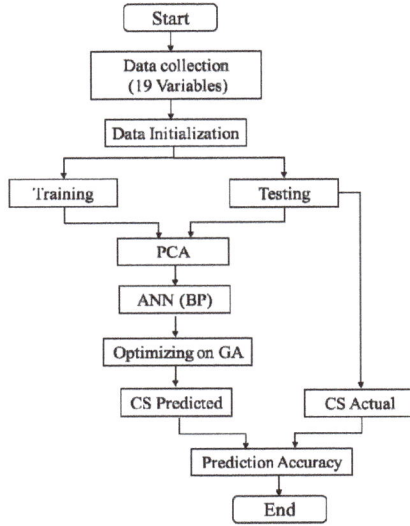

Figure 3. Flow chart of ANN model.

2.3.1. Information Forward Propagation of BP Network

In the BP network mentioned here, a three-layer network structure, $m\sim(m+1)\sim1$ (where, m is the number of new PCA variables obtained by the PCA operation), used in the present modeling, has been shown in Figure 4. The numbers of input variables here were m. The number of input neurons in the hidden layer, designed according a serial of model validations, is assured at $m+1$, and the neuron in output layer is 1, which represents the CS.

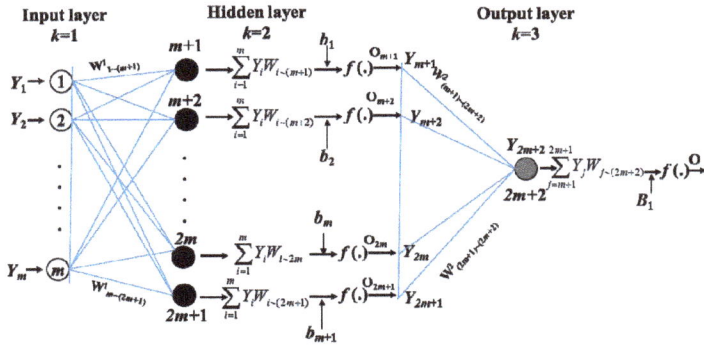

Figure 4. Schematic diagram of BP neural network.

In Figure 4, the input to the neurons in the hidden layer and the output layer is the sum of all the input neurons as shown in Equation (7):

$$Y_i = \sum Y_j W_{j\sim i} + b_i \tag{7}$$

where Y_i: input sum of ith neuron in the k layer, $k = 2, 3$.
Y_j: input sum of jth neuron in the $k-1$ layer.

i: ith neuron in k layer (k = 2, 3), for k = 2, i = 1, 2, ... , m+1; k = 3, i = 1.

j: jth neutron in $k-1$ layer, j = 1, 2, ... , m for input layer, j = m+1, m+2, ... , $2m$+1 for the hidden layer.

W_{j-i}: weight of between jth neuron in $k-1$ layer and ith neuron in k layer.

b_i: threshold of ith neuron in k layer(k = 2), (b_i = B_i, k = 3).

Basing on preliminary experiments, it was found that the Tan-Sigmoid function [19] was accurate to obtain target results; thus, it was arranged as a candidate transfer function of adjacent neurons (Equation (8)):

$$O_i = f(.) = \frac{2}{1 + e^{-2Y_i}} - 1 \tag{8}$$

where O_i: output of ith in k layer (k = 2, 3).

Therefore, using the weights and thresholds of the neurons in hidden, signals transferred by each neuron to the proceeding neuron can be estimated by Equation (9):

$$Y_{m+1} = Y_1 W_{1\sim(m+1)} + Y_2 W_{2\sim(m+1)} + \cdots Y_m W_{m\sim(m+1)} + b_1 \tag{9}$$

Thus, using transfer function O_{m+1} output Y_{m+1}. The total input of neuron in the output layer can be given by Equation (10):

$$Y_{2m+2} = Y_{m+1} W_{(m+1)\sim(2m+2)} + Y_{m+2} W_{(m+2)\sim(2m+2)} + \cdots Y_{2m+1} W_{(2m+1)\sim(2m+2)} + b_{m+2} \tag{10}$$

Again, using transfer function O_{2m+2}, Y_{2m+2} can be converted to output CS.

2.3.2. Error Inverse Propagation of BP Network

All the weights (W) and thresholds (b) are randomized at the first training step. The network generates weights, while training and producing output. Its output is then compared with the target output of all the neurons of the output layer; the error is back propagated by changing the interconnect weights (W) and thresholds (b), in case of any discrepancy. In the BP method here, the least square error (E_k^2) is given by Equation (11):

$$E_k^2 = 0.5 \sum (t_i - O_i)^2 \tag{11}$$

where t_i: targeted output in k layer; O_i: predicted output in k layer.

2.3.3. Optimize the BP Network with GA

It is acknowledged that the weights and thresholds of the BP network may stay at a local minimum in the training process, which may limit network convergence. In order to make the model accurate, the GA was conducted to optimize the initial weights and thresholds of the BP network. Generally, the GA includes selection operation, crossover operation, and mutation operation [20]. The details are listed as follows:

(1) The weights (W) and thresholds (b) of the whole neural network are regarded as a group of chromosomes, as shown in Equation (12):

$$W_{\text{whole}} = \{W_{j\sim i}, b_i\} \tag{12}$$

where W_{whole} is an assemblage of weights and thresholds.

(2) Obtain the fitness of the chromosomes (f_i) in Equation (13):

$$f_i = \frac{M}{\Delta E_i} \tag{13}$$

where M is a constant, ΔE_i is the absolute error in the training model.

(3) Select some individuals whose fitness is larger. The selection probability (P_k) is shown in Equation (14):

$$P_k = \frac{f_i}{\sum f_i} \tag{14}$$

For the sake of preventing degradation of the best sample, individuals with the maximum f_i are put to the next generation directly, without any genetic operations such as crossover and mutation.

(4) A new generation of sample will be generated after crossover and mutation operation.

(5) According to the GA operation, the weights and thresholds given the minimum error are picked up as the new weights and thresholds for the ANN network.

3. Results and Discussion

3.1. PCA

The 19 new components shown in Table 3 are produced basing on the PCA operation. It should be noted that the new 19 components here are different from the initial 19 variables, and the new components are numbered base on the eigenvalue (λ), from large to small. Besides, each new component is independent and contains one or more correlative initial variables information. In addition, the contribution rate and cumulative contribution rate of the new components are listed in Table 3. The contribution rate is shown in Figure 5.

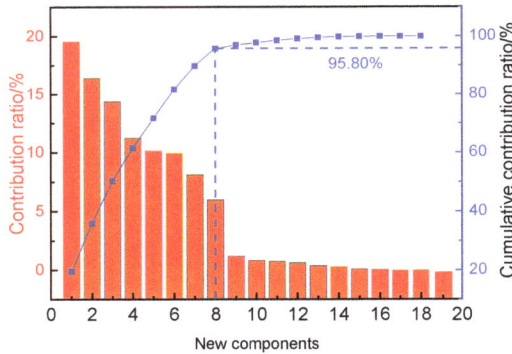

Figure 5. Contribution rate and cumulative contribution rate of new components.

In this paper, PCA variables are assured when the cumulative contribution rate of the new components is more than 90% (as mentioned in Equation (5)). From Figure 5, it can be seen that the cumulative contribution rate of the first eight components is 95.80% (>90%), which means that the first eight components can adequately describe the information existing in the 19 initial variables. Other contribution rates are too small and therefore omitted. We chose the first eight components as the new PCA variables to serve the BP network. Thus, $m = 8$ in the following BP network.

Table 2. λ and contribution rate of each new component.

New Components	λ	Contribution Rate (%)	Cumulative Contribution Rate (%)
1	2.12	19.548	19.55
2	1.78	16.413	35.96
3	1.56	14.385	50.35
4	1.22	11.249	61.60
5	1.1	10.143	71.74

Table 3. λ and contribution rate of each new component.

New Components	λ	Contribution Rate (%)	Cumulative Contribution Rate (%)
6	1.08	9.959	81.70
7	0.88	8.114	89.81
8	0.65	5.994	95.80
9	0.13	1.199	97.00
10	0.09	0.830	97.83
11	0.08	0.738	98.57
12	0.07	0.645	99.22
13	0.04	0.369	99.59
14	0.03	0.277	99.86
15	0.008	0.074	99.94
16	0.005	0.046	99.97
17	0.001	0.009	99.98
18	0.0009	0.007	99.99
19	0.0001	0.001	100.00

The score coefficient matrix of the eight components (PCA_i (i = 1~8) variables) is listed in Table 4. It can be seen that (1) PCA_1 contains the initial variable information: bentonite%, CS of green pellet, dropping strength, and CS of drying pellet; (2) PCA_2 contains the initial variable information: FeO%, firing temperature, and gas consumption; (3) PCA_3 contains the initial variable information: pellet diameter, bed depth, firing time, and charging rate; (4) PCA_4 contains the initial variable information: particle size (-74 μm%), and porosity of green pellet; (5) PCA_5 contains the initial variable information: MgO%; (6) PCA_6 contains the initial variable information: CaO%, and basicity; (7) PCA_7 contains the initial variable information: moisture, and burst temperature; (8) PCA_8 contains the initial variable information: Al_2O_3%.

Table 4. Score coefficient between the PCA_i (i = 1~8) and initial variables.

Initial Variables:	Score Coefficient							
	PCA_1	PCA_2	PCA_3	PCA_4	PCA_5	PCA_6	PCA_7	PCA_8
Particle size	~	~	~	0.654	~	~	~	~
Bentonite	0.536	~	~	~	~	~	~	~
Moisture	~	~	~	~	~	~	0.859	~
CS of green pellet	0.639	~	~	~	~	~	~	~
Dropping strength	0.551	~	~	~	~	~	~	~
Pellet diameter	~	~	0.589	~	~	~	~	~
Porosity	~	~	~	0.789	~	~	~	~
FeO	~	-0.669	~	~	~	~	~	~
MgO	~	~	~	~	0.898	~	~	~
CaO	~	~	~	~	~	0.898	~	~
Al_2O_3	~	~	~	~	~	~	~	0.458
Basicity (CaO/SiO$_2$)	~	~	~	~	~	0.758	~	~
Burst temperature	~	~	~	~	~	~	-0.856	~
CS of drying pellet	0.655	~	~	~	~	~	~	~
Bed depth	~	~	0.985	~	~	~	~	~
Firing temperature	~	0.858	~	~	~	~	~	~
Firing time	~	~	0.854	~	~	~	~	~
Gas consumption	~	0.746	~	~	~	~	~	~
Charging rate	~	~	-0.850	~	~	~	~	~

~ the score coefficient value is <0.1.

3.2. Training and Testing of the BP Network

For modeling purposes, operational data for 365 days were collected from the pellet plant. After preprocessing the data to eliminate noise and considering stable operating data, 300 data sets were finalized for training and testing of the network. The data were split into 260 for training and 40 for testing. The GA was conducted to optimize the initial weights and thresholds of the BP network. The training performance of the PCA-BP network and PCA-GA-BP network are shown in Figure 6. One can conclude from Figure 6a that the training speed of the PCA-GA-BP network is faster than that of the PCA-BP. For the PCA-GA-BP network, it takes 60 training times to obtain the target, while 90 training times is needed to meet the target goal for the PCA-BP network. In addition, the output error of the PCA-GA-BP network is smaller than that of the PCA-BP network (Figure 6b). Consequently, the GA operation for optimizing the BP network is virtually effective.

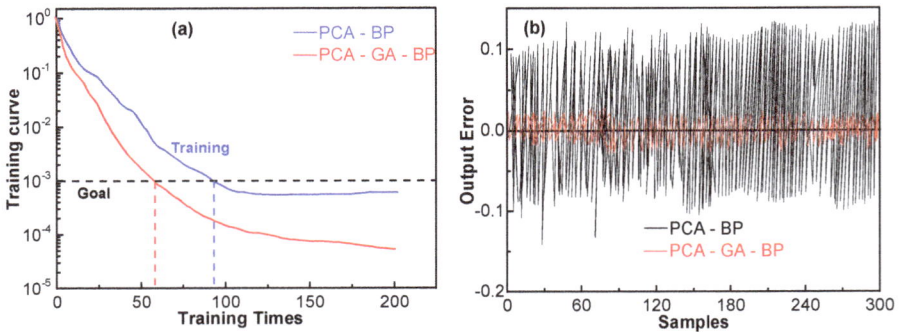

Figure 6. Training performance of PCA-BP network and PCA-GA-BP network (**a**) Training curve; (**b**) output error of the ANN network.

The network was trained till the minimum desired error was achieved. After the training operation, the PCA-GA-BP neural network was used to predict the testing samples; the testing results of the CS are shown in Figure 7. It has been found in Figure 7 that the prediction accuracy of the PCA-GA-BP network mentioned here is 96.4% and the regression coefficient (R^2) is 0.85, indicating that the ANN network can beneficially predict CS in the pelletizing process.

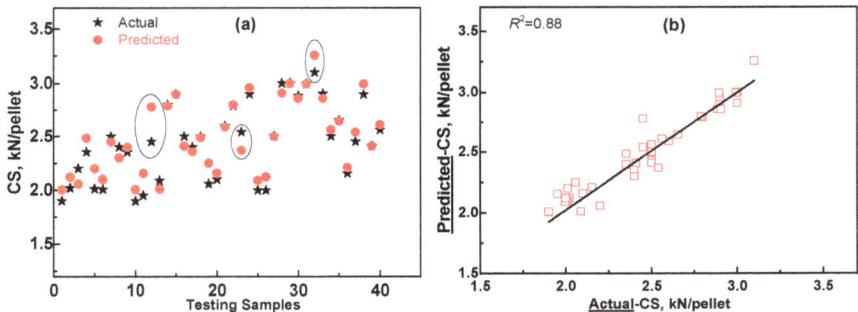

Figure 7. Testing results of PCA-GA-BP network for CS (**a**) Actual Vs Predicted results through ANN; (**b**) Actual Vs Predicted results for CS.

3.3. Sensitivity Analysis of the Influence Variables of CS

Obtaining the sensitivity of influence variables is a target for the PCA-GA-BP network referred to here. Therefore, the sensitivity analyses of some variables were carried out to understand their effect. Based on these results, we can obtain the detailed influence of an input variable on CS. Therefore, it is useful to form a CS control strategy in advance for the pelletizing process.

In the present study, the PCA process revealed that some input variables have a correlative relationship and those variables can comprise of a set of correlative variables. Therefore, only one or two variables in a correlative variable set were chosen to carry out the sensitivity analysis. The variables using sensitivity analysis in the present work include CS green pellet, bentonite%, firing temperature, charging rate, MgO%, basicity, −74 μm%, and burst temperature. The sensitivity analysis results are shown in Figure 8.

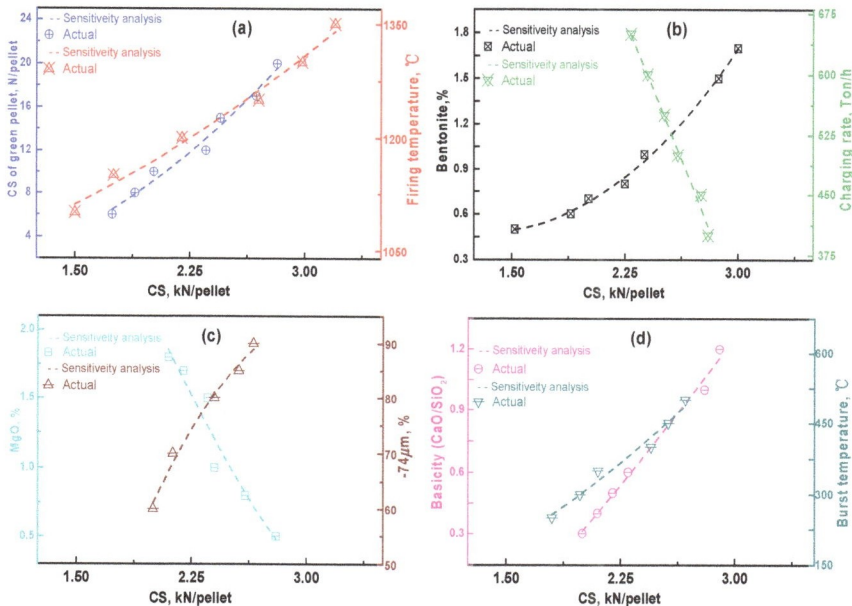

Figure 8. Sensitivity analysis of variables on CS. (**a**) CS of green pellet, firing temperature; (**b**) bentonite%, charging rate; (**c**) MgO%, particle size (−74 μm%); (**d**) basicity, burst temperature.

Figure 8a,b prove the following: (I) Increasing the bentonite dosage of green pellets can contribute to growth of the CS; the sensitivity analysis is also consistent with actual testing. This can be attributed to the addition of a dosage of bentonite (an efficient binder), which can highly enhance the CS of green pellets, leading to a higher CS. (II) High firing temperature improves CS because the induration process of pellet includes the oxidation of Fe_3O_4 and re-crystallization of Fe_2O_3, while high firing temperature simultaneously accelerates the oxidation of Fe_3O_4 and re-crystallization of Fe_2O_3, and obtains high strength [21]. The pellet samples in different firing temperatures (1150 °C, 1250 °C, and 1350 °C) were observed using a metallographic microscope, and the results are shown in Figure 9. For pellets indurated at higher temperatures (1250 °C, 1350 °C), the re-crystallization of Fe_2O_3 was completed; the crystalline grain of Fe_2O_3 compactly connected with each other and were evenly distributed. For pellets in lower indurated temperatures (at 1150 °C, Figure 9a), the re-crystallization of Fe_2O_3 was not complete. (III) A higher charging rate is found to decrease pellet strength. This can be attributed to a high charging rate, which can reduce the residence time of pellets in high temperature firing regions, resulting in low strength.

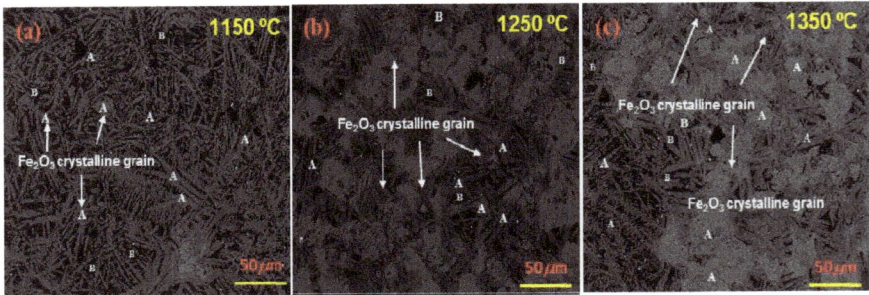

Figure 9. Micro-structure of pellets: (**a**) 1150 °C; (**b**) 1250 °C; (**c**) 1350 C. A is (Fe$_2$O$_3$ crystalline grain); B is (Un-oxidation Fe$_3$O$_4$).

From Figure 8c: (IV) High MgO content appears to attenuate pellet strength. It was pointed out by Shen et al. [22] that increasing MgO content has a negative effect on filling of Fe$_2$O$_3$ crystalline grain in the pore of the pellet.

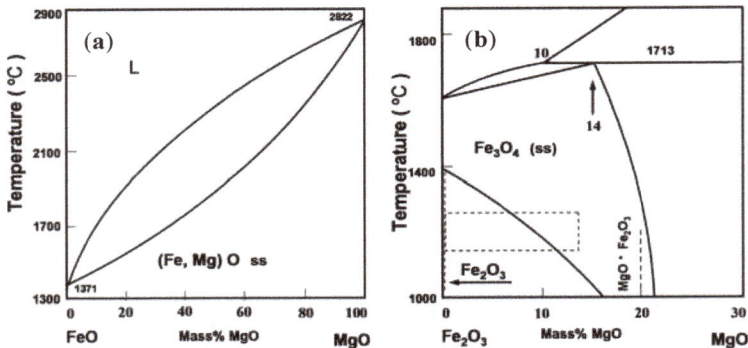

Figure 10. Phase diagram of FeO$_x$–MgO (**a**) FeO–MgO; (**b**) Fe$_2$O$_3$–MgO.

From the FeO$_x$–MgO phase diagram [23] (Figure 10), it is evident that FeO and MgO are unlimitedly mutually soluble (Figure 10a). Fe^{2+} and Mg^{2+} can be partially replaced with each other, and there is a large homogeneous zone meeting the induration of oxidized pellets (dotted line rectangle in Figure 10b). During the induration process, the reactions between MgO and Fe$_2$O$_3$ can produce MF minerals—{(Fe$_x$ Mg$_{1-x}$) O · Fe$_2$O$_3$}. Similar to the structure of magnetite (Fe$_3$O$_4$), MF is also a reverse spinel structure. Consequently, MgO can restrain the crystal transition from Fe$_3$O$_4$ to Fe$_2$O$_3$, resulting in incomplete re-crystallization and low strength. (V) The high strength of a large portion of -74 μm iron ore can be attributed to the fact that it is easier for iron ore with low particle size to be pulled together and indurated adequately.

In addition, from Figure 8d: (VI) Increase in pellet basicity has a major impact on CS. This could be attributed to the fact that higher CaO results in the formation of a stronger calcium ferrite phase by a reaction between calcium oxide and iron oxide [24]. The CaO–Fe$_2$O$_3$ phase diagram [23] is shown in Figure 11; the calcium ferrite phase (CF) is thus generated, as the firing temperature is higher than 1200 °C.

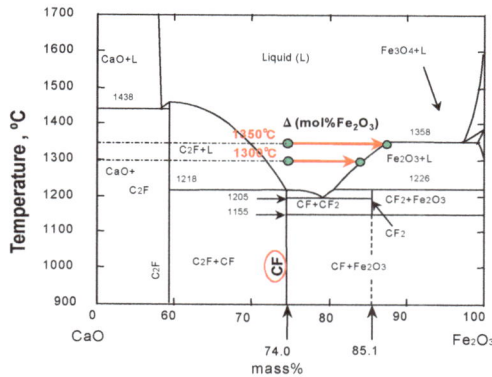

Figure 11. Phase diagram of CaO–Fe$_2$O$_3$.

(VII) Higher burst temperatures offer substantial improvement in pellet strength. Green pellets are susceptible to bursting during the drying process (Figure 12), if the burst temperature is too low. Therefore, enhancing the burst temperature (especially onset temperature) can help limit the same in the drying and firing processes. Similarly, the PCA operation has shown that the burst temperature and the moisture of green pellets have a negative correlation, and low moisture dosage is found to improve CS. Gao et al. [25] maintained that low moisture content in green pellets can improve burst temperature, thus increasing CS.

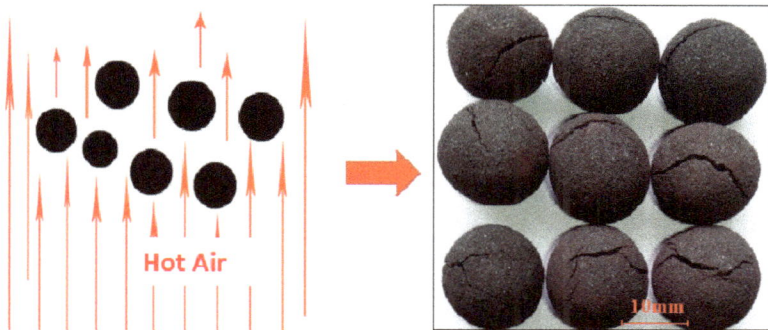

Figure 12. Burst process of green pellets in hot air.

3.4. Discussion

The CS of a pellet is of significance in the BF process. Therefore, the influence factors and control strategy of a pellet's CS should be explored with comprehensive consideration. The model mentioned here has a high accuracy and can be well used in the industry to predict and control pellet CS. Based on the different sets of input variables ($\leq$19, for the initial variables in Table 1), the model can predict pellet CS in advance. In addition, the model is available to optimize the existing pelletizing process, based on a target CS. For instance, if we set the target CS in 2500 N/pellet, the model will adjust the input variables to product pellets with desired CS. Also, the adjustment of the variables is permitted to a finite level to avoid operation fluctuations and maintain stable production. Moreover, the model can be used to deal with emergencies in production. Based on sensitivity analysis results of influence variables, we can tinker with high sensibility parameters to assure continuous production.

4. Conclusions

In order to provide a reliable control strategy in the production process and predict pellet CS, a prediction model based on the PCA-GA-BP network was proposed. The main findings are summarized as follows:

(1) The dimensionality influence factors of the CS were reduced by the PCA operation. The new independent PCA variables containing adequate information of the initial variables can also be brought out to substitute the initial variables.

(2) The GA was provided to offer accuracy and stability of the model and to optimize the initial weights and thresholds of the BP network. The prediction accuracy of the PCA-GA-BP network mentioned here is 96.4%, indicating that the ANN network was effective to predict CS in the pelletizing process.

(3) Sensitivity analyses of some variables were carried out to understand their effect on CS. Large bentonite dosage, high CS of green pellets, and high firing temperature increase the CS of pellets. Enhancing the charging rate and high MgO content of pellets are found to decrease pellet CS. Higher basicity, small particle size of iron ore (-74 μm%), and high burst temperature positively affect pellet CS.

Author Contributions: Q.G. and Y.Z. designed the experiments; Q.G. performed the experiments, analyzed the data, and compiled the text; X.J., H.Z. and F.S. supervised the experimental work and reviewed the manuscript.

Funding: This research was funded by the National Science Foundation of China (51604069 and 51604049), the Fundamental Research Funds for the Central Universities, China (N162504004) and the National Key R&D Program of China (2017YFB0603800 and 2017YFB0603802).

Acknowledgments: The authors wish to acknowledge the contributions of associates and colleagues at Northeastern University of China. The financial support of the National Science Foundation of China (51604069 and 51604049), the Fundamental Research Funds for the Central Universities, China (N162504004) and the National Key R&D Program of China (2017YFB0603800 and 2017YFB0603802) are also appreciated.

Conflicts of Interest: The authors declare no conflict of interest.

References

1. Biswas, A.K. *Principles of Blast Furnace Ironmaking*; Cootha Publishing House: Brisbane, Australia, 1981; pp. 33–39. ISBN 10:0949917001.
2. Fu, J.Y.; Zhu, D.Q. *Basic Principles, Techniques and Equipment of the Iron Ore Oxidized Pellets*; Central South University Press: Changsha, China, 2005; pp. 323–336. ISBN 9787811050516.
3. Gao, Q.; Shen, F.; Jiang, X.; Wei, G.; Zheng, H.Y. Gas-solid reduction kinetic model of MgO-fluxed pellets. *Int. J. Miner. Metall. Mater.* **2014**, *21*, 12–17. [CrossRef]
4. Nabeel, M.; Karasev, A.; Jönsson, P.G. Evaluation of dust generation during mechanical wear of iron ore pellets. *ISIJ Int.* **2016**, *56*, 960–966. [CrossRef]
5. China Metallurgical Construction Association. *Code for Design of Iron Pelletizing Engineering*; China Planning Press: Beijing, China, 2009; pp. 20–25. ISBN 1580177244.
6. Shen, F.; Gao, Q.; Jiang, X.; Wei, G.; Zheng, H. Effect of magnesia on the compressive strength of pellets. *Int. J. Miner. Metall. Mater.* **2014**, *21*, 431–437. [CrossRef]
7. Sadrnezhaad, S.K.; Ferdowsi, A.; Payab, H. Mathematical model for a straight grate iron ore pellet induration process of industrial scale. *Comp. Mater. Sci.* **2009**, *44*, 296–302. [CrossRef]
8. Batterham, R.J. Modeling the development of strength in pellets. *Metall. Mater. Trans. B.* **1986**, *17*, 479–485. [CrossRef]
9. Gardner, M.; Dorling, S.R. Artificial neural network (Multilayer Perceptron)—A review of applications in atmospheric sciences. *Atmos. Environ.* **1998**, *32*, 2627–2636. [CrossRef]
10. Shuang, C.; Wei, X. Design and selection of construction, parameters and training method of BP network. *Comput. Eng.* **2001**, *92*, 336–337.
11. Markward, S.W.; Lu, Y.Z. Integrated Neural System for coating weight control of a hot Dip Galvanizing Lina. *Iron Steel Eng.* **1995**, *72*, 45–49.
12. Portman, N.F.; Lindhoft, D. Application of neural networks in rolling mill automation. *Iron Steel Eng.* **1995**, *72*, 33–36.

13. Pomerleau, D.; Hodouin, D.; Poulin, É. A first principle simulator of an iron oxide pellet induration furnace–an application to optimal tuning. *Can. Metall. Quart.* **2013**, *44*, 571–582. [CrossRef]

14. Fan, X.; Yang, G.; Chen, X.; Gao, L.; Huang, X.; Li, X. Predictive models and operation guidance system for iron ore pellet induration in traveling grate-rotary kiln process. *Comp. Chem. Eng.* **2015**, *79*, 80–90. [CrossRef]

15. Im, H.J.; Song, B.C.; Park, Y.J.; Song, K. Classification of materials for explosives from prompt gamma spectra by using principal component analysis. *Appl. Radiat. Isotopes.* **2009**, *67*, 1458–1462. [CrossRef] [PubMed]

16. Wang, B.; Ma, J.H.; Wu, Y.P. Application of artificial neural network in prediction of abrasion of rubber composites. *Mater. Des.* **2013**, *49*, 802–807. [CrossRef]

17. Hartmann, S. A competitive genetic algorithm for resource-constrained project scheduling. *Nav. Res. Log.* **2015**, *45*, 733–750. [CrossRef]

18. El Kadi, H.; Al-Assaf, Y. Prediction of the fatigue life of unidirectional glass fiber/epoxy composite lamina using different neural network paradigms. *Compos. Struct.* **2002**, *55*, 239–246. [CrossRef]

19. Selvakumar, S.; Arulshri, K.P.; Padmanaban, K.P.; Sasikumar, K.S.K. Design and optimization of machining fixture layout using ANN and DOE. *Int. J. Adv. Manuf. Technol.* **2013**, *65*, 1573–1586. [CrossRef]

20. Brauer, M.J.; Holder, M.T.; Dries, L.A.; Zwickl, D.J.; Lewis, P.O.; Hillis, D.M. Genetic algorithms and parallel processing in maximum-likelihood phylogeny inference. *Mol. Biol. Evol.* **2002**, *19*, 1717–1726. [CrossRef] [PubMed]

21. Chen, Y.; Li, J. Crystal rule of Fe_2O_3 in oxidized pellet. *J. Cent. South Univ. Technol.* **2007**, *38*, 70–73.

22. Shen, F.; Gao, Q.; Wei, G.; Jiang, X.; Shen, Y. Densification process of MgO bearing pellets. *Steel Res. Int.* **2015**, *86*, 644–650. [CrossRef]

23. Eisenhüttenleute, V.D.; Allibert, M. *Slag Atlas*, 2nd ed.; Verlag Stahleisen GmbH: Düsseldorf, Germany, 1995; p. 128. ISBN 3514004579.

24. Matsumura, M.; Hoshi, M.; Kawaguchi, T. Improve of sinter softening property and reducibility by controlling chemical compositions. *ISIJ Int.* **2005**, *45*, 598–607. [CrossRef]

25. Gao, Q.; Wei, G.; Jiang, X.; Shen, F. Effect of calcinated magnesite on burst temperature of green pellet. *J. Northeastern Univ. Nat. Sci.* **2013**, *34*, 542–544.

Article

Phosphorus-Containing Mineral Evolution and Thermodynamics of Phosphorus Vaporization during Carbothermal Reduction of High-Phosphorus Iron Ore

Yuanyuan Zhang, Qingguo Xue, Guang Wang and Jingsong Wang *

State Key Laboratory of Advanced Metallurgy, University of Science and Technology Beijing, Beijing 100083, China; 13070116832@163.com (Y.Z.); xueqingguo@ustb.edu.cn (Q.X.); wangguang@ustb.edu.cn (G.W.)
* Correspondence: wangjingsong@ustb.edu.cn

Received: 29 May 2018; Accepted: 11 June 2018; Published: 13 June 2018

Abstract: High-phosphorus iron ore is not used because of its high phosphorus content. Phosphorus is mainly present in fluorapatite. In this work, the phosphorus vaporization that occurs during the carbothermal reduction of fluorapatite was investigated. The thermodynamic principle of vaporization, which removes phosphorus during carbothermal reduction, was elucidated, and the mineral evolution of high-phosphorus iron ore was summarized. The results demonstrate that it was difficult to reduce fluorapatite when only carbon was added. When Al_2O_3, SiO_2, and Fe_2O_3 were added, the dephosphorization of fluorapatite was stimulated, and the dephosphorization temperature decreased. A phosphorus-containing gas was generated during this process. SiO_2 had the strongest effect on the dephosphorization of fluorapatite. The carbothermal reduction rate of fluorapatite accelerated when SiO_2, Al_2O_3, and Fe_2O_3 were concurrently added. These oxides were advantageous for vaporization dephosphorization. The gas-phase volatiles were detected through gas-phase mass spectrometry. The volatiles were primarily P_2 or PO. The temperature range of 1000–1100 °C was the optimum for vaporization dephosphorization. This article provides a theoretical and experimental basis for the development and utilization of high-phosphorus iron ore through vaporization dephosphorization.

Keywords: high-phosphorus iron ore; fluorapatite; carbothermal reduction; vaporization dephosphorization

---

1. Introduction

High-phosphorus oolitic hematite constitutes an important iron ore resource and is commonly found in central and northern Western Europe [1], Ukraine, and Canada. The high-phosphorus iron ore processing characteristics are based on the following: (1) The distribution of hematite and fluorapatite is embedded and finely grained. Both minerals are surrounded by layers that form a ring. This leads to difficulty during sorting and grinding. (2) The phosphorus-containing mineral and hematite interface are closely related. Harmful minerals of P could easily enter the iron, leading to difficult dephosphorization. (3) Oolite is a sedimentary iron deposit of low hardness and that contains many clay minerals. During the fine grinding, the oolite produces higher iron-bearing slime and increases the sorting difficulty. The development of smelting technology that is suitable for high-phosphorus iron ore is required in order to guarantee the supply of iron ore resources in China.

Rotary hearth coal-based direct reduction has potential for the efficient utilization of high-phosphorus iron ore [2]. At present, many studies exist regarding the direct reduction of high-phosphorus iron ore [3–7]. The key to the direct reduction of high-phosphorus ore is to increase the metallization rate while inhibiting phosphorus from entering the iron. Only the carbothermal reduction mechanism of iron oxides and fluorapatite in high-phosphorus iron ore can be determined. Consequently, the causes of the metallization degree increase during the iron and phosphorus content reduction in the iron for high-phosphorus ore can be achieved. Much research exists on the reduction mechanism of iron oxides. Adversely, as a result of the difficulty in the preparation of fluorapatite and incomplete thermodynamic data, only little research exists on the carbothermal mechanisms of fluorapatite [8–16]. The mechanism of fluorapatite reduction is not yet clear.

The mineral evolution of high-phosphorus iron ore and the migration of phosphorus have been studied during carbothermal reduction [17,18]. It was found that approximately 15–30% of phosphorus was volatilized as a gas during the carbothermal reduction, but the phosphorus gasification was never studied. In order to understand the phosphorus vaporization and to improve the thermodynamic data on fluorapatite, the thermodynamics of fluorapatite during carbothermal reduction are discussed here in detail. The effects of the main gangue phases (SiO_2, Al_2O_3, and Fe_2O_3) on the gasification of phosphorus were analyzed. The phosphorus gasification was detected through quadrupole mass spectrometry. The purpose of this study was to provide a theoretical basis for the dephosphorization of high-phosphorus iron ore through vaporization, as well as to provide new insights for the utilization of high-phosphorus iron ore.

2. Experiments

2.1. Experimental Materials

The migration of phosphorus in high-phosphorus iron ore was studied in the authors' laboratory. It was discovered that approximately 15–30% of the phosphorus was volatilized as a gas [19]. The purpose of this study was to define the phosphorus gasification and to determine the thermodynamic principles of fluorapatite dephosphorization through vaporization during the carbothermal reduction of high-phosphorous iron ore. The chemical composition of high-phosphorus iron ore is presented in Table 1. The mass fraction of materials constituting below 1% (CaO and MgO) was ignored in order to reduce the influencing factors. The mechanism of phosphorus vaporization during the carbothermal reduction of fluorapatite was studied through thermodynamic calculations and experiments. Therefore, the experimental raw-material ratio scheme was based on the molar ratio of the reactants in the reaction equation. The experimental materials were as follows: analytically pure Fe_2O_3, SiO_2, and Al_2O_3, as well as custom-made high-purity fluorapatite. The raw-material ratio scheme is presented in Table 2.

Table 1. Chemical composition of high-phosphorus iron ore (wt %).

Composition	Fe_2O_3	$Ca_{10}(PO_4)_6F_2$	SiO_2	Al_2O_3	CaO	MgO
Chemical composition of raw ore	73.46	6.23	7.77	5.07	0.79	0.74

Table 2. Substance mixing ratio of fluorapatite carbothermal reduction experiments (molar ratio).

Experiment No.	$Ca_{10}(PO_4)_6F_2$	Al_2O_3	SiO_2	Fe_2O_3	C
1	1	0	0	0	15
2	1	9	0	0	15
3	3	31	0	0	45
4	1	0	9	0	15
5	2	0	21	0	30
6	2	20	41	0	30
7	2	20	41	0	18
8	1	0	0	1	18
9	2	20	41	2	36

2.2. Experimental Procedure

The raw materials were mixed uniformly according to Table 2, and 1 g of the mixture was pressed into cylindrical pellets. The pellets were consequently dried for experimental use. The samples were placed in a tube furnace under an Ar gas atmosphere. The temperature was increased to 1200 °C at a rate of 10 °C/min. Beyond 400 °C, the reactor was connected to the quadrupole mass spectrometer (QMS). Figure 1 presents the schematic diagram of the experimental apparatus. The generated gas was detected. The thermodynamics of the carbofuran reduction of fluorapatite was calculated with FactSage7.0 software [20].

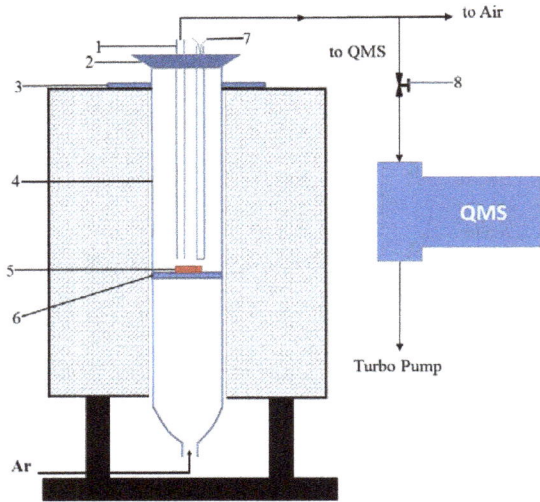

Figure 1. Schematic diagram of experimental apparatus (1: quartz tube; 2: sealing cover; 3: fixed bracket; 4: quartz reaction tube; 5: sample; 6: quartz sieve; 7: thermocouple; 8: variable leak valve).

3. Results and Discussion

3.1. Mineral Reaction Summary and Thermodynamics Discussion of High-Phosphorus Iron Ore Reduction

In order to clarify the phosphorus vaporization, the thermodynamics of carbothermal reduction of fluorapatite is discussed in detail.

(1) Carbothermal reduction of $Ca_{10}(PO_4)_6F_2$: The possible dephosphorization reaction equations are the following:

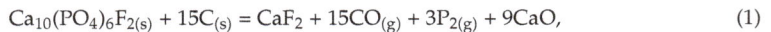

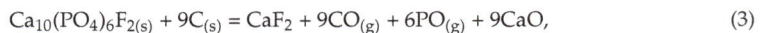

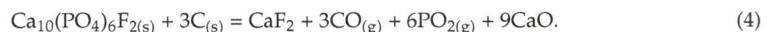

$$Ca_{10}(PO_4)_6F_{2(s)} + 15C_{(s)} = CaF_2 + 15CO_{(g)} + 3P_{2(g)} + 9CaO, \tag{1}$$

$$2Ca_{10}(PO_4)_6F_{2(s)} + 30C_{(s)} = 2CaF_2 + 30CO_{(g)} + 3P_{4(g)} + 18CaO, \tag{2}$$

$$Ca_{10}(PO_4)_6F_{2(s)} + 9C_{(s)} = CaF_2 + 9CO_{(g)} + 6PO_{(g)} + 9CaO, \tag{3}$$

$$Ca_{10}(PO_4)_6F_{2(s)} + 3C_{(s)} = CaF_2 + 3CO_{(g)} + 6PO_{2(g)} + 9CaO. \tag{4}$$

Figure 2 presents the relationship between the standard Gibbs free energy and temperature for the aforementioned four equations. The thermodynamic calculations indicated that fluorapatite could be reduced beyond 1400 °C when carbon alone was added. According to the standard Gibbs free energy, the products of phosphorus vaporization were most likely P_2 and P_4. In the case of high temperature and no oxygen, P_4 could be converted into P_2. P_2 is a small-sized molecule. Consequently, it was more likely to be volatilized. Therefore, the vaporization product of fluorapatite was mainly P_2.

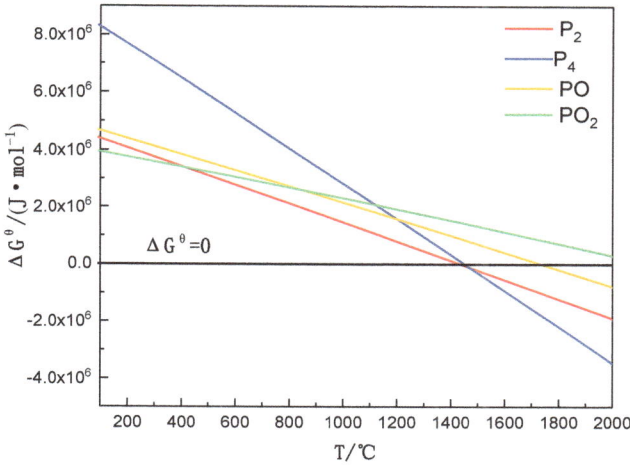

Figure 2. Relationship between standard Gibbs free energy and temperature when only carbon was added.

(2) Carbothermal reduction of $Ca_{10}(PO_4)_6F_2$ with gangue Al_2O_3: The possible dephosphorization reaction equations are the following:

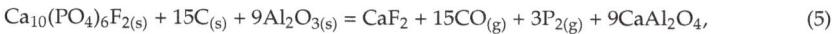

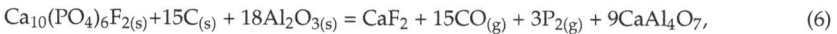

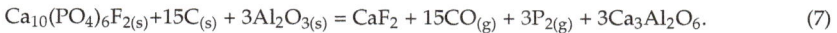

$$Ca_{10}(PO_4)_6F_{2(s)} + 15C_{(s)} + 9Al_2O_{3(s)} = CaF_2 + 15CO_{(g)} + 3P_{2(g)} + 9CaAl_2O_4, \qquad (5)$$

$$Ca_{10}(PO_4)_6F_{2(s)} + 15C_{(s)} + 18Al_2O_{3(s)} = CaF_2 + 15CO_{(g)} + 3P_{2(g)} + 9CaAl_4O_7, \qquad (6)$$

$$Ca_{10}(PO_4)_6F_{2(s)} + 15C_{(s)} + 3Al_2O_{3(s)} = CaF_2 + 15CO_{(g)} + 3P_{2(g)} + 3Ca_3Al_2O_6. \qquad (7)$$

Figure 3 presents the relationship between the standard Gibbs free energy and temperature for different Ca–Al–O products. When Al_2O_3 was added, the $Ca_{10}(PO_4)_6F_2$ was reduced to form Ca–Al–O gangue to promote the dephosphorization of fluorapatite. According to thermodynamic calculations, the standard Gibbs free energy required for the formation of $CaAl_2O_4$ was the lowest. Therefore, when Al_2O_3 was present, the dephosphorization product of $Ca_{10}(PO_4)_6F_2$ was most likely $CaAl_2O_4$. $CaAl_2O_4$ as the product below is used to discuss the possible reduction of fluorapatite when additional Al_2O_3 (molar ratio of $Al_2O_3/Ca_{10}(PO_4)_6F_2 > 10.3$) was added:

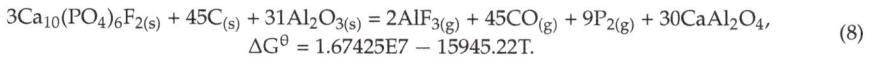

$$3Ca_{10}(PO_4)_6F_{2(s)} + 45C_{(s)} + 31Al_2O_{3(s)} = 2AlF_{3(g)} + 45CO_{(g)} + 9P_{2(g)} + 30CaAl_2O_4,$$
$$\Delta G^\theta = 1.67425E7 - 15945.22T. \qquad (8)$$

From the aforementioned calculations, $CaAl_2O_4$ was observed to promote the dephosphorization of fluorapatite, particularly when the Al_2O_3 was in excess (molar ratio of $Al_2O_3/Ca_{10}(PO_4)_6F_2 > 10.3$); fluorapatite was defluorinated to produce more easily reduced $Ca_3(PO_4)_2$, which promoted the dephosphorization of fluorapatite.

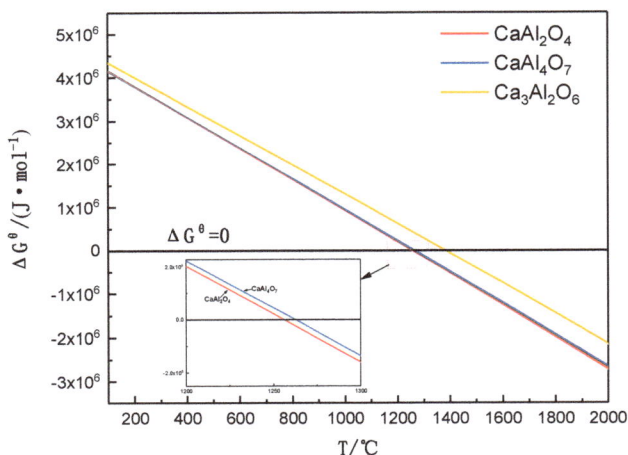

Figure 3. Relationship between standard Gibbs free energy and temperature following Al_2O_3 addition.

(3) Carbothermal reduction reaction of $Ca_{10}(PO_4)_6F_2$ with gangue SiO_2: The possible dephosphorization reaction equations are the following:

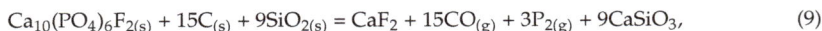

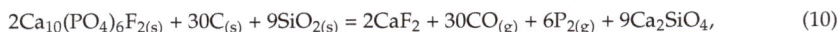

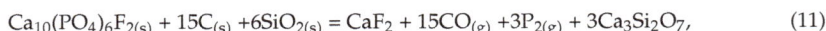

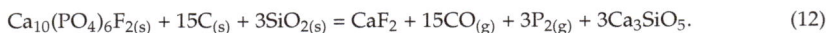

$$Ca_{10}(PO_4)_6F_{2(s)} + 15C_{(s)} + 9SiO_{2(s)} = CaF_2 + 15CO_{(g)} + 3P_{2(g)} + 9CaSiO_3, \qquad (9)$$

$$2Ca_{10}(PO_4)_6F_{2(s)} + 30C_{(s)} + 9SiO_{2(s)} = 2CaF_2 + 30CO_{(g)} + 6P_{2(g)} + 9Ca_2SiO_4, \qquad (10)$$

$$Ca_{10}(PO_4)_6F_{2(s)} + 15C_{(s)} + 6SiO_{2(s)} = CaF_2 + 15CO_{(g)} + 3P_{2(g)} + 3Ca_3Si_2O_7, \qquad (11)$$

$$Ca_{10}(PO_4)_6F_{2(s)} + 15C_{(s)} + 3SiO_{2(s)} = CaF_2 + 15CO_{(g)} + 3P_{2(g)} + 3Ca_3SiO_5. \qquad (12)$$

Figure 4 presents the relationship between the standard Gibbs free energy and temperature for different Ca–Si–O products. When SiO_2 was added, $Ca_{10}(PO_4)_6F_2$ was reduced to form Ca–Si–O gangue to promote the dephosphorization of fluorapatite. According to thermodynamic calculations, the standard Gibbs free energy required for the formation of $CaSiO_3$ was the lowest. Therefore, when SiO_2 was present, the dephosphorization product of $Ca_{10}(PO_4)_6F_2$ was most likely $CaSiO_3$. $CaSiO_3$ as the product below is used to discuss the possible reduction of fluorapatite when additional SiO_2 (molar ratio of $SiO_2/Ca_{10}(PO_4)_6F_2 > 10.5$) was added:

$$2Ca_{10}(PO_4)_6F_{2(s)} + 30C_{(s)} + 21SiO_{2(s)} = SiF_{4(g)} + 30CO_{(g)} + 6P_{2(g)} + 20CaSiO_3,$$
$$\Delta G^\theta = 7.14216E6 - 7002.12T. \qquad (13)$$

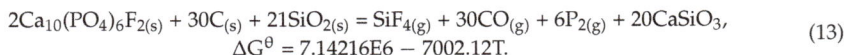

From the calculation results of Equation (13), it could be observed that when the $SiO_2/Ca_{10}(PO_4)_6F_2$ molar ratio was greater than 10.5, fluorapatite was more easily reduced. This occurred because the defluorination of fluorapatite was promoted by SiO_2 to generate more easily reduced $Ca_3(PO_4)_2$, which promoted the reduction of fluorapatite. The partial pressure of the phosphorus-containing gas was decreased by the generated SiF_4 gas, and the dephosphorization of fluorapatite was further promoted.

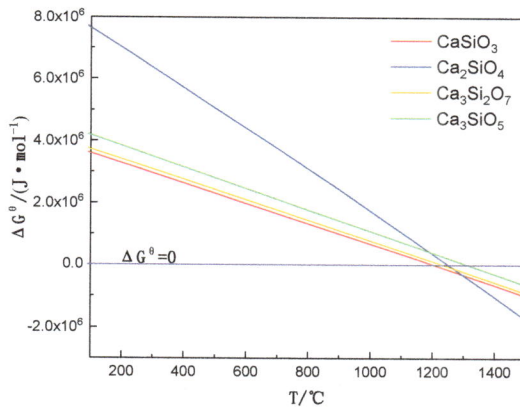

Figure 4. Relationship between standard Gibbs free energy and temperature following SiO_2 addition.

(4) Carbothermal reduction reaction of $Ca_{10}(PO_4)_6F_2$ with gangue Al_2O_3 and SiO_2: The possible dephosphorization reaction equations are the following:

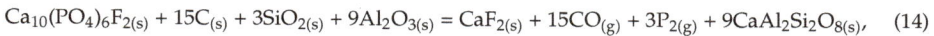

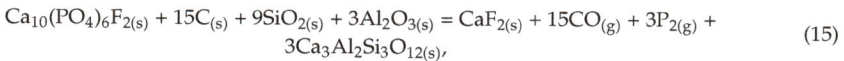

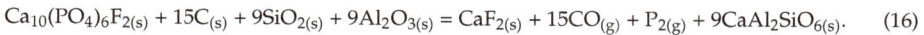

$$Ca_{10}(PO_4)_6F_{2(s)} + 15C_{(s)} + 3SiO_{2(s)} + 9Al_2O_{3(s)} = CaF_{2(s)} + 15CO_{(g)} + 3P_{2(g)} + 9CaAl_2Si_2O_{8(s)}, \quad (14)$$

$$Ca_{10}(PO_4)_6F_{2(s)} + 15C_{(s)} + 9SiO_{2(s)} + 3Al_2O_{3(s)} = CaF_{2(s)} + 15CO_{(g)} + 3P_{2(g)} + \\ 3Ca_3Al_2Si_3O_{12(s)}, \quad (15)$$

$$Ca_{10}(PO_4)_6F_{2(s)} + 15C_{(s)} + 9SiO_{2(s)} + 9Al_2O_{3(s)} = CaF_{2(s)} + 15CO_{(g)} + P_{2(g)} + 9CaAl_2SiO_{6(s)}. \quad (16)$$

Figure 5 presents the relationship between the standard Gibbs free energy and temperature for different Ca–Al–Si–O products. The thermodynamic calculations demonstrated that when Al_2O_3 and SiO_2 were simultaneously added, the standard Gibbs free energy required for the reduction of fluorapatite was further decreased. When the product was $CaAl_2Si_2O_8$, the required standard Gibbs free energy was the lowest. Therefore, when Al_2O_3 was present along with SiO_2, fluorapatite was more easily reduced to produce $CaAl_2Si_2O_8$.

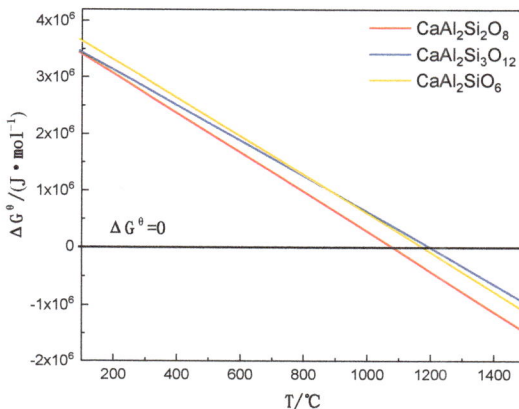

Figure 5. Relationship between standard Gibbs free energy and temperature following Al_2O_3 and SiO_2 addition.

(5) Carbothermal reduction reaction of $Ca_{10}(PO_4)_6F_2$ with gangue Fe_2O_3: The possible dephosphorization reaction equations are the following:

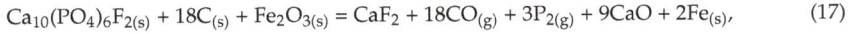

$$Ca_{10}(PO_4)_6F_{2(s)} + 18C_{(s)} + Fe_2O_{3(s)} = CaF_2 + 18CO_{(g)} + 3P_{2(g)} + 9CaO + 2Fe_{(s)}, \quad (17)$$

$$Ca_{10}(PO_4)_6F_{2(s)} + 18C_{(s)} + Fe_2O_{3(s)} = CaF_2 + 45CO_{(g)} + 2P_{2(g)} + 9CaO + 2Fe_3P + 14Fe_{(s)}, \quad (18)$$

$$Ca_{10}(PO_4)_6F_{2(s)} + 45C_{(s)} + Fe_2O_{3(s)} = CaF_2 + 45CO_{(g)} + 9CaO + 6Fe_3P + 2Fe_{(s)}. \quad (19)$$

Figure 6 presents the relationship between the standard Gibbs free energy and temperature for the previous three reaction equations. It could be observed that the standard Gibbs free energy required for the reduction of fluorapatite was decreased by Fe_2O_3. In contrast, the mechanism promoted by Fe_2O_3 was different compared to the Al_2O_3 and SiO_2 mechanisms. The iron formed by reduction had a strong absorption capacity for phosphorus to generate Fe_3P. Therefore, the dephosphorization of fluorapatite was promoted, but it was not favorable for the separation of iron and phosphorus. For the high-phosphorus iron ore, to achieve dephosphorization, the diffusion of phosphorus into iron should be inhibited.

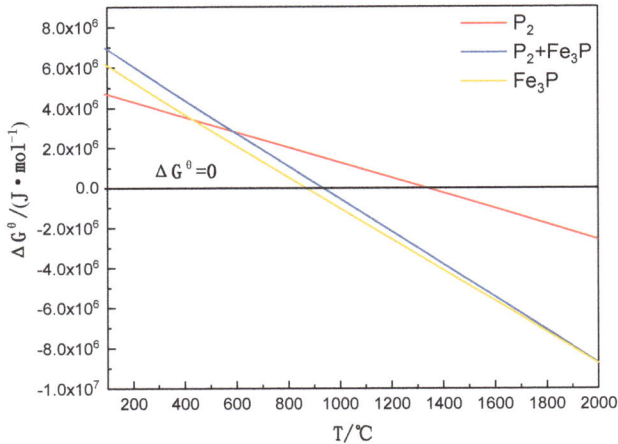

Figure 6. Relationship between standard Gibbs free energy and temperature following Fe_2O_3 addition.

The process of the carbothermal reduction of $Ca_{10}(PO_4)_6F_2$ with three types of gangue was complex. Taking into account the previous analysis and the composition of high-phosphorus iron ore, the possible overall reaction equation is the following:

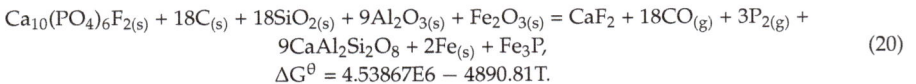

$$Ca_{10}(PO_4)_6F_{2(s)} + 18C_{(s)} + 18SiO_{2(s)} + 9Al_2O_{3(s)} + Fe_2O_{3(s)} = CaF_2 + 18CO_{(g)} + 3P_{2(g)} +$$
$$9CaAl_2Si_2O_8 + 2Fe_{(s)} + Fe_3P, \quad (20)$$
$$\Delta G^\theta = 4.53867E6 - 4890.81T.$$

Through the thermodynamic analysis, it was found that when only carbon was present, fluorapatite dephosphorization was difficult to achieve. The starting temperature (temperature at $\Delta G^\theta = 0$) for the dephosphorization of fluorapatite exceeded 1400 °C. It was also discovered that the higher the C/O (O from $Ca_{10}(PO_4)_6F_2$), the lower the dephosphorization temperature of fluorapatite. Following the SiO_2 or Al_2O_3 addition, the oxides reacted with $Ca_{10}(PO_4)_6F_2$ to form $CaAl_2O_4$ and $CaSiO_3$, which promoted the vaporization dephosphorization reaction. The effect of SiO_2 was stronger than that of Al_2O_3. When the amount of SiO_2 was sufficient, fluorapatite was defluorinated to $Ca_3(PO_4)_2$, and it was easier to reduce $Ca_3(PO_4)_2$ than $Ca_{10}(PO_4)_6F_2$. In addition, the partial pressure of the phosphorus-containing gas was decreased by the generation of fluorine-containing gas, which improved the thermodynamic conditions of dephosphorization. This promoted the decomposition of fluorapatite and resulted in an acceleration of the vaporization dephosphorization rate of fluorapatite. As the gangue-phase amount increased, the dephosphorization temperature of fluorapatite was gradually reduced. When three gangue oxides were simultaneously present, the gangue phase first transformed into the new gangue phase, and fluorapatite was dephosphorized through gasification in the role of the new gangue phase. The starting temperature (temperature at $\Delta G^{\theta} = 0$) for dephosphorization of fluorapatite was 928 °C.

According to the previous thermodynamic analysis, the evolution of fluorapatite during the carbothermal reduction was summarized and is presented in Figure 7. When only carbon was added, fluorapatite was not easily dephosphorized by vaporization. This occurred because fluorapatite was not easily decomposed. The starting temperature (temperature at $\Delta G^{\theta} = 0$) for dephosphorization of fluorapatite was 1442 °C. When Al_2O_3, SiO_2, and Fe_2O_3 were added, the dephosphorization temperature of fluorapatite was decreased. When low amounts of Al_2O_3 (molar ratio of $Al_2O_3/Ca_{10}(PO_4)_6F_2 < 10.3$) were added, fluorapatite began to be dephosphorized at 1306 °C. In this time frame, the products were mainly $CaAl_2O_4$, CaF_2, P_2, and CO. When a sufficient amount of Al_2O_3 (molar ratio of $Al_2O_3/Ca_{10}(PO_4)_6F_2 > 10.3$) was added, fluorapatite was defluorinated to form AlF_3 gas, which improved the thermodynamic conditions of the reaction and caused fluorapatite to be dephosphorized at approximately 1050 °C. The resulting products were mainly $CaAl_2O_4$, P_2, CO, and AlF_3. When low amounts of SiO_2 (molar ratio of $SiO_2/Ca_{10}(PO_4)_6F_2 < 10.5$) were added, fluorapatite was dephosphorized at 1200 °C. The subsequent products were mainly $CaSiO_3$, CaF_2, P_2, and CO. When a sufficient amount of SiO_2 (molar ratio of $SiO_2/Ca_{10}(PO_4)_6F_2 > 10.5$) was added, the melting point of the system was reduced, which promoted the mass transfer conditions to accelerate the dephosphorization of fluorapatite. Fluorapatite was dephosphorized at 1019 °C. At this point, the decomposed CaF_2 was consumed. The main products were $CaSiO_3$, CO, P_2, and SiF_4. When only Fe_2O_3 was added, the phosphorus was absorbed by the iron phase to form Fe_3P, which promoted the dephosphorization of fluorapatite. When SiO_2, Al_2O_3, and Fe_2O_3 were added, the reduction route of the iron oxide was changed. Certain amounts of FeO and gangue generated intermediate products (Fe_2SiO_4 and $FeAl_2O_4$), which were difficult to reduce. Fluorapatite was dephosphorized under the combined action of Fe_2SiO_4 and $FeAl_2O_4$. When a sufficient amount of carbon was available (molar ratio of $C/O > 1$, where O was from fluoroapatite and Fe_2O_3), fluorapatite was completely reduced, and the phosphorus vaporization product was P_2. When the amount of carbon was insufficient (molar ratio of $C/O < 1$, where O was from fluorapatite and Fe_2O_3), fluorapatite was not completely reduced and was primarily volatilized as phosphorus oxides.

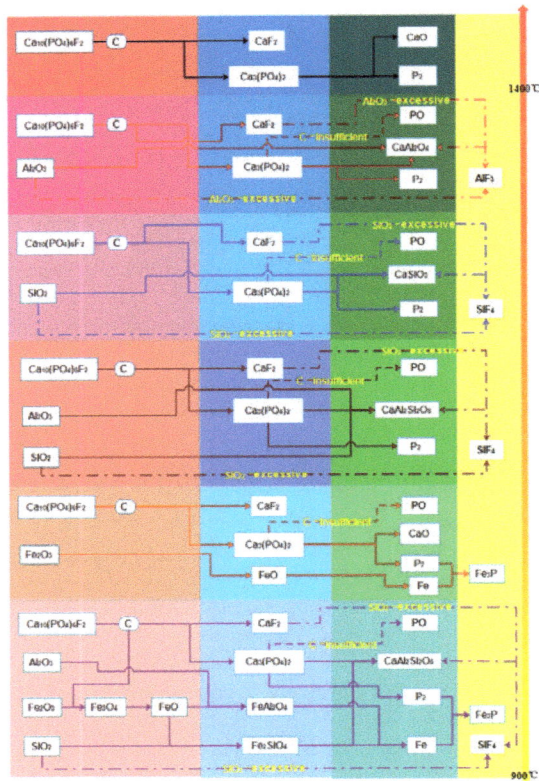

Figure 7. Schematic diagram of mineral evolution during fluorapatite direct reduction.

3.2. Phosphorus Vaporization in Carbothermal Reduction under Different Conditions

From thermodynamic analysis, it is known that fluorapatite is dephosphorized by gasification. The gasification of fluorapatite has not been studied. In order to clarify the gasification process of phosphorus, the gas in the carbothermal reduction was detected by quadrupole mass spectrometry. The results are presented in Figure 8. Figure 8a presents that no gas was generated within the reduction temperature range (900–1200 °C). The result confirmed that fluorapatite was difficult to reduce when only carbon was present (experiment 1).

Figure 8b presents that when the molar ratio of Al_2O_3 to $Ca_{10}(PO_4)_6F_2$ was below 10.3 (experiment 2), the thermodynamic conditions of reduction were not obtained at 1200 °C. Fluorapatite was not reduced. When the molar ratio of Al_2O_3 to $Ca_{10}(PO_4)_6F_2$ exceeded 10.3 (experiment 3), CO was generated at approximately 1100 °C. This meant that fluorapatite was reduced. Phosphorus was volatilized in the form of P_2 and PO, but the amount of P_2 was higher compared to PO. AlF_3 gas was also detected. The reaction equation is presented as Equation (8). As the temperature increased, the amount of phosphorous gas constantly increased. The limiting factor of the reaction at this time was the temperature. When the temperature increased beyond 1150 °C, the phosphorus gas was generated slower. The limiting factor of the reaction at this time was the number of reactants. Additional amounts of Al_2O_3 reduced the melting point of the system, improving its mass transfer conditions. Fluorapatite was reduced at a lower temperature. During the reduction, fluorapatite was defluorinated to form AlF_3 gas, which improved the thermodynamic conditions of the reaction.

Figure 8c presents that PO, P_2, CO, and SiF_4 gases were detected when the molar ratio of SiO_2 to $Ca_{10}(PO_4)_6F_2$ was higher than 10.5 (experiment 5) and the temperature exceeded 1000 °C. The reaction equation is presented as Equation (11). There was no gas generation detected when a low amount of SiO_2 (molar ratio of $SiO_2/Ca_{10}(PO_4)_6F_2 < 10.5$) was added (experiment 4), indicating that fluorapatite was not substantially reduced. The contact area with fluorapatite increased, and the melting point of the gangue phase decreased to form a liquid phase when the SiO_2 amount was sufficient, which improved the diffusion conditions. When the molar ratio of SiO_2 to $Ca_{10}(PO_4)_6F_2$ exceeded 10.5, fluorapatite was defluorinated to produce SiF_4 and $Ca_3(PO_4)_2$, which were more easily reduced. The thermodynamic conditions of dephosphorization of fluorapatite were optimized, and the dephosphorization temperature was reduced.

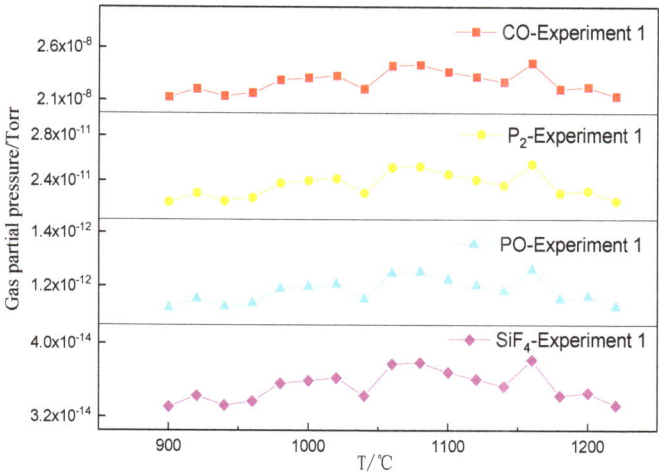

(a)

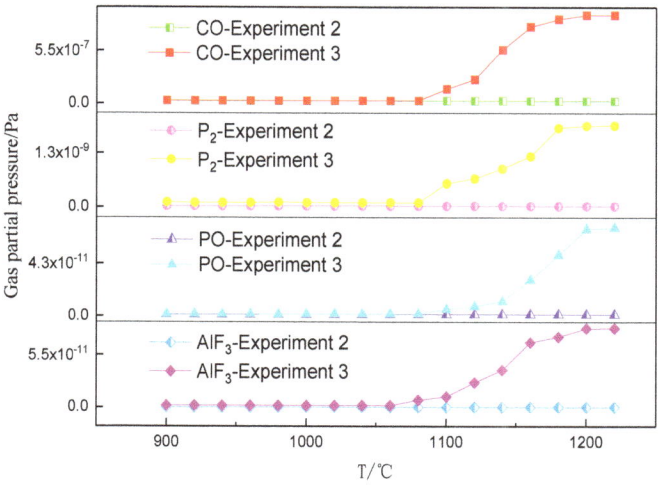

(b)

Figure 8. *Cont.*

143

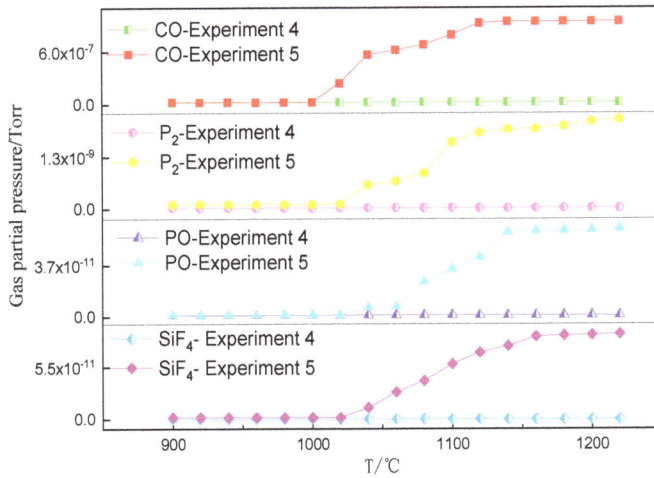

Figure 8. Gas analysis for gas species that were vaporized during solid-state carbothermic reduction reaction (**a**) without additives, (**b**) with Al_2O_3, and (**c**) with SiO_2.

Through the thermodynamic calculation results' comparison of reaction Equations (1)–(4), it was found that the C/O molar ratio had a high effect on the dephosphorization of fluorapatite. The effect of C/O on the dephosphorization of fluorapatite was studied through experiments 6 and 7. Figure 9 presents that when the amount of carbon was sufficient (experiment 6; C/O = 1), fluorapatite was reduced at 980 °C. The reduced phosphorus was primarily volatilized in the form of P_2. When the carbon amount was insufficient (experiment 7; C/O < 1), fluorapatite was reduced beyond 1080 °C. The reduced phosphorus was primarily volatilized as PO. This was consistent with the above thermodynamic calculations. When an adequate amount carbon was present, fluorapatite came into closer contact with carbon, which provided good reduction conditions for fluorapatite. The dephosphorization of fluorapatite was accompanied by electron transfer, while the conditions favoring electron transfer were provided by the sufficient amount of carbon, which allowed P^{5+} to obtain enough electrons to form P_2. When the carbon amount was insufficient, the number of electrons was not sufficient to be supplied to P^{5+} to form P_2. Consequently, the dephosphorization product was phosphorus oxide.

Figure 10 presents the results of the quadrupole mass spectrometry detection following the addition of Fe_2O_3 during the carbothermal reduction (experiment 8). It was discovered that when the temperature exceeded 950 °C, CO was generated, which indicated that Fe_2O_3 began to be reduced. P_2 was generated beyond 1050 °C as the temperature increased. At this point, fluorapatite began to be dephosphorized. The reduction order of the iron oxide was $Fe_2O_3 \rightarrow Fe_3O_4 \rightarrow FeO \rightarrow Fe$. The generated iron was wrapped on the surface of fluorapatite, and P_2 was easily absorbed by the iron phase, which presented an uneven distribution. The phosphorus distribution in the iron phase following a carbothermal reduction is depicted in Figure 11. The corresponding reaction is given as Equation (21). As the temperature increased, liquid iron was gradually formed. The mass transfer kinetic conditions were improved by liquid iron. The unreacted graphite particles were adhered to the fluorapatite surface, while the contact area of graphite and fluorapatite increased, which prompted the dephosphorization of fluorapatite. Adversely, when the temperature exceeded 1150 °C, the amount of volatilized P_2 began to decrease, which may have been caused by the increase in liquid iron and the high amount of P_2 entering the iron phase. In order to prevent liquid iron from inhibiting phosphorus to enter the iron phase, the reduction temperature should be reduced.

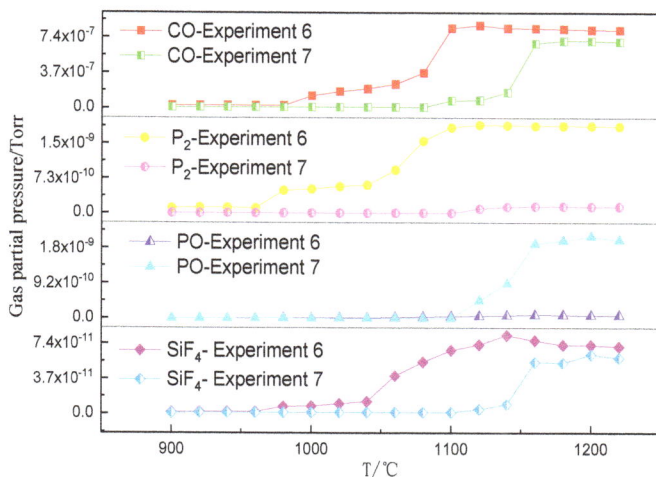

$$Fe_{(s)} + P_{2(g)} = Fe_3P_{(s)} \tag{21}$$

Figure 9. Effect of carbon content on phosphorus-containing gas products.

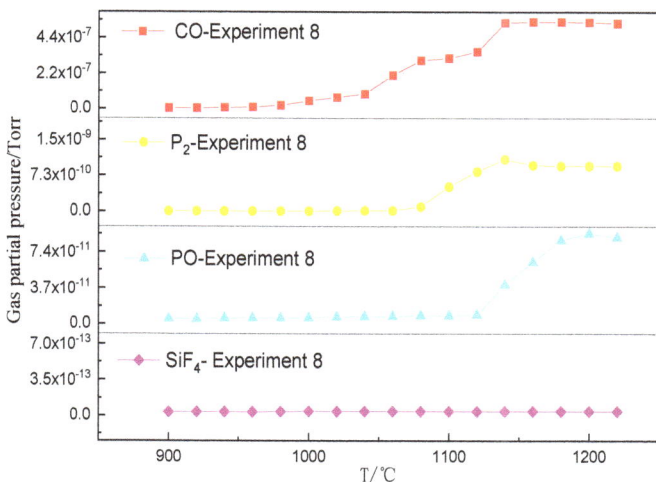

Figure 10. Gas-phase products during direct reduction with Fe_2O_3.

Figure 11. Distribution of Fe and P in the DRI phase microstructure.

Figure 12 presents the gas formation during the carbothermal reduction when three gangue oxides were added. It was observed that the CO, P_2, PO, and SiF_4 gases were generated initially at 900 °C, which indicated that fluorapatite had begun to be dephosphorized through vaporization. When Al_2O_3, SiO_2, and Fe_2O_3 (experiment 9) were added, the FeO reduction product reacted with SiO_2 and Al_2O_3 to form Fe_2SiO_4 and $FeAl_2O_4$. Fe_2SiO_4 is a low-melting-point material; it combined with SiO_2 to form $Fe_2SiO_4.SiO_2$ with a lower melting point, resulting in the pellets melting. A high amount of Fe–Si–Al gangue phase was formed to improve the mass transfer kinetics conditions. This resulted in a further decrease in the dephosphorization temperature of fluorapatite.

When the temperature was below 1000 °C, only a low amount of fluorapatite was dephosphorized, and less phosphorus-containing gas was produced. As the temperature increased, the generated amount of phosphorus-containing gas gradually increased. The latter analysis demonstrated that the dephosphorization of fluorapatite underwent two stages. The first stage occurred during the carbothermal reduction, in which the dephosphorization of fluorapatite was promoted by SiO_2 and Al_2O_3. Only a low amount of fluorapatite was dephosphorized at this stage, and the dephosphorization ratio was low. As the reaction progressed, high amounts of Fe_2SiO_4 and $FeAl_2O_4$ were formed from SiO_2, Al_2O_3, and FeO, leading to the second phase initiation. At this stage, the reaction thermodynamics and the kinetic conditions were improved, and the dephosphorization of fluorapatite primarily occurred during this stage.

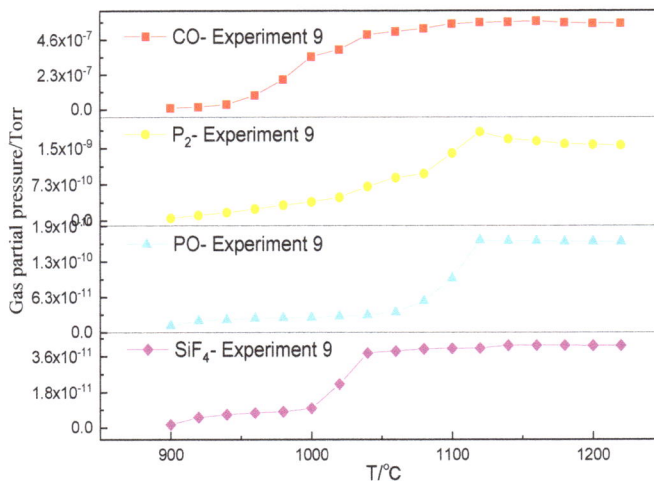

Figure 12. Gas-phase products present during direct reduction with three additives.

3.3. Effect of Temperature on Dephosphorization Extent and Distribution of Phosphorus in High-Phosphorus Iron Ore

The effect of gangue oxides on the dephosphorization of fluorapatite was investigated through the aforementioned experiments. When the composition of the sample was fixed, the temperature became a factor that limited the dephosphorization of fluorapatite. Consequently, the effect of temperature on the dephosphorization extent was studied. In order to study the effect of temperature on the dephosphorization of high-phosphorus iron ore, the pure substance was used to simulate the chemical composition of high-phosphorus iron ore. In order to reduce the influencing factors, the substances with a mass fraction of <1 (CaO and MgO) were ignored. The raw-material ratio is presented in Table 1 (C/O = 1). The sample was reduced at a temperature range of 800–1200 °C, and the phosphorus content of the reduced sample at different temperatures was determined through chemical analysis. Figure 13 presents the dephosphorization extent of the sample subsequent to reduction at different temperatures. The dephosphorization extent was very low in the temperature range of 800–1000 °C. When the temperature was in the range of 1000–1100 °C, the dephosphorization extent rapidly increased. Beyond 1100 °C, the dephosphorization extent was essentially unchanged or even slightly decreased. When the temperature was lower than 1100 °C, liquid iron was not formed, and the solubility of phosphorus in the solid-phase iron was low. Additionally, almost all generated P_2 was volatilized. As the temperature increased, the iron phase appeared in the liquid phase. Because the solubility of phosphorus in the liquid iron was relatively high, generated P_2 was absorbed by the liquid iron, resulting in the stability or even decrease of the dephosphorization extent. In the analysis, the 1000–1100 °C temperature range was considered the best period for the vaporization dephosphorization.

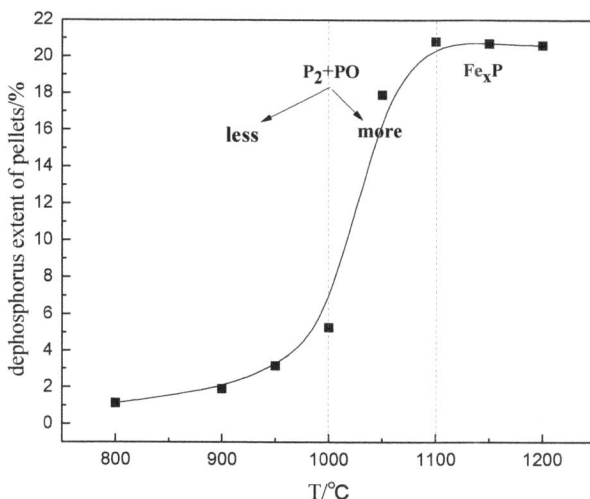

Figure 13. Effect of reduction temperature on dephosphorization extent.

To further understand the migration of phosphorus, the evolution of fluorapatite at different temperatures was studied through SEM (Scanning Electron Microscope). Figure 14 presents the distribution of elements following reduction at different temperatures. At 1000 °C, the distributions of Ca and P were similar. The distributions of Ca, Si, and Al overlapped to a low degree, but no overlapping region existed between Fe and P. It was revealed that most fluorapatite particles were still intact, while a low amount of fluorapatite was dephosphorized to form a Ca–Al–Si gangue phase. The entire reduced phosphorus amount was volatilized at 1000 °C. The distributions of Fe, Si, and Al also overlapped. This was because the Al_2O_3 and SiO_2 gangue phases reacted with iron oxides. Additionally, Fe_2SiO_4 and $FeAl_2O_4$ were formed on the surfaces of Al_2O_3 and SiO_2. As the temperature increased to 1050 °C, the Ca–Al–Si gangue phase grew in size, and no overlapping region existed between Fe and P. The reduced phosphorus was still discharged as a gas, and still high amounts of $FeAl_2O_4$ and Fe_2SiO_4 existed. At 1100 °C, a high amount of iron was generated and gathered. Phosphorus was present around the iron phase. Adversely, phosphorus was inhibited from entering the metal iron, and no phosphorus was detected inside the metal iron. As the temperature continued to increase, the reduced iron was carburized and the metallic iron began to melt. At 1150 °C, a high amount of fluorapatite was dephosphorized, and generated P_2 was rapidly absorbed by the liquid iron and consequently distributed within the iron phase in a network. At 1200 °C, the amount of liquid iron increased, the phosphorus in the iron phase increased, and the distribution tended to be uniform. In this time frame, the gangue phase was uniform $CaAl_2Si_2O_8$. The study revealed that when the reduction temperature reached 1150 °C, the phosphorus started to enter the iron phase, and a reduction temperature below 1150 °C was favorable for the dephosphorization of the high-phosphorus iron ore.

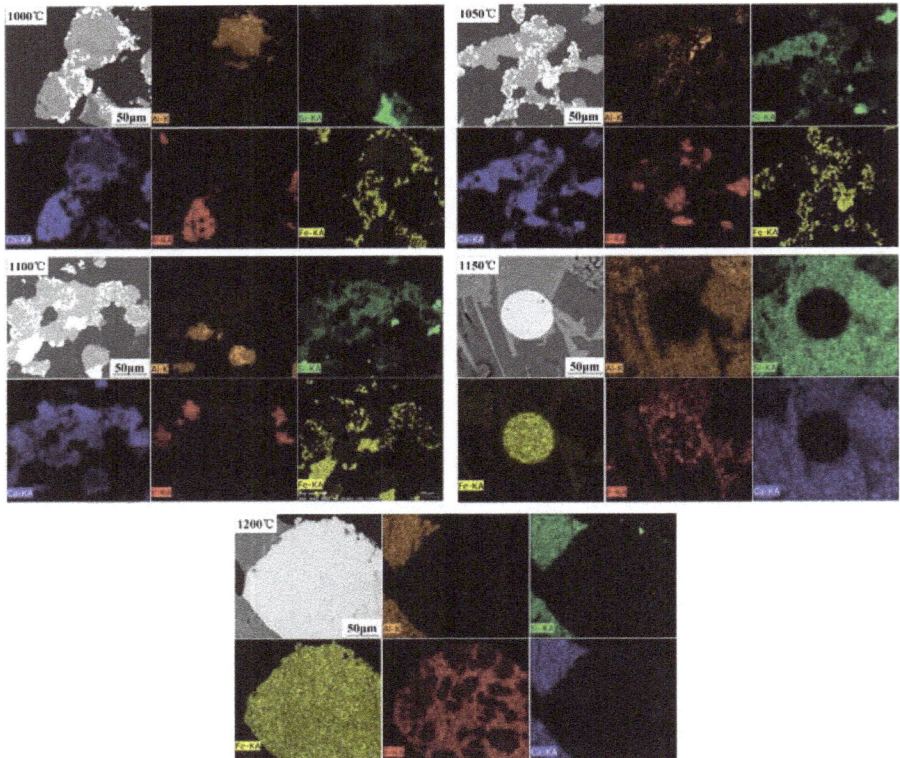

Figure 14. SEM-EDS(Scanning Electron Microscope—Energy Dispersive Spectrometer analysis of reduced pellets at different temperatures.

4. Conclusions

In summary, in the carbothermal reduction process, fluorapatite was dephosphorized, and the dephosphorization product was mainly P_2 gas. In contrast, PO was generated when the carbon amount was insufficient. All gangue oxides (Al_2O_3, SiO_2, and Fe_2O_3) in the high-phosphorus iron ore promoted the dephosphorization of fluorapatite, while the promotion effect of SiO_2 was the strongest. Although the P_2 could be absorbed by the iron phase, this mainly occurred at high temperatures (>1150 °C). In the 1000–1100 °C range, gasification dephosphorization was possible. In future studies, the feasibility of gasification dephosphorization will be studied in detail. The results of the study are summarized as follows:

1. Without the additives' contribution, the fluoroapatite reduction was difficult to achieve. The theoretical reduction temperature exceeded 1400 °C. When Al_2O_3 or SiO_2 were added, the dephosphorization of fluorapatite was promoted, and the phosphorus-containing gas was generated. The effect of SiO_2 was stronger compared to Al_2O_3. When a sufficient amount of Al_2O_3 or SiO_2 was added, the defluorination reaction of fluorapatite occurred, which accelerated the vaporization dephosphorization of fluorapatite.

2. Fluorapatite was reduced when the molar ratio of Al_2O_3 to $Ca_{10}(PO_4)_6F_2$ was below 10.3 and the reduction temperature exceeded 1350 °C. Fluorapatite was reduced at 1050 °C when the molar ratio of Al_2O_3 to $Ca_{10}(PO_4)_6F_2$ exceeded 10.3. Fluorapatite was reduced at temperatures beyond 1250 °C when the molar ratio of SiO_2 to $Ca_{10}(PO_4)_6F_2$ was below 10.5. Fluorapatite was reduced at 1019 °C when the molar ratio of SiO_2 to $Ca_{10}(PO_4)_6F_2$ exceeded 10.5, which reduced the reduction temperature of fluorapatite. These results indicated that it was possible for gasification dephosphorization to occur at lower temperatures.

3. The amount of carbonhad a significant effect on the gas-phase products of phosphorus. When a sufficient amount of carbon was added, the phosphorus-containing gas was primarily produced as P_2, and the dephosphorization temperature was low. When inadequate amounts of carbon were present, the phosphorus-containing gas primarily produced PO, and the dephosphorization temperature was high.

4. The dephosphorization ratio was very low within the temperature range of 800–1000 °C. When the temperature was in the range of 1000–1100 °C, the dephosphorization ratio rapidly increased. Beyond 1100 °C, the dephosphorization ratio was essentially unchanged or even slightly decreased. The temperature range of 1000–1100 °C was the optimum for gasification dephosphorization.

5. During the carbothermal reduction process of high-phosphorus iron ore, the most effective means of dephosphorization was to volatilize phosphorus as a gas, which could prevent phosphorus from entering the iron in the subsequent process. This paper provides a theoretical basis for the dephosphorization of high-phosphorus iron ore by vaporization and provides new insights for the utilization of high-phosphorus iron ore.

Author Contributions: Conceptualization: J.W.; data curation: Y.Z.; formal analysis: Y.Z.; funding acquisition: J.W., Q.X., and G.W.; investigation: Y.Z.; methodology: Y.Z.; project administration: J.W., Q.X., and G.W.; resources: J.W., Q.X., and G.W.; supervision: J.W., Q.X., and G.W.; validation: Y.Z.; writing of original draft: Y.Z.; writing of review and editing: J.W.

Acknowledgments: The authors acknowledge the support by the National Natural Science Foundation of China (51374024), Fundamental Research Funds for the Central Universities (FRF-TP-16-019A1), and the China Postdoctoral Science Foundation (2016M600919).

Conflicts of Interest: The authors declare no conflict of interest.

References

1. Baioumy, H.; Omran, M.; Fabritius, T. Mineralogy, geochemistry and the origin of high-phosphorus oolitic iron ores of Aswan, Egypt. *Ore Geol. Rev.* **2017**, *80*, 185–199. [CrossRef]
2. Zhu, D.; Guo, Z.; Xue, Q.G.; Zhang, F. Synchronous upgrading iron and phosphorus removal from high phosphorus oolitic hematite ore by high temperature flash reduction. *Metals* **2016**, *6*, 123. [CrossRef]
3. Matinde, E.; Hino, M. Dephosphorization treatment of high phosphorus iron ore by pre-reduction, air jet milling and screening methods. *Int. ISIJ* **2011**, *51*, 544–551. [CrossRef]

4. Huang, D.B.; Zong, Y.B.; Wei, R.F.; Gao, W.; Liu, X.M. Direct reduction of high phosphorus oolitic hematite ore based on biomass pyrolysis. *J. Iron Steel Res. Int.* **2016**, *23*, 874–883. [CrossRef]

5. Sun, Y.S.; Han, Y.X.; Gao, P.; Wang, Z.H.; Ren, D.Z. Recovery of iron from high phosphorus oolitic iron ore using coal-based reduction followed by magnetic separation. *Int. J. Miner. Metall. Mater.* **2013**, *20*, 411–419. [CrossRef]

6. Lei, Y.; Li, Y.; Chen, W. Microwave carbothermic reduction of oolitic hematite. *Int. ISIJ* **2017**, *57*, 791–794. [CrossRef]

7. Matinde, E.; Hino, M. Dephosphorization treatment of high phosphorus iron ore by pre-reduction, mechanical crushing and screening methods. *Int. ISIJ* **2011**, *51*, 220–227. [CrossRef]

8. Yu, W.; Tang, Q.Y.; Chen, J.A.; Sun, T.C. T Thermodynamic analysis of the carbothermic reduction of a high-phosphorusoolitic iron ore by FactSage. *Int. J. Miner. Metall. Mater.* **2016**, *23*, 1126–1132. [CrossRef]

9. Mu, J.; Leder, F.; Park, W.C. Reduction of phosphate ores by carbon: Part 1 process variables for design of rotary kiln system. *Metall. Trans. B* **1986**, *17*, 861–868. [CrossRef]

10. Mu, J.; Leder, F.; Park, W.C. Reduction of phosphate ores by carbon: Part 2 rate limiting steps. *Metall. Trans. B* **1986**, *17*, 869–877. [CrossRef]

11. L'vov, B.V. Mechanism of carbothermal reduction of iron, cobalt, nickel and copper oxides. *Thermochim. Acta* **2000**, *360*, 109–120. [CrossRef]

12. Jiang, L.K.; Qiu, L.Y.; Liang, B. Solid reaction mechanism for the thermal reduction of fluorapatite by carbon. *J. Chengdu Univ. Sci. Technol.* **1995**, *5*, 1–7.

13. Tang, X.L.; Zhang, Z.T.; Guo, M. Viscosities behavior of CaO-SiO$_2$-MgO-alumina slag with low mass ratio of CaO to SiO$_2$ and wide range of alumina content. *J. Iron Steel Res. Int.* **2011**, *18*, 1–6. [CrossRef]

14. Williams, Q.; Knittle, E. Infrared and Raman spectra of Ca$_5$(PO$_4$)$_3$F$_2$-fluorapatite at high pressures: Compression-induced changes in phosphate site and Davydov splittings. *Phys. Chem. Solids* **1996**, *57*, 417–422. [CrossRef]

15. Ning, X.Y.; Xue, Q.G.; Wang, G. Mechanism of direct reduction carbon and melting separation for carbon-bearing pellets. *Chin. J. Eng.* **2014**, *36*, 1166–1173.

16. Liu, Y.C.; Li, Q.X.; Liu, Y.C. Preparation of phosphorus by carbothermal reduction mechanism in vacuum. *Adv. Mater. Res.* **2012**, *49*, 268–274. [CrossRef]

17. Cheng, C.; Xue, Q.G.; Wang, G.; Zhang, Y.Y.; Wang, J.S. Phosphorus migration during direct reduction of coal composite high-phosphorus iron ore pellets. *Metall. Mater. Trans. B* **2016**, *47*, 154–163. [CrossRef]

18. Zhang, Y.Y.; Xue, Q.G.; Wang, G.; Wang, J.S. Intermittent microscopic observation of structure change and mineral reactions of high phosphorus oolitic hematite in carbothermic reduction. *ISIJ Int.* **2017**, *57*, 1149–1155. [CrossRef]

19. Cheng, C.; Xue, Q.G.; Zhang, Y.Y. Dynamic migration process and mechanism of phosphorus permeating into metallic iron with carburizing in coal-based direct reduction. *ISIJ Int.* **2015**, *55*, 2576–2581. [CrossRef]

20. Bale, C.W.; Bélisle, E.; Chartrand, P. FactSage thermochemical software and databases, 2010–2016. *Calphad-Comput. Coupling Phase Diagr. Thermochem.* **2016**, *54*, 35–53. [CrossRef]

metals

MDPI

Article

Modeling and Experimental Study of Ore-Carbon Briquette Reduction under CO–CO$_2$ Atmosphere

Huiqing Tang *, Zhiwei Yun, Xiufeng Fu and Shen Du

State Key Laboratory of Advanced Metallurgy, University of Science and Technology Beijing, Beijing 100083, China; ustbyzwei@gmail.com (Z.Y.); apt-fu@outlook.com (X.F.); dushen@ustb.edu.cn (S.D.)
* Correspondence: hqtang@ustb.edu.cn; Tel.: +86-10-8237-7180

Received: 26 February 2018; Accepted: 21 March 2018; Published: 23 March 2018

Abstract: Iron ore-carbon briquette is often used as the feed material in the production of sponge iron via coal-based direct reduction processes. In this article, an experimental and simulation study on the reduction behavior of a briquette that is made by hematite and devolatilized biochar fines under CO–CO$_2$ atmosphere was carried out. The reaction model was validated against the corresponding experimental measurements and observations. Modeling predictions and experimental results indicated that the CO–CO$_2$ atmosphere significantly influences the final reduction degree of the briquette. Increasing the reduction temperature did not increase the final reduction degree but was shown to increase the carbon that was consumed by the oxidative atmosphere. The influence of the CO–CO$_2$ atmosphere on the briquette reduction behavior was found to be insignificant in the early stage but became considerable in the later stage; near the time of the briquette reaching its maximum reduction degree, both iron oxide reduction and metallic iron re-oxidation were able to occur.

Keywords: ore-carbon briquette; CO–CO$_2$ atmosphere; simulation; re-oxidation; reduction

1. Introduction

An ore-carbon briquette is a composite briquette consisting of iron-bearing oxide and carbonaceous materials that were used as feed material in some coal-based direct reduction processes, such as FASTMET® (FASTMET is a trade mark of MIDREX Co., USA.) and ITMK3® (ITMK3 is a trademark of KOBE Steel Co., Japan.) [1–4]. The use of these briquettes offers advantages, such as a high reduction rate, utilization of non-coking coal, and biochar for producing sponge iron economically. The ore-carbon briquette reduction technology is often used in treating various metallurgical dust and sludge [5–7], recovering valuable metals, such as nickel and titanium, from complicated minerals [8–11], and upgrading refractory iron ores by removing detrimental minerals, such as quartz and alumina [12–14]. The reduction of ore-carbon briquettes is usually carried out at the industrial level using rotary hearth furnace (RHF) reduction technology [15]. In an RHF process, the briquette is fed onto the rotating hearth of the RHF and is reduced into sponge iron in a high-temperature environment. The flue gas from the combustion of coal gas, composed of CO, CO$_2$, H$_2$, and H$_2$O, forms a weakly oxidative atmosphere on the surface of the briquette [16–18]. The CO$_2$ or H$_2$O levels may allow for the metallic iron oxidation of the ore-carbon briquette during the RHF process, which would have a negative effect on the quality of the products. Therefore, studies are required to address the reduction behavior of ore-carbon briquettes under the RHF reactive atmosphere.

The reaction kinetics of iron ore-carbon briquettes under inert atmosphere (nitrogen or argon) have been extensively studied, and their major features are well established [19–22]. The reduction of the briquette proceeds rapidly under high temperatures and generates a large amount of gas inside the briquette, making it more complicated than the reaction behavior of an iron ore or coal briquette. The reaction kinetics depend significantly on the chemical composition and physical properties of the briquette. Some studies on ore-carbon reduction have been conducted under the oxidative atmosphere by Singh et al. [23] and by Ghosh et al. [24], but such studies are scarce. There are several mathematical descriptions of the reduction phenomenon of the iron ore-carbon briquette. For example, Moon et al. [25] developed a model with an assumption of uniform conversion of iron oxide and carbon particles in the briquette, Sun and Lu [26] and Shi et al. [27] developed models including the expressions of chemical kinetics, equations of mass transfer, and equations of heat transfer, and Donskoi et al. [28] developed a model when considering the swelling/shrinkage of the briquette. Although these models attempt to give a comprehensive understanding of the reduction behavior of the briquette, the interaction between the briquette and the oxidative atmosphere has not been included. Therefore, the effect of the oxidative atmosphere on the reduction behavior of ore-carbon briquette has been overlooked in the existing studies.

The first aim of this study was to conduct kinetic experiments of the isothermal reduction behavior of the ore-carbon briquette under a simulated RHF atmosphere (CO–CO_2 atmosphere with $P_{CO}/P_{CO2} = 1.0$). The second aim was to establish and develop a reaction model of ore-carbon briquette reduction that includes the reaction of metallic iron with CO_2. The simulation results were compared to the experimental results in respect to briquette mass change, briquette carbon conversion, briquette reduction degree, and briquette reduction progress. The reduction behavior of the briquette under the CO–CO_2 atmosphere was also analyzed.

2. Experiments

2.1. Materials and Briquette Preparation

The iron ore sample was from Tangshan Iron and Steel Company (Tangshan), China. The carbonaceous reductant sample was prepared by carbonizing the biochar under 1273 K for 1 h. The chemical composition of the employed biochar is given in [29]. Chemical composition of the ore sample is listed in Table 1. Fe_2O_3 content in Table 1 was analyzed by chemical analysis (iron chloride method) and contents of other components in Table 1 were examined by energy-dispersive X-ray fluorescence spectrometry (XRF) using an XRF 1800 spectrometer (Shimadzu Co., Kyoto, Japan), and the proximate analysis of the carbonaceous reductant sample is listed in Table 2. Both of the samples were ground using a F-P400 ball mill (Focucy Co., Changsha, China), and the average sizes of the ore fines and reductant fines were 100 and 80 μm, respectively. The mixture was thoroughly mixed with an addition of 2% cellulose binder (Dingshengxin Co., Tianjin, China), 5% reagent-grade CaO powder (Xilong Co., Shantou, China), and 10% distilled water. Molar ratio of fixed carbon in the reductant fines to oxygen in the iron oxide of the ore fines was 1.0. The moistened fines were pressed with a die under a pressure of 40 MPa to make the briquettes. The briquettes were then air-dried for 24 h, followed by drying at 473 K for 2 h. The prepared briquettes had a cylindrical shape with a diameter (D) of 20 mm, a height (H) of 10 mm, and a mass of approximately 6.0 g.

Table 1. Chemical composition of the iron ore sample (wt %).

Fe_2O_3	SiO_2	CaO	Al_2O_3	MgO	MnO	LOI
91.77	2.9	0.1	4.05	0.56	0.14	0.48

LOI: loss on ignition.

Table 2. Proximate analysis of the carbonaceous reductant sample (wt %).

Volatile	Fixed Carbon	Ash
0.41	96.30	3.29

2.2. Experimental Setup and Procedures

The experimental device, schematically presented in Figure 1, includes a gas supply system, an electronic scale with an accuracy of ±0.001 g, and a temperature-controlled furnace with an accuracy of ±2 K. The furnace was heated by $MoSi_2$ elements, producing a 50 mm hot zone in the reaction tube, with an inner diameter of 60 mm. The sample holder was made of a heat-resistant alloy wire (Fe–Cr–Al).

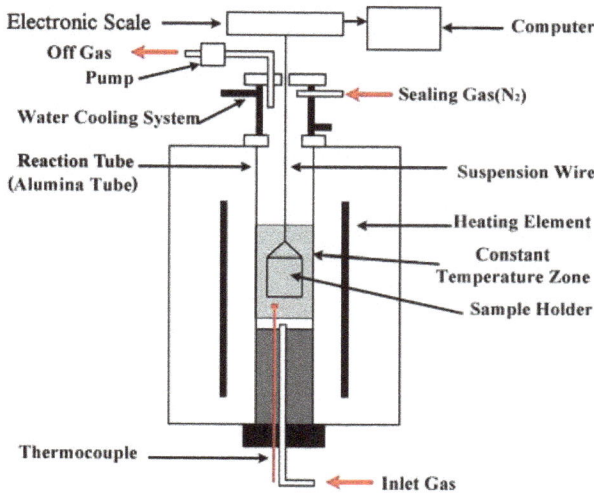

Figure 1. Schematic diagram of the experimental setup.

The furnace was preheated to the desired temperature under N_2. One briquette was then loaded into the sample holder, preheated at 773 K for 5 min in the upper part of the reaction tube, and then introduced into the hot zone. At this time, the N_2 feed was replaced with a $CO–CO_2$ gas mixture, $P_{CO}/P_{CO2} = 1.0$, and the mass loss/gain of the briquette was measured by an electronic scale and was recorded by a computer at an interval of 2 s. After the predetermined time, the briquette was withdrawn from the reaction tube and quenched by a N_2 stream. In all of the individual tests, a constant gas flow rate of 1600 cm^3/min (Standard Temperature and Pressure) at the gas inlet was maintained. In addition, some reduced briquettes were subjected to carbon content analysis, scanning electron microscopy (SEM), and energy dispersive spectrometry (EDS) examinations. The carbon analysis was conducted using a CS-2800 infrared carbon sulfur analyzer (NCS, Beijing, China); SEM and EDS were performed using a Quanta-250 scanning electron microscope (FEI, Hillsboro, OR, USA). The definitions and calculation methods of mass-loss fraction (f_m), reduction degree (f_O) and carbon conversion (f_C) of the tested briquette are $f_m = \Delta m_t/(m_C + m_O)$, $f_C = \Delta m_C/m_C = 1.0 - (m_b - \Delta m_t)[C]_t/m_C$, and $f_O = \Delta m_O/m_O = f_m + m_C/m_O(f_m - f_C)$, respectively [30], where, Δm_t, Δm_C, and Δm_O are the mass loss, carbon mass loss, and oxygen mass loss of the briquette at time t, respectively; m_b, m_C, and m_O are the initial mass, initial carbon mass, and initial iron-oxide oxygen mass of the preheated briquette, respectively; and $[C]_t$ is the carbon content of the sample at time t. m_b, m_C, and m_O are available, according to the preparation procedure of the briquette.

3. Mathematical Model

A mathematical model was established for the reduction process on a single cylindrical briquette. According to the symmetry of the geometry and the experimental conditions, a simplified geometrical model is shown in Figure 2a. The computational domain was chosen as 0.5 radian, and three types of boundary conditions, including wall, symmetry, and axis, were used in the simulations. Figure 2b schematically shows the structure's grid system, which used a grid of 40 × 20.

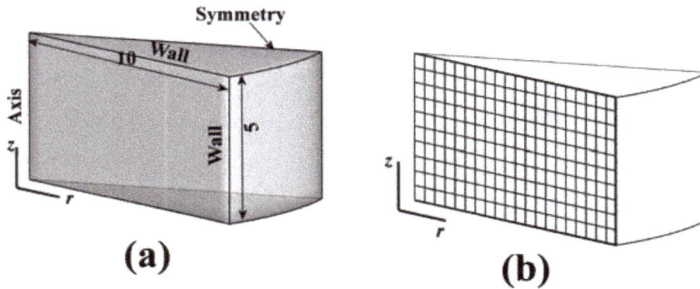

Figure 2. Geometrical and mathematical model for the modeling of a single briquette: (a) 0.5 radian computational domain; and, (b) grid system.

Several assumptions were made during model development to express the behavior of a briquette. Briquette reduction was considered under isothermal conditions, swelling/shrinking of the briquette was not considered, and the porosity of the briquette was assumed to be constant. Similar assumptions can be found in many studies [7,25–27] involving the modeling of the reduction of the ore-carbon briquette, and therefore it is considered that these assumptions would not cause considerable errors.

The overall reduction process is believed to occur through an intermediate gas phase. The reactions that take place in the briquette are the stage-wise reductions of iron oxide by CO (Equations (1)–(3)), the Boudouard reaction of carbon particles (Equation (4)), and metallic iron oxidation (Equation (5)). The reduction of iron oxide by solid carbon may also occur; however, it is difficult to estimate the extent of this reaction, and the share of the solid-state reduction is minimal when considering the much larger gas–solid contact area. Therefore, no solid–solid reaction was taken into account.

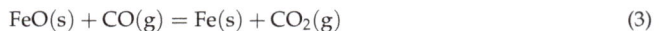

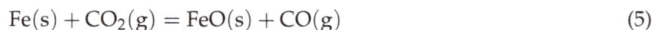

$$3\ Fe_2O_3(s) + CO(g) = 2\ Fe_3O_4(s) + CO_2(g) \tag{1}$$

$$Fe_3O_4(s) + CO(g) = 3\ FeO(s) + CO_2(g) \tag{2}$$

$$FeO(s) + CO(g) = Fe(s) + CO_2(g) \tag{3}$$

$$C(s) + CO_2(g) = 2CO(g) \tag{4}$$

$$Fe(s) + CO_2(g) = FeO(s) + CO(g) \tag{5}$$

The gas phase in the briquette was considered to be ideal and to consist of CO and CO_2. Dependencies of the gas properties on local temperature and composition were calculated according to the ideal gas law in the FLUNET material database [31].

Gas transfer in the reduction of an ore-carbon briquette is driven by two factors: the concentration gradient and the pressure gradient. Gas transfer by pressure gradient is negligible in most cases; therefore, the total gas pressure on the computational domain was fixed at atmospheric pressure, and it is shown below in Equation (6).

$$P_{CO} + P_{CO_2} = 1.01 \times 10^5 \tag{6}$$

The equation of species CO_2 on the computational domain was then given by Equation (7) using the effective coefficient with Fick's law:

$$\frac{\partial(\alpha\rho_g y_{CO_2})}{\partial t} = \frac{1}{r}\frac{\partial}{\partial r}(r\rho_g D_{eff}\frac{\partial y_{CO_2}}{\partial r}) + \frac{\partial}{\partial z}(\rho_g D_{eff}\frac{\partial y_{CO_2}}{\partial z}) + S_{CO_2} \tag{7}$$

where $S_{CO_2} = M_{CO_2}/M_O(R_1 + R_2 + R_3) - M_{CO_2}/M_C R_4 - M_{CO_2}/M_O R_5$. The equation of species CO is shown below in Equation (8).

$$y_{CO} + y_{CO_2} = 1.0 \tag{8}$$

D_{eff} in Equation (7) depends on the porous structure of the briquette, and it was determined using the Weisz–Schwartz relationship, given by Equation (9) [32].

$$D_{eff} = \alpha^2 D_{CO-CO_2}/\sqrt{3} \tag{9}$$

Equations (6)–(8) formed the governing equations of the gas phase. The boundary and initial conditions for Equation (7) were given by Equations (10)–(13). Under experimental conditions, the velocity and other gas properties of the furnace atmosphere in Equations (10) and (11) were considered as equal to their respective values at the gas inlet of the furnace. Additionally, the equivalent diameter of the cylindrical briquette was assumed as the overall diameter.

$$r = D/2, \rho_g D_{eff} \frac{\partial y_{CO_2}}{\partial r} = k_g(y_{CO_2,f}\rho_{g,f} - y_{CO_2}\rho_g) \tag{10}$$

$$z = H/2, \rho_g D_{eff} \frac{\partial y_{CO_2}}{\partial z} = k_g(y_{CO_2,f}\rho_{g,f} - y_{CO_2}\rho_g) \tag{11}$$

where $k_g = D_{CO-CO_2,f}\left(2.0 + 0.6Re_f^{1/2}Sc_f^{1/3}\right)/D$ [33].

$$z = 0, \frac{\partial y_{CO_2}}{\partial r} = 0; \ r = 0, \frac{\partial y_{CO_2}}{\partial z} = 0 \tag{12}$$

$$r \in (0, D/2), z \in (0, H/2), \ y_{CO_2} = y_{CO_2,f} \tag{13}$$

The governing equations used for the solid phase are given below as Equations (14)–(18).

$$\partial \rho_{Fe_2O_3}/\partial t = (3M_{Fe_2O_3}/M_O)(-R_1) \tag{14}$$

$$\partial \rho_{Fe_3O_4}/\partial t = (M_{Fe_3O_4}/M_O)(2R_1 - R_2) \tag{15}$$

$$\partial \rho_{FeO}/\partial t = (M_{FeO}/M_O)(3R_2 - R_3 + R_5) \tag{16}$$

$$\partial \rho_{Fe}/\partial t = (M_{Fe}/M_O)(R_3 - R_5) \tag{17}$$

$$\partial \rho_C/\partial t = -R_4 \tag{18}$$

The initial conditions for Equations (15)–(18) are $r \in (0, D/2), z \in (0, H/2), \rho_{Fe_2O_3} = \rho_{Fe_2O_3,0}, \rho_{Fe_3O_4} = 0.0, \rho_{FeO} = 0.0, \rho_{Fe} = 0.0$, and, $\rho_C = \rho_{C,O}$.

4. Solution Method

The FLUENT CFD package (v6.3, Fluent Inc., Lebanon, NH, USA) [31] was used for conducting the numerical simulations. Equation (7) was spatially and temporally discretized using a fully implicit first-order upwind scheme. The time step was 0.01 s, the under-relaxation factor was 0.1, and the convergence criterion used was 1.0×10^{-5}. An explicit time integration method was adopted for solid-phase equations.

5. Results and Discussion

5.1. Determination of Model Parameters and Reaction Rates of Involved Reactions

Before performing simulations, some parameters and rate expressions of the involved reactions must be determined. Lu and Sun [21] reported that variation of the briquette porosity varied from

0.40 to 0.68 in the reduction process of the ore-carbon briquette, and the porosity of the briquette was thus assumed to be $\alpha = 0.50$ in the present study.

Gaseous reduction of iron ore particles has been extensively studied in its thermodynamics and kinetics. Generally, shrink unreacted core model is consistent with the step-wise of iron oxides. As the hematite particles are very small, the reactions that are given by Equations (1)–(3) were considered to proceed independently [34]. Their reduction rates were thus described using a one-interface unreacted shrinking core model, as Equation (19).

$$R_i = \gamma 4\pi d_{ore}^2 M_O N_P (P_{CO} - \frac{P_{CO_2}}{K_i}) / (RT) / ((\frac{d_{ore}}{2D_{e,i}}((1-f_i)^{-\frac{1}{3}} - 1) + (\frac{K_i}{k_i(1+K_i)})(1-f_i)^{-\frac{2}{3}})) \quad (19)$$

where $k_1 = \exp(-1.445 - 6038/T)$, $K_1 = \exp(7.255 + 3720/T)$, and $D_{e,1} = \infty$ for R_1; $k_2 = 1.70\exp(2.515 - 4811/T)$, $K_2 = \exp(5.289 - 4711/T)$ and $D_{e,2} = \exp(-1.835 - 7180/T)/P_g$ for R_2; and, $k_3 = \exp(0.805 - 7385/T)$, $K_3 = \exp(-2.946 + 2744.63/T)$, and $D_{e,3} = \exp(0.485 - 8770/T)/P_g$ for R_3 [35]. Definition of f_i in Equation (19) is given in [32]. The internal gas diffusion resistance of R_1 was not considered because the transformation of Fe_2O_3 to Fe_3O_4 (Equation (1)) proceeds very quickly in the initial stage of the briquette reduction.

The reaction rate of the reaction given in Equation (4) is Equation (20) [36].

$$R_4 = k_4(1-f_4)^{0.44}(P_{CO_2}/1.01 \times 10^5)\rho_{C,0} \quad (20)$$

where $f_4 = 1.0 - \rho_C/\rho_{C,0}$ and $k_4 = 1.8 \times 10^3 \exp(-139000/RT)$ [29].

The rate of the reaction given in Equation (5) is Equation (21) [37].

$$R_5 = \gamma 4\pi d_{ore}^2 N_P k_5(P_{CO_2} - P_{CO}/K_5)/(1.01 \times 10^5) \quad (21)$$

where $k_5 = 0.011\exp(-42611/RT)$, and $K_5 = 1/K_3$.

In estimating N_P in Equations (19)–(21), the true density of the hematite particles was assumed to be 5000 kg/m^3. γ in Equations (19)–(21) is a coefficient for adjusting the specific area of ore particles due to their irregular geometry, and it was determined to be 0.4 by a trial-and-error method.

5.2. Briquette Mass Change

Mass change in the briquette during reduction was caused by several reactions; the reactions given by Equations (1)–(4) led to a decrease in mass, whereas the reaction given by Equation (5) led to an increase in mass. Measured and model-predicted mass-loss fraction curves under different temperatures were compared, and the results are shown in Figure 3. In simulations, the briquette mass-loss fraction at time t was calculated using Equation (22).

$$f_m = 1.0 - \sum V_{cell}((3\rho_{Fe_2O_3}/M_{Fe_2O_3} + 4\rho_{Fe_3O_4}/M_{Fe_3O_4} + \rho_{FeO}/M_{FeO})M_O + \rho_C)/\sum(V_{cell}((3\rho_{Fe_2O_3}/M_{Fe_2O_3})M_O + \rho_{C,0})) \quad (22)$$

The model predictions closely match the experimental measurements at 1273 K and 1373 K, as can be seen in Figure 3a,b; however, under 1473 K, some deviation occurs in the later reduction stage (Figure 3c). In view of the assumptions that are made in the model and errors in the measurements, the agreement between them is considered to be satisfying.

The shape of all of the mass-loss fraction curves presents some common features. The briquette reduction can be divided into three fairly distinguishable stages. The first stage comprises the mass loss, the second stage corresponds to the mass loss reaching its maximum value, and the third stage includes the mass increase. The mass loss/gain characteristics became more evident with an increasing temperature.

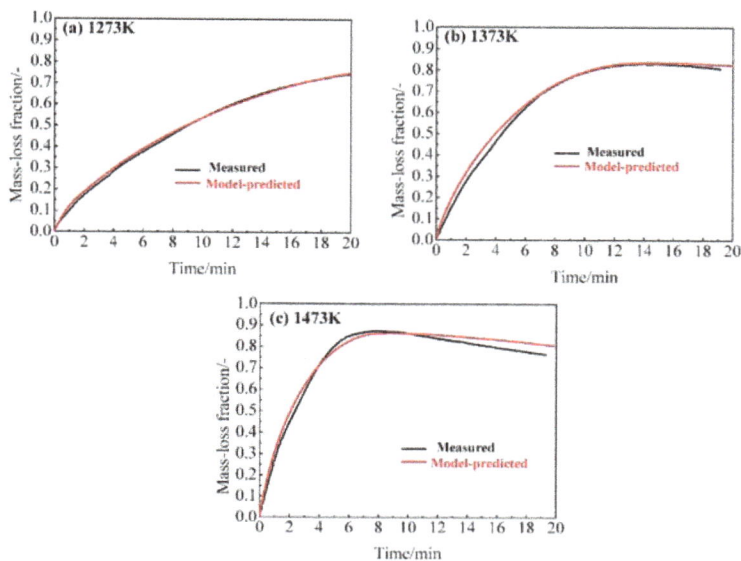

Figure 3. Model-predicted and experimental mass-loss fraction curves under different temperatures: (a) 1273 K; (b) 1373 K; and, (c) 1473 K.

5.3. Briquette Reduction Degree and Briquette Carbon Conversion

Because the mass change of the briquette is affected by several reactions, it cannot adequately reflect the briquette reduction behavior. However, the briquette reduction degree and carbon conversion are important parameters for evaluating the quality of the reduced briquette. In the simulations, the briquette reduction degree at time t was calculated using Equation (23), and the briquette carbon conversion at time t using Equation (24).

$$f_O = 1.0 - \sum V_{cell}(3\rho_{Fe_2O_3}/M_{Fe_2O_3} + 4\rho_{Fe_3O_4}/M_{Fe_3O_4} + \rho_{FeO}/M_{FeO})/\sum V_{cell}(3\rho_{Fe_2O_3}/M_{Fe_2O_3}) \quad (23)$$

$$f_C = 1.0 - \sum(V_{cell}\rho_C)/(V_{Cell}\rho_{C,0}) \quad (24)$$

The model-predicted and measured briquette reduction degrees under three temperatures are shown in Figure 4. The average difference in reduction degree between model predictions and experimental measurements was less than 0.03; therefore, it can be concluded that the developed model is applicable to the reduction of the ore-carbon briquette. Model predictions at 1273 K present a successive increase of reduction degree throughout the reduction period. At 1373 K, the reduction degree increased in the early stage and reached its maximum reduction degree of 0.75 at approximately 12 min; thereafter, it decreased. Briquette reduction behavior under 1473 K was similar to that under 1373 K, except that it reached a maximum reduction degree of 0.80 at approximately 6 min. After 15 min, the reduction degrees of the briquette under 1473 K and under 1373 K are quite close, and by 20 min, the reduction degrees under the three temperatures are nearly the same. These findings indicate that, under the oxidative atmosphere, increasing the temperature does not increase the final reduction degree of the briquette.

The model-predicted briquette carbon conversion is compared to the corresponding experimental measurements in Figure 5. The briquette carbon conversion increased with time at all three temperatures. Model-predicted curves at 1373 K and 1473 K show that, after reaching their maximum reduction degree (at 12 min under 1373 K and at 7 min under 1473 K), the carbon conversion increased much more slowly. By 20 min, the final carbon conversion was lower at 1273 K than at 1373 K and 1473 K. Combing the results from Figures 4 and 5 indicates that more biochar could be consumed by the atmosphere with the increase of the temperature.

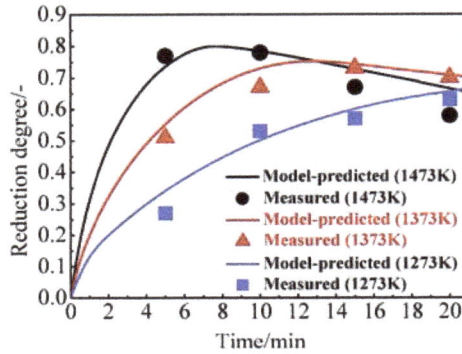

Figure 4. Model-predicted and measured briquette reduction degrees.

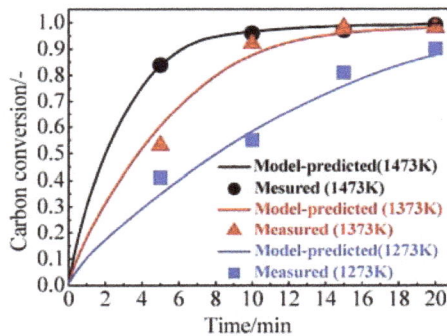

Figure 5. Model-predicted and measured briquette carbon conversions.

5.4. Briquette Reduction Progress

Briquette reduction under 1473 K was selected for further study of the briquette reduction progress because it displayed the major characteristics of hematite-biochar reduction under the oxidative atmosphere: fast reduction in the early stage and obvious metallic iron oxidation in the later stage.

The simulation results of the development of local P_{CO}/P_{CO2}, the local reduction degree profile, and the local carbon conversion profile on the briquette cross section are displayed in Figure 6. In Figure 6, the local reduction degree and carbon conversion at time t were calculated by $1.0 - (3\rho_{Fe_2O_3}/M_{Fe_2O_3} + 4\rho_{Fe_3O_4}/M_{Fe_3O_4} + \rho_{FeO}/M_{FeO})/(3\rho_{Fe_2O_3,0}/M_{Fe_2O_3})$ and $1.0 - \rho_C/\rho_{C,0}$, respectively. These simulation results were then used to assume the reduction progress of the briquette. At the beginning stage, the dominant reactions were assumed as the reactions that are given by Equations (1), (2) and (4). Owing to the intense gas generation, the briquette reduction progress was assumed to not be influenced by the atmosphere or gas diffusion. At 1 min, P_{CO}/P_{CO2} was nearly uniformly profiled at a level of approximately 4.0; both the reduction degree and the carbon conversion profiles were uniform and at a degree of approximately 0.3. As time proceeded, the reactions that were given by Equations (3) and (4) became dominant. As the rate of generated product gas was reduced, the effect of gas diffusion increased. At 5 min, the atmosphere began to influence the briquette reduction progress, and the P_{CO}/P_{CO2} profile in the briquette became uneven: at the core, it increased to 10.28, whereas at the surface, it only increased to 3.26. Correspondingly, the reduction degree increased to 0.88 at the core and to 0.60 at the surface, and carbon conversion increased to 0.76 at the core and to 0.90 at the surface. By 10 min, P_{CO}/P_{CO2} remained higher than 2.95 in the internal part of the briquette ($K_5 = 2.95$ under 1473 K), and reached 110 at the core, so that the reaction given by Equation (3) could still proceed there. However, P_{CO}/P_{CO2} decreased to less than 2.95 at the surface of the briquette, and consequently, the reaction that was given by Equation (5) took place and became a dominant reaction. At 10 min, reduction degree increased in the internal area and reached 0.98 at the core; at the surface, it decreased to 0.50. By the end of the reduction, the dominant reactions changed to those given by Equations (4) and (5). The contour line of $P_{CO}/P_{CO2} = 2.95$ became closer to the core, and P_{CO}/P_{CO2} at the core decreased to 20.0, which reflected that the atmosphere was extending its influence toward the core. At 15 min, the reaction that was given by Equation (3) ceased. In the region with $P_{CO}/P_{CO2} > 2.95$, the reduction degree remained higher than 0.90. However, the reaction given by Equation (5) became significant as the region with $P_{CO}/P_{CO2} < 2.95$ was enlarged. The reduction degree at the surface further decreased to 0.40, and the weak gasification of the remaining carbon particles was then attributed to CO_2 from the atmosphere.

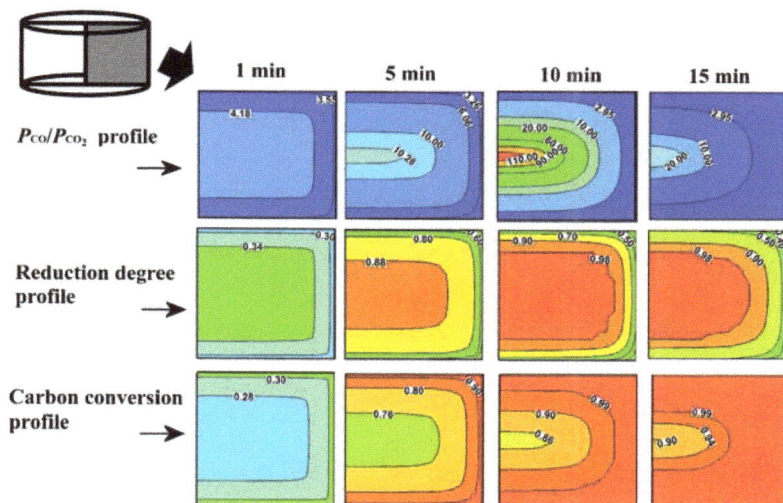

Figure 6. Simulation results on the development of the profile of local P_{CO}/P_{CO2}, the profile of local reduction degree, and the profile of local carbon conversion.

Thus, the reduction of the briquette presented a mainly homogeneous reaction system in the period from 1 to 5 min, whereas the gas transfer and the oxidative atmosphere became considerable after 5 min. Both metallic re-oxidation and iron oxide reduction appeared as the briquette neared its maximum reduction degree.

SEM–EDS results on intermittent morphologies at the briquette side surface in the reduction course at 1473 K are shown in Figure 7. The ore particles were uniformly distributed at 5 min (Figure 7a), and a mixture of tiny metallic iron grains (Point 1 in Figure 7b), wustite grains (Point 2 in Figure 7b), and gangue grains (Point 3 in Figure 7b) was presented within ore particles (Figure 7b). The deformation and the sintering of ore particles had occurred by 10 min (Figure 7c), and iron grains became scarce in the particles (Figure 7d). The decrease of iron grains was due to the re-oxidation by the atmosphere, and the ore particle sintering and deforming were attributed to wustite reacting with gangue components (CaO, SiO_2, and Al_2O_3) to form low-melting-point compounds (glass phase).

By 15 min, the sintering degree of ore particles had increased (Figure 7e), and the size of some remaining iron grains in ore particles was enlarged (Figure 7f). Iron grain growth was attributed to an agglomeration of tiny iron grains that was facilitated by the glass phase. Overall, the SEM–EDS results indicate that the briquette side surface underwent oxidation after 5 min in the reduction course, which was in accordance with the simulated reduction process. The intense self-reduction in the early reduction stages lead to crack generation at the briquette surface so that iron oxidation near the surface was accelerated in the later stage as the porosity near the surface was increased. Therefore, crack formation near the surface could be the main reason for the deviation between model predictions and the experimental measurements in the later stage of reduction, as shown in Figure 3c.

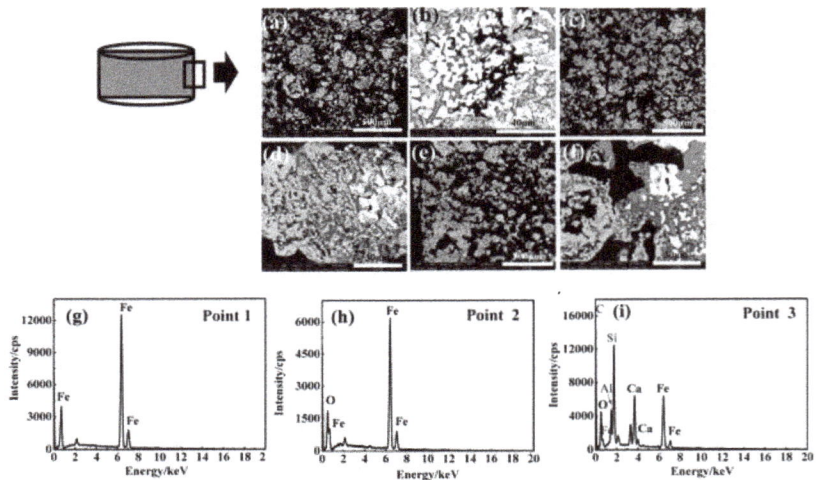

Figure 7. Intermittent scanning electron microscopy–energy dispersive spectrometry (SEM–EDS) results at the briquette surface: (**a**,**b**) 5 min; (**c**,**d**) 10 min; (**e**,**f**) 15 min; and (**g–i**) EDS results of Points 1, 2 and 3 in Figure 7b, respectively.

6. Conclusions

1. A model to predict the reduction behavior of the ore-carbon briquette under $CO-CO_2$ atmosphere was developed. The model included the kinetics of the stage-wise reduction of iron oxide, carbon gasification and metallic iron oxidation, and it was with the assumptions of constant porosity and size of the briquette. The simulation results were validated by the experimental measurements and observations and the model was found to be reliable.

2. The $CO-CO_2$ atmosphere can significantly influence the final reduction degree of the briquette, and the briquette cannot reach higher final reduction degree by further increasing the temperature. Under higher temperatures, more carbon is consumed by the reactive atmosphere.

3. In the briquette reduction progress, the briquette reduction behavior is not initially influenced by the $CO-CO_2$ atmosphere; however, near the maximum reduction degree, both iron oxide reduction and metallic iron re-oxidation can occur in the briquette.

Acknowledgments: The authors thank the National Natural Science Foundation of China (No. 51144010), and the State Key Laboratory of Advance Metallurgy USTB for the financial support of this work.

Author Contributions: Huiqing Tang conceived and designed the experiments; Zhiwei Yun and Xiufeng Fu performed the experiments; Zhiwei Yun and Xiufeng Fu analyzed the data; Shen Du contributed reagents/materials/analysis tools; Huiqing Tang wrote the paper.

Conflicts of Interest: The authors declare no conflict of interest.

Abbreviations

Table of Symbols

d_{ore}	diameter of ore particle, (m)
$D_{e,i}$	interior diffusion coefficient of the reaction given by Equation (i), $(m \cdot s^{-1})$
D_{CO-CO_2}, D_{eff}	gas diffusivity, effective gas diffusivity, $(m^2 \cdot s^{-1})$
f_i	reaction fraction of the reaction given by Equation (i)
k_i	reaction rate constant of the reaction given by Equation (i), (unit vary)
k_g	external mass transfer coefficient, $(m \cdot s^{-1})$
K_i	equilibrium constant of the reaction given by Equation (i)
M	molar weight, $(kg \cdot mol^{-1})$
N_P	number density, (m^{-3})
Re	Reynolds number
R	gas constant, $(8.314 \, J \cdot mol^{-1} \cdot K^{-1})$
R_i	reaction rate of the reaction given by Equation (i), $(kg \cdot m^{-3} \cdot s^{-1})$
Sc	Schemidt number
T	temperature, (K)
V_{cell}	cell volume, (m^3)
t	time, (s)
y	mass fraction
α	porosity
ρ	local density, $(kg \cdot m^{-3})$
Subscripts	
0	initial
g	gas
f	furnace
Species or element name	variable of assigned species or element

References

1. Ahmed, H.M.; Viswanathan, N.; Bjorkman, B. Composite pellets—A potential raw material for iron-making. *Steel Res. Int.* **2014**, *85*, 293–306. [CrossRef]
2. Nikai, I.; Garbers-Craig, A.M. Use of iron ore fines in cold-bonded self-reducing composite pellets. *Miner. tProcess. Extr. Metall. Rev.* **2016**, *37*, 42–48. [CrossRef]
3. Chukwuleke, O.P.; Cai, J.J.; Chukwujekwu, S.; Song, X. Shift from coke to coal using direct reduction method and challenges. *J. Iron Steel Res. Int.* **2009**, *16*, 1–5. [CrossRef]
4. Manning, C.P.; Fruehan, R.J. Emerging technologies for iron and steelmaking. *JOM* **2001**, *53*, 36–39. [CrossRef]
5. Mae, K.; Inaba, A.; Hanaki, K. Production of iron/carbon composite from low rank coal as a recycle material for steel industry. *Fuel* **2005**, *84*, 227–233. [CrossRef]
6. El-Hussiny, N.A.; Shalabi, M.E.H.A. Self-reduced intermediate product from iron and steel plants waste materials using a briquetting process. *Powder Technol.* **2011**, *205*, 217–223. [CrossRef]
7. Kuwauchi, Y.; Barati, M. A mathematical model for carbothermic reduction of dust-carbon composite agglomerates. *ISIJ Int.* **2013**, *53*, 1097–1105. [CrossRef]
8. Li, G.; Shi, T.; Rao, M.; Jiang, T.; Zhang, Y. Beneficiation of nickeliferous laterite by reduction roasting in the presence of sodium sulfate. *Miner. Eng.* **2012**, *32*, 19–26. [CrossRef]
9. Cheng, G.; Gao, Z.; Yang, H.; Xue, X. Effect of calcium oxide on the crushing strength, reduction, and smelting performance of high-chromium vanadium–titanium magnetite pellets. *Metals* **2017**, *7*, 181. [CrossRef]
10. Zhu, D.; Guo, Z.; Pan, J.; Zhang, F. Synchronous upgrading iron and phosphorus removal from high phosphorus Oolitic hematite ore by high temperature flash reduction. *Metals* **2016**, *6*, 123. [CrossRef]
11. Wang, Z.; Chu, M.; Liu, Z.; Wang, H.; Gao, L. Preparing Ferro-Nickel alloy from low-grade laterite nickel ore based on metallized reduction–magnetic separation. *Metals* **2017**, *7*, 313. [CrossRef]

12. Li, G.; Luo, J.; Jiang, T.; Peng, Z.; Zhang, Y. Digestion of alumina from non-magnetic material obtained from magnetic separation of reduced iron-rich diasporic bauxite with sodium salts. *Metals* **2016**, *6*, 294. [CrossRef]
13. Tang, H.; Fu, X.; Qin, Y. Production of low-silicon molten iron from high-silica hematite using biochar. *J. Iron Steel Res. Int.* **2017**, *24*, 27–33. [CrossRef]
14. Zhou, X.; Zhu, D.; Pan, J.; Luo, Y.; Liu, X. Upgrading of high-aluminum hematite-limonite ore by high temperature reduction-wet magnetic separation process. *Metals* **2016**, *6*, 57. [CrossRef]
15. McClelland, J.M.; Metius, G.E. Recycling ferrous and nonferrous waste streams with FASTMET. *JOM* **2003**, *55*, 30–34. [CrossRef]
16. Wu, Y.; Jiang, Z.; Zhang, X.; Xue, Q.; Yu, A.; Shen, Y. Modeling of thermochemical behavior in an industrial-scale rotary hearth furnace for metallurgical dust recycling. *Metall. Mater. Trans. B* **2017**, *48*, 2403–2418. [CrossRef]
17. Liu, Y.; Su, F.Y.; Wen, Z.; Li, Z.; Yong, H.Q. CFD modeling of flow, temperature, and concentration fields in a pilot-scale rotary hearth furnace. *Metall. Mater. Trans. B* **2014**, *45*, 251–261. [CrossRef]
18. Liu, Y.; Zhi, W.; Lou, G.; Li, Z.; Yong, H.; Feng, X. Numerical investigation of the Effect of C/O mole ratio on the performance of rotary hearth furnace using a combined model. *Metall. Mater. Trans. B* **2014**, *45*, 2370–2381. [CrossRef]
19. Dutta, S.K.; Ghosh, A. Study of nonisothermal reduction of iron ore-coal/char composite pellet. *Metall. Mater. Trans. B* **1994**, *25*, 15–26. [CrossRef]
20. Murao, M.B.; Takano, C. Self-reducing pellets for ironmaking: Reaction rate and processing. *Miner. Process. Extr. Metall. Rev.* **2003**, *24*, 183–202. [CrossRef]
21. Sun, S.; Lu, W.K. A theoretical investigation of kinetics and mechanisms of iron ore reduction in an ore/coal composite. *ISIJ Int.* **1999**, *39*, 123–129. [CrossRef]
22. Park, H.; Sohn, I.; Freislich, M. Investigation on the reduction behavior of coal composite pellet at temperatures between 1373 and 1573 K. *Steel Res. Int.* **2017**, *88*, 1–13. [CrossRef]
23. Singh, A.; Deo, K.; Ghosh, A. Reduction behavior of powder mixtures of iron oxide and carbon in reactive atmospheres. *Steel Res. Int.* **2001**, *72*, 136–140. [CrossRef]
24. Ghosh, A.; Mungole, M.N.; Gupta, G.; Tiwari, S. A preliminary study of influence of atmosphere on reduction behavior of iron ore-coal composite pellets. *ISIJ Int.* **1999**, *39*, 829–831. [CrossRef]
25. Moon, J.; Sahajwalla, V. Kinetic model for the uniform conversion of self-reducing iron oxide and carbon briquettes. *ISIJ Int.* **2003**, *43*, 1136–1142. [CrossRef]
26. Sun, S.; Lu, W.K. Building of a mathematical model for the reduction of iron ore in ore/coal composites. *ISIJ Int.* **1999**, *39*, 130–138. [CrossRef]
27. Shi, J.; Donskoi, E.; McElwain, D.L.S.; Wibberley, L.J. Modelling the reduction of an iron ore-coal composite pellet with conduction and convection in an axisymmetric temperature field. *Math. Comput. Model.* **2005**, *42*, 45–60. [CrossRef]
28. Donskoi, E.; McElwain, D.L.S. Mathematical modelling of non-isothermal reduction in highly swelling iron ore–coal char composite pellet. *Ironmak. Steelmak.* **2001**, *28*, 384–389. [CrossRef]
29. Tang, H.; Qi, T.; Qin, Y. Production of low-phosphorus molten iron from high-phosphorus oolitic hematite using biomass char. *JOM* **2015**, *67*, 1956–1965. [CrossRef]
30. Iguchi, Y.; Takada, Y. Rate of direct reactions measured in vacuum of iron ore-carbon composite pellets heated at high temperatures: Influence of carbonaceous materials, oxidation degree of iron oxides and temperature. *ISIJ Int.* **2004**, *44*, 673–681. [CrossRef]
31. FLUENT Inc. *FLUENT User Guide*; FLUENT Inc.: Lebanon, NH, USA, 2006.
32. Leffler, A.J. Determination of effective diffusivities of catalysts by gas chromatography. *J. Catal.* **1966**, *5*, 22–26. [CrossRef]
33. Ge, Q. *Kinetics of Gas-Solid Reactions*, 1st ed.; Nuclear Energy Press: Beijing, China, 1991; pp. 9–12. ISBN 7502204008.
34. Tang, H.; Guo, Z.; Kitagawa, K. Simulation study on performance of z-path moving-fluidized bed for gaseous reduction of iron ore fines. *ISIJ Int.* **2012**, *52*, 1241–1249. [CrossRef]
35. Natsui, S.; Kikuchi, T.; Suzuki, R.O. Numerical analysis of carbon monoxide–hydrogen gas reduction of iron ore in a packed bed by an Euler–Lagrange approach. *Metall. Mater. Trans. B* **2014**, *45*, 2395–2413. [CrossRef]

36. Wang, L.; Sandquist, J.; Varhegyi, G.; Guell, B.M. CO_2 gasification of chars prepared from wood and forest residue: A kinetic study. *Energy Fuel* **2013**, *27*, 6098–6107. [CrossRef]

37. Kaushik, P.; Fruehan, R.J. Behavior of direct reduced iron and hot briquetted iron in the upper blast furnace shaft: Part I. Fundamentals of kinetics and mechanism of oxidation. *Metall. Mater. Trans. B* **2006**, *37*, 715–725. [CrossRef]

metals

MDPI

Article

Damage Mechanism of Copper Staves in a 3200 m³ Blast Furnace

Haibin Zuo [1,*] , Yajie Wang [1] and Xuebin Wang [2]

[1] State Key Laboratory of Advanced Metallurgy, University of Science and Technology Beijing, Beijing 100083, China; 18810634675@163.com

[2] Shandong Iron and Steel Co., Ltd. Laiwu Branch, Laiwu 271104, China; erli2000@126.com

* Correspondence: zuohaibin@ustb.edu.cn; Tel.: +86-139-1094-9735

Received: 24 October 2018; Accepted: 8 November 2018; Published: 13 November 2018

Abstract: Copper staves have been widely applied in large blast furnaces especially those whose inner volumes exceed 2000 m³ due to high cooling capacity. In the past decade, copper staves suffered severe damages in some blast furnaces, which not only shortened their campaign lives, but also caused huge economic losses. In order to make out this phenomenon, the damage mechanism of copper staves was investigated via analyzing the chemical composition, thermal conductivity, metallographic aspects and microstructure in this paper. As a result, the working state was more likely to damage copper staves instead of their materials. At the beginning, the poor quality of the coke and the large bosh angle promoted the development of edge airflow, which intensified the erosion of refractory materials, resulting in the fall-off of slag crusts and damage of cooling water pipes. After repair, the cooling capacity of copper staves still declined, causing the temperature to rise easily; consequently, hydrogen attack happened when the temperature reached 370 °C, which degraded the performance of copper staves. Therefore, copper staves were worn too quickly to form slag crusts, which finally failed under the hydrogen attack and the scouring of the edge airflow at high temperatures.

Keywords: blast furnace; copper stave; hydrogen attack; slag crust

1. Introduction

In the long-term smelting practice, iron-smelting workers gradually realized that slag crusts are the best furnace lining to protect blast furnaces instead of refractory materials [1–4]. The super cooling capacity of cooling staves is the key to the formation of stable slag crusts [5]. Compared to previously used cast iron cooling staves, copper cooling staves are better thermal conductors because of their good thermal conductivity, thermal shock properties, and heat carrying capacity [6–8]. Therefore, copper staves have become a major development model to decrease the cost of iron making and extend the campaign lives [9–11]. Because they can reduce the thermal stress and further prevent material deformation and cracking of materials [12–15]. Copper staves have been used in many newly built and overhauled large-scale blast furnaces including more than 200 blast furnaces in China and over 45 blast furnaces abroad. The spread of copper staves has not only saved many refractory materials, but also greatly promoted the stable operation of blast furnaces. However, some blast furnaces have experienced major failures in popularizing and applying copper staves since 2010, which caused catastrophic consequences [16].

Three types of copper staves are used in blast furnaces: (1) calendering (including rolling or forging) copper plate welding copper staves; (2) welding copper staves from continuous casting slabs; and (3) buried pipe cast copper staves [17]. Currently, the most widely used is the first type. There are mainly three damaged forms of copper staves: (1) the damage of copper staves in the tuyere area caused by the unreasonable design of the furnace structure; (2) the damage caused by the manufacturing defects; and (3) the large area damage of the local hot surface caused by the "severe shortage of cooling strength". What is commonly found in blast furnaces is the first two damaged forms. However, the third damaged form occurred frequently recently in China, shown as follows: (1) Copper staves were burnt out in a large area, whose shape was destroyed. (2) The local part of copper staves was seriously worn, losing its original shape and weakening its function. (3) Cooling water pipes were exposed because severe wear occurred at both ends of copper staves. (4) Copper staves were bent and deformed, which were convex toward the hot surface. To solve these problems, the damage mechanism of the third damaged form is studied in depth in this paper.

2. Materials and Methods

Damaged copper staves were gathered from No. 3 blast furnace of Laiwu Steel, which was 3200 m^3 in inner volume and put into operation on 16 March 2010. From December 2015, the eighth story-cooling pipes began to experience damage. Despite remedial measures, such as installing a point cooler on the damaged part, spraying water and covering the tank outside the shell, multiple cooling pipes in bosh (No. 7 and No. 8 story) and belly (No. 9 story) were damaged one after another in the following half year, especially in the group 35–39.

In order to analyze the damage mechanism of copper staves, five copper columns drilled from different positions of the damaged stave and the adhesive material on the hot surface of stave were collected. The chemical composition was detected according to the Chinese standards of GB/T and NACIS/C, respectively. Using the LFA427 laser thermal conductivity meter, the thermal conductivity was determined according to the method of ASTM E 1461. The metallographic aspects of copper samples, which were coarsely ground, polished, and then pickled with a solution of ferric chloride aqueous hydrochloric acid, were observed under a JX-4R metallurgical microscope (PDV, Beijing, China). Using the Quanta FEG 250 scanning electron microscope (FEI, Hillsboro, OR, USA) and energy spectrum analysis, changes in the microstructure and the chemical composition at each location of the cooling stave after service were studied.

3. Results

3.1. Chemical Composition

Table 1 shows the chemical composition of TU2 (An oxygen-free copper, in line with GB/T 5231-2001) the material of copper staves that we investigated. Tables 2 and 3 list the chemical composition of damaged copper stave at different sampling point and the harmful element composition of the adhesive material, respectively. The chemical composition of copper samples was basically consistent with oxygen-free copper TU2 except Zn of No. 5 sample. Zn was also high in the adherent slag, while K, Na, As, P, and S were at normal level compared to common slag. Although Zn can greatly reduce the thermal conductivity of the copper, its content was not enough to have an impact. Therefore, the chemical composition of the cooling stave was qualified.

Table 1. Chemical composition of TU2 (wt. %).

Si	P	S	Ni	Ag	As	Bi
0.001	0.002	0.004	0.002	–	0.002	0.001
Cu + Ag	O	Sb	Fe	Pb	Sn	Zn
99.95	0.003	0.002	0.004	0.004	0.002	0.003

Table 2. Chemical composition of test samples (wt. %).

Sample	Si	P	S	Ni	Ag	As	Bi
1	<0.0005	<0.0005	0.0013	<0.0005	0.0003	<0.0005	<0.00001
2	<0.0005	<0.0005	0.0014	<0.0005	0.0005	<0.0005	<0.00001
3	<0.0005	<0.0005	0.0013	<0.0005	0.0008	<0.0005	<0.00001
4	<0.0005	<0.0005	0.0014	<0.0005	0.0008	<0.0005	<0.00001
5	<0.0005	<0.0005	0.0012	<0.0005	0.0008	<0.0005	0.00003

Sample	Cu + Ag	O	Sb	Fe	Pb	Sn	Zn
1	99.98	0.0005	<0.0001	0.003	0.0002	<0.0001	<0.0005
2	99.97	0.0001	<0.0001	0.003	0.0002	<0.0001	0.0015
3	99.96	0.0004	<0.0001	0.002	0.0002	<0.0001	<0.0005
4	99.96	0.0003	<0.0001	0.006	0.0003	<0.0001	0.0015
5	99.96	<0.0001	<0.0001	0.002	0.0015	<0.0001	0.013

Table 3. Harmful element composition of the adherent slag (wt. %).

C	Mn	P	S	Ti	As
2.54	0.21	0.12	0.43	0.18	<0.005
Ca	K	Na	Pb	SiO$_2$	Zn
12.80	0.54	0.24	0.0026	18.72	0.34

3.2. Thermal Conductivity

The thermal conductivity of copper samples was tested after service via sampling the hot surface, as shown in Table 4. Under normal circumstances, the thermal conductivity of the copper is 340–385 W/(m·K) at room temperature, which decreases gradually with the increase of temperature. Therefore, overheating should be avoided as much as possible.

Table 4. Thermal conductivity of copper samples (W/(m·K)).

Sample	Room Temperature	100 °C	200 °C
1	380.83	358.08	347.97
2	394.22	372.46	357.48
3	382.02	359.06	349.26
4	379.84	362.41	350.60
5	364.82	354.23	348.46

3.3. Metallographic Aspects

Thermal shock occurred repeatedly on the hot surface during the service of copper staves. As shown in Figure 1, the metallographic structure was observed under optical microscope to study the influence of temperature fluctuations on the hot surface. Figure 2 shows the metallographic structure of the pure copper. Compared to the pure copper, the metallographic structure of the copper stave had little change after service as the grain growth did not happen, thus, the performance of copper staves was stable.

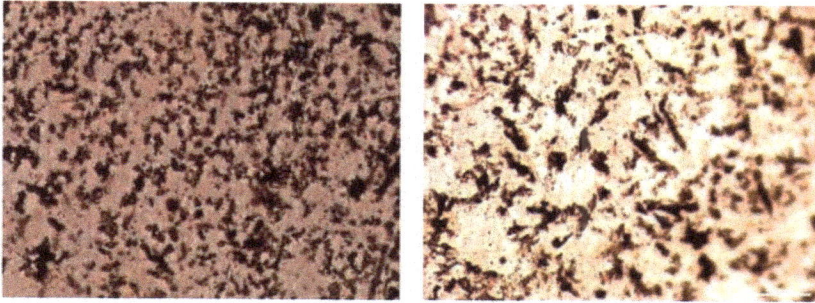

Figure 1. Metallographic structure of copper staves: (**a**) amplification factor, 60; (**b**) amplification factor, 150.

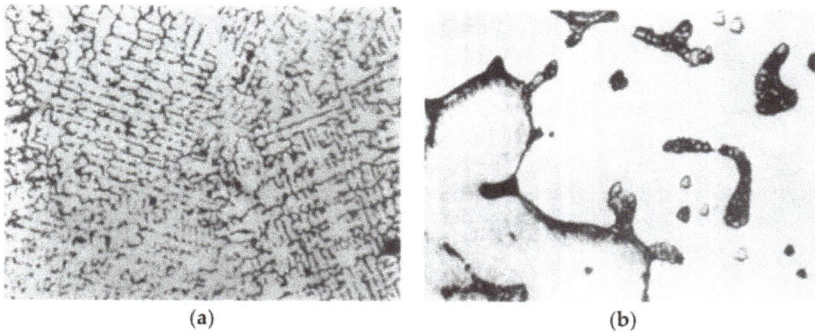

(**a**) (**b**)

Figure 2. Metallographic structure of the pure copper: (**a**) amplification factor, 60; (**b**) amplification factor, 150.

3.4. Microstructure

Figure 3 shows the microstructure of copper staves observed via scanning electron microscopy. Some tiny cracks (about 40–60 μm in length) distributed on the image of copper staves, which had a tendency to develop further as their ends were tapered. Therefore, these cracks would grow to larger cracks if they were stretched or compressed under the force of thermal stress, resulting in rupture and leak of copper staves.

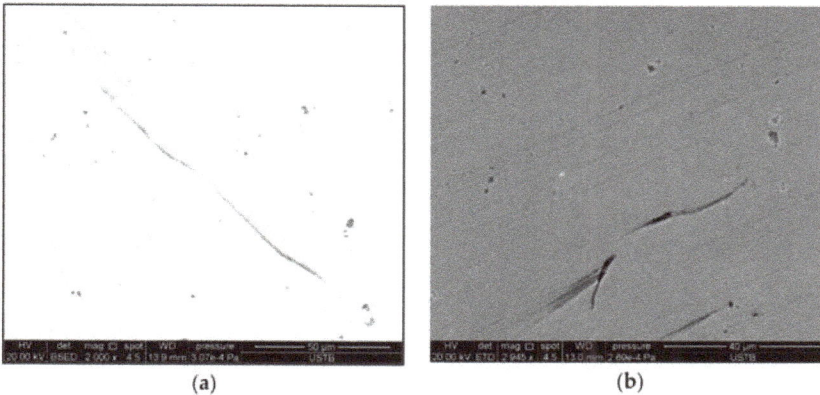

(**a**) (**b**)

Figure 3. Microstructure of copper staves: (**a**) Backscatter imaging; (**b**) Secondary electron imaging.

The energy spectrum analysis was used to determine the chemical composition. As shown in Figure 4, there were Cu, O, C, and Si at the crack, while there was no O at the crack-free location, as shown in Figure 5. Figure 6 shows the energy spectrum analysis of the spot-like inclusion, where Si, C, and Cu were found. As the same as Figure 4, C and Si might come from the polishing paste. Figure 7 shows the energy spectrum analysis at the interface between copper staves and slag crusts. As a result, copper staves were seriously eroded as Cu, O, C, Si, and Fe were found at the interface.

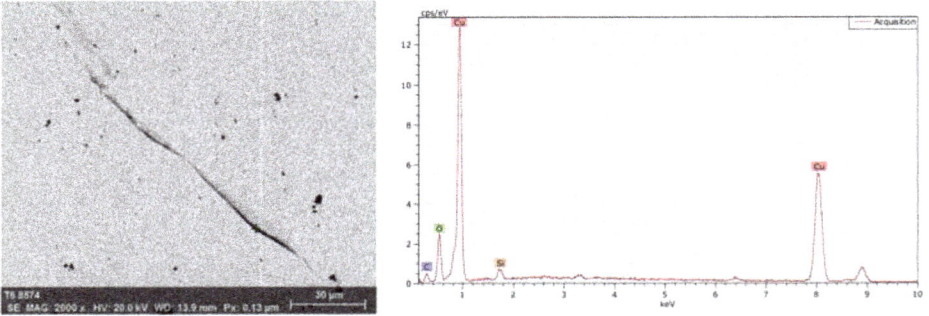

Figure 4. Energy spectrum analysis at the crack.

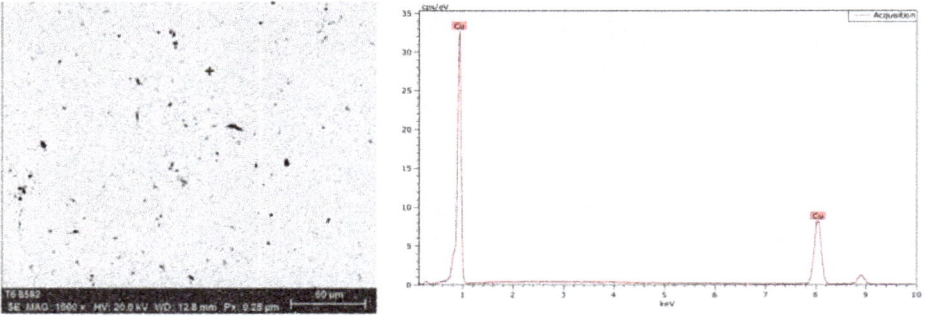

Figure 5. Energy spectrum analysis at the crack-free location.

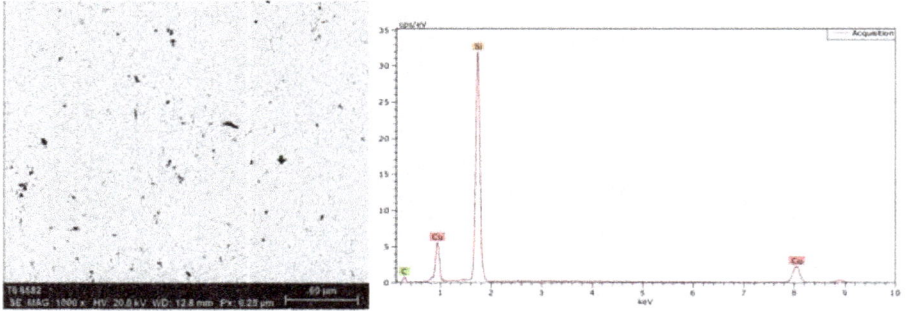

Figure 6. Energy spectrum analysis of the spot-like inclusion.

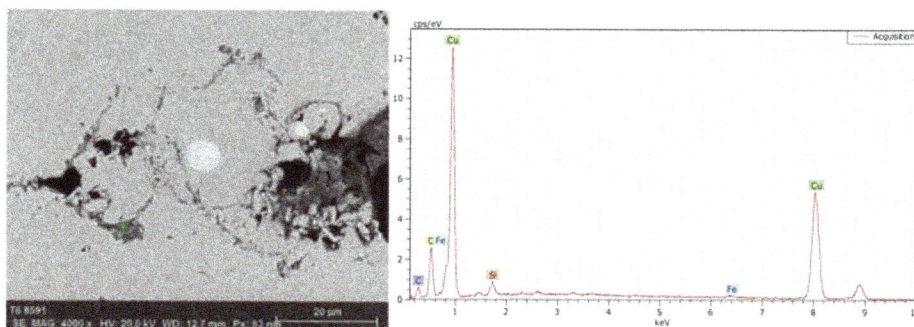

Figure 7. Energy spectrum analysis at the interface.

4. Discussion

4.1. Materials

With extremely low impurity elements, copper staves have the content of O less than 0.0005% and the content of Cu + Ag higher than 99.96%, so the chemical composition of copper staves completely meets industrial requirements. The thermal conductivity is almost above 350 W/(m·K) at different temperatures, which is consistent with the performance of pure copper. Also, the mechanical properties were not deteriorated as no grain growth was observed in the metallographic structure. Therefore, there are no defects in the materials. In addition, the rolled copper drilling cooling stave is used in the blast furnace, so the air gap thermal resistance caused by the poor integration between water pipes and stave body does not exist. Welding problems were also not found in service. Therefore, the wear caused by the materials can be eliminated.

4.2. Working State

The damage mostly occurred in the belly and the waist of the blast furnace, especially in the interface. Copper staves need to withstand the scouring of the edge airflow and the erosion of slag and iron, so that refractory materials cannot work for a long time without the protection of slag crusts, the formation of which is mainly affected by temperature and operational stability. The thickness of slag crusts can be calculated by the following formula [18]:

$$\delta = \lambda(t_2 - t_1)/q, \tag{1}$$

δ, thickness; λ, thermal conductivity; t_2, high temperature; t_1, low temperature; q, heat flow intensity.

As shown in formula (1), a lower temperature of copper staves is beneficial to the formation of slag crusts, but the development of the edge airflow is harmful because it increases the heat flow intensity. When the temperature of copper staves rises, the slag crusts will fall off frequently and the generation cycle will be prolonged, so the time exposed to the scour of the edge airflow becomes longer, resulting in wear of copper staves, as shown in Figure 8.

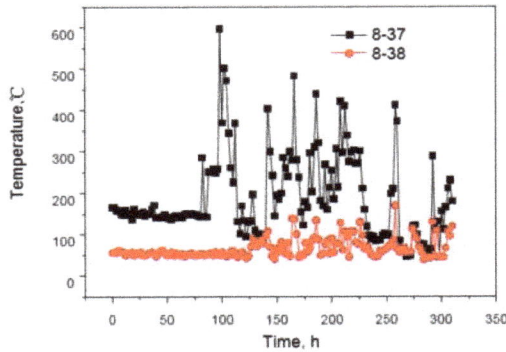

Figure 8. Changes in temperature under normal and abnormal conditions.

Figure 9 shows the changes in temperature after copper staves were damaged. The cooling capacity decreased due to the loss of cooling water after cooling water pipes damaged, therefore, the temperature of copper staves rose and fluctuated rapidly [19]. The average temperature of the 37th and 38th groups exceeded 550 °C and 400 °C in the later period, much higher than the allowable working temperature (180 °C) in the long term. Consequently, slag crusts were difficult to form at this temperature. Also, the hot surface became smooth because of the wear of copper staves, which was also not conductive to the stability of slag crusts.

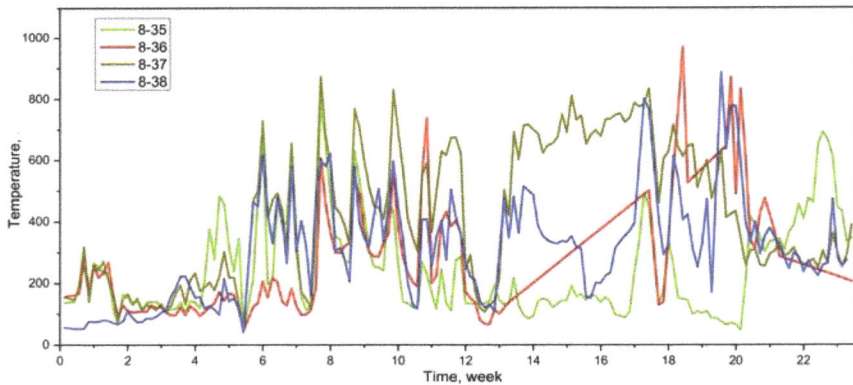

Figure 9. Changes in temperature after copper staves were damaged.

The blast furnace utilization coefficient was 2.5–2.7 t/(m^3·d) for a long time, indicating the high working strength, thus, the amount of gas and the gas flow rate were both large in the furnace. The edge airflow would develop when the raw materials were bad. It is mainly the coke that determined the permeability of the charge. The average ash content and sulfur content of coke for the blast furnace were 12.77% and 0.818%, respectively, which were all at a high level in China. In addition, the indicators of coke quality such as M40, M10, CRI, and CSR were also poor compared to the same size blast furnaces in China. Poor coke quality would degrade the permeability of the burden, resulting in the development of edge airflow. A large amount of coke breeze was generated during the production due to the low thermal intensity, which had a strong scouring effect on the refractory materials and stave body under the condition of high-speed airflow.

With the development of thin-walled blast furnaces, the furnace type has undergone tremendous changes, resulting in the disappearance of the difference between the design furnace type and the working furnace type. Therefore, many of problems occur with the continuation of the previous bosh angle for the thin-walled blast furnaces, especially for use of copper staves. One of the problems is that the bosh angle is too large as the abrasion of the refractory bricks increases with the increase of the bosh angle [20]. The bosh angle is greater than 78 deg for the blast furnace in Laiwu Steel, however, it is 72–74 deg for the advanced blast furnaces in the world, which have long campaign lives and low fuel ratios. Therefore, the large bosh angle creates conditions for the development of the edge airflow. Figure 10 shows the radial distribution of the airflow velocity under different bosh angles simulated by Fluent 14.5 (ANSYS, Pittsburgh, PA, USA).

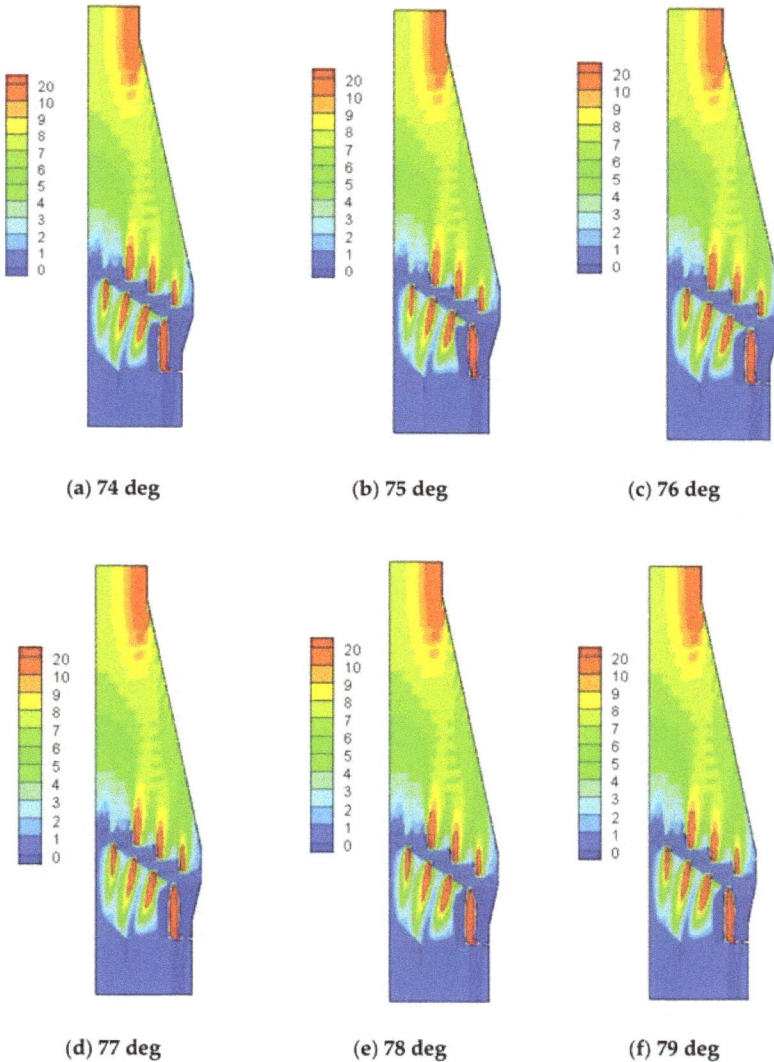

(a) 74 deg	(b) 75 deg	(c) 76 deg
(d) 77 deg	(e) 78 deg	(f) 79 deg

Figure 10. Radial distribution of the airflow velocity (m/s) under different bosh angles.

As shown in Figure 10, the velocity of the edge airflow increased as the bosh angle increased, especially in the boundary between the belly and the waist, where the light-colored area increased, and the color became lighter, indicating the increase of the velocity of the edge airflow. Figure 11 shows the velocity distribution of the edge airflow under different bosh angles. The velocity of the edge airflow increased exponentially as the bosh angle increased from 74 deg to 79 deg, which intensified the wear of copper cooling stave.

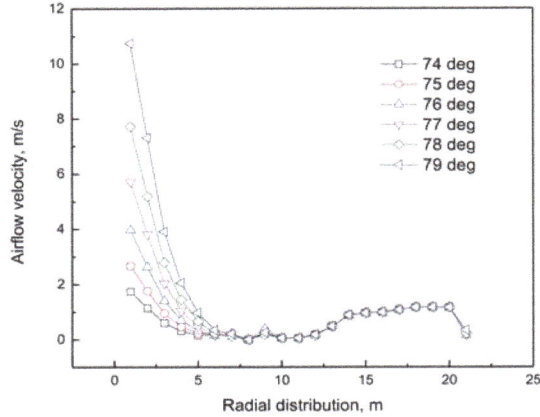

Figure 11. Radial distribution of edge airflow velocity under different bosh angles.

4.3. Damage Process

The damage of copper staves has gone through the following stages based on above analysis:

The first stage was the erosion and wear of refractory bricks. A certain thickness of refractory bricks was embedded in the dovetail groove of the hot surface. When the refractory bricks were intact, copper staves were well protected, and the temperature and thermal stress were very low at this time, therefore, the refractory bricks resisted the initial erosion and wear. Figure 12 shows the temperature distribution of refractory bricks under different thicknesses simulated by ANSYS 14.5 (ANSYS, Pittsburgh, PA, USA).

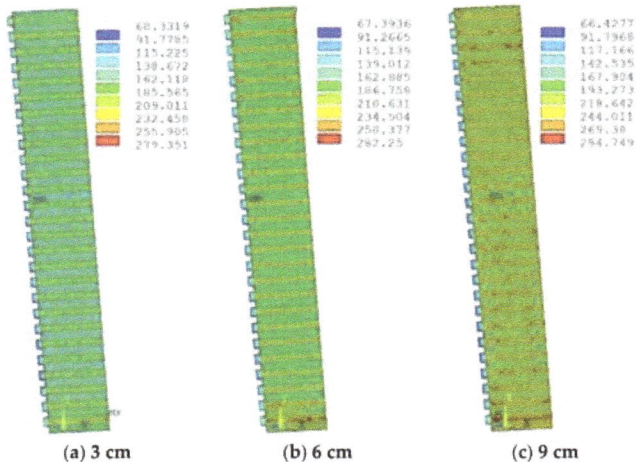

(a) 3 cm   (b) 6 cm   (c) 9 cm

Figure 12. *Cont.*

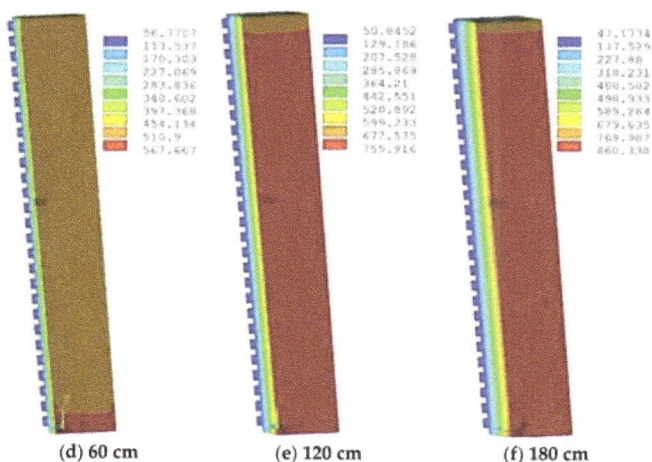

(d) 60 cm   (e) 120 cm   (f) 180 cm

Figure 12. Temperature (°C) distribution of refractory bricks under different thicknesses.

The increase in the thickness led to an increase in temperature, because it increased the total thermal resistance, leading to the decrease in heat transfer capacity. The highest temperature appeared in the lowermost corner position, which was prone to damage. The slag crust was difficult to form when the temperature exceeded 800 °C, resulting in the ablation and thinning of refractory bricks. Figure 13 shows the stress distribution of refractory bricks under different thicknesses. The maximum stress underwent a process of decreasing first and then increasing as the thickness of refractory bricks increased. Under the action of high thermal stress and high temperature, the refractory bricks were continuously eroded by the liquid slag and iron, and scoured by the high-temperature airflow carrying a large number of coke breeze without the protection of slag crusts [21].

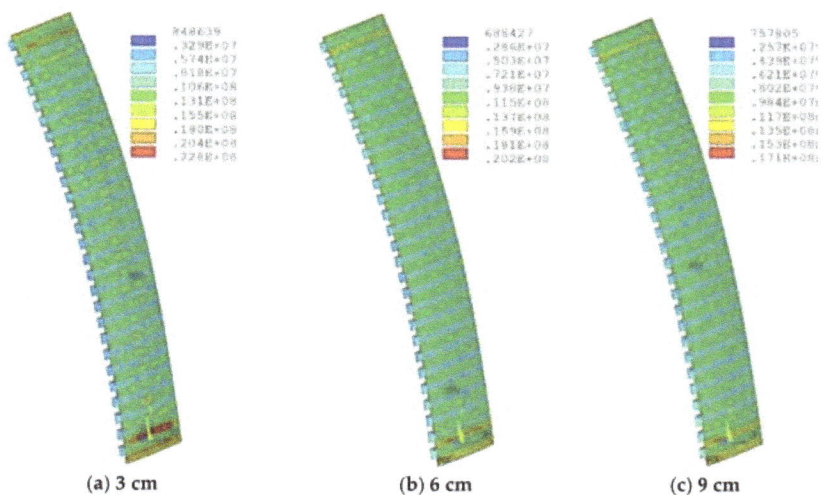

(a) 3 cm   (b) 6 cm   (c) 9 cm

Figure 13. *Cont.*

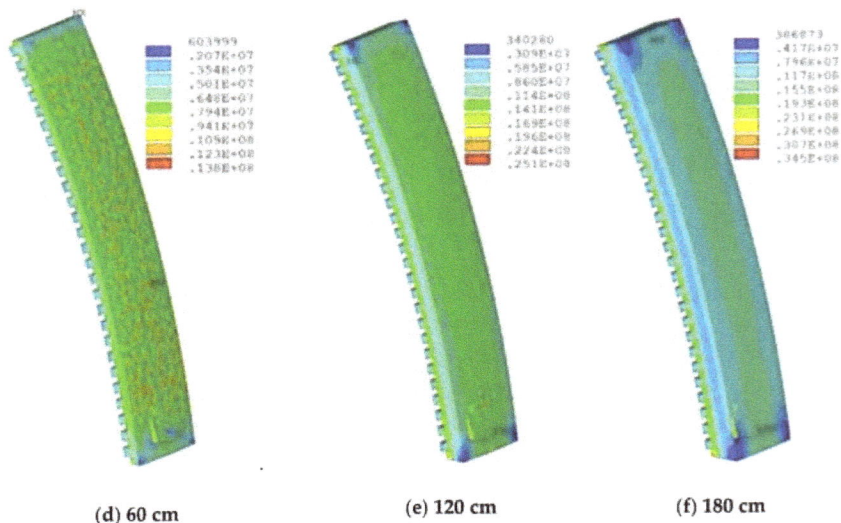

(d) 60 cm **(e) 120 cm** **(f) 180 cm**

Figure 13. Stress (Pa) distribution of refractory bricks under different thicknesses.

The second stage was the damage of copper staves after the refractory bricks disappeared. As a result, copper staves were directly exposed to high-speed airflow, but the maximum temperature would not exceed 300 °C as long as the cooling capacity was sufficient. Figure 14 shows the thermocouple temperature under different working conditions. The stave body would not be ablated even if there was no protection of refractory bricks. The major damage was due to the scouring of the high-temperature airflow carrying the coke breeze. As the wear of copper staves intensified, it was forced to break the water and block the water pipe when a part was worn to the cooling water pipe. The reduction in cooling capacity and the increase in temperature of copper staves were inevitable results regardless of the remedy.

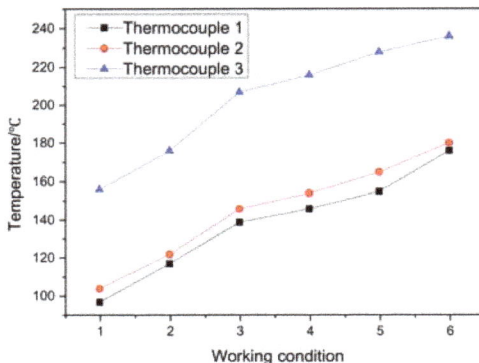

Figure 14. Thermocouple temperature under different working conditions (Working conditions 1–3: flue gas temperature, 1000 °C, 1100 °C, and 1200 °C, respectively; cooling water flow rate, 16 t/h. Working conditions 4–6: flue gas temperature, 1200 °C; cooling water flow, 12 t/h, 8 t/h, and 4 t/h, respectively. Thermocouples 1–3: thermocouple temperature of 10 mm from the hot surface at the upper, middle and lower parts of copper staves.).

The third stage was the hydrogen attack after the temperature rose because of the damage of cooling water pipes [16]. The contents of O and C were high at the crack and interface. Generally, the solid solution of oxygen in the copper is very low, but there is a strong affinity between the oxygen and copper; therefore, excess oxygen and copper will form Cu_2O and distribute on the grain boundaries or in the dendrite network, and the tendency increases as the temperature rises. The harm of hydrogen on oxygen-containing copper is related to the temperature. Hydrogen does not harm the oxygen-containing copper at 150 °C because the water vapor is in the condensed state. Under a hydrogen atmosphere, the oxygen-containing copper will not crack within 10 years at 150 °C, but it will crack after no more than 1.5 years at 200 °C or 70 hours at 400 °C [22]. Ordinary copper (oxygen content, 0.02–0.1%) is prone to have a hydrogen attack in a reducing medium with a temperature greater than 370 °C. Unlike ordinary copper, oxygen-free copper (oxygen content, ≤0.002%) can be processed and used under reducing atmosphere conditions, but it is also likely to have a hydrogen attack. The occurrence of hydrogen attack must meet the following conditions at the same time: the first is the reducing atmosphere, which is available in the blast furnace, the second is the oxygen content, and the third is the temperature of copper staves. The temperature of the thermocouple installed at the damaged copper cooling stave exceeded 400 °C for a long time according to the monitoring results, which created conditions for cracking and damage caused by hydrogen attack.

The final stage was the wear of copper staves under high temperatures. First, the hot surface of copper staves damaged severely due to hydrogen attack; meanwhile, the poor raw materials and large bosh angle created conditions for the development of edge airflow. Therefore, the high-speed and high-temperature airflow carrying a large amount of coke breeze intensified the wear of copper staves. Thus, the surface of copper staves was worn smooth, and the temperature rose rapidly due to the damage of cooling pipes. Consequently, the stable slag crusts were difficult to form, leading to the loss of protection to copper staves. Finally, the temperature of the hot surface was continuously increased, resulting in a decrease in the hardness and wear resistance of the copper, thus, copper staves were completely damaged under high-speed wear.

5. Conclusions

The damage process of copper staves is as follows:

(1) The poor raw materials, especially poor coke quality, and the large bosh angle caused the development of the edge airflow to maintain the high smelting strength. The high-temperature and high-speed airflow carrying the coke breeze generated by the deterioration had a strong scouring effect on the refractory bricks, which were continuously eroded by the slag, iron, and airflow under thermal stress and high temperature.

(2) Slag crusts were difficult to be stable due to the increase of temperature, causing damage to the water pipes. The cooling strength inevitably declined after remedy, resulting in an increase in the temperature of copper staves and a decrease in the capacity to form slag crusts.

(3) The hydrogen attack happened under the reducing atmosphere when the temperature was greater than 370 °C, leading to the development of cracks at the hot surface. As a result, the performance of copper staves such as the hardness at the high temperature and the wear resistance decreased. Copper staves were quickly worn to form a smooth surface under continuous scouring of the edge airflow, thus, it became more difficult to form slag crusts. Under the hydrogen attack and the scouring of the edge airflow at the high temperature, copper staves finally failed.

Author Contributions: Conceptualization, H.Z.; Data curation, H.Z.; Resources, X.W.; Writing—original draft, Y.W.; Writing—review and editing, Y.W.

Funding: This research was funded by National Key Research and Development Program (No. 2016YFB0601304).

References

1. Frolova, I.B.; Minkin, V.M.; Frolov, V.A.; Yukhimenko, V.I. The slag crust in high-temperature iron ore smelting. *Refract. Ind. Ceram.* **1976**, *17*, 429–432. [CrossRef]
2. Zamoskovtsev, D.E.; Lednov, A.Y.; Kishchuk, V.D.; Chaplouskii, A.A. Ultrasonic diagnosis in control of the slag crust. *Metallurgist* **1997**, *41*, 436. [CrossRef]
3. An, J.Q.; Zhang, J.L.; Wu, M.; She, J.H.; Terano, T. Soft-sensing method for slag-crust state of blast furnace based on two-dimensional decision fusion. *Neurocomputing* **2018**, *315*, 405–411. [CrossRef]
4. Jiao, K.X.; Zhang, J.L.; Liu, Z.J.; Liu, F.; Liang, L.S. Formation mechanism of the graphite-rich protective layer in blast furnace hearths. *Int. J. Miner. Metall. Mater.* **2016**, *23*, 16–24. [CrossRef]
5. Zhang, H.; Jiao, K.X.; Zhang, J.L.; Chen, Y.B. A new method for evaluating cooling capacity of blast furnace cooling stave. *Ironmak. Steelmak.* 2018. [CrossRef]
6. Mohanty, T.R.; Sahoo, S.K.; Moharana, M.K. Computational modeling of blast furnace save cooler based on steady state heat transfer analysis. *Procedia Eng.* **2015**, *127*, 940–946. [CrossRef]
7. Xu, X.; Wu, L.J.; Lu, Z.A. Performance optimization criterion of blast furnace stave. *Int. J. Heat Mass Transf.* **2017**, *105*, 102–108. [CrossRef]
8. Wu, L.J.; Zhou, W.G.; Su, Y.L.; Li, X.J. Experimental and operational thermal studies on blast furnace cast steel staves. *Ironmak. Steelmak.* **2008**, *35*, 179–182. [CrossRef]
9. Yeh, C.P.; Ho, C.K.; Yang, R.J. Conjugate heat transfer analysis of copper staves and sensor bars in a blast furnace for various refractory lining thickness. *Int. Commun. Heat Mass Transf.* **2012**, *39*, 58–65. [CrossRef]
10. Balamurugan, S.; Shunmugasundaram, R.; Patra, M.; Pani, S.; Dutta, M. Evaluation of copper stave remnant thickness in blast furnace using ultrasonic method. *ISIJ Int.* **2015**, *55*, 605–610. [CrossRef]
11. Liu, Z.J.; Zhang, J.L.; Zuo, H.B.; Yang, T.J. Recent progress on long service life design of Chinese blast furnace hearth. *ISIJ Int.* **2012**, *52*, 1713–1723. [CrossRef]
12. Chen, W.C.; Cheng, W.T. Numerical simulation on forced convective heat transfer of titanium dioxide/water nanofluid in the cooling stave of blast furnace. *Int. Commun. Heat Mass Transf.* **2016**, *71*, 208–215. [CrossRef]
13. Liu, Q.; Cheng, S.S. Heat transfer and thermal deformation analyses of a copper stave used in the belly and lower shaft area of a blast furnace. *Inter. J. Therm. Sci.* **2016**, *100*, 202–212. [CrossRef]
14. Wu, T.; Cheng, S.S. Model of forming-accretion on blast furnace copper stave and industrial application. *J. Iron Steel Res. Int.* **2012**, *19*, 1–5. [CrossRef]
15. Xie, N.Q.; Cheng, S.S. Analysis of effect of gas temperature on cooling stave of blast furnace. *J. Iron Steel Res. Int.* **2010**, *17*, 1–6. [CrossRef]
16. Deng, Y.; Jiao, K.X.; Wu, Q.C.; Zhang, J.L.; Yang, T.J. Damage mechanism of copper stave used in blast furnace. *Ironmak. Steelmak.* 2017. [CrossRef]
17. Wang, X.L. *Iron and Steel Metallurgy (Ironmaking)*, 3rd ed.; Metallurgical Industry Press: Beijing, China, 2013; p. 391. ISBN 978-7-5024-6130-0.
18. Hua, J.S.; Zhu, J.; Li, X.M.; Ma, Y.P. *Principles of Transfer in Metallurgy*, 1st ed.; Northwestern Polytechnical University Press: Xi'an, China, 2005; p. 88. ISBN 7-5612-1904-0.
19. Li, Y.L.; Cheng, S.S. Cooling capacity recovery of copper stave based on heat transfer. *J. Iron Steel Res.* **2012**, *24*, 5. [CrossRef]
20. Zhang, J.L.; Chen, Y.X.; Fan, Z.Y.; Hu, Z.W.; Yang, T.J.; Tatsuro, A. Influence of profile of blast furnace on motion and stress of burden by 3D-DEM. *J. Iron Steel Res. Int.* **2011**, *18*, 1–6. [CrossRef]
21. Li, F.G.; Zhang, J.L. Stress distribution law and adherent dross stability of the copper cooling stave with variable slag coating thickness. *Chin. J. Eng.* **2017**, *39*, 389–398. [CrossRef]
22. Zhong, W.J. *Practical Manual for Processing Technology of Copper*, 1st ed.; Metallurgical Industry Press: Beijing, China, 2007; p. 78. ISBN 978-7-5024-4100-5.

![metals logo] *metals*

MDPI

Article

Effects of Fe₂O₃ on Reduction Process of Cr-Containing Solid Waste Self-Reduction Briquette and Relevant Mechanism

Tuo Wu, Yanling Zhang *, Zheng Zhao and Fang Yuan

State Key Laboratory of Advanced Metallurgy, University of Science and Technology Beijing, Beijing 100083, China; wutuo90@163.com (T.W.); ext_zheng@163.com (Z.Z.); 408995357@163.com (F.Y.)
* Correspondence: zhangyanling@metall.ustb.edu.cn; Tel.: +86-139-1189-1432

Received: 21 November 2018; Accepted: 29 December 2018; Published: 7 January 2019

Abstract: High-temperature quench method, scanning electron microscope-energy dispersive spectroscopy (SEM-EDS), and thermodynamic analysis were adopted to study the effects of Fe₂O₃ on reduction process of Cr-containing solid waste self-reduction briquette (Cr-RB). Moreover, the relevant mechanism was also studied. The results clearly showed that the addition of Fe₂O₃ decreased the chromium-iron ratio (Cr/(Fe + Cr)) of Cr-RB itself and promoted the reduction of chrome oxide in the Cr-containing solid wastes such as stainless steel slag and dust. A large number of Fe-C alloy droplets generated in the lower temperature could decrease the activity of reduced chromium by in situ dissolution and the reduction of Cr-oxide was accelerated. Rapid separation of metal and slag could be achieved at a relatively lower temperature, which was very beneficial to the efficient recovery of Cr. Finally, the corresponding mechanism diagram was presented.

Keywords: Cr recovery; self-reduction briquette; reaction mechanism

---

1. Introduction

In the entire process of stainless steel smelting, chromium is inevitably oxidized into stainless steel slag (SSS) [1,2], especially for the EAF (electric-arc furnace) smelting process, stainless steel dust (SSD) [3,4], and mill scales [5] to form Cr-containing solid wastes. These solid wastes are valuable secondary resources, especially containing a considerable amount of chromium metal. Fully recycling chromium in these solid wastes will not only improve enterprises' economic efficiency, but also reduce the risk of heavy metal pollution caused by Cr in solid wastes after being leached in the natural acidic or alkaline environment [6]. In all of Cr-containing solid wastes treatment technologies, high-temperature reduction technology using carbonaceous reductants has been paid extensive attention due to its unique characteristics compatible with metallurgical enterprises [5,7–10]. Due to its excellent reactivity, self-reduction briquette bearing C was mostly adopted [11–14]. The composition design of self-reduction briquette is critical to efficient recovery of Cr as the high temperature reduction technology is applied. Studies have shown that FeO$_x$ (x = 0, 1, 4/3) can significantly affect the carbothermal reduction process of chromium oxides which mainly was the bulk phase in the Cr-containing solid wastes. Görnerup et al. [15,16] studied the reduction behavior of FeO and Cr₂O₃ in stainless steel slag and found that the reduction of Cr₂O₃ by C had a pronounced incubation period and the reduction of FeO did not have any incubation period. The reduction of Cr₂O₃ did not start until some liquid Fe-C alloys had formed and the product was Fe-Cr-C alloy. Abundant FeO in the slag was beneficial to rapid reduction of Cr₂O₃. Chakraborty et al. [17] investigated the carbothermal reduction of chrome ore ((Fe, Mg)O·(Fe, Cr, Al)₂O₃) and found that iron had been fully reduced before the reduction of Cr. Hu et al. [18–20] studied the alloying effect of special precursor made up of chromite ore, mill scale, and carbon, and pointed out that the Cr/(Cr + Fe) of the alloy precursor had important influence on the yield of Cr. Therefore, the introduction of FeO_x into Cr-containing solid waste self-reduction briquette (Cr-RB, for short) is very necessary. However, the intrinsic behaviour of reduction and melting-separation processes of Cr-RB with the addition of FeO_x has not been fully investigated. Accordingly, high temperature reduction-rapid quench, SEM-EDS, and thermodynamic analysis methods were adopted to clarify this problem for the Cr-RB with the addition of Fe_2O_3 or not (A: Fe_2O_3 + SSS + SSD + C, B: SSS + SSD + C). A detailed analysis of reduction results at different temperatures and times would contribute to understand the relevant mechanism of reduction process of Cr-containing solid waste self-reduction briquette.

2. Materials and Methods

2.1. Materials

Figure 1 shows the three experimental materials, which are SSD (AOD dust), SSS (EAF slag), and Fe_2O_3 respectively. The SSS was coarsely crushed by a jaw crusher and finely ground by an ore crusher. Then, metallic iron particles contained in it were removed by a magnet. The powdery slag and the SSD were all passed through a 60-mesh steel screen in order to eliminate the effect caused by the uneven particle size. Reductant used in the experiment was high-purity graphite powder (C $\geq$99.85%) and the flux was analytical reagent SiO_2. All materials were dried at 200 °C for 24 h before experiment. The composition of three experimental materials was shown in Table 1.

Figure 1. Experimental material (**a**) stainless steel dust (SSD) (**b**) stainless steel slag (SSS) (**c**) Fe_2O_3.

Table 1. Composition of experimental materials wt.%.

Mater.	TFe	Cr	CaO	SiO_2	MgO	Al_2O_3	NiO	MnO	MoO_3	V_2O_5	F	Others
SSD	20.18	35.80	20.67	3.10	3.49	1.70	0.11		0.54	0.10		Bal.
SSS	3.20	9.12	41.01	28.22	9.43	2.14		1.50		0.28	0.64	Bal.
Fe_2O_3	$\geq$99%											Bal.

Figure 2 shows the XRD spectrum of SSD and SSS. It can be seen that the SSD mainly contains CaO, $FeCr_2O_4$, Fe_3O_4, $MgFe_2O_4$, and small amounts of $Ca_3Mg(SiO_4)_2$ and metal Fe. Among them, the valuable elements Fe and Cr mainly existed in the spinel phase. Due to the isomorphism phenomenon of spinel phase, Fe and Cr in the dust actually were in a solid solution similar to (Fe, Mg)O·(Fe, Cr)$_2$O$_3$. The SSS mainly contained $MgCr_2O_4$, $Ca_3Mg(SiO_4)_2$ and Fe, in which Cr was mainly in the Magnesiochromite ($MgCr_2O_4$) with a high melting point (2390 °C). In summary, most of the valuable metals Fe and Cr in the Cr-containing solid wastes were in their corresponding spinel solid solution phase.

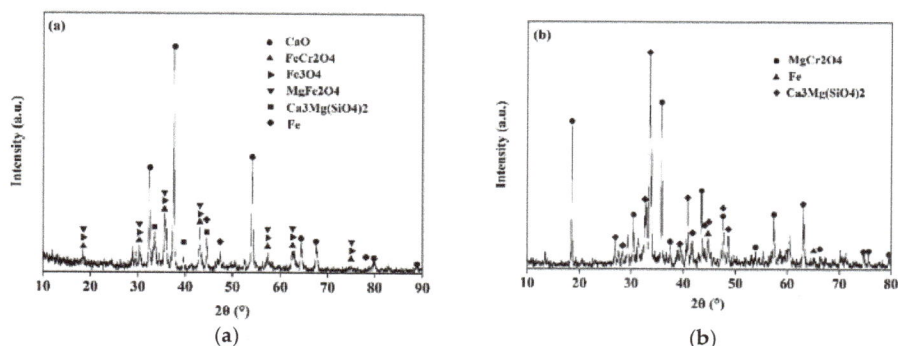

Figure 2. XRD spectrum of SSD (**a**) and SSS (**b**).

2.2. Apparatus and Procedure

According to the proportion scheme shown in Table 2, the experimental materials were accurately weighed and thoroughly mixed in a mixer. Then, for A and B groups, nine parts of mixture with a quantity of about 10 g were weighed. The powder mixture was made into a self-reduction briquette by a constant pressure of 30 MPa for 2 min in a steel mold (Inner diameter: 20 mm), and then put in a high-purity MgO crucible (Outside diameter: 28 mm, Inner diameter: 21 mm, Height: 68 mm). A vertical tube furnace (Version: BLMT-1700 °C) heated by six U-shaped MoSi$_2$ heating elements was used for the reduction experiment. The schematic of the experimental set-up was shown in Figure 3. Its constant temperature zone with fluctuation of ±2 °C was 6 cm, and the temperature control accuracy was ±1 °C.

Table 2. Experiment scheme.

Proportion Scheme	Temperature	Holding Time/Min			Cr/(Cr + Fe)
A	L: 1350	5	20	40	
(30%Dust+20%SSS+50%Fe$_2$O$_3$): 100 g Graphite: n_C:n_O = 1.2	M: 1450	5	20	40	0.23
SiO$_2$: R = 1.2	H: 1550	5	20	40	
B	L: 1350	5	20	40	
(60%Dust+40%SSS): 100 g Graphite: n_C:n_O = 1.2	M: 1450	5	20	40	0.65
SiO$_2$: R = 1.2	H: 1550	5	20	40	

n_C is the number of C moles for graphite, and n_O is the number of O moles in FeO$_x$ and Cr$_2$O$_3$ from slag, dust and Fe$_2$O$_3$. R is the basicity of briquette, (wt.% CaO)/(wt.% SiO$_2$).

First, high-purity Ar (1 L/min, ≥99.999%) was introduced into the furnace tube throughout the entire experiment before the temperature control program was run. When the temperature was raised to the target (1350 °C, 1450 °C, 1550 °C) and stabilized, the crucible containing the sample was preheated in the upper low temperature zone for 1.5 min and then placed in the constant temperature zone for 5, 20, 40 min as shown in Table 2. Once the holding time was finished, the hot sample was rapidly taken out from furnace tube and quenched by the high-purity Ar. The quench time was about 2 min. When the crucible cooled down to room temperature, its weight loss ratio was calculated. Finally, the crucible containing the reaction product was integrally inlaid by resin and longitudinally cut. One half of it was ground, polished, and sprayed with gold to meet demand of SEM. For convenience, each sample was numbered. For example, AM20 represented a sample of A which was heated at 1450 °C for 20 min.

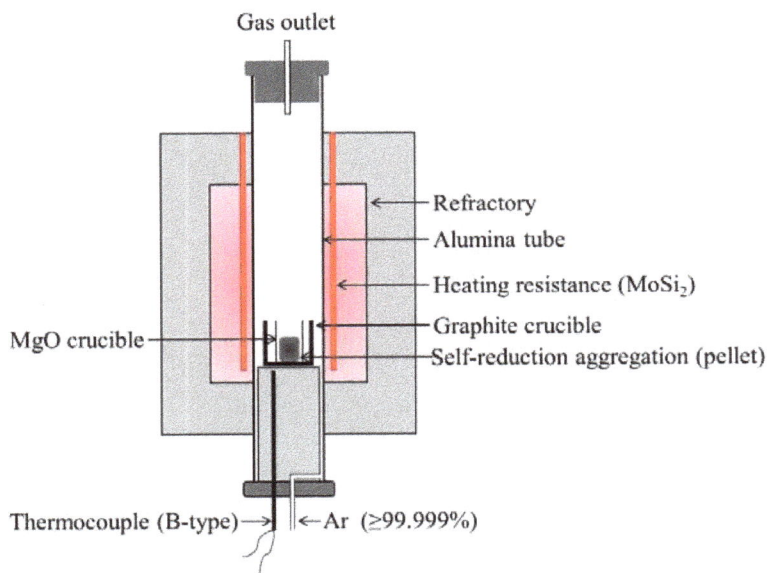

Figure 3. Schematic of tube furnace.

3. Results and Discussion

3.1. Reduction Yield

Based on the mass variation of reaction system before and after high temperature reduction, the corresponding weight loss could be obtained, and the reduction yield (φ) could be calculated according to Equation (1), where $M_{(weight\ loss)}$ is actual mass loss and $M_{(theory\ loss)}$ is theoretical mass loss calculated as the maximum mass loss due to assumed emission of CO produced after the complete reduction of Cr_2O_3 and FeO_x in the self-reduction briquette. Since reduction experiments were carried out at relatively high temperatures ($\geq 1350\ °C$), in which CO was more stable than CO_2 and high-purity argon would take the gas product continuously, and an excessive amount of C could react with CO_2 produced by indirect reduction (Equation (2)) to form CO based on Boudouard reaction (Equation (3)), it was assumed that the gas product is mainly CO (Equation (4)) when the Equations (2) and (3) were combined.

$$\varphi = \frac{M_{(weight\ loss)}}{M_{(theory\ loss)}} \times 100\% \tag{1}$$

$$Me_xO_y(s) + CO(g) = Me(s) + y/2CO_2 \text{ (Me: Fe and Cr)} \tag{2}$$

$$y/2CO_2(g) + yC(s) = 2yCO(g) \tag{3}$$

$$yC(s) + Me_xO_y(s) = xMe(s) + yCO(g) \text{ (Me: Fe and Cr)} \tag{4}$$

Figure 4 shows the time and temperature dependences of reduction yield of A and B-type self-reduction briquettes. As shown that the reduction yield of them increases rapidly before 5 min and then its increasing rate slows down with the extension of time. In addition, higher temperature results with higher reduction yield for the A-type Cr-RB after 5 min, which reflects the superiority of carbothermal reduction reaction at higher temperature. However, the reduction yield of the B-type Cr-RB does not show the same trend after 5 min, which may be due to the samples' easier reoxidation caused by quench operation. Compared with the low-Cr metal product ($Cr/(Cr + Fe) = 0.23$) in the A-type self-reduction briquette, high-Cr metal product ($Cr/(Cr + Fe) = 0.65$) in the B-type self-reduction briquette has higher Cr activity coefficient especially at higher temperature, which leads to the increase of ΔG of Reaction (4) and the decrease of reduction yield. In general, the reduction yield of the A-type Cr-RB is higher than that of the B-type self-reduction briquette, which reflects that Fe_2O_3 can significantly improve the reducibility of B-type Cr-RB.

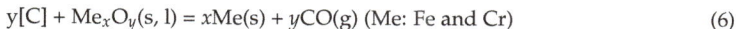

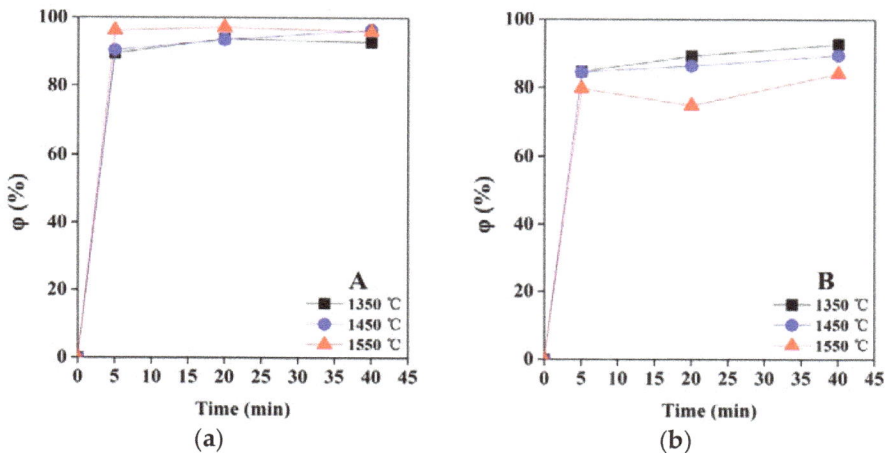

$$(Fe, Cr)(s) + C(s) = [Fe, Cr, C](l) \tag{5}$$

$$y[C] + Me_xO_y(s, l) = xMe(s) + yCO(g) \text{ (Me: Fe and Cr)} \tag{6}$$

Figure 4. Time and temperature dependences of reduction yield of A (**a**) and B-type (**b**) self-reduction briquettes.

Obviously, the self-reduction reaction of briquettes was basically finished within 5 min according to Reaction (4). The carburization reaction (Reaction (5)) also occurred subsequently. After that, the dissolved [C] would react with the Fe or Cr oxides by Reaction (6). However, due to the decrease in activities of C and oxides, the reduction yield increased slowly after 5 min.

3.2. Morphology of Products

Figure 5 shows the longitudinal section overview of the reduction products at different temperatures and times. It can be seen that as the temperature keeps constant, the metal products in every samples gradually contact each other and grow up, and the metal-slag interface are clearer which means the separation of them are more complete as the reaction time prolongs. The same phenomenon also occurred for the increasing temperature when reaction time remained constant. Compared with the sample BL5, sample AL5 has better reactivity, and many small metal particles are clearly visible in the reduction products after holding at 1350 °C for 5 min. Although BL5 has a violent reaction seen from many holes caused by gas product, no clear metal particles are formed, and slag and metal mixes together. As seen in AM20 sample, a good separation between slag and metal is achieved after holding at 1450 °C for 20 min. But for the corresponding BM20 sample, a holding of 20 min at 1550 °C must be needed. It is observed that the addition of Fe_2O_3 was very effective for rapid slag-metal separation from the Cr-RB at a relatively low temperature.

Figure 5. Longitudinal section overview of the reduction product.

3.3. Variation of Product Morphology Observed by SEM

Figures 6 and 7 are the morphologies of the reaction products of Cr-RB under different conditions observed by SEM. It can be further seen that the increase of reaction temperature and the prolongation of reaction time facilitate the effective separation of the metal product from the slag. Higher temperature increases carbothermal reduction rate of Fe or Cr oxides [21,22]. In addition, the increase in temperature enhances the fluidity of metal and slag, and accelerates the separation between them, which make metal aggregate, grow, and deposit more easily. The prolongation of reaction time and the increase of temperature are in favor of metal recovery from Cr-containing solid wastes, especially for the Cr-RB with some Fe_2O_3. Moreover, the metal produced by A is mostly spherical before the metal particles completely aggregates, and its contact surface with the slag is small to separate; the metals in B are mostly sharp-edged blocks or strips, and they are mutually overlapped with slag and is not easy to separate. The above results indicate that the rapid reduction of chromium oxides in Cr-containing solid waste can be achieved while accompanying the reduction of iron oxides.

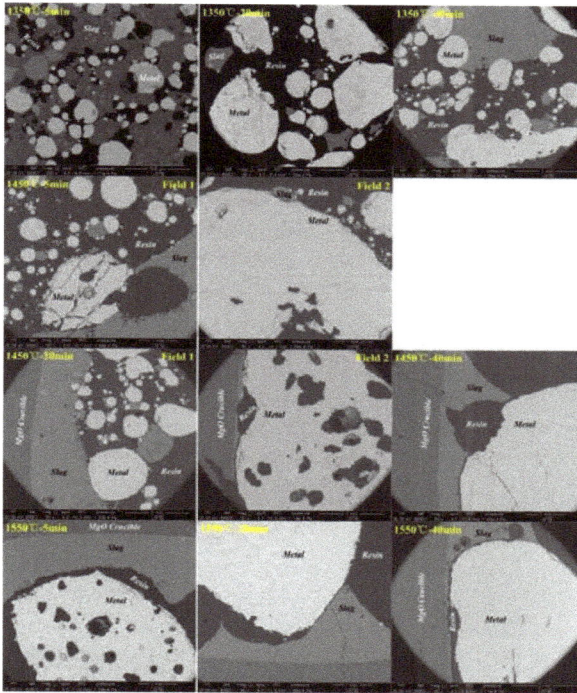

Figure 6. Morphology of the reduction product of A-type Cr-containing solid waste self-reduction briquette (Cr-RB) under different conditions observed by SEM.

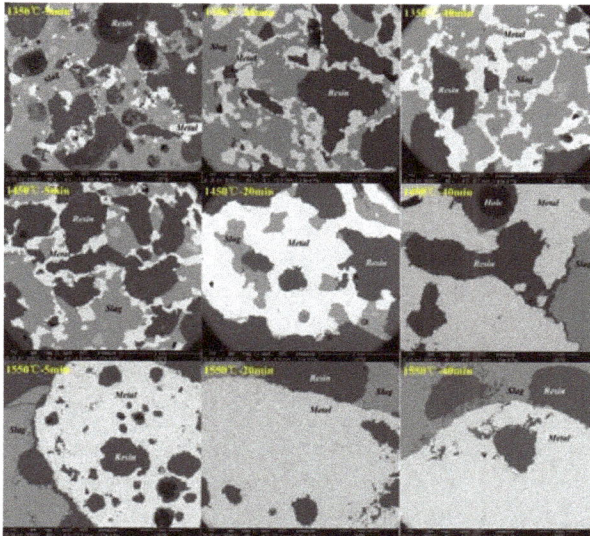

Figure 7. Morphology of the reduction product of B-type Cr-RB under different conditions observed by SEM.

Figures 8 and 9 show the morphology, element distribution, and EDS spectra of typical phase in AL5 and BL5 samples. It can be found that a large number of Fe-Cr-C metal particles were formed in AL5 (Figure 8). When the Cr concentration is low, it is mostly spherical (point 1). When the Cr concentration is high, the shape is irregular (point 9). There are also (Fe, Mg)O·(Cr, Al)$_2$O$_3$ spinels (point 13), which are from the raw material, not yet sufficiently reduced, and a slag phase (point 11) where the Cr concentration reaches 3.15%. Similar to AL5, a large number of irregular Cr-Fe-C alloys with high Cr concentration (point 1), slag (point 11), and insufficient reduction of (Fe, Mg)O·(Cr, Al)$_2$O$_3$ spinel (point 13) were generated in BL5 (Figure 9). The element distribution results showed that in the reduction process, a large number of spherical Fe-based metal droplets were generated in AL5, and a certain concentration of Cr and C dissolved in them; a large amount of Cr-based metals were produced in BL5, and a certain concentration of Fe and C dissolved in them too. This further reflects that the relative proportion of Fe and Cr in the Cr-RB would affect the speed of the entire reduction progress and the recovery of Cr element. Studies [23] have pointed out that chromium ore with higher iron oxide content can achieve higher reduction at relatively low temperatures. In the study of the carbothermal reduction process of the stainless steel slag in the electric furnace, Görnerup et al. [15,16] obtained that the increase of the liquid metal containing C in the slag will promote the recovery of Cr and suggested to appropriately increase the FeO content in the slag. Hu et al. [20] pointed out that Cr/(Fe + Cr) was a critical parameter affecting the recovery of Cr in alloying agents when studying the direct alloying of steel by chromium ore, and thought that the alloying precursor with low Cr/(Fe + Cr) were more likely to obtain high Cr recovery. In this study, the Cr/(Fe + Cr) of A and B-Type self-reduction aggregation are 0.23 and 0.65, respectively. So the former is more efficient for the recovery of Cr.

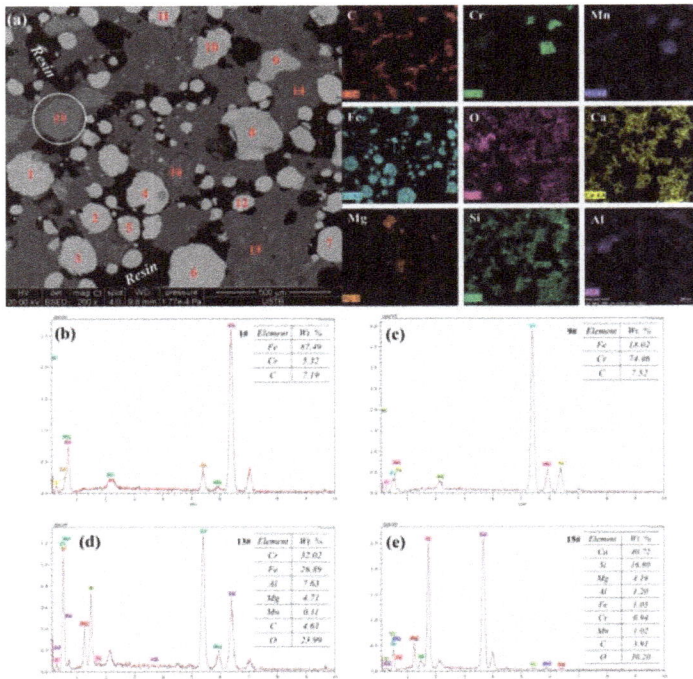

Figure 8. Element distribution of AL5 sample (**a**), EDS spectra of typical phases Point 1 (**b**), Point 9 (**c**), Point 13 (**d**) and Point 15 (**e**).

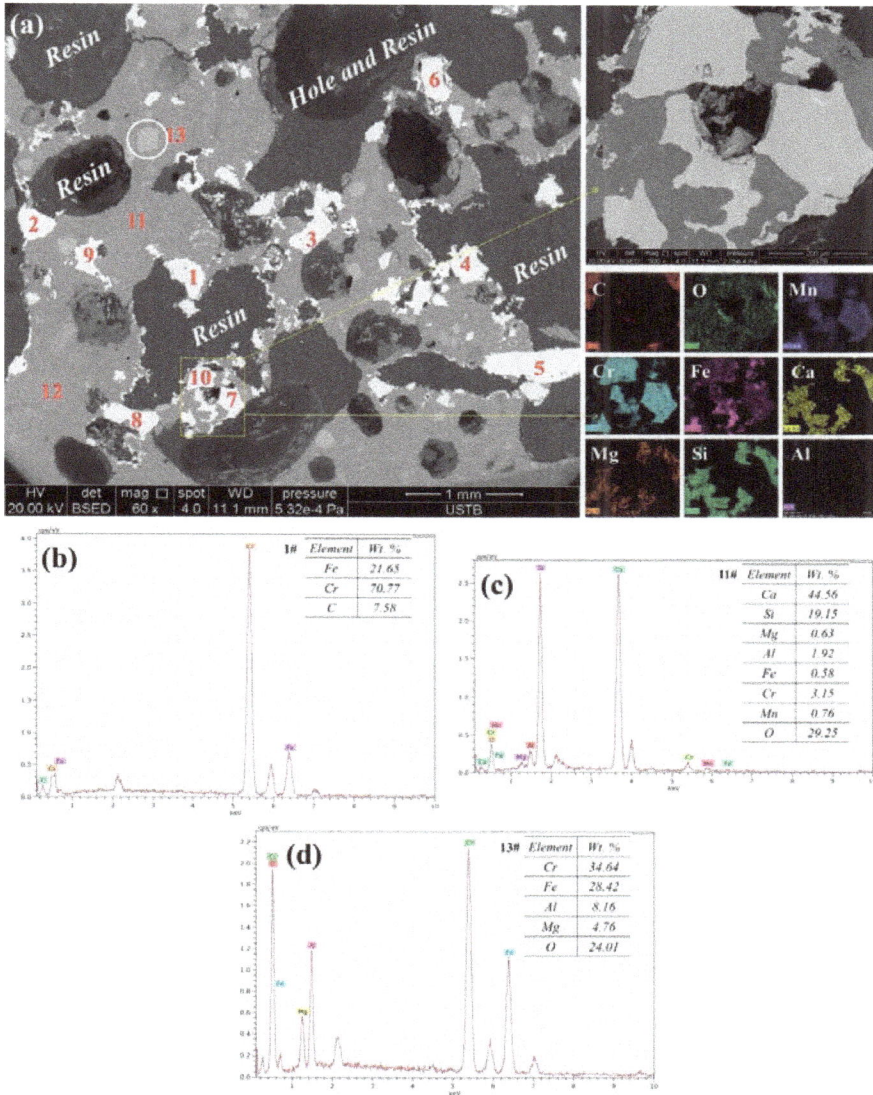

Figure 9. Element distribution of BL5 sample (**a**), EDS spectra of typical phases Point 1 (**b**), Point 11 (**c**), and Point 13 (**d**).

Further comparison of the two slags after holding at 1450 °C for 5 min was made, and the results revealed that there were residual Cr-containing spinel phases ((Fe, Mg)O·(Cr, Al)$_2$O$_3$) remaining in the BM5 sample as shown in Figure 10, which came from SSS or SSD. However, this phase was rarely seen in the slag of the AM5 sample. This indicated that the Cr-containing phase could not be rapidly reduced in the B-type Cr-RB without Fe$_2$O$_3$, which corresponded to that the reduction yield of the A-type Cr-RB was higher than that of the B-type Cr-RB at 5 min.

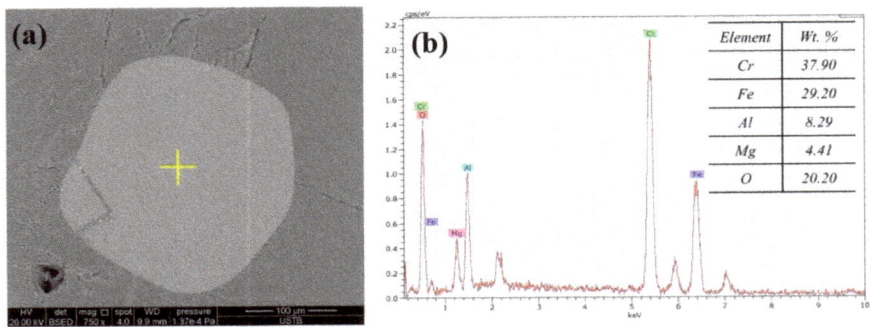

Figure 10. Residual (Fe, Mg)O·(Cr, Al)$_2$O$_3$ Spinel in BM5 sample, (**a**) morphology and (**b**) EDS.

3.4. Variation of Metal Composition

Base on EDS analysis, the composition of different small metal particles and large metal product in the observation field of SEM could be obtained. By means of Fe-Cr-C ternary phase diagram, the variation of metal composition in the samples at different conditions could be clearly observed, which was shown in Figure 11. As shown that the liquidus temperature increased with increasing concentration of C and Cr. For example, the liquidus temperature increased to 1575 °C from 1314 °C when the Cr concentration increased to 70 wt.% from 10 wt.% for the Fe-6 wt.%C-Cr system. Specifically for the AL5 sample, the metal products mostly distributed near the liquidus of 1350 °C, and C in the metals was in a saturated state. A few metal products were in the M7C3 ((Cr, Fe)$_7$C$_3$) and liquid phase coexistence zone, which was away from the liquidus. The Cr concentration in the metals increased with increasing reaction time, which indicated that the metal with a high Cr concentration and the metal with a low Cr concentration contacted and dissolved each other. When the temperature was 1450 °C and 1550 °C, the bulk metal component was in the liquid phase zone and had good fluidity and reactivity. For BL5 sample, the metal composition was far away from the liquid zone, and its melting point was as high as 1650–1750 °C. The metals contained a large amount of high-melting phase M7C3, which increased the viscosity of the metal [24] and was not conducive to the separation of metal and slag. The increase of the temperature to 1450 °C and 1550 °C accelerated the metal agglomeration, but its composition was still outside the liquid zone of the corresponding temperature, and M7C3 still existed.

Figure 12 is a Fe-Cr-C ternary phase diagram with iso-activity line of Cr at 1450 °C, in which the metal product composition of A and B-type Cr-RB at different temperature and holding time were projected. As shown that the Cr activity depended intensely on the Cr concentration once the C concentration was relatively constant. For example, the Cr activity increased to 0.5 from 0.02 when the Cr concentration increased to 70 wt.% from 10 wt.% for the Fe-6 wt.%C-Cr system. Obviously, the Cr activity of the metals in the samples of group A was significantly lower than that of the metals in the samples of group B. Higher Cr activity in the metals was not conducive to Cr transmit from slag into metal. Thus, the Cr in the A-type Cr-RB tended to move toward the metal faster.

In conclusion, due to the relatively low chromium-iron ratio (Cr/(Fe + Cr)) of the A-type Cr-RB, the metal products are more likely to rapidly aggregate and separate from the slag. The low Cr concentration of the metal products in the A-type Cr-RB is more thermodynamically beneficial for Cr recovery.

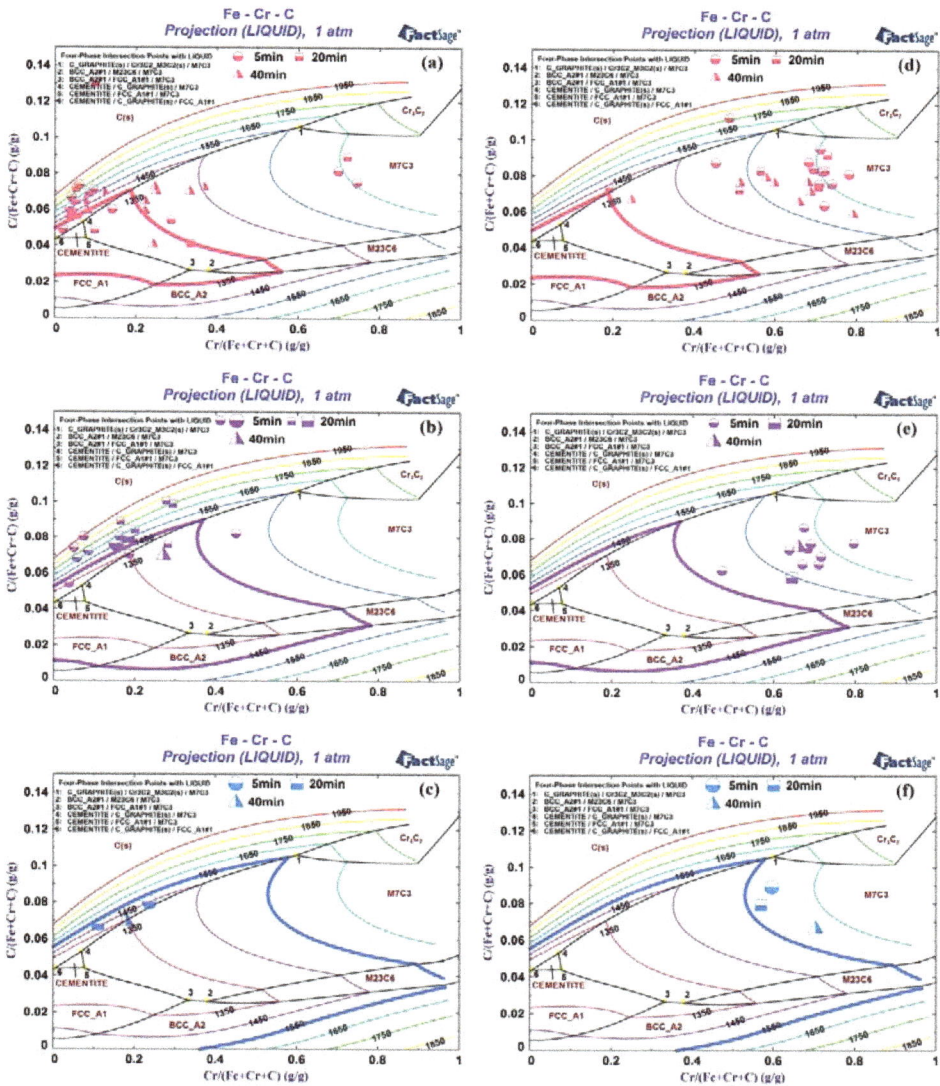

Figure 11. Metal composition variation in A and B-type Cr-RBs at different temperatures and times, (**a**) A-1350 °C, (**b**) A-1450 °C, (**c**) A-1550 °C, (**d**) B-1350 °C, (**e**) B-1450 °C, (**f**) B-1550 °C (Calculated by FactSage7.0, FSstel Database).

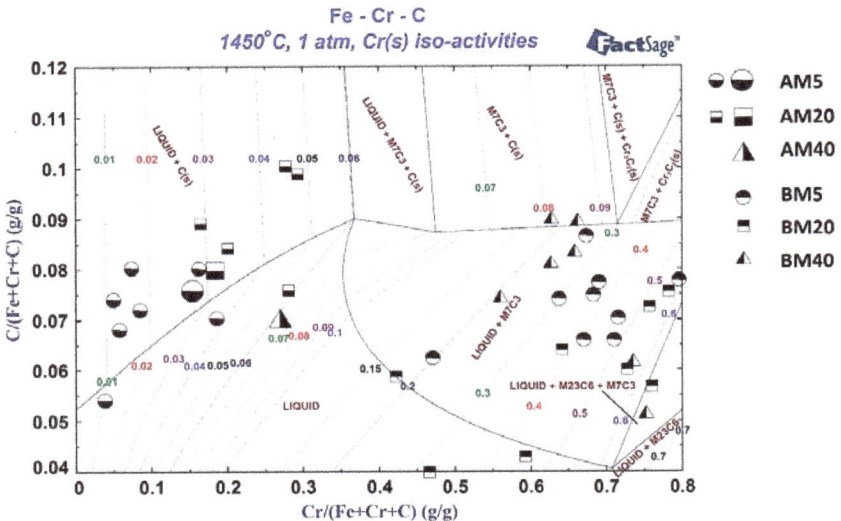

Figure 12. Activity of Cr in metals at 1450 °C (pure Cr as standard state, calculated by FactSage 7.0, FSstel Database).

3.5. Reaction Mechanism

Figure 13 shows that the initial reaction temperature ($T_{initial}$) of Reactions (7) to (14) in the standard state are 712 °C, 655 °C, 653 °C, 1053 °C, 1248 °C, 1141 °C, 1253 °C, 1710 °C, respectively. Obviously, the initial reduction temperatures of iron-containing oxides are below 712 °C. Once the Fe was first reduced, it could be carburized by C or CO to produce a large amount of liquid Fe-C alloys (Reaction (15)) [25,26]. The initial reduction temperatures of the Cr-containing oxides are all above 1000 °C, and especially for the reaction between Cr_2O_3 and Cr_7C_3, the $T_{initial}$ reaches to 1710 °C. Research [27] has pointed out that the carbothermal reduction of Cr_2O_3 was carried out according to $Cr_2O_3 \rightarrow Cr_3C_2 \rightarrow Cr_7C_3 \rightarrow Cr_{23}C_6 \rightarrow Cr$, and the stability range of different chromium carbides was different [23] (Cr_7C_3: 1250–1600 °C). Therefore, Cr_7C_3 produced at the experimental temperature would delay the transition of Cr_2O_3 to Cr. When the Cr-RB had a lower Cr/(Fe + Cr), a large amount of Fe-C alloy droplets would be generated before the reduction of Cr_2O_3. The pure Cr or Cr_7C_3 produced later would in-situ dissolve into the Fe-C alloys, and the liquid Fe-Cr-C alloy was formed (Reaction (16)). Meanwhile, the activity of the reduced product Cr or Cr_7C_3 decreased. As shown in Equations (17) and (18), the decrease in Cr activity was favorable for the decrease in ΔG of the reactions, which accelerated Cr recovery. Therefore, the reduction efficiency of Cr in A-type Cr-RB with Fe_2O_3 was higher than that of B-type Cr-RB. Based on the above experimental results and analysis, a schematic diagram of reaction mechanism was shown in Figure 14.

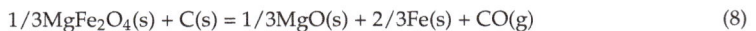

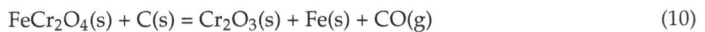

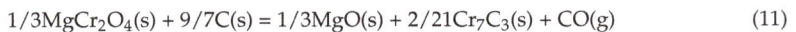

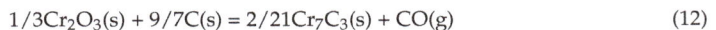

$$1/4Fe_3O_4(s) + C(s) = 3/4Fe(s) + CO(g) \tag{7}$$

$$1/3MgFe_2O_4(s) + C(s) = 1/3MgO(s) + 2/3Fe(s) + CO(g) \tag{8}$$

$$1/3Fe_2O_3(s) + C(s) = 2/3Fe(s) + CO(g) \tag{9}$$

$$FeCr_2O_4(s) + C(s) = Cr_2O_3(s) + Fe(s) + CO(g) \tag{10}$$

$$1/3MgCr_2O_4(s) + 9/7C(s) = 1/3MgO(s) + 2/21Cr_7C_3(s) + CO(g) \tag{11}$$

$$1/3Cr_2O_3(s) + 9/7C(s) = 2/21Cr_7C_3(s) + CO(g) \tag{12}$$

$$1/3Cr_2O_3(s) + C(s) = 2/3Cr(s) + CO(g) \tag{13}$$

$$1/3Cr_2O_3(s) + 1/3Cr_7C_3(s) = 3Cr(s) + CO(g) \tag{14}$$

$$Fe(s) + C(s) = (Fe\text{-}C)_{alloy} \tag{15}$$

$$Fe\text{-}C_{alloy} + Cr + Cr_7C_3 = (Fe\text{-}Cr\text{-}C)_{alloy} \tag{16}$$

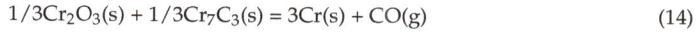

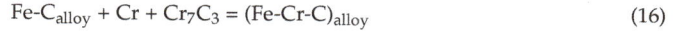

$$\Delta G_{13} = \Delta G_{13}^0 + RT \ln \frac{\frac{P_{CO}}{P^0} \cdot a_{Cr}^{2/3}}{a_C \cdot a_{Cr_2O_3}^{1/3}} \tag{17}$$

$$\Delta G_{14} = \Delta G_{14}^0 + RT \ln \frac{\frac{P_{CO}}{P^0} \cdot a_{Cr}^3}{a_{Cr_7C_3}^{1/3} \cdot a_{Cr_2O_3}^{1/3}} \tag{18}$$

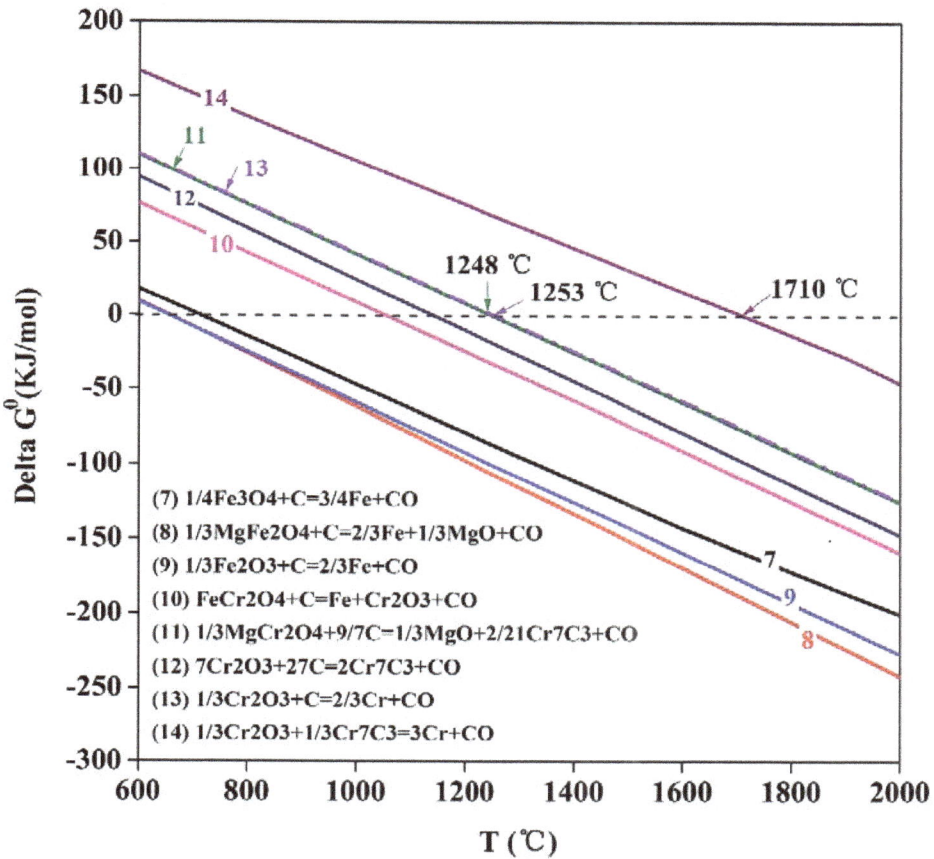

Figure 13. Variation of standard Gibbs free energy with temperature for each reaction (Calculated by FactSage 7.0, FactPS Database).

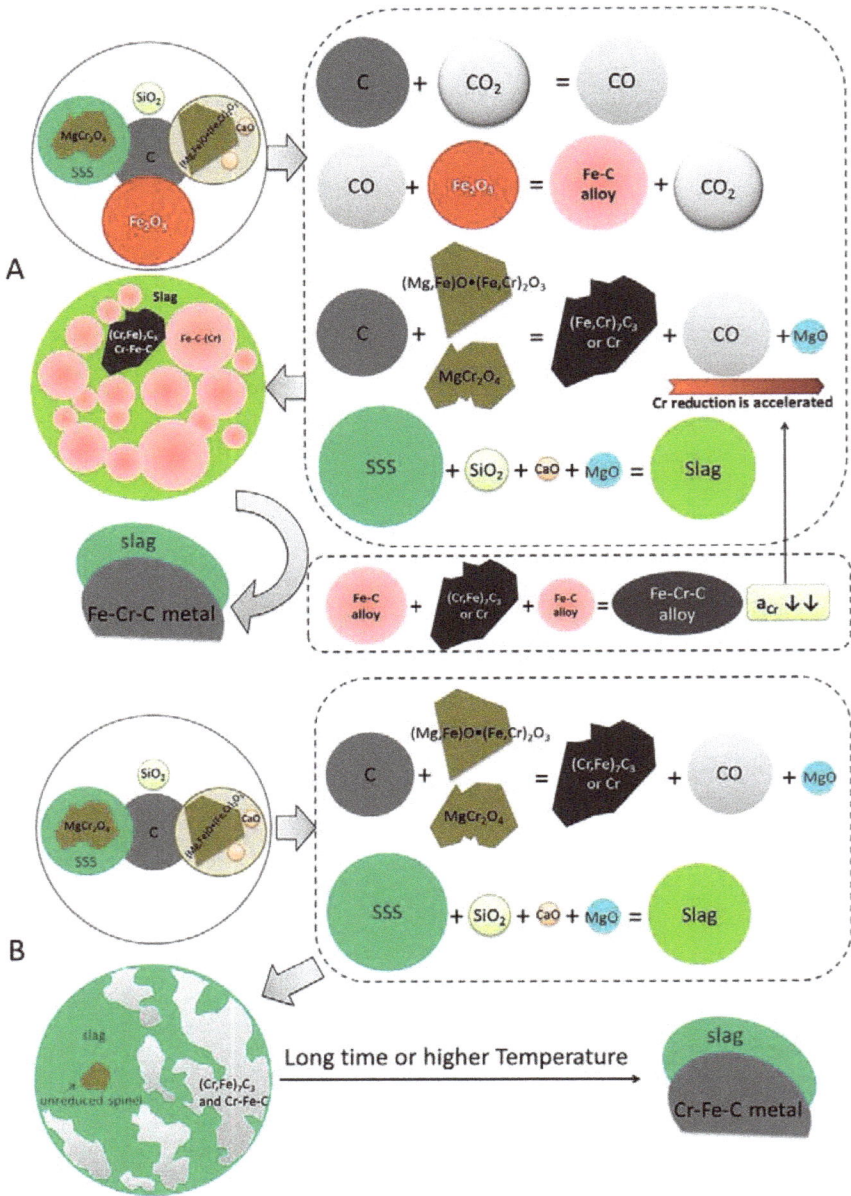

Figure 14. Mechanism diagram of the reduction-melting process of two Cr-RBs.

4. Conclusions

(1) The reduction yield of the two self-reduction briquettes A and B increased rapidly within 5 min, and then the growth rate became slower. The introduction of Fe_2O_3 to the Cr-containing solid waste self-reduction briquettes increased its reduction yield at a certain temperature.

(2) The addition of Fe_2O_3 to the Cr-containing solid waste self-reduction briquettes decreased its $Cr/(Cr + Fe)$, and a large amount of spherical Fe-C alloys with low Cr concentration was produced at a relatively low temperature. They could in-situ dilute the pure Cr or Cr_7C_3 to decrease the activity of Cr and promote the rapid reduction of Cr_2O_3, improving the recovery efficiency of Cr. Therefore, when the composition of Cr-containing solid waste self-reducing aggregation was designed for the furnaces with strong reducing atmosphere such as rotary hearth furnace, shaft furnace, or EAF furnace et al., much attention should be paid for its chromium-iron ratio $(Cr/(Fe + Cr))$.

Author Contributions: conceptualization, Y.Z.; methodology, T.W.; software, T.W.; validation, Y.Z.; formal analysis, T.W.; investigation, T.W., Z.Z., and F.Y.; resources, Y.Z.; data curation, T.W.; writing—original draft preparation, T.W.; writing—review and editing, Y.Z.; visualization, T.W.; supervision, Y.Z.; project administration, Y.Z.; funding acquisition, Y.Z.

Funding: This research was funded by National Natural Science Foundation, grant number 51674022.

Conflicts of Interest: The authors declare no conflict of interest.

References

1. Guo, M.; Durinck, D.; Jones, P.T.; Heylen, G.; Hendrickx, R.; Beaten, R.; Blanpain, B.; Wollants, P. EAF stainless steel refining-Part I: Observational study on chromium recovery in an eccentric bottom tapping furnace and a spout tapping furnace. *Steel Res. Int.* **2007**, *78*, 117–124. [CrossRef]
2. Durinck, D.; Jones, P.T.; Guo, M.; Verhaeghe, F.; Heylen, G.; Hendrickx, R.; Beaten, R.; Blanpain, B.; Wollants, P. EAF stainless steel refining-Part II: Microstructural slag evolution and its implications for slag foaming and chromium recovery. *Steel Res. Int.* **2007**, *78*, 125–135. [CrossRef]
3. Ma, G.; Garbers-Craig, A.M. Cr (VI) containing electric furnace dusts and filter cake from a stainless steel waste treatment plant: Part 1-Characteristics and microstructure. *Ironmak. Steelmak.* **2006**, *33*, 229–237. [CrossRef]
4. Ma, G.; Garbers-Craig, A.M. Cr (VI) containing electric furnace dusts and filter cake from a stainless steel waste treatment plant: Part 2-Formation mechanisms and leachability. *Ironmak. Steelmak.* **2006**, *33*, 238–244. [CrossRef]
5. Cho, S.; Lee, J. Metal recovery from stainless steel mill scale by microwave heating. *Met. Mater. Int.* **2008**, *14*, 193–196. [CrossRef]
6. Zhang, H.W.; Hong, X. An overview for the utilization of wastes from stainless steel industries. *Resour. Conserv. Recycl.* **2011**, *55*, 745–754.
7. Takano, C.; Cavallante, F.L.; Santos, D.M.D.; Mourão, M.B. Recovery of Cr, Ni and Fe from dust generated in stainless steelmaking. *Miner. Process Extr. Metall.* **2013**, *114*, 201–206. [CrossRef]
8. Zhang, Y.; Guo, W.; Jia, X. Reduction of chromium oxides in stainless steel dust. *Int. J. Min. Met. Mater.* **2015**, *22*, 573–581. [CrossRef]
9. Qi, Q.S.; Xu, A.J.; He, D.F.; Pan, J.T. Carbothermic reduction for recycling of iron and chromium from stainless steel slag. *Iron Steel.* **2017**, *52*, 82–87.
10. Adamczyk, B.; Brenneis, R.; Adam, C.; Mudersbach, D. Recovery of chromium from AOD-Converter slags. *Steel Res. Int.* **2010**, *81*, 1078–1083. [CrossRef]
11. Duan, J.; Zhang, Y.; Li, H.; Wang, J. Experimental analysis on direct recycling Cr-Ni stainless steel making dust in EAF. *Iron Steel.* **2009**, *44*, 76–80.
12. McClelland, J.M.; Metius, G.E. Recycling ferrous and nonferrous waste streams with FASTMET. *JOM* **2003**, *55*, 30–34. [CrossRef]
13. Ye, G.; Burström, E.; Kuhn, M.; Piret, J. Reduction of steel-making slags for recovery of valuable metals and oxide materials. *Scand. J. Metall.* **2010**, *32*, 7–14. [CrossRef]
14. Zhao, H.; Qi, Y.; Shi, Y.; Feng, H.; Na, X. A test research on treatment of stainless steel dust by Oxycup. *J. Mat. Metal.* **2017**, *16*, 58–62.
15. Görnerup, M.; Lahiri, A.K. Reduction of electric arc furnace slags in stainless steelmaking Part 1 Observations. *Ironmak. Steelmak.* **1998**, *25*, 317–322.
16. Görnerup, M.; Lahiri, A.K. Reduction of electric arc furnace slags in stainless steelmaking: Part 2 Mechanism of CrO_x reduction. *Ironmak. Steelmak.* **1998**, *25*, 382–386.

17. Chakraborty, D.; Ranganathan, S.; Sinha, S.N. Investigations on the carbothermic reduction of chromite ores. *Metall. Mater. Trans. B* **2005**, *36*, 437–444. [CrossRef]
18. Hu, X.; Teng, L.; Wang, H.; Sundqvist Ökvist, L.; Yang, Q.; Björkma, B.; Seetharaman, S. Carbothermic reduction of synthetic chromite with/without the addition of iron powder. *ISIJ Int.* **2016**, *56*, 2147–2155. [CrossRef]
19. Hu, X.; Yang, Q.; Sundqvist Ökvist, L.; Björkma, B. Thermal analysis study on the carbothermic reduction of chromite ore with the addition of mill scale. *Steel Res. Int.* **2016**, *87*, 562–570. [CrossRef]
20. Hu, X.; Sundqvist Ökvist, L.; Eriksson, J.; Yang, Q.; Björkma, B. Direct alloying steel with chromium by briquettes made from chromite ore, mill Scale, and petroleum coke. *Steel Res. Int.* **2017**, *88*, 1–10. [CrossRef]
21. Zhang, Y.; Liu, Y.; Wei, W. Products of carbothermic reduction of Fe-Cr-O and Fe-Cr-Ni-O systems. *Trans. Nonferr. Met. Soc. China* **2014**, *24*, 1210–1219. [CrossRef]
22. Zhang, Y.; Liu, Y.; Wei, W. Carbothermal reduction process of the Fe-Cr-O system. *Int. J. Min. Met. Mater.* **2013**, *20*, 931–940. [CrossRef]
23. Eric, R.H. Chapter 1.10-Production of Ferroalloys. In *Treatise on Process Metallurgy*; Elsevier: Amsterdam, The Netherlands, 2014; Volume 3, pp. 477–532.
24. Mori, Y.; Ototake, N. Viscosity of solid-liquid mixture. *Chem. Eng.* **1956**, *20*, 488. [CrossRef]
25. Ohno, K.; Miki, T.; Hino, M. Kinetic analysis of iron carburizaiton during smelting reduciton. *ISIJ Int.* **2004**, *44*, 2033–2039. [CrossRef]
26. Ohno, K.; Miki, T.; Sasaki, Y.; Hino, M. Carburization degree of iron nugget produced by rapid heating of powdery iron, iron oxide in slag and carbon mixture. *ISIJ Int.* **2008**, *48*, 1368–1372. [CrossRef]
27. Mori, T.; Yang, J.; Kuwabara, M. Mechanism of carbothermic reduction of chromium oxide. *ISIJ Int.* **2007**, *47*, 1387–1393. [CrossRef]

![metals logo] *metals*

MDPI

Article

Thermal Behavior of Hydrated Iron Sulfate in Various Atmospheres

Ndue Kanari [1,]*[ⓘ], Nour-Eddine Menad [2], Etleva Ostrosi [3], Seit Shallari [4], Frederic Diot [1], Eric Allain [1] and Jacques Yvon [1]

[1] GeoRessources Laboratory, UMR 7359 CNRS, CREGU, Université de Lorraine, 2, rue du doyen Roubault, BP 10162, 54505 Vandoeuvre-lès-Nancy, France; frederic.diot@univ-lorraine.fr (F.D.); ericgallain@gmail.com (E.A.); jacques.yvon@univ-lorraine.fr (J.Y.)
[2] BRGM, 3 av. C. Guillemin, BP 36009, CEDEX 2, 45060 Orléans, France; N.Menad@brgm.fr
[3] Ville de Montréal, Direction de l'Environnement, Division de la Planification et du Suivi Environnemental, 801, rue Brennan, Montréal, QC H3C 0G4, Canada; etlevao@yahoo.com
[4] Faculty of Agriculture and Environment, Agriculture University of Tirana, 1029 Tirana, Albania; seitshallari@gmail.com
* Correspondence: ndue.kanari@univ-lorraine.fr; Tel.: +33-372-744-530

Received: 19 November 2018; Accepted: 13 December 2018; Published: 19 December 2018

Abstract: Iron sulfate, in particular $FeSO_4 \cdot 7H_2O$, is derived from titanium dioxide production and the steel pickling process. Regarding TiO_2 manufacturing, the amount of the resultant $FeSO_4 \cdot 7H_2O$ can be as high as 6 tons per ton of produced TiO_2, leading to a huge amount of ferrous sulfate heptahydrate, which is considered an environmental and economic concern for the titanium dioxide industry in European countries. The present paper focuses on the thermal treatment of ferrous sulfate (heptahydrate and monohydrate) samples under different conditions. Nonisothermal thermogravimetric (TG) analysis was used to study the behavior of iron sulfate samples at temperatures of up to 1000 °C in $Cl_2 + O_2$, O_2, and N_2 atmospheres. Results showed that the dehydration of iron sulfate heptahydrate in nitrogen started at room temperature and resulted in iron sulfate tetrahydrate ($FeSO_4 \cdot 4H_2O$). The ferrous sulfate monohydrate ($FeSO_4 \cdot H_2O$) was formed at temperatures close to 150 °C, while the anhydrous ferrous sulfate ($FeSO_4$) was obtained when the samples were heated in nitrogen at over 225 °C. The kinetic features of $FeSO_4$ decomposition into Fe_2O_3 were revealed under isothermal conditions at temperatures ranging from 500 to 575 °C. The decomposition of iron sulfate was characterized by an apparent activation energy of around 250 kJ/mol, indicating a significant temperature effect on the decomposition process. The obtained powder iron oxide could be directed to the agglomeration unit of iron and the steelmaking process.

Keywords: iron sulfate; TG analysis; thermal treatment; iron oxide; kinetics; activation energy

---

1. Introduction

Titanium oxide (TiO_2) is manufactured from materials such as ilmenite, rutile, anatase, and slags using sulfate or chloride processes. The simplified schemes for the industrial processes that are currently used are presented in Figure 1 [1]. Ilmenite and titanium slags are the raw materials used for TiO_2 manufacturing through the sulfate process. The ilmenite is digested in sulfuric acid, generating a solution that contains titanyl sulfate ($TiOSO_4$) and iron sulfate. The solution is treated with scrap iron to reduce the ferric ions into a ferrous state to avoid the precipitation of ferric hydroxide. The ferrous sulfate then crystallizes into $FeSO_4 \cdot 7H_2O$ (melanterite) and is separated from the liquor. Additional steps (see Figure 1) are necessary to obtain the TiO_2 base pigment. Depending on the quality of the raw materials used in the TiO_2 production, the amount of iron sulfate produced can reach up to 6 tons of $FeSO_4 \cdot 7H_2O$ per ton of produced TiO_2 when ilmenite is used as the raw material. The sulfate route, which is mainly used in the production of TiO_2 in European countries, generates a huge amount of wasted ferrous sulfate heptahydrate. On the contrary, North American countries produce TiO_2 through the chlorine route, the main steps of which are also shown in Figure 1.

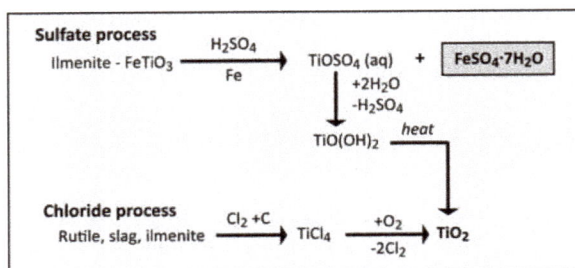

Figure 1. Schematic representation of the main steps in titanium oxide manufacturing using sulfate and chloride processes, adapted from Reference [1].

An extensive overview of the titanium metallurgical processes was recently conducted in Reference [2]. This report compares the main characteristics of the classical and emerging processes for TiO_2 manufacturing from economic and environmental viewpoints. As mentioned in earlier works [3,4], one drawback of the sulfate process is the amount of iron sulfate heptahydrate and the spent acid generated during TiO_2 production. Only a small part of the iron sulfate is reused, and the remaining part must be disposed of as waste. It must be noted that the amount of $FeSO_4 \cdot 7H_2O$ generated from the surface treatment of steel is decreasing due to the use of hydrochloric acid (HCl) instead of sulfuric acid (H_2SO_4).

Several investigations previously conducted in our laboratory were focused on the use of industrial iron sulfate for the synthesis of alkali ferrates [3,5,6]. The potassium ferrate synthesis (K_2FeO_4) from spent steel pickling liquid was also reported by Wei et al. [7]. A recent work applied the reductive decomposition reaction of iron sulfate with pyrite into Fe_3O_4 at a relatively low temperature [4].

In this context, the present work dealt with the dehydration and the decomposition of various iron sulfate samples in oxidizing and neutral atmospheres. Thermogravimetric (TG) analysis was used as an appropriate method to continuously follow the reaction kinetics in the decomposition of $FeSO_4$ into iron oxide under isothermal conditions. The iron oxide obtained can be used as a raw material in the ironmaking sectors. Many recent reports are available on the reduction of iron-oxide-bearing materials from various known agents [8–16].

2. Materials and Methods

Several samples of the iron sulfate heptahydrate generated from industrial operators were collected, the physicochemical characterization of which has been previously given in Reference [3]. For this investigation, two samples of iron sulfate were selected. The first sample—iron sulfate monohydrate—was provided by an industrial operator and was named the IND sample. The second sample—LAB sample—was obtained through the two-step dehydration process of analytical-grade iron sulfate heptahydrate ($FeSO_4 \cdot 7H_2O$) in a laboratory oven. The heating of the $FeSO_4 \cdot 7H_2O$ at about 60–70 °C led to the loss of 3 mol of water, resulting in the formation of $FeSO_4 \cdot 4H_2O$. An increase in the temperature to about 150 °C provoked the dehydration of the iron sulfate tetrahydrate into $FeSO_4 \cdot H_2O$. However, in the presence of air, the oxidation of Fe(II) into Fe(III) may occur. Both the IND and LAB samples were subjected to a variety of analyses to determine their composition.

The total iron and Fe(II) contents of the samples were determined using chemical analysis. After sample digestion, the Fe(II) was determined using potassium dichromate titration. Table 1 gives the average values of iron in both the samples. The IND sample contained about 31% Fe, and the whole iron was in a divalent state. X-Ray diffraction (XRD) analysis only showed the presence of $Fe^{II}SO_4 \cdot H_2O$ in the crystallized phase. The total iron content of the LAB sample was 32.8% in which 14.8% was in a divalent state and 18.0% was in a trivalent state. In other words, in the 100% Fe_{total} LAB sample, about 45% was Fe(II) and about 55% was Fe(III). Concerning the results of XRD (see Table 1), they revealed the presence of $Fe^{II}SO_4 \cdot H_2O$ and $Fe^{III}SO_4 \cdot OH$ in the LAB sample. This confirmed that the dehydration of $FeSO_4 \cdot 4H_2O$ into $FeSO_4 \cdot H_2O$ and partial oxidation of Fe(II) to Fe(III) occurred by 150 °C.

Table 1. Results of chemical and XRD analysis of two iron sulfate samples.

Sample	Chemical Analysis (%)			XRD
	Fe_{total}	Fe(II)	Fe(III)	
IND	30.9	30.9	trace	$Fe^{II}SO_4 \cdot H_2O$
LAB	32.8	14.8	18.0 [1]	$Fe^{II}SO_4 \cdot H_2O$ and $Fe^{III}SO_4 \cdot OH$

[1] By difference.

Experimental tests of thermogravimetric analysis were performed using a CAHN 1000 microbalance capable of resisting corrosive atmospheres to check the thermal behavior of $FeSO_4 \cdot 7H_2O$ under different atmospheres ($Cl_2 + O_2$, Cl_2, O_2, and N_2). Furthermore, a TG 2171 Cahn balance was used to study the dehydration/decomposition kinetics of iron sulfate samples under N_2 by simultaneous TG and differential thermal (DT) measurements. Solid reaction products were examined by X-ray diffraction, scanning electron microscopy, and Mössbauer spectroscopy.

3. Results

3.1. Nonisothermal TG Analysis of $FeSO_4 \cdot 7H_2O$ under Different Atmospheres

The thermal behavior of a $FeSO_4 \cdot 4H_2O$ sample in various gaseous atmospheres ($Cl_2 + O_2$, O_2, and N_2) was investigated by TG analysis utilizing nonisothermal conditions [3]. The results are drawn in Figure 2 as the evolution of the percent mass loss (% ML) of the sample versus temperature up to 300 °C. The calculated limits corresponding to different hydrated states of ferrous sulfate are also shown in Figure 2. Results of XRD and Mössbauer analyses [1] showed that $FeSO_4 \cdot 4H_2O$, $FeSO_4 \cdot H_2O$, and $FeSO_4$ are the main crystallized phases in the solid product obtained at 75 °C, 150 °C and 300 °C, respectively, during the treatment of iron sulfate heptahydrate sample under nitrogen atmosphere. Conversely, the treatment of the sample under oxidizing atmosphere ($Cl_2 + O_2$, O_2) led to the transformation of Fe(II) into Fe(III), and the product obtained at 300 °C was mainly composed of $FeSO_4 \cdot OH$.

As revealed by Mössbauer analysis, the product resulting from the treatment of $FeSO_4 \cdot 7H_2O$ in N_2 at 150 °C was composed of Fe(II) in totality, while the product generated by the treatment in $Cl_2 + O_2$ contained iron, mostly in a three-valent state. To observe the reactivity of iron sulfate toward O_2 and $Cl_2 + O_2$, TG tests were performed at temperatures up to 1000 °C, and the corresponding data is plotted in Figure 3. As can be seen, the curves for both oxidizing gas mixtures have roughly similar shapes for temperatures up to 675 °C. This observation suggests that chlorine reacted with the sample only after the decomposition of iron sulfate into ferric oxide (hematite), producing ferric chloride ($FeCl_3$) as a final reaction product [3]. The kinetics of the reaction of Fe_2O_3 with Cl_2 and $Cl_2 + O_2$ were further discussed in earlier articles [17,18].

The treatment of $FeSO_4 \cdot 7H_2O$ in nitrogen at different heating rates was followed by DT analysis, and a data summary is given in Table 2. These results, combined with those of TG analysis and XRD analysis, show that the sequence of sample transformation was the following: $FeSO_4 \cdot 7H_2O \rightarrow FeSO_4 \cdot 4H_2O \rightarrow FeSO_4 \cdot H_2O \rightarrow FeSO_4 \rightarrow Fe_2O_3$.

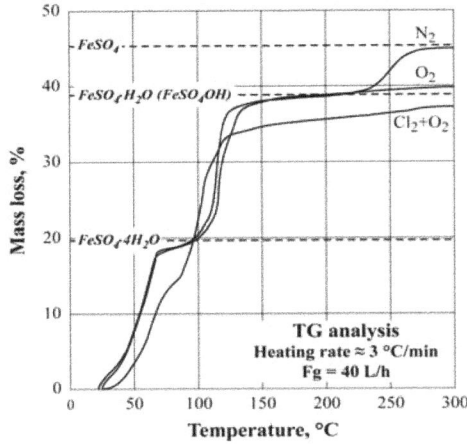

Figure 2. Thermogravimetric (TG) analysis of a $FeSO_4 \cdot 7H_2O$ sample under different atmospheres.

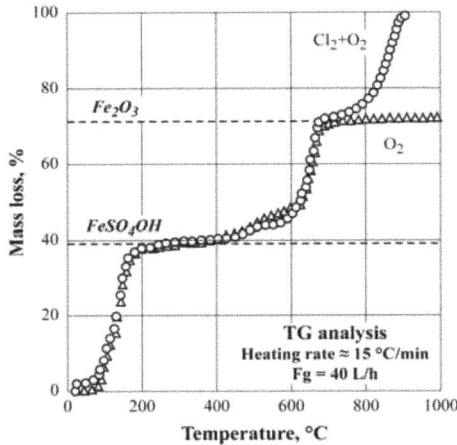

Figure 3. TG analysis of a $FeSO_4 \cdot 7H_2O$ sample in O_2 and $Cl_2 + O_2$ atmospheres.

Table 2. Endothermic peaks (°C) revealed by differential thermal (DT) analysis for the treatment of $FeSO_4 \cdot 7H_2O$ under nitrogen at various heating rates.

Heating Rate, °C/min			Possible Reaction Steps		
2.5	5.0	10.0			
70	80	98	$FeSO_4 \cdot 7H_2O$	$\rightarrow$	$FeSO_4 \cdot 4H_2O$
86	133	159	$FeSO_4 \cdot 4H_2O$	$\rightarrow$	$FeSO_4 \cdot H_2O$
227	250	283	$FeSO_4 \cdot H_2O$	$\rightarrow$	$FeSO_4$
653	687	716	$FeSO_4$	$\rightarrow$	Fe_2O_3

3.2. Nonisothermal TG Analysis of $FeSO_4 \cdot H_2O$ under Nitrogen

Nonisothermal TG tests up to 1000 °C in nitrogen were performed for both IND and LAB samples. The furnace heating rates were fixed at 2.5 and 20.0 °C/min, and the data is plotted as the evolution of the % ML as a function of temperature. Figure 4 shows the results for the IND sample. The % ML obtained between 200 and 400 °C corresponds to the dehydration of $FeSO_4 \cdot H_2O$

into $FeSO_4$. The theoretical % ML for the dehydration of iron sulfate monohydrate into iron sulfate (dashed horizontal line) matches well with the experimental % ML of this sample. As shown in Figure 5, the behavior of the LAB sample seems to be somewhat different when compared with the IND sample. Only 7% ML was observed at temperatures less than or equal to 400 °C. This could be attributed to the dehydration of $FeSO_4 \cdot H_2O$ into $FeSO_4$. The continuous mass loss of the LAB sample between 400 and 550 °C was probably due to the transformation of $FeSO_4 \cdot OH$ into ferric oxysulfate ($Fe_2O(SO_4)_2$). The decomposition of iron sulfates producing ferric oxides takes place at temperatures higher than 575 °C, and the curve shapes for both samples seem to be similar. The XRD analysis of the decomposition product showed the presence of Fe_2O_3 in the main crystallized phase.

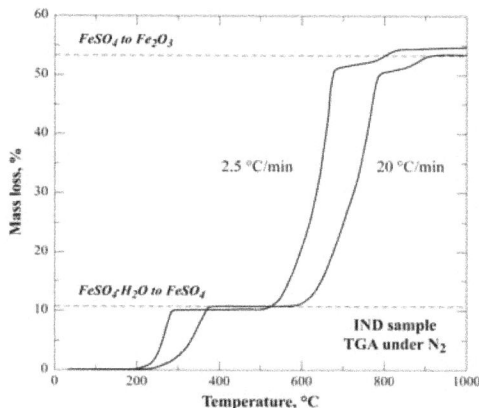

Figure 4. TG analysis in N_2 of the IND sample.

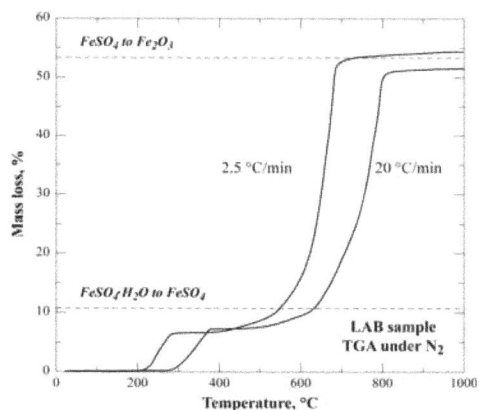

Figure 5. TG analysis in N_2 of the LAB sample.

Based on these TG results, the decomposition kinetics in nitrogen atmosphere of both iron sulfate samples into iron oxide were studied under isothermal conditions at temperatures higher than or equal to 500 °C.

3.3. Isothermal Decomposition of $FeSO_4 \cdot H_2O$ Samples

This decomposition process was tested at low temperatures (between 500 and 575 °C) to minimize the iron sulfate decomposition during the nonisothermal temperature rise. Thus, the initial temperature increased linearly (heating rate = 5.0 °C/min) up to a fixed value, then the temperature remained constant, and the decomposition extent followed a function of time. A typical example of experimental conditions and results is given in Figure 6. The temperature profile was programmed to increase up to 560 °C; from this point, the decomposition rate (% ML vs. time) was measured at 560 °C. It should be noted that the % ML observed during nonisothermal test ($\leq$12% ML) corresponds to the dehydration step of $FeSO_4 \cdot H_2O$ to $FeSO_4$.

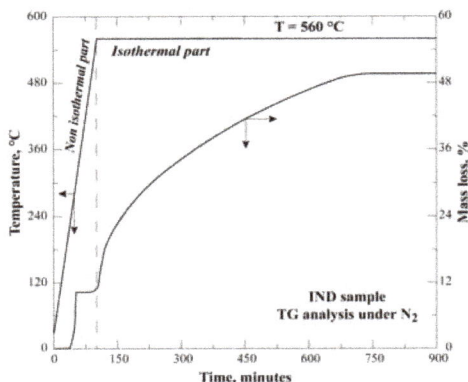

Figure 6. Typical example of TG analysis results of iron sulfate decomposition.

The behavior of both IND and LAB samples under isothermal treatment was monitored in order to check the eventual impact of Fe(III) sulfate on the transformation kinetics. The results obtained from the treatment of the IND sample in nitrogen are represented in Figure 7. About 50 h were necessary for the half decomposition of iron sulfate at 500 °C, while full decomposition was achieved in less than 7 h at 575 °C. This result seems to indicate that the decomposition process depends highly on the temperature.

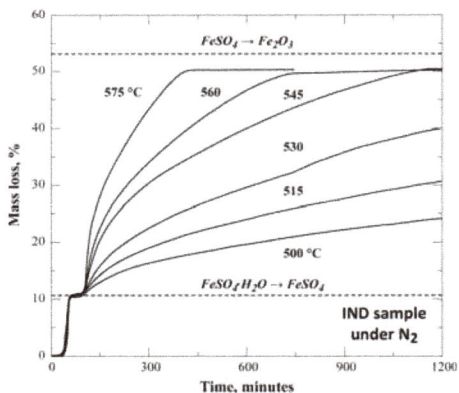

Figure 7. Evolution of the percent mass loss (% ML) as a function of time during the isothermal treatment of an industrial sample under nitrogen between 500 and 575 °C.

Experimental results corresponding to the treatment of LAB sample are illustrated in Figure 8 as % ML versus time for temperatures ranging from 500 to 575 °C. The isothermal part ($\geq$12% ML) of the curves has a similar shape to that observed in the IND sample. When the temperature was increased from 500 to 575 °C, the initial decomposition rate was multiplied by a factor of about 26.

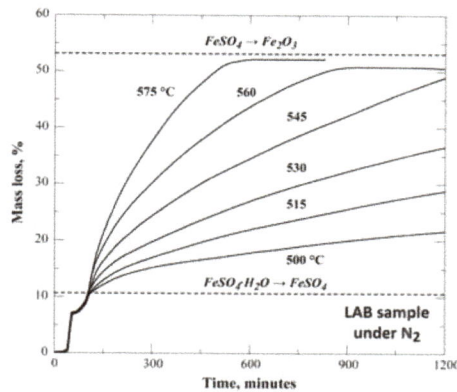

Figure 8. Evolution of the % ML as a function of time during the isothermal treatment of a laboratory sample under nitrogen between 500 and 575 °C.

Arrhenius diagrams for both samples were established in order to evaluate the temperature effect on the initial decomposition rate of iron sulfates. The mean decomposition rate was calculated by linearization of the isothermal data corresponding to 12.5% $\leq$ ML $\leq$ 22.5%. As shown in Figure 9, the decomposition rate of the IND sample was higher than that for the LAB sample over the complete range of temperatures. The values of the apparent activation energy for the treatment of industrial and laboratory sample were about 262 and 238 kJ/mol, respectively, showing the strong effect of temperature on the decomposition process of iron sulfate between 500 and 575 °C. A similar value of apparent activation energy (average value of 244 kJ/mol) was obtained in the work reported by Huang et al. [4].

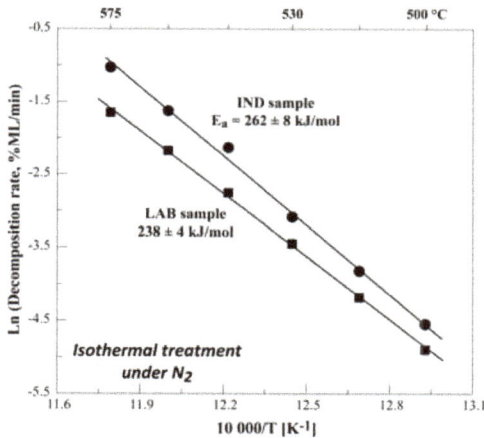

Figure 9. Arrhenius diagram for the decomposition of the iron sulfate samples under nitrogen for temperatures ranging from 500 to 575 °C.

The reaction product generated by the thermal treatment of iron sulfates was mainly composed of pure iron (III) oxide. Such a material, in powder state, must be agglomerated before use for pig iron production and/or for more valuable end uses, such as high-grade pigments for cosmetics.

4. Conclusions

The dehydration of iron sulfate heptahydrate ($FeSO_4 \cdot 7H_2O$) in nitrogen occurred through at least three steps: $FeSO_4 \cdot 7H_2O \rightarrow FeSO_4 \cdot 4H_2O \rightarrow FeSO_4 \cdot H_2O \rightarrow FeSO_4$. Complete dehydration of $FeSO_4 \cdot 7H_2O$ occurred at a temperature lower than 300 °C. The treatment of $FeSO_4 \cdot 7H_2O$ under oxidizing atmosphere led to the formation of $Fe^{III}SO_4 \cdot OH$ as a final stable product at 300 °C.

The decomposition of iron sulfates ($FeSO_4$ and $Fe^{III}SO_4 \cdot OH$) generating ferric oxide, in nonisothermal conditions, started at T > 500 °C, and the final temperature of their full decomposition depended on the heating rate in the furnace.

The isothermal decomposition of selected iron sulfate samples was strongly affected by temperature as it proceeded with a value of apparent activation energy of 250 kJ/mol between 500 and 575 °C. The obtained iron oxide could be used as an appropriate raw material in the ferrous metallurgy sector and/or for more noble end uses.

Author Contributions: Conceptualization, N.K., N.-E.M. and E.O.; Formal analysis, N.-E.M. and E.O.; Investigation, N.K., E.O. and F.D.; Visualization, S.S. and F.D.; Resources, F.D. and J.Y.; Writing—original draft, N.K., E.O., S.S. and E.A.; Writing—review and editing, N.K., E.O, E.A. and J.Y.

Funding: A part of this work was performed in the frame of contract No. BRPR-CT97-0392 of the European Union. Another part of this development work was supported by the French National Research Agency through the program "Investissements d'avenir" with the reference ANR-10-LABX-21-01/LABEX RESSOURCES21.

Conflicts of Interest: The authors declare no conflict of interest.

References

1. Kanari, N. *Contribution to Chlorine Chemistry and its Applications: Synthesis of Alkali Ferrates (VI). A Study of the Kinetics of Chlorine-Solid Reactions*; Defense of the Habilitation Diploma (HDR), Institut National Polytechnique de Lorraine: Nancy, France, 2000.
2. Zhang, W.; Zhu, Z.; Cheng, C.Y. A literature review of titanium metallurgical processes. *Hydrometallurgy* **2011**, *108*, 177–188. [CrossRef]
3. Kanari, N.; Filippova, I.; Diot, F.; Mochón, J.; Ruiz-Bustinza, I.; Allain, E.; Yvon, J. Utilization of a waste from titanium oxide industry for the synthesis of sodium ferrate by gas-solid reactions. *Thermochim. Acta* **2014**, *575*, 219–225. [CrossRef]
4. Huang, P.; Deng, S.; Zhang, Z.; Wang, X.; Chen, X.; Yang, X.; Yang, L. A sustainable process to utilize ferrous sulfate waste from titanium oxide industry by reductive decomposition reaction with pyrite. *Thermochim. Acta* **2015**, *620*, 18–27. [CrossRef]
5. Kanari, N.; Ostrosi, O.; Ninane, N.; Neveux, N.; Evrard, O. Synthesizing alkali ferrates using a waste as a raw material. *JOM* **2005**, *57*, 39–42. [CrossRef]
6. Kanari, N. Method of Producing Ferrates (VI). French Patent n° 2 905 609, 14 March 2008.
7. Wei, Y.-L.; Wang, Y.-S.; Liu, C.-H. Preparation of potassium ferrate from spent steel pickling liquid. *Metals* **2015**, *5*, 1770–1787. [CrossRef]
8. Pineau, A.; Kanari, N.; Gaballah, I. Kinetics of reduction of iron oxides by H_2: Part I: Low temperature reduction of hematite. *Thermochim. Acta* **2006**, *447*, 89–100. [CrossRef]
9. Jozwiak, W.K.; Kaczmarek, E.; Maniecki, T.P.; Ignaczak, W.; Maniukiewicz, W. Reduction behavior of iron oxides in hydrogen and carbon monoxide atmospheres. *Appl. Catal. A* **2007**, *326*, 17–27. [CrossRef]
10. Pineau, A.; Kanari, N.; Gaballah, I. Kinetics of reduction of iron oxides by H_2: Part II. Low temperature reduction of magnetite. *Thermochim. Acta* **2007**, *456*, 75–88. [CrossRef]
11. Tang, J.; Chu, M.S.; Ying, Z.W.; Li, F.; Feng, C.; Liu, Z.G. Non-isothermal gas-based direct reduction behavior of high chromium vanadium-titanium magnetite pellets and the melting separation of metallized pellets. *Metals* **2017**, *7*, 153. [CrossRef]
12. Cao, Y.; Zhang, Y.; Sun, T. Dephosphorization behavior of high-phosphorus oolitic hematite-solid waste containing carbon briquettes during the process of direct reduction-magnetic separation. *Metals* **2018**, *8*, 897. [CrossRef]
13. Oh, J.; Noh, D. The reduction kinetics of hematite particles in H_2 and CO atmospheres. *Fuel* **2017**, *196*, 144–153. [CrossRef]

14. Chen, Z.; Dang, J.; Hu, X.; Yan, H. Reduction kinetics of hematite powder in hydrogen atmosphere at moderate temperatures. *Metals* **2018**, *8*, 751. [CrossRef]

15. Tang, H.; Yun, Z.; Fu, X.; Du, S. Modeling and experimental study of ore-carbon briquette reduction under CO–CO_2 atmosphere. *Metals* **2018**, *8*, 205. [CrossRef]

16. Fukushima, J.; Takizawa, H. In situ spectroscopic analysis of the carbothermal reduction process of iron oxides during microwave irradiation. *Metals* **2018**, *8*, 49. [CrossRef]

17. Kanari, N.; Mishra, D.; Filippov, L.; Diot, F.; Mochón, J.; Allain, E. Kinetics of hematite chlorination with Cl_2 and Cl_2+O_2: Part I. Chlorination with Cl_2. *Thermochim. Acta* **2010**, *497*, 52–59. [CrossRef]

18. Kanari, N.; Mishra, D.; Filippov, L.; Diot, F.; Mochón, J.; Allain, E. Kinetics of hematite chlorination with Cl_2 and Cl_2+O_2. Part II. Chlorination with Cl_2+O_2. *Thermochim. Acta* **2010**, *506*, 34–40. [CrossRef]

metals

MDPI

Article

Crystallization Behaviors of Anosovite and Silicate Crystals in High CaO and MgO Titanium Slag

Helin Fan [1,2], Dengfu Chen [1,2,*], Tao Liu [1,2], Huamei Duan [1,2,*], Yunwei Huang [1,2], Mujun Long [1,2] and Wenjie He [1,2]

[1] College of Materials Science and Engineering, Chongqing University, Chongqing 400044, China; fanhelin@cqu.edu.cn (H.F.); cqltao@cqu.edu.cn (T.L.); cquyunwei@cqu.edu.cn (Y.H.); longmujun@cqu.edu.cn (M.L.); hewenjie@cqu.edu.cn (W.H.)

[2] Chongqing Key Laboratory of Vanadium-Titanium Metallurgy and Advanced Materials, Chongqing University, Chongqing 400044, China

* Correspondence: chendfu@cqu.edu.cn (D.C.); duanhuamei@cqu.edu.cn (H.D.); Tel.: +86-23-6510-2467 (D.C. & H.D.)

Received: 31 August 2018; Accepted: 21 September 2018; Published: 24 September 2018

Abstract: Electric-furnace smelting has become the dominant process for the production of the titanium slag from ilmenite in China. The crystallization behaviors of anosovite and silicate crystals in the high CaO and MgO titanium slag were studied to insure smooth operation of the smelting process and the efficient separation of titanium slag and metallic iron. The crystallization behaviors were studied by a mathematical model established in this work. Results show that the crystallization order of anosovite and silicate crystals in high CaO and MgO titanium slag during cooling is: Al_2TiO_5 > Ti_3O_5 > $MgTi_2O_5$ > $MgSiO_3$ > $CaSiO_3$ > $FeTi_2O_5$ > Mn_2SiO_4 > Fe_2SiO_4. Al_2TiO_5 and Ti_3O_5 have higher crystallization priority and should be responsible for the sharp increase in viscosity of titanium slag during cooling. The total crystallization rates of anosovite and silicate crystals are mainly controlled by Al_2TiO_5 and $MgSiO_3$, respectively. The mass ratio of $Ti_2O_3/\Sigma TiO_2$ has a prominent influence on the total crystallization rate of anosovite crystals while the mass ratio of MgO/FeO has a slight influence on the total crystallization rate of anosovite crystals.

Keywords: crystallization behaviors; crystallization rate; anosovite crystals; silicate crystals; titanium slag

1. Introduction

Titanium dioxide and metallic titanium are the main products of titanium metallurgy. Titanium dioxide is widely employed in the fields of coating, painting, paper, plastics, rubber, and ceramic because of its non-toxicity, opacity, whiteness and brightness [1–3]. Metallic titanium has various applications such as aviation, aerospace, biomedical, marine, and nuclear waste storage due to its high corrosion resistance, high specific strength, light weight, high melting point, and high chemical/heat stability [3–5]. A large amount of high-quality titanium-rich materials are increasing required due to the wider and wider application of titanium dioxide and metallic titanium. Titanium-rich materials mainly include two types: Rutile and titanium slag. With the depletion of rutile resource, titanium slag has become the main titanium-rich material to produce the metallic titanium and titanium white in China. The electric-furnace smelting has become the dominant process to produce titanium slag from ilmenite in China. During the process of the electric-furnace smelting, iron oxide in ilmenite concentrate is reduced to metallic iron and titanium oxide in ilmenite is left in the slag (called titanium slag). The flow behavior of titanium slag at high temperature is a crucially significant factor, influencing, for example, the ability to tap the titanium slag from the electric furnace, the efficiency of separation of titanium slag from metal iron, the foaming extent of titanium slag, and the kinetics of titanium smelting reactions [6]. There is some research about the flow behavior of titanium slag at high temperature in recent decades [7–10]. It has been concluded from the aforementioned research that the titanium slag crystallizes much more easily during tapping from the electric furnace and separation is more difficult from liquid iron than other metallurgical slags. Therefore, the investigation on crystallization behavior of titanium slag at a high temperature will contribute to insuring the smooth operation of the smelting and the efficient separation of titanium slag and metallic iron.

The crystallization of melts consists of two steps: Nucleation and growth [11,12]. The crystallization behavior of melts can be studied by three kinds of methods: Experimental techniques, molecular dynamic simulation, and classical crystallization theory. For example, the crystallization behaviors of mold flux and Ti-bearing blast furnace slag were intensively investigated by the differential scanning calorimeter [13,14], the single hot thermocouple technique/double hot thermocouple technique [15–17], the confocal laser scanning microscope [18–20], and the viscosity-temperature curve method. Watson et al. [21] studied the crystal nucleation and growth in Pd-Ni alloys by the molecular dynamic simulation, which was only suitable for a relatively simple system. Turnbull [22] interpreted convincingly the nucleation phenomena on the basis of the classical crystallization theory. In contrast, there is almost no effective experimental technique to investigate the crystallization behaviors of many different crystals in the relatively complex system of titanium slag. Furthermore, based on the fact that the titanium slag can corrode all refractory oxide containers [9], studies of the crystallization behavior of titanium slag using experiment techniques are extremely difficult and dangerous to carry out. Therefore, few research is focused on the crystallization behavior of titanium slag at a high temperature.

Panzhihua, located in southwestern China, holds ilmenite reserves of 870 million tons, accounting for 35% of total ilmenite reserves in the world [23]. The composition of mineral phases in titanium slag produced from Panzhihua ilmenite is complex due to its high content of CaO and MgO. According to the previous study by Dong et al. [24], the mineral phases in the high CaO and MgO titanium slag consisted of anosovite ($FeTi_2O_5$, $MgTi_2O_5$, Ti_3O_5, and Al_2TiO_5), olivine (Fe_2SiO_4 and Mn_2SiO_4), and augite ($CaSiO_3$ and $MgSiO_3$) as determined by X-ray diffraction and Energy Dispersive Spectrometer. Fan et al. [25] also confirmed the complex composition of mineral phases in the high CaO and MgO titanium slag. Therefore, the crystallization behaviors of crystals in the high CaO and MgO titanium slag would become more complex.

In this work, a mathematical model was established to describe the crystallization behaviors of anosovite and silicate crystals in the high CaO and MgO titanium slag. Specifically, two main issues will be addressed. First, the crystallization behaviors of anosovite and silicate crystals in the high CaO and MgO titanium slag will be determined. The crystallization ability, crystallization order, and optimum temperature ranges of different crystals in the high CaO and MgO titanium slag will be analyzed and discussed in detail. Second, the effect of the composition of titanium slag on the crystallization behaviors of anosovite crystals will be investigated. Based on the results, the smelting parameters of titanium slag will be adjusted in industrial production in further works to enhance the smooth operation of the electric furnace and to improve the separation of titanium slag and metallic iron.

2. Mathematical Model

According to the classical crystallization kinetics from D. Turnbull et al. [26], the nucleation rate I of a crystal and the formation energy of a crystal nucleus could be expressed as Equations (1) and (2) respectively.

$$I = \frac{nD}{a_0^2} \exp\left(-\frac{\Delta G^*}{kT}\right) \tag{1}$$

$$\Delta G^* = \frac{16\pi\sigma^3}{3(\Delta g_V)^2} \tag{2}$$

$$n = \frac{1}{a_0^3} \tag{3}$$

where I is the nucleation rate of a crystal in $m^{-3} \cdot s^{-1}$, n is the atom number per unit volume, D is the self-diffusion coefficient of atoms in the melt in $m^2 \cdot s^{-1}$, a_0 is the lattice parameter in m, ΔG^* is the formation energy of crystal nucleus in $J \cdot mol^{-1}$, k is the Boltzmann constant, T is the temperature of the melt in K, π is the pi constant, σ is the interfacial energy between the melt and crystal in $J \cdot m^{-2}$, and Δg_V is the change of Gibbs free energy per unit volume in $J \cdot mol^{-1} \cdot m^{-3}$.

The relationship between the self-diffusion coefficient D and the viscosity η could be expressed as Equation (4) [27,28].

$$D = \frac{kT}{3\pi a_0 \eta} \tag{4}$$

$$\Delta g_V = \frac{\Delta H_m^g}{V} \Delta T_r \tag{5}$$

where η is the viscosity of the melt in Pa·s, ΔH_m^g is the fusion heat per mole melt in $J \cdot mol^{-1}$, V is the mole volume of the crystal in $m^3 \cdot mole^{-1}$, and $\Delta T_r = 1 - T_r$ ($T_r = T/T_m$, T_m is the melting temperature of a crystal in K).

The viscosity involves the physicochemical property of the melt and can be estimated by the National Physical Laboratory model [29]. The optical basicity of the slag component for the calculation of titanium slag is from references [29–31]. The relationship between the viscosity and temperature was expressed as Equation (6).

$$\ln \eta = \ln A + B/T \tag{6}$$

The relationship between optical basicity and constants A and B could be expressed as Equations (7) and (8) [29].

$$\ln \frac{B}{1000} = -1.77 + \frac{2.88}{\Lambda} \tag{7}$$

$$\ln A = -232.69(\Lambda)^2 + 357.32\Lambda - 144.17 \tag{8}$$

The optical basicity of the titanium slag could be calculated by the following Equation (9).

$$\Lambda = \frac{\sum \chi_i n_i \Lambda_i}{\sum \chi_i n_i} \tag{9}$$

where χ_i and n_i is the mole fraction and the number of oxygen atoms of an independent component respectively and Λ_i is the optical basicity of the corresponding independent component.

Taking Equations (2)–(5) into Equation (1), the nucleation rate I of a crystal can be deduced as Equation (10).

$$I = \frac{kT}{3\pi a_0^6 \eta} \exp \left\{ -\frac{16\pi}{3T_r(\Delta T_r)^2} \left[\frac{(N_0 V^2)^{1/3} \sigma}{\Delta H_m^g} \right]^3 (\frac{\Delta H_m^g}{RT_m}) \right\} \tag{10}$$

where α and β are the two dimensionless parameters which have the important influence on the crystal melting process. α and β could be expressed as following Equations (11) and (12) [32].

$$\alpha = \frac{(N_0 V^2)^{1/3} \sigma}{\Delta H_m^g} \tag{11}$$

$$\beta = \frac{\Delta H_m^g}{RT_m} \tag{12}$$

where N_0 is the Avogadro constant.

Combined with Equations (11) and (12), the nucleation rate I of a crystal could be expressed as Equation (13).

$$I = \frac{kT}{3\pi a_0^6 \eta} \exp \left[-\frac{16\pi \alpha^3 \beta}{3T_r(\Delta T_r)^2} \right] \tag{13}$$

According to the classical kinetics of crystallization, the growth rate I_L of a crystal could be expressed as Equation (14).

$$I_L = \frac{f_s D}{a_0} \left[1 - \exp(\frac{\Delta G_g}{RT}) \right] \tag{14}$$

where f_s is the proportionality coefficient of the position benefiting the atomic adsorption on the crystal at interface between melt and crystal and ΔG_g is the change of Gibbs free energy per mole between the melt and crystal. *fs* and ΔG_g could be expressed as Equations (15) and (16) [33].

$$f_s = \begin{cases} 1 & \Delta H_m 2RT_m \\ 0.2\frac{T_m - T}{T_m} & \Delta H_m 4RT_m \end{cases} \tag{15}$$

$$\Delta G_g = \Delta H_m^g \frac{T_m - T}{T_m} \tag{16}$$

Combined with Equations (14)–(16), the growth rate of a crystal could be expressed as Equation (17).

$$I_L = \frac{f_s kT}{3\pi a_0^2 \eta} \left\{ 1 - \exp\left[\left(\frac{\Delta H_m^g}{RT_m}\right)\left(\frac{\Delta T_r}{T_r}\right) \right] \right\} \tag{17}$$

Combined with Equations (11), (12) and (17), the growth rate of a crystal could be expressed as Equation (18) [34].

$$I_L = \frac{f_s kT}{3\pi a_0^2 \eta} \left[1 - \exp\left(-\beta \frac{\Delta T}{T_r} \right) \right] \tag{18}$$

According to a previous research by Uhlmann et al. [33], the volume fraction (*x*) of a crystal could be expressed as a function with the nucleation rate (*I*), growth rate (I_L) and time *t*.

$$x = \frac{1}{3}\pi I I_L^3 t^4 \tag{19}$$

$$r_i = \frac{1}{3}\pi I I_L^3 \tag{20}$$

The total crystallization rate of a crystal r_{total} could be expressed as a function with mole fraction (ω_i) and crystallization rate (r_i) of a given crystal as Equation (21).

$$r_{total} = \sum_i \omega_i r_i \tag{21}$$

The chemical composition of titanium slag is as follows (wt. %): TiO_2 = 55.20, FeO = 12.21, SiO_2 = 5.89, MnO = 1.53, MgO = 2.51, Al_2O_3 = 2.83, CaO = 1.00, and Ti_2O_3 = 18.92. The structural parameters and melting temperatures of these crystals in titanium slag are listed in Table 1 [35–37]. $a_0 = (a \times b \times c)^{1/3}$, in which *a*, *b* and *c* are lattice parameters of crystals in the titanium slag. Because the content of Mn-anosovite is small, Mn-anosovite is neglected in this work.

Table 1. Structural parameters and melting temperature of crystals in titanium slag.

Mineral Phase	Chemical Formula	a_0 (Å)	T_m (K)
Anosovite	$FeTi_2O_5$	7.192 [35]	1767 [36]
	$MgTi_2O_5$	7.145 [35]	1930 [37]
	Ti_3O_5	7.186 [35]	1991 [37]
	Al_2TiO_5	6.858 [35]	2133 [36]
Olivine	Fe_2SiO_4	6.752 [35]	1478 [36]
	Mn_2SiO_4	6.878 [35]	1618 [36]
Augite	$CaSiO_3$	7.351 [35]	1817 [36]
	$MgSiO_3$	7.464 [35]	1830 [36]

3. Results and Discussion

3.1. Nucleation and Growth of Crystals in Titanium Slag

3.1.1. Nucleation and Growth of Anosovite Crystals

From the viewpoint of the classical crystallization theory, the crystallization process includes two stages: Nucleation and growth. The nucleation refers to the generation of a crystal nucleus from a mother phase. The growth of the crystal is realized by the growth of the crystal nucleus. Although the nucleation process is transitory compared with the subsequent growth process, the nucleation is completely different from the growth process.

Figure 1 displays the nucleation and growth rates of different anosovite crystals in the titanium slag. It can be observed from Figure 1 that the nucleation rates first increase from zero to the maximum value and then decrease to zero with the degree of undercooling increasing from zero. The profiles of nucleation rates should be attributed to the interaction of two contradictory factors related to the crystallization process: The degree of undercooling and diffusion ability of atoms. With a too high or too low degree of undercooling, the small nucleation rates of crystals are formed. With a reasonable degree of undercooling, the maximum nucleation rates of crystals are formed. The profiles of growth rates have the similar characteristics to the profiles of nucleation rates. The maximum nucleation rates and maximum growth rates of anosovite crystals follow the same order: $Al_2TiO_5 > Ti_3O_5 > MgTi_2O_5 > FeTi_2O_5$. The maximum nucleation rates (maximum growth rates) of different anosovite crystals are very different, indicating that the crystallization abilities of different anosovite crystals may differ remarkably.

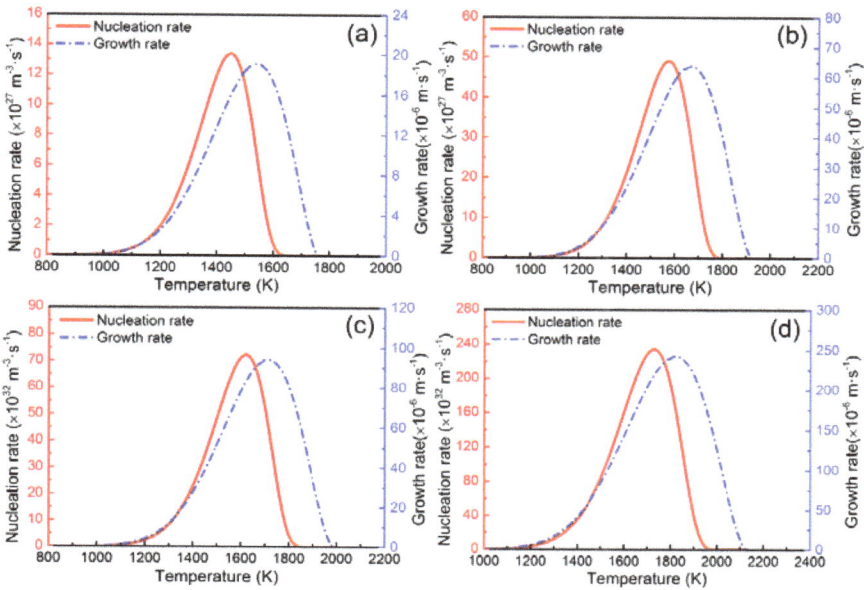

Figure 1. Nucleation and growth rates of different anosovite crystals: (**a**) $FeTi_2O_5$, (**b**) $MgTi_2O_5$, (**c**) Ti_3O_5, and (**d**) Al_2TiO_5.

3.1.2. Nucleation and Growth of Silicate Crystals

Figure 2 displays the nucleation and growth rates of different silicate crystals in titanium slag. It can be observed from Figure 2 that the profiles of nucleation rates of silicate crystals have the similar characteristics to the profiles of nucleation rates of anosovite crystals. The profiles of growth rates of silicate crystals have the similar characteristics to the profiles of growth rates of anosovite crystals. Both the maximum nucleation rates and the maximum growth rates of silicate crystals follow the same order: $MgSiO_3 > CaSiO_3 > Mn_2SiO_4 > Fe_2SiO_4$. The maximum nucleation rates (maximum growth rates) of different silicate crystals are similar in value, indicating that silicate crystals have the similar crystallization ability. The maximum nucleation rates (maximum growth rates) of silicate crystals are much smaller than those of anosovite crystals, demonstrating that silicate crystals may have much a weaker crystallization ability than the anosovite crystals.

The driving force of the crystallization and diffusion ability of atoms is large enough when the temperature is between that of the maximum nucleation rate and the maximum growth rate. Therefore, the range from the temperature of maximum nucleation rate to the temperature of maximum growth rate is specified as the characteristic temperature range of crystallization in this work. It is easy to nucleate, grow and obtain coarse grain within the characteristic temperature range. Figure 3 displays the characteristic temperature ranges of crystallization of anosovite and silicate crystals in titanium slag. It can be observed from Figure 3 that the characteristic temperature ranges of crystallization of crystals in titanium slag follow the sequence of $Al_2TiO_5 > Ti_3O_5 > MgTi_2O_5 > MgSiO_3 > CaSiO_3 > FeTi_2O_5 > Mn_2SiO_4 > Fe_2SiO_4$. The characteristic temperature ranges of crystallization of these crystals are in sequence as follows: 1728–1819 K, 1620–1712 K, 1573–1666 K, 1497–1589 K, 1487–1579 K, 1448–1540 K, 1333–1424 K, 1225–1312 K. The characteristic temperature ranges are consistent with the tapping temperature of smelting ilmenite to titanium slag in an electric furnace.

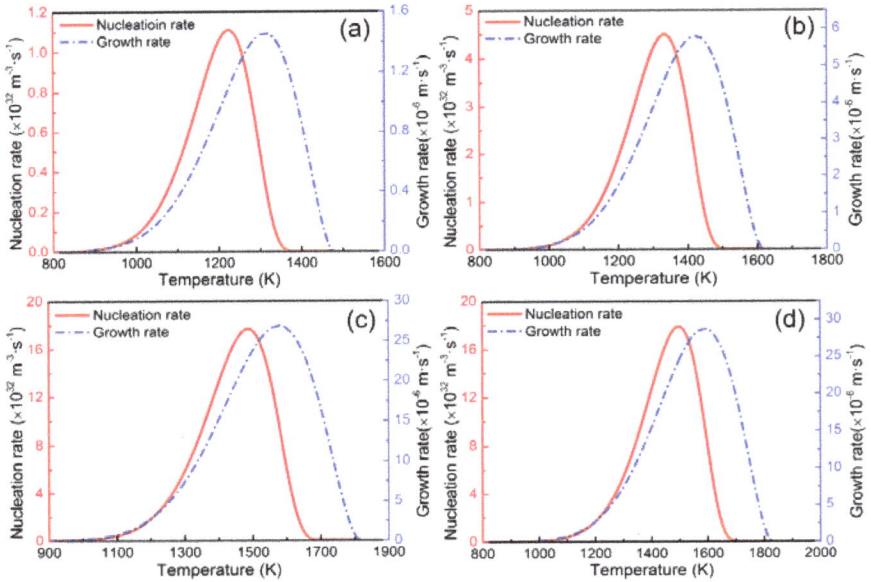

Figure 2. Nucleation and growth rates of different silicate crystals: (**a**) Fe_2SiO_4, (**b**) Mn_2SiO_4, (**c**) $CaSiO_3$ and (**d**) $MgSiO_3$.

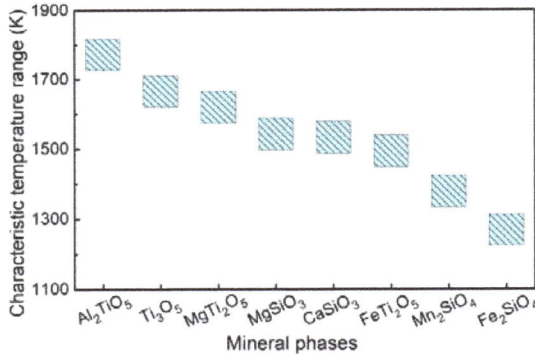

Figure 3. Characteristic temperature ranges for crystallization of anosovite and silicate crystals in titanium slag.

3.2. Crystallization Behavior of Anosovite and Silicate Crystals

The crystallization rate of crystals is the main and direct indicator for evaluating the crystallization behavior. Therefore, the optimum temperature range of crystallization can be determined by the crystallization rate of crystals. Figure 4 shows the crystallization rates of different anosovite and silicate crystals in titanium slag. It can be seen from Figure 4a that the maximum crystallization rate of Al_2TiO_5 has two higher orders of magnitude than Ti_3O_5. The maximum crystallization rate of $MgTi_2O_5$ has five higher orders of magnitude than $FeTi_2O_5$. It is notable that the total crystallization rate of anosovite crystals is mainly controlled by Al_2TiO_5. The optimum temperature ranges of anosovite crystals are as the following: Al_2TiO_5 (1744–1804 K) > Ti_3O_5 (1636–1696 K) > $MgTi_2O_5$ (1589–1649 K) > $FeTi_2O_5$ (1346–1406 K). It can be seen from Figure 4b that the maximum crystallization rates of

MgSiO$_3$ and CaSiO$_3$ have three higher orders of magnitude than Mn$_2$SiO$_4$ and five higher orders of magnitude Fe$_2$SiO$_4$. The total crystallization rate of silicate crystals is mainly controlled by MgSiO$_3$. The optimum temperature ranges of silicate crystals are as the following: MgSiO$_3$ (1512–1572K) > CaSiO$_3$ (1502–1562 K) > Mn$_2$SiO$_4$ (1346–1406 K) > Fe$_2$SiO$_4$ (1236–1296 K).

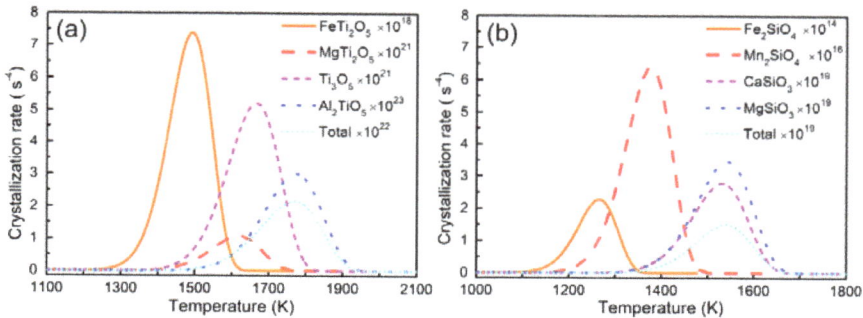

Figure 4. Crystallization rate of different crystals in titanium slag: (**a**) anosovite and (**b**) silicate.

Notably, both the maximum crystallization rate and the optimum temperature ranges of Al$_2$TiO$_5$ and Ti$_3$O$_5$ are much higher than other crystals in high CaO and MgO titanium slag. The crystallization priority of Al$_2$TiO$_5$ and Ti$_3$O$_5$ is consistent with the report from Zhao et al. [7] that the sharp increase in viscosity of titanium slag may be caused by the high Al$_2$O$_3$ and Ti$_2$O$_3$ content in titanium slag. The accuracy of the mathematical model can be preliminarily demonstrated by the consistency between the crystallization priority of Al$_2$TiO$_5$ and Ti$_3$O$_5$ in titanium slag and the remarkable effect of Al$_2$O$_3$ and Ti$_3$O$_5$ on viscosity of titanium slag. This result implies that the variation of Al$_2$O$_3$ and Ti$_3$O$_5$ content in titanium slag should be paid close attention to during the electric-furnace smelting process. Furthermore, Song et al. [38] and Handfield et al. [9] reported that the melting temperature of the titanium slag, whose composition was similar with the titanium slag in this work, was approximately 1873 K. Al$_2$TiO$_5$ in this work has the crystallization rate of 1.03×10^{23} (s^{-4}) at the temperature of 1873 K. The accuracy of the mathematical model can be further demonstrated by the crystallization rate of Al$_2$TiO$_5$ at the melting temperature of 1873 K.

T_c is the peak temperature of maximum crystallization rate of crystals. T_m is the melting point of the crystal. $(T_m - T_c)/T_m$ represents the crystallization process during cooling. Figure 5a shows the relationship between melting temperature and the crystallization process. It can be seen from Figure 5a that the crystal with higher melting temperature crystallizes earlier during cooling. This crystallization order agrees with the work from Diao et al. [39] that the mineral phases of vanadium-chromium slag crystallized in the order of their melt points during cooling. Figure 5b shows the relationship between the maximum crystallization rate and the peak temperature. It can be seen from Figure 5b that the crystallization order of crystals in titanium slag is obtained during cooling as follows: Al$_2$TiO$_5$ > Ti$_3$O$_5$ > MgTi$_2$O$_5$ > MgSiO$_3$ > CaSiO$_3$ > FeTi$_2$O$_5$ > Mn$_2$SiO$_4$ > Fe$_2$SiO$_4$. The crystal with a higher maximum crystallization rate has a higher peak temperature. This relationship may be ascribed to the strong diffusion ability of atoms with the higher temperature. Obviously, anosovite crystals crystallize overall earlier than silicate crystals during cooling. The crystallization order in this work is consistent with the reports from Pistorius et al. [40] and Wang et al. [41] that anosovite and silicate crystals occupied the center and outer layer of titanium slag particle, respectively. The accuracy of the mathematical model can be demonstrated once again by the consistency between the overall crystallization order and spatial distribution of anosovite and silicate crystals in high CaO and MgO titanium slag.

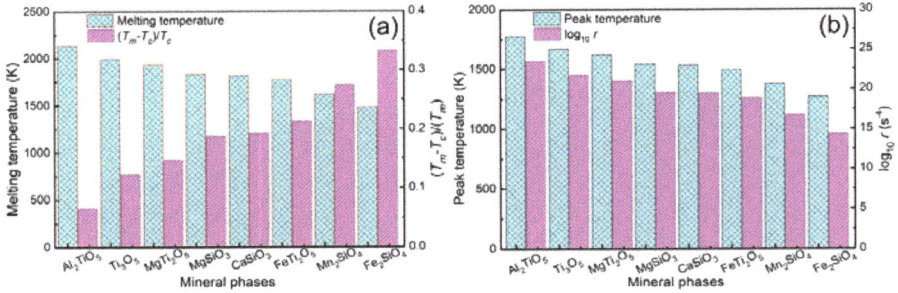

Figure 5. (**a**) Relationship between melting temperature and $(T_m - T_c)/T_c$ and (**b**) Relationship between maximum crystallization rate and peak temperature.

3.3. Effects of Slag Composition on Crystallization Behaviors of Anosovite Crystals

To investigate effects of the slag composition on crystallization behaviors of anosovite crystals, the mass ratio of $Ti_2O_3/\Sigma TiO_2$ and the mass ratio of MgO/FeO are selected as two variables. The chemical compositions of slag are shown in Table 2. The mass ratio of $Ti_2O_3/\Sigma TiO_2$ is from 0.05 (CS1) to 0.45 (CS5) and the mass ratio of MgO/FeO is from 0.10 (CS6) to 0.90 (CS10).

Table 2. Compositions of titanium slag (wt. %).

Samples	TiO$_2$	FeO	SiO$_2$	MnO	MgO	Al$_2$O$_3$	CaO	Ti$_2$O$_3$	Mass Ratio
CS1	72.46	12.21	5.89	1.53	2.51	2.83	1.00	3.78	0.05
CS2	63.62	12.21	5.89	1.53	2.51	2.83	1.00	11.35	0.15
CS3	55.20	12.21	5.89	1.53	2.51	2.83	1.00	18.92	0.25
CS4	46.79	12.21	5.89	1.53	2.51	2.83	1.00	26.49	0.35
CS5	38.37	12.21	5.89	1.53	2.51	2.83	1.00	34.06	0.45
CS6	55.20	13.38	5.89	1.53	1.34	2.83	1.00	18.92	0.10
CS7	55.20	11.32	5.89	1.53	3.40	2.83	1.00	18.92	0.30
CS8	55.20	9.81	5.89	1.53	4.91	2.83	1.00	18.92	0.50
CS9	55.20	8.66	5.89	1.53	6.06	2.83	1.00	18.92	0.70
CS10	55.20	7.75	5.89	1.53	6.97	2.83	1.00	18.92	0.90

Figure 6 shows the effects of mass ratio of $Ti_2O_3/\Sigma TiO_2$ and the mass ratio of MgO/FeO on the total crystallization rates of anosovite crystals. It can be observed that the total crystallization rates of anosovite crystals increases one order of magnitude with the mass ratio of $Ti_2O_3/\Sigma TiO_2$ increasing from 0.05 to 0.45 and increases only three times with the mass ratio of MgO/FeO increasing from 0.10 to 0.90. It can be deduced that the mass ratio of $Ti_2O_3/\Sigma TiO_2$ has a remarkable influence on the total crystallization rate of anosovite crystals while the mass ratio of MgO/FeO has a slight influence on the total crystallization rate of anosovite crystals.

Figure 7 shows the effects of the mass ratio of $Ti_2O_3/\Sigma TiO_2$ and the mass ratio of MgO/FeO on the maximum total crystallization rate and peak temperature of anosovite crystals. It can be seen from Figure 7 that the maximum crystallization rate of anosovite crystals decreases with the mass ratio of $Ti_2O_3/\Sigma TiO_2$ increasing, while the maximum crystallization rate of anosovite crystals increases with the mass ratio of MgO/FeO increasing. This phenomenon can be attributed to the variation of the viscosity of titanium slag resulting from the composition change. The peak temperatures of total crystallization rates of anosovite crystals increase with the mass ratio of $Ti_2O_3/\Sigma TiO_2$ increasing, while the peak temperatures of total crystallization rates of anosovite crystals decrease with the mass ratio of MgO/FeO increasing. This phenomenon demonstrates that the increase in the viscosity of titanium slag leads to the diffusion difficulty of atoms and the larger degree of undercooling is needed to crystallize.

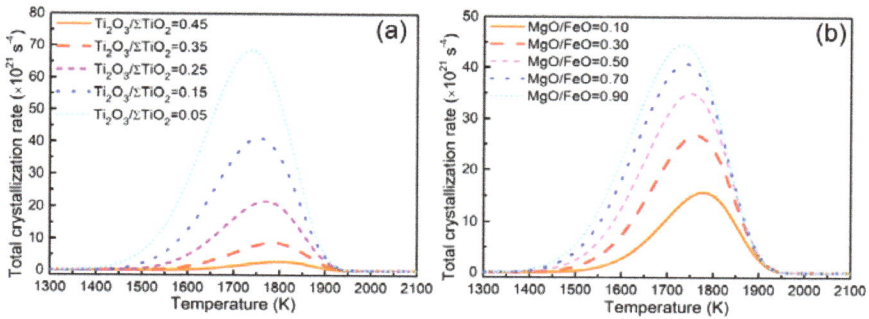

Figure 6. Effects of slag composition on total crystallization rates of anosovite crystals: (**a**) $Ti_2O_3/\Sigma TiO_2$ and (**b**) MgO/FeO.

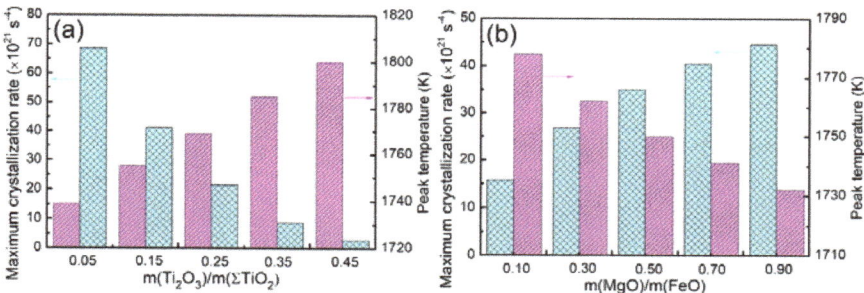

Figure 7. Effects of slag composition on maximum crystallization rates and peak temperatures of anosovite crystals: (**a**) $Ti_2O_3/\Sigma TiO_2$ and (**b**) MgO/FeO.

4. Conclusions

A mathematical model was established to describe the crystallization behaviors of anosovite and silicate crystals in high CaO and MgO titanium slag. The main results are shown as follows:

(1) The crystallization order of anosovite and silicate crystals in high CaO and MgO titanium slag is obtained during cooling as follows: $Al_2TiO_5 > Ti_3O_5 > MgTi_2O_5 > MgSiO_3 > CaSiO_3 > FeTi_2O_5 > Mn_2SiO_4 > Fe_2SiO_4$.

(2) The optimum temperature ranges of crystallization of these crystals in titanium slag are in their crystallization order as follows: 1744–1804 K, 1636–1696 K, 1589–1649 K, 1512–1572 K, 1502–1562 K, 1463–1523 K, 1346–1406 K, 1236–1296 K.

(3) The total crystallization rates of anosovite and silicate crystals are mainly controlled by Al_2TiO_5 and $MgSiO_3$, respectively. Al_2TiO_5 and Ti_3O_5 have a higher crystallization priority and should be responsible for the sharp increase in viscosity of titanium slag during cooling.

(4) The mass ratio of $Ti_2O_3/\Sigma TiO_2$ has a remarkable influence on the total crystallization rate of anosovite crystals while the mass ratio of MgO/FeO has a slight influence on the total crystallization rate of anosovite crystals.

Author Contributions: H.F., D.C., T.L., and H.D. deduce and validate the model. H.F., T.L., Y.H., M.L. and W.H. performed the calculation. All of the authors analyzed and discussed the data; H.F. wrote the paper.

Funding: This research received no external funding.

Acknowledgments: Special appreciation is extended to PhD candidates Huabiao Chen and Sheng Yu at Chongqing University for the help on the graphic design.

Conflicts of Interest: The authors declare no conflict interest.

References

1. Liu, Y.; Meng, F.; Fang, F.; Wang, W.; Chu, J.; Qi, T. Preparation of rutile titanium dioxide pigment from low-grade titanium slag pretreated by the naoh molten salt method. *Dyes Pigment.* **2016**, *125*, 384–391.

2. Liu, W.; Lü, L.; Yue, H.; Liang, B.; Li, C. Combined production of synthetic rutile in the sulfate TiO_2 process. *J. Alloys Compd.* **2017**, *705*, 572–580. [CrossRef]

3. Fan, H.; Chen, D.; Liu, P.; Duan, H.; Huang, Y.; Long, M.; Liu, T. Structural and transport properties of feo-TiO_2 system through molecular dynamics simulations. *J. Non-Cryst. Solids* **2018**, *493*, 57–64. [CrossRef]

4. Cho, J.; Roy, S.; Sathyapalan, A.; Free, M.L.; Fang, Z.Z.; Zeng, W. Purification of reduced upgraded titania slag by iron removal using mild acids. *Hydrometallurgy* **2016**, *161*, 7–13. [CrossRef]

5. Chen, G.; Song, Z.; Chen, J.; Srinivasakannan, C.; Peng, J. Investigation on phase transformation of titania slag using microwave irradiation. *J. Alloys Compd.* **2013**, *579*, 612–616. [CrossRef]

6. Kondratiev, A.; Jak, E.; Hayes, P. Predicting slag viscosities in metallurgical systems. *JOM* **2002**, *54*, 41–45. [CrossRef]

7. Zhao, Z.; Ma, E.; Lian, Y. Behavior of Al_2O_3 in titianium slag. *Iron Steel Vanadium Titan.* **2002**, *23*, 36–38.

8. Li, S.; Lv, X.; Song, B.; Miu, H.; Han, K. Rheological property of TiO_2-FeO-(SiO_2, CaO, MgO) ternary slag. *Chin. J. Nonferrous Metals* **2016**, *26*, 2015–2022.

9. Handfield, G.; Charette, G. Viscosity and structure of industrial high TiO_2 slags. *Can. Metall. Q.* **1971**, *10*, 235–243. [CrossRef]

10. Gao, G.; Yang, Y.; Yang, D. Viscosity and melting point determination of titanium slag. *Iron Steel Vanadium Titan.* **1987**, *9*, 51–55.

11. Komatsu, T. Design and control of crystallization in oxide glasses. *J. Non-Cryst. Solids* **2015**, *428*, 156–175. [CrossRef]

12. Iqbal, N.; Van Dijk, N.; Offerman, S.; Moret, M.; Katgerman, L.; Kearley, G. Real-time observation of grain nucleation and growth during solidification of aluminium alloys. *Acta Mater.* **2005**, *53*, 2875–2880. [CrossRef]

13. Gan, L.; Zhang, C.; Shangguan, F.; Li, X. A differential scanning calorimetry method for construction of continuous cooling transformation diagram of blast furnace slag. *Metall. Mater. Trans. B* **2012**, *43*, 460–467. [CrossRef]

14. Gan, L.; Zhang, C.; Zhou, J.; Shangguan, F. Continuous cooling crystallization kinetics of a molten blast furnace slag. *J. Non-Cryst. Solids* **2012**, *358*, 20–24. [CrossRef]

15. Li, J.; Wang, X.; Zhang, Z. Crystallization behavior of rutile in the synthesized Ti-bearing blast furnace slag using single hot thermocouple technique. *ISIJ Int.* **2011**, *51*, 1396–1402. [CrossRef]

16. Klug, J.L.; Hagemann, R.; Heck, N.C.; Vilela, A.C.; Heller, H.P.; Scheller, P.R. Crystallization control in metallurgical slags using the single hot thermocouple technique. *Steel Res. Int.* **2013**, *84*, 344–351. [CrossRef]

17. Zhou, L.; Wang, W.; Huang, D.; Wei, J.; Li, J. In situ observation and investigation of mold flux crystallization by using double hot thermocouple technology. *Metall. Mater. Trans. B* **2012**, *43*, 925–936. [CrossRef]

18. Hu, M.; Liu, L.; Lv, X.; Bai, C.; Zhang, S. Crystallization behavior of perovskite in the synthesized high-titanium-bearing blast furnace slag using confocal scanning laser microscope. *Metall. Mater. Trans. B* **2014**, *45*, 76–85. [CrossRef]

19. Liu, J.; Verhaeghe, F.; Guo, M.; Blanpain, B.; Wollants, P. In situ observation of the dissolution of spherical alumina particles in CaO–Al_2O_3–SiO_2 melts. *J. Am. Ceram. Soc.* **2007**, *90*, 3818–3824.

20. Ruvalcaba, D.; Mathiesen, R.; Eskin, D.; Arnberg, L.; Katgerman, L. In situ observations of dendritic fragmentation due to local solute-enrichment during directional solidification of an aluminum alloy. *Acta Mater.* **2007**, *55*, 4287–4292. [CrossRef]

21. Watson, K.D.; Nguelo, S.T.; Desgranges, C.; Delhommelle, J. Crystal nucleation and growth in Pd–Ni alloys: A molecular simulation study. *CrystEngComm* **2011**, *13*, 1132–1140. [CrossRef]

22. Turnbull, D. Formation of crystal nuclei in liquid metals. *J. Appl. Phys.* **1950**, *21*, 1022–1028. [CrossRef]

23. Wu, F.; Li, X.; Wang, Z.; Wu, L.; Guo, H.; Xiong, X.; Zhang, X.; Wang, X. Hydrogen peroxide leaching of hydrolyzed titania residue prepared from mechanically activated panzhihua ilmenite leached by hydrochloric acid. *Int. J. Miner. Process.* **2011**, *98*, 106–112. [CrossRef]

24. Dong, H. Study on Production of High-Quality Synthetic Rutile from Electric Furnace Titanium Slag with High Content of Calcium and Magnesium. Ph.D., Thesis, Central South University, Changsha, China, 2010.

25. Fan, H.; Duan, H.; Tan, K.; Li, Y.; Chen, D.; Long, M.; Liu, T. Production of synthetic rutile from molten titanium slag with the addition of B_2O_3. *JOM* **2017**, *69*, 1914–1919. [CrossRef]

26. Seitz, F.; Turnbull, D. *Solid State Physics*; Academic Press: New York, NY, USA, 1958; Volume 7.

27. Meyer, R.E. Self-diffusion of liquid mercury. *J. Phys. Chem.* **1961**, *65*, 567–568. [CrossRef]

28. Hoffman, R.E. The self-diffusion of liquid mercury. *J. Chem. Phys.* **1952**, *20*, 1567–1570. [CrossRef]

29. Mills, K.C.; Sridhar, S. Viscosities of ironmaking and steelmaking slags. *Ironmak. Steelmak.* **1999**, *26*, 262–268. [CrossRef]

30. Zhang, Y. Electronegativities of elements in valence states and their applications. 1. Electronegativities of elements in valence states. *Inorg. Chem.* **1982**, *21*, 3886–3889.

31. Dimitrov, V.; Komatsu, T. Correlation among electronegativity, cation polarizability, optical basicity and single bond strength of simple oxides. *J. Solid State Chem.* **2012**, *196*, 574–578. [CrossRef]

32. Turnbull, D. Under what conditions can a glass be formed? *Contemp. Phys.* **2006**, *10*, 473–488. [CrossRef]

33. Uhlmann, D.R. A kinetic treatment of glass formation. *J. Non-Crystall. Solids* **1972**, *7*, 337–348. [CrossRef]

34. Dai, D. *Amorphous Physics*; Electronic Industry Press: Beijing, China, 1989.

35. Powder Diffraction File. *International Centre for Diffraction Data*; Swarthmore: Newtown Square, PA, USA, 2000.

36. Atlas, S. (Ed.) *By Vdeh*; Verlag Stahleisen GmbH: Düsseldorf, Germany, 1995; pp. 48–90.

37. Eriksson, G.; Pelton, A.D. Critical evaluation and optimization of the thermodynamic properties and phase diagrams of the MnO-TiO_2, MgO-TiO_2, FeO-TiO_2, Ti_2O_3-TiO_2, Na_2O-TiO_2, and K_2O-TiO_2 systems. *Metall. Mater. Trans. B* **1993**, *24*, 795–805. [CrossRef]

38. Song, B. *Rheological Property and Melt Structure of High Titania Slag*; Chongqing University: Chongqing, China, 2015.

39. Diao, J.; Zhou, W.; Gu, P.; Ke, Z.; Qiao, Y.; Xie, B. Competitive growth of crystals in vanadium–chromium slag. *CrystEngComm* **2016**, *18*, 6272–6281. [CrossRef]

40. Pistorius, P.C.; Kotzé, H. Role of silicate phases during comminution of titania slag. *Miner. Eng.* **2009**, *22*, 182–189. [CrossRef]

41. Wang, D.; Wang, Z.; Qi, T.; Wang, L.; Xue, T. Decomposition kinetics of titania slag in eutectic naoh-nano 3 system. *Metall. Mater. Trans. B* **2016**, *47*, 1–9.

metals **MDPI**

Article

Model Development for Refining Rates in Oxygen Steelmaking: Impact and Slag-Metal Bulk Zones

Ameya Kadrolkar * and Neslihan Dogan

McMaster Steel Research Centre, Department of Materials Science and Engineering, McMaster University, Hamilton, ON L8S 4L8, Canada; dogann@mcmaster.ca
* Correspondence: kadrola@mcmaster.ca

Received: 22 December 2018; Accepted: 2 March 2019; Published: 8 March 2019

Abstract: A new approach has been adopted to predict the contribution of the impact and slag-metal bulk zones to the refining rates of impurities in a top blown oxygen steelmaking process. The knowledge pertaining to the behavior of top-jets and bottom stirring plumes (water model and industrial studies) was adapted. For the impact zone, the surface renewal generated by the top jet as well as bottom stirring plumes is incorporated in the current model, whereas in the case of slag-metal bulk zones the surface renewal is caused solely by the bottom stirring plumes. This approach helped in achieving a more explicit use of process parameters in quantifying the slag formation. The results suggest a minor contribution of these two zones to the overall refining of impurities throughout the oxygen blow.

Keywords: oxygen steelmaking; refining kinetics; slag formation; penetration theory

---

1. Introduction

The slag formation in oxygen steelmaking consists of oxidation reactions (Si, Mn, Fe, P) and the dissolution of flux additions such as CaO and MgO. At the start of the blow, there is rapid oxidation of Si and Mn due to thermodynamic favorability [1]. This is followed by the main decarburization period, in which a majority of the carbon removal takes place. Deo and Boom [2] claimed that [Si] > 0.05 wt % suppresses the CO formation/decarburization and the effective estimation of desiliconization rate is essential to predict the start of main decarburization period. Phosphorus removal takes place primarily within the metal droplets in the emulsion zone. With the dissolution of fluxes, the slag basicity increases and the oxides of P and Mn are reduced to a certain extent in the emulsion zone. Therefore, there is an increase in P and Mn contents in the bath in the middle of the blow and this phenomenon is termed "reversion" [2,3]. Although these reactions and their order of events are mostly understood through the sampling studies of the bath, the exact contribution of various reaction zones to these reactions has not been well understood.

Most oxidation reactions are extremely favorable at high temperatures and the thermodynamic aspect of these reactions is known under oxygen steelmaking conditions. The progress of these reactions is limited by the kinetics of the refining reactions within individual zones. The path of slag evolution has been reported based on plant trials [4–8] and previous modeling attempts [9–16] in literature. Early models developed by Asai and Muchi [9] and Jalkanen et al. [10] assumed that the reactions take place in a single zone. Asai and Muchi [9] suggested that slag is formed solely on the surface of the cavity, by incorporating the absorption of oxygen and the simultaneous oxidation of carbon, silicon and manganese. The rate constants (and their relative magnitudes) for these oxidation reactions were considered as input data. The mean residence time of steel at the surface of the cavity was extremely small ~ 10^{-5} s, which does not reflect the actual circulation of metal underneath the cavity. Jalkanen et al. [10,11] described refining with a generalized reaction zone. They incorporated the reaction affinities and diffusivity of the impurities, and the transport of the impurities to the reaction zone due to energy dissipation from top blown and bottom stirring gases. An energy dissipation model developed by Nakanishi et al. [17] was applied. The mass transfer of impurities is correlated with top gas and bottom stirred gas flow rates by introducing two fitting parameters. Even though their approach is an effective way to describe the oxidation rate of impurities and the slag formation phenomena, an assessment of these two parameters is not available. These fitted parameters may vary from one vessel to another or with operational conditions. Therefore, it is difficult to further assess the significance of these parameters on the calculation of mass transfer rates.

Since the distinct role of impact and emulsion zones as reaction zones has been established [7,8,18], there have been recent attempts to model refining reactions in these zones separately [12–16,19]. Sarkar et al. [12] assumed that a stoichiometric equivalent quantity of hot-metal to the O_2 jet gets oxidized at the impact zone, which acts as a precursor for further refining in the emulsion zone. Their modeling predictions for [Si] and [Mn] did not correlate well with those reported by Cicutti et al. [4,5]. Rout et al. [13,14] presented a three zone model in which they applied a first order rate equation to predict the refining rates of Si and Mn at the impact zone. The mass transfer of impurities in the metal phase was assumed to be the rate limiting step. The mass transfer coefficient was calculated using an empirical correlation suggested by Kitamura et al. [20] which incorporated induced stirring energy (by bottom stirring gas) in the bath and geometrical parameters of the furnace. As this correlation [20] was based on measuring the oxidation rates of impurities in hot metal in contact with a layer of FeO containing slag for experimental and industrial ladles systems, the fluid flow dynamics of these systems are different from those applicable to the impact zone in an oxygen steelmaking furnace. At the impact zone intense turbulence is generated due to the impingement of the oxygen jet and there is a relatively small amount of slag in contact with the metal bath, as the slag is displaced by the O_2 jet in lateral direction. Some multi-zone models [15,16] provide a reasonable approach to predict slag formation, however the details of this model are not available in open literature. Knoop et al. [15] presented a "slag-droplet model" based on the multi-component mixed transport control (MMTC) theory [21]. It was assumed that FeO is formed in the impact zone, followed by an FeO reduction with dissolved carbon in metal droplets. The refining reactions were assumed to occur in the emulsion zone, and oxygen was supplied through the formation of FeO at the impact zone. Jung et al. [16] presented a thermodynamic model to represent various phenomena on the oxygen steelmaking such as slag formation, scrap and flux dissolution. The eight phenomena were located in a bulk metal bath (1.scrap dissolution and 2.metal homogenization), impact zone (3.surface and 4.hot-spot volume) and slag (5.slag-metal bulk reaction, 6.emulsion, 7.flux dissolution, 8.slag homogenization). The kinetics associated with these phenomena were incorporated by varying the volume of reaction zones as a function of the blowing conditions. At the impact zone, only the surface oxidation of metal is assumed to occur during the soft blow period, while during the medium and hard blow period oxidation is assumed to occur at a depth beneath the impact zone. However, the criterion for a variation of reaction zone volume is unclear and the underlying empirical reactions used to determine the volume of the reaction zones were not described. Dogan et al. [19] developed a model to predict the decarburization rate at the impact and emulsion zones separately. This work was able to predict the end point carbon content of liquid metal however it does not include other important refining reactions. Further, some recent findings on the bloating behavior of droplets were not included in these models. Coley and coworkers [22–27] conducted numerous high temperature experiments using an X-ray fluoroscopy technique to quantify the nucleation, growth and escape of CO gas bubbles within droplets and the interplay between decarburization and dephosphorization kinetics of bloated droplets in steelmaking slags. They studied the effect of temperature, metal chemistry and FeO content in slag. It was concluded that the refining rates within droplets are extremely fast and the droplet generation rate is a limiting factor to extend the refining rates for the oxygen steelmaking process. Developing a comprehensive model for top blown oxygen steelmaking that incorporates and critically investigates these new findings is currently the focus of a study by the authors of this paper. The central thesis of the model is that the kinetics of oxygen steelmaking is dominated by changes in the motion of iron droplets from the moment they are ejected from the surface of the metal bath to the moment they return to metal bath.

The process model focusses on the refining rates of major impurities such as carbon, silicon and manganese in different reaction zones to predict the metal and slag chemistry throughout the blow. The overarching aim of the current work is to provide better knowledge on the contribution of the refining rates at the impact zone and slag-metal bulk solely based on operational parameters using a mechanistic approach. This study is an attempt to use the theoretical findings from the experimental studies to the full-scale operating conditions by minimizing the use of empirical/fitting parameters. This would make the application of the model to different steelmaking furnaces straightforward. The conceptual model developed by Dogan et al. [19,28–30] will be used in this work. This model consists of various sub-modules to describe scrap and flux dissolution, emulsion and impact zone decarburization. In this study, two reaction zones are considered, namely reactions at the impact zone and at the slag-bulk metal interface, while the contribution of other reaction zones will be described in the subsequent work. Only few studies [4,5,31] have reported the path of slag evolution using industrial data. In the current study, the measured data of Cicutti et al. [4,5] is used to analyze the importance of refining rates for an industrial practice since the slag path was described based on the steel and slag samples taken at various times of the blow.

2. Model Development

The authors suggest that the refining rates of impurities are controlled by the mass transfer of impurities in the metal at the gas-metal and the slag-metal interfaces. The refining rate of solutes [Si, Mn] can then be calculated using the following equation.

$$W_{[X]} = \frac{J_{[X]} M_X}{1000} = \begin{pmatrix} \frac{k_{X-gm} A_{gm} \rho_m}{100} \left\{ [wt \% X]_b - [wt \% X]_{i-gm} \right\} \\ + \left(\frac{k_{X-sm} A_{sm} \rho_m}{100} \left\{ [wt \% X]_b - [wt \% X]_{i-sm} \right\} \right) \end{pmatrix} \tag{1}$$

where $W_{[X]}$ is the weight of solute removed (kg/s), X represents solutes such as Si and Mn in the liquid metal, $J_{[X]}$ is the moles of solute removed per unit time, (mol/s), M_X is the molecular weight of solute, and the subscripts gm and sm represent the gas-metal and slag-metal interface, respectively. A is the contact area/interfacial area, ρ_m is the density of hot-metal, k_X is the mass transfer coefficient of solute (m/s), $[wt \% X]_b$ and $[wt \%X]_i$ are the solute contents in the bath and at the interface, respectively. It is assumed that all silicon and manganese brought to the impact zone (gas-metal interface) are oxidized since oxidation of these elements is highly favorable at steelmaking temperatures, and hence $[wt \% X]_{i-gm} \approx 0$, whereas the interfacial equilibrium concentration at the slag-metal bulk is determined by the distribution coefficient between metal and slag.

$$[wt \% X]_{i-sm} = L_X (wt \% X) \tag{2}$$

where [] indicates the element dissolved in iron and () indicates the compound dissolved in slag. L_X is the distribution coefficient of solute between metal and slag. The L_{Si} [32] and L_{Mn} [33] values are calculated through the approach adopted by Rout et al. [13,14,34].

2.1. Description of Fluid Flow at the Impact Zone and Slag-Metal Bulk Due to Top-Oxygen Jet

Figure 1 schematically depicts phenomena at the impact zone of the oxygen steelmaking furnace [3,35]. The top gas jet impinging on the metal bath surface forms a cavity and causes the displacement of liquid metal and hence leads to the continuous renewal of the reaction area. The supersonic oxygen jet is obstructed by the metal bath and its velocity is reduced to impingement point velocity, u_j. The jet emerges in a radially outward direction with a further reduced velocity called tangential gas velocity, u_g. A fraction of jet momentum is used to generate metal droplets from the bath while the residual jet momentum induces circular eddy flows in the bath, which bring the elements like C, Si and Mn to the surface of the cavity. The velocity of the bath underneath the cavity surface due to top-jets is termed the surface renewal velocity, u_l.

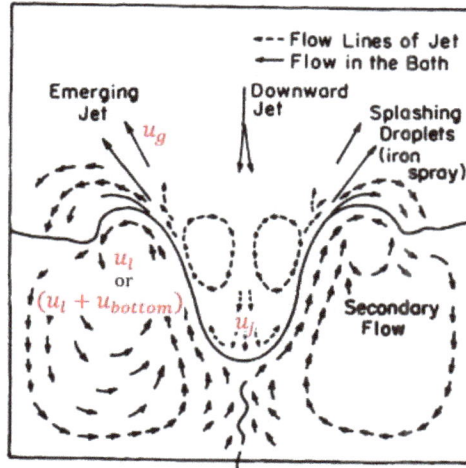

Figure 1. Fluid flow behavior at impact zone by gas jet impingement. Reproduced from [3,35], with permission of Springer, 1980. u_j, u_g, u_l, u_{bottom} are vertical velocity at the impingement point, tangential velocity of gas-jet, and surface renewal velocity of the metal bath due to oxygen jet and bottom stirring plumes, respectively. $u_j > u_g >> u_l$. In the current model the surface renewal velocity due to top jet bottom stirring plumes is $(u_l + u_{bottom})$.

Observation of actual fluid flow behavior in oxygen steelmaking furnace is impossible due to extreme conditions. Sharma, Hlinka and Kern [36] observed the flow behavior of metal due to the interaction of an oxygen jet with a 200 lb steel bath using an X-ray adjacent to a quartz window in order to establish the direction of fluid flow at the jet-metal impingement point. It is important to note that they didn't propose any correlation to predict the velocity of liquid using this technique. Davenport et al. [37] took images to track circulation of plastic beads in water induced by the impinging gas jet. The density of plastic beads was equal to the water. They were able to observe liquid behavior at a rapid speed in a radially outward direction close to the surface of the bath. They found that momentum gained from the gas jet was sufficient to carry this liquid metal stream down to the sides and back to the center. The evaluation of the surface velocity of liquid metal is very critical to the calculation of mass transfer coefficient. Even though previous studies [38–42] based on CFD simulations provide the values of surface velocity for a certain time step, a correlation incorporating the effects of blowing profile of oxygen steelmaking on surface velocity is necessary. Recently Hwang and Irons [43] performed water modelling studies to evaluate the velocity of surface renewal of a water bath due to the impinging gas jet. They measured the cavity dimensions using high speed imaging and local and bulk liquid velocities using the particle image velocimetry (PIV) technique as a function of various lance heights. They simplified stress balance and suggested the following correlation,

$$u_l = Au_g + B \tag{3}$$

where u_g and u_l are tangential velocities of gas and liquid (due to momentum of top-jet), respectively. A and B are constant values. Upon employment of the assumption that a linear relationship exists between the impact and tangential gas velocities, i.e., $u_j = \eta.u_g$ and application of local modified

Froude number similarity to the Equation (3), the following correlation was obtained between liquid velocity and depth of cavity;

$$u_l = A'\sqrt{n_o} + B \tag{4}$$

where n_o is the depth of the cavity created due to impingement of the jet on the bath surface. Based upon their experimental observation, they suggested that this correlation was valid for varying lance heights relevant to steelmaking conditions. The effect of cavity shape [43]:

$$\theta = tan^{-1}\left(\frac{d_c}{2n_o}\right) \tag{5}$$

where θ is the cavity angle (°) and d_c is the cavity diameter. The contact distance (between jet and cavity) increases as the cavity angle θ increases. This correlation is given by [43].

$$u_l \times 100 \times cos\theta = \left((0.026 \pm 0.004) \times \sqrt{n_o \times 100}\right) - (0.02 \pm 0.006) \tag{6}$$

This correlation incorporates the cavity dimension: shape (θ) and depth n_o and explicitly correlates jet parameters with liquid metal velocity at the impact zone. In this study, the methodology by Dogan et al. [19] was adapted to calculate the diameter and depth of the cavity. Thus, the knowledge of cavity dimensions allows us to calculate the velocity of metal displaced underneath the cavity due to the impact of the oxygen jet.

2.2. Description of Fluid Flow at the Impact Zone and the Slag-Metal Bulk Interface Due to Bottom Stirring

Bottom stirring by gases like Ar and N_2 is widely used to homogenize the metal bath in the oxygen steelmaking process [2]. As the stirring gas is introduced from the bottom of the vessel (through either porous plugs or tuyeres), gas-metal plumes are formed, and the amount of metal reaching the surface of the metal bath increases. This increases splashing and the amount of metal in contact with the oxygen jet [44–47]. Therefore, a quantification of the amount of metal being brought to the impact zone is necessary. Krishnapishrody and Irons [48] developed a correlation between various plume parameters on a fundamental basis. They characterized the gas-metal plumes using scaled parameters to evaluate the operating variables such as metal and gas velocities in plume and metal circulation rate. Since their model results were validated against a wide variety of industrial data, this model was preferred in comparison to other models [49,50], and used in the current study to quantify the amount of metal being brought to the impact zone. A brief description of the model is as follows.

Non-dimensional gas flow rate Q^* and height z^* are defined by the following two equations, respectively.

$$Q^* = \frac{Q}{g^{0.5}H^{2.5}} \tag{7}$$

$$z^* = \frac{z}{H} \tag{8}$$

The superscript * is used to indicate non-dimensional quantity. The functional relationship between the non-dimensional liquid velocity, u^*_{bottom} and Q^* and z^* is given by [48]

$$u^*_{bottom} = 1.16 \times (Q^*)^{0.32}(z^*)^{-0.28} \tag{9}$$

The actual liquid plume velocity, u_{bottom} can be calculated using the following non-dimensional relationship:

$$u_{bottom} = u^*_{bottom}\sqrt{gH} \tag{10}$$

In ladles, single plumes are used to increase the mass transfer between slag and metal [51–53]. Contrary to a single gas-metal plume in a ladle, the emergence of bottom stirring plumes on a free surface is more complex. Based on locations of the bottom stirring plugs the plume may emerge to different locations on a free surface. In this study, the bottom stirred plug configuration is applied based on the data reported by Bertezzolo et al. [54] for the 200-t oxygen steelmaking furnace. The top view of this interaction between 8-bottom stirring plumes and a 6-holed lance for a 200-t oxygen steelmaking furnace is shown in Figure 2a,b. It should be noted that the intention of this figure is a schematic diagram to represent the assumption related to the plume-cavity interaction. The fluid flow profile is not computed by the authors. The interaction of the plumes with the cavities and the slag metal bulk is characterized on the basis of the following assumptions:

1. The eight plumes are represented by a sub-sector of 45 degrees each. The downward circulating plumes do not affect the flow beyond their respective sub-sector.
2. Each plume has a significant momentum and by virtue of that, undergoes complete radial expansion in its subsector. This leads to surface renewal and supply of metal to the gas-metal interface (cavities) and slag-metal interface (slag-metal bulk).
3. Since only six cavities are created (by 6 holed lance), in contrast to the 8 bottom stirring plumes, the plumes are classified in two sets, namely:

   a. (Plume set A) 4 Partial expanded plumes: Two plumes are responsible for bringing liquid metal in contact with cavity. This leads to surface renewal of single cavity as shown in Figure 2a.
   b. (Plume set B) 4 Total expanded plumes: Each of the plumes causes surface renewal of the single cavity, as shown in Figure 2b

4. The behavior of the plumes in the annular region surrounding the cavities is uniform in each sub-sector.
5. The instantaneous dimension of the cavities can be calculated as a function of the lance parameters and from that the width of the annulus is calculated. These values are used to calculate the instantaneous refining in the respective zones.
6. The metal flow resulting from the top-jet and the bottom stirring plumes is assumed to be additive, hence the surface renewal velocity is the sum of the top-jet and the bottom-stirring surface renewal velocities.

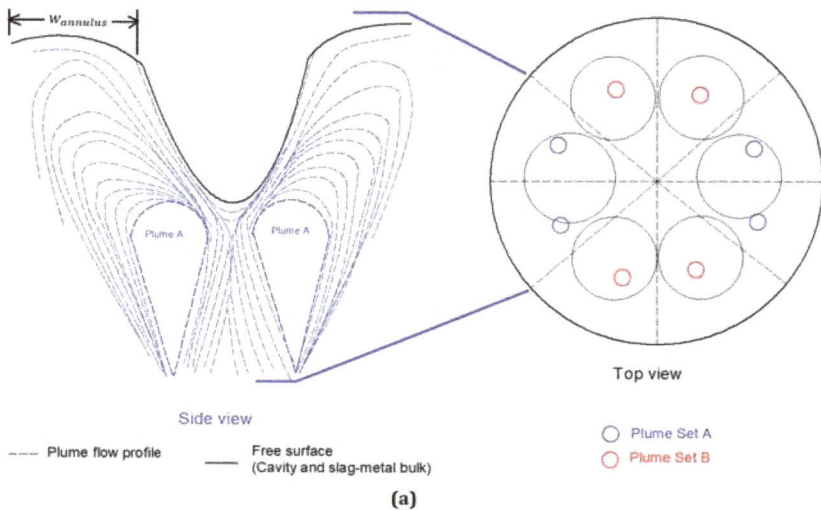

Figure 2. *Cont.*

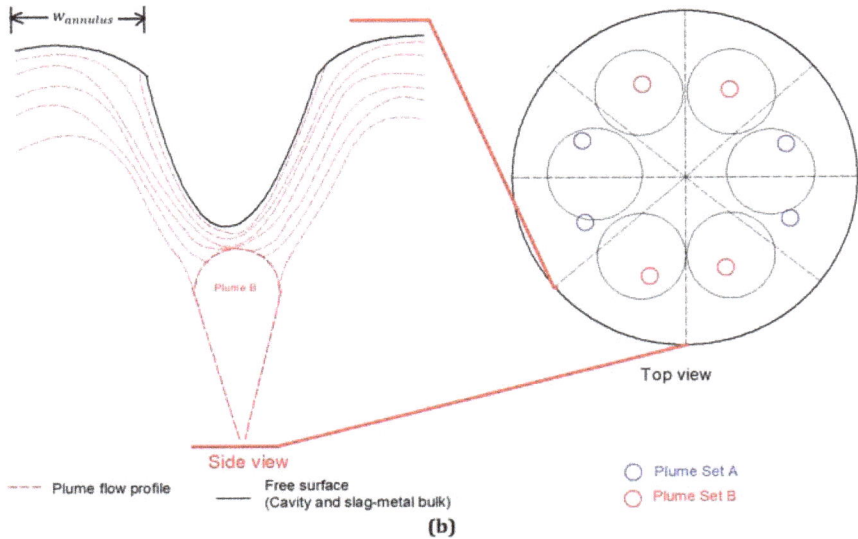

Figure 2. Schematic representation of the interaction between cavities and plumes (**a**) Expansion of plume set A underneath cavities (Side view), (**b**) Expansion of plume set B underneath cavities (Side view).

The solid lines in the figure represent the free surfaces (cavities, slag-metal bulk and vessel wall) whereas the dotted lines represent the plume flow profiles. The exact location of the porous plugs at the bottom is currently unknown and is based on bottom stirring configuration of the 200-t furnace being modeled in the study of Bertezzolo et al. [53].

The instantaneous density and composition differences between upper and lower baths are quite possible, but a large circulation of metal throughout the bath would reduce these gradients to a significant extent. A high bath circulation rate of 126 t/min (refer Appendix A) calculated in the current work for a 170–190 t bath indicates a high turnover of bath, thus decreasing these (density and composition) gradients and their effect on the metal flow behavior in a relatively short period of time. The concentration and temperature gradients can be neglected under the defined stirring conditions.

2.3. Determination of Mass Transfer at the Impact Zone

The mass transfer coefficient, k, can be defined according to Higbie's penetration theory [55],

$$k = 2 \times \sqrt{\frac{D}{\pi t_c}} \tag{11}$$

where D is the diffusion coefficient of the reacting element X, t_c is the residence time of the reacting element at the impact zone/reaction interface and is defined by

$$t_c = \frac{l_c}{u} \tag{12}$$

where u is the velocity of surface renewal, l_c represents the 'characteristic length' and it is the half of the circumference (i.e., arc length) of the paraboloid cavity, since the surface renewal is assumed to be symmetric about the axis of the cavity. Based on the above stated assumptions the mass transfer coefficients at the impact zone and slag-metal bulk are calculated as shown in Tables 1 and 2.

Table 1. Evaluation of mass transfer parameters at impact zone (gas-metal interface).

Plume Set	Reaction Area, m²	Characteristic Length, m	Time of Contact, s	Mass Transfer Coefficient, m/s
A	$\frac{1}{2} \times \left(\frac{A_{gm}}{n_{cav}} \right)$	$l_{c-sm} = \frac{1}{2} \times C_{cavity}$	$t_c = \frac{l_{c-gm}}{(u_l + u_{bottom})}$	$k_{gm} = 2 \times \sqrt{\frac{D}{\pi t_c}}$
B	$\left(\frac{A_{gm}}{n_{cav}} \right)$			

where A_{gm} = Total area of cavities (gas-metal interfacial area), (A_{gm}/n_{cav}) represents area of a single cavity and C_{cavity} = Circumference of the cavity. It should be noted that u_l and u_{bottom} are calculated using Equations (6) and (10).

Table 2. Evaluation of mass transfer parameters at slag-metal bulk (slag-metal interface).

Reaction Area, m²	Characteristic Length, m	Time of Contact, s	Mass Transfer Coefficient, m/s
$A_{sm} = A_{vessel} - A_{gm}$	$l_{c-sm} = w_{annulus}$	$t_{c-sm} = \frac{l_{c-sm}}{(u_l + u_{bottom})}$	$k_{sm} = 2 \times \sqrt{\frac{D}{\pi t_{c-sm}}}$

where A_{sm} = Area of slag-metal bulk, m², $w_{annulus}$ = Width of annulus between cavities and wall of vessel.

The values for the diffusion coefficients of Si and Mn in the liquid iron, which are determined from experimental studies [56–58] and compiled by Kawai and Shiraishi [59], are used in the current study. They range from 4×10^{-9} to 5×10^{-9} m²/s for Si (at 1550 °C–1725 °C) and 1.77×10^{-9} to 2.5×10^{-9} m²/s for Mn (at 1550 °C–1700 °C). However, another study by Grace and Derge [60] suggests that the values are ranging from 1.78×10^{-8} to 2.11×10^{-8} m²/s for Si and 8.8×10^{-9} to 1.05×10^{-8} m²/s for Mn in carbon saturated liquid iron. It is important to note that these values are one order of magnitude higher than those obtained from Calderon et al. [56], Majdic et al. [57] and Kawai et al. [58]. Since these sets of experimental values were arrived at independently, under different experimental conditions the values were used in the calculations contrary to those suggested by Grace and Derge [60].

2.4. Determination of Impact (Reaction) Area

In this study, the impact zone is defined as the smooth surface of the cavities formed by the supersonic jets from the lance, where the oxygen comes in contact with the metal bath. Previous studies [61–64] indicated that the cavity surface is rough, resulting in the generation of "splash sheets" [64] or "metal-bath spraying effect" [61]. However, the surface area enhancement due to splash sheet formation is difficult to estimate. Therefore, surface roughness is not included in the model development. In this study, the methodology by Dogan et al. [19] was adapted to calculate the impact area as follows.

The depth n_o and diameter d_c of the cavity are calculated using the correlation developed by Koria and Lange [65], for the penetrability of impinging oxygen jets in molten pig-iron baths. They found that these parameters were mainly affected by the oxygen supply pressure, nozzle diameter and the lance height which contribute to the momentum of gas jet. The following equations are used to calculate the depth and diameter of the cavity.

$$n_o = 4.469 \times h \times \left(0.7854 \times 10^5 d_{th}^2 P_a \left(1.27 \frac{P_o}{P_a} - 1 \right) \cos \alpha \frac{1}{g \rho_m h^3} \right)^{0.66} \tag{13}$$

$$d_c = 2.813 \times h \times \left(0.7854 \times 10^5 d_{th}^2 P_a \left(1.27 \frac{P_o}{P_a} - 1 \right) (1 + \sin \alpha) \frac{1}{g \rho_m h^3} \right)^{0.282} \tag{14}$$

where h is the lance height, d_t is the throat diameter of lance's nozzle, P_o is the supply pressure of oxygen, P_a is the ambient pressure inside the vessel, g is the acceleration due to gravity, α is the nozzle inclination angle. Then the area of single cavity, A_c is calculated using the correlation

$$A_c = \frac{\pi r_c}{6n_o^2}\left(\left(r_c^2 + 4n_o^2\right)^{\frac{3}{2}} - r_c^3\right) \tag{15}$$

where r_c is the radius of the cavity.

The total impact area of jets can be calculated by summation of individual cavities for multi-nozzle lances [2].

$$A_{gm} = \sum_{c=1}^{n_{cav}} A_c \tag{16}$$

The slag-metal bulk is defined by the interface between the bath and slag at the annular region between the cavities and the wall of the furnace. It is evaluated by subtracting the total area of cavities from the cross-sectional area of the vessel.

The sequence of calculation for the refining model at the impact and slag-metal bulk zones is represented in Figure 3. For every time step, the data from hot metal and scrap composition such as manganese and silicon, metal-bath height, H, lance height, h, velocity of the oxygen jet at the impact point u_j and bottom gas flow rate, Q_b with time are taken as inputs. The resultant cavity parameters and reaction areas (impact zone and slag-metal bulk) are then evaluated, followed by the calculation of the surface renewal velocities due to the top-jets and bottom-stirring separately as described in the previous section. The mass transfer coefficients at the two zones are calculated using Equation (11) while the refining rate at the two zones is calculated using Equation (1). Subsequently the composition of the bath is updated by taking into account the instantaneous bath weight W_b, melted scrap weight W_{sc}, weight of solute removed W_X and the sequence of calculation is repeated for the next time step.

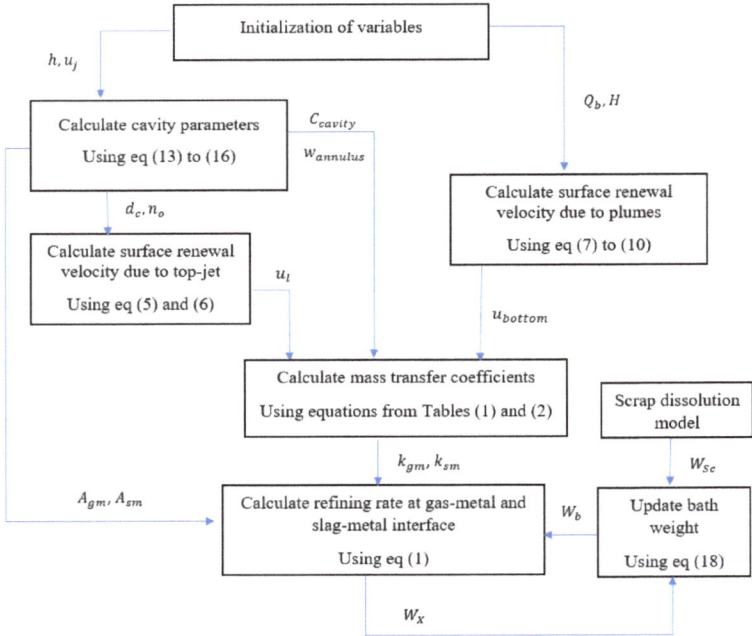

Figure 3. Schematic representation of calculation procedure for refining model.

3. Results and Discussion

3.1. Liquid Velocity

The surface renewal velocities at the impact zone were calculated with respect to time, considering the instantaneous blowing parameters for the operation of Cicutti et al. [4,5]. Figure 4 illustrates the variation of the surface renewal velocity due to the top-jet with respect to cavity dimensions at various lance heights prevalent during the blow. The decrease in lance height decreases the surface renewal velocity and increases the cavity depth and radius. The decrease in lance height would increase the momentum transferred by the oxygen jet to the cavity, which is expected, but this increased momentum supply is consumed in droplet generation and does not translate into an increase in surface renewal rate. This result is consistent with the literature. Hwang and Irons [43] also stated that the kinetic energy transfer (from the jet to the bath) was more efficient at higher lance heights based on their observation. The Energy Transfer Index (ETI) values for higher lance heights were higher than those for lower lance heights. The ETI is defined as a ratio between the kinetic energy of the bath and the input kinetic energy of the jet [43]. Similar observations were made in recent study by Zhou et al. [66] on kinetic energy dissipation by metal bath and slag in oxygen steelmaking vessels. For a 100-t oxygen steelmaking vessel they calculated the ETI value to decrease by 36%, when the lance height was lowered from 1.6 m to 1 m (for O_2 flow rate of 3.76 $\frac{Nm3}{t \times min}$).

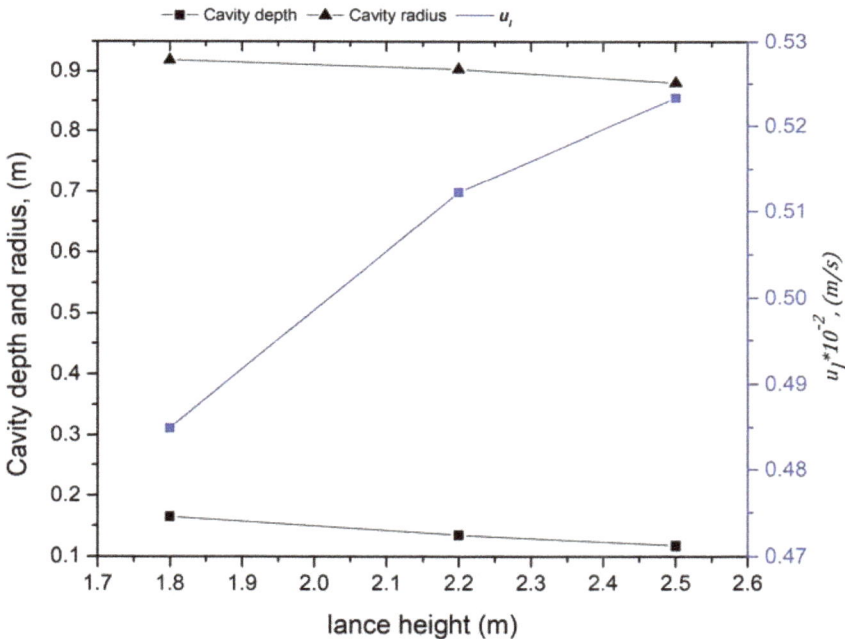

Figure 4. Change of cavity dimensions and surface velocity at various lance heights.

The surface renewal velocities due to the top jet, u_l is compared with the surface renewal velocity due to bottom stirring, u_{bottom} in Figure 5. The velocity of the metal circulation due to top jet varies from 0.0048 m/s to 0.0052 m/s and it is two orders of magnitude lower than the metal velocity at the top of the plume (0.45 to 0.64 m/s). Even though the characterization of metal flow is complex due to the overlap of the varying circulation patterns with respect to top-jet and plumes, this result shows that the plumes dominate the supply of metal to the cavities and the slag-metal bulk.

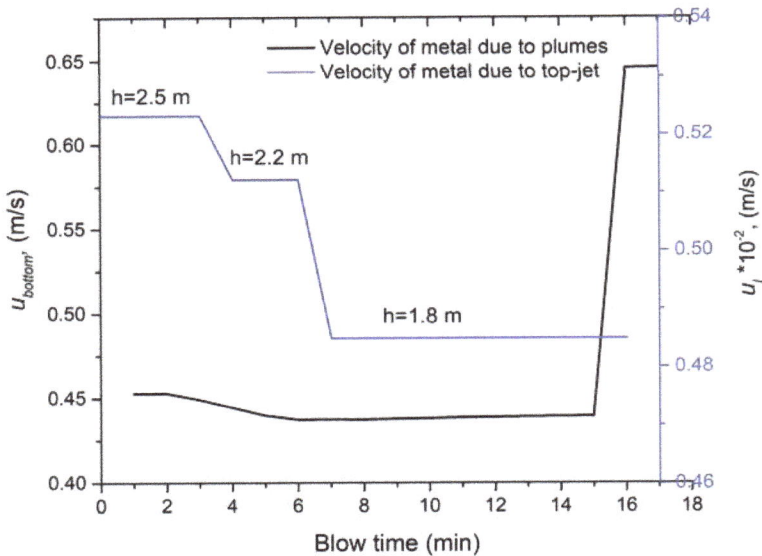

Figure 5. Comparison of surface renewal velocities due to top-jet and bottom stirring, (h = lance height, m).

It should be noted that this model predicts a large quantity of metal (~126 t/min: procedure for the calculation described in Appendix A) is circulated by the bottom stirring plumes for a 200-t oxygen steelmaking furnace described by Cicutti et al. [4,5]. However, it is worth considering exactly how much of this metal comes into contact with oxygen from the jet or the slag.

3.2. Mass Transfer Coefficients

The values of mass transfer coefficients are calculated with respect to the process parameters such as lance height and bottom stirring gas flow rate. These parameters affect the cavity dimensions and the metal circulated coming in contact with the oxygen jet. The mass transfer coefficient at the cavities k_{cav}, related to liquid velocity due to the top jet varies from 6.1×10^{-6} to 5.8×10^{-6} m/s from start to end of the blow. If the contribution of both top-jet and bottom stirring plumes is considered, k_{gm} varies from 5.77×10^{-5} to 6.70×10^{-5} m/s. Similarly, the mass transfer coefficient at the slag-metal bulk due to the top jet varies from 6.3×10^{-6} to 6.5×10^{-6} m/s and for a combined contribution of top jet and bottom stirring plumes it varies from 5.94×10^{-5} to 7.45×10^{-5} m/s. This study shows that the magnitude of the mass transfer coefficients for the top jet is considerably lower than combined effect of top blowing and bottom stirring. This finding is consistent with literature [17,53,67,68].

It should be noted that the magnitudes are lower as compared with those reported in the literature. The values of mass transfer coefficients which have been reported by Ohguchi et al. [21], $k_m = 4 \times 10^{-4}$ m/s and $k_s = 2 \times 10^{-4}$ m/s (metal and slag phase mass transfer coefficients respectively, for gas stirred mass transfer between the two phases).

In Figure 6, the predictions of mass transfer coefficients at impact zone, k_{cav} from current study are compared with the values calculated though Kitamura et al.'s [20] approach for Cicutti et al.'s steelmaking operation (wherein the mass transfer coefficient varies as a function of stirring energy imparted on the bath through top-jet and bottom stirring gas). The mass transfer coefficients have been evaluated at two distinct bottom stirring flow rates, namely the base case of 2.5 Nm³/min and 5 Nm³/min. If the bottom stirring rate is doubled, an 11% increase in mass transfer coefficient is observed whereas in Kitamura et al.'s case the increase is 23%. The increase in bottom stirring rate would indeed increase the metal re-circulated per unit time (thus reducing the mixing time) but does not proportionately increase the metal reaching the interface. There is a limit to the metal coming in contact with the interfaces which is discussed in Section 3.3.

The values predicted by Kitamura et al. [20] are at-least one order of magnitude higher than those predicted in the current study. In Kitamura et al.'s [20] case the decrease in lance height translates to an increase in momentum transfer to the bath and a corresponding increase in mass transfer coefficient is observed. In the current study the magnitude of mass transfer coefficient decreases due to a decrease in lance height as discussed in the previous section. The increased supply of momentum is expended in droplet generation rather than agitation of bath. The sudden rise in mass transfer coefficients in the last 2 min of the blow is due to threefold increase in the bottom stirring rate (for bath homogenization).

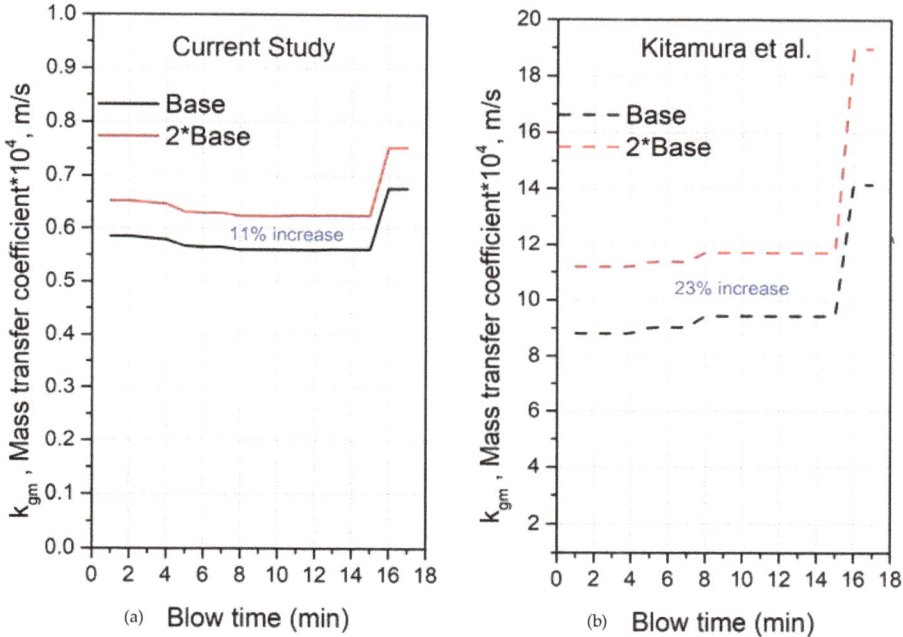

Figure 6. Comparison of metal phase mass transfer coefficients calculated by (**a**) Current model and (**b**) correlation obtained from the study of Kitamura et al. [20].

3.3. Effect of Lance Height and Bottom Stirring on the Metal Circulation Rate at the Interfaces

As discussed in the previous section the metal circulating at the interfaces, MCRI, (cavities and slag-metal bulk) is calculated separately for each time step, based on the instantaneous mass transfer coefficient and areas and is given by the following equation.

$$MCRI = \left\{ \left(k_{gm} A_{gm} \rho_{metal} \right) + \left(k_{sm} A_{sm} \rho_{metal} \right) \right\} \qquad (17)$$

Figure 7a shows the variation of total metal circulating at the interfaces and the instantaneous lance height and bottom stirring rates. The metal circulation rate at interfaces (MCRI) varies between 388 kg/min and 468 kg/min. As expected the MCRI shows a similar dependence on the lance height as shown by the mass transfer coefficient. The bottom stirring rate affects this parameter significantly. The lowered lance height represents a harder blow as shown in Figure 7b. This figure indicates that there is a marginal change as compared to the base case. On the other hand, doubling the bottom stirring rate increases the MCRI and it varies from 446 kg/min to 523 kg/min as shown in Figure 7c. The cross-sectional area of the plume increases successively as it rises towards the top of the metal bath. However, regardless of the mode of emergence of the plume on the free surface, the plume is able to circulate the metal across a larger area, where metal (and consequently the impurities like C, Si and Mn) comes in contact with oxygen from either the jet or FeO in slag.

It should be noted that the un-melted scrap most likely interacts with the plumes depending on scrap dimensions. Consequently, fluid flow at the impact and slag-metal bulk zones may also be affected by the presence of un-melted scrap. However, to the best of authors' knowledge, there is no study available for the effects of un-melted scrap on the fluid flow behavior as well as mass transfer of solutes in the open literature. Therefore, the interaction of un-melted scrap with the rising plume is not considered in the current study.

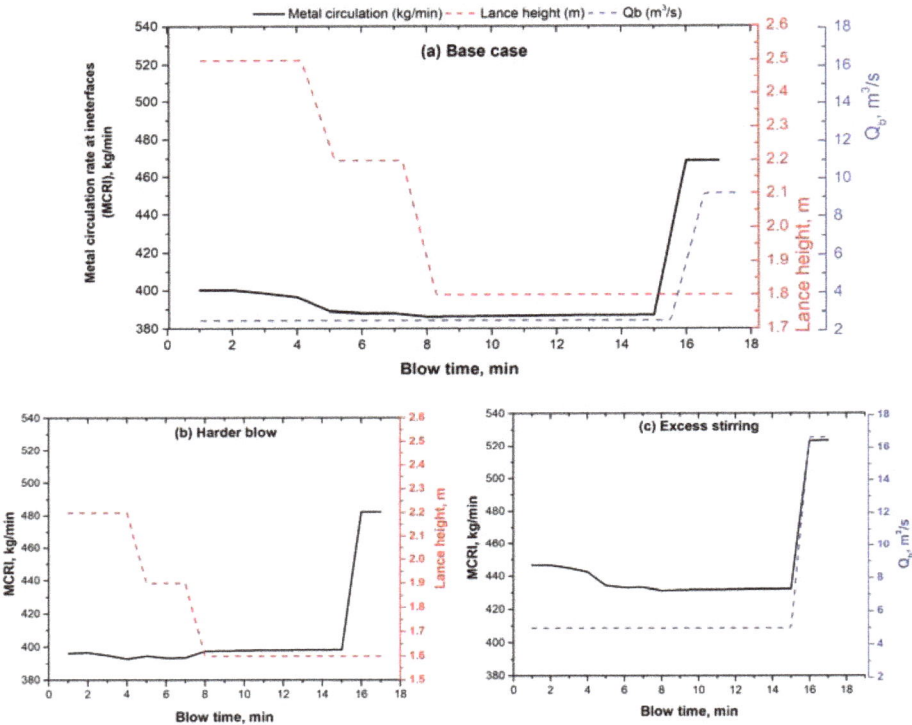

Figure 7. Effect of lance height and stirring rate on metal circulated at cavities and slag-metal bulk. (a) Base case (for operation of Cicutti et al., (b) Harder blow i.e., consistently lower lance height (2.2 m/1.9 m/1.6 m), (c) Excess stirring rate: 5 m³/s (0–15 min), 16.66 m³/s (16–17 min).

3.4. Refining Rates at the Interface

The change in bath weight during the blow was calculated by an approach previously suggested by Dogan et al. [28,29]. This involves accounting for the instantaneous amount of scrap freezing/melting W_{sc}^t, droplet ejected $W_{md}^{ejection}$ and droplet returned W_{md}^{return} to the bath and metal consumed in slag formation W_{ox}, as indicated in Equation (18)

$$W_b^t = W_b^{t-\Delta t} + W_{sc}^t - W_{md}^{ejection} + W_{md}^{return} - W_{ox} \tag{18}$$

The change in bath and scrap masses for Cicutti et al.'s operation is shown in Figure 8. There is a minor decrease in metal bath weight in the first 2 min of the blow due to the freezing of metal on the surface of cold scrap and a corresponding rise in scrap weight. Subsequently the scrap melts steadily and the bath weight increases linearly. There is a minor change in the bath weight after scrap is fully melted.

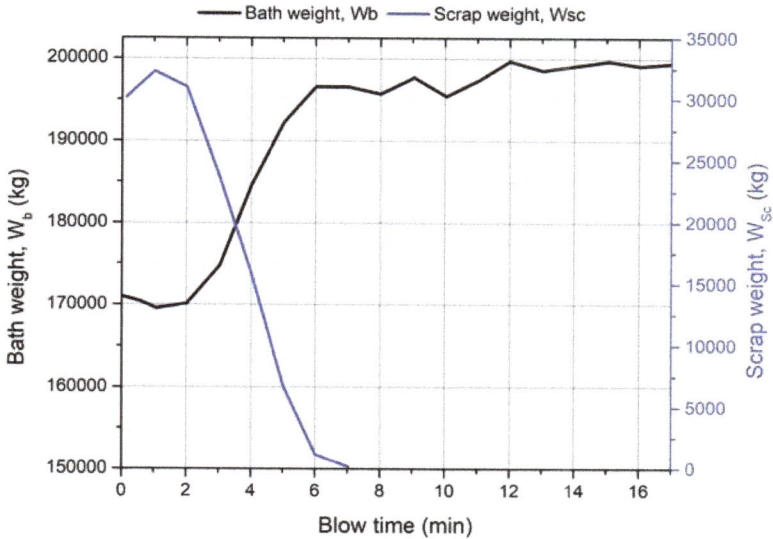

Figure 8. Bath and scrap weight changes for Cicutti et al.'s [4,5] operation.

Figure 9 compares the effect of top oxygen jets and combined blow (top oxygen jets + bottom stirring plumes) on the oxidation rates of silicon and manganese in the liquid metal at the cavities and the slag-metal bulk (as described in the combined refining model in Figure 3). The refining model predicts oxidation throughout the blow. Si content in the metal is observed to decrease in the early part of the blow, primarily due to a reduction in bath weight (due to freezing of metal on scrap). During the later part of the blow, there is a negligible change due to an increase in bath weight, even though oxidation continues to occur. A similar trend is observed for refining of Mn from liquid metal. There is hardly any oxidation of silicon and manganese taking place due to top oxygen jets. This is most likely due to the fact that surface renewal velocity is very small and hence only a marginal momentum of the oxygen jets is transferred to the bath.

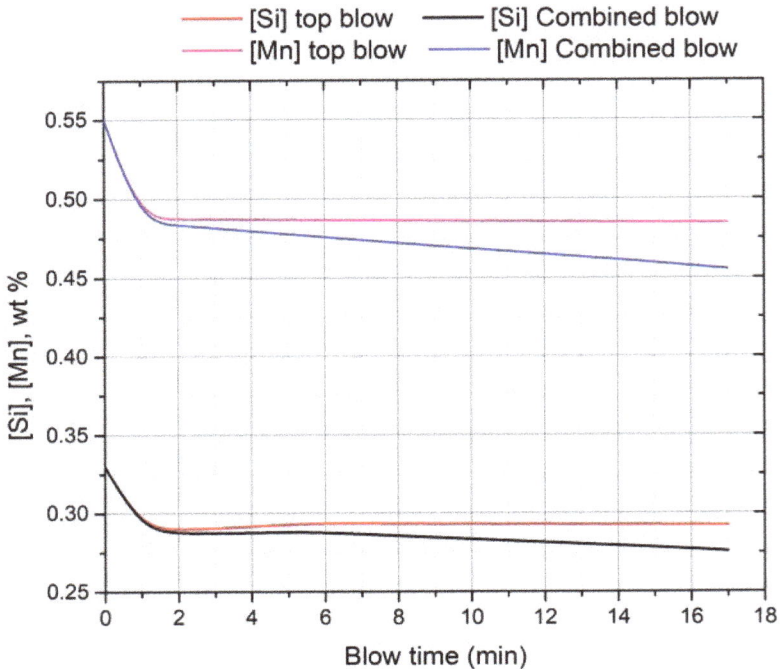

Figure 9. Effect of top and combined blows on the silicon and manganese oxidation rates during oxygen blow.

Figure 10 compares the predicted concentration of silicon and manganese in the metal to the measured values reported in the study of Cicutti et al. [4,5]. The calculated values are significantly higher than the measured values due to low refining rates at the impact and slag-metal bulk zones. The value of product of mass transfer coefficient and reaction area ($k * A$) should be 50–100 times higher if the impact and slag-metal bulk zones are considered as major refining zones. This indicates that the contribution of these reaction zones is insignificant. The authors are currently working on the calculation of refining rates in other possible reaction zones. It is still important to identify the possible sources of such a low prediction. It is worth considering how the selection of diffusion coefficient values from literature would affect the refining rate predictions. The lower diffusion coefficient values by Grace and Derge [54] (discussed in Section 2.3) will affect the mass transfer coefficients to an extent, however the mass transfer rates don't increase significantly and there is no appreciable increase in the refining rates (the predicted final silicon content decreases from 0.2758 wt % to 0.25 wt %.

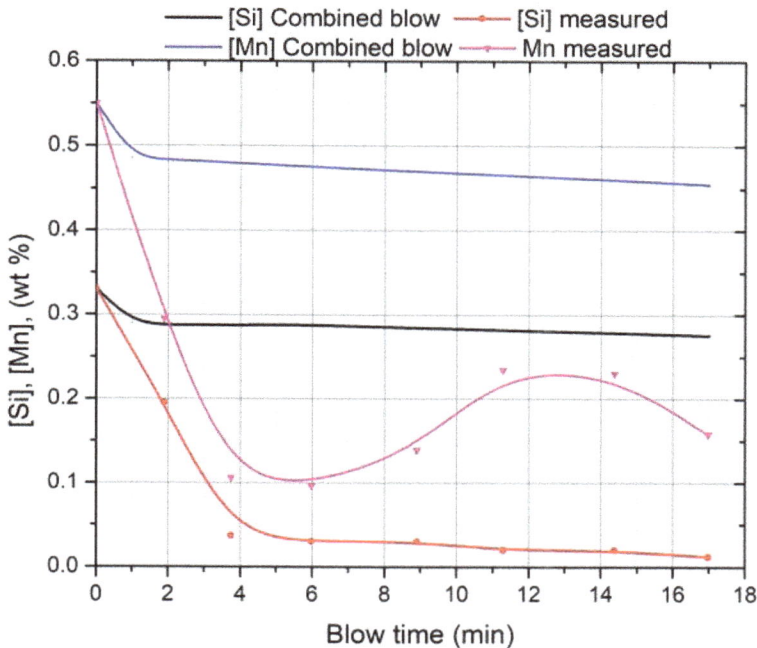

Figure 10. Comparison of measured and predictions for silicon and manganese in the liquid metal.

Figure 11 shows the variations in the area of the impact zone and the slag-metal bulk zone and the individual contributions to the refinement of silicon. The area of the impact zone increases as the lance height decreases. Therefore, the slag-metal bulk area decreases. The refining rate of silicon is affected proportionately. The silicon refining rate falls between minutes 2 and 4 due to a decrease in the weight of the metal bath. It can be inferred from this figure that a change in lance height will vary the areas of cavities and slag-metal bulk, but this will not significantly increase the refining in the corresponding reaction zones.

Another possible reason for the low predictions might be a small gas-metal reaction interface which is related to the assumption that the cavity surface is smooth (approximately 15 m²). If the surface area of the impact zone rises by a factor of 10 (by assuming roughness due to splash sheet formation), the corresponding increase in the silicon refining rate is shown in Figure 12. The final silicon content goes down to 0.20 wt % if the surface roughness is considered, but it still cannot explain the rapid oxidation of silicon during the initial part of the blow. It is worth mentioning about the hot-model study by Koch et al. [61]. They conducted experiments which involved the blowing of an oxygen jet on a 50 kg molten Fe-C bath to simulate the top-blown oxygen steelmaking process. The generated droplets due to "metal-bath spraying" at the cavity were sampled at various points of the cross section of the metal sampler during a blow. The authors observed that at any given instant, the droplets had lower carbon content than the metal bath. Further the "inner" droplets (ejected into the jet) experienced much higher decarburization as compared to the "outer" droplets (ejected away from jet). This indicates a possibility of significant contribution of decarburization and other refining reactions via the gas-metal droplet interface rather than the smooth cavity surfaces.

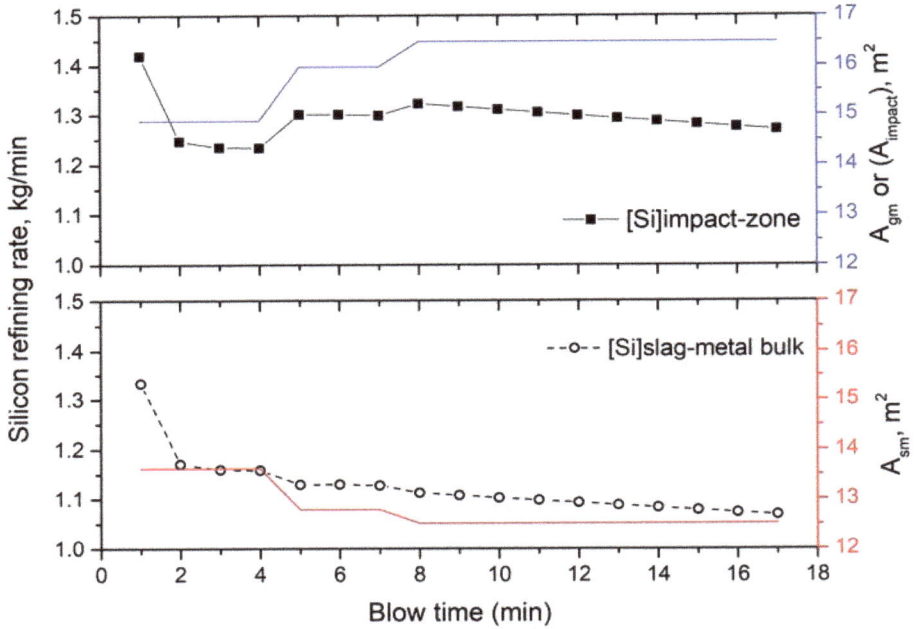

Figure 11. Variation of area and contribution to refining by impact and slag-metal bulk zones.

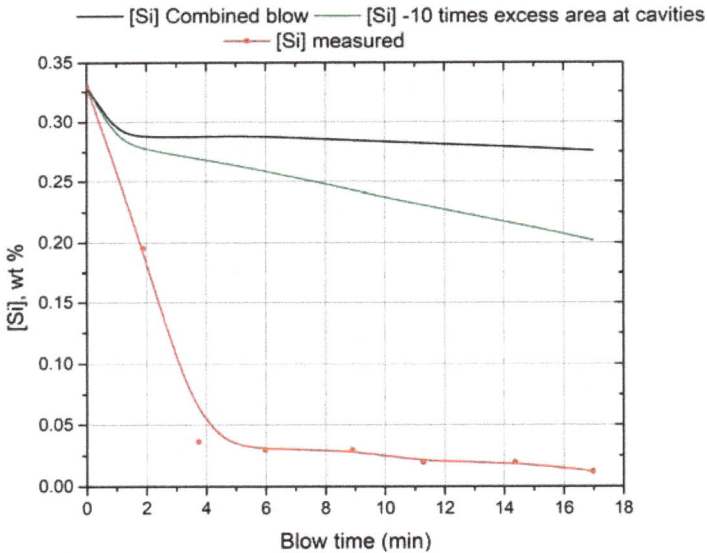

Figure 12. Effect of surface roughness on oxidation rate of silicon.

As discussed in the section above, the flow of bottom stirring plumes to the top causes the circulation of metal. Additionally there is an escape of argon bubbles through the slag-metal interface as argon solubility in the metal is very low. During the passage of the bubble into the slag, a thin film of metal is carried over, and the drainage of this film causes the suspension of metal droplets in slag. This phenomena of the passage of argon bubbles through the liquid iron-slag interface has been investigated using an in situ X-ray transmission technique [69,70] as well as mathematical modelling [71]. These studies described the effect of bubble size and interfacial tension between slag-metal on the metal suspended in slag. However, the typical observed [69] mass of ejection/entrainment of metal in the slag phase was extremely low (0.0065 g/bubble for surface tension of 1.8 N/m and bubble diameter of 11.5 mm). This would entail an ejection of 2.55 kg metal/min for Cicutti et al.'s [4] bottom stirring rate of 2.5 Nm³/min. This indicates a much lower contribution of refining through this mechanism.

4. Conclusions

A mechanistic description of the refining phenomena at the impact and the slag-metal bulk zones is presented. This description includes the role of the oxygen jets and bottom stirring plumes in bringing the metal into contact with oxygen, using independent models for each.

1. Top-blown jets appear to cause a negligible renewal of the surface at the impact zone despite their high momentum. This leads to the conclusion that this momentum is expended in the generation of droplets.
2. The bottom stirring plumes cause a significant circulation of metal (~125 tonne/min for 2.5 m³ stirring gas/min in 200-t furnace) but do not aid the refining reactions at the impact and the slag-metal bulk zones.
3. The contribution of the impact and slag-metal bulk zones appear to be negligible in the refining reactions while the emulsion zone appears to be a significant contributor to the refining reactions.

Author Contributions: Supervision, N.D.; Formulation of model, A.K. and N.D.; Review, N.D.; Editing, A.K. and N.D.; A.K. wrote the original draft of paper.

Funding: This research was funded by Natural Sciences and Engineering Research Council of Canada (NSERC), project number 20007117 and the McMaster Steel Research Centre (SRC).

Acknowledgments: The authors would like to thank Gordon Irons and Anand Senguttuvan for the valuable discussions regarding this work.

Conflicts of Interest: The authors declare no conflict of interest.

Nomenclature

A_c	Area of single cavity, m²
A_{gm}	Total area of cavities/impact zone, m²
A_{sm}	Area between bulk slag and metal bath, m²
C_{cavity}	Circumference of cavity, m
D	Diffusion coefficient of impurity in hot-metal, m²/s
d_c	Diameter of cavity, m
d_{th}	Throat diameter of nozzle, m
H	Height of metal bath, m
h	Lance height, m
J_X	Moles of solute X transferred to interface, $\frac{moles}{s}$
k	Mass transfer coefficient, $\frac{m}{s}$
k_{X-gm}	Mass transfer coefficient of solute X at the impact zone/gas-metal interface, $\frac{m}{s}$
k_{X-sm}	Mass transfer coefficient of solute X at the slag-metal interface, $\frac{m}{s}$
L_X	Distribution coefficient of silicon between hot-metal and slag
l_{c-gm}	Characteristic length of cavity, m
l_{c-sm}	Characteristic length of slag-metal interface, m
$MCRI$	Metal circulating at the interfaces, $\frac{kg}{min}$
M_X	Molecular weight of solute X, $\frac{kg}{mole}$
n_{bse}	Number of bottom stirring elements i.e., number of plumes
n_{cav}	Number of cavities
n_o	Depth of cavity, *m*

P_a	Ambient pressure inside vessel, $\frac{kg}{m.s^2}$
P_o	Oxygen supply pressure, $\frac{kg}{m.s^2}$
Q_b	Bottom stirring gas flow rate, $\frac{m3}{s}$
r_c	Radius of cavity, m
$t, \Delta t$	Time instant and time step, respectively
t_c	Residence time of an element at interface, m/s
u_{bottom}	surface renewal velocity due to bottom stirring, m/s
u_g	Tangential velocity of oxygen jet, m/s
u_j	Velocity of oxygen jet at impact point, m/s,
u_l	Surface renewal velocity of hot-metal due to oxygen jet, m/s
V_m	Volume of metal circulated, m^3
W_b	Weight of metal bath, kg
W_{sc}	Weight of melted scrap, kg
$W_{[X]}$	Weight of solute X (silicon and manganese) removed, kg/min
$W_{md}^{ejection}$	Weight of droplets ejected from the bath, kg
W_{md}^{return}	Weight of droplets returning to the bath, kg
W_{ox}	Weight of hot metal oxidized, kg
$w_{annulus}$	Width of annular region between cavity and vessel walls, m
X	Solute in hot-metal like Si or Mn
α	Inclination angle of the nozzle, ($°$)
θ	Cavity angle (cavity slope), ($°$)
ρ_m	Density of hot-metal, $\frac{kg}{m^3}$

Appendix A

The metal circulation for Cicutti et al. data is calculated as follows:
For 8 bottom stirring plug.
Bottom stirring gas flow rate = 150 Nm^3/h (for entire blow except last two minutes).
Bottom stirring gas flow rate, Q = 150/3600/8 = 0.005208 Nm^3/s/plug.

Table A1. Process parameters.

Q, m^3/s	0.005208
H (height of metal bath), m	0.89

The value of the metal circulated by the plume (bottom stirred gas) is calculated by Equations (7) through (10).

Table A2. Calculation of metal circulation by single plume.

Q^*	z^*	u_{bottom}^*	u_{bottom}, m/s	A_p, m^2	Metal Circulated by Plume V_m, m^3/s	Metal Circulated by Single Plume, kg/min
0.002225	1	0.1642	0.4854	0.0773	0.0375	$V_m\rho_m$ = 15774.98

So total metal circulated by 8 bottom stirring plumes (MCRI) = 15774 × 8 = 126,199.8 kg/min.

References

1. Miller, T.W.; Jimenez, J.; Sharan, A.; Goldstein, D.A. *The Making, Shaping and Treating of Steel*, 11th ed.; Carnegie Steel Company: Pittsburgh, PA, USA, 1998.
2. Deo, B.; Boom, R. *Fundamentals of Steel Making Metallurgy*; Pretince Hall International: Upper Saddle River, NJ, USA, 1993.
3. Pehlke, R.D. Steelmaking—The jet age. *Metall. Trans. B* **1980**, *11*, 539–562. [CrossRef]
4. Cicutti, C.; Valdez, M.; Pérez, T.; Petroni, J.; Gomez, A.; Donayo, R.; Ferro, L. Study of slag-metal reactions in an LD-LBE converter. In Proceedings of the 6th International Conference on Molten Slags, Fluxes and Salts, Helsinki, Finland, 12–17 June 2000; p. 367.

5. Cicutti, C.; Valdez, M.; Pérez, T.; Donayo, R.; Petroni, J. Analysis of slag foaming during the operation of an industrial converter. *Lat. Am. Appl. Res.* **2002**, *32*, 237–240.
6. van Hoorn, A.I.; van Konynenburg, J.T.; Kreyger, P.J. Evolution of slag composition and weight during the blow. In *The Role of Slag in Basic Oxygen Steelmaking Processes, McMaster Symposium on Iron and Steelmaking No.4*; McMaster University: Hamilton, ON, Canada, 1976.
7. Meyer, H.W.; Porter, W.F.; Smith, G.; Szekely, J. Slag-Metal Emulsions and Their Importance in BOF Steelmaking. *J. Met.* **1968**, *20*, 35–42. [CrossRef]
8. Schoop, J.; Resch, W.; Mahn, G. Reactions Occuring During the Oxygen Top-Blown Process and The Calculation of Metallurgical Control Parameters. *Ironmak. Steelmak.* **1978**, *2*, 72–79.
9. Asai, S.; Muchi, I. Theoretical Analysis by the Use of Mathematical Model in LD Converter Operation. *Trans. ISIJ* **1970**, *10*, 250.
10. Jalkanen, H. Experiences in physicochemical modelling of oxygen converter process(BOF). *Sohn Int. Symp. Adv. Process. Met. Mater.* **2006**, *2*, 541–554.
11. Jalkanen, H.; Holappa, L. On the role of slag in the oxygen converter process. In Proceedings of the VII International Conference on Molten Slags Fluxes and Salts, Cape Town, South Africa, 25–28 January 2004; pp. 71–76.
12. Sarkar, R.; Gupta, P.; Basu, S.; Ballal, N.B. Dynamic Modeling of LD Converter Steelmaking: Reaction Modeling Using Gibbs' Free Energy Minimization. *Metall. Mater. Trans. B* **2015**, *46*, 961–976. [CrossRef]
13. Rout, B.K.; Brooks, G.A.; Li, Z.; Rhamdhani, A. Dynamic Modeling of Oxygen Steelmaking Process: A Multi-Zone Kinetic Approach. *AISTech* **2017**, *2017*, 1315–1326.
14. Rout, B.K.; Brooks, G.A.; Li, Z.; Rhamdhani, M.A. Analysis of Desiliconization Reaction Kinetics in a BOF. *AISTech* **2016**, *2016*, 1019–1026.
15. Knoop, W.V.D.; Deo, B.; Snoijer, A.B.; Unen, G.V.; Boom, R. A Dynamic Slag-Droplet Model for the Steelmaking Process. In Proceedings of the 4th International Conference on Molten Slags and Fluxes, Sendai, Japan, 8–11 June 1992; pp. 302–307.
16. Jung, I.H.; Hudon, P.; van Ende, M.A.; Kim, W.Y. Thermodynamic Database for P_2O_5 Containing Slags and Its Application to the Dephosphorisaing process. *AISTech Proc.* **2014**, *2014*, 1257–1268.
17. Nakanishi, K.; Saito, K.; Nozaki, T.; Kato, Y.; Suzuki, K.; Emi, T. Physical and Metallurgical Characteristics of Combined Processes. In Proceedings of the 65th Steelmaking Conference Proceedings, Pittsburgh, PA, USA, 28–31 March 1982; pp. 101–108.
18. Price, D.J. Steelmaking: Significance of the emulsion in carbon removal. In Proceedings of the Process Engineering of Pyrometallurgy Symposium, London, UK, 8–15 January 1974; pp. 8–15.
19. Dogan, N.; Brooks, G.A.; Rhamdhani, M.A. Comprehensive Model of Oxygen Steelmaking Part 3: Decarburization in Impact Zone. *ISIJ Int.* **2011**, *51*, 1102–1109. [CrossRef]
20. Kitamura, S.; Kitamura, T.; Shibata, K.; Mizukami, Y.; Mukawa, S.; Nakagawa, J. Effect of Stirring Energy, Temperature and Flux Composition on Hot Metal Dephosphorization Kinetics. *ISIJ Int.* **1991**, *31*, 1322–1328. [CrossRef]
21. Ohguchi, S.; Robertson, D.G.; Deo, B.; Grieveson, P.; Jeffes, J.H. Simultaneous dephosphorization and desulfurization of molten pig iron. *Ironmak. Steelmak.* **1984**, *11*, 202–213.
22. Chen, E. Kinetic study of Droplet Swelling in BOF steelmaking. Ph.D Thesis, McMaster University, Hamilton, ON, Canada, 2011.
23. Chen, E.; Coley, K.S. Kinetic study of droplet swelling in BOF steelmaking. *Ironmak. Steelmak.* **2010**, *37*, 541–545. [CrossRef]
24. Coley, K.S.; Chen, E.; Pomeroy, M. Kinetics of reaction important in oxygen steelmaking. In Proceedings of the Extraction and Processing Division Symposium on Pyrometallurgy, San Diego, CA, USA, 16–20 June 2014; pp. 289–302.
25. Pomeroy, M.D. Decarburization Kinetics of Fe-C-S Droplets in Oxygen Steelmaking Slags. Master's Thesis, McMaster University, Hamilton, ON, Canada, 2011.
26. Gu, K.; Dogan, N.; Coley, K.S. The Influence of Sulfur on Dephosphorization Kinetics Between Bloated Metal Droplets and Slag Containing FeO. *Metall. Mater. Trans. B* **2017**, *48*, 2343–2353. [CrossRef]
27. Gu, K.; Dogan, N.; Coley, K.S. An Assessment of the General Applicability of the Relationship Between Nucleation of CO Bubbles and Mass Transfer of Phosphorus in Liquid Iron Alloys. *Metall. Mater. Trans. B* **2018**, *49*, 1119–1135. [CrossRef]

28. Dogan, N. Mathematical Modelling of Oxygen Steelmaking. Ph.D Thesis, Swinburne University, Hawthorn, VIC, Australia, 2011.

29. Dogan, N.; Brooks, G.A.; Rhamdhani, M.A. Comprehensive Model of Oxygen Steelmaking Part 1: Model Development and Validation. *ISIJ Int.* **2011**, *51*, 1086–1092. [CrossRef]

30. Dogan, N.; Brooks, G.A.; Rhamdhani, M.A. Comprehensive Model of Oxygen Steelmaking Part 2: Application of Bloated Droplet Theory for Decarburization in Emulsion Zone. *ISIJ Int.* **2011**, *51*, 1093–1101. [CrossRef]

31. Masui, A.; Yamada, K.; Takahashi, K. Slagmaking, Slag/metal reactions and their sites in BOF refining processes. In *The Role of Slag in Basic Oxygen Steelmaking Processes, McMaster Symposium on Iron and Steelmaking No.4*; McMaster University: Hamilton, ON, Canada, 1976.

32. Narita, K.; Makino, T.; Matsumoto, H.; Hikosaka, A.; Katsuda, J. Oxidation Mechanism of Silicon in Hot Metal. *Tetsu-to-Hagane* **1983**, *69*, 1722–1733. [CrossRef]

33. Suito, H.; Inoue, R. Thermodynamic Assessment of Manganese Distribution in Hot Metal and Steel. *ISIJ Int.* **1995**, *35*, 266–271. [CrossRef]

34. Rout, B.; Brooks, G.; Rhamdhani, M.A.; Li, Z.; Schrama, F.N.; Sun, J. Dynamic Model of Basic Oxygen Steelmaking Process Based on Multi-zone Reaction Kinetics: Model Derivation and Validation. *Metall. Mater. Trans. B* **2018**, *49*, 537–557. [CrossRef]

35. Simotsuma, T.; Sano, K. The Influence of Spouting by Bath Motion. *Tetsu-to-Hagane* **1965**, *51*, 1909.

36. Sharma, S.K.; Hlinka, J.W.; Kern, D.W. The Bath Circulation, Jet Penetration and High Temperature Reaction Zone in BOF Steelmaking. In Proceedings of the InOpen Hearth and Basic Oxygen Steel Conference, Pittsburgh, PA, USA, 17–20 April 1977; pp. 187–197.

37. Davenport, W.G.; Wakelin, D.; Bradshaw, A. Interaction of both bubbles and gass jets with liquids. *Heat Mass Transf. Process Metall.* **1967**, *24*, 207–245.

38. Odenthal, H.; Falkenreck, U.; Schlüter, J. CFD Simulation of Multiphase Melt Flows in Steelmaking Converters. In Proceedings of the European Conference on Computational Fluid Dynamics, Egmond aan Zee, The Netherlands, 5–8 September 2006.

39. Odenthal, H.; Kempken, J.; Schlüter, J.; Emling, W.H. Advantageous numerical simulation of the converter blowing process. *Iron Steel Technol.* **2007**, *4*, 71–89.

40. Odenthal, H. Latest Developments for the BOF Converter. In Proceedings of the 6th International Congress on the Science and Technology of Steelmaking, Beijing, China, 12–14 May 2015.

41. Li, Y.; Lou, W.T.; Zhu, M.Y. Numerical simulation of gas and liquid flow in steelmaking converter with top and bottom combined blowing. *Ironmak. Steelmak.* **2013**, *40*, 505–514. [CrossRef]

42. Ersson, M.; Höglund, L.; Tilliander, A.; Jonsson, L.; Jönsson, P. Dynamic Coupling of Computational Fluid Dynamics and Thermodynamics Software: Applied on a Top Blown Converter. *ISIJ Int.* **2008**, *48*, 147–153. [CrossRef]

43. Hwang, H.Y.; Irons, G.A. A water model study of impinging gas jets on liquid surfaces. *Metall. Mater. Trans. B Process Metall. Mater. Process. Sci.* **2012**, *43*, 302–315. [CrossRef]

44. He, Q.L.; Standish, N. A Model Study of Droplet Generation in the BOFSteelmaking. *ISIJ Int.* **1990**, *30*, 305–309. [CrossRef]

45. Luomala, M.J.; Fabritius, T.M.J.; Härkki, J.J. The Effect of Bottom Nozzle Configuration on the Bath Behaviour in the BOF. *ISIJ Int.* **2004**, *44*, 809–816. [CrossRef]

46. Maia, B.T.; Imagawa, R.K.; Tavares, R.P. Cold Model Bath Behavior Study in LD Converter With Bottom Blowing. *AISTech* **2016**, *55*, 1083–1094.

47. Maia, B.T.; Diniz, C.N.A.; Carvalho, D.A.; Souza, D.L.D.; Guimarães, J.A.; Raissa, S. TBM Tuyeres Arrangements and Flow—Comparison between BOF thyssenkrupp CSA and Cold Model. *AISTech* **2017**, *2017*, 1335–1346.

48. Krishnapisharody, K.; Irons, G.A. An Analysis of Recirculatory Flow in Gas-Stirred Ladles. *Steel Res. Int.* **2010**, *81*, 880–885. [CrossRef]

49. Sano, M.; Mori, K. Fluid Flow and Mixing Characteristics in a Gas-stirred Molten Bath. *Trans. Iron Steel Inst. Jpn.* **1983**, *23*, 169–175. [CrossRef]

50. Murthy, G.G.K.; Ghosh, A.; Mehrotra, S.P. Characterization of two-phase axisymmetric plume in a gas stirred liquid bath-A water model study. *Metall. Trans. B* **1988**, *19*, 885–892. [CrossRef]

51. Szekely, J.; Lehner, T.; Chang, C. Flow phenomena, mixing, and mass transfer in argon stirred ladles. *Ironmak. Steelmak.* **1979**, *6*, 285.

52. Hsiao, C.; Lehner, T. Fluid Flow in Ladles-Experimental Results. *Scand. J. Metall.* **1980**, *9*, 105–110.

53. Nakanishi, K.; Fujii, T.; Szekely, J. Possible relationship between energy dissipation and agitation in steel processing operations. *Ironmak. Steelmak.* **1975**, *2*, 193.

54. Bertezzolo, U.; Donayo, R.; Gomez, A.; Denier, G.; Stomp, H. The LBE process at Siderar. In Proceedings of the 2nd European Oxygen Steelmaking Congress, Taranto, Italy, 13–15 October 1997.

55. Higbie, R. The rate of absorption of a pure gas into a still liquid during short periods of exposure. *Trans. Am. Inst. Chem. Eng.* **1935**, *35*, 36–60.

56. Calderon, F.; Sano, N. Diffusion of Mn and Si in liquid Fe over the whole range of composition. *Metall. Trans. B* **1971**, *2*, 3325. [CrossRef]

57. Majdic, A.; Graf, D.; Schenk, J. Diffusion of Si,P,S and Mn in molten Fe. *Arch. für das Eisenhüttenwes* **1969**, *40*, 627.

58. Saito, T.; Kawai, Y.; Maruya, K. Diffusion of Some Alloying Elements in Liquid Iron. *Tohuku Daigaku Senk.* **1960**, *16*, 15.

59. Kawai, Y.; Shiraishi, Y. *Handbook of Physico-Chemical Properties at High Temperatures*; Iron and Steel Institute of Japan: Tokyo, Japan, 1988.

60. Grace, R.; Derge, G. Diffusion of Third Elements in Liquid Iron Saturated with Carbon. *Trans. Metall. Soc.* **1958**, *212*, 331–337.

61. Koch, K.; Falkus, J.; Ralf, B. Hot model experiments of the metal bath spraying effect during the decarburization of Fe-C melts through oxygen top blowing. *Steel Res. Int.* **1993**, *64*, 15–21. [CrossRef]

62. Lee, M.; Whitney, V.; Molloy, N. Jet-liquid interaction in a steelmaking electric arc furnace. *Scand. J. Metall.* **2001**, *30*, 330–336. [CrossRef]

63. Lee, M.S.; O'Rourke, L.; Molloy, N.A. Oscillatory flow in the steelmaking vessel. *Scand. J. Metall.* **2003**, *32*, 281–288. [CrossRef]

64. Sabah, S.; Brooks, G.A. Splash Distribution in Oxygen Steelmaking. *Metall. Mater. Trans. B* **2014**, *46*, 863–872. [CrossRef]

65. Koria, S.; Lange, K.W. Penetrability of impinging gas jets in molten steel bath. *Steel Res.* **1987**, *58*, 421–426. [CrossRef]

66. Zhou, X.; Ersson, M.; Zhong, L.; Jönsson, P. Numerical Simulations of the Kinetic Energy Transfer in the Bath of a BOF Converter. *Metall. Mater. Trans. B Process Metall. Mater. Process. Sci.* **2016**, *47*, 434–445. [CrossRef]

67. Koria, S. Studies of the Bath Mixing Intensity in converter Steelmaking Processes. *Can. Metall. Q.* **1992**, *31*, 105–112. [CrossRef]

68. Koria, S.; Pal, S. Experimental study of the effect of gas injection parameters on bath mixing intensity induced during steelmaking. *Steel Res.* **1991**, *2*, 47–53. [CrossRef]

69. Han, Z.; Holappa, L. Bubble Bursting Phenomenon in Gas/Metal/Slag Systems. *Metall. Mater. Trans. B* **2003**, *34*, 525–532. [CrossRef]

70. Han, Z.; Holappa, L. Mechanisms of Iron Entrainment into Slag due to Rising Gas Bubbles. *ISIJ Int.* **2003**, *43*, 292–297. [CrossRef]

71. Kobayashi, S. Iron Droplet Formation Due to Bubbles Passing through Molten Iron/Slag Interface. *ISIJ Int.* **1993**, *33*, 577–582. [CrossRef]

![metals logo] *metals*

MDPI

Article

Desiliconisation and Dephosphorisation Behaviours of Various Oxygen Sources in Hot Metal Pre-Treatment

Youngjo Kang

Department of Material Science and Engineering, Dong-A University, Saha-gu, Busan 48315, Korea;
youngjok@dau.ac.kr; Tel.: +82-51-200-7758

Received: 25 January 2019; Accepted: 17 February 2019; Published: 20 February 2019

Abstract: In order to obtain a better understanding of the efficiencies of desiliconisation and dephosphorisation reactions during hot metal pretreatment in an open ladle, a number of simulation experiments were carried out with various oxygen sources. Three types of solid oxygen materials (sintered return ore, scale briquette and fine mill scale) were carefully investigated as hot metal pre-treatment agents, evaluating their desiliconisation and dephosphorisation efficiencies. The method applied for supplying gaseous oxygen was also assessed. The comparison between top blowing and injection methods indicated that injected oxygen gas is more advantageous for desiliconisation, while top-blown oxygen gas is favourable for dephosphorisation. The obtained information on the characteristics of gaseous oxygen can be used for the optimisation of blowing patterns, in order to improve the efficiency of the hot metal pre-treatment.

Keywords: hot metal pre-treatment; desiliconisation; dephosphorisation; solid and gaseous oxygen

1. Introduction

Steelmaking based on blast furnaces has been dominant for many years, due to its high productivity and low cost. However, due to the presence of considerable amounts of impurities (such as C, S, P and Si) in the hot metal from blast furnaces, a careful refinement must be carried out to obtain final products with acceptable properties. In order to reduce the burden in the refining of the converter in the conventional process, the hot metal pre-treatment technique (originally developed in Japan) has been improved over several decades, in order to achieve optimum balance across the whole refining process. Hot metal pre-treatment generally includes desiliconisation, dephosphorisation and desulfurisation processes. Because each process has different favourable conditions in operational as well as thermodynamic terms, iron- and steelmaking companies have been adopting their own sequences and procedures.

The development of pre-treatment technologies has been accelerated by many valuable experimental, industrial and even theoretical studies. This is particularly true for the changes made in the dephosphorisation process [1,2], in terms of flux composition [3,4] and in thermodynamic and kinetic modelling [5,6]. Simultaneous desulfurisation and dephosphorisation of the hot metal was suggested by careful analyses of the oxygen potential and flux composition [7–11]. Based on the high oxygen supply rates and lime injection, the dephosphorisation in the converter became advantageous, and, currently, the process is widely used in many steelmaking mills [12–15]. Nonetheless, a preliminary dephosphorisation in the transfer vessel may be still required in the production of ultra-low phosphorus grade steel. Moreover, the dephosphorisation within the converter cannot be applied to the production of high-alloy steel using hot metal, due to the oxidation of the alloy elements.

However, hot metal transfer vessels such as open ladles or torpedoes generally have low stirring capability and insufficient free board. To overcome these limitations, solid and gaseous oxygen sources can be used, by supplying them onto the surface or injecting them into the hot metal. Dephosphorisation can be carried out by either injected oxygen during flotation or slag-metal reaction with a top flux. While the phosphorus removal in the top flux is a 'permanent' reaction, the dephosphorisation with injected oxygen is described as 'transitory' reaction, because the phosphorus oxidised by the injected oxygen may be readily reduced by carbon in the hot metal. An optimal control of both reactions is of primary importance for effective dephosphorisation [16].

Various kinds of oxygen sources can be supplied in different ways to perform the dephosphorisation in the transfer vessel; moreover, desiliconisation should be carried out in advance and also at the same time as the dephosphorisation. Therefore, it is necessary to understand in detail the properties of the solid and gaseous oxygen sources, as well as the mechanisms of the transitory and permanent reactions during desiliconisation and dephosphorisation. In this study, the influences of the oxygen source type and supply method on the desiliconisation and dephosphorisation processes are investigated by laboratory-scale experiments; moreover, we discuss the reaction mechanisms involved in these processes.

2. Materials and Methods

Three series of laboratory experiments were carried out to investigate the effects and roles of various oxygen sources in the desiliconisation and dephosphorisation processes.

Dephosphorisation experiments using various solid oxygen sources were carried out in a high-frequency induction furnace equipped with a vacuum chamber (Hyundai Electronics, Seoul, Korea) as shown in Figure 1. Several types of raw materials (sintered return ore, scale briquette and fine mill scale) were used as solid oxygen materials. Their composition and sizes are summarised in Table 1; each material was prepared for the laboratory-scale hot metal dephosphorisation by crushing it into similar size (~1 mm) and drying at 373 K for 6 h. Electrolytic steel, graphite and reagent-grade iron phosphide were charged in a MgO crucible with inner diameter of 55 mm, to prepare a hot metal with composition C 4.2 wt. %, Si 0.15 wt. %, P 0.1 wt. %. After evacuating the chamber at least three times, deoxidised Ar gas, previously purified with sodium hydroxide and magnesium perchlorate, was introduced into the chamber and injected into the hot metal through an immersed alumina lance, at a rate of 0.4 SLPM (standard litres per minute at 273.15 K and 101.325 kPa). The temperature was then increased to 1623 K, using a predetermined pattern and controlled by a thermocouple placed under a carbon crucible. Once about 0.8 kg of hot metal was completely molten, an initial metal sample ($t = 0$) was collected; the purified Ar was then injected into the hot metal through an immersed alumina lance, at a rate of 0.4 SLPM. Thereafter, the solid oxygen material was added, together with CaO powder, from the top of the furnace. The addition was carried out every 10 min for a total of four times over 40 min. The total amount of solid oxygen was estimated in such a way as to achieve an acceptable composition of the hot metal (Si, trace; P, 0.02 wt. %), assuming constant desiliconisation and dephosphorisation efficiencies during the experiment. Various operational data from hot metal pre-treatment plants were used to estimate the values of the desiliconisation and dephosphorisation efficiencies (60% and 20%, respectively). Moreover, the amount of CaO was determined to adjust % CaO/% SiO_2 to be 3.0.

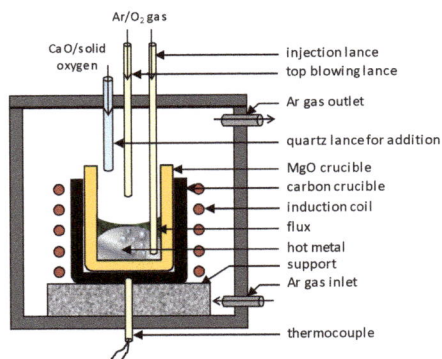

Figure 1. Schematic illustration of experimental apparatus.

Table 1. Characteristic of various solid oxygen sources considered in the present study.

Oxygen Source	Composition (wt. %)									Original Size (mm)
	T-Fe*	M-Fe*	O	FeO	Fe₂O₃	SiO₂	CaO	P	Others	
Sintered return ore	58.3	-	24.1	7.8	74.7	5.51	9.57	0.05	2.46	2–10
Scale briquette	73.6	12.8	22.4	33.7	49.4	0.80	0.10	0.02	0.21	40–50
Fine mill scale	74.5	-	24.9	63.7	35.8	0.15	0.18	0.01	0.12	5–8

* T-Fe and M-Fe stand for total Fe and metallic Fe contents, respectively.

Similar experiments were carried out to study the influence of gaseous oxygen, blown either from the top or into the hot metal, on desiliconisation and dephosphorisation. About 0.8 kg of hot metal in an MgO crucible was melted in a high-frequency induction furnace, to yield the same composition described earlier. Ar and O_2 gases were blown through two alumina lances, which were placed on the top of and into the hot metal for top blowing and injection, respectively. The specific blowing conditions used with the two lances are listed in Table 2. CaO was added intermittently from the top of the hot metal to achieve a stable dephosphorisation reaction.

Table 2. Gas blowing conditions for hot metal desiliconisation and dephosphorisation.

Blowing Conditions	Top Lance	Bottom Lance
1	O_2 0.4 SLPM*	Ar 0.4 SLPM
2	O_2 0.2 SLPM, Ar 0.2 SLPM	O_2 0.2 SLPM, Ar 0.2 SLPM
3	Ar 0.4 SLPM	O_2 0.4 SLPM

* SLPM: standard litres per minute.

In another set of experiments in a larger furnace scale, the effects of the top-charged solid oxygen and injected gaseous oxygen on desiliconisation and dephosphorisation were compared upon injecting CaO into the hot metal. The injection of CaO powder was performed with an injection lance especially designed for upward and downward movement, as shown in Figure 2. About 200 kg of hot metal with the same composition was melted in a high-frequency induction furnace. In the experiments with solid oxygen, sintered return ore was dropped onto the hot metal, while CaO was injected with Ar as a carrier gas. On the other hand, in the gaseous oxygen experiments CaO was injected into the hot metal with dehydrated air, which played the roles of oxygen source and carrier gas. The amount of each oxygen source was adjusted to yield an equivalent amount of oxygen.

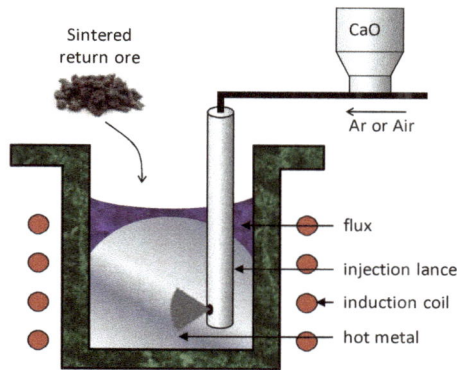

Figure 2. Schematic illustration of experimental apparatus with CaO injection.

In each experiment, metal samples were collected by suction every 5 min to investigate the changes in composition with time. While the carbon content in the samples was determined by combustion infrared detection using an elemental analyser (CS600, LECO Corp., Miami, FL, USA), the silicon and phosphorus contents were measured with an inductively coupled plasma optical emission spectrometer (ARCOS EOP, Spectro, Kleve, Germany).

3. Results and Discussion

3.1. Desiliconisation and Dephosphorisation Behaviours of Various Solid Oxygen Materials

Figure 3 shows the changes in the C, Si and P contents in the hot metal when various types of solid oxygen material were added. Regardless of the kind of solid oxygen source, the C, Si and P contents in the hot metal were found to decrease with time. Compared to the phosphorus content in the hot metal, however, the Si content sharply decreased to a value as low as 0.02 wt. % in 35 min. Dephosphorisation appeared to be negligible in the early stages and, notably, took place when the Si content was less than 0.02 wt. %. This suggests the high importance of performing the desiliconisation as rapidly as possible, in order to achieve a satisfactory dephosphorisation efficiency.

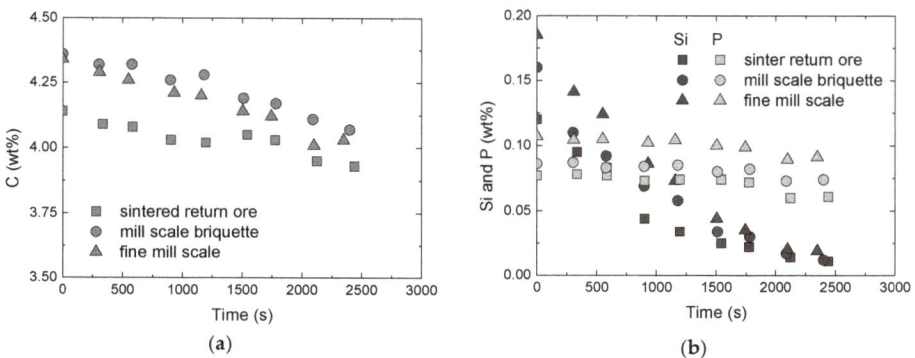

Figure 3. Changes in (**a**) C and (**b**) Si/P contents of hot metal upon addition of various solid oxygen materials.

In order to quantitatively compare the effects of different solid oxygen sources, the amount of oxygen consumed for desiliconisation/dephosphorisation and the corresponding efficiencies were estimated by considering the initial contents and their preferential oxidation tendencies. Based on the total amount of oxygen consumed by each reaction between solid oxygen and hot metal, decarburisation occurs first, followed by desiliconisation, while dephosphorisation takes place last. Therefore, the amount of oxygen consumed for the preceding reactions should be taken into account when estimating the desiliconisation and dephosphorisation efficiencies, as shown in Equations (1) and (2).

$$De - Si \ efficiency = \frac{(O \ for \ De - Si)}{(Total \ O) - (O \ for \ De - C)} \times 100\% \quad (1)$$

$$De - P \ efficiency = \frac{(O \ for \ De - P)}{(Total \ O) - (O \ for \ De - C) - (O \ forDe - Si)} \times 100\% \quad (2)$$

The estimated decarburisation, desiliconisation and dephosphorisation efficiencies of the various solid oxygen materials in the whole experiment are shown in Figure 4. Compared to the changes in the C, Si and P contents in Figure 3, the reaction efficiencies of each solid oxygen material showed a significantly distinct behaviour during the experiments. Among the three types of solid oxygen materials, the highest efficiencies in all reactions were obtained for fine mill scale. Although the amounts of C, Si and P removed by sintered return ore and by the other materials were comparable, sintered return ore showed the lowest efficiencies for decarburisation, desiliconisation and dephosphorisation. This difference may be attributed to the presence of higher contents of Fe_2O_3 than FeO, which is relatively more reducible at equivalent oxygen contents in sintered return ore. A higher efficiency of fine mill scale may result in a significant reduction in the required amount of refining agents and processing time.

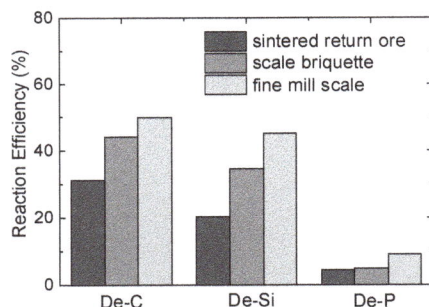

Figure 4. Decarburisation, desiliconisation and dephosphorisation reaction efficiencies of various solid oxygen materials.

3.2. Effect of Gaseous Oxygen Supply Method on Desiliconisation and Dephosphorisation of Hot Metal

To further investigate the effect of gaseous oxygen on the desiliconisation and dephosphorisation of the hot metal, a series of experiments with 0.8 kg of hot metal were carried out using different supply methods of gaseous oxygen. The detailed conditions are described in Table 2. Similar to solid oxygen, the contents of C, Si and P in the hot metal decreased with time, as shown in Figure 5, with gaseous oxygen supplied by either top blowing or injection. Regardless of how gaseous oxygen was supplied, the carbon content showed a gradual decrease. While extensive oxidation of silicon took place in the early stages of oxygen blowing, the dephosphorisation of the hot metal was found to be far from a satisfactory level. In particular, when gaseous oxygen was bottom-blown by injection, phosphorus removal was negligible compared to the removal observed with top-blown oxygen. Based on the different dephosphorisation behaviour shown in Figure 5, it could be concluded that the contribution of the transitory reaction to the hot metal dephosphorisation was relatively small, compared to that of the permanent reaction between hot metal and top slag.

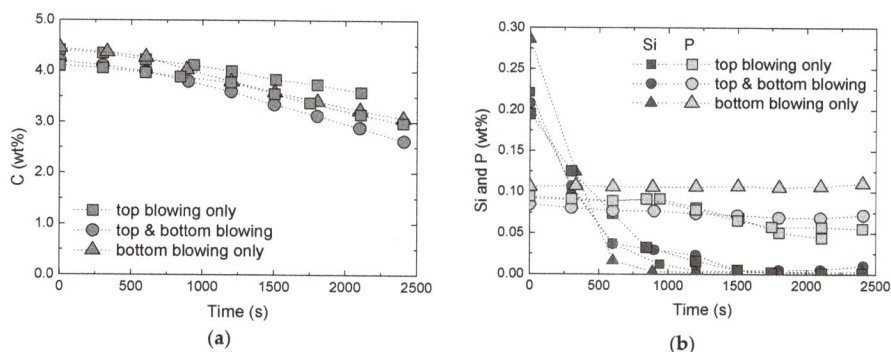

Figure 5. Changes in (**a**) C and (**b**) Si/P contents of hot metal with different supply methods of gaseous oxygen.

A number of studies have focused on the permanent and transitory reactions in dephosphorisation since the 1980s [16–20]. Although in the transitory reaction the injected oxygen may oxidise phosphorus in the hot metal, phosphorus oxides are readily reduced during flotation by C or Si in the hot metal. This also implies that simultaneous CaO injection is necessary for P_2O_5 stabilisation. In contrast, top oxygen blowing enables active Fe oxidation, leading to the formation of a top slag with high basicity and high Fe oxide content. The higher dephosphorisation capability of top-blown oxygen compared to that of injected oxygen may be due to the permanent reaction between hot metal and top slag. When oxygen was blown onto the top surface of the hot metal, dephosphorisation became significant after about 900 s, which corresponds to a Si content lower than 0.025 wt. %. This confirms that a rapid desiliconisation is critical for achieving a more effective dephosphorisation.

In terms of desiliconisation of the hot metal, top- and bottom-blown oxygen showed an opposite tendency to that found for dephosphorisation. The desiliconisation rate of bottom-blown oxygen was much faster than that of top-blown oxygen. Moreover, the desiliconisation efficiency of total supplied oxygen showed a marked increase to 32.9% with injected oxygen, compared to 18.9% value obtained in the experiments with top-blown oxygen. It can be expected that the injected oxygen has a higher probability and a larger interface to react with the hot metal; furthermore, the formed Si oxide could float up to the top slag without being reduced during flotation. Therefore, it can be recommended that the gaseous oxygen is injected at the first step of the dephosphorisation (further desiliconisation of pre-desiliconised hot metal), followed by switching over to the top charging of solid oxygen for deep dephosphorisation. Of course, effective countermeasures should be considered to overcome the low dephosphorisation efficiency at higher temperature, which may result from the desiliconisation following injection of gaseous oxygen.

The desiliconisation and dephosphorisation efficiencies were compared more systematically by higher-scale experiments also involving the injection of CaO powder. The oxygen for desiliconisation and dephosphorisation was supplied by either top charging of iron ore or injection of dehydrated air with an equivalent oxygen amount. The changes in the C, Si and P contents of the hot metal under the different conditions are presented in Figure 6. For both types of oxygen sources, the carbon and silicon contents showed a monotonic decrease, which denotes a similar behaviour to that observed in the previous experiments. However, the dephosphorisation behaviours of the two types of oxygen were significantly different from each other. In contrast to the behaviour of top-charged solid oxygen, the dephosphorisation by injected oxygen was found to be insufficient, indicating that the CaO injection was essentially inadequate. To improve the dephosphorisation efficiency, direct contact with CaO may be required, in order to enable P_2O_5 to reach the top slag unreduced.

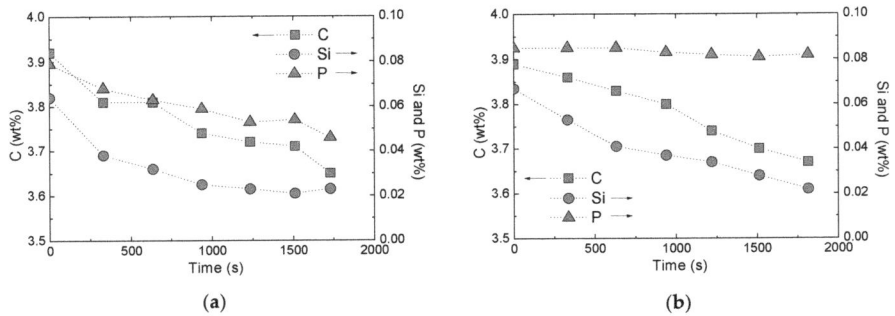

Figure 6. Changes in C, Si and P contents of hot metals subjected to different oxygen supply methods: (a) iron ore top charging; (b) dehydrated air injection.

In order to understand the relatively low dephosphorisation ability of injected oxygen, the reaction efficiencies of desiliconisation and dephosphorisation during the experiments were evaluated, as shown in Figure 7. As mentioned earlier, the overall efficiencies of desiliconisation and dephosphorisation were higher with iron ore top charging than those with dehydrated air injection. Besides the difference in overall efficiency, it should be noted that the desiliconisation and dephosphorisation efficiencies of top-charged solid oxygen drastically decreased with time in a very similar manner. This drop of the efficiency is likely due to the increase in the top slag amount, which may hinder the active contact between top-charged solid oxygen and hot metal. On the other hand, in the experiments using air injection, the desiliconisation and dephosphorisation efficiencies behaved in the opposite way. These different behaviours reflect well the characteristics of the permanent and transitory reactions during hot metal treatment. In the transitory reaction, dephosphorisation is only feasible when phosphorus oxide can float upward without being reduced by Si in the hot metal. Therefore, a more effective dephosphorisation can be achieved by optimising the type and supply method of the oxygen material, based on the relationship between desiliconisation and dephosphorisation.

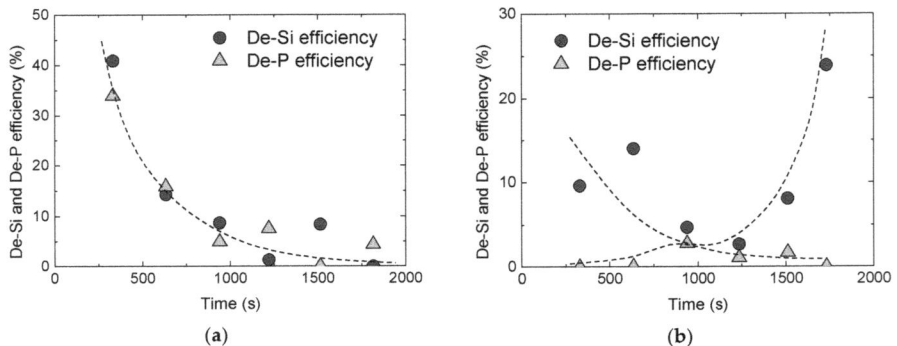

Figure 7. Changes in desiliconisation and dephosphorisation efficiencies with different oxygen supply methods: (a) iron ore top charging; (b) dehydrated air injection.

4. Conclusions

The effects of the oxygen source type and supply method on the hot metal desiliconisation and dephosphorisation were experimentally investigated. The reaction efficiencies achieved with injected gaseous oxygen and other oxygen sources were compared via a series of experiments. Among various types of solid oxygen materials, scale materials turned out to be more favourable than iron ore, mainly because of the higher FeO content. The experiments comparing the top blowing and injection methods of gaseous oxygen revealed that the injected gaseous oxygen hardly contributes to the dephosphorisation, due to the low dephosphorisation efficiency of the transitory reaction. However, desiliconisation by injected oxygen was found to be considerably effective in the early stages. These findings suggest the possibility to optimise the supply method of gaseous oxygen according to the reaction steps. In larger-scale experiments with controlled CaO injection, it was also observed that the dephosphorisation ability of injected gaseous oxygen was still unsatisfactory, compared to that of top-charged solid oxygen. Moreover, the permanent and transitory reactions involved in the desiliconisation and dephosphorisation processes were illustrated by the changes in their respective efficiencies. In particular, the desiliconisation and dephosphorisation reactions by injected oxygen turned out to be in a competitive relationship during the transitory reaction.

Funding: This work was supported by the Dong-A University research fund.

Acknowledgments: The author is grateful to Lee Heeho for his assistance with very complicated experiments. Valuable and stimulating advice from Suk Min-Oh is also acknowledged.

Conflicts of Interest: The author declares no conflicting interest.

References

1. Ogawa, Y.; Maruoka, N. Progress of Hot Metal Treatment Technology and Future Outlook. *Tetsu Hagane J. Iron Steel Inst. Jpn.* **2014**, *100*, 434–444. [CrossRef]
2. Iwasaki, M.; Matsuo, M. Change and Development of Steelmaking Technology. *Nippon Steel Tech. Rep.* **2012**, *101*, 89–94.
3. Guo, S.; Dong, Y.; Chen, E.; Wang, H. Dephosphorization and rephosphorization of liquid steel by lime based fluxes. *Gangtie* **2000**, *3*, 5.
4. Yue, K.; Li, J. Dephosphorization of Hot Metal With Slag Containing CaO-Fe$_2$O$_3$. *J. Iron Steel Res.* **2006**, *9*, 2.
5. Yang, X.; Li, J.; Chai, J.; Duan, D.; Zhang, J. A Thermodynamic Model for Predicting Phosphorus Partition between CaO-based Slags and Hot Metal during Hot Metal Dephosphorization Pretreatment Process Based on the Ion and Molecule Coexistence Theory. *Metall. Mater. Trans. B* **2016**, *47*, 2279–2301. [CrossRef]
6. Pak, J.J.; Fruehan, R.J. Dynamics of the hot metal dephosphorization with Na$_2$O slags. *Metall. Trans. B* **1987**, *18B*, 687–693. [CrossRef]
7. Nakamura, Y.; Harashima, K.; Fukuda, Y. Dephosphorization and Desulfurization of Molten 4%C-Fe Alloy with CaO-based flux Containing Halide. *Tetsu-to-Hagané* **1981**, *67*, 2138–2144. [CrossRef]
8. Inoue, H.; Shigeno, Y.; Tokuda, M.; Ohtani, M. On Simultaneous Dephosphorization and Desulphurization of Pig Iron with CaO-CaCl$_2$ Fluxes. *Tetsu-to-Hagané* **1983**, *69*, 210–219. [CrossRef]
9. Taskinen, A.; Janke, D. Slag Pretreatment for Simultaneous Dephosphorization and Desulphurization of Hot Metal. *Stahl Eisen* **1983**, *103*, 491–496.
10. Umezawa, K.; Matsunaga, H.; Arima, R.; Tonomura, S.; Furugaki, I. The Influence of Operating Condition on Dephosphorization and Desulphurization Reactions of Hot Metal with Lime-based Flux. *Tetsu-to-Hagané* **1983**, *69*, 1810–1817. [CrossRef]
11. Hernandez, A.; Romero, A.; Chavez, F.; Angeles, M. Dephosphorization and desulfurization pretreatment of molten iron with CaO-SiO$_2$-CaF$_2$-FeO-Na$_2$O slags. *ISIJ Int.* **1998**, *38*, 126–131. [CrossRef]
12. Nozaki, T.; Nakanishi, K.; Morishita, H.; Yamada, S.; Sudo, F. Characteristics of Dephosphorization in a Bottom Blown Converter and Its Application to the Preliminary Treatment of Hot Metal. *Tetsu-to-Hagané* **1982**, *68*, 1737–1743. [CrossRef]
13. Ono, H.; Masui, T.; Mori, H. Dephosphorization Kinetics of Hot Metal by Lime Injection with Oxygen Gas. *Tetsu-to-Hagané* **1983**, *69*, 1763–1770. [CrossRef]
14. Ogawa, Y.; Yano, M.; Kitamura, S.; Hirata, H. Development of the Continuous Dephosphorization and Decarburization Process Using BOF. *Tetsu-to-Hagané* **2001**, *87*, 21–28. [CrossRef]
15. Mukawa, S.; Mizukami, Y. Effect of Stirring Energy and Rate of Oxygen Supply on the Rate of Hot Metal Dephosphorization. *ISIJ Int.* **1995**, *35*, 1374–1380. [CrossRef]
16. Haida, O.; Takeuchi, S.; Nozaki, T.; Emi, T.; Sudo, F. Mechanism of Hot Metal Dephosphorization by Injecting Lime Base Fluxes into Bottom Blown Converter. *Tetsu-to-Hagané* **1982**, *68*, 1744–1753. [CrossRef]

17. Kawauchi, Y.; Maede, H.; Kamisaka, E.; Satoh, S.; Inoue, T.; Naki, M. Metallurgical Characteristics of Hot Metal Desiliconization by Injecting Gaseous Oxygen. *Tetsu-to-Hagané* **1983**, *69*, 1730–1737. [CrossRef]

18. Saito, K.; Nakanishi, K.; Misaki, N.; Nakai, K.; Onishi, M. Dephosphorization of Hot Metal with Injection of Lime Bearing Fluxes in a Ladle. *Tetsu-to-Hagané* **1983**, *69*, 1802–1809. [CrossRef]

19. Inoue, T.; Yoshida, M.; Sato, H.; Yonenaka, E. De-siliconization and de-phosphorization treatment of hot metal by gaseous oxygen injection in TPC. *Tetsu-to-Hagané* **1985**, *71*, S943.

20. Roy, G.G.; Chaudhary, P.N.; Minj, R.K.; Goel, R.P. Dephosphorization of ferromanganese using BaCO$_3$-based fluxes by submerged injection of powders: A preliminary kinetic study. *Metall. Mater. Trans. B* **2001**, *32B*, 558–561. [CrossRef]

![metals logo] *metals*

MDPI

Article

A Further Evaluation of the Coupling Relationship between Dephosphorization and Desulfurization Abilities or Potentials for CaO-based Slags: Influence of Slag Chemical Composition

Xue Min Yang [1,*], Jin Yan Li [2], Meng Zhang [2], Fang Jia Yan [1], Dong Ping Duan [1] and Jian Zhang [3]

[1] CAS Key Laboratory of Green Process and Engineering, Institute of Process Engineering, Chinese Academy of Sciences, Beijing 100190, China; yanfangjia@gmail.com (F.J.Y.); douglass@ipe.ac.cn (D.P.D.)
[2] Department of Metallurgy and Raw Materials, China Metallurgical Industry Planning and Research Institute, Beijing 100711, China; lijinyan@mpi1972.com (J.Y.L.); zhangmeng@mpi1972.com (M.Z.)
[3] School of Metallurgical and Ecological Engineering, University of Science and Technology Beijing, Beijing 100083, China; zhangjian_27@126.com
*   Correspondence: yangxm71@ipe.ac.cn; Tel.: +86-10-82544930

Received: 20 November 2018; Accepted: 17 December 2018; Published: 19 December 2018

Abstract: The coupling relationships between dephosphorization and desulfurization abilities or potentials for $CaO–FeO–Fe_2O_3–Al_2O_3–P_2O_5$ slags over a large variation range of slag oxidization ability during the secondary refining process of molten steel have been proposed by the present authors as $\log L_P + 5 \log L_S$ or $\log C_{PO_4^{3-}} + \log C_{S^{2-}}$ in the reducing zone and as $\log L_P + \log L_S - 5 \log N_{Fe_tO}$ or $\log C_{PO_4^{3-}} + \log C_{S^{2-}} - \log N_{FeO}$ in the oxidizing zone based on the ion and molecule coexistence theory (IMCT). In order to further verify the validation and feasibility of the proposed coupling relationships, the effects of chemical composition of the CaO-based slags are provided. The chemical composition of slags was described by three group parameters including reaction abilities of components represented by the mass action concentrations N_i, two kinds of slag basicity as simplified complex basicity $(\% \, CaO) / [(\% \, P_2O_5) + (\% \, Al_2O_3)]$ and optical basicity Λ, and the comprehensive effect of iron oxides Fe_tO and basic oxide CaO. Comparing with the strong effects of chemical composition of the CaO-based slags on dephosphorization and desulfurization abilities or potentials, the proposed coupling relationships have been confirmed to not only be independent of slag oxidization ability as expected but also be irrelevant to the aforementioned three groups of parameters for representing the chemical composition of the CaO-based slags. Increasing temperature from 1811 to 1927 K (1538 to 1654 °C) can result in a decreasing tendency of the proposed coupling relationships. In terms of the proposed coupling relationships, chemical composition of slags or fluxes with assigned dephosphorization ability or potential can be theoretically designed or optimized from its desulfurization ability or potential, and vice versa. Considering the large difference of magnitude between phosphate capacity $C_{PO_4^{3-}}$ and sulfide capacity $C_{S^{2-}}$, the proposed coupling relationships between dephosphorization and desulfurization abilities for CaO-based slags are recommended to design or optimize chemical composition of slags.

Keywords: phosphate capacity; sulfide capacity; phosphorus distribution ratio; sulfur distribution ratio; evaluation of coupling relationship; secondary refining process, CaO–based slags

---

1. Introduction

For the purpose of refining low or ultra-low phosphorus and sulfur steel products with high mechanical properties, simultaneous dephosphorization and desulfurization of iron-based melts has been widely applied as a routine sub-process during hot metal pretreatment operation and the secondary refining process of molten steel in most metallurgical companies. It is well known that the greater oxygen potential of slags or iron-based melts, higher content of basic oxides in slags, and lower temperature at dephosphorization zone are three preferred operation conditions for promoting dephosphorization reactions under a fixed mass ratio of slags to iron-based melts from the viewpoint of dephosphorization thermodynamics. However, the corresponding three preferred operation conditions for promoting desulfurization reactions can be summarized as smaller oxygen potential of slags or iron-based melts, higher content of basic oxides in slags, and higher temperature at the desulfurization zone. Evidently, conditions for promoting dephosphorization reactions are to some degree opposite to those for enhancing desulfurization reactions for an assigned slag system. Moreover, a larger amount of slags or fluxes is also beneficial for promoting dephosphorization as well as desulfurization from the viewpoint of kinetics. However, a reasonable mass ratio of slags to iron-based melts should be controlled in order to decrease production cost in industrial plants. It can be concluded that besides the easily controlled temperature and content of basic oxides in slags or fluxes, controlling the optimal range of slag oxidization ability is a challenging task to successfully maintain ideal dephosphorization ability and acceptable desulfurization ability during simultaneous dephosphorization and desulfurization processes of iron-based melts.

CaO–Fe_tO–Al_2O_3 slag system was recommended by Ban–ya et al. [1] for simultaneous dephosphorization and desulfurization during the secondary refining process of molten steel. The recommended CaO–FeO–Fe_2O_3–Al_2O_3–P_2O_5 slags by Ban–ya et al. [1] exhibited a large variation range of slag oxidization ability with the mass percentage of Fe_tO varying from 1.88 to 55.50. Nevertheless, no conclusions or results on the linkage between dephosphorization and desulfurization abilities or potentials of the slags were provided by Ban–ya et al. [1]. The coupling relationships between dephosphorization and desulfurization abilities or potentials for CaO–FeO–Fe_2O_3–Al_2O_3–P_2O_5 slags [1] over a large range of slag oxidization ability during the secondary refining process of molten steel have been recently proposed by Yang et al. [2] as $\log L_P + 5 \log L_S$ or $\log C_{PO_4^{3-}} + \log C_{S^{2-}}$ in the reducing zone and as $\log L_P + \log L_S - 5 \log N_{Fe_tO}$ or $\log C_{PO_4^{3-}} + \log C_{S^{2-}} - \log N_{FeO}$ in the oxidizing zone through deleting or omitting the term of slag oxidization ability represented by the comprehensive mass action concentration N_{Fe_tO} of iron oxides Fe_tO based on the ion and molecule coexistence theory (IMCT) [2–18]. The proposed coupling relationships [2] for the CaO-based slags have been verified to be independent of slag oxidization ability as expected.

It should be specially mentioned that the linkage between phosphate capacity $C_{PO_4^{3-}}$ and sulfide capacity $C_{S^{2-}}$ for slags was first correlated by Sano et al. [19] in 1990 as $\log C_{PO_4^{3-}} = 1.5 \log C_{S^{2-}} +$ const. through deleting activity $a_{O^{2-}}$ of oxygen ion O^{2-} in slags. The defined constant term by Sano et al. [19] as $\log\left\{ \left[K_{PO_4^{3-}}^{\ominus} \gamma_{S^{2-}}^{3/2} (\Sigma n_i^0)^{1/2} M_{PO_4^{3-}} \right] / \left[(K_{S^{2-}}^{\ominus})^{3/2} \gamma_{PO_4^{3-}} M_S^{3/2} \right] \right\}$ can hold constant only under conditions that both ratios of $K_{PO_4^{3-}}^{\ominus} / \gamma_{PO_4^{3-}}$ and $\gamma_{S^{2-}}^{3/2} / (K_{S^{2-}}^{\ominus})^{3/2}$ keep constants simultaneously, which is also derived in details by Yang et al. [2] However, Ban–ya et al. [20,21] clearly proved and argued that the ratio of the activity coefficient to the standard equilibrium constant, i.e., $f_{\%, i} / K_i^{\ominus}$ or $\gamma_i / K_i^{\ominus}$, of dephosphorization and desulfurization products cannot hold constant in multi-components solutions like molten slags, fluxes and salts. Thus, the assumption of the term on the right-hand side of the relationship between $C_{PO_4^{3-}}$ and $C_{S^{2-}}$ by Sano et al. [19] being constant is not a theoretically correct conclusion. Furthermore, the intrinsic relationship between the phosphorus distribution ratio $L_P = (\% P_2O_5) / [\% P]^2$ and sulfur distribution ratio $L_S = (\% S) / [\% S]$ for slags with fixed chemical compositions has scarcely been investigated.

Under these circumstances, the proposed coupling relationships for the CaO-based slags should be further verified and evaluated from the viewpoint of whether or not they are also independent of slag chemical composition. The slag chemical composition was described in this contribution by three group parameters including the reaction abilities of components described by the mass action concentrations N_i, activity $a_{R,i}$ relative to pure liquid or solid matters as standard state, slag basicity containing simplified complex basicity $(\% \, CaO)/[(\% \, P_2O_5) + (\% \, Al_2O_3)]$ or optical basicity Λ, and the comprehensive influence of iron oxides Fe_tO and basic oxide CaO.

The ultimate objectives of this study can be summarized as (1) to further verify the linkage between dephosphorization and desulfurization abilities or potentials for a fixed flux or slags not only regardless of slag oxidization ability as expected but also independent of slag chemical composition; (2) to provide fundamental information for optimizing the chemical composition of slags or fluxes with the aim of enhancing simultaneous dephosphorization and desulfurization abilities or potentials of iron-based melts by a fixed flux or slags; (3) to enrich the foundations of the reaction mechanism during the simultaneous dephosphorization and desulfurization process of iron-based melts by a fixed flux or slags over a large variation range of slag oxidization ability; (4) moreover, to open new application fields of the IMCT [2–18] for metallurgical slags.

2. Influence of Slag Chemical Composition on Proposed Coupling Relationships between Dephosphorization and Desulfurization Abilities and Potentials for CaO–based Slags

The chemical compositions of $CaO–FeO–Fe_2O_3–Al_2O_3–P_2O_5$ slags equilibrated with liquid iron by Ban-ya et al. [1] and three parameters for representing slag oxidization ability as the mass percentage of Fe_tO through $(\% \, Fe_tO) = (\% \, FeO) + 0.9(\% \, Fe_2O_3)$; calculated [3] comprehensive mass action concentration N_{Fe_tO} of Fe_tO and oxygen potential p_{O_2} of the CaO-based slags over a temperature range from 1811 to 1927 K (1538 to 1654 °C) are summarized in Table 1. In addition, the determined [2] and calculated [2] coupling relationship terms as $\log L_P + 5 \log L_S$ or $\log C_{PO_4^{3-}} + \log C_{S^{2-}}$ in the reducing zone and as $\log L_P + \log L_S - 5 \log N_{Fe_tO}$ or $\log C_{PO_4^{3-}} + \log C_{S^{2-}} - \log N_{FeO}$ in the oxidizing zone based on measured $\log L_{P,\,measured}$ or measured $\log L_{S,\,measured}$ by Ban-ya et al. [1], predicted $\log L_{P,\,calculated}^{IMCT}$ [3] or $\log L_{S,\,calculated}^{IMCT}$ [4] by the IMCT models, determined $\log C_{PO_4^{3-},\,determined}$ [3] or determined $\log C_{S^{2-},\,determined}$ [5] after Ban–ya et al. [1], and predicted $\log C_{PO_4^{3-},\,calculated}^{IMCT}$ [3] or $\log C_{S^{2-},\,calculated}^{IMCT}$ [5] by the IMCT models are also tabulated in Table 2 for comparison. Thus, the effects of chemical composition of slags described by three group parameters including the reaction abilities of components represented by the mass action concentrations N_i, two kinds of slag basicity as simplified complex basicity $(\% \, CaO)/[(\% \, P_2O_5) + (\% \, Al_2O_3)]$ and optical basicity Λ, and the comprehensive effect of iron oxides Fe_tO and basic oxide CaO on proposed coupling relationships [2] for the CaO-based slags are further evaluated in the next section.

Table 1. Chemical compositions of CaO–FeO–Fe$_2$O$_3$–Al$_2$O$_3$–P$_2$O$_5$ slags over a large range of slag oxidization ability equilibrated with liquid iron after Ban-ya et al. [1] and three parameters for representing slag oxidization ability as the mass percentage of Fe$_t$O, calculated [3] comprehensive mass action concentration N_{Fe_tO} of Fe$_t$O, and oxygen potential [3] pO_2 of the slags over a temperature range from 1811 to 1927 K (1538 to 1654 °C).

New Test No. [2]	Old Test No. [1]	Chemical Composition of Slags [1] (mass %)						Chemical Composition of Liquid Iron [1] (mass %)			T [K (°C)]	Slag Oxidization Ability		
		(CaO)	(FeO)	(Fe$_2$O$_3$)	(Al$_2$O$_3$)	(P$_2$O$_5$)	(S)	[P]	[S]	[O]		(Fe$_t$O) (mass %)	N_{Fe_tO}[3] (–)	pO_2[3] (Pa)
1	18	55.19	1.61	0.23	39.65	2.5048	1.079	1.8766	0.04	0.0039	1822 (1549)	1.88	0.019	8.18 × 10⁻⁸
2	19	56.69	2.36	0.39	40.34	0.5629	0.862	0.1973	0.072	0.0051	1818 (1545)	2.72	0.026	1.54 × 10⁻⁷
3	17	56.14	2.67	0.76	37.35	2.1138	1.364	0.5017	0.079	0.0077	1821 (1548)	3.46	0.031	2.18 × 10⁻⁷
4	20	55.64	3.74	0.16	38.25	1.348	0.944	0.1296	0.107	0.0069	1820 (1547)	3.97	0.042	4.02 × 10⁻⁷
5	10	53.69	2.63	0.19	39.89	0.2418	0.181	0.0580	0.017	0.0087	1876 (1603)	2.91	0.031	5.38 × 10⁻⁷
6	27	58.31	2.78	0.00	36.81	0.5744	0.802	0.2766	0.075	0.0127	1918 (1645)	2.84	0.030	1.03 × 10⁻⁶
7	11	53.58	4.79	1.01	35.63	0.4165	0.216	0.0275	0.029	0.0165	1873 (1600)	6.00	0.056	1.72 × 10⁻⁶
8	14	56.84	3.71	0.42	35.71	2.3081	0.920	0.4648	0.083	0.0150	1927 (1654)	4.23	0.042	2.20 × 10⁻⁶
9*	1	52.67	5.47	1.24	38.13	0.3284	0.154	0.0120	0.026	0.0179	1874 (1601)	6.75*	0.064*	2.27 × 10⁻⁶*
10	21	54.47	12.01	3.25	30.36	1.5259	0.986	0.0093	0.158	0.0242	1822 (1549)	14.92	0.133	4.18 × 10⁻⁶
11	2	53.13	9.28	2.83	32.44	0.5394	0.214	0.0060	0.035	0.0275	1874 (1601)	12.11	0.107	6.38 × 10⁻⁶
12	22	51.69	18.59	3.92	23.61	1.4744	0.899	0.0047	0.145	0.0315	1811 (1538)	22.61	0.205	8.27 × 10⁻⁶
13	3	51.28	11.09	5.26	29.78	0.3168	0.134	0.0016	0.037	0.0356	1874 (1601)	16.24	0.133	9.93 × 10⁻⁶
14	28	53.35	10.19	0.93	30.35	1.7682	1.010	0.0283	0.168	0.0267	1922 (1649)	11.63	0.114	1.54 × 10⁻⁵
15	4	50.58	15.86	6.58	25.05	0.3575	0.167	0.0017	0.028	0.0448	1876 (1603)	22.21	0.187	2.03 × 10⁻⁵
16	23	49.27	25.15	7.22	17.78	1.3705	0.952	0.0026	0.125	0.0448	1828 (1555)	31.83	0.281	2.08 × 10⁻⁵
17	29	55.46	15.10	3.60	24.06	1.472	0.983	0.0085	0.154	0.0407	1924 (1651)	18.67	0.164	3.27 × 10⁻⁵
18	24	45.27	35.25	7.72	10.96	1.6423	0.958	0.0022	0.096	0.0530	1824 (1551)	42.54	0.387	3.68 × 10⁻⁵
19	5	47.42	20.83	10.83	18.09	0.3611	0.181	0.0004	0.024	0.0521	1874 (1601)	31.47	0.261	3.82 × 10⁻⁵
20	9	45.40	21.92	16.63	16.48	0.4327	0.211	0.0006	0.029	0.0541	1875 (1602)	36.73	0.301	5.15 × 10⁻⁵
21	6	46.35	27.70	11.89	11.59	0.352	0.175	0.0006	0.021	0.0584	1873 (1600)	39.37	0.338	6.29 × 10⁻⁵
22	15	51.99	20.69	3.17	17.92	2.9425	0.896	0.0126	0.129	0.0555	1927 (1654)	25.11	0.226	6.51 × 10⁻⁵
23	25	37.11	40.14	11.61	5.10	1.5949	1.027	0.0011	0.084	0.0612	1827 (1554)	53.84	0.506	6.64 × 10⁻⁵
24	26	38.61	47.49	13.02	0.00	1.5678	1.029	0.0021	0.083	0.0624	1821 (1548)	59.73	0.558	7.28 × 10⁻⁵
25	7	42.12	33.57	13.99	8.40	0.3142	0.172	0.0002	0.018	0.0666	1870 (1597)	47.06	0.425	9.49 × 10⁻⁵
26	8	43.63	33.64	15.20	5.25	0.3065	0.168	0.0002	0.017	0.0694	1873 (1600)	48.42	0.426	9.97 × 10⁻⁵
27	16	49.13	26.21	8.98	11.50	2.9552	1.067	0.0062	0.111	0.0690	1925 (1652)	35.79	0.304	1.14 × 10⁻⁴
28	12	43.52	38.77	13.86	2.30	0.4959	0.237	0.0008	0.021	0.0713	1873 (1600)	52.05	0.463	1.18 × 10⁻⁴
29	13	39.23	40.00	12.78	2.67	0.5014	0.241	0.0007	0.020	0.0768	1874 (1601)	54.40	0.500	1.40 × 10⁻⁴
30	30	45.80	37.65	10.16	5.34	1.5302	1.083	0.0025	0.108	n/a†	1927 (1654)	47.29	0.420	2.25 × 10⁻⁴
31	31	42.61	45.79	9.49	0.00	1.5405	1.139	0.0021	0.092	n/a†	1927 (1654)	55.50	0.503	3.23 × 10⁻⁴

* The No. 9 test run in the new test run number corresponds to the obtained [2–5] criterion for distinguishing the reducing and oxidizing zones of the slags that correspond to (% Fe$_t$O) as 6.75 or calculated [3] N_{Fe_tO} as 0.0637 or oxygen potential pO_2 of the slags as 2.27×10^{-6}. † n/a (not applicable) means that the oxygen content in the new or original test runs No. 30 and No.31 was not reported by Ban-ya et al. [1].

Table 2. Comparison between determined and calculated coupling relationship terms between dephosphorization and desulfurization abilities or potentials for CaO–FeO–Fe$_2$O$_3$–Al$_2$O$_3$–P$_2$O$_5$ slags over a large range of slag oxidization ability equilibrated with liquid iron during the secondary refining process of molten steel based on measured log L_P, measured or log L_S, measured by Ban-ya et al. [1], predicted log $L_{P,calculated}^{IMCT}$ [3] or log $L_{S,calculated}^{IMCT}$ [4] by the IMCT models, determined log $C_{PO_4^{3-}}$, determined [3] or determined log $C_{S^{2-}}$, determined [5] after Ban-ya et al. [1], and predicted log $C_{PO_4^{3-}}^{IMCT}$, calculated [3] or log $C_{S^{2-}}^{IMCT}$, calculated [5] by the IMCT models over a temperature range from 1811 to 1927 K (1538 to 1654 °C).

New Test No. [2]	Old Test No. [1]	De-P and De-S Abilities (–)				De-P and De-S Potentials (–)				Coupling Relationship Term between De-P and De-S Abilities (–)				Coupling Relationship Term between De-P and De-S Potentials (–)			
		logL_P		logL_S		log$C_{PO_4^{3-}}$		log$C_{S^{2-}}$		reducing zone logL_P+5logL_S		oxidizing zone logL_P+logL_S−5logN_{FeO}		reducing zone log$C_{PO_4^{3-}}$+log$C_{S^{2-}}$		oxidizing zone log$C_{PO_4^{3-}}$+log$C_{S^{2-}}$−logN_{FeO}	
		Meas.† [1]	Cal.† [3]	Meas.† [1]	Cal.† [4]	Detd.† [3]	Cal.† [3]	Detd.† [5]	Cal.† [5]	Detd.† [2]	Cal.† [2]	Detd.† [2]	Cal.† [2]	Detd.† [2]	Cal.† [2]	Detd.† [2]	Cal.† [2]
1	18	-0.148	1.294	1.431	1.880	18.921	20.912	-2.258	-2.028	7.007	10.696	—	—	16.663	18.884	—	—
2	19	1.160	1.848	1.078	1.218	19.398	20.258	-2.127	-2.057	6.551	7.936	—	—	17.271	18.201	—	—
3	17	0.924	2.133	1.237	1.231	19.030	20.734	-1.859	-2.063	7.110	8.286	—	—	17.172	18.671	—	—
4	20	1.904	2.772	0.946	0.983	19.647	20.479	-2.118	-2.067	6.632	7.688	—	—	17.528	18.412	—	—
5	10	1.857	1.344	1.027	1.109	18.597	18.408	-1.826	-1.975	6.993	6.886	—	—	16.671	16.433	—	—
6	27	0.876	0.885	1.029	1.209	17.580	18.164	-1.799	-1.877	6.021	6.928	—	—	15.753	16.287	—	—
7	11	2.741	2.649	0.872	0.822	18.497	18.765	-1.783	-1.993	7.101	6.758	—	—	16.698	16.772	—	—
8	14	1.029	1.454	1.045	1.121	17.680	18.514	-1.871	-1.871	6.252	7.058	—	—	15.896	16.643	—	—
9*	1	3.358	2.874	0.773	0.741	18.662	18.479	-1.860	-2.012	7.221	6.577	—	—	16.802	16.467	—	—
10	21	4.247	5.068	0.795	0.849	19.521	20.362	-1.707	-1.660	—	—	9.442	10.317	—	—	18.692	19.579
11	2	4.176	3.960	0.786	0.764	18.721	18.702	-1.661	-1.763	—	—	9.831	9.593	—	—	18.031	17.911
12	22	4.824	6.079	0.792	0.898	19.594	20.734	-1.591	-1.440	—	—	9.075	10.435	—	—	18.692	19.983
13	3	5.093	4.389	0.559	0.716	18.788	18.316	-1.778	-1.713	—	—	10.050	9.503	—	—	17.896	17.478
14	28	3.344	3.553	0.779	0.932	18.323	18.379	-1.713	-1.498	—	—	8.837	9.199	—	—	17.553	17.823
15	4	5.092	5.051	0.776	0.762	18.611	20.288	-1.463	-1.516	—	—	9.528	9.473	—	—	17.816	17.821
16	23	5.307	6.438	0.882	0.882	19.315	20.288	-1.354	-1.291	—	—	8.967	10.098	—	—	18.512	19.548
17	29	4.309	4.273	0.805	0.882	18.301	18.163	-1.501	-1.383	—	—	9.054	9.096	—	—	17.585	17.565
18	24	5.531	7.041	0.999	0.941	19.304	20.518	-1.161	-1.100	—	—	8.620	10.072	—	—	18.556	19.830
19	5	6.354	5.698	0.877	0.808	19.035	18.289	-1.295	-1.328	—	—	10.184	9.460	—	—	18.323	17.543
20	9	6.080	5.892	0.862	0.797	18.892	18.695	-1.296	-1.277	—	—	9.616	9.363	—	—	18.118	17.739
21	6	5.990	6.197	0.921	0.876	18.732	18.560	-1.203	-1.151	—	—	9.309	9.470	—	—	18.000	18.015
22	15	4.268	4.882	0.842	0.898	18.070	20.225	-1.332	-1.225	—	—	8.350	9.021	—	—	17.384	17.981
23	25	6.120	7.378	1.087	0.955	19.414	20.711	-1.011	-0.964	—	—	8.737	9.862	—	—	18.699	19.556
24	26	5.551	7.618	1.093	0.986	19.147	20.711	-0.995	-0.901	—	—	7.966	9.927	—	—	18.405	20.063
25	7	6.895	6.587	0.980	0.980	19.039	18.390	-1.087	-1.042	—	—	9.788	9.388	—	—	18.323	17.719
26	8	6.884	6.584	0.983	0.878	18.964	18.496	-1.056	-1.048	—	—	9.792	9.376	—	—	18.279	17.676
27	16	4.886	5.429	0.983	0.883	18.156	19.014	-1.095	-1.113	—	—	8.486	8.929	—	—	17.579	17.899
28	12	5.889	6.735	1.053	0.926	18.544	18.935	-0.987	-0.964	—	—	8.666	9.385	—	—	17.891	18.385
29	13	6.010	6.809	1.081	0.924	18.521	18.221	-0.928	-0.931	—	—	8.650	9.292	—	—	17.894	18.305
30	30	5.389	5.914	1.001	0.980	17.697	18.208	-0.851	-0.872	—	—	8.310	8.814	—	—	17.222	17.726
31	31	5.543	6.180	1.093	1.051	17.572	18.208	-0.685	-0.726	—	—	8.181	8.776	—	—	17.189	17.784

* The No. 9 test run in the new test run number corresponds to the obtained [2–5] criterion for distinguishing the reducing and oxidizing zones of the slags that correspond to (% Fe$_t$O) as 6.75 or calculated [3] N_{Fe_tO} as 0.0637 or oxygen potential p_{O_2} of the slags as 2.27 × 10^{-6}. † Cal. = Calculated, Meas. = Measured, Detd. = Determined.

2.1. Influence of Reaction Abilities of Components on Coupling Relationships between Dephosphorization and Desulfurization Abilities or Potentials for CaO-based Slags

It was verified [3] that good corresponding relationships between mass percentages of components and calculated [3] mass action concentrations N_i of CaO, FeO, Fe_2O_3, and Al_2O_3 as components are established for the CaO-based slags. Thus, the calculated [3] mass action concentrations N_i of components based on the IMCT [2–18] can be applied to the representation of the chemical composition of the CaO-based slags, like the mass percentage (% *i*) of components. As a newly-formed structural unit $FeO \cdot Fe_2O_3$, according to the IMCT [2–18], the calculated [3–5] mass action concentration $N_{FeO \cdot Fe_2O_3}$ of $FeO \cdot Fe_2O_3$ can be used to describe the reaction ability of $FeO \cdot Fe_2O_3$. In addition, the calculated [3–5] comprehensive mass action concentration N_{Fe_tO} of iron oxides Fe_tO can also be applied to the description of reaction ability of Fe_tO. Thus, the calculated [3–5] mass action concentrations N_i of CaO, FeO, Fe_2O_3, Al_2O_3, $FeO \cdot Fe_2O_3$, and Fe_tO are used to represent reaction abilities of components in the CaO-based slags, like the traditional applied activities $a_{R, i}$ of components in the classical metallurgical physicochemistry.

2.1.1. Influences of Reaction Abilities of Components on Coupling Relationships between Dephosphorization and Desulfurization Abilities for CaO-based Slags

The relationships of calculated [3–5] mass action concentrations, N_i, of six components CaO, FeO, Fe_2O_3, Al_2O_3, $FeO \cdot Fe_2O_3$, and Fe_tO against the calculated [3] phosphorus distribution ratio $\log L_{P, \text{calculated}}^{IMCT}$ using the IMCT–L_P model for the CaO-based slags equilibrated with liquid iron are shown in the first layers of Figure 1. Likewise, the relationships of the N_i of six components against the calculated [4] sulfur distribution ratio $\log L_{S, \text{calculated}}^{IMCT}$ using the IMCT–L_S model for the CaO-based slags are also illustrated in the second layers of Figure 1. Meanwhile, the relationship of the aforementioned N_i of the six components against determined $\log L_{P, \text{measured}} + 5 \log L_{S, \text{measured}}$ or $\log L_{P, \text{measured}} + \log L_{S, \text{measured}} - 5 \log N_{Fe_tO}$ after original data from Ban-ya et al. [1], or calculated $\log L_{P, \text{calculated}}^{IMCT} + 5 \log L_{S, \text{calculated}}^{IMCT}$ or $\log L_{P, \text{calculated}}^{IMCT} + \log L_{S, \text{calculated}}^{IMCT} - 5 \log N_{Fe_tO}$ based on results by Yang et al. [3,4] for the CaO-based slags is displayed in the first and second layers of Figure 2, respectively. The average values of term $\log L_P + 5 \log L_S$ or $\log L_P + \log L_S - 5 \log N_{Fe_tO}$ after original data from Ban-ya et al. [1] or based on results from Yang et al. [3,4] over three sub-divided temperature ranges are also exhibited in Figure 2 by lines. The distinguishing lines between CaS and FeS for representing the reducing and oxidizing zones are added on the horizontal ordinates in the sub-figures of Figure 1 and other figures in the following text if necessary. To improve the display resolution, the horizontal ordinates in the sub-figures of Figure 1 and other figures in the following text, if necessary, are also split into two zones through adding break symbols for describing the reducing and oxidizing zones. It should be emphasized that the exponential growing tendency of the phosphorus distribution ratio $\log L_P = \log \left\{ (\% \, P_2O_5) / [\% \, P]^2 \right\}$ against the slag oxidization ability expressed by N_{Fe_tO} of iron oxides in Figure 1(f1) as well as the backward tick-shaped or asymmetrical relationship of the sulfur distribution ratio $\log L_S = \log \{ (\% \, S) / [\% \, S] \}$ against N_{Fe_tO} of iron oxides in Figure 1(f2) based on the normal scale of the horizontal ordinates cannot normally be displayed. However, the intrinsic relationships of the mass action concentrations N_i of six components against $\log L_P$ or $\log L_S$ cannot be changed by adding break symbols on the horizontal ordinates in the sub-figures of Figure 1. It can be observed in Figures 1 and 2 that the criterion for distinguishing the reducing and oxidizing zones corresponds to N_{CaO} in 0.778, N_{FeO} in 0.0626, $N_{Fe_2O_3}$ in 1.00×10^{-4}, $N_{Al_2O_3}$ in 1.50×10^{-3}, $N_{FeO \cdot Fe_2O_3}$ in 4.10×10^{-5}, and N_{Fe_tO} in 0.0637, respectively.

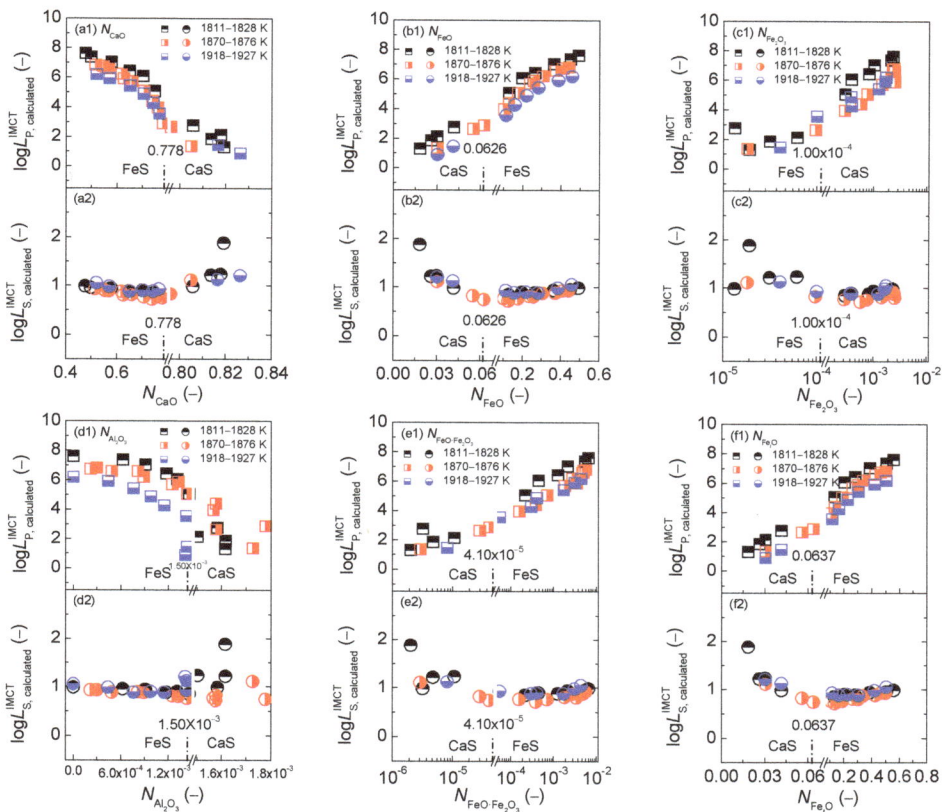

Figure 1. Relationship of the calculated [3] mass action concentration N_i of CaO (**a**), FeO (**b**), Fe_2O_3 (**c**), Al_2O_3 (**d**), $FeO \cdot Fe_2O_3$ (**e**), and Fe_tO (**f**) against the calculated [3] phosphorus distribution ratio $\log L_{P, calculated}^{IMCT}$ by IMCT–L_P model in the first layer or calculated [4] sulfur distribution ratio $\log L_{S, calculated}^{IMCT}$ by IMCT–L_S model in the second layer for $CaO–FeO–Fe_2O_3–Al_2O_3–P_2O_5$ slags equilibrated with liquid iron over a temperature range of 1811 to 1927 K (1538 to 1654 °C), respectively.

With regard to the dephosphorization ability of the CaO-based slags, it can be observed in the first layers of Figure 1 that increasing N_{CaO} or $N_{Al_2O_3}$ can result in a significantly decreasing tendency of $\log L_P$ as shown in Figure 1(a1,d1); however, increasing N_{FeO}, $N_{Fe_2O_3}$ or $N_{FeO \cdot Fe_2O_3}$ can lead to an increasing trend of $\log L_P$ as shown in Figure 1(b1,c1,e1). Certainly, the result in Figure 1(a1) is not consistent with the widely-accepted consensus that basic oxide CaO can promote dephosphorization reactions. It can be obtained from the relationship of calculated N_{Fe_tO} against N_{CaO} or N_{FeO} or $N_{Fe_2O_3}$ or $N_{Al_2O_3}$ or $N_{FeO \cdot Fe_2O_3}$ for the CaO-basedslags as illustrated in Figure 3 that increasing N_{FeO} or $N_{Fe_2O_3}$ or $N_{FeO \cdot Fe_2O_3}$ can result in an increasing tendency of N_{Fe_tO} as shown in Figure 3(b,c,e); however, increasing N_{Fe_tO} can lead to a decreasing trend of N_{CaO} or $N_{Al_2O_3}$ as illustrated in Figure 3(a,d). The promotive effect of increasing N_{Fe_tO} on $\log L_P$ can be counteracted by the decrease of N_{CaO} for the CaO-based slags. There are some extreme proofs to support this finding as the CaO-based slags with high CaO but very low Fe_tO, which are widely applied at reduction stage during electric arc furnace (EAF) steelmaking process or used during the blast furnace (BF) ironmaking process, can only extract sulphur, rather than phosphorus, from iron–based melts. Thus, not only the independent effect of iron oxides Fe_tO and basic oxide CaO, but also the comprehensive effect of iron oxides Fe_tO and basic oxide CaO plays a decisive role in $\log L_P$ between the CaO-based slags and liquid iron.

This finding is in good agreement with the well-known conclusion that the CaO-based slags with middle Fe$_t$O and high CaO, which are commonly used in a top–bottom combined oxygen blowing converter for the steelmaking process or dephosphorization pretreatment of hot metal, indicates a greater dephosphorization ability coupling with limited desulfurization ability for iron-based melts. This result can be applied to the explanation of reason that the promotive effect of increasing N_{FeO} or $N_{Fe_2O_3}$ or $N_{FeO \cdot Fe_2O_3}$ or N_{Fe_tO} on $\log L_P$ can be counteracted by a decrease of N_{CaO} for the CaO-based slags as shown in Figure 1(a1).

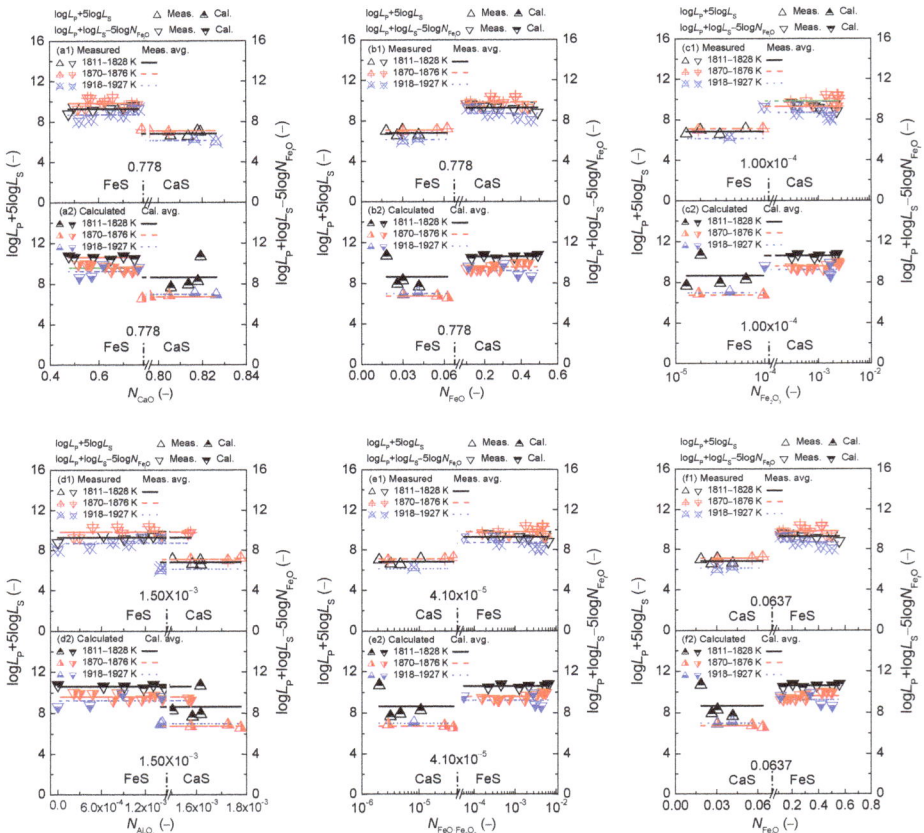

Figure 2. Relationship of calculated [3] mass action concentration N_i of CaO (**a**), FeO (**b**), Fe$_2$O$_3$ (**c**), Al$_2$O$_3$ (**d**), FeO·Fe$_2$O$_3$ (**e**), and Fe$_t$O (**f**) against determined term $\log L_{P,\,measured} + 5\log L_{S,\,measured}$ or $\log L_{P,\,measured} + \log L_{S,\,measured} - 5\log N_{Fe_tO}$ after Ban-ya et al. [1] in the first layer or calculated term $\log L_{P,\,calculated}^{IMCT} + 5\log L_{S,\,calculated}^{IMCT}$ or $\log L_{P,\,calculated}^{IMCT} + \log L_{S,\,calculated}^{IMCT} - 5\log N_{Fe_tO}$ based on the results from Yang et al. [3,4] in the second layer for CaO–FeO–Fe$_2$O$_3$–Al$_2$O$_3$–P$_2$O$_5$ slags equilibrated with liquid iron over a temperature range of 1811 to 1927 K (1538 to 1654 °C), respectively.

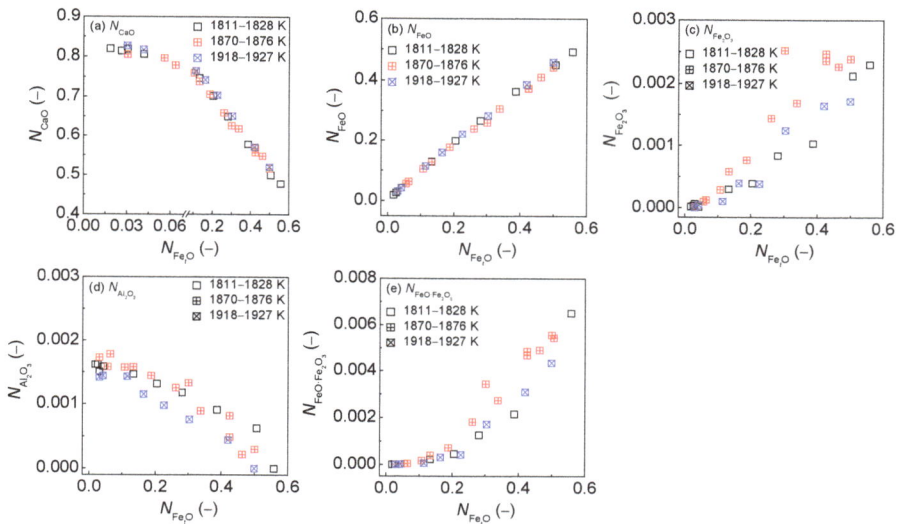

Figure 3. Relationship of calculated [3] comprehensive mass action concentration N_{Fe_tO} of iron oxides against calculated [3] mass action concentration N_{CaO} (**a**) or N_{FeO} (**b**) or $N_{Fe_2O_3}$ (**c**) or $N_{Al_2O_3}$ (**d**) or $N_{FeO \cdot Fe_2O_3}$ (**e**) for CaO–FeO–Fe$_2$O$_3$–Al$_2$O$_3$–P$_2$O$_5$ slags equilibrated with liquid iron over a temperature range of 1811 to 1927 K (1538 to 1654 °C), respectively.

With respect to the desulfurization ability of the CaO-based slags, it can be obtained from the second layers of Figure 1 that increasing N_{CaO} or $N_{Al_2O_3}$ accompanied with a decrease in N_{FeO} or $N_{Fe_2O_3}$ or $N_{FeO \cdot Fe_2O_3}$ or N_{Fe_tO} can result in an obviously increasing tendency of the sulfur distribution ratio $\log L_S$ of the CaO-based slags in the reducing zone. This result can be reasonably explained by the IMCT–L_S model [4] for the CaO-based slags in the reducing zone that basic oxide CaO expressed by N_{CaO} shows a promoting effect on desulfurization ability, while iron oxides Fe$_t$O expressed by N_{Fe_tO} exhibit a decaying influence on the desulfurization ability of the CaO-based slags in the reducing zone. However, increasing slag oxidization ability, i.e., increasing N_{FeO} or $N_{Fe_2O_3}$ or $N_{FeO \cdot Fe_2O_3}$ or N_{Fe_tO}, can lead to a slightly increasing trend of $\log L_S$ of the CaO-based slags in the oxidizing zone. This result can be explained by the IMCT–L_S model [4] for the CaO-based slags in the oxidizing zone that only ferrous oxide FeO expressed by N_{FeO} influences $\log L_S$ of the CaO-based slags in the oxidizing zone.

On the proposed coupling relationships [2] between L_P and L_S for the CaO–based slags, it can be observed in Figure 2 that the proposed terms $\log L_P + 5 \log L_S$ in the reducing zone and $\log L_P + \log L_S - 5 \log N_{Fe_tO}$ in the oxidizing zone are doubtlessly independent of variation of N_{CaO} or $N_{Al_2O_3}$ as well as N_{FeO} or $N_{Fe_2O_3}$ or $N_{FeO \cdot Fe_2O_3}$ or N_{Fe_tO}. Increasing temperature from 1811 to 1927 K (1538 to 1654 °C) can result in a slightly decreasing tendency of proposed coupling relationships between L_P and L_S for the CaO-based slags. Thus, the proposed coupling relationships between L_P and L_S are not only independent of the slag oxidization ability as shown in Figure 1(f1,f2) but is also irrelevant to the mass action concentrations N_i of six components over a narrow temperature range.

2.1.2. Influences of Reaction Abilities of Components on Coupling Relationships between Dephosphorization and Desulfurization Potentials for CaO–based Slags

The relationship of calculated [3–5] mass action concentrations N_i of six components as CaO, FeO, Fe$_2$O$_3$, Al$_2$O$_3$, FeO·Fe$_2$O$_3$, and Fe$_t$O against the calculated [3] phosphate capacity $\log C_{PO_4^{3-}}^{IMCT}$, calculated using the IMCT–$C_{PO_4^{3-}}$ model for the CaO-based slags are shown in the first layers of Figure 4, respectively. Similarly, the relationship of N_i of six components against the calculated [5] sulfide capacity $\log C^{IMCT}_{S^{2-}, \text{calculated}}$ using the IMCT–$C_{S^{2-}}$ model for the CaO-based slags are also illustrated in the second layers of Figure 4, respectively. The relationship of N_i of six components against determined $\log C_{PO_4^{3-}, \text{determined}} + \log C_{S^{2-}, \text{determined}}$ or $\log C_{PO_4^{3-}, \text{determined}} + \log C_{S^{2-}, \text{determined}} - \log N_{FeO}$ after original data from Ban-ya et al. [1], or calculated $\log C^{IMCT}_{PO_4^{3-}, \text{calculated}} + \log C^{IMCT}_{S^{2-}, \text{calculated}}$ or $\log C^{IMCT}_{PO_4^{3-}, \text{calculated}} + \log C^{IMCT}_{S^{2-}, \text{calculated}} - \log N_{FeO}$ based on results by Yang et al. [3,5] for the CaO-based slags are displayed in the first and second layers of Figure 5, respectively. In addition, the average values of term $\log C_{PO_4^{3-}} + \log C_{S^{2-}}$ or $\log C_{PO_4^{3-}} + \log C_{S^{2-}} - \log N_{FeO}$ after original data from Ban-ya et al. [1] or based on results from Yang et al. [3,5] over three sub-divided temperature ranges are also exhibited in Figure 5 by horizontal lines, respectively.

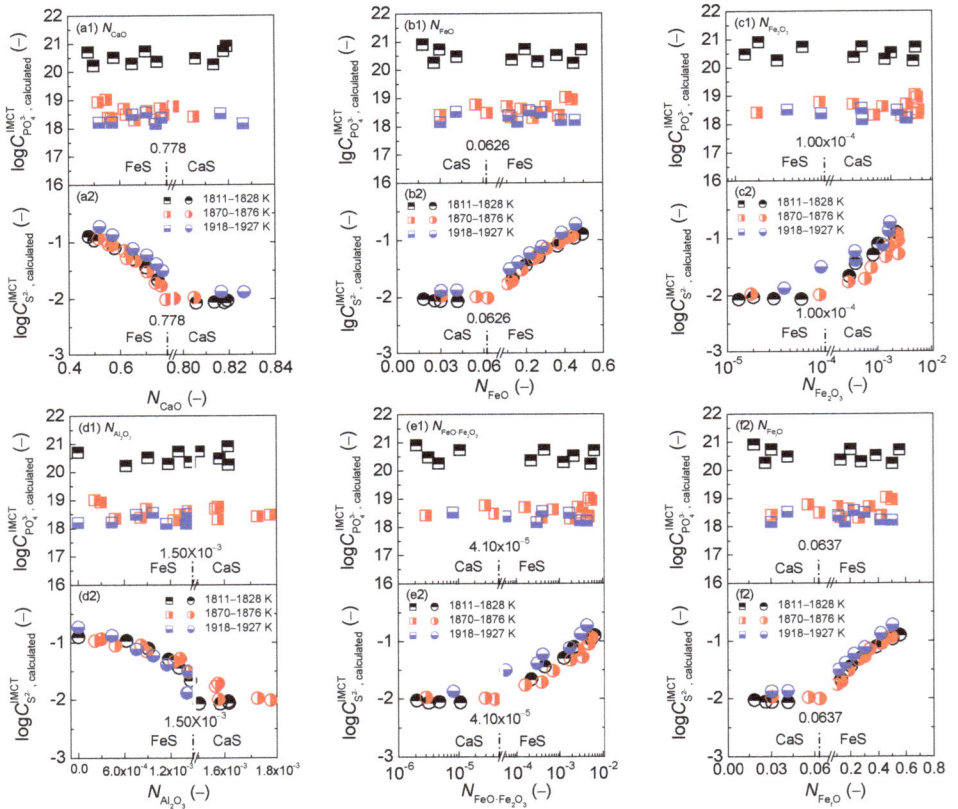

Figure 4. Relationship of calculated [3] mass action concentration N_i of CaO (**a**), FeO (**b**), Fe$_2$O$_3$ (**c**), Al$_2$O$_3$ (**d**), FeO·Fe$_2$O$_3$ (**e**), and Fe$_t$O (**f**) against calculated [3] phosphate capacity $\log C^{IMCT}_{PO_4^{3-}, \text{calculated}}$ by the IMCT–$C_{PO_4^{3-}}$ model in the first layer or calculated [5] sulfide capacity $\log C^{IMCT}_{S^{2-}, \text{calculated}}$ by the IMCT–$C_{S^{2-}}$ model in the second layer for CaO–FeO–Fe$_2$O$_3$–Al$_2$O$_3$–P$_2$O$_5$ slags equilibrated with liquid iron over a temperature range of 1811 to 1927 K (1538 to 1654 °C), respectively.

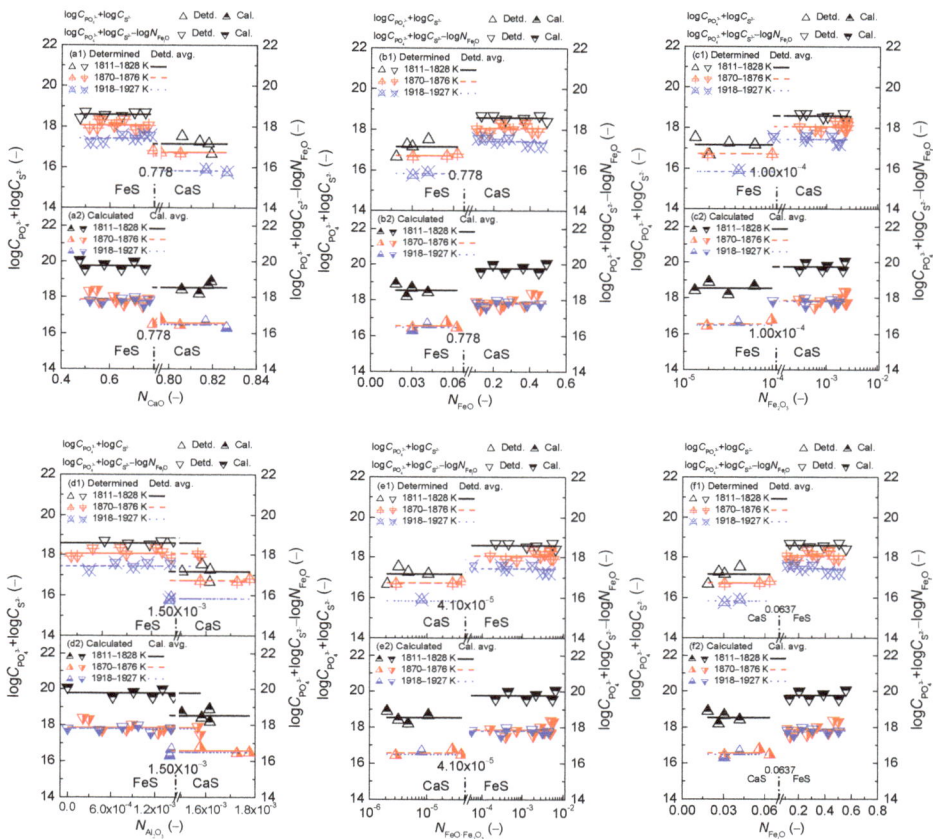

Figure 5. Relationship of calculated [3] mass action concentration N_i of CaO (**a**), FeO (**b**), Fe$_2$O$_3$ (**c**), Al$_2$O$_3$ (**d**), FeO·Fe$_2$O$_3$ (**e**), and Fe$_t$O (**f**) against determined term $\log C_{PO_4^{3-}, \text{determined}}$ + $\log C_{S^{2-}, \text{determined}}$ or $\log C_{PO_4^{3-}, \text{determined}}$ + $\log C_{S^{2-}, \text{determined}}$ − $\log N_{FeO}$ after Ban-ya et al. [1] in the first layer or calculated term $\log C_{PO_4^{3-}, \text{calculated}}^{IMCT}$ + $\log C_{S^{2-}, \text{calculated}}^{IMCT}$ or $\log C_{PO_4^{3-}, \text{calculated}}^{IMCT}$ + $\log C_{S^{2-}, \text{calculated}}^{IMCT}$ − $\log N_{FeO}$ based on results from Yang et al. [3,5] in the second layer for CaO–FeO–Fe$_2$O$_3$–Al$_2$O$_3$–P$_2$O$_5$ slags equilibrated with liquid iron over a temperature range from 1811 to 1927 K (1538 to 1654 °C), respectively.

With respect to the dephosphorization potential of the CaO-based slags, it can be observed in the first layers of Figure 4 that phosphate capacity $C_{PO_4^{3-}}$ of the CaO-based slags is almost unchangeable with the increase of N_i of six components over a narrow temperature range because the comprehensive effect of iron oxides Fe$_t$O and basic oxide CaO plays the key role in dephosphorization potential of the CaO-based slags as discussed in Section 2.3.2 and elsewhere [3].

With regard to the desulfurization potential [5] of the CaO-based slags, it can be observed in the second layers of Figure 4 that sulfide capacity $C_{S^{2-}}$ of the CaO-based slags in the reducing zone also keeps almost constant with the increase of N_i of six components. However, sulfide capacity $C_{S^{2-}}$ of the CaO-based slags in the oxidizing zone displays an obviously increasing tendency with the increase of N_{FeO} or $N_{Fe_2O_3}$ or $N_{FeO·Fe_2O_3}$ or N_{Fe_tO} over a narrow temperature range, but exhibits a largely decreasing trend with the increase of N_{CaO} or $N_{Al_2O_3}$, which has been explained elsewhere [5]. Basic oxides CaO expressed by N_{CaO} largely the affect desulfurization potential of the CaO-based slags in the reducing zone from the IMCT–$C_{S^{2-}}$ model [5], while ferrous oxides expressed by N_{FeO} significantly influence the desulfurization potential of the CaO-based slags in the oxidizing zone from the IMCT–$C_{S^{2-}}$ model [5]. The very small decreasing tendency of N_{CaO} in the reducing zone in Figure 3(a) cannot cause an obvious increasing tendency of desulfurization potential of the CaO-based slags in the reducing zone. Furthermore, sulfide capacity $C_{S^{2-}}$ of the CaO-based slags in the reducing zone is also independent of N_{FeO} or $N_{Fe_2O_3}$ or $N_{FeO \cdot Fe_2O_3}$ or N_{Fe_tO}. The decreasing tendency of the sulfide capacity $C_{S^{2-}}$ of the CaO-based slags in the oxidizing zone with the increase of N_{CaO} or $N_{Al_2O_3}$ can be attributed to the largely decreasing trend of N_{Fe_tO} as shown in Figure 3. Increasing N_{FeO} or $N_{Fe_2O_3}$ or $N_{FeO \cdot Fe_2O_3}$ or N_{Fe_tO} can significantly promote the sulfide capacity $C_{S^{2-}}$ of the CaO-based slags in oxidizing zone.

With regards to the proposed coupling relationships between $C_{PO_4^{3-}}$ and $C_{S^{2-}}$ for the CaO-based slags, it can be observed in Figure 5 that the proposed terms [2] $\log C_{PO_4^{3-}} + \log C_{S^{2-}}$ and $\log C_{PO_4^{3-}} + \log C_{S^{2-}} - \log N_{FeO}$ are independent of N_i of six components. Increasing temperature from 1811 to 1927 K (1538 to 1654 °C) can result in a decreasing tendency of proposed coupling relationships between $C_{PO_4^{3-}}$ and $C_{S^{2-}}$ for the CaO-based slags. Thus, the proposed coupling relationships between $C_{PO_4^{3-}}$ and $C_{S^{2-}}$ for the CaO-based slags are not only independent of slag oxidization ability as shown in Figure 5(f) but are also irrelevant to the mass action concentrations N_i of six components over a narrow temperature range.

However, small discrepancies of proposed terms [2] $\log C_{PO_4^{3-}} + \log C_{S^{2-}}$ and $\log C_{PO_4^{3-}} + \log C_{S^{2-}} - \log N_{FeO}$ based on results by Yang et al. [3,5] and that based on determined ones after original data from Ban-ya et al. [1] for the CaO-based slags can be observed in each sub-figure of Figure 5. It is widely accepted that the phosphate capacity $C_{PO_4^{3-}}$ of slags can be determined or calculated from the corresponding phosphorus distribution ratio L_P through the relationship [3,11] between L_P and $C_{PO_4^{3-}}$ for slags, meanwhile, sulfide capacity $C_{S^{2-}}$ of slags can also be determined or calculated from the sulfur distribution ratio L_S through the relationship [5,7,9] between L_S and $C_{S^{2-}}$ for slags. It was verified by Yang et al. [4,5] that the calculated [4] results of $\log L_{S, \text{calculated}}^{IMCT}$ by the IMCT–L_S model are in good consistency with measured [1] ones by Ban-ya et al., meanwhile the calculated [5] results of $\log C_{S^{2-}, \text{calculated}}^{IMCT}$ by the IMCT–$C_{S^{2-}}$ model are in good accord with determined [5] $\log C_{S^{2-}, \text{determined}}$ after the original data from Ban–ya et al. [1]. However, it was also verified by Yang et al. [3] that the calculated results of $\log L_{P, \text{calculated}}^{IMCT}$ by the IMCT–L_P model are not in good agreement with the measured $\log L_{P, \text{measured}}$ by Ban-ya et al. [1], especially over the lower temperature range of 1811 to 1828 K (1538 to 1555 °C). Thus, the large deviation between calculated [3] $\log C_{PO_4^{3-}, \text{calculated}}^{IMCT}$ by the IMCT–$C_{PO_4^{3-}}$ model and determined [3] results of $\log C_{PO_4^{3-}, \text{determined}}$ after the original data from Ban–ya et al. [1], especially over the lower temperature range, is caused by the relationship [3,11] between L_P and $C_{PO_4^{3-}}$ for slags. Evidently, the accuracy of the phosphorus distribution ratio L_P is very important to obtain the precise phosphate capacity $C_{PO_4^{3-}}$ of slags through the relationship [3,11] between L_P and $C_{PO_4^{3-}}$ for slags. This means that the experimental uncertainties for dephosphorization reactions by Ban-ya et al. [1] can be effectively relieved by the predicted results of $\log L_{P, \text{calculated}}^{IMCT}$ by the IMCT–L_P [3] and $\log C_{PO_4^{3-}, \text{calculated}}^{IMCT}$ by the IMCT–$C_{PO_4^{3-}}$ model [3]. It can be deduced that the relationships of the mass action concentrations N_i of six components against calculated coupling relationships between $C_{PO_4^{3-}}$ and $C_{S^{2-}}$ based on results by Yang et al. [3,5] are more accurate than those against determined ones after original data from Ban-ya et al. [1] for the CaO–based slags.

2.2. Influence of Slag Basicity on Coupling Relationships between Dephosphorization and Desulfurization Abilities or Potentials for CaO-based Slags

For the purpose of investigating the influence of slag basicity on proposed coupling relationships, two kinds of slag basicity as simplified complex basicity and optical basicity Λ are applied in this study. The commonly applied complex basicity $[(\% \, CaO) + 1.4(\% \, MgO)] / [(\% \, SiO_2) + (\% \, P_2O_5) + (\% \, Al_2O_3)]$ [15,22–24] can be simplified as $(\% \, CaO) / [(\% \, P_2O_5) + (\% \, Al_2O_3)]$ due to no SiO_2 in the CaO-based slags.

Three group values of optical basicity for FeO and Fe_2O_3 have been recommended as (1) Λ_{FeO} = 0.51 and $\Lambda_{Fe_2O_3}$ = 0.48 from Pauling electronegativity [25]; (2) Λ_{FeO} = 0.93 and $\Lambda_{Fe_2O_3}$ = 0.69 from average electron density [26]; (3) Λ_{FeO} = 1.0 and $\Lambda_{Fe_2O_3}$ = 0.75 based on mathematical regression [27] from numerous experimental data. According to the evaluation results [3–5] of the aforementioned three group values of optical basicity for FeO and Fe_2O_3, the obtained Λ_{FeO} = 1.0 and $\Lambda_{Fe_2O_3}$ = 0.75 from mathematical regression [27] are recommended to represent optical basicity for FeO and Fe_2O_3, which are similar to those recommended Λ_{FeO} = 1.0 and $\Lambda_{Fe_2O_3}$ = 0.77 by Young et al. [28]

2.2.1. Influence of Slag Basicity on Coupling Relationships between Dephosphorization and Desulfurization Abilities for CaO-based Slags

The relationship between the simplified complex basicity $(\% \, CaO) / [(\% \, P_2O_5) + (\% \, Al_2O_3)]$ or optical basicity Λ and calculated [3] phosphorus distribution ratio $\log L_{P, \, calculated}^{IMCT}$ for the CaO-based slags is shown in the first layers of Figure 6, respectively. Similarly, the relationship between the two kinds of slag basicity and the calculated [4] sulfur distribution ratio $\log L_{S, \, calculated}^{IMCT}$ for the CaO-based slags is also illustrated in the second layers of Figure 6, respectively. The relationship between two kinds of slag basicity and determined term $\log L_{P, \, measured} + 5 \log L_{S, \, measured}$ or $\log L_{P, \, measured} + \log L_{S, \, measured} - 5 \log N_{Fe_tO}$ after original data from Ban-ya et al. [1], or calculated $\log L_{P, \, calculated}^{IMCT} + 5 \log L_{S, \, calculated}^{IMCT}$ or $\log L_{P, \, calculated}^{IMCT} + \log L_{S, \, calculated}^{IMCT} - 5 \log N_{Fe_tO}$ based on results by Yang et al. [3,4] for the CaO-based slags are displayed in the first and second layers of Figure 7, respectively. It can be obtained from Figures 6 and 7 that the criterion for distinguishing, reducing and oxidizing zones corresponds to simplified complex basicity $(\% \, CaO) / [(\% \, P_2O_5) + (\% \, Al_2O_3)]$ in 1.56 or optical basicity Λ in 0.80, respectively. It should be pointed out that the obtained criterion of optical basicity Λ as 0.80 for separating the reducing and oxidizing zones in this study is in good agreement with that by Young et al. [25] for developing the sulfide capacity $C_{S^{2-}}$ prediction model of $CaO–SiO_2–MgO–FeO–MnO–Al_2O_3$ slags.

With respect to the dephosphorization ability of the CaO-based slags, it can be observed in the first layers of Figure 6 that increasing two kinds of slag basicity can result in an exponentially growing tendency of the phosphorus distribution ratio $\log L_P$, which has been explained by Yang et al. elsewhere [3]. Regarding the desulfurization ability of the CaO-based slags, it can be obtained from the second layers of Figure 6 that increasing two kinds of slag basicity can lead to a backward tick-shaped variation trend of sulfur distribution ratio $\log L_S$, which has been explained by Yang et al. elsewhere [4]. Certainly, adding break symbols on the horizontal ordinates in the sub-figures of Figure 6 can only destroy the apparent relationship of the exponentially growing tendency of $\log L_P$ against two kinds of slag basicity in the first layers of Figure 6 as well as the backward tick-shaped or asymmetrical relationship of $\log L_S$ against two kinds of slag basicity in the second layers of Figure 6. The intrinsic relationships of two kinds of slag basicity against $\log L_P$ or $\log L_S$ cannot be changed through adding break symbols on the horizontal ordinates in the sub-figures of Figure 6.

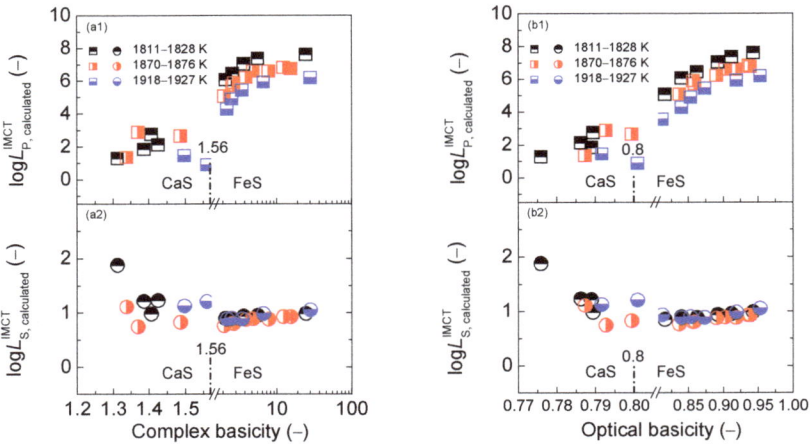

Figure 6. Relationship of simplified complex basicity $(\% \, CaO)/[(\% \, P_2O_5) + (\% \, Al_2O_3)]$ (**a**) or optical basicity Λ (**b**) against the calculated [3] phosphorus distribution ratio $\log L_{P, calculated}^{IMCT}$ by the IMCT–L_P model in the first layer or calculated [4] sulfur distribution ratio $\log L_{S, calculated}^{IMCT}$ by the IMCT–L_S model in the second layer for CaO–FeO–Fe_2O_3–Al_2O_3–P_2O_5 slags equilibrated with liquid iron over a temperature range of 1811 to 1927 K (1538 to 1654 °C), respectively.

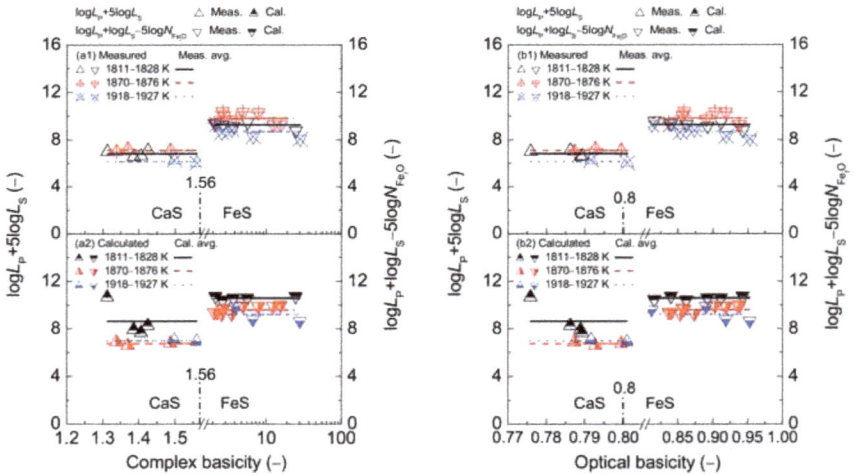

Figure 7. Relationship of simplified complex basicity $(\% \, CaO)/[(\% \, P_2O_5) + (\% \, Al_2O_3)]$ (**a**) or optical basicity Λ (**b**) against the determined term $\log L_{P, measured} + 5 \log L_{S, measured}$ or $\log L_{P, measured} + \log L_{S, measured} - 5 \log N_{Fe_tO}$ after Ban-ya et al. [1] in the first layer or calculated term $\log L_{P, calculated}^{IMCT} + 5 \log L_{S, calculated}^{IMCT}$ or $\log L_{P, calculated}^{IMCT} + \log L_{S, calculated}^{IMCT} - 5 \log N_{Fe_tO}$ based on results from Yang et al. [3,4] in the second layer for CaO–FeO–Fe_2O_3–Al_2O_3–P_2O_5 slags equilibrated with liquid iron over a temperature range of 1811 to 1927 K (1538 to 1654 °C), respectively.

With respect to the proposed coupling relationships [2] between L_P and L_S for the CaO-based slags, it can be observed in Figure 7 that increasing two kinds of slag basicity cannot cause a visible variation of the proposed term $\log L_P + 5 \log L_S$ or $\log L_P + \log L_S - 5 \log N_{Fe_tO}$ for the CaO-based slags over a narrow temperature range. Thus, the proposed coupling relationships [2] between L_P and L_S for the CaO-based slags are also independent of simplified complex basicity $(\% \text{ CaO})/[(\% \text{ P}_2\text{O}_5) + (\% \text{ Al}_2\text{O}_3)]$ or optical basicity Λ.

2.2.2. Influence of Slag Basicity on Coupling Relationship between Dephosphorization and Desulfurization Potentials for CaO-based Slags

The relationships of two kinds of slag basicity against calculated [3] phosphate capacity $\log C_{\text{PO}_4^{3-}, \text{ calculated}}^{\text{IMCT}}$ for the CaO-based slags are shown in the first layers of Figure 8, respectively. Likewise, the relationships of two kinds of slag basicity against calculated [5] sulfide capacity $\log C_{\text{S}^{2-}, \text{ calculated}}^{\text{IMCT}}$ for the CaO-based slags are also illustrated in the second layers of Figure 8, respectively. The relationships of two kinds of slag basicity against determined term $\log C_{\text{PO}_4^{3-}, \text{ determined}} + \log C_{\text{S}^{2-}, \text{ determined}}$ or $\log C_{\text{PO}_4^{3-}, \text{ determined}} + \log C_{\text{S}^{2-}, \text{ determined}} - \log N_{\text{FeO}}$ after original data from Ban-ya et al. [1], or calculated term $\log C_{\text{PO}_4^{3-}, \text{ calculated}}^{\text{IMCT}} + \log C_{\text{S}^{2-}, \text{ calculated}}^{\text{IMCT}}$ or $\log C_{\text{PO}_4^{3-}, \text{ calculated}}^{\text{IMCT}} + \log C_{\text{S}^{2-}, \text{ calculated}}^{\text{IMCT}} - \log N_{\text{FeO}}$ based on results by Yang et al. [3,5] for the CaO-based slags are displayed in the first and second layers of Figure 9, respectively.

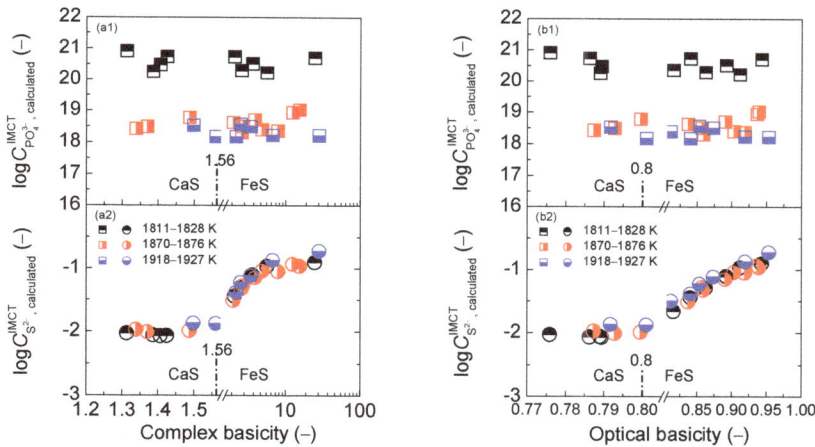

Figure 8. Relationship of simplified complex basicity $(\% \text{ CaO})/[(\% \text{ P}_2\text{O}_5) + (\% \text{ Al}_2\text{O}_3)]$ **(a)** or optical basicity Λ **(b)** against the calculated [3] phosphate capacity $\log C_{\text{PO}_4^{3-}, \text{ calculated}}^{\text{IMCT}}$ by the IMCT–$C_{\text{PO}_4^{3-}}$ model in first layer or calculated [5] sulfide capacity $\log C_{\text{S}^{2-}, \text{ calculated}}^{\text{IMCT}}$ by the IMCT–$C_{\text{S}^{2-}}$ model in second layer for CaO–FeO–Fe$_2$O$_3$–Al$_2$O$_3$–P$_2$O$_5$ slags equilibrated with liquid iron over a temperature range of 1811 to 1927 K (1538 to 1654 °C), respectively.

With respect to the dephosphorization potential of the CaO-based slags, it can be observed in the first layers of Figure 8 that increasing two kinds of slag basicity cannot result in an obvious influence on dephosphorization potential over a narrow temperature range, which has been explained by Yang et al. elsewhere [3]. With regard to the desulfurization potential of the CaO-based slags, it can be observed in the second layers of Figure 8 that the desulfurization potential of the CaO-based slags in the reducing zone keeps almost constant with the increase of two kinds of slag basicity as illustrated in the left regions of Figure 8(a2,b2); however, sulfide capacity $C_{\text{S}^{2-}}$ of the CaO-based slags in oxidizing zone displays an obviously increasing tendency with the increase of two kinds of slag basicity as illustrated in the right regions of Figure 8(a2,b2), which has been explained by Yang et al. elsewhere [5].

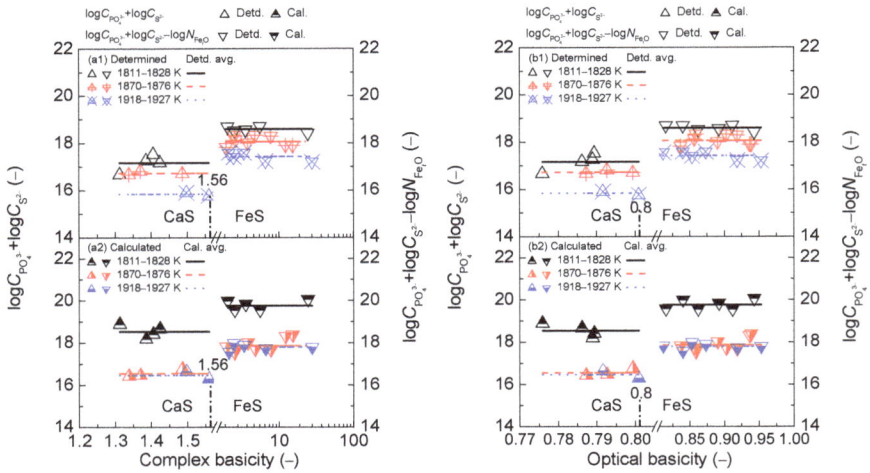

Figure 9. Relationship of simplified complex basicity $(\% \text{CaO})/[(\% \text{P}_2\text{O}_5) + (\% \text{Al}_2\text{O}_3)]$ (**a**) or optical basicity Λ (**b**) against determined term $\log C_{\text{PO}_4^{3-}, \text{determined}} + \log C_{\text{S}^{2-}, \text{determined}}$ or $\log C_{\text{PO}_4^{3-}, \text{determined}} + \log C_{\text{S}^{2-}, \text{determined}} - \log N_{\text{FeO}}$ after Ban-ya et al. [1] in the first layer or calculated term $\log C_{\text{PO}_4^{3-}, \text{calculated}}^{\text{IMCT}} + \log C_{\text{S}^{2-}, \text{calculated}}^{\text{IMCT}}$ or $\log C_{\text{PO}_4^{3-}, \text{calculated}}^{\text{IMCT}} + \log C_{\text{S}^{2-}, \text{calculated}}^{\text{IMCT}} - \log N_{\text{FeO}}$ based on results from Yang et al. [3,5] in the second layer for CaO–FeO–Fe$_2$O$_3$–Al$_2$O$_3$–P$_2$O$_5$ slags equilibrated with liquid iron over a temperature range of 1811 to 1927 K (1538 to 1654 °C), respectively.

Regarding the proposed coupling relationships [2] between $C_{\text{PO}_4^{3-}}$ and $C_{\text{S}^{2-}}$ for the CaO-based slags, it can be observed in Figure 9 that increasing two kinds of slag basicity cannot cause a visible variation of proposed coupling relationships between $C_{\text{PO}_4^{3-}}$ and $C_{\text{S}^{2-}}$ for the CaO-based slags over a narrow temperature range. The relationships of two kinds of slag basicity against calculated [2] term $\log C_{\text{PO}_4^{3-}, \text{calculated}}^{\text{IMCT}} + \log C_{\text{S}^{2-}, \text{calculated}}^{\text{IMCT}}$ or $\log C_{\text{PO}_4^{3-}, \text{calculated}}^{\text{IMCT}} + \log C_{\text{S}^{2-}, \text{calculated}}^{\text{IMCT}} - \log N_{\text{FeO}}$ based on the results by Yang et al. [3,4] are more accurate than those against determined ones after original data from Ban-ya et al. [1] for the CaO-based slags in the reducing or oxidizing zone. Thus, the proposed coupling relationships [2] between $C_{\text{PO}_4^{3-}}$ and $C_{\text{S}^{2-}}$ for the CaO-based slags are also independent of two kinds of slag basicity over a narrow temperature range.

2.3. Comprehensive Effect of Fe$_t$O and CaO on Coupling Relationships between Dephosphorization and Desulfurization Abilities or Potentials for CaO-based Slags

It was verified by Yang et al. [3–5] that the mass percentage ratios or the mass action concentration ratios of various iron oxides to basic oxide CaO can be applied to the elucidation of the comprehensive effect of iron oxides Fe$_t$O and basic oxide CaO on dephosphorization and desulfurization reactions of the CaO-based slags. It was also verified by Yang et al. [4] that the mass percentage ratio (% FeO)/(% CaO) or (% Fe$_2$O$_3$)/(% CaO) or (% Fe$_t$O)/(% CaO) can correlate a good linear relationship with the mass action concentration ratio $N_{\text{FeO}}/N_{\text{CaO}}$ or $N_{\text{Fe}_2\text{O}_3}/N_{\text{CaO}}$ or $N_{\text{Fe}_t\text{O}}/N_{\text{CaO}}$ for the CaO-based slags, respectively. Thus, the aforementioned mass action concentration ratios of various iron oxides to basic oxide CaO can be reliably substituted by the corresponding mass percentage ratios. As the newly formed structural unit FeO·Fe$_2$O$_3$ in the CaO-based slags in terms of the IMCT [2–18], the mass action concentration ratio $N_{\text{FeO·Fe}_2\text{O}_3}/N_{\text{CaO}}$ is also applied to the evaluation of the comprehensive influence of FeO·Fe$_2$O$_3$ and basic oxide CaO on the proposed coupling relationships [2].

2.3.1. Comprehensive Effect of Fe$_t$O and CaO on Coupling Relationships between Dephosphorization and Desulfurization Abilities for CaO-based Slags

The relationship between the mass action concentration ratios of various iron oxides to basic oxide CaO, i.e., N_{FeO}/N_{CaO} or $N_{Fe_2O_3}/N_{CaO}$ or $N_{FeO \cdot Fe_2O_3}/N_{CaO}$ or N_{Fe_tO}/N_{CaO} and calculated [3] phosphorus distribution ratio $\log L_{P,\,calculated}^{IMCT}$ for the CaO-based slags is shown in the first layers of Figure 10, respectively. Similarly, the relationship between aforementioned four mass action concentration ratios and calculated [4] sulfur distribution ratio $\log L_{S,\,calculated}^{IMCT}$ for the CaO-based slags is also illustrated in the second layers of Figure 10, respectively. The relationships between four mass action concentration ratios and determined term $\log L_{P,\,measured} + 5\log L_{S,\,measured}$ or $\log L_{P,\,measured} + \log L_{S,\,measured} - 5\log N_{Fe_tO}$ after original data from Ban-ya et al. [1], or calculated term $\log L_{P,\,calculated}^{IMCT} + 5\log L_{S,\,calculated}^{IMCT}$ or $\log L_{P,\,calculated}^{IMCT} + \log L_{S,\,calculated}^{IMCT} - 5\log N_{Fe_tO}$ based on results by Yang et al. [3,4] for the CaO-based slags are displayed in the first and second layers of Figure 11, respectively. It can be observed in Figures 10 and 11 that the criterion for distinguishing the reducing and oxidizing zones corresponds to N_{FeO}/N_{CaO} in 0.08, $N_{Fe_2O_3}/N_{CaO}$ in 1.59×10^{-4}, $N_{FeO \cdot Fe_2O_3}/N_{CaO}$ in 5.27×10^{-5}, and N_{Fe_tO}/N_{CaO} in 0.082, respectively.

With respect to the dephosphorization ability of the CaO-based slags, it can be observed in the first layers of Figure 10 that increasing four mass action concentration ratios can result in an exponentially growing tendency of the phosphorus distribution ratio $\log L_P$, which has been discussed by Yang et al. elsewhere [3]. With regard to the desulfurization ability of the CaO-based slags, it can be obtained from the second layers of Figure 10 that increasing four mass action concentration ratios can lead to a backward tick-shaped variation trend of sulfur distribution ratio $\log L_S$, which has been discussed by Yang et al. elsewhere [4].

On the proposed coupling relationships [2] between L_P and L_S for the CaO-based slags, it can be observed in Figure 11 that the proposed term $\log L_P + 5\log L_S$ or $\log L_P + \log L_S - 5\log N_{Fe_tO}$ after original data from Ban-ya et al. [1] or based on results by Yang et al. [3,4] keeps almost constant with the increase of four mass action concentration ratios for the CaO-based slags in reducing and oxidizing zones over a narrow temperature range. Thus, the proposed coupling relationships [2] between L_P and L_S for the CaO-based slags are independent of the comprehensive influence of iron oxides Fe$_t$O and basic oxides expressed by the mass action concentration ratio N_{FeO}/N_{CaO} or $N_{Fe_2O_3}/N_{CaO}$ or $N_{FeO \cdot Fe_2O_3}/N_{CaO}$ or N_{Fe_tO}/N_{CaO}, or the mass percentage ratio (% FeO)/(% CaO) or (% Fe$_2$O$_3$)/(% CaO) or (% Fe$_t$O)/(% CaO).

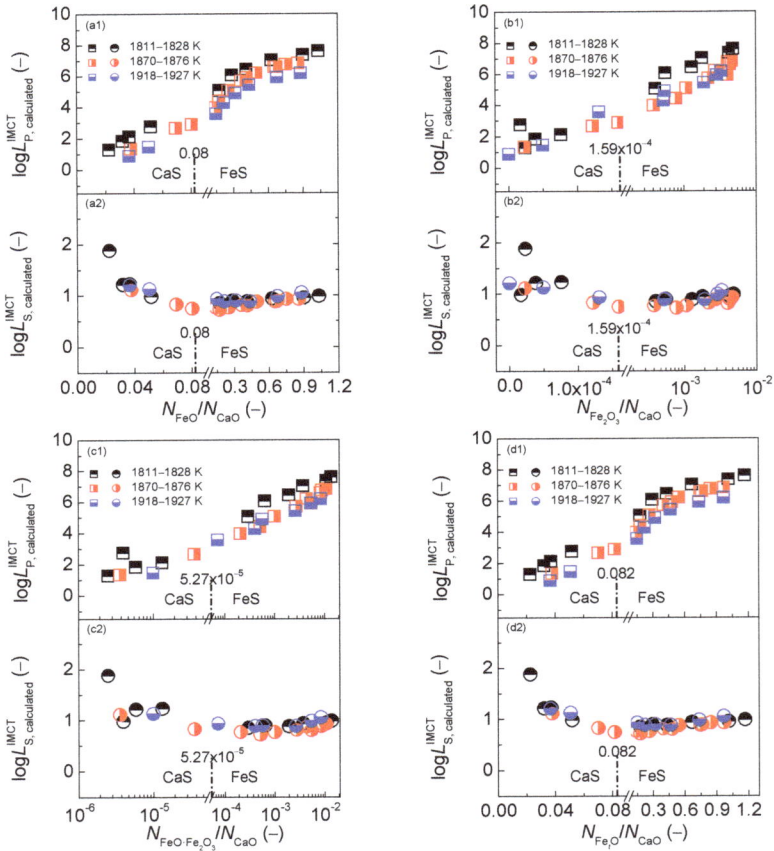

Figure 10. Relationship of the mass action concentration ratio N_{FeO}/N_{CaO} (**a**) or $N_{Fe_2O_3}/N_{CaO}$ (**b**) or $N_{FeO \cdot Fe_2O_3}/N_{CaO}$ (**c**) or N_{Fe_tO}/N_{CaO} (**d**) against the calculated [3] phosphorus distribution ratio $\log L_{P, calculated}^{IMCT}$ by the IMCT–L_P model in first layer or the calculated [4] sulfur distribution ratio $\log L_{S, calculated}^{IMCT}$ by IMCT–L_S model in the second layer for CaO–FeO–Fe$_2$O$_3$–Al$_2$O$_3$–P$_2$O$_5$ slags equilibrated with liquid iron over a temperature range of 1811 to 1927 K (1538 to 1654 °C), respectively.

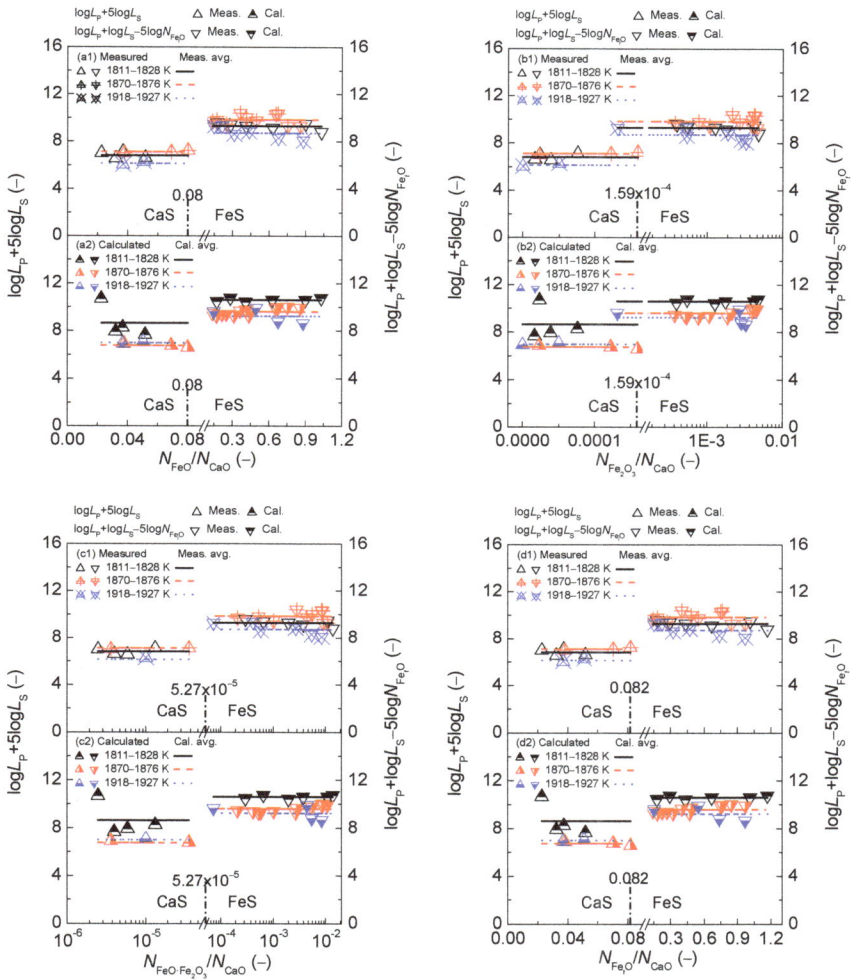

Figure 11. Relationship of the mass action concentration ratio N_{FeO}/N_{CaO} (**a**) or $N_{Fe_2O_3}/N_{CaO}$ (**b**) or $N_{FeO·Fe_2O_3}/N_{CaO}$ (**c**) or N_{Fe_tO}/N_{CaO} (**d**) against the determined term $\log L_{P,\ measured} + 5\log L_{S,\ measured}$ or $\log L_{P,\ measured} + \log L_{S,\ measured} - 5\log N_{Fe_tO}$ after Ban-ya et al. [1] in the first layer or calculated term $\log L_{P,\ calculated}^{IMCT} + 5\log L_{S,\ calculated}^{IMCT}$ or $\log L_{P,\ calculated}^{IMCT} + \log L_{S,\ calculated}^{IMCT} - 5\log N_{Fe_tO}$ based on the results from Yang et al. [3,4] in the second layer for CaO–FeO–Fe$_2$O$_3$–Al$_2$O$_3$–P$_2$O$_5$ slags equilibrated with liquid iron over a temperature range of 1811 to 1927 K (1538 to 1654 °C), respectively.

2.3.2. Comprehensive Effect of Fe$_t$O and CaO on Coupling Relationships between Dephosphorization and Desulfurization Potentials for CaO-based Slags

The relationship between the aforementioned four mass action concentration ratios of various iron oxides to basic oxide CaO and calculated [3] phosphate capacity $\log C_{PO_4^{3-},\ calculated}^{IMCT}$ for the CaO-based slags is shown in the first layers of Figure 12, respectively. Likewise, the relationship between four mass action concentration ratios and the calculated [4] sulfide capacity $\log C_{S^{2-},\ calculated}^{IMCT}$ for the CaO-based slags is also illustrated in the second layers of Figure 12, respectively. The relationship between four mass action concentration ratios and the determined term $\log C_{PO_4^{3-},\ determined} + \log C_{S^{2-},\ determined}$

or $\log C_{PO_4^{3-},\,\text{determined}} + \log C_{S^{2-},\,\text{determined}} - \log N_{FeO}$ after the original data from Ban-ya et al. [1], or calculated term $\log C_{PO_4^{3-},\,\text{calculated}}^{IMCT} + \log C_{S^{2-},\,\text{calculated}}^{IMCT}$ or $\log C_{PO_4^{3-},\,\text{calculated}}^{IMCT} + \log C_{S^{2-},\,\text{calculated}}^{IMCT} - \log N_{FeO}$ based on results by Yang et al. [3,5] for the CaO-based slags is also illustrated in the first and second layers of Figure 13, respectively.

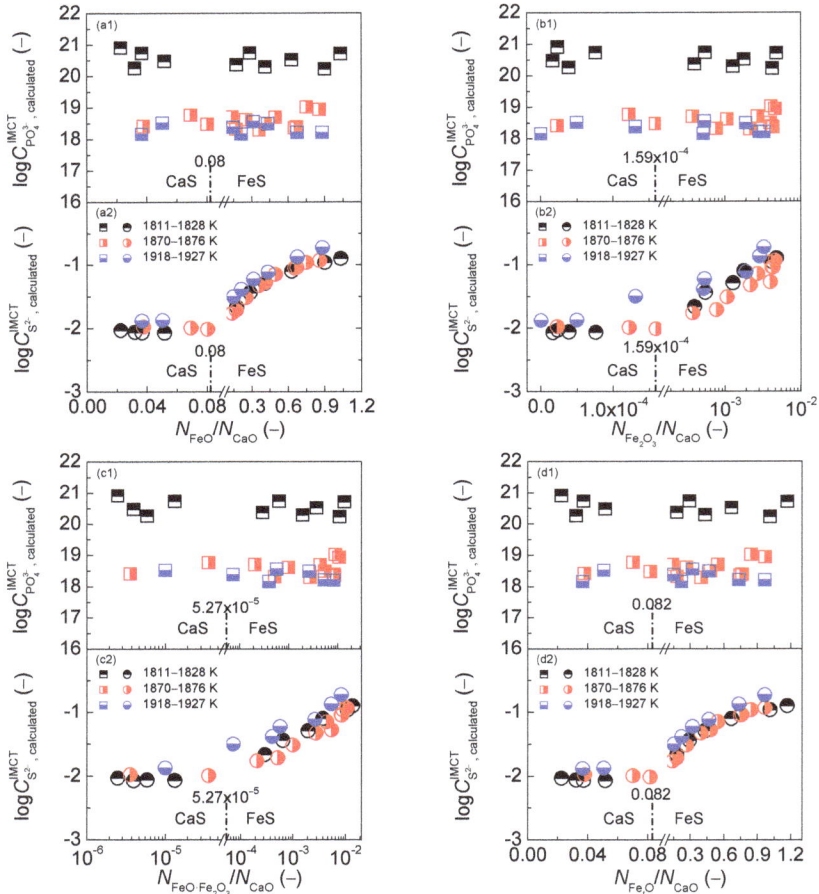

Figure 12. Relationship of mass action concentration ratio N_{FeO}/N_{CaO} (**a**) or $N_{Fe_2O_3}/N_{CaO}$ (**b**) or $N_{FeO\cdot Fe_2O_3}/N_{CaO}$ (**c**) or N_{Fe_tO}/N_{CaO} (**d**) against the calculated [3] phosphate capacity $\log C_{PO_4^{3-},\,\text{calculated}}^{IMCT}$ by the IMCT–$C_{PO_4^{3-}}$ model in the first layer or calculated [5] sulfide capacity $\log C_{S^{2-},\,\text{calculated}}^{IMCT}$ by the IMCT–$C_{S^{2-}}$ model in the second layer for CaO–FeO–Fe$_2$O$_3$–Al$_2$O$_3$–P$_2$O$_5$ slags equilibrated with liquid iron over a temperature range of 1811 to 1927 K (1538 to 1654 °C), respectively.

With respect to the dephosphorization potential of the CaO-based slags, it can be observed in the first layers of Figure 12 that increasing the four mass action concentration ratios cannot result in an obvious variation of dephosphorization potential over a narrow temperature range. This result was explained by Yang et al. [3] as that greater values of calculated dephosphorization ability are the reason for larger ones of dephosphorization potential by the IMCT$-C_{PO_4^{3-}}$ model over the lower temperature range of 1811 to 1828 K (1538 to 1555 °C). With regard to the desulfurization potential of the CaO-based slags, it can be obtained from the second layers of Figure 12 that increasing four mass action concentration ratios can lead to the similar variation trend of sulfide capacity $C_{S^{2-}}$ against slag oxidization ability expressed by N_{Fe_tO} of iron oxides in Figure 4(f2), which has been explained by Yang et al. elsewhere [5].

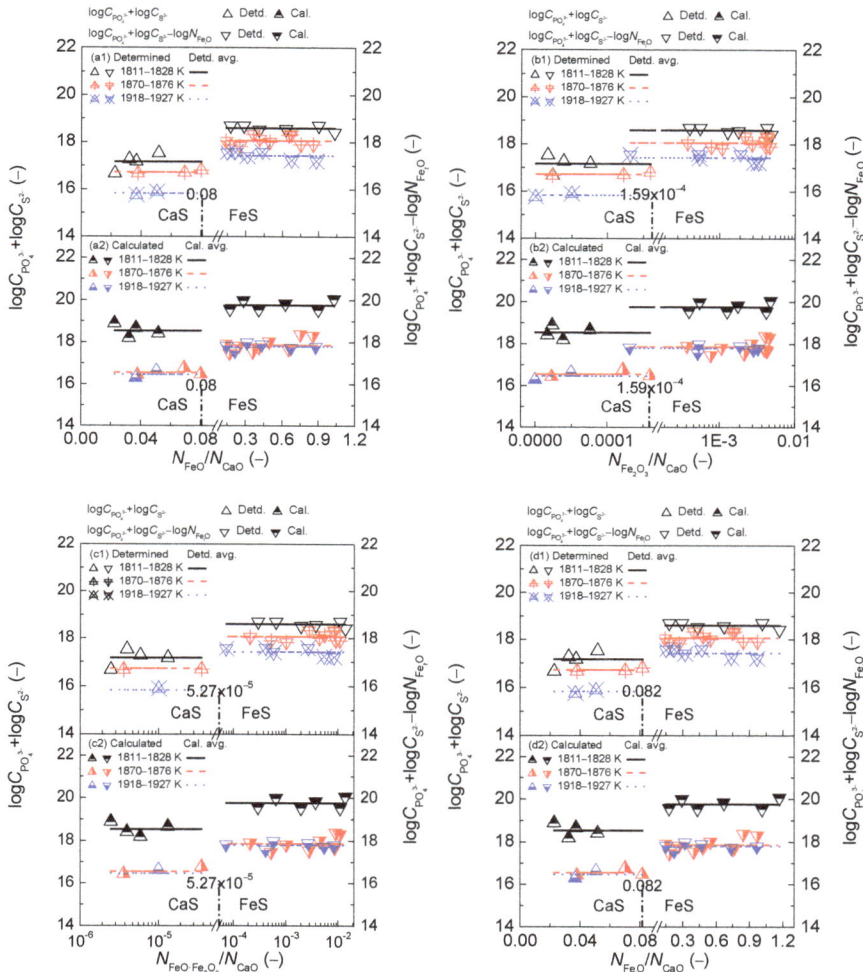

Figure 13. Relationship of the mass action concentration ratio N_{FeO}/N_{CaO} (**a**) or $N_{Fe_2O_3}/N_{CaO}$ (**b**) or $N_{FeO·Fe_2O_3}/N_{CaO}$ (**c**) or N_{Fe_tO}/N_{CaO} (**d**) against the determined term $\log C_{PO_4^{3-}, determined} + \log C_{S^{2-}, determined}$ or $\log C_{PO_4^{3-}, determined} + \log C_{S^{2-}, determined} - \log N_{FeO}$ after Ban-ya et al. [1] in the first layer or calculated term $\log C_{PO_4^{3-}, calculated}^{IMCT} + \log C_{S^{2-}, calculated}^{IMCT}$ or $\log C_{PO_4^{3-}, calculated}^{IMCT} + \log C_{S^{2-}, calculated}^{IMCT} - \log N_{FeO}$ based on results from Yang et al. [3,5] in the second layer for CaO–FeO–Fe$_2$O$_3$–Al$_2$O$_3$–P$_2$O$_5$ slags equilibrated with liquid iron over a temperature range of 1811 to 1927 K (1538 to 1654 °C), respectively.

On the proposed coupling relationships [2] between $C_{PO_4^{3-}}$ and $C_{S^{2-}}$ for the CaO-based slags, it can be observed in Figure 13 that the proposed term $\log C_{PO_4^{3-}} + \log C_{S^{2-}}$ or $\log C_{PO_4^{3-}} + \log C_{S^{2-}} - \log N_{FeO}$ after the original data from Ban-ya et al. [1] or based on results by Yang et al. [3,4] keeps almost constant with the increase of aforementioned four mass action concentration ratios for the CaO-based slags in the reducing and oxidizing zones over a narrow temperature range. The relationships of four mass action concentration ratios against calculated term $\log C^{IMCT}_{PO_4^{3-},\,calculated} + \log C^{IMCT}_{S^{2-},\,calculated}$ or $\log C^{IMCT}_{PO_4^{3-},\,calculated} + \log C^{IMCT}_{S^{2-},\,calculated} - \log N_{FeO}$ based on results by Yang et al. [3,4] are more accurate than those against determined terms after original data from Ban-ya et al. [1] for the CaO-based slags in the reducing and oxidizing zones. Thus, the proposed coupling relationships [2] between $C_{PO_4^{3-}}$ and $C_{S^{2-}}$ for the CaO-based slags are certainly independent of the comprehensive influence of iron oxides Fe_tO and basic oxides expressed by the mass action concentration ratio N_{FeO}/N_{CaO} or $N_{Fe_2O_3}/N_{CaO}$ or $N_{FeO\cdot Fe_2O_3}/N_{CaO}$ or N_{Fe_tO}/N_{CaO}, or the mass percentage ratio $(\% \, FeO)/(\% \, CaO)$ or $(\% \, Fe_2O_3)/(\% \, CaO)$ or $(\% \, Fe_tO)/(\% \, CaO)$.

3. Discussion on Proposed Coupling Relationships between Dephosphorization and Desulfurization Abilities or Potentials for CaO-based Slags

3.1. Magnitude of Proposed Coupling Relationships between Dephosphorization and Desulfurization Abilities for CaO-based Slags

Values of the phosphorus distribution ratio $\log L_P$ for the CaO-based slags in the reducing zone as shown in the first layers of Figures 1, 6 and 10 vary from 1.0 to 3.0, while data of the sulfur distribution ratio $\log L_S$ for the CaO-based slags in the reducing zone, as illustrated in the second layers of Figures 1, 6 and 10, change from 1.7 to 0.9. However, results of the proposed term $\log L_P + 5 \log L_S$ for the CaO-based slags in the reducing zone, as displayed in Figures 2, 7 and 11, fluctuate from 6.7 to 8.5. Thus, the magnitude of proposed term $\log L_P + 5 \log L_S$ for the CaO-based slags in the reducing zone is mainly decided by that of $\log L_S$ because the desulfurization ability $\log L_S$ indicates a five-time contribution to the proposed term $\log L_P + 5 \log L_S$ compared with the one-time dephosphorization ability of $\log L_P$. It is a well-known viewpoint that reducing slags exhibits good desulfurization ability with limited dephosphorization ability. This means that the magnitude of the proposed term $\log L_P + 5 \log L_S$ for the CaO-based slags in the reducing zone is decided by desulfurization ability. Theoretically, higher temperature can promote the desulfurization reaction of the CaO-based slags. However, increasing the temperature from 1811 to 1927 K (1538 to 1654 °C) cannot cause an obvious increase of desulfurization ability for the CaO-based slags as shown in the second layers of Figures 1, 6 and 10. Furthermore, higher temperature can inhibit dephosphorization reactions of the CaO-based slags. Thus, increasing the temperature from 1811 to 1927 K (1538 to 1654 °C) can result in an effectively decreasing influence on dephosphorization ability of the CaO-based slags as illustrated in the first layers of Figures 1, 6 and 10. Evidently, increasing temperature from 1811 to 1927 K (1538 to 1654 °C) can lead to a slightly decreasing tendency of the proposed term $\log L_P + 5 \log L_S$ for the CaO-based slags in the reducing zone in Figures 2, 7 and 11.

Values of the phosphorus distribution ratio $\log L_P$ for the CaO-based slags in the oxidizing zone, as shown in the first layers of Figures 1, 6 and 10 vary from 3.0 to 8.0, while the data of the sulfur distribution ratio $\log L_S$ for the CaO-based slags in the oxidizing zone as illustrated in the second layers of Figures 1, 6 and 10 change from 0.7 to 1.0. However, the results of the proposed term $\log L_P + \log L_S - 5 \log N_{Fe_tO}$ for the CaO-based slags in the oxidizing zone, as displayed in Figures 2, 7 and 11, fluctuate from 9.2 to 10.6. Thus, the magnitude of proposed term $\log L_P + \log L_S - 5 \log N_{Fe_tO}$ for the CaO-based slags in the oxidizing zone is mainly decided by that of $\log L_P$. It is a widely-accepted viewpoint that oxidizing slags exhibit good dephosphorization ability with limited desulfurization ability. This indicates that the proposed term $\log L_P + \log L_S - 5 \log N_{Fe_tO}$ for the CaO-based slags in the oxidizing zone is controlled by the dephosphorization ability. This means that the temperature effect on the dephosphorization ability of the CaO-based slags can decide the influence of the increasing temperature from 1811 to 1927 K (1538 to 1654 °C) on the proposed term $\log L_P + \log L_S - 5 \log N_{Fe_tO}$ for the CaO-based slags in the oxidizing zone in Figures 2, 7 and 11.

3.2. Magnitude of Proposed Coupling Relationships between Dephosphorization and Desulfurization Potentials for CaO-based Slags

Values of phosphate capacity $\log C_{PO_4^{3-}}$ for the CaO-based slags in the reducing zone, as shown in the first layers of Figures 4, 8 and 10, vary from 18.0 to 20.0, while the data of sulfide capacity $\log C_{S^{2-}}$ for the CaO-based slags in the reducing zone, as illustrated in the second layers of Figures 4, 8 and 10, keep almost constant at -2.0. However, the results of the proposed term $\log C_{PO_4^{3-}} + \log C_{S^{2-}}$ for the CaO-based slags in the reducing zone, as shown Figures 5, 9 and 13 fluctuate from 16.5 to 18.5. Thus, the magnitude of the proposed term $\log C_{PO_4^{3-}} + \log C_{S^{2-}}$ for the CaO-based slags in the reducing zone is mainly decided by that of $\log C_{PO_4^{3-}}$ because the dephosphorization potential of the CaO-based slags in the reducing zone with smaller oxygen partial potential p_{O_2} can produce a large value of $C_{PO_4^{3-}}$ according to the defined phosphate capacity $C_{PO_4^{3-}}$ by Wagner [29]. Thus, using inaccurate values of phosphate capacity $C_{PO_4^{3-}}$ in the proposed term $\log C_{PO_4^{3-}} + \log C_{S^{2-}}$ for the CaO-based slags in the reducing zone can cause some degree of risk for designing or optimizing slag chemical composition, as described in Section 3.3. As pointed out in Section 2.3.2 that the calculated [3] $\log C_{PO_4^{3-}, \text{ calculated}}^{\text{IMCT}}$ can relieve the experimental uncertainties. Therefore, the calculated [3] $\log C_{PO_4^{3-}, \text{ calculated}}^{\text{IMCT}}$ for the CaO-based slags, rather than the determined [3] $\log C_{PO_4^{3-}, \text{ determined}}$ after the original data from Ban-ya et al. [1], is applied in this study.

Values of phosphate capacity $\log C_{PO_4^{3-}}$ for the CaO-based slags in the oxidizing zone, as shown in the first layers of Figures 4, 8 and 10, also vary from 18.0 to 20, while the data of sulfide capacity $\log C_{S^{2-}}$ for the CaO-based slags in the oxidizing zone, as illustrated in the second layers of Figures 4, 8 and 10 change from -2.0 to -0.75. However, the results of proposed term $\log C_{PO_4^{3-}} + \log C_{S^{2-}} - \log N_{\text{FeO}}$ for the CaO-based slags in the oxidizing zone, as displayed in Figures 5, 9 and 13 fluctuate from 17.5 to 20.0. Thus, the magnitude of the proposed term $\log C_{PO_4^{3-}} + \log C_{S^{2-}} - \log N_{\text{FeO}}$ for the CaO-based slags in the oxidizing zone is mainly decided by that of $\log C_{PO_4^{3-}}$. Oxidizing slags have good dephosphorization ability with limited desulfurization ability. This means that the proposed term $\log C_{PO_4^{3-}} + \log C_{S^{2-}}$ or $\log C_{PO_4^{3-}} + \log C_{S^{2-}} - \log N_{\text{FeO}}$ for the CaO-based slags in reducing and oxidizing zones includes the key factor of dephosphorization potential. In addition, higher temperature can restrain the dephosphorization ability and potential of the CaO–based slags. Increasing the temperature from 1811 to 1927 K (1538 to 1654 °C) can effectively decrease the dephosphorization potential of the CaO-based slags as illustrated in the first layers of Figures 4, 8 and 12. Thus, increasing the temperature from 1811 to 1927 K (1538 to 1654 °C) can result in a slightly decreasing tendency of the proposed term $\log C_{PO_4^{3-}} + \log C_{S^{2-}}$ or $\log C_{PO_4^{3-}} + \log C_{S^{2-}} - \log N_{\text{FeO}}$ for the CaO-based slags in the reducing and oxidizing zones in Figures 5, 9 and 13.

Considering the large difference of magnitude between $C_{PO_4^{3-}}$ and $C_{S^{2-}}$, the proposed coupling relationships [2] as $\log L_P + 5 \log L_S$ and $\log L_P + \log L_S - 5 \log N_{\text{Fe}_t\text{O}}$, rather than $\log C_{PO_4^{3-}} + \log C_{S^{2-}}$ and $\log C_{PO_4^{3-}} + \log C_{S^{2-}} - \log N_{\text{FeO}}$, are recommended to design or optimize the chemical composition of slags under the fixed experimental uncertainties as described in Section 3.3.

3.3. Prospect and application for Proposed Coupling Relationship between Dephosphorization and Desulfurization Abilities or Potentials for CaO–based Slags

The proposed coupling relationships [2] for CaO–FeO–Fe$_2$O$_3$–Al$_2$O$_3$–P$_2$O$_5$ slags are not only independent of slag oxidization ability as expected but are also irrelevant to slag chemical composition represented by the reaction abilities of components, two kinds of slag basicity as simplified complex basicity (% CaO)/[(% P$_2$O$_5$) + (% Al$_2$O$_3$)] and optical basicity Λ, and the comprehensive effect of iron oxides Fe$_t$O and basic oxide CaO. Thus, the proposed coupling relationships [2] for the CaO-based slags remain almost constant over a narrow temperature range although changing slag chemical composition can significantly affect its dephosphorization and desulfurization abilities or potentials. This means that the maximum values of the sum of dephosphorization and desulfurization abilities or potentials for the assigned slags in reducing and oxidizing zones can be determined by the proposed coupling relationships [2]. Additionally, the counteraction characteristics between the dephosphorization and desulfurization abilities or potentials for reducing slags can be theoretically explained and quantitatively expressed as $\log L_P + 5 \log L_S$ or $\log C_{PO_4^{3-}} + \log C_{S^{2-}}$. The promotive effect of slag oxidization ability described by the comprehensive mass action concentration N_{Fe_tO} of iron oxides on the maximum values of the sum of dephosphorization and desulfurization abilities or potentials for oxidizing slags can be reasonably explained and quantitatively described as $\log L_P + \log L_S - 5 \log N_{Fe_tO}$ or $\log C_{PO_4^{3-}} + \log C_{S^{2-}} - \log N_{FeO}$.

It has been verified by Yang et al. [3–5] that the IMCT–L_P [3] or IMCT–$C_{PO_4^{3-}}$ [3] or IMCT–L_S [4] or IMCT–$C_{S^{2-}}$ [5] models can be accurately applied to the prediction of dephosphorization and desulfurization abilities or potentials of the assigned slags. Thus, the dephosphorization abilities or potentials of the assigned slags or fluxes can be theoretically predicted from its desulfurization abilities or potentials based on the proposed coupling relationships [2], and vice versa. This means that a new method of designing or optimizing chemical composition of slags or fluxes with the assigned dephosphorization abilities or potentials can be developed based on the proposed coupling relationships [2].

The proposed coupling relationships [2] between dephosphorization and desulfurization abilities as $\log L_P + 5 \log L_S$ in the reducing zone and as $\log L_P + \log L_S - 5 \log N_{Fe_tO}$ in the oxidizing zone have been verified to be valid based on the reported equilibrium experiments in laboratory scale by Ban-ya et al. [1] Actually, reactions of dephosphorization and desulfurization at the final stage of many refining processes such as the dephosphorization pretreatment process of hot metal [15–17], the simultaneous dephosphorization and desulfurization operation of iron-based melts during secondary refining process [1–5], desulfurization reaction during the ladle furnace (LF) refining process [8,9], dephosphorization reaction at the blowing end-point during top–bottom combined blown converter steelmaking process [10,11], and so on can be considered to reach quasi-equilibrium at the interface between the slags and metal. It can be deduced that the proposed coupling relationships [2] are also suitable to industrial operations during the dephosphorization and desulfurization processes.

4. Conclusions

The proposed coupling relationships between the dephosphorization and desulfurization abilities or potentials for CaO–FeO–Fe$_2$O$_3$–Al$_2$O$_3$–P$_2$O$_5$ slags over a large variation range of slag oxidization ability during the secondary refining process of molten steel as $\log L_P + 5 \log L_S$ or $\log C_{PO_4^{3-}} + \log C_{S^{2-}}$ in the reducing zone and as $\log L_P + \log L_S - 5 \log N_{Fe_tO}$ or $\log C_{PO_4^{3-}} + \log C_{S^{2-}} - \log N_{FeO}$ in the oxidizing zone have been further verified as valid and feasible through investigating the influence of slag chemical composition. The main summary remarks can be obtained as follows:

(1) The proposed coupling relationships for the CaO-based slags in both the reducing and oxidizing zones are not only independent of slag oxidization ability described by the comprehensive mass action concentration N_{Fe_tO} of iron oxides but are also irrelevant to the reaction abilities of components expressed by the mass action concentrations N_i over a narrow temperature range in comparison with significant influences of slag oxidization ability as well as reaction abilities of components on dephosphorization and desulfurization abilities or potentials.

(2) The proposed coupling relationships for the CaO-based slags in both the reducing and oxidizing zones keep almost constant with the variation of two kinds of slag basicity as the simplified complex basicity (% CaO)/[(% P$_2$O$_5$) + (% Al$_2$O$_3$)] and optical basicity Λ over a narrow temperature range compared with the strong effects of two kinds of slags basicity on dephosphorization and desulfurization abilities or potentials.

(3) The proposed coupling relationships for the CaO-based slags in both reducing and oxidizing zones are independent of the comprehensive effect of iron oxides Fe_tO and basic oxide CaO described by the mass action concentration ratio N_{FeO}/N_{CaO} or $N_{Fe_2O_3}/N_{CaO}$ or $N_{FeO·Fe_2O_3}/N_{CaO}$ or N_{Fe_tO}/N_{CaO}, or the mass percentage ratio (% FeO)/(% CaO) or (% Fe_2O_3)/(% CaO) or (% Fe_tO)/(% CaO) in comparison with the large influences of the aforementioned comprehensive effect of iron oxides Fe_tO and basic oxide CaO on dephosphorization and desulfurization abilities or potentials.

(4) Increasing the temperature from 1811 to 1927 K (1538 to 1654 °C) can result in a slightly decreasing tendency of the proposed coupling relationships for the CaO-based slags in reducing and oxidizing zones.

(5) Chemical composition of slags or fluxes with the assigned dephosphorization ability or potential can be theoretically designed or optimized by its desulfurization ability or potential, and vice versa, in terms of the obtained maximum values of dephosphorization and desulfurization abilities or potentials for the CaO-based slags in both reducing and oxidizing zones.

(6) The proposed coupling relationships between L_P and L_S for the CaO-based slags as $\log L_P + 5\log L_S$ and $\log L_P + \log L_S - 5\log N_{Fe_tO}$ in reducing and oxidizing zones are recommended to design or optimize the chemical composition of slags or fluxes due to a large difference of magnitude between phosphate capacity $C_{PO_4^{3-}}$ and sulfide capacity $C_{S^{2-}}$.

Author Contributions: X.M.Y. conceived and designed the study. J.Y.L. and M.Z. performed the simulations. X.M.Y. and J.Y.L. wrote the main draft of the manuscript. X.M.Y., J.Y.L. and F.J.Y. revised the manuscript. All authors contributed to the discussion of the results, and commented on the manuscript.

Funding: This research was funded by the Beijing Natural Science Foundation [Grant No. 2182069] and the National Natural Science Foundation of China [Grant No. 51174186].

Conflicts of Interest: The authors declare no conflict of interest.

Nomenclatures

$a_{R, i}$	Activity of components i in slags or element i in liquid iron relative to pure solid or liquid component i or element i as standard state with mole fraction x_i as concentration unit and following Raoult's law under the condition of taking ideal solution as reference state, i.e., $a_{R, i} = x_i\gamma_i$, (–);
$C_{PO_4^{3-}}$	Phosphate capacity of slags based on gas–slag equilibrium, (–);
$C_{S^{2-}}$	Sulfide capacity of slags based on gas–slag equilibrium, (–);
$f_{\%, i}$	Activity coefficient of element i in liquid iron related with activity $a_{\%, i}$, (–);
$K_i^{\ominus}$	Standard equilibrium constant of chemical reaction for forming component i or structural unit i, (–);
L_P	Phosphorus distribution ratio between slags and liquid iron, defined as $L_P = (\% P_2O_5)/[\% P]^2$, (–);
L_S	Sulphur distribution ratio between slags and liquid iron, defined as $L_S = (\% S)/[\% S]$, (–);
M_i	Relative atomic mass of element i or relative molecular mass of component i, (–);
Σn_i^0	Total mole number of all components in 100 g slags, (mol).

Greek symbols

γ_i	Activity coefficient of component i in slags related with activity $a_{R, i}$, (–);
Λ	Optical basicity of slags, (–).

References

1. Ban–ya, S.; Hino, M.; Sato, A.; Terayama, A.O. O, P and S Distribution Equilibria between Liquid Iron and CaO–Al$_2$O$_3$–FetO Slag Saturated with CaO. *Tetsu–to–Hagané* **1991**, *77*, 361–368.

2. Yang, X.M.; Li, J.Y.; Zhang, M.; Chai, G.M.; Duan, D.P.; Zhang, J. Coupling Relationship between Dephosphorization and Desulfurization Abilities or Potentials for CaO–FeO–Fe$_2$O$_3$–Al$_2$O$_3$–P$_2$O$_5$ Slags over a Large Variation Range of Slag Oxidization Ability Based on the Ion and Molecule Coexistence Theory. *Ironmak. Steelmak.* **2018**, *45*, 25–43. [CrossRef]

3. Yang, X.M.; Chai, G.M.; Zhang, M.; Li, J.Y.; Liang, Q.; Zhang, J. Thermodynamic Models for Predicting Dephosphorization Ability and Potential of CaO–FeO–Fe$_2$O$_3$–Al$_2$O$_3$–P$_2$O$_5$ Slags during Secondary Refining Process of Molten Steel Based on the Ion and Molecule Coexistence Theory. *Ironmak. Steelmak.* **2016**, *43*, 663–687. [CrossRef]

4. Yang, X.M.; Li, J.Y.; Zhang, M.; Zhang, J. Prediction Model of Sulfur Distribution Ratio between CaO–FeO–Fe$_2$O$_3$–Al$_2$O$_3$–P$_2$O$_5$ Slags and Liquid Iron in a Large Variation Range of Oxygen Potential during Secondary Refining Process of Molten Steel Based on the Ion and Molecule Coexistence Theory. *Ironmak. Steelmak.* **2016**, *43*, 39–55. [CrossRef]

5. Yang, X.M.; Li, J.Y.; Chai, G.M.; Zhang, M.; Zhang, J. Prediction Model of Sulfide Capacity for CaO–FeO–Fe$_2$O$_3$–Al$_2$O$_3$–P$_2$O$_5$ Slags in a Large Variation Range of Oxygen Potential Based on the Ion and Molecule Coexistence Theory. *Metall. Mater. Trans. B* **2014**, *45*, 2118–2137. [CrossRef]

6. Yang, X.M.; Jiao, J.S.; Ding, R.C.; Shi, C.B.; Guo, H.J. A Thermodynamic Model for Calculating Sulphur Distribution Ratio between CaO–SiO$_2$–MgO–Al$_2$O$_3$ Ironmaking Slags and Carbon Saturated Hot Metal Based on the Ion and Molecule Coexistence Theory. *ISIJ Int.* **2009**, *49*, 1828–1837. [CrossRef]

7. Shi, C.B.; Yang, X.M.; Jiao, J.S.; Li, C.; Guo, H.J. A Sulphide Capacity Prediction Model of CaO–SiO$_2$–MgO–Al$_2$O$_3$ Ironmaking Slags Based on the Ion and Molecule Coexistence Theory. *ISIJ Int.* **2010**, *50*, 1362–1372. [CrossRef]

8. Yang, X.M.; Shi, C.B.; Zhang, M.; Chai, G.M.; Wang, F. A Thermodynamic Model of Sulfur Distribution Ratio between CaO–SiO$_2$–MgO–FeO–MnO–Al$_2$O$_3$ Slags and Molten Steel during LF Refining Process Based on the Ion and Molecule Coexistence Theory. *Metall. Mater. Trans. B* **2011**, *42*, 1150–1180. [CrossRef]

9. Yang, X.M.; Shi, C.B.; Zhang, M.; Chai, G.M.; Zhang, J. A Sulfide Capacity Prediction Model of CaO–SiO$_2$–MgO–FeO–MnO–Al$_2$O$_3$ Slags during LF Refining Process Based on the Ion and Molecule Coexistence Theory. *Metall. Mater. Trans. B* **2012**, *43*, 241–266. [CrossRef]

10. Yang, X.M.; Duan, J.P.; Shi, C.B.; Zhang, M.; Zhang, Y.L.; Wang, J.C. A Thermodynamic Model of Phosphorus Distribution Ratio between CaO–SiO$_2$–MgO–FeO–Fe$_2$O$_3$–MnO–Al$_2$O$_3$–P$_2$O$_5$ Slags and Molten Steel during Top–Bottom Combined Blown Converter Steelmaking Process Based on the Ion and Molecule Coexistence Theory. *Metall. Mater. Trans. B* **2011**, *42*, 738–770. [CrossRef]

11. Yang, X.M.; Shi, C.B.; Zhang, M.; Duan, J.P.; Zhang, J. A Thermodynamic Model of Phosphate Capacity for CaO–SiO$_2$–MgO–FeO–Fe$_2$O$_3$–MnO–Al$_2$O$_3$–P$_2$O$_5$ Slags Equilibrated with Molten Steel during a Top–Bottom Combined Blown Converter Steelmaking Process Based on the Ion and Molecule Coexistence Theory. *Metall. Mater. Trans. B* **2011**, *42*, 951–976. [CrossRef]

12. Yang, X.M.; Shi, C.B.; Zhang, M.; Zhang, J. A Thermodynamic Model for Prediction of Iron Oxide Activity in Some FeO–Containing Slag Systems. *Steel Res. Int.* **2012**, *83*, 244–258. [CrossRef]

13. Yang, X.M.; Zhang, M.; Zhang, J.L.; Li, P.C.; Li, J.Y.; Zhang, J. Representation of Oxidation Ability for Metallurgical Slags Based on the Ion and Molecule Coexistence Theory. *Steel Res. Int.* **2014**, *85*, 347–375. [CrossRef]

14. Li, J.Y.; Zhang, M.; Guo, M.; Yang, X.M. Enrichment Mechanism of Phosphate in CaO–SiO$_2$–FeO–Fe$_2$O$_3$–P$_2$O$_5$ Steelmaking Slags. *Metall. Mater. Trans. B* **2014**, *45*, 1666–1682. [CrossRef]

15. Yang, X.M.; Li, J.Y.; Chai, G.M.; Duan, D.P.; Zhang, J. A Thermodynamic Model for Predicting Phosphorus Partition between CaO–based Slags and Hot Metal during Hot Metal Dephosphorization Pretreatment Process Based on the Ion and Molecule Coexistence Theory. *Metall. Mater. Trans. B* **2016**, *47*, 2279–2301. [CrossRef]

16. Yang, X.M.; Li, J.Y.; Chai, G.M.; Duan, D.P.; Zhang, J. Critical Evaluation of Prediction Models for Phosphorus Partition between CaO–based Slags and Iron–based Melts during Dephosphorization Processes. *Metall. Mater. Trans. B* **2016**, *47*, 2302–2329. [CrossRef]

17. Yang, X.M.; Li, J.Y.; Chai, G.M.; Duan, D.P.; Zhang, J. A Thermodynamic Model for Predicting Phosphate Capacity of CaO−based Slags during Hot Metal Dephosphorization Pretreatment Process. *Ironmak. Steelmak.* **2017**, *44*, 437–454. [CrossRef]

18. Zhang, J. *Computational Thermodynamics of Metallurgical Melts and Solutions*; Metallurgical Industry Press: Beijing, China, 2007.

19. Tsukihashi, F.; Nakamura, M.; Orimoto, T.; Sano, N. Thermodynamics of Phosphorus for the CaO–BaO–CaF$_2$–SiO$_2$ and CaO–Al$_2$O$_3$ Systems. *Tetsu–to–Hagané* **1990**, *76*, 1664–1671. [CrossRef]

20. Ban-ya, S.; Hobo, M.; Kaji, T.; Itoh, T.; Hino, M. Sulphide Capacity and Sulphur Solubility in $CaO-Al_2O_3$ and $CaO-Al_2O_3-CaF_2$ Slags. *ISIJ Int.* **2004**, *44*, 1810–1816. [CrossRef]
21. Ban-ya, S.; Hino, M. Comments on "Evaluation of Thermodynamic Activity of Metallic Oxide in a Ternary Slag from the Sulphide Capacity of the Slag". *ISIJ Int.* **2005**, *45*, 1754–1756. [CrossRef]
22. Wei, S.K. *Thermodynamics of Metallurgical Processes*; Science Press: Beijing, China, 2010.
23. Zhang, J.Y. *Metallurgical Physicochemistry*; Metallurgical Industry Press: Beijing, China, 2004.
24. Chen, J.X. *Handbook of Common Figures, Tables and Data for Steelmaking*, 2nd ed.; Metallurgical Industry Press: Beijing, China, 2010.
25. Sosinsky, D.J.; Sommerville, I.D. The Composition and Temperature Dependence of the Sulfide Capacity of Metallurgical Slags. *Metall. Trans. B* **1986**, *17*, 331–337. [CrossRef]
26. Nakamura, T.; Ueda, Y.; Toguri, J.M. A Critical Review of Optical Basicity on Metallurgical application. In Proceedings of the Third International Conference on Metallurgical Slags and Fluxes, University of Strathclyde, Glasgow, Scotland, 27–29 June 1988; The Institute of Metals: London, UK; pp. 146–149.
27. Mills, K.C.; Sridhar, S. Viscosities of Ironmaking and Steelmaking Slags. *Ironmak. Steelmak.* **1999**, *26*, 262–268. [CrossRef]
28. Young, R.W.; Duffy, J.A.; Hassall, G.J.; Xu, Z. Use of Optical Basicity Concept for Determining Phosphorus and Sulfur Slag–Metal Partitions. *Ironmak. Steelmak.* **1992**, *19*, 201–219.
29. Wagner, C. The Concept of the Basicity of Slags. *Metall. Trans. B* **1975**, *6*, 405–409. [CrossRef]

metals

MDPI

Article

Inclusions Control and Refining Slag Optimization for Fork Flat Steel

Yangyang Ge [1], Shuo Zhao [1,2,*], Liang Ma [1], Tao Yan [1], Zushu Li [2] and Bin Yang [3]

[1] Department of Materials Science and Engineering, Hebei University of Engineering, Handan 056000, China; g514477239@gmail.com (Y.G.); Ma_liang@hebeu.edu.cn (L.M.); yantao@hebeu.edu.cn (T.Y.)
[2] Warwick Manufacturing Group, University of Warwick, Coventry CV4 7AL, UK; z.li.19@warwick.ac.uk
[3] Hebei Yongyang Special Steel Group Co., Ltd., Handan 056000, China; yangbin2027@126.com
* Correspondence: zhaos418@hebeu.edu.cn; Tel.: +86-310-857-7969

Received: 20 December 2018; Accepted: 16 February 2019; Published: 20 February 2019

Abstract: In order to investigate the causes of the large number of cracks and porosities formed in 33MnCrTiB fork flat steel produced by a special steel plant, scanning electron microscopy (SEM), energy dispersive spectrometer (EDS) analysis, and large sample electrolysis of the obtained steel samples were carried out in different steps of the steelmaking processes. The main micro-inclusions in the fork flat steel samples were Al_2O_3, CaO-MgO-Al_2O_3-SiO_2, and TiN, and the macro-inclusions were mainly Al_2O_3, CaO-Al_2O_3, CaO-Al_2O_3-SiO_2-TiO_2, and CaO-MgO-Al_2O_3-SiO_2-TiO_2-(K_2O) systems which originated from the ladle slag and mold flux in the production process. In order to reduce the number of micro-inclusions effectively, the control range of components in the refining slag was confirmed by the thermodynamic calculation, where the mass ratio of CaO/Al_2O_3 should be in the range of 1.85–1.92, and the mass fraction of SiO_2 and MgO should be controlled to 7.5–20% and 6–8%, respectively. In addition, the numbers of macro-inclusions in the flat steel should be effectively reduced by optimizing the flow field of mold and preventing the secondary oxidation, and the flat steel quality problems caused by the inclusions can be improved by the optimization process above.

Keywords: fork; flat steel; inclusions; 33MnCrTiB; slag

---

1. Introduction

Forklift trucks are common production vehicles, which convey raw materials to the production line, scrap, and final products to the storage areas or to the transportation trucks. The fork is one of the most important parts of a forklift. It not only bears a large external force, but also needs to be a small cross section and lightweight. The strength and stiffness of the fork must meet the safety requirements, but they also require a good plasticity index and certain low-temperature impact properties. The research shows that the load-support parts of forklift forks are subjected to repetitive stress conditions of a variety of load and frequency spectra [1,2], including loading and unloading cycles and vibrations coming from moving on irregular terrain conditions [3,4]. The raw materials' quality is one of the main causes inducing fatigue failure of forklift forks.

Common forks generally use 33MnCrTiB, 35CrMo, and 40Cr flat steel as raw materials [5]. The 33MnCrTiB fork flat steel, produced by a special steel enterprise in China, has high strength, is lightweight, and has long service life, thus making it a relatively high-quality fork flat steel. Recently, cracks, porosity, and other defects have often been found in the flat steel produced by steel mills. Many inclusions have been found to be the source of the cracks. Therefore, to find out the sources of inclusions in steel and make relevant optimization measures, steel samples obtained from steelmaking plants are analyzed in this study by means of large sample electrolysis, metallographic observation, and scanning electron microscopy.

2. Materials and Experimental Methods

2.1. Fork Flat Steel Production Process

The production process of a 33MnCrTiB low-alloy flat steel plant includes: Scrap + Hot Metal → Consteel Electric Arc Furnace (EAF) → Ladle Furnace (LF) → 160 × 225 mm Billet Continuous Casting → Rolling → Heat Treatment → Product. The chemical compositions of defective samples in the steel plant were analyzed by X-ray, and the results are shown in Table 1. The internal control requirements were satisfied according to the steel standards. In order to reduce the nitrides in the continuous casting process, it was necessary to precisely control the trace elements Ti and B components to the lower limit, and the alloy was then pre-heated to prevent an increase of nitrogen in the molten steel.

Table 1. Chemical compositions of 33MnCrTiB, wt %.

C	Si	Mn	P	S	Cr	Al	Ti	B
0.33	0.26	1.36	0.011	0.005	0.40	0.009	0.034	0.0013
0.30–0.35	0.17–0.37	1.25–1.50	≤0.035	≤0.035	0.30–0.60	-	≥0.015	0.0005–0.0030

2.2. Experimental Methods

In order to study the influence of inclusions on the straightening cracks of fork flat steel, we analyzed the inclusions in steel according to two sources of inclusions—exogenous and endogenous inclusions. The research objects were 33MnCrTiB steel samples obtained from EAF production processes.

Firstly, the billets were cut and carried out the detection of large inclusions with steel sample electrolysis experiments, where mainly the types, sizes, and microstructures of exogenous inclusions were analyzed [6,7]. In this study, 33MnCrTiB samples were obtained from steel mills, which were cut and surface-cleaned to meet the sample preparation requirements for electrolysis of large inclusions. Then, a cylinder steel sample (Y) was machined as the electrolytic sample. Finally, the electrolysis experiment was carried out at the State Key Laboratory of Advanced Metallurgy in the University of Science and Technology, Beijing. The sample analysis process of large sample electrolysis is shown in Figure 1.

Then, the composition and morphology of endogenous inclusions in steel samples taken during the refining and casting processes were analyzed by the JEOL JSM-6480LV scanning electron microscope (JEOL Ltd., Akishima, Japan) with energy spectrum analysis. Finally, combined with the thermodynamic software Factsage 7.1, the slag compositions and typical inclusions in steel were effectively controlled by using the calculation and experiment results.

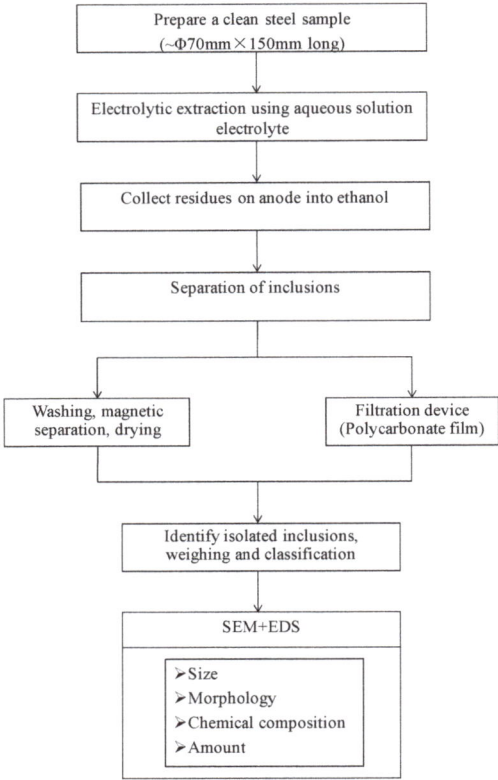

Figure 1. Large sample electrolysis process.

3. Analysis and Discussion

3.1. Exogenous Inclusions in Steel

The size and morphology of large inclusions was observed by SEM, respectively, and the inclusions morphology is shown in Figure 2. The types of large inclusions were analyzed by EDS, and Table 2 shows the composition of each inclusion labelled in Figure 2.

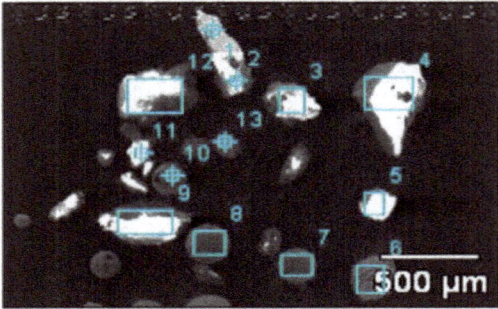

Figure 2. Morphology of large inclusions in Sample Y.

Table 2. Analysis results of inclusions compositions in Figure 2, wt %.

No.	O	Mg	Al	Si	S	K	Ca	Ti	Cr	Mn	Fe	Cu	Zr
1	50.85	1.33	12.76	23.22	-	3.22	3.05	1.85	0.52	-	3.2	-	-
2	60.39	-	22.58	3.21	-	-	5.03	1.71	-	-	7.09	-	-
3	50.12	0.85	23.33	12.08	-	1.62	3.87	3.06	-	-	5.09	-	-
4	45.98	-	34.43	7.55	-	0.98	0.95	2.31	0.51	-	7.28	-	-
5	60.89	-	21.2	2.61	-	-	9.83	-	-	-	5.46	-	-
6	20.85	2.03	7.18	1.03	1.11	-	6.03	6.56	3.46	5.01	46.74	-	-
7	29.42	8.6	24.96	0.78	2.15	-	11.16	12.71	-	-	7.78	2.45	-
8	24.45	3.47	25.24	0.86	-	-	18.2	18.15	-	-	2.19	-	7.44
9	48.79	-	51.21	-	-	-	-	-	-	-	-	-	-
10	14.14	1.09	23.22	3.56	5.47	-	32.27	14.55	-	-	5.7	-	-
11	44.09	-	55.91	-	-	-	-	-	-	-	-	-	-
12	60.67	-	20.78	-	-	-	10.03	-	-	-	8.52	-	-
13	42.75	-	32.08	-	-	-	25.17	-	-	-	-	-	-

The weighting and classification of large oxide inclusions obtained after the sample electrolysis can be seen in Table 3. The total amount of inclusions is 6.21 mg/10 kg steel.

Table 3. Analysis results of large inclusions in steel.

Sample	Original Weight	Remaining Weight	Electrolytic Weight	Total Inclusions		Inclusion Particle Size Classification, µm			
						<80	80~140	140~300	>300
	kg	kg	kg	mg	mg/10 kg	mg	mg	mg	mg
Y	4.438	0.251	4.178	2.60	6.21	0.10	0.20	1.20	1.10

Figure 3 is the morphology of large inclusions in Sample Y, which was observed at 20× magnification. Through an observation by scanning electron microscope of Sample Y, it can be concluded that the representative inclusions in the electrolysis sample were mainly $CaO-MgO-Al_2O_3-SiO_2-TiO_2$, $CaO-Al_2O_3-SiO_2-TiO_2$, Al_2O_3, and $CaO-Al_2O_3$ oxides, where some quaternary inclusions also included less ZrO_2, K_2O, etc. There was also a small amount of sulfide inclusions, such as MnS and CuS in the sample. The size of the large inclusions was mostly more than 140 µm in Sample Y, and only small amounts were below 80 µm (Figure 4). Such large oxide inclusions are harmful to the cutting properties, weldability, and corrosion resistance of the steel, and their existence destroys the continuity of the steel matrix and affects the relevant properties of the steel [8,9]. Most of the surface defects on the steel, such as the turnover skull, scabs, concavo-convex, and cracks, are related to large oxide inclusions. The hot brittleness of the steel can be caused by sulfide inclusions in steel, and will affect the plasticity and toughness of the steel.

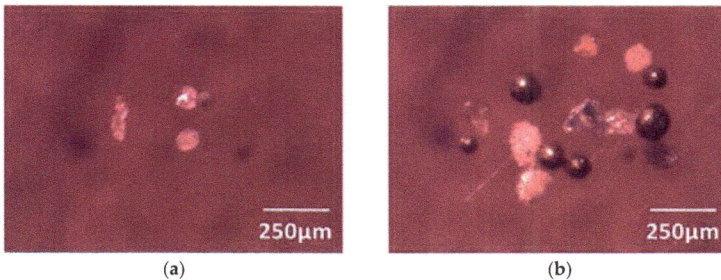

(a)

(b)

Figure 3. *Cont.*

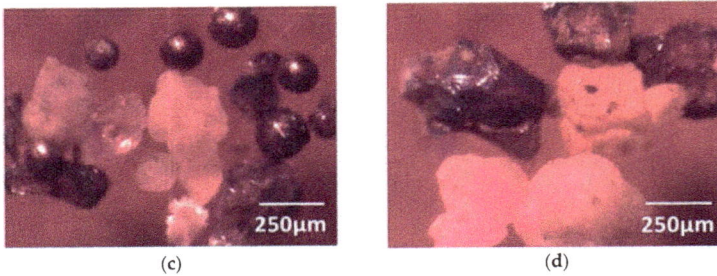

(c)   (d)

Figure 3. Large inclusions in Sample Y at 20× magnification: (**a**) <80 µm; (**b**) 80~140 µm; (**c**) 140~300 µm; (**d**) >300 µm.

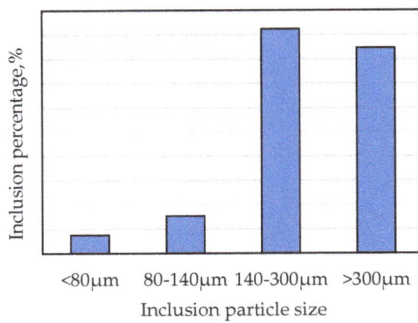

Figure 4. Size distribution of inclusion particles in Sample Y.

3.2. Endogenous Inclusions in Steel

The origin of endogenous inclusions is closely related to steel quality. Because such inclusion is mainly generated inside the molten steel during steelmaking or continuous casting, endogenous inclusions are more difficult to remove than exogenous inclusions. Next, to identify the components and types of inclusions in steel during the different process, the microscopic analysis of each sample was conducted separately.

3.2.1. Pre-Refining of LF Furnace

After the EAF furnace tapping, the steel samples were obtained from the ladle furnace in the steel mill, and metallographic observations were performed in the laboratory. The inclusion morphology and compositions were also analyzed by SEM and EDS. The experimental results are shown in Figure 5.

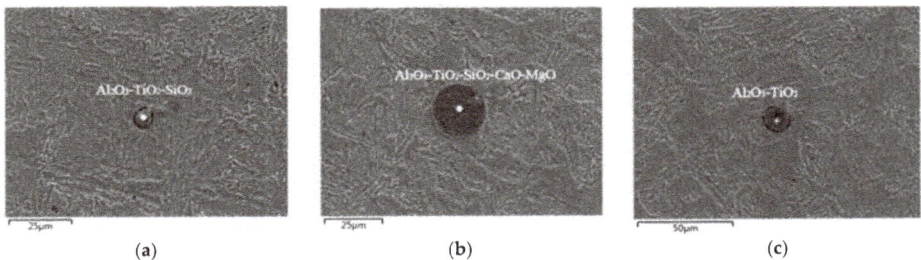

(a)   (b)   (c)

Figure 5. Types and compositions analysis of inclusions at the beginning of refining (Ladle Furnace (LF)). (**a**) Al_2O_3-TiO_2-SiO_2; (**b**) Al_2O_3-TiO_2-SiO_2-CaO-MgO; (**c**) Al_2O_3-TiO_2.

According to the analysis results of the steel sample at the beginning of LF refining, the typical inclusions in the steel sample are mainly Al_2O_3-TiO_2, Al_2O_3-TiO_2-SiO_2, and Al_2O_3-TiO_2-SiO_2-CaO-MgO composite oxides when the ladle enters into working position; there is also a small amount of MnS wrapped around the oxide. Such inclusions mainly come from alloying deoxidation products and the top slag modifier, where they will affect the slagging function in the early stage of refining. The size range of the inclusions is relatively small, the maximum size not exceeding 25 μm and the smallest one only being about 5 μm. The welding performance, corrosion resistance, and fatigue resistance of the material will be seriously destroyed by such complex inclusions existing in the steel. However, the generation of such inclusions can be effectively reduced by the composition's adjustment of refining slag [10].

3.2.2. Wire Feeding Process

At the end of ladle furnace refining, the steel needs to be further treated with feeding the cored wire, and the pure Ca or CaSi wire is mainly used in the wire feeding process. The metallographic observations were also conducted on steel samples from the steel plant, and SEM and EDS analyses were performed before and after the wire feeding. The inclusions morphology and compositions are shown in Figure 6 (before wire feeding) and Figure 7 (after wire feeding), respectively. The following types are the relevant experimental results in the steel after wire feeding (Ca cored wire feeding in this trial).

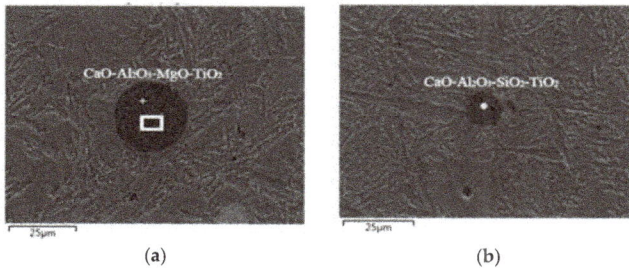

(a) (b)

Figure 6. SEM and EDS analysis of inclusions before wire feeding. (a) CaO-Al_2O_3-MgO-TiO_2; (b) CaO-Al_2O_3-SiO_2-TiO_2.

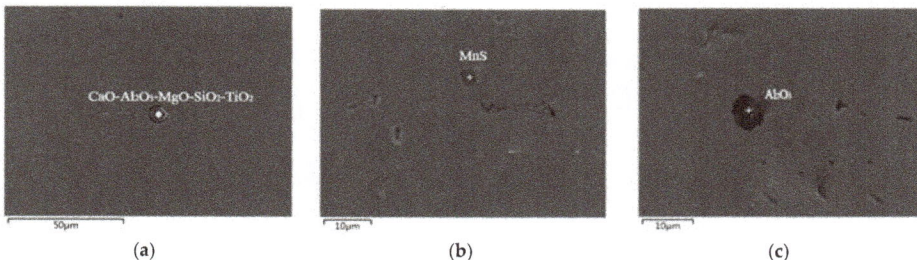

(a) (b) (c)

Figure 7. SEM and EDS analysis of inclusions after wire feeding. (a) CaO-Al_2O_3-MgO-SiO_2-TiO_2; (b) MnS; (c) Al_2O_3.

By analyzing the above SEM diagrams of inclusions in steel before and after wire feeding, it can be concluded that the typical inclusions in steel before wire feeding mainly include CaO-Al_2O_3-MgO-TiO_2 and CaO-Al_2O_3-SiO_2-TiO_2 inclusions. In addition, there is also a small amount of MnO and a very small amount of sulphide, and the size of the inclusion particles is small, which is between 10 μm and 20 μm. After the wire feeding, the typical inclusions in steel are mainly Al_2O_3 and CaO-Al_2O_3-MgO-SiO_2-TiO_2 with a particle size below 10 μm, and there is also less MnS. Compared with the inclusions in steel before wire feeding, it was found that the TiO_2 content was greatly reduced, and only a very small amount of titanium oxide was contained. Al_2O_3, SiO_2, and other components were also reduced after wire feeding. This is because feeding Ca wire led to a significant increase in CaO content in the inclusions in steel and consequently formed large-size inclusions that were easier to float out of the steel. As a result, only a few simple inclusions remained in the steel and their sizes were also significantly reduced compared to those before wire feeding.

3.2.3. Tundish Casting Process

After the refining process, the molten steel is transferred to the tundish and mold, and it keeps casting until it is solidified into the final billet. At the same time, the steel samples taken from the tundish were subjected to metallographic and energy dispersive analysis. The results are shown in Figure 8.

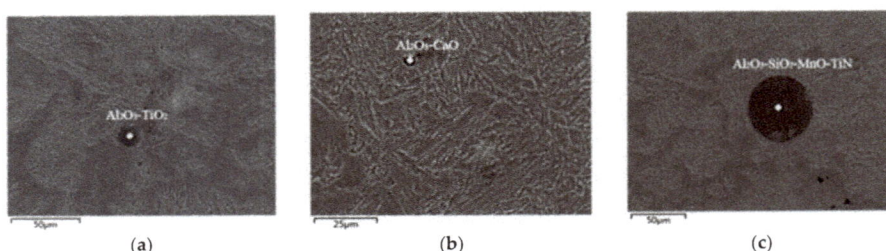

Figure 8. SEM and EDS analysis of inclusions in tundish. (**a**) Al_2O_3-TiO_2; (**b**) Al_2O_3-CaO; (**c**) Al_2O_3-SiO_2-MnO-TiN.

Through the SEM and EDS analysis of the inclusions in the tundish, it can be shown that the main types of the inclusions in steel samples are Al_2O_3-CaO, Al_2O_3-TiO_2, Al_2O_3-SiO_2-MnO-TiN composite inclusions, and the size difference of the inclusions is huge—the larger is about 50 μm and the smaller is less than 10 μm. It is not desirable for these composite inclusions which exist in steel to be seen in actual production. They mainly come from the cooling and solidification process of continuous casting, due to the decrease in solubility and element interactions. Oxygen, nitrogen, and other impure elements dissolved in molten steel were precipitated as the compound from the liquid phase or solid solution during the cooling and solidification process, and finally, they remained in the steel to form inclusions [11]. This kind of inclusion is harmful to the welding performance, corrosion resistance, and fatigue resistance of steel. However, by analyzing the solubility product of related compounds, we can find ways to reduce the generation of such inclusions.

3.2.4. Casting Billet

After the casting flame-cutting, the fixed-length billet was obtained, and its relevant components were analyzed by scanning electron microscopy and an energy spectrum analyzer. Figure 9 shows the results of the experimental analysis.

By analyzing the SEM results of the billet, the main inclusions in the obtained slab were TiN, SiO_2-Al_2O_3-MgO-CaO, and a small amount of TiO_2 inclusions. The inclusions size is generally small, with the largest being about 25 μm and the smallest being less than 10 μm. The casting billet contains many irregular TiN, and such inclusions can easily cause fatigue fracture during the processing of the billet or rolled materials. In order to reduce the number of TiN inclusions, it is necessary to avoid the reoxidation of molten steel and reduce the precipitation amounts during the cooling and solidification process.

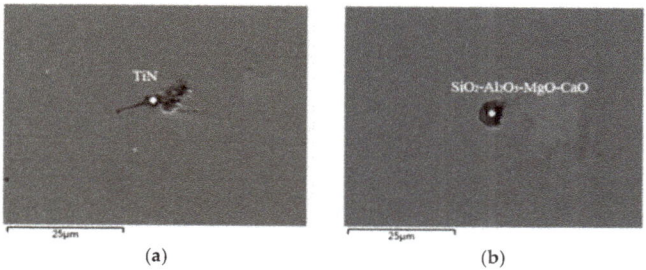

Figure 9. SEM and EDS analysis of inclusions in the casting billet. (a) TiN; (b) SiO_2-Al_2O_3-MgO-CaO.

4. Control of Refining Slag and Precipitation Inclusions

4.1. Compositions Control of Refining Slag

It can be concluded that there are mainly Al_2O_3 and CaO-MgO-Al_2O_3-SiO_2 oxide inclusions in flat steel during the entire production process. Through the composition determination of the micro-inclusions, it can be concluded that the oxide inclusions in steel mainly come from the slag and deoxidation agents, so it is very important to improve the compositions of refining slag for absorbing the inclusions. In view of the reasonable composition control of final refining slag, the following four points need be followed [12–14]:

(1) The melting-point of the final slag should be appropriate, i.e., lower than the temperature of molten steel at the tundish nozzle (about casting temperature $-10\ ^\circ$C);
(2) The refining slag should be fully reacted with the Al_2O_3 deoxidation product so that it can be assimilated and absorbed as much as possible. That is, the viscosity of refining slag and the initial activity of Al_2O_3 in the slag cannot be too large;
(3) The slag should avoid chemical reactions with the steel, and the sulfur capacity should be as high as possible;
(4) To ensure that the lining is not eroded by slag, the MgO content should be reduced as much as possible to reduce the melting point and viscosity of the slag. Wang et al. pointed out that the continuous increase of MgO content will reduce the saturation solubility of CaO, which will not only increase the melting point of the slag, but also reduce its desulfurization capacity when w(MgO) < 8% in the slag, its melting point and desulfurization ability are better [15]. Therefore, the MgO content is generally controlled at about 6% to 8%.

Next, the compositions of the refining slag are optimized by the thermodynamic calculations, and then the compositions of inclusions and steel can be controlled effectively due to the slag-steel balanced reaction. The following slag-steel reactions can occur during the refining process:

Steel-slag oxidation reaction: [Al] + 0.75(SiO_2) = 0.5(Al_2O_3) + 0.75[Si]

$$\lg K^\theta_{\text{Al-Si}} = \frac{8595.52}{T} - 1.40 \tag{1}$$

Slag-steel desulphurization reaction: 3[S] + 3(CaO) + 2[Al] = 3(CaS) + (Al_2O_3)

$$\lg K^\theta_{\text{CaS}} = \frac{44,279}{T} - 15.12 \tag{2}$$

According to the system, the Gibbs free energy isothermal equation is $\Delta G = \Delta G^0 + RT \ln J$. In the above formula, ΔG^0 is the Gibbs free energy in a standard state; J is the activity ratio of the substance of above reactions under the actual condition; and R, T is the ideal gas constant and temperature, respectively. When $\Delta G > 0$, the forward reaction is inhibited, favoring the reverse reaction;

when $\Delta G < 0$, it favors the forward reaction and the reverse reaction is inhibited; and when $\Delta G = 0$, the chemical reaction reaches equilibrium. Therefore, the steel-slag reaction in this study needs to suppress the oxidation reaction ($\Delta G > 0$) and promote the desulfurization reaction ($\Delta G < 0$).

Therefore, the suppression conditions for the above two reactions at the refining temperature of 1873 K by calculation are as follows:

Inhibiting the steel-slag oxidation reaction was:

$$\frac{(f[\%Si])^{0.75} \cdot a_{Al_2O_3}^{0.5}}{f_{Al} \cdot [\%Al] \cdot a_{SiO_2}^{0.75}} > 1.55 \times 10^3 \tag{3}$$

Promoting the desulfurization reaction was:

$$\frac{a_{CaS}^3 \cdot a_{Al_2O_3}}{(f_S[\%S])^3 \cdot (f_{Al}[\%Al])^2 \cdot a_{CaO}^3} < 3.32 \times 10^8 \tag{4}$$

The activities (a_i) and activity coefficients (f_i) of each element in the steel can be calculated by the following formula:

$$a_i = f_i \times [\%i], \ \lg f_i = \sum e_i^j[\%j] \tag{5}$$

The chemical compositions of 33MnCrTiB steel are listed in Table 1, and the element interaction coefficients (e_i^j) in liquid iron are presented in Table 4.

Table 4. First-order interaction coefficients in liquid iron at 1873 K.

e_i^j	Al	B	C	Cu	Ni	N	Mn	H	P	S	Si	Ti	Cr
e_{Al}^j	0.05	-	0.09	0.01	-	-0.1	-	0.24	-	0.03	0.01	-	-
e_S^j	0.04	0.13	0.11	0	0	0.01	-0.026	0.12	0.03	0	0.06	-0.1	-0.011
e_{Si}^j	0.06	0.2	0.18	0.01	0.01	0.09	0.002	0.64	0.11	0.06	0.11	-	-0.0003
e_{Ti}^j	-	-	-0.165	-	-	-1.8	0.0043	-	-0.0064	-0.11	0.05	0.013	0.055
e_N^j	-0.3	0.094	0.13	0.009	0.01	0	-0.021	-	0.045	0.007	0.047	-0.53	-0.047

The activities of [%S], [%Si], and [%Al] can be calculated according to the interaction coefficients in Table 4, where a_{CaS} should be chosen as 1 because of its pure solid state, and then the following calculation results can be drawn: For the inhibiting of steel-slag oxidation, $\frac{a_{Al_2O_3}^{0.5}}{a_{SiO_2}^{0.75}} > 36.89$; for the promoting of slag-steel desulfurization, $\frac{a_{Al_2O_3}}{a_{CaO}^3} < 3.6965$.

Based on the refining data of the thermodynamic balancing calculation, the corresponding activity values and mass fractions of CaO, Al_2O_3, and SiO_2 which satisfied the above two inequality formulas can firstly be found in the phase diagram database of Factsage; then, the mass fractions of three oxides that satisfy the preconditions are marked in the quaternary phase diagram, and it was found that most points were concentrated in one area. This red area can be seen in Figure 10. Thus, the refining slag range can be obtained, where the CaO content should be controlled in the range of 44.0% to 57.0%, the Al_2O_3 content between 19.5% and 40.0%, and the SiO_2 content between 7.5% and 20.0%.

In addition to the above-mentioned requirements of refining slag, it is also necessary to consider the influence of its own viscosity, melting point, and other factors. Therefore, in order to fully absorb the Al_2O_3 in molten steel, it is necessary that the refining slag has a suitable melting point and viscosity. Previous studies [14] pointed out that the slag viscosity increases as the MgO content continues to increase, and that when $63\% < w(CaO + MgO) < 65\%$ and $w(MgO) = 4\%~8\%$, the viscosity of refining slag is the smallest (0.05~0.06 Pa·s) in the CaO-SiO_2-Al_2O_3-MgO system. In other words, the Al_2O_3 assimilation and absorption ability of the slag are relatively strong. Meanwhile, the higher CaO/Al_2O_3 ratio in the slag should be maintained, as this will help reduce the activity of Al_2O_3 and improve the absorption ability of refining slag to the inclusions with high melting points.

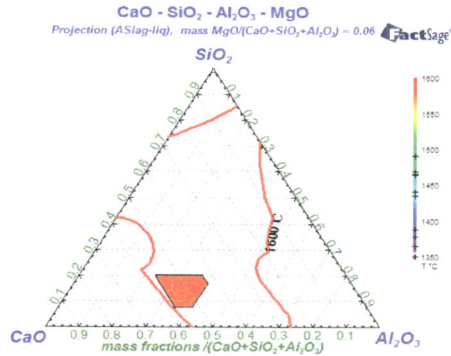

Figure 10. $T = 1873$ K, quaternary phase diagram of CaO-SiO$_2$-Al$_2$O$_3$-6%MgO slag.

In order to obtain a lower melting point, the three oxides' components should be chosen in the area below 1600 °C, and the red region in Figure 10 can meet the low melting-point requirement. It is also known that when 63% < w(CaO + MgO) < 65%, the viscosity of the quaternary slag system is particularly suitable—thus, the following control range of refining slag could be obtained: the CaO/Al$_2$O$_3$ ratio being between 1.85 and 1.92, 7.5% < w(SiO$_2$) < 20%, and 6% < w(MgO) < 8%.

4.2. Thermodynamic Calculation of TiN Precipitations

According to the above analysis, TiN is the typical nitride and mainly formed during the cooling and solidification stage in the continuous casting process. In order to effectively reduce the amount of precipitation, the following calculations were carried out. The chemical reaction equations in molten steel are as follows [16,17]:

$$[\text{Ti}] + [\text{N}] = \text{TiN}_{(s)} \quad K_{\text{TiN}} = \frac{a_{\text{TiN}}}{a_{[\text{Ti}]} \cdot a_{[\text{N}]}} = \frac{1}{f_{\text{Ti}}[\%\text{Ti}] f_{\text{N}}[\%\text{N}]} \tag{6}$$

$$\lg K_{\text{TiN}} = \frac{-\Delta G^{\ominus}}{RT/\lg(e)} = \frac{-\Delta G^{\ominus}}{2.3RT} = \frac{291,000 - 107.91T}{2.3RT} = \frac{15,217}{T} - 5.643 \tag{7}$$

The following expression of the logarithmic equation was derived:

$$\lg f_{\text{Ti}} + \lg f_{\text{N}} + \lg([\%\text{Ti}][\%\text{N}]) = 5.643 - \frac{15,217}{T} \tag{8}$$

It can be known from the above formulas that the equilibrium activity product of TiN inclusions is mainly affected by the temperature T of the molten steel, and the calculation formula of activity coefficients of the elements in molten steel at 1873 K can refer to Formula (5). $f_{\text{Ti}}, f_{\text{N}}$ are the activity coefficients of Ti and N in the molten steel, respectively, and the activity coefficients of Ti and N have the following relationship with the temperature:

$$\lg f_{\text{Ti}} = \left(\frac{2557}{T} - 0.365\right) \sum \left(e^j_{\text{Ti}(1873)}[\%j]\right) \tag{9}$$

$$\lg f_{\text{N}} = \left(\frac{3280}{T} - 0.75\right) \sum \left(e^j_{\text{N}(1873)}[\%j]\right) \tag{10}$$

From the chemical compositions of the target steel in Table 1 and the interaction coefficient data in Table 4, it can be calculated:

$$\lg([\%\text{Ti}][\%\text{N}]) = 5.627 - \frac{15,216}{T} \tag{11}$$

Under equilibrium conditions, the relationship between nitrogen and titanium content in molten steel under different temperature conditions is shown in Figure 11:

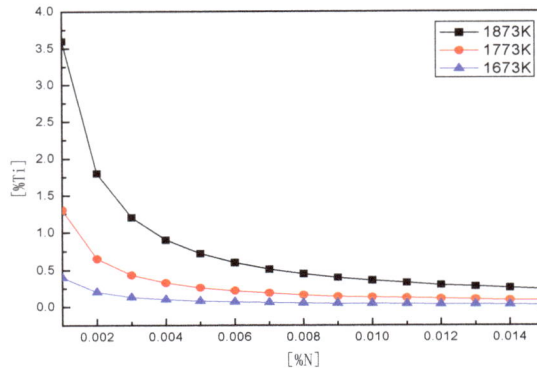

Figure 11. [Ti] and [N]-balanced concentrations in molten steel under different temperatures.

As shown in Figure 11, the concentration product curves of [Ti] and [N] at three different temperatures under equilibrium conditions were obtained by theoretical calculations. According to the changing trend of different temperature curves, the concentration of TiN in steel also decreases as the temperature continues to decrease. The solubility product of TiN in steel is the smallest when the temperature is 1673 K. [Ti] content in steel is known to be 0.034%, and [N] content is 0.0053%. At this time, the generation concentration of TiN is basically at the temperature curve of 1773 K. It can be concluded that when the molten steel is at 1773 K, TiN in molten steel can be precipitated in the liquid–solid two-phase region—that is, controlling the [Ti], [N] content in the steel and rapid solidification, which can effectively reduce the amount of TiN inclusions in molten steel.

5. Conclusions

According to the above analysis and calculations, the following conclusions can be drawn:

(1) There are lots of exogenous large inclusions in the billet, including: Al_2O_3, CaO-Al_2O_3, CaO-Al_2O_3-SiO_2-TiO_2, and CaO-MgO-Al_2O_3-SiO_2-TiO_2. Such large-scale inclusions mainly come from refractory materials, mold flux, and covering slag. The large-scale inclusions can generally be removed by optimizing the flow-field of mold and tundish, and adjusting the components of metallurgy auxiliary materials.

(2) The typical endogenous oxides in molten steel are the Al_2O_3 and CaO-MgO-Al_2O_3-SiO_2 systems. The size of endogenous inclusions are smaller than that of exogenous inclusions. By optimizing the composition of top slags in the refining process, the amount of Al_2O_3 and CaO-MgO-Al_2O_3-SiO_2 inclusions can be effectively reduced. According to the thermodynamic calculations, the control range of refining slag should be: CaO/Al_2O_3 between 1.85 and 1.92, $7.5\% < w(SiO_2) < 20\%$, and $6\% < w(MgO) < 8\%$.

(3) The precipitation amount of TiN could be reduced during the cooling process of continuous casting by controlling the [Ti], [N] content in steel and rapid solidification.

Author Contributions: Methodology and Investigation, Y.G. and S.Z.; Data Curation, L.M.; Funding Acquisition, S.Z. and T.Y.; Writing-Original Draft Preparation, Y.G.; Results Discussion, Writing-Review and Editing, Z.L.; Project Administration, B.Y.

Funding: The project was supported by Natural Science Foundation—Steel and Iron Foundation of Hebei Province [E2016402096, E2016402111], National Natural Science Foundation of China [51501052, 51804094] and Higher Education Teaching Reform Project of Hebei Province [2017GJJG129].

Metals **2019**, *9*, 253

Acknowledgments: The authors would like to appreciate the support from Advanced Manufacturing and Materials Centre, University of Warwick.

Conflicts of Interest: The authors declare no conflict of interest.

References

1.   Figueiredo, M.V.; Oliveira, F.M.F.; Goncalves, J.P.M.; de Castro, P.M.S.T.; Fernandes, A.A. Fracture analysis of a heavy duty lift truck. *Eng. Fail. Anal.* **2001**, *8*, 411–421. [CrossRef]
2.   Massone, J.M.; Boeri, R.E. Failure of forklift forks. *Eng. Fail. Anal.* **2010**, *17*, 1062–1068. [CrossRef]
3.   Wang, S.L. Strength structure and manufacture of forks for forklifts. *Sci. Technol. Enterp.* **2016**, *9*, 215–217. (In Chinese)
4.   Yin, X.G.; Wang, G.; Zhang, X.H. Development of low-alloy high-strength fork flat steel. *Mod. Metall.* **2009**, *37*, 13–15. (In Chinese)
5.   Wu, X.B.; Pang, B.H.; Wang, B.H.; Zhang, L. Analysis on the development of the material for forks. *Heat Treat. Tech. Equip.* **2006**, *27*, 20–22. (In Chinese)
6.   Yoshida, Y.; Funahashi, Y. On the extraction of large inclusions in steel by slime method and classification according to the size. *J. Iron Steel Inst. Jpn.* **1975**, *61*, 2489–2500. (In Japan) [CrossRef]
7.   Tunde, I.O.; Taiwo, E.A.; Augusta, I.E. Investigation of mechanical properties and parametric optimization of the dissimilar GTAW of AISI 304 stainless steel and low carbon steel. *World Eng.* **2018**, *15*, 584–591.
8.   Boya, O.; Danjuma, Y.; Ibraheem, S.; Dagwa, I.; David, O. Effects of electrode type on the mechanical properties of weldments of some steel samples produced in Nigeria. *World Eng.* **2014**, *11*, 95–106.
9.   Li, D.Z. *Non-metal Inclusions in Steel*, 1st ed.; Science China Press: Beijing, China, 1983; pp. 57–76.
10. Zhao, S.; Ge, Y.Y.; Ma, L.; Yan, T.; Lyu, J.C.; Li, Z.S. Formation analysis of edge cracks of 33MnCrTiB fork steel. *Metals* **2018**, *8*, 587. [CrossRef]
11. Huang, X.H. *Metallurgical Principle of Iron and steel*, 3rd ed.; Metallurgical Industry Press: Beijing, China, 2002; pp. 23–52. (In Chinese)
12. Wang, Q.; He, S.P. Optimization of LF refining process and slag for low carbon aluminum containing steel. *J. Univ. Sci. Technol. Beijing* **2007**, *29*, 15–17. (In Chinese)
13. Zhao, S.; He, S.P.; Chen, G.J.; Peng, M.M.; Wang, Q. Castability of molten steel and cleanliness of slab for high strength low alloy steel without calcium treatment. *Ironmaking Steelmaking* **2014**, *41*, 153–160. [CrossRef]
14. Cui, L.Z. Control non-metal inclusion in steel by LF refining furnace. *Hebei Metall.* **2010**, *179*, 30–31. (In Chinese)
15. Hao, N.; Wang, X.H.; Liu, J.G.; Wang, W.J. Effect of MgO content on desulphurization of CaO-Al$_2$O$_3$-SiO$_2$-MgO slag. *Steelmaking* **2009**, *25*, 16–19. (In Chinese)
16. Yang, J.; Wang, X.H.; Gong, Z.X.; Jiang, M.; Wang, G.C.; Huang, F.X. Precipitation thermodynamics analysis and control of titanium nitride inclusions in extra- low oxygen wheel steel. *J. Univ. Sci. Technol. Beijing* **2010**, *32*, 1139–1143. (In Chinese)
17. Kang, Y.B.; Lee, H.G. Thermodynamic analysis of Mn-depleted zone near Ti oxide inclusions for intragranular nucleation of ferrite in steel. *ISIJ Int.* **2010**, *50*, 501–508. [CrossRef]

MDPI

Article

Modification of Non-Metallic Inclusions in Oil-Pipeline Steels by Ca-Treatment

Elena Sidorova [1,2], Andrey V. Karasev [1,*], Denis Kuznetsov [2] and Pär G. Jönsson [1]

[1]   Department of Materials Science and Engineering, KTH Royal Institute of Technology, Brinellvägen 23,
     10044 Stockholm, Sweden; elena.sidorova91@gmail.com (E.S.); parj@kth.se (P.G.J.)
[2]   Department of Functional Nanosystems and Hightemperature Materials, National University of Science and
     Technology (MISIS), Leninsky Prospect 4, 119049 Moscow, Russia; dk@misis.ru
*   Correspondence: karasev@kth.se; Tel.: +46-(0)8-790-8357

Received: 18 January 2019; Accepted: 25 March 2019; Published: 28 March 2019

Abstract: Corrosion rate in different steel grades (including oilfield pipeline steels) is determined by the presence of non-metallic inclusions (NMI) in steels. Specifically, the effect of different inclusions on the quality of steels depends on their characteristics such as size, number, morphology, composition, and physical properties, as well as their location in the steel matrix. Therefore, the optimization and control of NMI in steels are very important today to obtain an improvement of the material properties of the final steel products. It is well known that a Ca-treatment of liquid steels in ladle before casting is an effective method for modification of non-metallic inclusions for improvement of the steel properties. Therefore, the NMI characteristics were evaluated in industrial steel samples of low carbon Ca-treated steel used for production of oil-pipelines. An electrolytic extraction technique was used for extraction of NMI from the steel samples followed by three-dimensional investigations of different inclusions and clusters by using SEM in combination with EDS. Moreover, the number and compositions of corrosion active non-metallic inclusions were estimated in hot rolled steel samples from two different heats. Finally, the corrosion resistance of these steels can be discussed depending on the characteristics of non-metallic inclusions present in the steel.

Keywords: oil-pipeline steel; Ca-treatment; non-metallic inclusions; electrolytic extraction; corrosion

1. Introduction

The increasing energy consumption and demands for oil and natural gas requires safe and effective possibilities to transport them under high pressure for long distances to customers. Therefore, the requirements to the material properties of steels, which are used for oil- or gas-pipelines, increase year by year. In previous studies [1–3] it was found that the hydrogen induced cracking (HIC) and sulfide stress cracking (SSC) are main reasons of corrosion and damages in pipeline steels during transportation of oil and natural gas containing H_2S and H_2O. Specifically, the corrosion is induced by the penetration of hydrogen atoms from oil or gas into the steel, their accumulation on surfaces of microdefects in the steel matrix (such as grain boundaries, pores, and non- metallic inclusions), and initiations of cracks due to a high internal pressure from the formed H_2 gas [4–6].

Generally, the corrosion resistance of steels depends mostly on the following steel characteristics: (1) the content of alloying elements (chromium, nickel and copper), which are involved in the formation of protective films of corrosion products on the steel surface, (2) the steel microstructure, and (3) the presence of components in the steel structure that cause an increased levels of stress as well as contribute to the destruction of protective films. Such components of the structure include, in particular, non-metallic inclusions (NMI) of unfavorable chemical compositions, and isolation of excess phases, including nanoscale phases. The content of chromium, nickel, and copper, which is necessary to ensure a high corrosion resistance, can be reduced in the absence of unfavorable structural components in the steel bar.

Based on detailed investigations of the HIC and SSC defects in different pipeline steels [7,8], it was found that the non-metallic inclusions (such as MnS and Al_2O_3, CaO-Al_2O_3 or complex inclusions containing Al_2O_3 and CaO) are one of the major reasons for hydrogen induced corrosions. An addition of Ca during ladle treatment of the liquid steel is a commonly used technique for modification of MnS inclusions (due to reduction of size and aspect ratio of inclusions) and improvement of the corrosion resistance of steels [3,9].

It is known that the main cause of high corrosion rates of oilfield pipelines is contaminations of steel by certain harmful non-metallic inclusions, which are precipitated in steel during the ladle treatment and casting process. Such inclusions are called corrosion-active non-metallic inclusions [10,11]. However, these non-metallic inclusions in modern steels have usually a complex chemical composition, where often the influence of the composition on the corrosion is not fully known. Previous authors have reported negative effects on the corrosion resistance of oil-field pipeline steels due to two types of non-metallic inclusions [10,11]. The first type is non-metallic inclusions based on calcium aluminates, which sometimes contain magnesium and silicon oxides. The second type is complex inclusions, which have a core of calcium aluminate (at different ratios of CaO and Al_2O_3), manganese sulfide or another inclusion, but surrounded by a calcium sulfide shell. It should be noted that non-metallic inclusions, regardless of their type, affect the resistance of steel to local corrosion, according to the same mechanism as any other heterogeneity.

Today, various methods can be used to assess non-metallic inclusions in different steels and alloys with a high accuracy. The conventional method for evaluation of NMI is the two-dimensional (2-D method) investigations of non-metallic inclusions on polished surfaces of steel samples by using light optical microscopy (LOM) or scanning electron microscopy (SEM). However, three-dimensional (3-D method) investigations of NMI in various steels by using an electrolytic extraction technique combined with SEM investigation have been applied over the last 10 to 20 years. The latter method shows a number of significant advantages compared to the conventional 2-D method [12–17].

In this study, the electrolytic extraction technique was applied for the 3-D investigations of non-metallic inclusions in industrial steels. In addition, the number and compositions of corrosion active NMI were investigated in hot rolled steel samples to evaluate the corrosion resistance of these steels depending on the characteristics of non-metallic inclusions present in the steels.

2. Materials and Methods

2.1. Steel Production and Sampling

In this study, steel samples from two industrial heats (Heat A and Heat B) of low-carbon steels for production of oil-pipeline were used for the evaluation of the non-metallic inclusion characteristics as well as their effect on the corrosion properties of steels. The production technology of the steels included the converter, primary ladle treatment, DH-vacuum treatment, final treatment in the ladle furnace, continuous casting, and hot rolling processes. The main technological parameters and alloy contents in both heats were very similar. Furthermore, the same amount of aluminum wire was added in both heats, but in Heat A it was added during the vacuum treatment of the melt while in Heat B it was added during the final ladle treatment before casting. During the ladle refining, a modification of the non-metallic inclusions was done by an addition of calcium carbide to the liquid steel during the DH-vacuum treatment. However, the amount of calcium carbide added in Heat A (0.25 kg/ton) was significantly larger than that in Heat B (0.23 kg/ton). Moreover, the addition of the calcium carbide in Heat A was done 10 min earlier than that in Heat B. The slabs after continuous casting were rolled under similar conditions, according to the required mechanical and structural properties of the sheet with a thickness of 8 mm. A schematic illustration of production technology and steel sampling for two investigated heats (Heat A and Heat B) for oil-pipeline steel are shown in Figure 1. It should be pointed out that the modification of inclusion characteristics were investigated only in the followed steel samples: samples A2 and B2—initial conditions of non-metallic inclusions before Ca-treatment; samples A4, B4 and A5, B5—modified NMI after Ca-treatment before and during casting, respectively; samples A6 and B6—NMI in the final steel product after hot rolling.

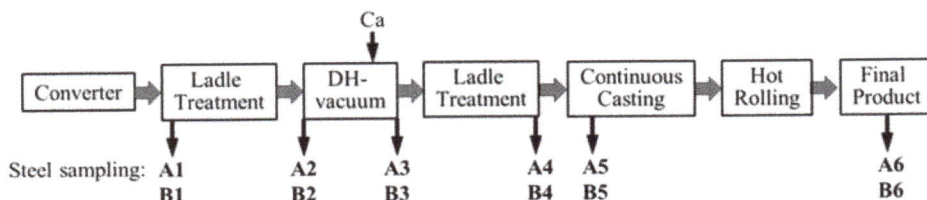

Figure 1. Schematic illustration of the process steps and sampling times.

2.2. Evaluation of the Non-Metallic Inclusions and Microstructure in Steel Samples

As mentioned earlier, non-metallic inclusions were investigated in steel samples taken from two heats (Heats A and B) of low carbon steels, which are used for the production of oil-pipelines. The characteristics of non-metallic inclusions in steel samples (such as composition, morphology, size and number) were evaluated by using the electrolytic extraction (EE) technique in combination with SEM, which has successfully been applied for precise 3-D investigations of inclusions and clusters in different steel grades, as was reported in separate articles [17]. A 10% AA electrolyte (10% acetylacetone-1% tetramethylammonium chloride-methanol) was used for dissolution of the metal matrix. The non-metallic inclusions, which were more stable and did not dissolved in the electrolyte, were collected on a surface of a membrane polycarbonate (PC) film filter (having a 0.4 μm open-pore diameter) by filtration of the electrolyte after the completed extraction. Figure 2 shows a schematic illustration of the electrolytic extraction process and typical SEM images of non-metallic inclusions on film filters after an electrolytic extraction. The following electric parameters were used during the electrolytic extractions of steel samples: electric current—40–60 mA, voltage—2.9–4.5 V, electric charge—500 or 1000 Coulomb. Furthermore, the weight of the dissolved steel during EE process varied from 0.11 up to 0.23 g depending on electric parameters.

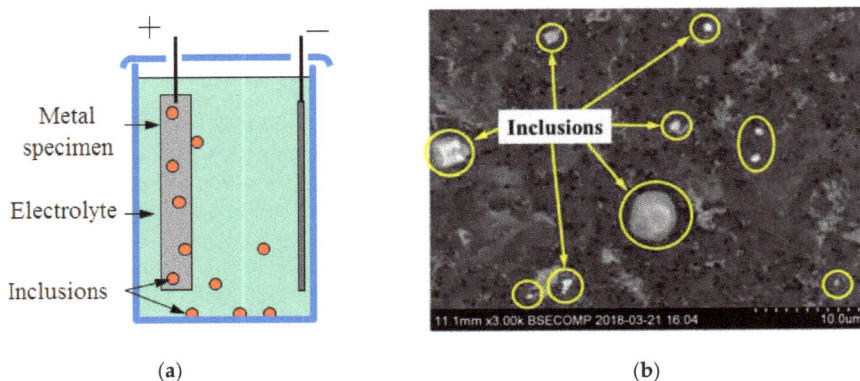

(a) (b)

Figure 2. (**a**) Schematic illustration of the electrolytic extraction process; (**b**) SEM image of some typical non-metallic inclusions present on a film filter after a completed electrolytic extraction.

The size, number, composition and morphology of different inclusions on surfaces of film filters were determined by using scanning electron microscope (SEM) in combination with energy dispersive spectroscopy (EDS). The compositions of 10–25 typical inclusions were determined by using the EDS and the data was recalculated to the main oxide and sulfide components of inclusions. Specifically, the lengths (L) and widths (W) of each inclusion or cluster were measured on the SEM images. The number of measured inclusions for each sample varied from 150 to 430 inclusions, depending on the sampling occasion and cleanness of steel. The equivalent size (d_{eq}) for each inclusion was determined as the average value between measured length and width: $d_{eq} = (L + W)/2$. The number of inclusions per unit volume of steel sample (N_V) for each size interval was calculated as follows:

$$N_V = n \cdot \frac{A_f}{A_{obs}} \cdot \frac{\rho_m}{W_{dis}} \tag{1}$$

where n is the number of inclusions in the given size interval. The parameters A_f and A_{obs} are the total area of the PC film filter containing inclusions after filtration (1200 mm^2) and the area of filter observed by SEM. Furthermore, ρ_m is the density of the steel (~0.0078 g/mm^3) and W_{dis} is the weight of the steel dissolved during the electrolytic extraction.

For the evaluation of corrosion active non-metallic inclusions, hot rolled steel samples A6 and B6 were polished and quickly etched (up to 60 s) in a reagent developed on the basis of the Obergoffer reagent, which is usually used to detect structural inhomogeneity in steels. Then, the active non-metallic inclusions causing corrosion were investigated quantitatively and qualitatively on the metal surfaces of the samples by using SEM in combination with EDS. Figure 3 shows typical SEM images including corrosion active non-metallic inclusion on steel surfaces of both heats after etching. Number and compositions of the corrosion active inclusions having diameter larger than 5 μm were evaluated in steel samples A6 and B6. The corrosion resistance and quality of steel products from Heats A and B were judged by the presence or quantitative characteristics of corrosion active non-metallic inclusions and areas affected by corrosion.

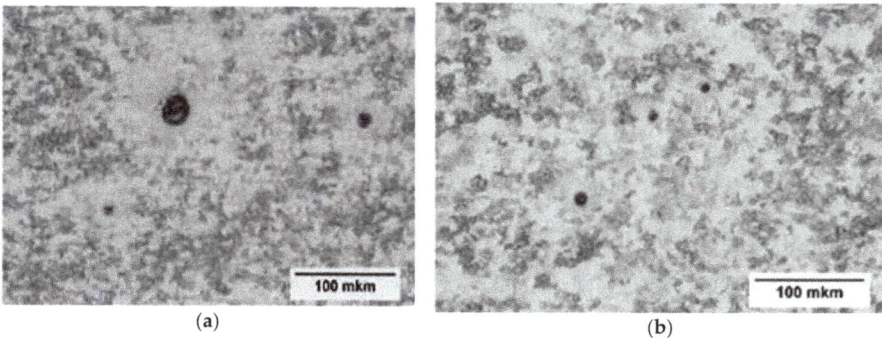

(a)　　　　　　　　　　　　　　　　　(b)

Figure 3. Typical SEM images of corrosion active non-metallic inclusion in steel samples after etching: (a) from Heat A; (b) from Heat B.

The microstructures of steel samples A6 and B6 have been revealed after using a common etching procedure for the given steel grade and investigated by using a light optical microscope at different magnifications.

3. Results and Discussions

3.1. Evaluation of Composition and Microstructure of Steels

The chemical compositions of steel samples taken from both heats after hot rolling are given in Table 1. It can be seen that the contents of main elements in steels of both heats are very similar. Only the contents of Al, Ca, and S are slightly higher and the N content is lower in Heat A compared to Heat B, which can effect on characteristics of non-metallic inclusions.

Table 1. Contents of main elements in steels from both heats (wt%).

Steel	C	Si	Mn	Cr	Ni	Cu	Al	Ti	Ca	S	N
A	0.06	0.24	0.63	0.43	0.17	0.33	0.025	0.019	0.0020	0.002	0.005
B	0.05	0.23	0.67	0.43	0.18	0.35	0.022	0.021	0.0014	0.001	0.007

Typical microstructures observed in both hot rolled steels (Samples A6 and B6) are shown in Figure 4. The results show that the microstructures of steel samples from both heats are very similar.

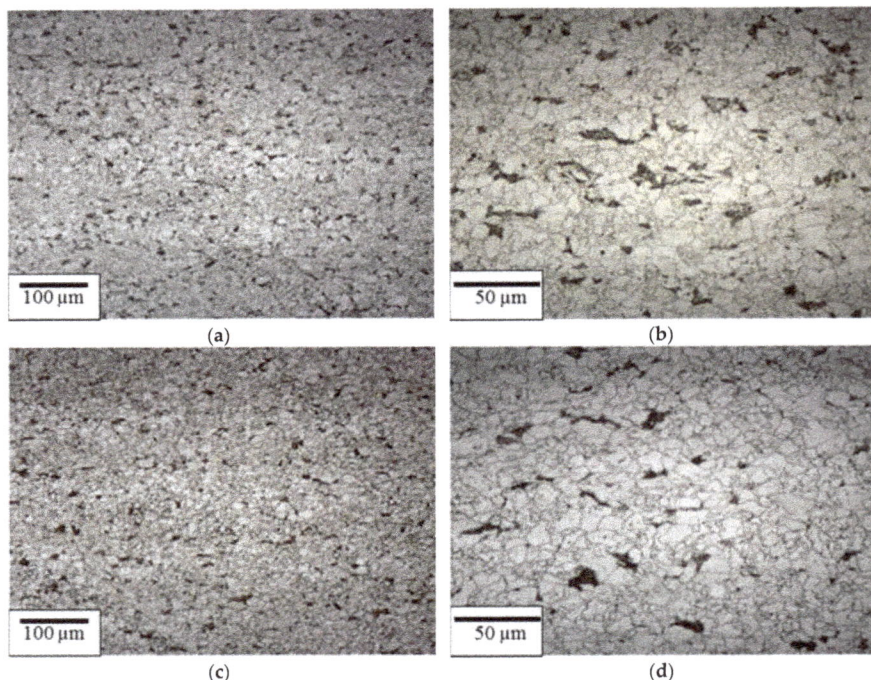

Figure 4. Typical microstructures observed in steel samples: (**a**) Heat A (magnification of ×200); (**b**) Heat A (magnification of ×500); (**c**) Heat B (magnification of ×200); (**d**) Heat B (magnification of ×500).

Based on the obtained results, it may be safely suggested that the chemical compositions and microstructures are very similar in both investigated heats. Thus, these parameters cannot explain the significant differences the corrosion resistances between these two heats.

3.2. Evaluation of NMI Characteristics after Electrolytic Extraction

The results show that despite that the production route and final steel compositions of both heats were very similar the characteristics of the non-metallic inclusions in these heats have significant differences.

The main characteristics of non-metallic inclusions (such as composition, morphology, number and size) were investigated on film filter after electrolytic extraction of steel samples from both heats, as explained earlier. The typical NMI were classified into six different groups, which are presented in Table 2. It can be seen that Type I inclusions (A, AM) are regular and irregular inclusions and clusters containing mostly pure Al_2O_3 or Al_2O_3-MgO. The content of other components such as CaO, CaS and

SiO_2 in these NMI are less than 17%. Also, the size of these observed NMI varied from 0.5 to 7.9 μm. The Type I inclusions and clusters were observed in samples A2 and B2 after alloying and deoxidation of the melt of both heats.

Table 2. Classification of typical non-metallic inclusions observed in different steel samples.

Type of NMI [1]	Composition	Size (μm)	Sample
Type I (A,AM)	Al_2O_3—75–84%, MgO—0–19%, CaO—0–17%, CaS—0–5%.	0.5–7.9	A2, B2
Type II (ASi)	Al_2O_3—5–78%, SiO_2—21–94%, MgO—0–2%, CaO—0–4%, CaS—0–5%.	0.7–5.6	B2
Type III (CAM+CS)	CaO—9–69%, Al_2O_3—2–54%, MgO—0–22%, SiO_2—0–7%, CaS—9–62%.	0.9–5.2	A4–A6, B4–B6
Type IV (CS)	CaS—97–100%, Al_2O_3—0–2%, MgO—0–1%, CaO—0–2%.	2.5–7.4	A6
Type V (CAM+CS+TN)	CaO—4–45%, Al_2O_3—4–45%, MgO—0–19%, SiO_2—1–5%, CaS—0–50%, TiN—1–53%.	1.1–5.0	A4–A6, B5–B6
Type VI (TN)	TiN—78–100%, Al_2O_3—0–5%, MgO—0–5%, CaO—0–5%, CaS—0–13%.	0.7–4.3	A4–A6, B4–B6

[1] A—Al_2O_3; M—MgO; Si—SiO_2; C—CaO; CS—CaS; TN—TiN.

Type II inclusions (ASi) can have spherical or irregular shapes and sizes in the range from 0.7 up to 5.6 μm. They contain mostly 21–94% of SiO_2 and 5–78% of Al_2O_3. It should be pointed out that these inclusions were only observed in samples B2 from Heat B.

Type III inclusions (CAM + CS) are spherical or irregular inclusions containing a complex oxide core (9–69% CaO, 2–54% Al_2O_3, 0–22% MgO and 0–7% SiO_2) and an outer layer consisting of CaS (9–62%). The size of these NMI varied from 0.9 to 5.2 μm. Also, this type of NMI was observed in all steel samples taken after a Ca addition as well as in the final product of both heats (samples A4–A6 and B4–B6).

Type IV inclusions (CS) also contain an oxide core and a larger outer layer consisting of CaS. The content of CaS in these inclusions during the EDS analysis could reach values up to between 97 to 100%. In addition, most of these inclusions have spherical or slightly elongated shapes. However, the NMI of Type IV were observed only in sample A6 from the final product of Heat A. The size of these inclusions was in the range from 2.5 to 7.4 μm. Moreover, it should be noted that the average value of aspect ratio for this type inclusions ($AR = L/N = 2.2$) is significantly larger than that for Types III, V, and VI (1.1–1.4).

Type V (CAM + CS + TN) and Type VI (TN) complex inclusions contain an oxide core and an outer precipitation of CaS and TiN. Therefore, these NMI have spherical/irregular or regular (cubic) shapes. The outer layer can contain 1–53% TiN and 0–50% CaS in the Type V inclusions and 78–100% TiN and only 0–13% CaS in the Type VI inclusions. These inclusions were observed to different extents in the steel samples A4–A6 and B4–B6 in both heats. The sizes varied from 1.1 to 5.0 μm for Type V inclusions and from 0.7 to 4.3 μm for Type VI inclusions.

The frequencies of NMI (in percentages) in various steel samples from both heats are shown in Figure 5 and sorted based on the type of inclusions and the stage of the steel production where the samples have been taken. It can be seen that the frequency of Type III inclusions (CAM + CS) decreases significantly from 98% (A4) to 36% (A6) in Heat A and from 89% (B4) to 20% (B6) in Heat B. The amount of Type IV inclusions (CS) in the final product of Heat A (sample A6) is only ~4%. However, the average size of these CaS inclusions (~5.3 μm) is considerable larger than the sizes for the other NMI. On the other hand, the frequency of NMI containing TiN (Type V and Type VI) increases from 2% (A4) to 59% (A6) in Heat A and from 11% (B4) to 80% (B6) in Heat B. This tendency may be explained by the precipitation of large amounts of small TiN, due to the cooling and solidification of the steel melt during sampling as well as during continuous casting.

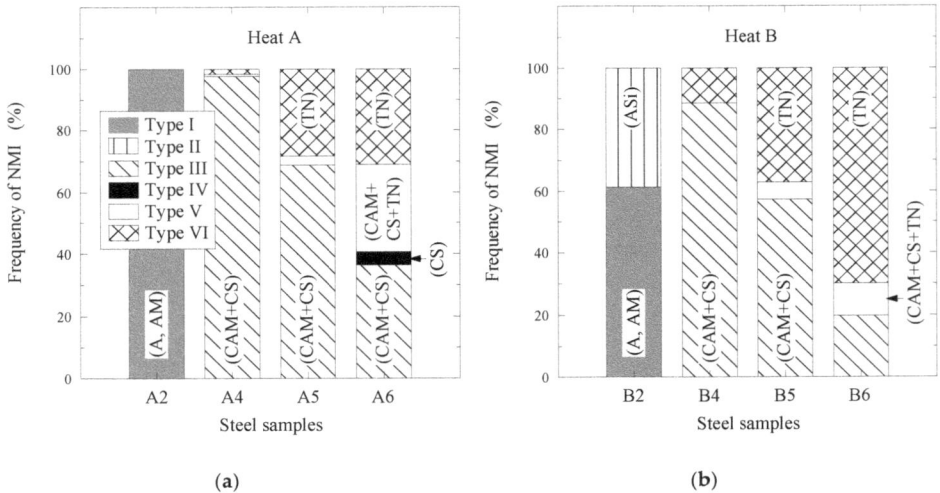

Figure 5. Different types of inclusions observed in steel samples: (**a**) Heat A; (**b**) Heat B.

It should be pointed out that the characteristics of TiN inclusions (size, number, concentration and morphology) precipitated during solidification of the steel melt in samples A4, A5 and B4, B5 can be significantly different compared to those in the final product after the completed hot rolling operation (samples A6 and B6). Therefore, the particle size distributions for the main inclusions (Types III, V and VI) as well as for all observed inclusions are compared in Figure 6 for samples A6 and B6 from both heats. It is apparent that the number and size of inclusions of Type III (CAM + CS) and Type V (CAM + CS + TN) inclusions are significantly smaller in Heat B compared to in Heat A, as shown in Figure 6a,b. However, the number of small size TiN inclusions (Type VI) in sample B6 is significantly larger compared to in sample A6 (Figure 6c), due to the higher content of N in steel in Heat B compared to Heat A. As a result, the total number of all inclusions is larger while the size of inclusions is smaller in steel sample B6 than those in sample A6.

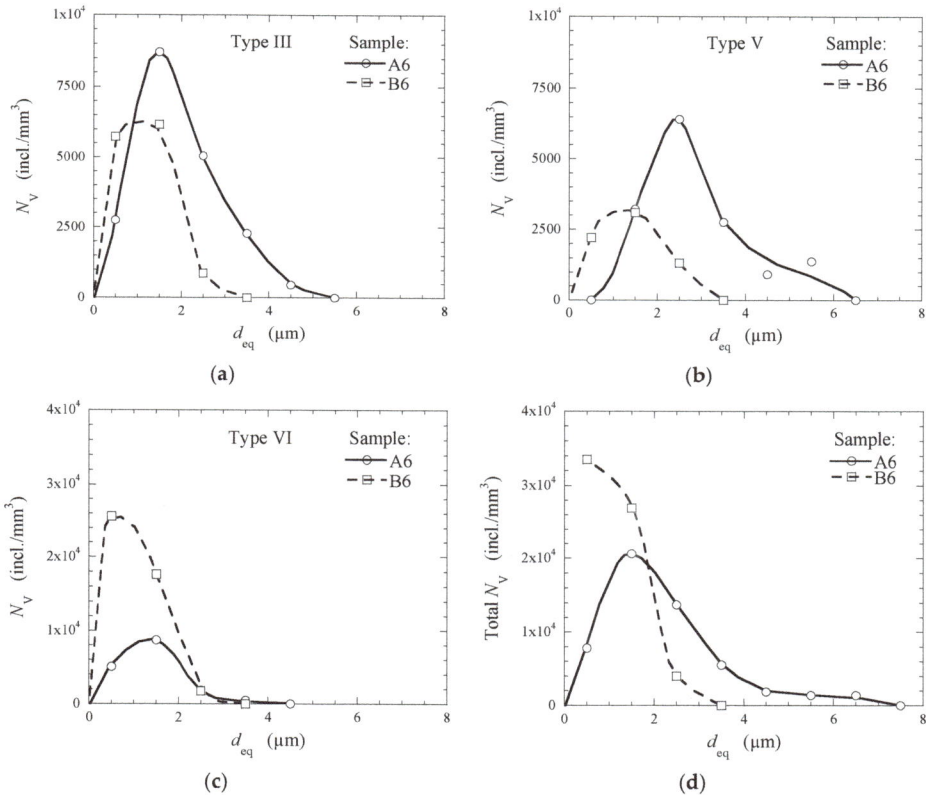

Figure 6. Particle size distributions for different non-metallic inclusions observed in the final products in Heat A and Heat B (samples A6 and B6): (a) Type III inclusions; (b) Type V inclusions; (c) Type VI inclusions; (d) all inclusions.

The modification of the NMI compositions in different steel samples is shown in the three phase CaO-Al$_2$O$_3$-CaS diagram in Figure 7 for both heats. It is apparent that the initial oxide inclusions of Type I and Type II (grey zones in diagrams) contained Al$_2$O$_3$, MgO and SiO$_2$ are modified in both heats after a Ca addition. As a result, the contents of CaO in the oxide core and CaS in the outer layer increases significantly. For instance, the Type III oxide inclusions in samples A4 and B4 (blue zones) can contain 40 to 90% of CaO. Then, the CaO content in Type III inclusions decreases, while the CaS content in these complex inclusions increases in steel samples A5, A6 and B5, B6. Especially, the CaS content increases up to the values between ~97 to 100% in outer layer of the Type IV inclusions in the A6 sample of Heat A (Figure 7a). However, most inclusions of Type III in the B6 sample of Heat B contain less than 45% of CaS (Figure 7b).

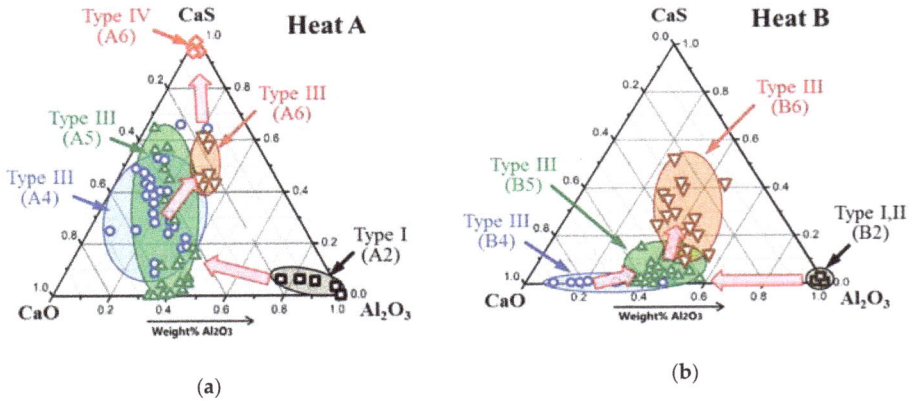

(a)

(b)

Figure 7. Modification of typical oxide and oxy-sulfide inclusions in steel samples: (**a**) Heat A; (**b**) Heat B.

3.3. Modification of Inclusions by Ca-Treatment

The modification of oxide inclusions by Ca-treatment of oil-pipeline steels in the ladle is shown in Figure 8.

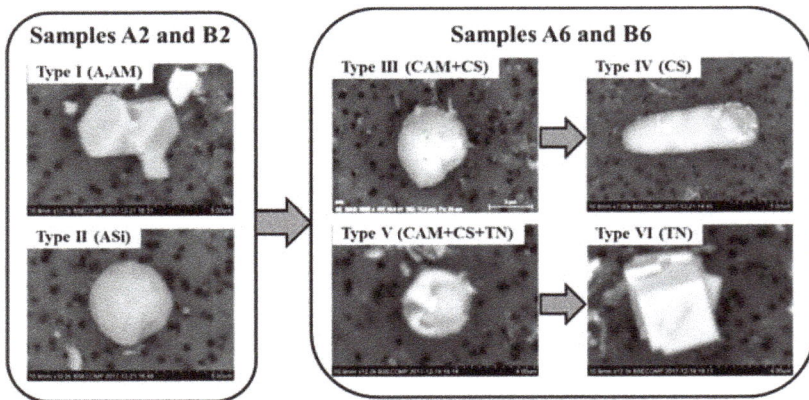

Figure 8. Modification of oxide inclusions by Ca-treatment of oil-pipeline steel.

It is apparent that the Ca-addition into the melt will firstly modify the Type I (Al_2O_3 and Al_2O_3-MgO) and Type II (Al_2O_3-SiO_2) oxide inclusions present in samples A2 and B2 into liquid or semi-liquid CaO-Al_2O_3-MgO inclusions of Types III and V, which are present in steel samples A4–A6 and B4–B6. Some amounts of CaS can also precipitate on the surface of these oxide inclusions depending on the concentration of S and Ca in the melt. A presence of the TiN phase on the surface of some Type V inclusions in samples A4, B4, A5, and B5 can be explained by the heterogeneous precipitation of TiN during the fast cooling and solidification of the melt in the samplers. Also, a slow solidification of the melt during the following continuous casting of steel provides larger segregations of S and N in some zones of the melt. As a result, the A6 and B6 steel samples taken from the final product contained Type IV (CS) and Type VI (TN) inclusions consisting of large outer layers of CaS and TiN, respectively. It should be noted that the Type VI inclusions were observed in both steel samples A6 and B6. However, the Type IV inclusions were only observed in the A6 sample, which had high S contents.

3.4. Evaluation of Corrosion Active Non-Metallic Inclusions

The corrosion resistance of two investigated hot rolled steels (Heats A and B) were evaluated and discussed based on the characteristics of the non-metallic inclusions observed in these steels. The numbers of corrosion active NMI per unit area (N_A) in the given steels are given in Table 3. It can be seen that a larger number of corrosion-active inclusions having diameters larger 5 µm were detected in steel samples from Heat A than from Heat B.

Table 3. Quantitative analysis of corrosion active non-metallic inclusions in hot rolled steels. NMI: non-metallic inclusions.

Steel	N_A of Corrosion Active NMI (incl./mm^2)
A	8.5
B	3.5

The results also showed that the morphology and compositions of the observed corrosion active NMI in dark pits on surfaces of steels A and B correspond to the large size inclusions observed in steel samples A6 and B6. Some typical corrosion active non-metallic inclusions in hot rolled steels after etching are shown in Figure 9.

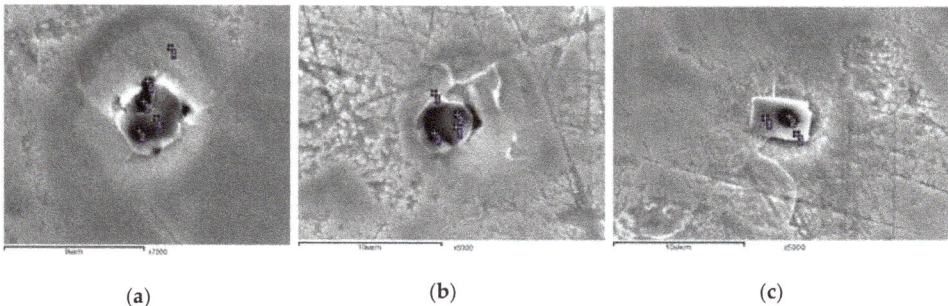

(a) (b) (c)

Figure 9. Typical corrosion active non-metallic inclusions observed after etching of hot rolled oil-pipeline steel samples: (**a**) Type III (CAM + CS), (**b**) Type V (CAM + CS + TN) and (**c**) Type VI (TN).

According to obtained results with respect to the characteristics of non-metallic inclusions and corrosion active inclusions in the investigated steels, it can be safely suggested that steel B will have a better corrosion resistance compared to steel A. The corrosion resistances of the given steels were evaluated according to the NACE Standards TM0284 ("Evaluation of Pipeline and Pressure Vessel Steels for Resistance to Hydrogen-Induced Cracking") and TM0177-2005 ("Laboratory Testing of Metals for Resistance to Sulfide Stress Cracking and Stress Corrosion Cracking in H$_2$S Environments"). The obtained results confirmed that steel B has a better corrosion resistance than steel A. However, detailed investigations of the effect of different non-metallic inclusions on the corrosion resistance of Ca treated hot rolled steels for oil-pipelines will be presented and discussed in a separate article.

4. Conclusions

The non-metallic inclusions were investigated in industrial steel samples taken from different stages of the steel production process and for two heats (Heat A and Heat B) of low carbon Ca-treated steel used for oil-pipelines. An electrolytic extraction (EE) technique was used for the extraction of NMI from the steel samples for the followed three-dimensional investigations of different inclusions and clusters by using SEM. Moreover, the number and compositions of corrosion active non-metallic inclusions were evaluated in hot rolled steel samples from both heats. The obtained results can be summarized as follows:

1. After Ca-treatment, the initial oxide NMI (Type I—Al_2O_3/Al_2O_3-MgO inclusions and clusters and Type II—Al_2O_3-SiO_2 inclusions) in steel samples A2 and B2 were modified to CaO-Al_2O_3-MgO inclusions. The oxide cores of Type III inclusions in steel samples taken after an addition of Ca (samples A4 and B4) contained about 40–90% of CaO in both heats. Then, the content of CaS significantly increased in the outer layer in Type III inclusions (up to 10–45% in the B6 sample and ~40–60% in the sample A6) and in the Type IV inclusions (up to 97–100% in A6 sample). Also, some amounts of TiN were found to precipitate in the outer layer of Type V (up to 53%) and Type VI (78–100%) inclusions during the solidification of steel.

2. The frequency of Type III oxy-sulfide inclusions in the steel samples after Ca-treatment decreases drastically from ~98% to 36% in Heat A and from ~89% to 20% in Heat B, while the frequency of Types V and VI inclusions containing TiN increased significantly up to values between 59% and 80% in the steel samples A6 and B6, respectively.

3. Although the total number of inclusions in the Heat B is larger than that in the Heat A, the average and maximum sizes of the observed inclusions in Heat B are significantly smaller than those in Heat A. Especially, the average size of the CaS inclusions of Type IV (~5.3 µm), which were only observed in the final product of Heat A, is significantly larger compared to the other types of NMI (1.1–2.8 µm) in both heats.

4. The final product of the Heat A has larger amounts (~2.4 times) of the discovered corrosion-active non-metallic inclusions in the size range of 5–7 µm compared to the level in Heat B. Also, most of the corrosion-active non-metallic inclusions correspond to large size inclusions of Types III, IV, V, and VI, which were observed in both heats.

Author Contributions: Conceptualization, A.V.K., and D.K.; methodology, E.S., and A.V.K.; investigation, E.S., and A.V.K.; resources, D.K., and P.G.J.; writing—original draft preparation, E.S., and A.V.K.; writing—review and editing, D.K., and P.G.J; supervision—D.K., and P.G.J.

Funding: This research received no external funding.

Conflicts of Interest: The authors declare no conflict of interest.

References

1. Carneiro, R.A.; Ratnapuli, R.C.; Lins, V.F.C. The Influence of Chemical Composition and Microstructure of API Linepipe Steels on Hydrogen Induced Cracking and Sulfide Stress Corrosion Cracking. *Mater. Sci. Eng. A* **2003**, *357*, 104–110. [CrossRef]

2. Xue, H.B.; Cheng, Y.F. Characterization of Inclusions of X80 Pipeline Steel and its Correlation with Hydrogen-induced Cracking. *Corros. Sci.* **2011**, *53*, 1201–1208. [CrossRef]

3. Yin, X.; Sun, Y.H.; Yang, Y.D.; Deng, X.X.; McLean, A. Effects of Non-Metallic Inclusions and Their Shape Modification on the Properties of Pipeline Steel. In Proceedings of the AISTech Conference Proceedings—7th International Conference on the Science and Technology of Ironmaking, Cleveland, OH, USA, 4–7 May 2015; PR-368-358-2015. pp. 3388–3406.

4. Tetelman, A.S.; Robertson, W.D. *The Mechanism of Hydrogen Embrittlement Observed in Iron-Silicon Single Crystals*; Ft. Belvoir Defense Technical Information Center: Fort Belvoir, VA, USA, 1961.

5. Oriani, R.A.; Josephic, P.H. Hydrogen-enhanced Load Relaxation in a Deformed Medium-carbon Steel. *Acta Metall.* **1979**, *27*, 997–1005. [CrossRef]

6. Herring, D.H. Hydrogen Embrittlement. Wire Forming Technology International. 2010. Available online: http://www.heat-treat-doctor.com/documents/Hydrogen%20Embrittlement.pdf (accessed on 17 December 2018).

7. Huang, F.; Liu, J.; Deng, Z.; Cheng, J.; Lu, Z.; Li, X. Effect of microstructure and inclusions on hydrogen induced cracking susceptibility and hydrogen trapping efficiency of X120 pipeline steel. *Mater. Sci. Eng. A* **2010**, *527*, 6997–7001. [CrossRef]

8. Kim, W.K.; Koh, S.U.; Yang, B.Y.; Kim, K.Y. Effect of Environmental and Metallurgical Factors on Hydrogen Induced Cracking of HSLA Steels. *Corros. Sci.* **2008**, *50*, 3336–3342. [CrossRef]

9. Moon, J.; Kim, S.J.; Lee, C. Role of Ca Treatment in Hydrogen Induced Cracking of Hot Rolled API Pipeline Steel in Acid Sour Media. *Met. Mater. Int.* **2013**, *19*, 45–48. [CrossRef]

10. Lube, I.I.; Pecheritsa, A.A.; Neklyudov, I.V.; Rodionova, I.G.; Zaitsev, A.I.; Marchenko, L.G.; Emel'yanov, A.V.; Stolyarov, V.I. Study of the effect of process parameters in steel production on the content of corrosion-active nonmetallic inclusions in corrosion-resistant pipes. *Metallurgist* **2005**, *49*, 269–275. [CrossRef]

11. Mitrofanov, A.V.; Petrova, M.V.; Kirillov, I.E.; Rodionova, I.G.; Udod, K.A.; Endel', N.I. Factors affecting housing and utilities pipe corrosion resistance. *Metallurgist* **2016**, *60*, 76–80. [CrossRef]

12. Karasev, A.V.; Suito, H. Analysis of Size Distributions of Primary Oxide Inclusions in Fe-10 mass pct Ni-M (M = Si, Ti, Al, Zr, and Ce) Alloy. *Metall. Mater. Trans. B* **1999**, *30*, 259–270. [CrossRef]

13. Ohta, H.; Suito, H. Characteristics of Particle Size Distribution of Deoxidation Products with Mg, Zr, Al, Ca, Si/Mn and Mg/Al in Fe-10mass% Ni Alloy. *ISIJ Int.* **2006**, *46*, 14–21. [CrossRef]

14. Doostmohammadi, H.; Karasev, A.; Jönsson, P.G. A Comparison of a Two-Dimensional and a Three-Dimensional Method for Inclusion Determinations in Tool Steel. *Steel Res. Int.* **2010**, *81*, 398–406. [CrossRef]

15. Kanbe, Y.; Karasev, A.; Todoroki, H.; Jönsson, P.G. Analysis of Largest Sulfide Inclusions in Low Carbon Steel by Using Statistics of Extreme Values. *Steel Res. Int.* **2011**, *82*, 313–322. [CrossRef]

16. Kanbe, Y.; Karasev, A.; Todoroki, H.; Jönsson, P.G. Application of Extreme Value Analysis for Two- and Three-Dimensional Determinations of the Largest Inclusion in Metal Samples. *ISIJ Int.* **2011**, *51*, 593–602. [CrossRef]

17. Karasev, A.; Jönsson, P.G. Assessment of Non-Metallic Inclusions in Different Industrial Steel Grades by Using the Electrolytic Extraction Method. In Proceedings of the 5th International Conference on Process Development in Iron and Steelmaking SCANMET-V, Luleå, Sweden, 12–15 June 2016; pp. 1–7.

![metals logo] *metals*

MDPI

Article

Characterisation of the Solidification of a Molten Steel Surface Using Infrared Thermography

Carl Slater [1,*] , Kateryna Hechu [2], Claire Davis [1] and Seetharaman Sridhar [3]

[1]   WMG, University of Warwick, CV7 4AL Coventry, UK; claire.davis@warwick.ac.uk
[2]   Tata Steel, Research & Development, 1970 CA IJmuiden, The Netherlands;
       Kateryna.Hechu@tatasteeleurope.com
[3]   George S. Ansell Department of Metallurgical and Materials Engineering, Colorado School of Mines, Golden,
       CO 80401, USA; sseetharaman@mines.edu
*   Correspondence: c.d.slater@warwick.ac.uk; Tel.: +442476151577

Received: 10 December 2018; Accepted: 23 January 2019; Published: 24 January 2019

Abstract: Infrared thermography provides an option for characterising surface reactions and their effects on the solidification of steel under different gas atmospheres. In this work, infrared thermography has been used during solidification of Twin Induced Plasticity (TWIP) steel in argon, carbon dioxide and nitrogen atmospheres using a confocal scanning laser microscope (CSLM). It was found that surface reactions resulted in a solid oxide film (in carbon dioxide) and decarburisation, along with surface graphite formation (in nitrogen). In both cases the emissivity and, hence, the cooling rate of the steel was affected in distinct ways. Differences in nucleation conditions (free surface in argon compared to surface oxide/graphite in carbon dioxide/nitrogen) as well as chemical composition changes (decarburisation) affected the liquidus and solidus temperatures, which were detected by thermal imaging from the thermal profile measured.

Keywords: liquid steel; non-contact measurement; oxides; steel-making

---

1. Introduction

With the ever-increasing demands on the steel industry to lower emissions and increase efficiency comes the equally demanding requirements for a higher quality product. This is leading to an increase in the number of sensors and monitoring devices used in the steelmaking process [1–3], whereby particular focus has been during the rolling process, where roll force, strip thickness, and even phase transformation after rolling (austenite to ferrite) can be monitored [4–8]. However, the conditions for higher temperature process monitoring are more challenging, for example, during casting or thermo-mechanical processing, but there is demand for real-time feedback to allow consistent and high quality steel to be produced. Some examples of where additional feedback related to steel surfaces would be useful for dynamic control include: the evolution of the solid shell during casting (morphology, phase, etc.), heat transfer during solidification and the rate/type of surface oxides forming in the ladle, which can be detrimental if entrained into the melt [9] or beneficial for lubrication during strip casting [10]. In addition, any ability to monitor scale formation during hot rolling of strip product would be beneficial, as the type, distribution and adhesion of the oxide scale influences the quality of the final product [11,12]. Identifying the type and distribution of molten and solid layers of slag on the steel surface in the ladle would be useful in controlling the steelmaking process, for example, if the slag location is detected correctly, the slag skimming process can be conducted with minimal steel losses, before pouring the molten metal into the tundish prior to casting [13]. Identifying if precipitation of solid phases occurs in the slag layer is also pertinent for controlling heat transfer during continuous casting [14] and for potential recovery of valuable metals from steelmaking slags [15].

This paper explores the potential for using infrared thermography for surface characterisation during solidification under different gas atmospheres, to detect and differentiate changes resulting from reactions (formation of surface films) and/or phase changes (solidification) due to the differences in emissivity and latent heat. In order to induce physical changes, resulting from phase transitions and chemical reactions, on the liquid steel surface and during solidification, controlled changes in temperatures and/or gas atmosphere (argon (Ar), nitrogen (N_2) or carbon dioxide (CO_2)) were used.

2. Experimental

High temperature confocal scanning laser microscopy (CSLM) has been increasingly utilised for metallographic investigation since it was pioneered by Yin et al. [16]. In more recent years its usage for steel processing, particularly in relation to oxides and slags, has gained momentum [17,18]. The work reported in this paper applied the technique, developed by Slater et al. [19], of using combined infrared thermography with CSLM. The CSLM used was a Yonekura VL2000DX-SVF17SP (Yonekura, Yokohama, Japan) with a He-Ne laser as the imaging source. Due to the wavelength of the halogen heat source in the CSLM, a TM160 (Micro Epsilon, Birkenhead, UK) 7–14 µm infrared thermographer was used to reduce the amount of background noise.

Prior to testing, the CSLM was calibrated to ensure the melting temperatures recorded were accurate using standard size (2 mm × 2 mm × 2 mm) samples from four pure metals (Al, Au, Ni and Fe). Before the test, the CSLM was evacuated and backfilled with N6 argon three times, to ensure an initial inert atmosphere in order to avoid any surface reactions. A typical thermography image of a hot sample in the heated crucible can be seen in Figure 1. Here, due to the curvature of the droplet surface, only the very top could be seen clearly without scaling. An R type thermocouple was situated on the underside of the crucible; this temperature reading was used to provide calibration for the thermographer. During testing, radiant heat values were determined from the crucible and from the sample. Radiant heat was used due to the uncertainty and variability in emissivity values. By using an emissivity value, and its corresponding apparent temperature, then this reduces the uncertainty in emissivity, because should an emissivity be used that is slightly off, then the apparent temperature would change, still providing the same radiant heat. The radiant heat (Q) can be calculated from the following equation:

$$Q = A\varepsilon\sigma T^4 \tag{1}$$

where σ is the Stefan–Boltzmann constant, T is the apparent absolute temperature given by the thermographer, A is the pixel area of the image and ε is the emissivity.

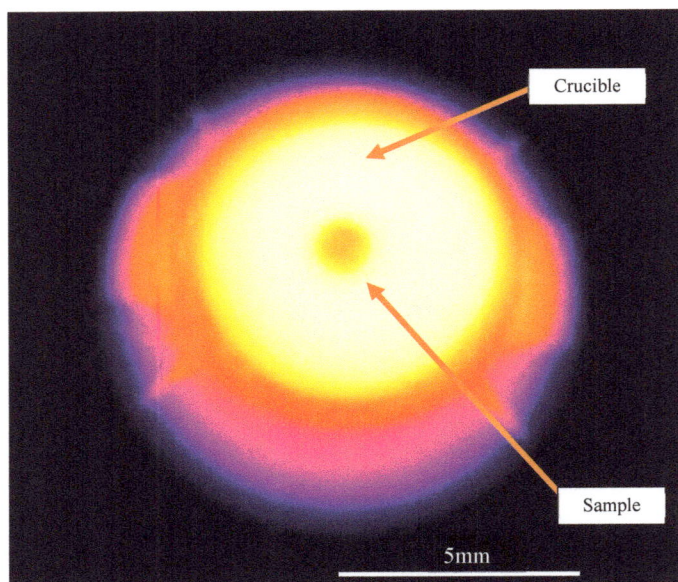

Figure 1. Typical thermography image taken from the confocal scanning laser microscope.

Three trials were carried out, all consisting of the same thermal schedule. A sample (2 mm ×
2 mm × 2 mm) of TWIP steel (15% Mn 1.5% Al, 3% Si, 0.7% C) was placed inside an alumina crucible.
The physical properties of the steel and crucible are given in Table 1, obtained using ThermoCalc
(version 2018b, Stockholm, Sweden). The specimen was heated to 200 °C for 2 min, to dry the samples,
before heating to 1500 °C at 10 °C/s and holding for 1 min. The heating was then turned off to allow
natural cooling of the sample. All the trials were conducted using argon during the initial heating
stage. However, during the dwell at 1500 °C, the atmosphere was either maintained as argon (Ar) or
switched to nitrogen (N_2) or carbon dioxide (CO_2) at a rate of 200 mL/min.

Table 1. Summary of expected properties of the materials used in this study.

Liquidus (°C)	Solidus (°C)	Emissivity of Solid Steel	Emissivity of Liquid Steel	Emissivity of Alumina
1380	1305	0.6	0.3	0.6–0.7

Once the CSLM reached 1500 °C, the crucible emissivity on the infrared thermographer was
calibrated to the CSLM thermocouple (the crucible calibration was taken 500 μm away from the steel
droplet). During dwell periods the temperature in the CSLM was much more homogenous and;
therefore, best suited for setting the initial conditions. The infrared thermographer reading during
cooling for all tests can be seen in Figure 2. Small differences near 1000 °C were due to small variations
in the mass of the sample, as well as the difference in cooling capacity of the different gases. However,
during the range of solidification (1380–1200 °C) very good repeatability can be seen. As the samples'
surfaces changed emissivity, both during solidification and in the presence of a surface product,
the stated temperatures in this paper refer to this calibrated point, and was, hence, why the distance
away from the sample was limited to 500 μm. This method has shown very good agreement previously,
where the impact of any thermal lag was shown to be negligible using differential calorimetry [19,20].
Therefore, the readings taken from the surface act to provide information on when events occur,
rather than providing a direct temperature reading itself.

In all cases readings were taken from a single pixel of size 10 μm × 10 μm. The sample reading was taken from the apex of the liquid droplet.

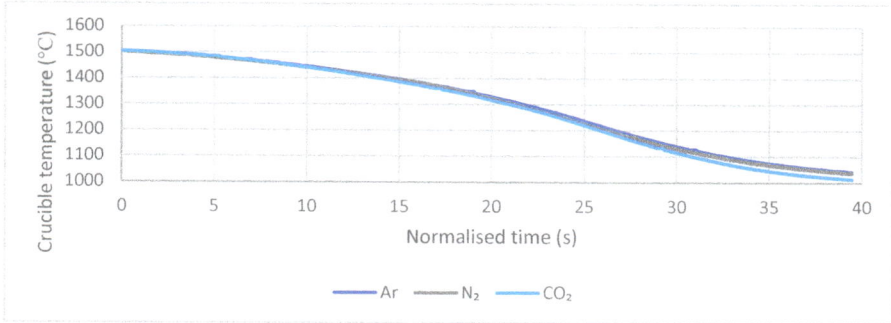

Figure 2. Temperature curves of the crucible during each test to ensure consistent conditions for each test. Normalised time refers to 0 s, being when the CLSM power was switched off.

3. Results and Discussion

In all three cases the radiant heat of the sample during the dwell at 1500 °C was 0.0033 W (±0.0002), an emissivity of around 0.08 was measured during the dwell. This was lower than seen in literature; however, the quality of vacuum in the CSLM was very high and likely to have less oxidation than a conventional exposed liquid steel surface.

The first condition observed was that of a continual argon atmosphere, which can be seen in Figure 3. Once the power of the heating was turned off (at about 5 s), a gradual decrease in the radiated heat could be seen due to the cooling of the sample. After around 20 s, the sample emittance dropped quickly (convection ripples on the liquid surface due to the formation of solid subsurface) before an increase could be seen, which was attributed to both the latent heat of fusion as well as the higher emissivity that the solid surface exhibited in comparison to the melt. After solidification, the slope of the radiated heat was steeper than prior to solidification, suggesting the higher emissivity was resulting in a greater cooling rate of the sample. As the thermal mass of the sample was much greater than that of the crucible (600 mg compared to 200 mg), then the cooling of the system was much more heavily dictated by the steel. Figure 4 shows CSLM images taken from the sample at specific time intervals from Figure 3, confirming that the sample was partially solidified at 23 s and full solidification had occurred at 30 s.

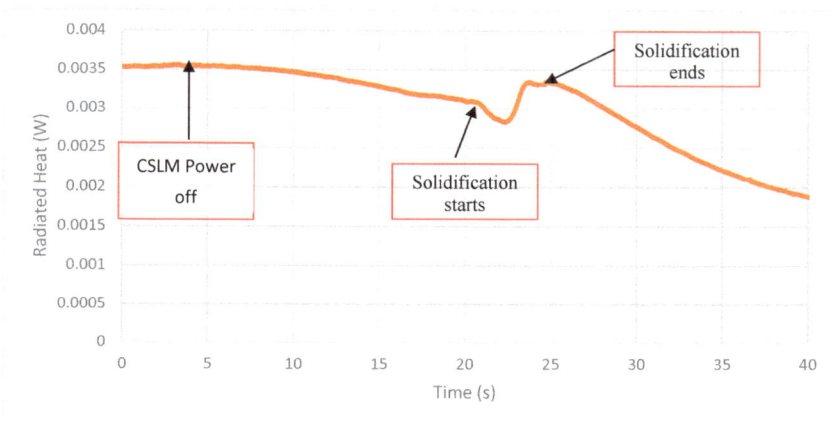

Figure 3. Radiated heat curve for the surface of the steel sample that was solidified in Ar.

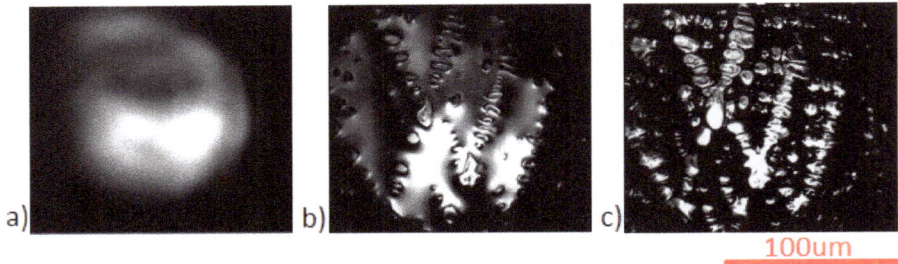

Figure 4. Images from the CSLM during cooling in Ar, taken at (**a**) 5 s (liquid steel), (**b**) 23 s (partial solidification) and (**c**) 30 s (fully solidified) according to Figure 3.

The second trial used a CO_2 atmosphere during the cooling of the steel (Figure 5). For this sample an initial fluctuation in the radiant heat could be seen in the sample as soon as the gas was changed (at approximately 5 s), this was followed by a gradual increase in radiated heat due to the formation of an oxide layer that had a higher emissivity. Turning the heating power off was followed by a gradual decrease in radiated heat as the sample cooled. A change in gradient was observed at approximately 33 s, when solidification started, but the effect of solidification on the curve was much smaller than observed during solidification under the argon atmosphere, Figure 3. This reduced signal from solidification was to be expected, as the increase in radiated heat was now only attributed to latent heat and not to a change in the emissivity of the surface, as there was already a solid oxide film present. Figure 6 shows the CSLM images taken from the CO_2 sample. Prior to the CO_2 introduction, the sample showed the reflective surface of the liquid steel droplet; however, once the CO_2 gas was introduced, then oxide formed, as can be seen in Figure 6b. It was seen from the continuous CLSM imaging that a complete oxide film formed over the surface within 0.5 s from the CO_2 gas being introduced. Therefore, the gradual increase in radiant heat, expected from an increase in emissivity, from the moment the CO_2 gas was switched on suggests that the thickness of the oxide was increasing with time.

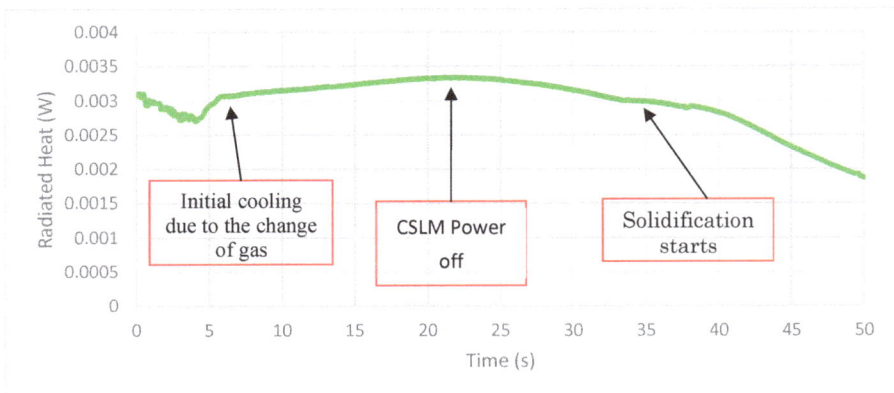

Figure 5. Radiated heat with time of the steel surface during cooling in a CO_2 atmosphere.

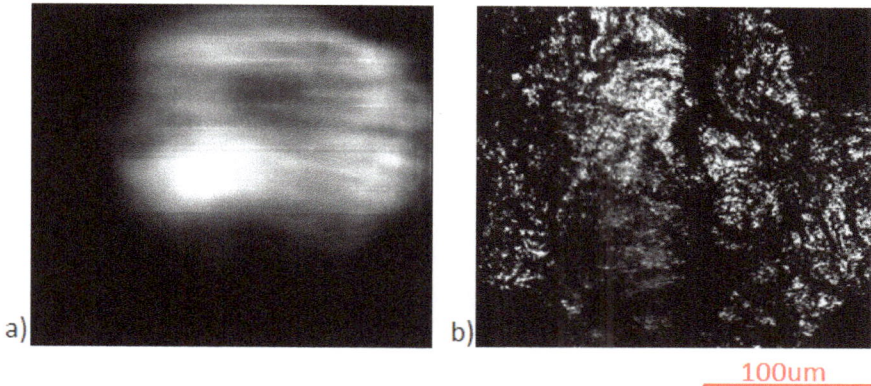

Figure 6. CSLM images taken from the sample cooled in a CO_2 atmosphere, taken at (**a**) 0 s (liquid steel) and (**b**) 25 s (oxide film) according to Figure 5.

The final sample was tested under a nitrogen atmosphere. Nitrogen has previously been seen to cause and stabilise the formation of graphite on the surface during decarburisation of the bulk steel [21]. It was suggested by Slater et al. [21] that this mechanism is a result of nitrogen reducing the solubility of carbon in the liquid metal, thus stabilizing graphite. Once graphite has formed, nitrogen can then further react with the graphite to produce a combination of C2N2 (cyanogen) and XCN (variable cyanides). This more dynamic reaction with the surface, compared to argon, is likely to impact the heat flux of the surface.

Figure 7 shows the radiant heat curve for this sample. It can be seen that there was a significant increase in emitted heat during the initial stage of gas interaction compared to the other samples, with the radiant heat increasing by more than $10\times$ that of the sample held in CO_2 during the same period. This was due to the formation of graphite on the surface, shown in Figure 8, which has a higher emissivity than the liquid steel and of an oxide, approaching that of a black body [1]. Incomplete coverage of the surface was seen, where the amount of graphite can be seen to increase from 0 to 30 s. Due to the scale of the graphite, the radiated heat in this case refers to the average within a pixel (approximately 100 μm^2); therefore, as surface coverage increased, the overall emissivity of a pixel increased also, thus giving feedback into the rate of reaction on the surface. At 30 s the reaction rate had reduced considerably (little change in graphite coverage of the sample), this may be due to importance of the Fe–N interaction in reducing the solubility of carbon, and thus could not reach a point where the whole surface was coated in graphite. Once the power of the CSLM was switched off, a large decrease in radiant heat was observed as the sample cooled quickly, and solidification could also be seen (at approximately 45 s), although the change in gradient was not as pronounced as for the sample in an argon atmosphere.

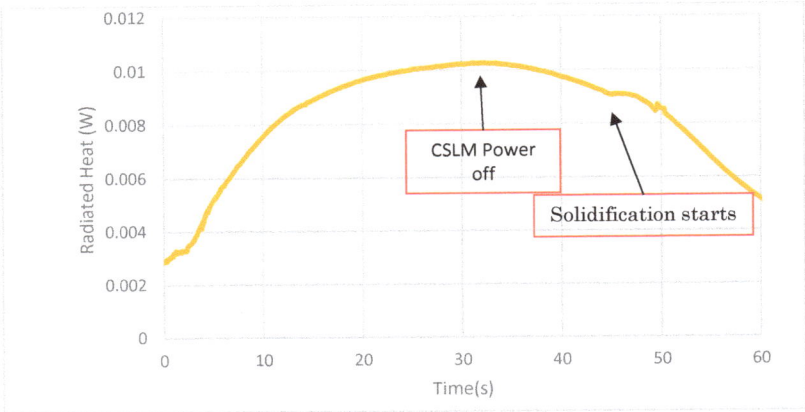

Figure 7. Radiant heat with time taken from the surface of a sample solidified in a N_2 atmosphere.

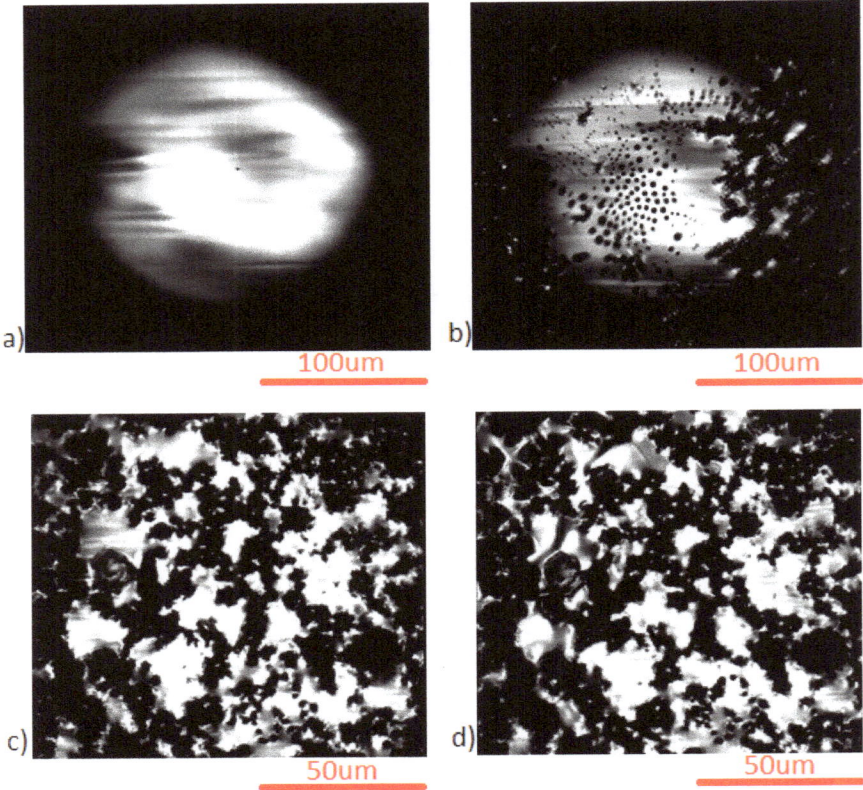

Figure 8. CSLM images taken during the solidification of steel under a N_2 atmosphere at (**a**) 0 s (liquid steel), (**b**) 10 s (graphite flake on steel surface), (**c**) 30 s (increase in graphite flakes) and (**d**) 45 s (start of solidification–dendrite interfaces arrowed).

Table 2 shows a summary of the solidus and liquidus temperatures of the steel solidified, under natural cooling, in the CLSM, in the three different atmospheres. It can be seen that the cooling under an argon atmosphere gave a liquidus much lower than that of predicted equilibrium solidification, which was expect as nucleation was limited from the free surface (the high purity of the atmosphere means that no inclusions or surface oxides were observed) and; therefore, a large undercooling was required locally before solidification was seen. This is consistent with previous work on the CSLM, where the smooth crucible surface and small sample size can reduce the chance of a nucleation event significantly [20]. Under a CO_2 atmosphere, the newly formed surface oxide can act as nucleation sites, and this minimises the undercooling required to a point where the liquidus matches very closely to the equilibrium predicted value (it is also possible some carbon was removed during oxidation, which would increase the liquidus; this is expected to be a minor influence as oxygen is predicted to more favourably bond with Al and Si, compared to forming CO (Factsage 7.2, Aachen, Germany)). Finally, the nitrogen atmosphere significantly increased both the liquidus and solidus temperatures, which can only be achieved through a dynamic change in the composition. As graphite forms on the surface, this decarburises the steel and thus increases the liquidus and solidus temperatures. If 0.5% carbon was removed from the bulk then the liquidus and solidus would have been increased to 1412 °C and 1355 °C, respectively. This level of decarburisation has been shown previously through a nitrogen atmosphere in a CSLM [21].

Table 2. Summary of the solidus and liquidus temperatures of the steel, determined from thermography in the three different atmospheres, and the predicted (from Thermo-Calc) values.

Equilibrium	Liquidus (°C)	Solidus (°C)
	1380	1305
Ar	1323	1279
CO_2	1373	1283
N_2	1402	1366

4. Conclusions

The work presented has highlighted that infrared thermography can be used to obtain information about surface reactions and solidification temperatures even when the liquid steel cannot be directly imaged. Specifically, TWIP steel samples were solidified under different gas (Ar, CO_2, N_2) atmospheres in a confocal laser scanning microscope, and infrared thermography was used to monitor the sample surface.

Different surface conditions were observed (simple liquid steel in Ar; oxidation in CO_2; decarburisation and graphite flake formation in N_2), which affected the emissivity and, hence, the radiated heat detected by thermography. The different surface reactions could be characterised by their different radiative properties.

Solidification of the steel could be detected from changes in the radiated heat signatures of the sample, even when a surface film (oxide or graphite) was present. Differences in the liquidus and solidus temperatures were related to changes in local chemistry (decarburisation in N_2) and nucleation (oxide and graphite) conditions.

The work has indicated that infrared thermography may be a suitable non-contact measurement method to provide information in areas such as the tundish in steel making (due to any emissivity changes of the slag material as it reacts with the steel) or the hot strip surface, as steel comes from the mould in a continuous caster in defining blowouts etc.

Author Contributions: C.S. and S.S. conceived and designed the experiments; C.S. performed the experiments; C.S., K.H., C.D. and S.S. analyzed the data; C.S. wrote the paper.

Funding: This research was funded by EPSRC, grant number EP/M014002/1.

Acknowledgments: The authors would like to thank EPSRC for funding and also WMG for their support and facilities.

Conflicts of Interest: The authors declare no conflict of interest.

References

1. Udayraj; Chakraborty, S.; Ganguly, S.; Chacko, E.Z.; Ajmani, S.K.; Talukdar, P. Estimation of surface heat flux in continuous casting mould with limited measurement of temperature. *Int. J. Therm. Sci.* **2017**, *118*, 435–447. [CrossRef]
2. Salah, B.; Zoheir, M.; Slimane, Z.; Jurgen, B. Inferential sensor-based adaptive principal components analysis of mould bath level for breakout defect detection and evaluation in continuous casting. *Appl. Soft Comput.* **2015**, *34*, 120–128. [CrossRef]
3. Takács, G.; Ondrejkovič, K.; Hulkó, G. A low-cost non-invasive slag detection system for continuous casting. *IFAC-PapersOnLine* **2017**, *50*, 438–445. [CrossRef]
4. Proxitron. Sensors for Steel Works Material Tracking in Steel and Rolling Mills Tube Manufacturing Sensors for Steel Works. Available online: http://www.elegance.co.za/download-catalogue/sensors-steel-works.pdf (accessed on 1 September 2018).
5. Molleda, J.; Usamentiaga, R.; García, D.F. On-line flatness measurement in the steelmaking industry. *Sensors* **2013**, *13*, 10245–10272. [CrossRef] [PubMed]
6. Zhou, L.; Liu, J.; Hao, X.J.; Strangwood, M.; Peyton, A.J.; Davis, C.L. Quantification of the phase fraction in steel using an electromagnetic sensor. *NDT E Int.* **2014**, *67*, 31–35. [CrossRef]
7. Legrand, N.; Labbe, N.; Weisz-Patrault, D.; Ehrlacher, A.; Luks, T.; Horský, J. Analysis of Roll Gap Heat Transfers in Hot Steel Strip Rolling through Roll Temperature Sensors and Heat Transfer Models. *Key Eng. Mater.* **2012**, *504–506*, 1043–1048. [CrossRef]
8. Legrand, N.; Lavalard, T.; Martins, A. New concept of friction sensor for strip rolling: Theoretical analysis. *Wear* **2012**, *286–287*, 8–18. [CrossRef]
9. Campbell, J.; Campbell, J. Casting Alloys. *Complet. Cast. Handb.* **2015**, 223–340.
10. Ebrill, N.; Durandet, Y.; Strezov, L. Influence of substrate oxidation on dynamic wetting, interfacial resistance and surface appearance of hot dip coatings. *Galvatech.* **2001**, 329–336.
11. Nilsonthi, T. Oxidation behaviour of hot-rolled Si-containing steel in water vapour between 600 and 900 °C. *Mater. Today Proc.* **2018**, *5*, 9552–9559. [CrossRef]
12. Na Kalasin, N.; Yenchum, S.; Nilsonthi, T. Adhesion behaviour of scales on hot-rolled steel strips produced from continuous casting slabs. *Mater. Today Proc.* **2018**, *5*, 9359–9367. [CrossRef]
13. ASM International. *ASM handbook-Casting. Materials Park*; ASM: Geauga County, OH, USA, 2008.
14. Zhang, Z.T.; Wen, G.H.; Liao, J.L.; Sridhar, S. Observations of Crystallization in Mold Slags with Varying Al_2O_3/SiO_2 Ratio. *Steel Res. Int.* **2010**, *81*, 516–528. [CrossRef]
15. Semykina, A.; Nakano, J.; Sridhar, S.; Shatokha, V.; Seetharaman, S. Confocal Scanning Laser Microscopy Studies of Crystal Growth During Oxidation of a Liquid FeO-CaO-SiO_2 Slag. *Metall. Mater. Trans. B* **2011**, *42*, 471–476. [CrossRef]
16. Yin, H.; Shibata, H.; Emi, H.; Suzuki, M. In-situ Observation of Collision, Agglomeration and Cluster Formation of Alumina Inclusion Particles on Steel Melts. *ISIJ Int.* **1997**, *37*, 936–945. [CrossRef]
17. Wang, Y.; Sridhar, S.; Valdez, M. Formation of CaS on Al_2O_3-CaO inclusions during solidification of steels. *Metall. Mater. Trans. B* **2002**, *33*, 625–632. [CrossRef]
18. Yang, L.; Webler, B.A.; Cheng, G. Precipitation behavior of titanium nitride on a primary inclusion particle during solidification of bearing steel. *J. Iron Steel Res. Int.* **2017**, *24*, 685–690. [CrossRef]
19. Slater, C.; Hechu, K.; Sridhar, S. Characterisation of solidification using combined confocal scanning laser microscopy with infrared thermography. *Mater. Charact.* **2017**, *126*, 144–148. [CrossRef]
20. Hechu, K.; Slater, C.; Santillana, B.; Clark, S.; Sridhar, S. A novel approach for interpreting the solidification behaviour of peritectic steels by combining CSLM and DSC. *Mater. Charact.* **2017**, *133*, 25–32. [CrossRef]
21. Slater, C.; Spooner, S.; Davis, C.; Sridhar, S. Chemically Induced Solidification: A New Way to Produce Thin Solid-Near-Net Shapes. *Metall. Mater. Trans. B* **2016**, *47*, 3221–3224. [CrossRef]

![metals logo] *metals*

MDPI

Article

Measurement of Molten Steel Velocity near the Surface and Modeling for Transient Fluid Flow in the Continuous Casting Mold

Tao Zhang [ID], Jian Yang * and Peng Jiang

State Key Laboratory of Advanced Special Steel, Shanghai University, Shanghai 200444, China;
zhang_tao@shu.edu.cn (T.Z.); jiangp@i.shu.edu.cn (P.J.)
*   Correspondence: yang_jian@t.shu.edu.cn; Tel.: + 86-21-6613-6580

Received: 11 December 2018; Accepted: 29 December 2018; Published: 4 January 2019

Abstract: In the current work, a rod deflection method (RDM) is conducted to measure the velocity of molten steel near the surface in continuous casting (CC) mold. With the experimental measurement, the flow velocity and direction of molten steel can be obtained. In addition, a mathematical model combining the computational fluid dynamics (CFD) and discrete phase method (DPM) has been developed to calculate the transient flow field in a CC mold. The simulation results are compared and validated with the plant measurement results. Reasonable agreements between the measured and simulated results are obtained, both in the trends and magnitudes for the flow velocities of molten steel near the mold surface. Based on the measured and calculated results, the velocity of molten steel near the surface in the mold increases with increasing casting speed and the casting speed can change the flow pattern in the mold. Furthermore, three different types of flow patterns of molten steel in the mold can be obtained. The pattern A is the single-roll-flow (SRF) and the pattern C is the double-roll-flow (DRF). The pattern B is a transition state between DRF and SRF, which is neither cause the vortices nor excessive surface velocity on the meniscus, so the slag entrainment rarely occurs. Argon gas injection can slow down the molten steel velocity and uplift the jet zone, due to the buoyancy of bubbles. Combination of the measurement and numerical simulation is an effective tool to investigate the transient flow behavior in the CC mold and optimize the actual operation parameters of continuous casting to avoid the surface defects of the automobile outer panels.

Keywords: flow velocity; casting speed; gas flow rate; flow pattern; continuous casting

---

1. Introduction

The surface defects on the cold rolled sheets for automobile outer panels related to the steelmaking process are mainly the large-sized Al_2O_3 clusters, "$Ar + Al_2O_3$" typed inclusions and the entrapped mold powder particles [1]. Generally, the non-metallic inclusions and the entrapped mold powder particles produced in the steelmaking and continuous casting (CC) process are firstly crushed down and then elongated along the rolling direction during the subsequent rolling process. Finally, they are discretely distributed along the rolling direction to form the line-shape defects, such as sliver, scab and blister.

Yasunaka H. [2] studied the surface defects in ULC steel and found that the defects were caused by the capture of argon bubbles whose diameter are 0.5–3 mm into the solidification structure 'hook.' Yang J. [3] researched the morphologies and chemical compositions for three kinds of surface defects on automobile outer panels and found that the stripe widths and lengths formed by three kinds of defects are different. Wang X.H. [4,5] studied the defects of IF steel in continuous casting slab and found that in the bottom slab of the cast, the number of inclusions from the mold powder entrapment is much more than that in the normal casting slabs. The number density of the large inclusions decreases with increasing casting speed when the casting speed is 1.0–2.0 m/min.

The formation of surface defects on cold rolled sheets is closely related to the flow behavior in the CC mold. Therefore, it is very important to investigate the flow field and flow pattern in the mold. The simulated calculation is an important means to research the flow behavior of molten steel in the CC mold. The transient flow field and the velocity near the surface in CC mold were investigated in many previous works [6–10]. The most popular numerical approach is the Reynolds time-averaged (RANS) model to predict steady-state and single-phase flow [11–14]. Liu C.L. [15] used a Eulerian-Lagrangian two-way coupled model to explore transient argon-steel system flow patterns in the mold. Wang Y.F. [16,17] investigated the complicated phenomena which are associated with transient stages during the continuous casting process using the Unsteady RANS model. They found that the acceleration rate for casting speed significantly affected the magnitude of velocity in the mold. In recent years, as the computer became more powerful than ever, the Large Eddy Simulation (LES) has been popular to investigate the transient flow and particles behavior in the CC mold [18–22].

The velocity of molten steel near the surface is one of the flow parameters which can be measured to predict the flow field and flow pattern in the CC mold. In many previous works, the water model experiments have been carried out to measure the flow velocity near the surface and investigate the flow field and flow pattern in the mold. Miki Y. [23] used a propeller-type velocity meter to measure the flow velocity at the 1/4 width position on the center plane of the thickness. The results showed that the fluctuations included a wide range of flow velocities as large as 0–0.6 m/s and the period of these fluctuations was estimated to be around 15 s based on the water model experiments. In Chaudhary R.'s study [24], the impeller velocity probes positioned below the top surface on both sides of the submerged entry nozzle (SEN) were used to investigate the horizontal velocity near the surface and instantaneous velocity data were collected at a sampling frequency of 1 Hz. The transient flow phenomena and the flow velocity near the surface in the water model of continuous casting process were also studied by use of the particle image velocity (PIV) technology [25–28].

However, the multi-phase flow behavior and related phenomena in the air-water system are quite different from that in the argon-molten steel system. Therefore, it is highly desirable to measure the velocity of molten steel directly and study the flow behavior of the molten steel in the CC mold. Up to now, the mainly reported method to measure the molten steel velocity near the mold surface is nail dipping method [29–31]. In this method, nails are inserted into the molten steel and a lump forms on the bottom of nails. Based on the lump shape and height difference between the two sides, the velocity near the surface and direction of flow can be measured in the mold. However, not only molten steel but also the molten layer of mold powder can form a lump on nails. The viscosity of the mold powder is about ten times higher than that of molten steel, which makes the flow behavior in the molten slag completely different from that in the molten steel. Therefore, the measurement error of molten steel velocity with the nail dipping method is obviously quite large. Several works about the dipped-in rod method were reported to measure the velocity near the surface [29,30,32,33]. But the information on how to build the relationship between the deflecting angle with the velocity near the surface and other technical details was not mentioned. As a consequence, it is of great significance to study an effective method to measure the velocity of molten steel near the surface in the CC mold.

In current work, a rod deflection method (RDM) is investigated to directly measure the flow velocity of molten steel near the surface in CC mold. In addition, a mathematical model combining the computational fluid dynamics (CFD) and discrete phase method (DPM) has been developed to calculate the transient flow behavior in the CC mold. The simulation results are compared and validated with the industrial experimental measurement results. Specifically, this work aims to investigate the effect of different casting speeds on the flow velocity of molten steel near the surface and the flow pattern in the CC mold from the numerical calculation and experimental measurements.

2. Experimental Apparatus and Method

Due to the high temperature up to about 1600 °C of molten steel, it is quite difficult to measure the flow velocity using the common devices and methods. Consequently, a speed measuring device and method is newly developed to measure the flow velocity of molten steel near the surface in an actual mold.

Figure 1 is the schematic diagram of the measurement method of molten steel velocity near the surface in the CC mold. The speed measuring device mainly consists of two parts: the deflecting part which can obtain the value of deflecting angle and the stainless steel detecting rod. By installing a counterweight as shown in Figure 2, the barycenter of the detecting device is adjusted to a position which is very close to the rotational pivot of the detecting rod. This design makes the detecting rod very sensitive to rotate and the accuracy of measurement can be significantly improved. Therefore, the flow direction of molten steel can be clearly known by the deflected direction and then the flow pattern in the mold can be obtained, which is helpful to control the flow behavior of molten steel in the mold. This method can be named the "Rod Deflecting Method" (RDM).

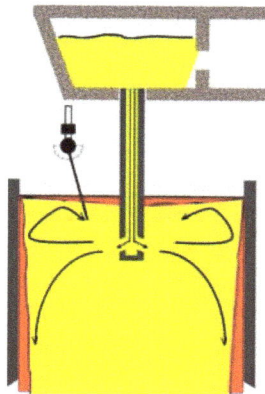

Figure 1. Schematic diagram of the measurement method of molten steel velocity near the surface in a mold.

Under each experimental condition, three detection rods were used to measure the velocity of molten steel near the surface. The detecting rod can stay in the molten steel for about 30 s, the data were recorded as many as possible. More than thirty measured velocity values were obtained for each operation condition, so the average value of these velocities could minimize measuring errors caused by various reasons, such as waves, level fluctuations and large argon bubbles.

Figure 2 is the analysis of the forces acted on a detecting rod. The detecting rod is dipped into the molten steel, while the other end is supported by a pivot where the rod can freely rotate with the molten steel flow in the mold. The detecting rod is subjected to three forces: gravity (G), buoyancy force (F_f) and the impact force (F_D). When the detecting rod leans to a certain angle (θ) and reaches a balanced state, the relationship of three forces can be described by the following equation:

$$GL_1 \sin\theta - F_f L_2 \sin\theta = F_D L_2 \cos\theta \qquad (1)$$

where G is the gravity of the deflecting rod, (N); F_f is the buoyancy force of the detecting rod immersion part, (N); F_D is the impact force acted on the detecting rod immersion part, (N); L_1 is the distance between the barycenter of detecting rod and the rotational pivot, (m); L_2 is the distance between the acting point of the impact force and the rotational pivot, (m); and θ is the rotational angle of the flow velocity detecting rod, (°).

Due to the impact force equivalent to the drag force on the detecting rod, the impact force can be expressed as the following equation:

$$F_D = C_D \frac{\rho U_0^2}{2} A \tag{2}$$

where C_D is the drag force coefficient which can be obtained from the relationship between the drag force coefficient and the Reynolds number (Re). U_0 is the velocity of the steel melt, (m·s^{-1}); ρ is the density of molten steel, (kg·m^{-3}); A is the projection area of the detecting rod immersion part in the vertical direction of the flowing steel melt, (m^2).

Equations (1) and (2) can be combined to give the following expression for calculating the velocity of the molten steel:

$$U_0 = \sqrt{\frac{2(GL_1 \tan\theta - F_f L_2 \tan\theta)}{L_2 C_D \rho A}} \tag{3}$$

As shown in Equation (3), G and L_1 are intrinsic parameters of the detecting rod, if the deflection angle (θ) and the immersion depth of the detecting rod (L_3) are measured, the velocity of molten steel can be obtained.

Figure 3 shows the relationship between the flow velocity of molten steel and deflection angle of detecting rod with different insertion depths. With the same insertion depth, the velocity of molten steel increases with increasing deflection angle. On the contrary, the velocity of molten steel decreases with increasing insertion depth for the same deflection angle. Thus all the experimental measurements are tried to be carried out with the same immersion depth in the present work.

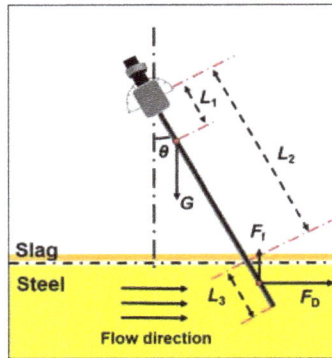

Figure 2. Analysis of the forces acted on a detecting rod.

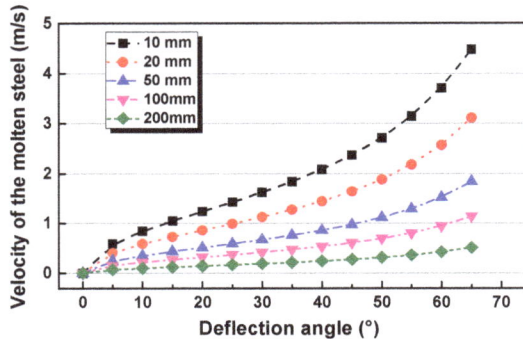

Figure 3. Relationship between flow velocity of molten steel and deflection angle of detecting rod with different insertion depths.

3. Numerical Simulation

A numerical model is developed to simulate the turbulence flow in the SEN and CC mold, using the standard k-ε turbulent model [34], coupled with the discrete phase model (DPM) [35]. To ease the complexity of the solution process, several reasonable assumptions are applied in the present work:

(1) The continuous phase is regarded as being an incompressible fluid;
(2) The effect of the slag layer on the flow behavior of molten steel is ignored;
(3) All bubbles are treated as the spherical shape and the effects of temperature and pressure on the bubble shape and size are ignored.
(4) The deformation of the solidified shell is ignored.
(5) The surface oscillations are not considered.

3.1. Fluid Flow Model

The continuity equation and momentum conservation for an incompressible fluid are given by the following equations:

$$\frac{\partial \rho}{\partial t} + \frac{\partial(\rho u_i)}{\partial x_i} = 0 \tag{4}$$

$$\frac{\partial}{\partial t}(\rho u_i) + \frac{\partial(\rho u_i u_j)}{\partial x_j} = -\frac{\partial p}{\partial x_i} + \frac{\partial}{\partial x_j}\left[(\mu_l + \mu_t)\left(\frac{\partial u_i}{\partial x_j} + \frac{\partial u_j}{\partial x_i}\right)\right] + \rho g_i + F \tag{5}$$

where ρ is the fluid-phase density; u is the fluid-phase average velocity; p is pressure; $(\mu_l + \mu_t)$ is the effective fluid-phase viscosity; F in Equation (5) is the source term for momentum exchange with the bubbles, representing the drag force, the lift force and virtual mass force respectively.

μ_t in Equation (5) is the turbulent viscosity, which is defined as

$$\mu_t = C_\mu \rho \frac{k^2}{\varepsilon} \tag{6}$$

The standard k-ε model is used to model turbulence, which means that the following transport equations of k and ε are solved.

$$\frac{\partial(\rho k)}{\partial t} + \frac{\partial(\rho u_i k)}{\partial x_i} = \frac{\partial}{\partial x_i}\left(\left(\mu_l + \frac{\mu_t}{\sigma_k}\right)\frac{\partial k}{\partial x_j}\right) + G_k - \rho \varepsilon \tag{7}$$

$$\frac{\partial(\rho \varepsilon)}{\partial t} + \frac{\partial(\rho u_i \varepsilon)}{\partial x_i} = \frac{\partial}{\partial x_i}\cdot\left(\left(\mu_l + \frac{\mu_t}{\sigma_\varepsilon}\right)\cdot\frac{\partial \varepsilon}{\partial x_j}\right) + C_1\frac{\varepsilon}{k}G_k - C_2\rho\frac{\varepsilon^2}{k} \tag{8}$$

$$G_k = \mu_t\left(\frac{\partial u_{i,j}}{\partial x_j} + \frac{\partial u_{i,j}}{\partial x_i}\right)\frac{\partial u_{i,j}}{\partial x_j} \tag{9}$$

The values of the constants are $C_\mu = 0.09$, $\sigma_k = 1.0$, $\sigma_\varepsilon = 1.3$, $C_1 = 1.44$, $C_2 = 1.92$.

3.2. Discrete Phase Model

In this study, the trajectories of argon gas bubbles can be simulated using an Euler-Lagrangian approach. The transport equation for bubbles are governed by Newton's second law and can be described as following:

$$m_b\frac{du_b}{dt} = F_d + F_g + F_f + F_x \tag{10}$$

The terms on the right-hand side of Equation (10) are drag force (F_d), gravitational force (F_g), buoyancy force (F_f) and other forces (F_x). F_x is an additional source term, mainly including virtual mass force (F_v), the pressure gradient force (F_p) and the lift force (F_L). All forces can be expressed as follows and details can be seen in previous work [9].

$$F_g = \frac{\pi d_b^3}{6} \rho_b g \tag{11}$$

$$F_f = \frac{\pi d_b^3}{6} \rho g \tag{12}$$

$$F_L = C_L \rho \frac{\pi}{6} d_b^3 (u_b - u)(\nabla \times u) \tag{13}$$

$$F_v = C_v \rho \frac{\pi}{6} d_b^3 \left(\frac{Du}{dt} - \frac{du_b}{dt} \right) \tag{14}$$

$$F_P = \rho_b \frac{\pi d_b^3}{6} \frac{Du}{Dt} \tag{15}$$

$$F_d = C_d \rho \frac{\pi}{8} d_b^2 (u_b - u)|u_b - u| \tag{16}$$

where u, u_b, ρ, ρ_b and d_b are the velocity of fluid, the velocity of bubbles, the density of molten steel, the density of bubbles and the diameter of bubbles, respectively. C_L, C_v and C_d are the coefficients of lift force, virtual mass force and drag force, respectively.

3.3. Simulation Details

Figure 4 shows the geometry (a) and grids (b) of the computational domain. In this paper, the three-dimensional calculation area is meshed by dividing some domains to reduce the false diffusion in the process of numerical simulation. Moreover, grid refinement is applied to the submerged entry nozzle (SEN) to improve the accuracy of the simulation.

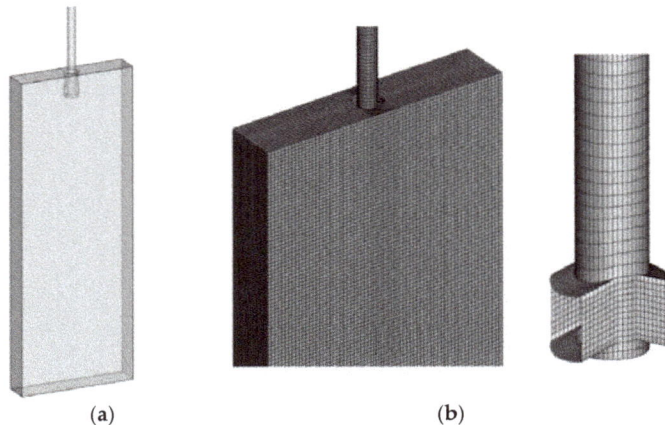

(a)   (b)

Figure 4. Geometry (a) and grids (b) of the computational domain.

The 3D turbulence fluid flow and the trajectories of gas bubbles in a CC mold are simulated. Gas bubbles are injected from the top of the SEN and follow the flow stream through the two nozzle outlet ports into the mold. The geometrical and process parameters are given in Table 1. The whole grid consists of about 570,000 cells. The time step for the simulation is 0.001 s.

Table 1. Geometrical and process parameters.

Parameters	Values	Parameters	Values
Slab width (mm)	1040, 1080, 1275	Fluid density (kg/m^3)	7020
Slab thickness (mm)	230	Gravity acceleration (m/s^2)	9.81
Slab length (mm)	3000	Gas density (kg/m^3)	0.27
SEN submerged depth (mm)	170	Gas bubble radius (mm)	0.5
Casting speed (m/min)	1.0, 1.3, 1.5, 1.7	Gas volume flow (l/min)	0, 4, 7
Fluid dynamic viscosity(N·s/m^2)	0.0056	SEN port angle (°)	20

For the molten steel, the inlet boundary condition is a constant velocity at the top of the SEN and the outlet condition is outflow at the bottom of the computational domain. The top surface of the mold is assumed to have a fixed and free-slip condition. And all the walls of the mold are assumed to be stationary and no-slip.

For argon gas bubbles, it is assumed that the bubbles remain spherical and their shape variations are neglected. Based on the previous works [36], the initial bubble radius is set to be 0.5 mm as a constant value for the convenience of calculation. An escape boundary condition is defined for the top surface and outlets of the mold and the trap boundary condition is defined for the walls of the mold where bubbles are predicted to be caught. However, the reflect boundary condition is defined for the walls of SEN where bubbles are predicted to return the computational domain.

4. Results and Discussion

4.1. The Comparison between the Experimental and Calculated Results

Figure 5 shows the distribution of molten steel velocities along the mold thickness direction under the conditions of (a) without and (b) with gas injection. From the central plane to the planes near the wall of the mold along the thickness direction, the velocities of molten steel gradually decrease. Compared the distribution of velocities with and without argon gas injection, it can be found that argon gas injection can slow down the molten steel velocity and uplift the jet zone. The central planes (Z = 0) in Figure 5 are chosen to make a detailed comparison of the different distances from SEN, as shown in Figure 6.

Figure 6 shows the comparison of calculated velocity profiles at the different distances from SEN on the central plane (Z = 0). It can be seen from Figure 6 how the jets spread as they move across the mold along the width direction. The velocity profiles of the molten steel at the position of 80 mm and 270 mm away from the nozzle center have convex shapes, due to the jet zone forming out of the nozzle outlet ports. The velocities of molten steel near the nozzle are larger than those far from the nozzle. When the molten steel arrives at the narrow wall, the shape of the velocity profile becomes a concave as shown in Figure 6c, due to there is a stagnation point at the zone of impingement. The velocity of molten steel is the largest at the nozzle outlet ports and it gradually decreases towards the narrow side. Compared the red lines and black lines in Figure 6, it is seen that the velocities of molten steel with gas blowing have a difference from the condition without gas injection. The reason is that argon gas injection can greatly slow down the velocity of molten steel and uplift the jet zone, due to the buoyancy of bubbles. The similar results were reported that the gas injection widened the jet slightly and diminished its peak velocity [36].

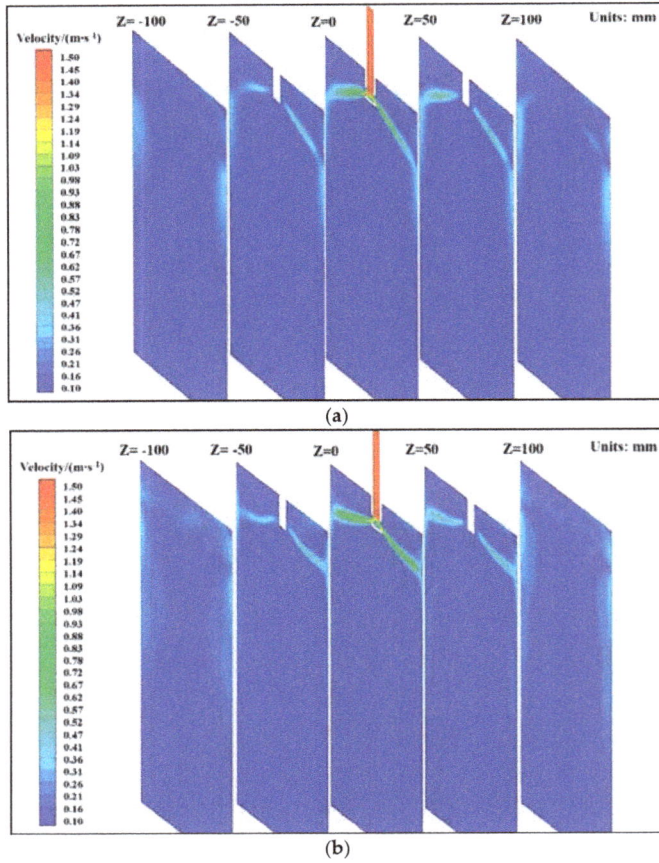

Figure 5. Distribution of molten steel velocity along the mold thickness direction. (**a**) Without gas injection; (**b**) With gas injection.

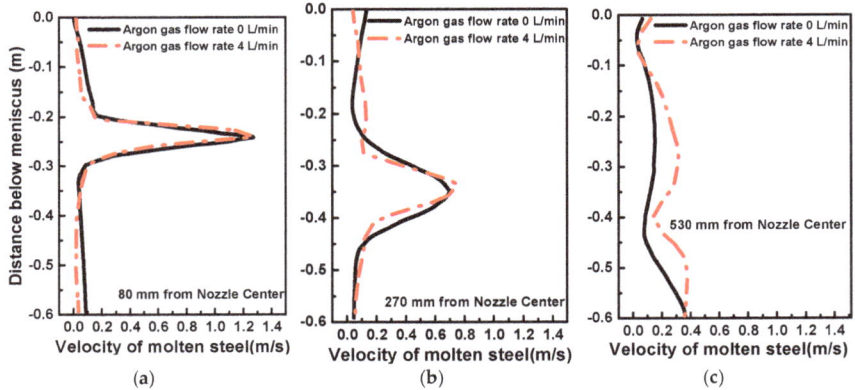

Figure 6. Comparison of calculated velocity profiles at the different distances from SEN. (**a**) 80 mm from nozzle center; (**b**) 270 mm from nozzle center; (**c**) 530 mm from nozzle center.

Figure 7 is the comparison of calculated and measured surface velocities with different casting speeds. Figure 7a shows the effect of different casting speeds on the molten steel velocities of 5 cm below the surface along the mold width direction on the center plane. The conditions are that the mold width is 1275 mm, the argon gas flow rate is 4 l/min and the immersion depth of SEN is 170 mm. The results show that velocities of molten steel increase with increasing casting speed, both from experimental measurement and the calculated results. The reason is that the large casting speed can increase the velocity of molten steel flowing out of the nozzle outlet ports and decrease the bubble size. On the one hand, the molten steel flowing out from the nozzle outlet ports has a large velocity to reach the narrow wall where the jet splits to flow upward and downward. On the other hand, the increased casting speed produces greater shear stress breaking large bubbles into smaller ones, which can reduce the effect of buoyancy of bubbles on slowing down the molten steel velocity and uplifting the jet zone [37]. When the casting speed is 1.0 m·s^{-1}, the flow velocity near the surface is negative at the 1/4 width of mold and the direction of flow is from the nozzle to the narrow wall. When the casting speeds are increased to 1.3 and 1.5 m·s^{-1}, the velocities of molten steel are positive and there is a reversal of flow direction.

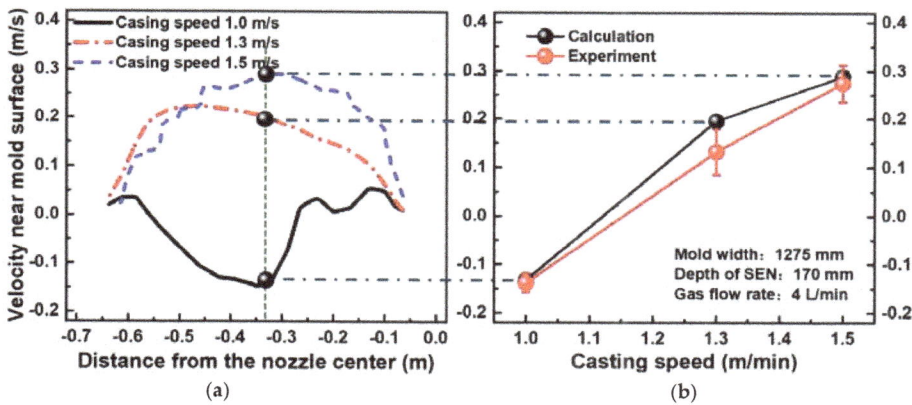

Figure 7. Comparison of calculated and measured velocities near the surface with different casting speeds. (**a**) Calculated results; (**b**). Comparison of experimental and calculated results

Figure 7b shows the comparison of experimental measurements and calculated results under the same conditions of Figure 7a, for the velocities of molten steel near the surface at the 1/4 width of mold. It gives a good agreement between the calculated and measured results, both in trends and magnitudes for the molten steel velocities. When the casting speed is increased from 1.0 m·s^{-1} to 1.5 m·s^{-1}, the velocity of molten steel near the surface increases from −0.13 m·s^{-1} to 0.28 m·s^{-1} and the flow direction also changes.

Figure 8 shows the comparison of calculated and measured velocities near the surface with different casting speeds under the conditions that the mold width is 1040 mm, the argon gas flow rate is 4 l/min and the immersion depth of SEN is 170 mm. The results show that velocities of molten steel near the surface increase with increasing casting speed, both from experimental measurement and the calculated results. There is a good agreement between the calculated and measured results in trends as shown in Figure 8b. When the mold width is decreased from 1275 mm to 1040 mm, it is noticed a slight less-prediction of the measured velocity. This may be because the narrow mold width decreases the volume flow rate of molten steel and enhances the influent of bubbles on non-steady flow phenomenon, as the previous work reported [13].

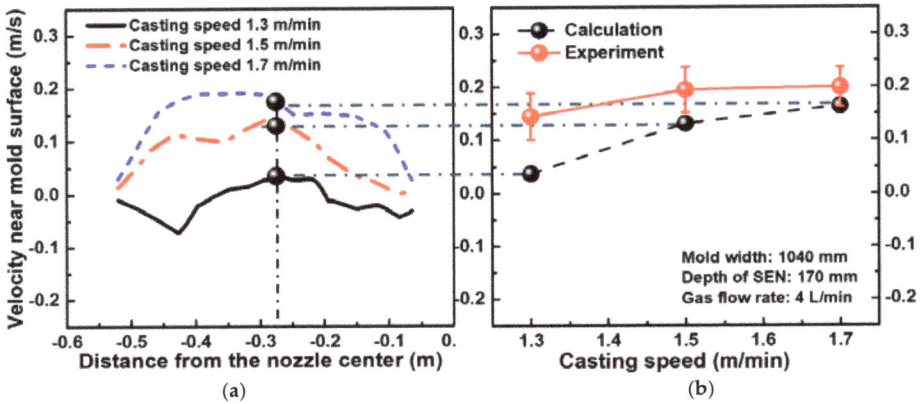

Figure 8. Comparison of calculated and measured velocities near the surface with different casting speeds. (**a**) Calculated results; (**b**). Comparison of experimental and calculated results

4.2. Flow Patterns in the Mold

Three different types of flow patterns of molten steel in the mold are shown in Figure 9, which can be obtained from the present calculation and measurement results. The pattern A is the single-roll-flow (SRF) which is formed with increasing argon gas flow rate when the casting speed is small in the wide slab mold. On the contrary, the pattern C is the double-roll-flow (DRF) which is formed with increasing casting speed in the narrow slab mold. The pattern B is a transition state between DRF and SRF and its formation conditions are the reasonable casting speed and argon gas flow rate in the mold.

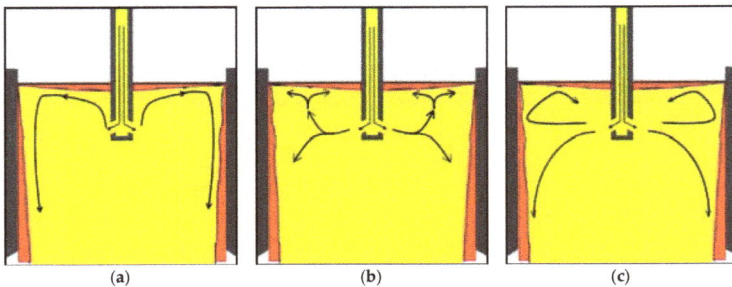

Figure 9. Different flow patterns of molten steel in the mold. (**a**) Pattern A (SRF); (**b**) Pattern B (Transition state); (**c**). Pattern C (DRF).

When the flow pattern is SRF, the strong upward flow and large argon bubbles near the SEN tend to result in the strong surface fluctuation of molten steel. Moreover, the large surface velocity is likely to lead to the longitudinal eddies near the narrow wall, which is one of the main reasons for the slag entrainment into the molten steel. For the flow pattern C, eddies and strong shear force seem to be easily produced by the excessive surface velocity of molten steel, which might result in the slag entrainment near the SEN. In comparison, the flow pattern B is neither cause the vortices nor excessive surface velocity on the meniscus, so the slag entrainment rarely occurs.

By means of the deflection angle measurement, not only the surface velocity but also the flow direction of molten steel can be measured and obtained. Thus, the flow pattern in the mold can be deduced from the flow directions at different positions of the mold. In the present study, the industrial experiments are performed by measuring the flow velocity at two different positions in the mold. One is at the location 10 cm from the narrow wall and the other at the 1/4 mold width.

Figure 10 shows the comparison between the calculated and measured velocities of molten steel under the same operation conditions. The black line presents the calculated velocity profiles of 5 cm below the surface along the mold width on the center plane. The experimental data in Figure 10a is measured corresponding to the simulation conditions which are as follows: the mold width is 1275 mm, the casting speed is 1.0 m·min^{-1}, the depth of SEN is 170 mm and the argon gas flow rate is 4 l·min^{-1}. Under these conditions, the velocities of molten steel near the surface show negative values at the 1/4 mold width, which indicate that the direction of molten steel flow is from the nozzle to the narrow wall. It is possible that the SRF pattern is formed in the mold. The SRF is an undesirable flow pattern which tends to cause more defects in the slab [38]. The strong upward flow and large argon bubbles near the SEN can result in the surface fluctuation of molten steel which is one of the main reasons for the slag entrainment. On the other hand, the downward flow near the mold narrow wall easily bring the bubbles and inclusions into the depth of mold which can also cause the surface defects. The flow velocity of molten steel is about 0.15 m·s^{-1} at the 1/4 width of the mold, which shows the good agreement between the measurement results of the three sets and the calculation result.

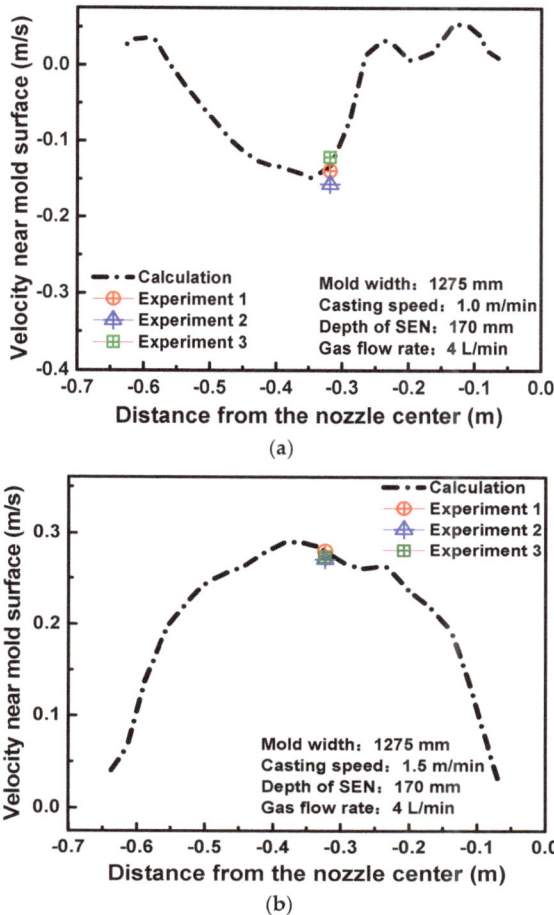

Figure 10. Comparison of the calculated and measured molten steel velocities near the surface. (**a**) Casting speed is 1.0 m/min; (**b**) Casting speed is 1.5 m/min.

In Figure 10b, when the casting speed is increased to 1.5 m/min with the same other parameters as those in Figure 10a, the velocities of molten steel near the surface dramatically change to the positive values, which means the direction of molten steel flow is from the narrow wall to the nozzle. The velocity of molten steel at the 1/4 width of the mold is about 0.28 m·s^{-1}, which agrees well with the measurement results. Compared with Figure 10a, the flow pattern changes from SRF to DRF with increasing the casting speed. The reason is that the molten steel out from the nozzle outlet ports has a large velocity to reach the narrow wall where the jet splits to flow more strongly upward and downward, which can form a typical DRF.

Figure 11 shows the measured velocities of molten steel near the surface at positions of 1/4 width and 10 cm from the narrow wall of the mold under the different casting speeds. The conditions of the experiment are as follows: the mold width is 1040 mm, the depth of SEN is 170 mm and the argon gas flow rate is 4 l·min^{-1}. No matter at 10 cm from the narrow wall or 1/4 width of the mold, the surface velocities increase with increasing the casting speed. It can be seen that when the casting speed is 1.3 m·min^{-1}, the flow direction at 1/4 width of the mold is from narrow wall to the nozzle but the flow direction is converse at 10 cm from the narrow wall of the mold. It is likely to indicate that the flow state in the mold is pattern B which is between SRF and DRF. When the casting speed is increased to 1.5 and 1.7 m·min^{-1}, the flow direction is from narrow wall to the nozzle both at 10 cm from the narrow wall and 1/4 width of the mold, which indicates that the flow pattern in the mold is pattern C of DRF.

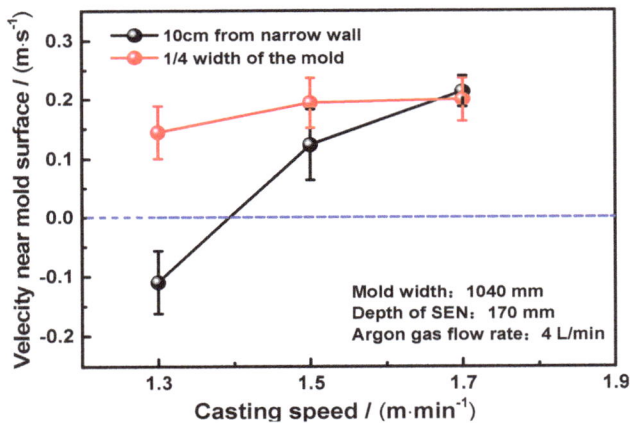

Figure 11. Measured velocities of molten steel near the surface at positions of 1/4 width and 10 cm from the narrow wall of the mold.

Figure 12 shows the predicted flow field of molten steel in the mold under different casting speeds. The calculated results show a good agreement with the measured results. When the casting speed is 1.3 m·min^{-1}, the flow direction of molten steel in the upper circulation zone is opposite between the 1/4 mold width and near the narrow face. When the casting speed is 1.7 m·min^{-1}, the flow pattern in the mold is a typical DRF. It strongly evidences that different casting speeds can change the flow pattern in the mold. In a consequence, the Rod Deflecting Method can be used to judge the flow pattern of the mold and optimize the operation conditions of continuous casting to improve the quality of the slab.

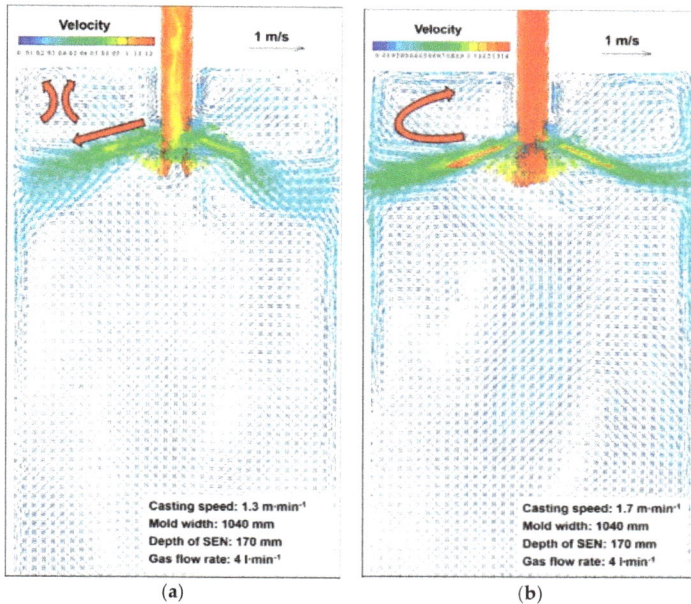

Figure 12. Flow field of molten steel in the mold under different casting speeds. (**a**) Casting speed is 1.3 m/min; (**b**) Casting speed is 1.7 m/min

5. Conclusions

In this study, the measurements of molten steel velocity near the mold surface are conducted using a newly developed speed measuring method under the different operation conditions of continuous casting. In addition, the mathematical model is adapted to simulate the flow field of molten steel in the mold. Based on the obtained results, the following conclusions are drawn:

1. The rod deflection method is conducted to measure the flow velocity of molten steel near the surface of the mold in the industrial experiment. With this measurement, the flow velocity and direction can be obtained.
2. Both experimental and calculated results show that the molten steel velocity near the mold surface increases with increasing casting speed. Furthermore, argon gas injection can slow down the molten steel velocity and uplift the jet zone, due to the buoyancy of bubbles.
3. Three different types of flow patterns of molten steel in the CC mold can be obtained from the present calculation and measurement results. The pattern A is the single-roll-flow (SRF) and the pattern C is the double-roll-flow (DRF). The pattern B is a transition state between DRF and SRF, which is neither cause the vortices nor excessive surface velocity on the meniscus, so the slag entrainment rarely occurs.
4. It is found both from the measurement results and calculated results of the molten steel velocity near the mold surface that the casting speed can change the flow pattern in the mold. When the argon flow rate is 4 l/min and the casting speed is 1.0 m/min, the SRF will be formed. However, the flow pattern becomes the DRF with increasing the casting speed to larger than 1.3 m/min.
5. Good and reasonable agreements have been obtained between the calculated and measured results. Combination of the measurement and calculation results is an effective tool to investigate the transient flow behavior in the CC mold and optimize the actual operation parameters of continuous casting to avoid the surface defects of the automobile outer panel.

Metals **2019**, *9*, 36

Author Contributions: Conceptualization, T.Z. and J.Y.; Methodology, T.Z. and J.Y; Software, T.Z. and P.J; Validation, T.Z., J.Y. and P.J.; Formal analysis, T.Z., J.Y. and P.J.; Investigation, T.Z. and J.Y.; Data curation, T.Z. and P.J.; Writing—original draft preparation, T.Z.; Writing—review and editing, J.Y. and T.Z.

Funding: This research received no external funding.

Conflicts of Interest: The authors declare no conflict of interest.

References

1. Wang, X.H. Non-metallic inclusion control technology for high quality cold rolled steel sheets. *Iron Steel* **2013**, *48*, 1–7.

2. Zeze, M.; Tanaka, A.; Tsujino, R. Formation Mechanism of Sliver-type Surface Defects with Oxide Scale on Sheet and Coil. *Tetsu-to-Hagane* **2001**, *87*, 15–22. [CrossRef]

3. Yang, J.; Zhi, J.J.; Wang, R.Z.; Zhu, K. Analysis of involved mold powders and inclusions for the surface defects on car body panels. In Proceedings of the 17th Chinese Steelmaking Conference, Hangzhou, China, 16 May 2013; pp. 809–814.

4. Wang, X.H. Possibility of producing high quality cold rolled coils with thin slab casting production route. *Iron Steel* **2004**, *39*, 18–25.

5. Yuan, F.M.; Wang, X.H.; Liu, X.M. Research of surface inclusive slag defects in interstitial-free steel slab. *Contin. Cast.* **2004**, *6*, 32–35.

6. Bai, H.; Thomas, B.G. Turbulent flow of liquid steel and argon bubbles in slide-gate tundish nozzles: Part I. Model development and validation. *Metall. Mater. Trans. B* **2001**, *32*, 253–267. [CrossRef]

7. Thomas, B.G. Review on modeling and simulation of continuous casting. *Steel Res. Int.* **2017**, *89*, 1700312. [CrossRef]

8. Zhang, T.; Luo, Z.G.; Liu, C.L.; Zhou, H.; Zou, Z.S. A mathematical model considering the interaction of bubbles in continuous casting mold of steel. *Powder Technol.* **2015**, *273*, 154–164. [CrossRef]

9. Zhu, M.Y.; Cai, Z.Z.; Yu, H.Q. Multiphase flow and thermo-mechanical behaviors of solidifying shell in continuous casting mold. *J. Iron Steel Res. Int.* **2013**, *20*, 6–17. [CrossRef]

10. Liu, Z.; Li, B.; Jiang, M. Transient asymmetric flow and bubble transport inside a slab continuous-casting mold. *Metall. Mater. Trans. B* **2013**, *45*, 675–697. [CrossRef]

11. Thomas, B.G.; Zhang, L. Mathematical modeling of fluid flow in continuous casting. *ISIJ Int.* **2001**, *41*, 1181–1193. [CrossRef]

12. Yuan, Q.; Zhao, B.; Vanka, S.P.; Thomas, B.G. Study of computational issues in simulation of transient flow in continuous casting. *Steel Res. Int.* **2005**, *76*, 33–43. [CrossRef]

13. Chaudhary, R.; Ji, C.; Thomas, B.G.; Vanka, S.P. Transient turbulent flow in a liquid-metal model of continuous casting, including comparison of six different methods. *Metall. Mater. Trans. B* **2011**, *42*, 987–1007. [CrossRef]

14. Kratzsch, C.; Timmel, K.; Eckert, S.; Schwarze, R. Urans simulation of continuous casting mold flow: Assessment of revised turbulence models. *Steel Res. Int.* **2015**, *86*, 400–410. [CrossRef]

15. Liu, C.L.; Luo, Z.G.; Zhang, T.; Deng, S.; Wang, N.; Zou, Z.S. Mathematical modeling of multi-sized argon gas bubbles motion and its impact on melt flow in continuous casting mold of steel. *J. Iron Steel Res. Int.* **2014**, *21*, 403–407. [CrossRef]

16. Wang, Y.F.; Zhang, L.F. Transient fluid flow phenomena during continuous casting: Part 1—cast start. *ISIJ Int.* **2010**, *50*, 1777–1782. [CrossRef]

17. Wang, Y.F.; Zhang, L.F. Transient fluid flow phenomena during continuous casting: Part 2—cast speed change, temperature fluctuation, and steel grade mixing. *ISIJ Int.* **2010**, *50*, 1783–1791. [CrossRef]

18. Thomas, B.G.; Yuan, Q.; Mahmood, S.; Liu, R.; Chaudhary, R. Transport and entrapment of particles in steel continuous casting. *Metall. Mater. Trans. B* **2014**, *45*, 22–35. [CrossRef]

19. Singh, R.; Thomas, B.G.; Vanka, S.P. Effects of a magnetic field on turbulent flow in the mold region of a steel caster. *Metall. Mater. Trans. B* **2013**, *44*, 1201–1221. [CrossRef]

20. Liu, Z.; Li, B.; Zhang, L.; Xu, G. Analysis of transient transport and entrapment of particle in continuous casting mold. *ISIJ Int.* **2014**, *54*, 2324–2333. [CrossRef]

21. Liu, Z.Q.; Li, L.; Li, B.; Jiang, M. Large eddy simulation of transient flow, solidification, and particle transport processes in continuous-casting mold. *JOM* **2014**, *66*, 1184–1196. [CrossRef]

22. Singh, R.; Thomas, B.G.; Vanka, S.P. Large Eddy Simulations of Double-Ruler Electromagnetic Field Effect on Transient Flow During Continuous Casting. *Mater. Trans. B* **2014**, *45*, 1098–1115. [CrossRef]

23. Miki, Y.; Takeuchi, S. Internal defects of continuous casting slabs caused by asymmetric unbalanced steel flow in mold. *ISIJ Int.* **2003**, *43*, 1548–1555. [CrossRef]

24. Chaudhary, R.; Lee, G.G.; Thomas, B.G.; Cho, S.M.; Kim, S.-H.; Kwon, O.D. Effect of stopper-rod misalignment on fluid flow in continuous casting of steel. *Metall. Mater. Trans. B* **2011**, *42*, 300–315. [CrossRef]

25. Yuan, Q.; Thomas, B.G.; Vanka, S.P. Study of transient flow and particle transport in continuous steel caster molds: Part I. Fluid flow. *Metall. Mater. Trans. B* **2004**, *35*, 685–702. [CrossRef]

26. Shen, B.; Shen, H.; Liu, B. Instability of fluid flow and level fluctuation in continuous thin slab casting mould. *ISIJ Int.* **2007**, *47*, 427–432. [CrossRef]

27. Thomas, B.G.; Yuan, Q.; Sivaramakrishnan, S.; Shi, T.; Vanka, S.P.; Assar, M.B. Comparison of four methods to evaluate fluid velocities in a continuous slab casting mold. *ISIJ Int.* **2001**, *41*, 1262–1271. [CrossRef]

28. Sanchez-Perez, R.; Morales, R.D.; Diaz-Cruz, M. A physical model for the two-phase flow in a continuous casting mold. *ISIJ Int.* **2003**, *43*, 637–646. [CrossRef]

29. Kubota, J.; Kubo, N.; Ishii, T. Steel flow control in continuous slab caster mold by traveling magnetic field. *NKK Tech. Rev.* **2001**, *85*, 1.

30. Liu, R.; Thomas, B.G.; Sengupta, J.; Chung, S.D.; Trinh, M. Measurements of molten steel surface velocity and effect of stopper-rod movement on transient multiphase fluid flow in continuous casting. *ISIJ Int.* **2014**, *54*, 2314–2323. [CrossRef]

31. Assar, M.B.; Dauby, P.H.; Lawson, G.D. *The 83rd Steelmaking Conference*; Iron and Steel Society: Pittsburgh, PA, USA, 2000.

32. Iguchi, M.; Kawabata, H.; Demoto, Y.; Morita, Z. Cold model experiments for developing a new velocimeter applicable to molten metal. *ISIJ Int.* **1994**, *34*, 461–467. [CrossRef]

33. Iguchi, M.; Terauchi, Y. Karman vortex probe for the detection of molten metal surface flow in low velocity range. *ISIJ Int.* **2002**, *42*, 939–943. [CrossRef]

34. Jones, W.P.; Launder, B.E. The prediction of laminarization with a two-equation model of turbulence. *Int. J. Heat Mass Transf.* **1972**, *15*, 301–314. [CrossRef]

35. Xu, Y.; Ersson, M.; Jönsson, P.G. A numerical study about the influence of a bubble wake flow on the removal of inclusions. *ISIJ Int.* **2016**, *56*, 1982–1988. [CrossRef]

36. Thomas, B.G.; Huang, X.; Sussman, R.C. Simulation of argon gas flow effects in a continuous slab caster. *Metall. Mater. Trans. B* **1994**, *25B*, 527. [CrossRef]

37. Zhang, T.; Luo, Z.; Zhou, H.; Ni, B.; Zou, Z. Analysis of two-phase flow and bubbles behavior in a continuous casting mold using a mathematical model considering the interaction of bubbles. *ISIJ Int.* **2016**, *56*, 116–125. [CrossRef]

38. Deng, X.; Ji, C.; Cui, Y.; Li, L.; Yin, X.; Yang, Y.; McLean, A. Flow pattern control in continuous slab casting moulds: Physical modelling and plant trials. *Ironmak. Steelmak.* **2016**, *44*, 461–471. [CrossRef]

metals

MDPI

Article

Formation of Surface Depression during Continuous Casting of High-Al TRIP Steel

Heng Cui [1,*], Kaitian Zhang [2], Zheng Wang [1], Bin Chen [3], Baisong Liu [3], Jing Qing [4] and Zhijun Li [4]

[1] Collaborative Innovation Center of Steel Technology, University of Science and Technology Beijing, Beijing 100083, China; wangzheng410@163.com
[2] Engineering Research Institute, University of Science and Technology Beijing, Beijing 100083, China; zhangkaitianbk@163.com
[3] Shougang Technology Research Center, Shougang Corporation, Beijing 100043, China; ustbchenbin@163.com (B.C.); baisongliu@163.com (B.L.)
[4] Beijing Shougang Co. Ltd., Beijing 100043, China; qingjing@sgqg.com (J.Q.); lizhijun@sgqg.com (Z.L.)
* Correspondence: cuiheng@ustb.edu.cn; Tel.: +86-136-7123-9796

Received: 31 December 2018; Accepted: 5 February 2019; Published: 9 February 2019

Abstract: High aluminum transformation-induced plasticity (TRIP) steels offer a unique combination of high tensile strength and ductility, high impact energy absorption and good formability. The surface of the slab is prone to depressions and longitudinal cracks during continuous casting due to the high Al content in steels. Surface depressions of the 1.35 wt.% Al-TRIP steel slab in a steel works were investigated by scanning electronic microscopy (SEM) and mold fluxes with different Al_2O_3/SiO_2 ratios were researched by thermodynamic calculations and high-temperature static balance experiments. The results show that some micro-cracks were distributed along the grain boundary in the surface depression of the slab. Inclusions containing K and Na, which were probably from mold flux, were found in the depression samples. Meanwhile, the components of reactive mold flux showed significant variation in their chemical composition during the continuous casting process of the Al-TRIP steel. A large number of depressions and irregular oscillation marks on the Al-TRIP steel slab surface were generated due to serious deterioration in the physical properties of the mold flux. Since the TRIP steel is a typical hypo-peritectic steel, the overly large thermal contraction and volume contraction during initial solidification is the intrinsic cause of surface depression. The change of mold flux properties during casting aggravates the formation of depressions.

Keywords: Al-TRIP steel; surface depression; cracks; non-metallic inclusion; mold flux; reactivity

1. Introduction

As automotive lightweighting and safety requirements increase, steels for automotive structural parts are required to have both high strength and ductility. Transformation-induced plasticity (TRIP) steels offer a unique combination of high tensile strength and ductility, high impact energy absorption and good formability [1,2]. Due to the poor wettability properties of silicon-alloyed TRIP steels during hot dip galvanizing, replacing all or part of the silicon with aluminum has been considered. The main advantage of aluminum-alloyed TRIP steels is that they do not form surface oxides, which improves their coating quality and galvanizing properties [3,4].

However, the high Al content in steel can easily lead to deterioration of the surface quality of continuous casting slabs. At present, the Al content of industrial aluminum-alloyed TRIP steel is generally 0.5–2 wt.%. The surface of the slab is prone to depressions and longitudinal cracks during continuous casting [5,6].

The depression formation during the continuous casting process is mainly determined by the chemical composition of steel grades [7–9]. Steels with 0.1 and 0.15 wt.% C tend to experience higher contraction forces, which build immediately upon initial solidification. This is the reason for the formation of surface depressions, as well as the higher sensitivity to the formation of longitudinal surface cracks [7]. Meanwhile, the addition of aluminum to TRIP steel causes the reaction between [Al] in molten steel and (SiO_2) in conventional lime-silica-based mold fluxes, resulting in many problems such as the variation of chemical composition, the instability of viscosity and the deterioration of other thermophysical properties of mold flux. These variations have been reported as the principal reasons for some problems during the continuous casting of high-Al steels such as breakout, uneven heat transfer across mold flux, inadequate lubrication and the poor surface quality of as-cast slabs [5,10–12].

In this present work, the metallographic samples at surface depressions of the high-Al TRIP slab were analyzed by scanning electron microscopy equipped with an energy dispersion spectrum (SEM/EDS), and the industrial mold flux was investigated by thermodynamic calculations and high-temperature static balance experiments. Furthermore, the formation of depressions and cracks was discussed from the two aspects of the chemical composition of the steel grade and the reactivity of the mold flux.

2. Experimental Materials and Methods

The 1.35 wt.% Al-TRIP steel used in this present work is produced by the 210t BOF-LF-RH-CC process at Shougang Qianan steel works. The main chemical composition of the Al-TRIP steel slab from the steel works is shown in Table 1. The trial was carried out at one heat with a two-strand continuous slab caster, and the slab section was 1400 mm × 230 mm with a casting speed of 0.9 m/min. The superheating of liquid steel within the tundish was controlled within the range of 20–35 °C.

Table 1. Main chemical compositions of Al-transformation-induced plasticity (TRIP) steel (wt.%).

C	Si	Mn	P	S	Alt	N
0.16	0.16	1.49	0.008	0.001	1.35	0.0016

A large amount of surface depression and irregular oscillation marks were found on the whole inner arc surface except within 100 mm from the narrow sides of the slab, which was the third slab of the #1 strand of the sequence, as shown in Figure 1. Samples were respectively cut with the size of 15 × 15 × 20 mm^3 from the surface depression and the normal surface of the near side and center positions of the slab width on the inner arc slab. For morphological observation, after immerging the material in acetone with ultrasonic oscillation to wash away the surface oil, the samples were subjected to cold acid pickling by dilute hydrochloric acid to remove mill scales. After this the depression morphologies were observed by SEM. For the non-metallic inclusion observation, grinding and polishing was performed on the surface of normal samples, and the thickness direction of both normal samples and depression samples, as well as the type, number and size of non-metallic inclusions of the samples' surface and subsurface (200 μm deep from slab surface), were observed by SEM/EDS. Two hundred and fifty grams of steel was added in a zirconia crucible (inner diameter = 45 mm, outer diameter = 50 mm, height = 75 mm), which was put in the constant temperature zone of an Si-Mo high-temperature furnace. Before heating, argon was piped into the closed furnace for 1 h with a flow rate of 5 l/min to eliminate internal air, preventing the oxidation of steel. The steel was then heated to 1550 °C by 10 °C/min and the temperature maintained after the steel melted completely. Sixty grams of pre-melted mold flux was added on the surface of the molten steel, and heat preservation continued for 3 h until the reaction tended to equilibrium. Finally, the crucible was removed, and after cooling, crushing, and grinding, the mold flux chemical compositions were analyzed by X-ray fluorescence (XRF). In addition, viscosity (1300 °C) was calculated by FactSage software (Thermfact/CRCT, Montreal, Canada and GTT-Technologies, Aachen, Germany) according to the chemical composition before and after the reaction of mold slag and steel.

Figure 1. Photograph of surface depressions of the 1.35 wt.% Al-TRIP steel slab.

3. Results

3.1. Metallographic Analysis

Figure 2 shows the typical morphology of micro-cracks on the depression samples observed by SEM. The crack length ranged from a few hundred microns to several millimeters and the gap width was tens of microns. The cracks were mainly star cracks along grain boundaries.

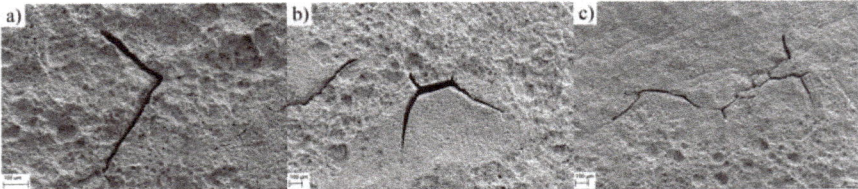

Figure 2. Different morphologies of star cracks on depression samples.

Figure 3 shows the statistic of the inclusion ($\geq$10 μm) comparison between normal samples and depression samples. Block Al_2O_3 inclusions were found on both normal and depression samples. A small amount of cluster Al_2O_3 was also found on normal samples, while inclusions containing K or Na were only found on depression samples. Moreover, whether on normal or depression samples, the number of inclusions near the side of the slab was more than that in the center of the slab.

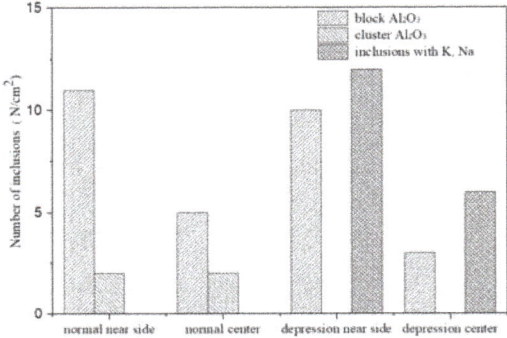

Figure 3. Distribution of inclusions in slab surface and subsurface.

Figure 4 shows typical inclusions in different positions. Table 2 shows the chemical composition of the inclusions. According to Figure 4 and Table 2, the inclusions in normal samples were mainly block (Figure 4a) and cluster (Figure 4b) Al_2O_3; the inclusions in depression samples were mainly Al_2O_3 (Figure 4c) and inclusions with the composition of Na and K (Figure 4d,e); the inclusion with the composition of K was found in the surface cracks of depression samples, as shown in Figure 4f. The Na and K elements in the inclusions were generally considered to be derived from the mold slag. As the inclusions in Figure 4d,f contained more than 40 wt.% Fe, the inclusions may have come from the mixture of mold slag and iron particles, which are probably attributed to argon bubbles scattering the iron particles over the meniscus [13]. Since the observed samples were from the surface of the slab, it was considered that the inclusions in Figure 4d,f may have resulted from the mold slag entrapping or pressing into the initial solidifying shell.

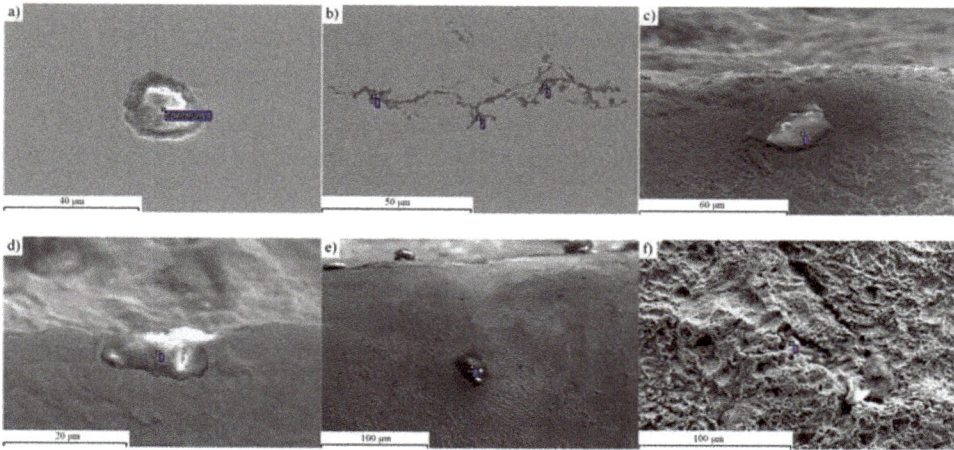

Figure 4. Morphology of non-metallic inclusions. (**a**,**b**) normal surface; (**c**–**f**) depression surface.

Table 2. Chemical compositions of the inclusions (a–f) (wt.%).

Samples	O	Na	Mg	Al	Si	Mn	P	S	K	Ca	Ti	Fe
a	38.19	-	-	57.21	-	-	-	-	-	-	-	4.60
b-1	55.90	-	1.90	42.21	-	-	-	-	-	-	-	-
b-2	51.09	-	4.30	42.55	-	2.07	-	-	-	-	-	-
b-3	55.25	-	5.93	38.82	-	-	-	-	-	-	-	-
c	62.14	-	-	37.31	-	-	-	-	-	-	-	0.55
d	43.26	2.51	0.64	0.53	1.30	-	0.42	5.39	0.57	2.49	-	42.89
e	7.56	4.55	0.64	1.76	0.59	1.60	-	0.57	0.62	0.56	-	81.56
f	36.43	-	-	5.26	0.99	0.54	-	-	1.15	0.48	-	0.79	54.37

3.2. Reactivity of Mold Fluxes for Al-TRIP Steel

The Al_2O_3 accumulation in the mold slag containing SiO_2 is unlikely to be avoided under the presence of Al in the steel by the following chemical reaction [12]:

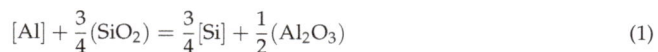

$$[Al] + \frac{3}{4}(SiO_2) = \frac{3}{4}[Si] + \frac{1}{2}(Al_2O_3) \tag{1}$$

$$\Delta G^{\theta}{}_1 = -164550 + 26.775T \tag{2}$$

$$\Delta G_1 = \Delta G^{\theta}{}_1 + RT \ln \frac{(a_{Al_2O_3})^{1/2}(a_{Si})^{3/4}}{(a_{SiO_2})^{3/4}(a_{Al})} \tag{3}$$

The activities of Al and Si (αAl and αSi) in liquid steel at 1550 °C were calculated by FactSage software and Wanger Model with $w[i] = 1\%$ solution as a standard. Due to the low content of other elements in liquid steel, the temperature has less influence on the activity interaction coefficient, and 1873 K was chosen for calculation. The results are as follows: $f_{Al} = 1.230$, $f_{Si} = 1.345$, $\alpha_{[Al]} = 1.660$, $\alpha_{[Si]} = 0.215$. The activities of Al_2O_3 and SiO_2 in the mold flux were calculated using the FToxid database in FactSage software. The results are as follows: $\alpha_{(SiO2)} = 5.25 \times 10^{-2}$, $\alpha_{(Al2O3)} = 5.50 \times 10^{-5}$. Finally, it was calculated that $\Delta G_1 = 181.7$ kJ/mol at 1550 °C.

A special mold flux for peritectic steel was used in the industrial test of the TRIP steel. The compositions of the industrial slag before and after the steel/slag reaction are shown in Table 3. This was investigated by high-temperature static balance experiments. The findings reveal that the industrial mold flux has a great tendency to react with liquid steel. After the equilibrium reaction, the w(SiO_2) in slag decreased from 40.35% to 28.90%; w(Al_2O_3) increased from 1.07% to 12.58%. This is similar to that reported in document [6]; after casting high-Al TRIP steel for 24 min, the mold flux composition became almost constant, the SiO_2 content was stable between 13 wt.% and 15 wt.% from 35.10 wt.% and the Al_2O_3 content was 40 wt.% from 1.50 wt.%. As a result, the properties of mold flux deteriorated; for example, there was a 79.74% increase in viscosity (from 0.232 Pa·s to 0.417 Pa·s), as shown in Table 3. Varying viscosity led to various casting problems, including sticking of the mold flux to copper mold, non-uniform heat transfer across the mold flux, reduced consumption of mold flux, reduced lubrication and poor as-cast slab surface quality, etc. [5,10].

Table 3. Compositions of industrial slag before and after steel/slag interfacial reaction (wt.%).

Slag Sample	CaO	SiO$_2$	Al$_2$O$_3$	CaF$_2$	MnO	BaO	Na$_2$O	Fe$_2$O$_3$	MgO	CaO/SiO$_2$	Viscosity (Pa·s)
Before reaction	36.30	40.35	1.07	13.66	0.21	0.50	6.64	0.58	0.68	0.90	0.232
After reaction	39.60	28.90	12.58	12.73	1.57	0.59	1.52	1.33	1.17	1.37	0.417

Non-reactive mold flux is expected to suppress reaction (1) by reducing $\alpha(SiO_2)$ and increasing $\alpha_{(Al2O3)}$. Therefore, an effective method is increasing the content of Al_2O_3 and reducing the content of SiO_2 in mold flux. The different ratios of Al_2O_3/SiO_2 mold flux samples were analyzed for their thermodynamics based on the industrial mold flux. The main components of slags are shown in Table 4. When the Al_2O_3/SiO_2 ratio increased, $\alpha_{(Al2O3)}$ increased while $\alpha_{(SiO2)}$ decreased and ΔG tended to 0, which means a significant reduction in the driving force of reaction (1). Specifically, the Al_2O_3/SiO_2 ratio increased from 0.16 to 2.08, and there was a 48.5% reduction of ΔG. In addition, ΔG changed sharply at the beginning while it tended to be stable until the Al_2O_3/SiO_2 ratio reached 1.18; after that, increasing Al_2O_3/SiO_2 had less effect on the reactivity.

Table 4. The mold flux for high aluminum steel with different Al_2O_3/SiO_2 ratios.

Samples	CaO	SiO$_2$	Al$_2$O$_3$	Al$_2$O$_3$/SiO$_2$	a(SiO$_2$)	a(Al$_2$O$_3$)	ΔG (kJ/mol)	Viscosity (Pa·s)
A-1	30	32	5	0.16	3.50×10^{-2}	1.10×10^{-3}	−154.4	0.293
A-2	30	27	10	0.37	2.24×10^{-3}	2.41×10^{-2}	−99.7	0.277
A-3	30	22	15	0.68	4.36×10^{-4}	1.55×10^{-2}	−84.5	0.258
A-4	30	17	20	1.18	2.76×10^{-4}	1.40×10^{-2}	−80.1	0.238
A-5	30	12	25	2.08	2.16×10^{-4}	1.06×10^{-2}	−79.4	0.219

All mold flux samples in Table 4 were tested for slag/steel reaction by the equilibrium experiments. The compositions and viscosities of slag samples with different Al_2O_3/SiO_2 ratios after reaction were analyzed, as shown in Table 5. Increasing the Al_2O_3/SiO_2 ratio can not only reduce the consumption of SiO_2 in mold slag, but it can also decrease the increment of Al_2O_3 significantly compared with that before the reaction. The above findings indicate that the properties of mold flux for high Al steel could be improved by increasing the Al_2O_3/SiO_2 ratio. It is possible to inhibit slag property deterioration due to reaction (1) during the continuous casting process.

Table 5. Compositions of mold slag with different Al_2O_3/SiO_2 ratios after steel/slag reaction (wt.%).

Sample	CaO	SiO_2	Al_2O_3	ΔSiO_2 (%)	ΔAl_2O_3 (%)	CaO/SiO_2	Viscosity (Pa·s)
A-1	34.25	20.14	17.53	−37.50	250.51	1.70	0.298
A-2	34.49	18.41	23.93	−31.82	139.29	1.87	0.286
A-3	31.94	15.18	25.42	−31.00	69.49	2.10	0.306
A-4	33.17	12.00	28.37	−29.41	41.85	2.76	0.245
A-5	32.33	8.58	34.00	−29.17	36.02	3.77	0.278

Figure 5 shows the viscosities of mold fluxes with different Al_2O_3/SiO_2 ratios calculated by FactSage software. With the increase of the Al_2O_3/SiO_2 ratio, the viscosity reduced for the samples before the reaction, and there was a significant reduction of viscosity variation (0.24–0.30 Pa·s) for the samples after the reaction compared with the industrial slag sample. Therefore, it is possible to reduce the mold flux reactivity significantly as long as the Al_2O_3/SiO_2 ratio can be adjusted to 1.18. In addition, some acidic oxides and fluxes which could adjust the basicity and crystallization properties such as B_2O_3 and LiO_2, should be added to slag because of the low content of SiO_2 [14,15].

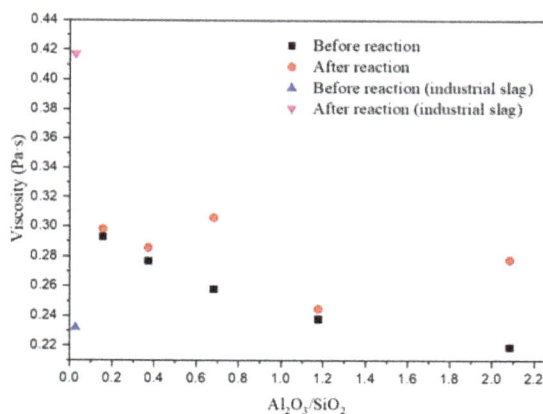

Figure 5. Comparison of the viscosity of the samples between before and after reaction.

4. Discussion

The thermal contraction of the solidifying shell in the meniscus region of a continuous casting mold is of great importance to the surface and subsurface quality of the cast products [7,16]. It is well known that peritectic steels with an equivalent carbon content (C_P) between 0.09 and 0.17 (the characteristic points C_A and C_B in the binary Fe–C system) are more prone to cracks and depressions than others. The binary Fe–C equilibrium phase diagram and the pseudo-binary Fe–C phase diagram of the 1.35 wt.% Al-TRIP steel were calculated by the commercial thermodynamic software ThermoCalc 2018b (database: TCFE8). Figure 6 shows the high-temperature ranges of the phase diagrams. The calculated chemical composition of the 1.35 wt.% Al-TRIP steel is shown in Table 1.

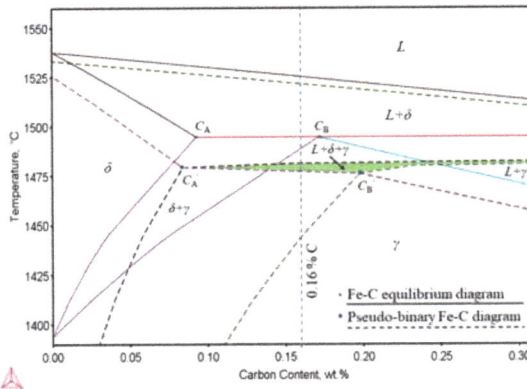

Figure 6. Fe–C equilibrium diagram and pseudo-binary Fe–C diagram of the 1.35 wt.% Al-TRIP steel.

According to the Fe–C phase diagram, C_A and C_B were 0.09 and 0.17 C wt.% at the peritectic phase transformation temperature of 1494.63 °C, respectively. The 1.35 wt.% Al-TRIP steel contains high Mn and Al elements, of which Mn is an austenite former and Al is a ferrite former. It can be seen that the presence of Mn and Al led to the formation of a peritectic ternary region ($L + \delta + \gamma$). In the phase diagram, C_A moved to 0.08 wt.% in the lower left at temperature of 1479.14 °C, while C_B moved to 0.20 wt.% in the lower right at temperature of 1475.92 °C. Compared with the Fe–C equilibrium phase diagram, the pseudo-binary Fe–C phase diagram has larger $L + \delta$ and $\delta + \gamma$ two-phase regions. The 1.35 wt.% Al-TRIP steel (0.16 wt.% C) is a typical hypo-peritectic steel, and its solidification sequence was $L \rightarrow L + \delta \rightarrow L + \delta + \gamma \rightarrow \delta + \gamma \rightarrow \gamma$ above 1400 °C, as shown in Figure 6. The liquid–solid phase transition of the TRIP steel was from 1521.80 to 1477.17 °C, while that of the Fe–C alloy was from 1525.41 to 1494.63 °C when the C content was 0.16 wt.%. Thus, the TRIP steel had larger thermal contraction due to the liquid–solid phase transition. Furthermore, the peritectic phase transition $L + \delta \rightarrow \gamma$ caused larger volume contraction during the liquid TRIP steel solidification, because the δ (body-centered cubic) transit to γ (face-centered cubic) was from 1481.32 to 1444.17 °C, while that of the Fe–C alloy was from 1494.63 to 1488.81 °C. As with the addition of Al and Mn in the Fe–C binary alloy (hypo-peritectic steel), the temperature range of the L to γ phase transition increased from 36.60 to 77.63 °C, and the solidifying shell of the TRIP steel was subjected to larger thermal and volumetric contraction forces than the normal hypo-peritectic steel during initial solidification. The contraction forces can reach maximum within a few seconds of solidification [7]. Therefore, a transverse air gap is generated between the solidifying shell and the mold, which prevents heat transfer. When the shell is pulled downward, it is moved close to the mold again by the heat flow and static pressure of the molten steel, thus forming a transverse depression. In order to reduce the risk of formation of surface depression, the C content of 1.35 wt.% Al-TRIP should move right to C_B' to reduce the region from the L phase to γ phase under the condition of satisfying the steel grade design.

Depression formation further results in uneven shell growth and coarse grains in the region of surface depression because the speed of solidification and cooling in depression parts is slower than that in other parts. Under the action of thermal stress and ferrostatic pressure, the stress at the depression is concentrated. When the stress exceeds the critical strength of the shell, cracks are generated at the weak grain boundary of the depression. As shown in Figure 2, depression was often accompanied by micro-cracks, which were cracked along the coarse grains in the depression surface or subsurface. The more serious depressions there are on the surface of the solidifying shell, the greater is the risk of crack formation. If there is too much additional external mechanical stress acting on the shell, the development of the micro-cracks will be further aggravated.

The two important functions of the mold flux are to lubricate the moving slab and to adjust the heat transfer from the solidifying shell to the mold wall. The stability of heat transfer in the mold is closely related to the uniform and stable inflow of mold slag during the continuous casting process. Because of the high requirement of heat transfer uniformity during the solidification and cooling process of peritectic steels, it is suitable to reduce the heat conduction of the mold flux and slow cool the slab under the condition of guaranteeing the thickness of the shell. For high-aluminum peritectic steels, special attention should also be paid to the reactivity of mold flux, as mentioned above.

The special mold flux for peritectic steel, which belongs to reactive mold flux, was used for this industrial test of the TRIP steel, as shown in Table 3. The slag chemical composition changed greatly during the high Al-TRIP steel continuous casting: $w(SiO_2)$ decreased by 28.38%; $w(Al_2O_3)$ increased by 1024.55%. These variations caused the serious deterioration of mold flux properties; for example, viscosity increased by 79.74%. The deterioration of mold flux properties during casting has serious effects on the liquid slag layer thickness, the mold slag inflow between the solidifying shell and mold wall and slag rim formation in the mold meniscus. The liquid level fluctuates acutely, resulting in slag entrainment. The heat transfer between the solidifying shell and mold wall becomes inhomogeneous, leading to the shell shrinking more inhomogeneously, and the risk of depression formation is increased. Furthermore, the slag rim aggravates the formation of depressions due to its obstacles to the infiltration of the slag film into the gap between the mold wall and the shell, and may press some slag film into the initial shell with mold oscillation. This is a reason why inclusions containing K and Na were found in the surface depression, as shown in Figure 4d,e.

It is possible to reduce the risk of surface depression of high-aluminum TRIP slab according to this present work and the literature using the following measures: (1) increasing the C content of steel, under the condition of satisfying the steel grade design, to reduce thermal and volume contraction during initial solidification; (2) adopting a non-reactive CaO–Al$_2$O$_3$-based mold flux with an appropriate melting point and viscosity, improving the mold flux viscosity and melting rate, controlling heat flux in the mold and liquid slag thickness to improve the heat transfer condition of the shell/mold and the homogeneous growth of the shell [5,10–12]; (3) feeding the molten mold flux into the casting mold to enhance the thermal insulation of the meniscus and, hence, the lubrication between the solidifying shell and the copper mold wall [17].

5. Conclusions

This research investigated the formation of depressions of 1.35 wt.% Al-TRIP steel. The following conclusions can be made:

(1) A large number of depressions and irregular oscillation marks were found on the slab surface of 1.35%Al-TRIP steel during the continuous casting process with reactive mold flux, and micro-cracks were found along the grain boundary on the slab surface and subsurface. Non-metallic inclusions containing K and Na were found in the depression surface samples, while the main inclusion in the normal surface sample was that of Al$_2$O$_3$;

(2) The 1.35 wt.% Al-TRIP steel is a typical hypo-peritectic steel; C_A moved to 0.08 wt.% in the lower left at a temperature of 1479.14 °C, while C_B moved to 0.20 wt.% in the lower right at a temperature of 1475.92 °C in the pseudo-binary Fe–C diagram. The overly large thermal contraction and volume contraction caused by the peritectic reaction during initial solidification is the intrinsic cause of surface depression;

(3) The change of mold flux properties during casting aggravates the formation of depressions due to the chemical reaction between the mold flux and the alloying element Al in Al-TRIP steel. Increasing the Al$_2$O$_3$/SiO$_2$ ratio of the mold flux could obviously reduce the reactivity of slag; however, there is no significant change in viscosity. As the Al$_2$O$_3$/SiO$_2$ ratio increased from 0.16 to 2.08, ΔG began to change rapidly and tended to stabilize when the ratio exceeded 1.18.

Author Contributions: Conceptualization, H.C.; Data curation, H.C.; Investigation, Z.W., B.L., J.Q. and Z.L.; Resources, B.C.; Writing—original draft, K.Z.; Writing—review and editing, H.C.

Funding: This research was funded by the National Natural Science Foundation of China (No. U1860106) and the China Scholarship Council (No. 201806465050).

Conflicts of Interest: The authors declare no conflict of interest.

References

1. Fonstein, N.; Pottore, N.; Lalam, S.H.; Bhattacharya, D. Phase transformation behavior during continuous cooling and isothermal holding of aluminum and silicon bearing TRIP steels. In Proceedings of the Materials Science and Technology, Chicago, IL, USA, 9–12 November 2003.

2. Tuling, A.; Banerjee, J.R.; Mintz, B. Influence of peritectic phase transformation on hot ductility of high aluminium TRIP steels containing Nb. *Mater. Sci. Tech.* **2011**, *27*, 1724–1731. [CrossRef]

3. Bellhouse, E.M.; Mertens, A.I.M.; McDermid, J.R. Development of the surface of TRIP steels prior to hot-dip galvanizing. *Mat. Sci. Eng. A* **2007**, *436*, 147–156. [CrossRef]

4. Bhattacharyya, T.; Singh, S.B.; Bhattacharyya, S.; Ray, R.K.; Bleck, W.; Bhattacharjee, D. An assessment on coatability of trans formation induced plasticity (TRIP)-aided steel. *Surf. Coat. Tech.* **2013**, *235*, 226–234. [CrossRef]

5. Shi, C.B.; Seo, M.D.; Cho, J.W.; Kim, S.H. Crystallization Characteristics of CaO- Al_2O_3-Based Mold Flux and Their Effects on In-Mold Performance during High-Aluminum TRIP Steels Continuous Casting. *Metall. Mater. Trans. B* **2014**, *45B*, 1081–1097. [CrossRef]

6. Ji, C.X.; Cui, Y.; Zeng, Z.; Tian, Z.H.; Zhao, C.L.; Zhu, G.S. Continuous casting of high-Al steel in Shougang Jingtang steel works. *J. Iron Steel Res. Int.* **2015**, *22*, 53–56. [CrossRef]

7. Bernhard, C.; Xia, G. Influence of alloying elements on the thermal contraction of peritectic steels during initial solidification. *Ironmak. Steelmak.* **2006**, *33*, 52–56. [CrossRef]

8. Pierer, R.; Bernhard, C. High temperature behavior during solidification of peritectic steels under continuous casting conditions. In Proceedings of the Materials Science and Technology, Cincinnati, OH, USA, 15–19 October 2006.

9. Presoly, P.; Pierer, R.; Bernhard, C. Identification of defect prone peritectic steel grades by analyzing high-temperature phase transformations. *Metall. Mater. Trans. A* **2013**, *44A*, 5377–5388. [CrossRef]

10. Wang, W.L.; Blazek, K.; Cramb, A. A study of the crystallization behavior of a new mold flux used in the casting of transformation-induced-plasticity steels. *Metall. Mater. Trans. B* **2008**, *39B*, 66–74. [CrossRef]

11. Cho, J.W.; Blazek, K.; Frazee, M.; Yin, H.B.; Park, J.H.; Moon, S.W. Assessment of CaO-Al_2O_3 based mold flux system for high aluminum TRIP casting. *ISIJ Int.* **2013**, *53*, 62–70. [CrossRef]

12. Liu, Q.; Wen, G.; Li, J.; Fu, X.; Tang, P.; Li, W. Development of mould fluxes based on lime-alumina slag system for casting high aluminium TRIP steel. *Ironmak. Steelmak.* **2014**, *41*, 292–297. [CrossRef]

13. Cui, H.; Wu, H.J.; Yue, F.; Wu, W.S.; Wang, M.; Bao, Y.P.; Chen, B.; Ji, C.X. Surface defects of cold-rolled Ti-IF steel sheets due to non-metallic inclusions. *J. Iron Steel Res. Int.* **2011**, *18*, 335–340.

14. Solek, K.; Korolczuk-Hejnak, M.; Slezak, W. Viscosity measurements for modeling of continuous steel casting. *Arch. Metall. Mater.* **2012**, *57*, 333–338.

15. Lu, B.X.; Chen, K.; Wang, W.L.; Jiang, B.B. Effects of Li_2O and Na_2O on the crystallization behavior of lime-alumina-based mold flux for casting high-Al steels. *Metall. Mater. Trans. B* **2014**, *45B*, 1496–1509. [CrossRef]

16. Thomas, B.G.; Storkman, W.R. Mathematical models of continuous slab casting to optimize mold taper. In Proceedings of the Modeling and Control of Casting and Welding Processes, Warrendale, PA, USA, 17–22 April 1988.

17. Cho, J.W.; Yoo, S.; Park, M.S.; Park, J.K.; Moon, K.H. Improvement of castability and surface quality of continuously cast TWIP slabs by molten mold flux feeding technology. *Metall. Mater. Trans. B* **2017**, *48B*, 187–196. [CrossRef]

metals

MDPI

Article

Effect of Substituting CaO with BaO and CaO/Al₂O₃ Ratio on the Viscosity of CaO–BaO–Al₂O₃–CaF₂–Li₂O Mold Flux System

Zhirong Li, Xinchen You, Min Li, Qian Wang, Shengping He * and Qiangqiang Wang

College of Materials Science and Engineering, Chongqing University, Chongqing 400044, China;
cqulzr@cqu.edu.cn (Z.L.); vanmeuen@cqu.edu.cn (X.Y.); sjzlimin@cqu.edu.cn (M.L.);
q_wang@cqu.edu.cn (Q.W.); wtfwawj@163.com (Q.W.)
* Correspondence: heshp@cqu.edu.cn; Tel.: +86-023-6510-2469

Received: 20 December 2018; Accepted: 26 January 2019; Published: 28 January 2019

Abstract: The effect of substituting CaO with BaO and CaO/Al₂O₃ ratio on the viscosity of CaO–BaO–Al₂O₃–CaF₂–Li₂O mold flux system was studied by rotational viscosity method. The results showed that the viscosity increased with increasing BaO as a substitute for CaO, while the viscosity decreased with the increase in CaO/Al₂O₃ ratio. The viscous activation energy of the slags is from 92.1 kJ·mol⁻¹ to 133.4 kJ·mol⁻¹. Either the Arhenius or the Weymann–Frenkel equation can be applied to establish the viscosity prediction model. In this paper, the Weymann–Frenkel equation and a new optical basicity with regard to Al₂O₃ as an acidic oxide were applied to the modified NPL model for predicting the viscosity of CaO–BaO–Al₂O₃–CaF₂–Li₂O mold flux system. The estimated viscosity is in good agreement with the measured viscosity.

Keywords: viscosity; BaO; CaO/Al₂O₃ ratio; modified NPL model

---

1. Introduction

High Mn-high Al steel, exemplified by twin-induced plasticity (TWIP) steel, has high strength and good ductility, as well as a smaller density compared with plain steels due to high contents of aluminum and manganese. This perfectly accords with next-generation steel for automobiles, the development of which is continuously devoted to pursuing safety and energy conservation and environmental protection [1,2]. However, the high aluminum content, is an unprecedented challenge to the production of high Mn-high Al steel using continuous casting, and the key to the solution for this is to select a proper mold flux [3]. It is well known that high Al content in molten steel will react with the traditional CaO–SiO₂-based mold flux, leading to a sharp reduction of SiO₂ and an increase of Al₂O₃ in molten slags, as shown in Equation (1), which will convert the original CaO–SiO₂-based mold flux into CaO–Al₂O₃-based mold flux [4,5]. With a substantial change of components, the physical properties of the slag, such as melting temperature and viscosity, will change dramatically, deteriorating the lubrication and heat transfer of slags and further resulting in all kinds of casting defects and even casting interruption [6–10].

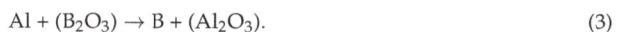

$$Al + (SiO_2) \rightarrow Si + (Al_2O_3), \tag{1}$$

$$Al + (Na_2O) \rightarrow Na(g) + (Al_2O_3), \tag{2}$$

$$Al + (B_2O_3) \rightarrow B + (Al_2O_3). \tag{3}$$

Significant efforts have been made to solve the casting problems by employing various kinds of mold fluxes. Strong oxidizing components, such as MnO and Fe₂O₃, which have priority over SiO₂ to react with Al in molten steel, have been added to mold fluxes so as to protect SiO₂ from oxidation [11], but this still led to a big increase in Al₂O₃ and altered the performance of the mold fluxes. Naturally, the $CaO-Al_2O_3$-based mold flux with low SiO_2 content was then introduced to cast high Mn-high Al steel in order to weaken or avoid interfacial reactions [7,12–14]. However, they still contain a small number of reactive components other than the low content of SiO_2, such as B_2O_3 and Na_2O, as shown in Equations (2) and (3). The components of the $CaO-BaO-Al_2O_3-CaF_2-Li_2O$ mold flux system studied in this paper have been verified not to react with Al in molten steel from thermodynamic calculation at the temperature of steelmaking [15]. Thus, the mold flux system can essentially avoid the reactivity problem and possess good application prospects in the casting of high Mn-high Al steel.

Viscosity is one of the most important indexes for evaluating the performance of mold fluxes, as it has a significant effect on the lubrication of casting blanks. In industrial production, the viscosity of mold fluxes at 1300 °C (η) and casting speed (υ) generally meet the correlation $\eta \cdot \upsilon = 1.0–3.5$ (Poise m·min^{-1}), so as to meet the demands for adequate lubrication and heat transfer [16]. A significant amount of research regarding the effect of Al_2O_3 on the viscosity of different slag systems has been conducted [17–22], while little research has been reported on mold fluxes in the absence of SiO_2. It has been concluded that Al_2O_3 exists as $[AlO_4]^{5-}$ ions with four coordinated oxygens and forms a tetrahedral network structure in basic melts [23,24]; however, more work needs to be done regarding how Al_2O_3 behaves in slags without SiO_2. It was also found that substituting CaO with BaO can reduce the viscosity and improve the vitrification rate of mold fluxes [25], which are conducive to the lubrication of the casting blank in continuous casting process. Thus, the viscosity of the $CaO-BaO-Al_2O_3-CaF_2-Li_2O$ mold flux system with various BaO/CaO and CaO/Al_2O_3 ratios in mass was studied in this paper.

Over the last few decades, a series of viscosity models has been developed to predict the viscosity of BF slags, refining slags and mold fluxes. The empirical and semi-empirical models, e.g., Urban model [26], Riboud model [27], Iida model [28,29] and NPL model [30], involve a numerical fit of viscosity to chemical composition based on Arrhenius or Weymann–Frenkel equation. A viscosity prediction model of the $CaO-BaO-Al_2O_3-CaF_2-Li_2O$ mold flux system according to the modified NPL model is established in this paper, and the estimated viscosity fits well with the experimental results.

2. Materials and Methods

2.1. Sample Preparation

Slag samples were prepared using reagent-grade Al_2O_3, CaF_2, CaO, $BaCO_3$ and Li_2CO_3 (Sinopharm Chemical Reagent Co., Ltd., Shanghai, China), with $BaCO_3$ and Li_2CO_3 being used as the sources for BaO and Li_2O, respectively. In a muffle furnace, Al_2O_3, CaF_2, $BaCO_3$ and Li_2CO_3 were calcined for 2 h at 500 °C to remove moisture, while CaO was calcined for 5 h at 1000 °C to decompose any carbonate and hydroxide before experiment. Table 1 gives the chemical compositions of samples. The prepared powders were then ground and weighed according to the designed compositions and mixed in a mortar for melting. Each fused sample was about 250 g.

Table 1. Chemical composition of experimental slags (wt %).

Slag No.	CaO	BaO	Al$_2$O$_3$	F$^-$	Li$_2$O	BaO/CaO	CaO/Al$_2$O$_3$
S-1	36	14	34	8	8	0.39	-
S-2	32	18	34	8	8	0.56	-
S-3	28	22	34	8	8	0.79	-
S-4	24	26	34	8	8	1.08	0.71
S-5	28	26	30	8	8	-	0.93
S-6	32	26	26	8	8	-	1.23
S-vali [1]	24	30	30	8	8	1.25	0.80

[1] S-vali is a slag for validation of the modified NPL (National Physical Laboratory, Teddington, UK) model.

2.2. Viscosity Measurement

Viscosity was measured using a rotation viscometer. The experimental apparatus (Chongqing Safety Production Scientific Research Co., Ltd., Chongqing, China) for viscosity measurement is shown in Figure 1. A high-temperature furnace with U-shape $MoSi_2$ heating elements is monitored by a B-type thermocouple and a proportional-integral-derivative (PID) controller to ensure that the deviation of temperature in the uniform temperature zone can be maintained within ±2 °C. Before the viscosity measurement, crucible and cylinder were set along the axis of the viscometer, and a calibration measurement was carried out at room temperature by using castor oil of known viscosity. Then, the furnace was heated to the experimental temperature (1400 °C) and held for 10 min to stabilize the temperature and homogenize the slag melt under an Ar gas atmosphere (flow rate: 0.5 L/min). After that the cylinder was carefully immersed into molten slag just 20mm above the bottom of graphite crucible and rotated at a speed of 12 rpm. When the signal from viscometer became stable, the furnace was cooled at a rate of 6 °C/min and the viscosity measurement started and lasted until the value of the measured viscosity reached nearly 3 Pa·S. In the end, the furnace was reheated to 1400 °C at a speed of 15 °C/min to pull out the cylinder and pour out the molten flux.

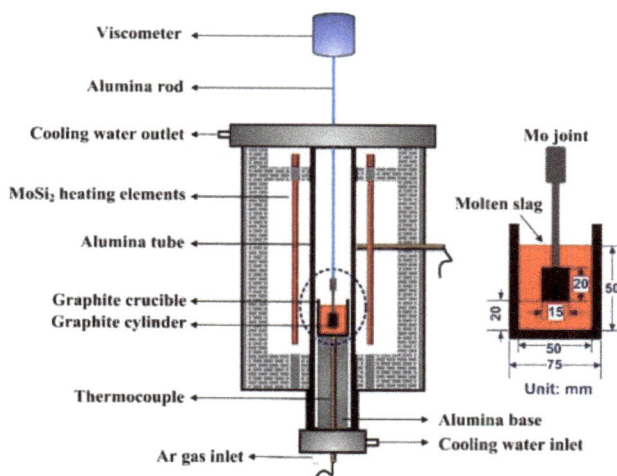

Figure 1. Schematic diagram for viscosity measurement.

3. Results and Discussion

3.1. The Effect of Substituting CaO with BaO on the Viscosity

Figure 2 shows the viscosity–temperature (η–T) curves of slags S-1 to S-4. Figure 3 shows the effect of substituting CaO with BaO on the viscosity of slags S-1 to S-4. It can be seen that the viscosity of the slags increases gradually with increasing BaO as a substitute for CaO. This is consistent with the trend observed by Wang et al. [31] for a $CaO-BaO-SiO_2-MgO-Al_2O_3$ slag system and Sukenaga et al. [32] for a $CaO-SiO_2-Al_2O_3-BaO$ slag system ($CaO/SiO_2 = 0.67$ in mass). Meanwhile, it is inconsistent with the results observed for a $BaO-CaO-Al_2O_3-MgO-B_2O_3-SiO_2-CaF_2-Li_2O$ slag system by Wu et al. [25], who found that the viscosity decreased with BaO substituting CaO. CaO and BaO, as alkaline-earth metallic oxides in the slags, have two completely different effects on the structure of the aluminates. On the one hand, as the Ba^{2+} ion radius (1.44 Å) is larger than the Ca^{2+} ion radius (1.08 Å) [33], the electrostatic potential of Ba^{2+} is smaller than that of Ca^{2+}. That is, it is easier for BaO to dissociate free oxygen (O^{2-}), and it possesses a greater depolymerization effect on network structure, which reduces the viscosity with BaO substituting CaO. On the other hand, in aluminosilicate systems, Al_2O_3

is a typical amphoteric oxide, and its behavior depends on the basicity of the melts [34]. When in an alkaline environment, it acts as a network former with charge compensation from cations and increases the complexity of the network structure. Since there is no SiO_2 in slags according to the composition in Table 1, Al_2O_3 serves as a network former and exists in the form of $[AlO_4]^{5-}$ tetrahedron with charge compensation from alkaline-earth metals [35]. As mentioned before, the electrostatic potential of Ba^{2+} is smaller than that of Ca^{2+}; that is, the capacity of charge compensation of Ba^{2+} is stronger than that of Ca^{2+}; therefore, substituting CaO with BaO can increase the structural stability of $[AlO_4]^{5-}$, which leads to the increase of the viscosity [31,36]. The experimental results indicate that the latter played a dominant role in affecting the network structure of the slags investigated in this study.

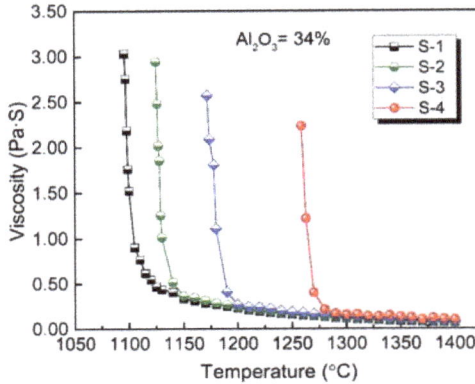

Figure 2. Viscosity–temperature curves of experimental slags with 34 wt % Al_2O_3 and various BaO/CaO ratios.

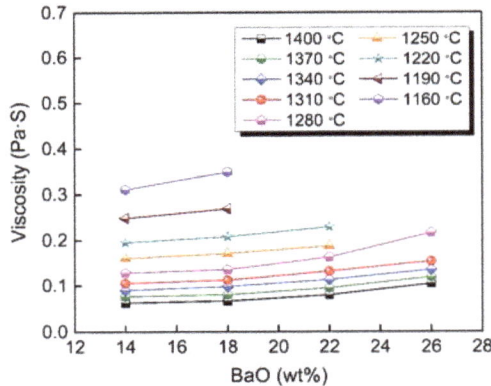

Figure 3. Viscosity of slags S-1 to S-4 in the fully fluid region with various BaO contents.

It should be noted that the substitution with BaO for CaO was performed by equal weight in mass. As the molecular weight of BaO is much greater than that of CaO, the free oxygen (O^{2-}) and charge compensation cations in the slags decreased after the substitution, which may also be a potential reason for the increase in viscosity.

3.2. Effect of CaO/Al$_2$O$_3$ Ratio on the Viscosity

Figure 4 shows the viscosity–temperature (η–T) curves of slags S-4 to S-6. Figure 5 shows the effect of CaO/Al_2O_3 ratio on the viscosity of slags S-4 to S-6. The results show that the viscosity of the slags decreased with the increase in CaO/Al_2O_3 ratio. This is consistent with the trend observed by Behera et al. [22] for an Al_2O_3–Cr_2O_3–CaO–CaF_2 slag system and Xu et al. [37] for a CaO–Al_2O_3–MgO slag system. Neuville et al. [38] studied the structure of crystals, glasses, and melts along CaO-Al_2O_3 join. The results show that Al is in octahedral coordination with high Al_2O_3 contents (>80 mol%) and essentially in fourfold coordination with 4 bridging O atoms (BOs) at Al_2O_3 contents of between 30 and 75 mol%. At around 25 mol% Al_2O_3, Al is in tetrahedral coordination with two BOs. The presence of higher-coordinated species at high Al_2O_3 contents and their absence at low Al_2O_3 contents imply different viscous flow mechanisms for high- and low-concentration Al_2O_3 networks. That is, a higher CaO/Al_2O_3 ratio tends to increase the concentration of free oxygen (O^{2-}), leading to the depolymerization of $[AlO_4]^{5-}$ tetrahedron converting into simple structure units, reducing the complexity of the network structure and decreasing the viscosity [35]. Therefore, the viscosity decreased with the increase of CaO/Al_2O_3 ratio from slags S-4 to S-6.

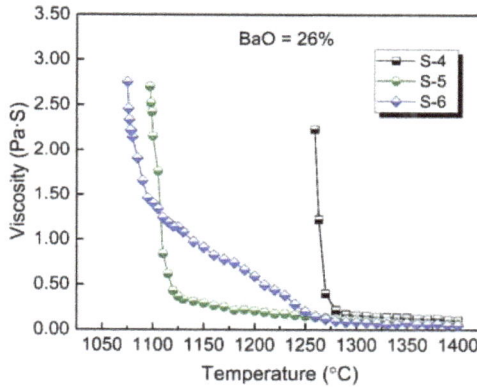

Figure 4. Viscosity–temperature curves of experimental slags with 26 wt % BaO and various CaO/Al_2O_3 ratios.

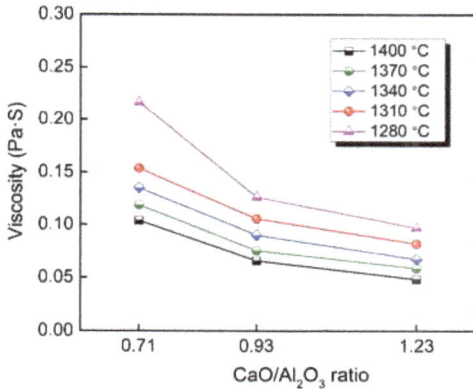

Figure 5. Viscosity of slags S-4 to S-6 in the fully fluid region with various CaO/Al_2O_3.

The optical basicity (Λ) can characterize the 'availability' of providing free oxygen ions, which indicates the degree of polymerization of the melts. A corrected optical basicity (Λ^{corr}) was proposed by Mills to charge balance the Al^{3+} ions incorporated into the Si^{4+} chain or ring [39]. Λ^{corr} is calculated by Equation (4), where X is the mole fraction, Λ is the optical basicity, and n refers to the number of oxygen atoms in the molecule, e.g., three for Al_2O_3, and one for CaF_2, etc., since two fluorine ions can be considered equivalent to one oxygen ion. A correction needs to be made in the calculation process: the component with the highest optical basicity will compensate the charge of Al_2O_3 first (if it is not enough, the one with the second-highest optical basicity will continue to compensate, and so on), and the remaining, together with the components not participating in charge compensation, contributes to the corrected optical basicity of the slag system. The values of the corrected optical basicity (Λ^{corr}) of slags S-1 to S-6 are listed in Table 2. For a series of slags with a similar structure in the silicate system, the viscosity of slags generally decreases with the increase in basicity. This is also applied to slags S-4 to S-6, the viscosity of which decreases with the increase of the corrected optical basicity (Λ^{corr}). Similarly, the viscosity of slags increases with the decrease of corrected optical basicity (Λ^{corr}) for slags S-1 to S-4.

$$\Lambda^{corr} = \frac{\sum(X_1 n_1 \Lambda_1 + X_2 n_2 \Lambda_2 + X_3 n_3 \Lambda_3 \cdots)}{\sum(X_1 n_1 + X_2 n_2 + X_3 n_3 \cdots)} \tag{4}$$

Moreover, Xu et al. found that the curves of viscosity versus CaO/Al_2O_3 ratio was a 'V' shape when studying the effect of CaO/Al_2O_3 ratio on the viscosity of a $CaO–Al_2O_3–MgO$ slag system [37]. They deemed that the formation of low melting compounds ($12CaO·7Al_2O_3$) was conducive to the viscosity reduction. It can be seen from Table 2 that the variation tendency of the melting temperature is correspondingly in agreement with that of the viscosity for slags S-4 to S-6 or slags S-1 to S-4.

Table 2. Melting temperature (T_m), break temperature (T_{br}), corrected optical basicity (Λ^{corr}), new optical basicity (Λ_{new}), and viscous activation energy of experimental slags.

Slag No.	T_m (°C)	T_{br} (°C)	Λ^{corr}	Λ_{new}	E_η (kJ·mol^{-1})	E_w (kJ·mol^{-1})
S-1	1087	1132	0.760	1.760	133.4	146.2
S-2	1096	1148	0.753	1.771	131.0	143.9
S-3	1114	1196	0.748	1.783	119.1	132.2
S-4	1119	1285	0.743	1.797	91.2	101.3
S-5	1114	1124	0.777	1.787	118.1	130.9
S-6	1080	1268	0.811	1.778	121.9	135.3

3.3. Viscous Flow Activation Energy

It can be observed from the viscosity–temperature curves in Figures 2 and 4 that there is a temperature point below which the viscosity of the slag increases sharply in a narrow range of temperature. This temperature point is known as the break temperature (T_{br}), which can be obtained from the graph of the natural logarithm of viscosity versus inverse absolute temperature, as shown in Figure 6, taking slag S-3 as an example. When the temperature is above the T_{br}, the molten slag is in the state of fully liquid phase, and is a Newtonian fluid, while it presents as a non-Newtonian fluid when the temperature is below the T_{br} as a result of crystallization or solidification [40,41]. The break temperature of the slags is also listed in Table 2.

The temperature dependence of the viscosity is usually expressed by Arrhenius equation (Equation (5)) or Weymann–Frenkel equation (Equation (6)):

$$\eta = A \exp(\frac{E_\eta}{RT}), \tag{5}$$

$$\eta = AT \exp(\frac{E_w}{RT}), \tag{6}$$

where A is a proportionality constant, E_η and E_w are the activation energies for viscous flow, R is the gas constant and T is the absolute temperature. Figure 7 shows a linear relationship between $\ln\eta$ and $1/T$ and Figure 8 shows a linear relationship between $\ln(\eta/T)$ and $1/T$ in the Newton fluid region, from which the activation energy is obtained. The deviations between E_η and E_w for slags in the present study are below 10%, indicating that either the Arrhenius or Weymann–Frenkel equation can be used to evaluate the activation energy for viscous flow. The activation energies according to Arrhenius listed in Table 2 are from 91.2 kJ·mol^{-1} to 133.4 kJ·mol^{-1}, which are lower than those by Xu et al. [37] and Kim et al. [42], with their results being in the range of 175–400 kJ·mol^{-1}. It may be that the slags in this study contained a higher content of fluxing agent—namely, Li_2O—and diluting agent—namely, CaF_2. Additionally, the difference in network structure between their slags and the authors' slags also leads to a difference in activation energy.

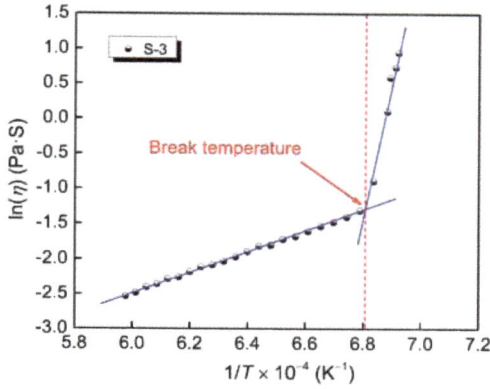

Figure 6. The natural logarithm of viscosity versus inverse absolute temperature of slag S-3.

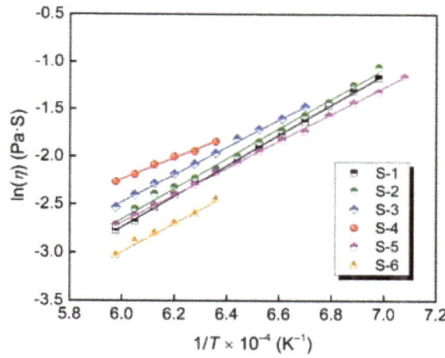

Figure 7. The natural logarithm of η as a function of $1/T$.

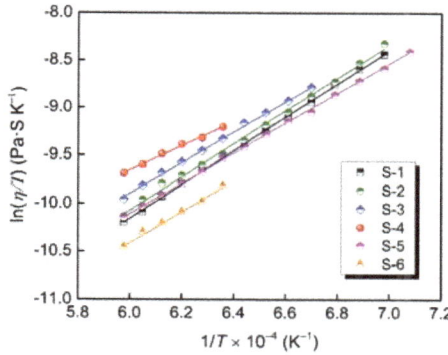

Figure 8. The natural logarithm of η/T as a function of $1/T$.

3.4. Industrial Application Prospect

To evaluate whether the viscosity of the slags in the present study is applicable for industrial application, 20Mn23AlV was selected as an instance. The casting speed of 20Mn23AlV is about 0.9 m/min, and the viscosities of the mold fluxes at 1300 °C in this paper are from 1.13 to 1.58 Poise; thus, the values of $\eta \cdot \upsilon$ are from 1.02 to 1.42, which meet the correlation $\eta \cdot \upsilon = 1.0$–3.5 (Poise m·min^{-1}). Moreover, according to the above empirical formula, the fluxes studied in this paper can be applied to industrial production with a maximum casting speed of approximately 3 m/min. Meanwhile, the casting speed of high Mn-high Al steel in industrial production is generally less than 2 m/min, due to the solidification characteristics of the steels and the limitations of the actual casting conditions. Therefore, from the perspective of slag viscosity, the fluxes studied in this paper have good prospects for industrial application. Nevertheless, other properties, such as break temperature, vitrification rate, etc., need to be taken into consideration, so as to select a proper mold flux.

4. Viscosity Model

Since viscosity measurement is time-consuming work, several researchers have tried to establish various viscosity models to predict the viscosity of slags, e.g., the Riboud model [27,43], the Urbain model [26,44], the Iida model [28], the KTH model [45], and the NPL model [30,46,47]. These viscosity models were mainly based on CaO–SiO$_2$-based slags or CaO–SiO$_2$–Al$_2$O$_3$-based slags with low Al$_2$O$_3$ content. Figure 9 shows the estimated viscosity against measured viscosity of slags in this study with the Urbain model, Riboud model, and NPL model. It can be observed that there is a large deviation between the viscosity predicted by the Urbain model and the Riboud model and the measured viscosity. However, the predicted results according to the NPL model are much better, with some estimated viscosity being in good agreement with the measured viscosity, while the rest does not have a large deviation. Therefore, a modified NPL model based on the Weymann–Frenkel equation and optical basicity to predict the viscosity of slags was established in this study.

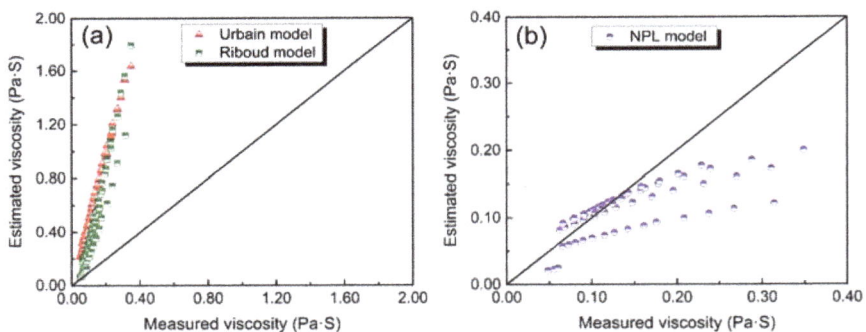

Figure 9. The estimated viscosity against the measured viscosity using different models: (**a**) Urbain and Riboud models, (**b**) NPL model.

The Weymann–Frenkel equation (Equation (6)) can be rewritten as follows:

$$\ln \frac{\eta}{T} = \ln A + \frac{1000B}{T},$$
(7)

where parameter B ($B = \frac{E_w}{1000R}$) is a quadratic expression of the optical basicity, parameters A and B are in a linear relationship [44,46]. Moreover, according to the intercept and slope of each fitted line in Figure 8, the values of parameters A and B of each mold flux (or each optical basicity) can be obtained, respectively.

Similar to the traditional definition of basicity, Shankar et al. [47] redefined the optical basicity, termed as the new optical basicity, and its expression is as follows:

$$\Lambda_{new} = \frac{\dfrac{\sum(X_B n_B \Lambda_B + \cdots)}{\sum(X_B n_B + \cdots)}}{\dfrac{\sum(X_A n_A \Lambda_A + \cdots)}{\sum(X_A n_A + \cdots)}}. \tag{8}$$

In Equation (8), the new optical basicity is defined as the ratio of the total basic oxides to the total acidic oxides in the slag. Al_2O_3 is regarded as an acidic oxide in the slags studied in this paper, as it is the only network former of the molten slags. The new optical basicity is employed to predict the parameters A and B in Equation (7).

Figure 10 shows the relationship between the new optical basicity and parameter B calculated from the measured viscosity by Equation (7). Their relation can be fitted by a quadratic polynomial with a determination coefficient of 0.972, as shown in Equation (9). The parameters B and $\ln A$ are in a linear relationship, as shown in Figure 11, which can be expressed by Equation (10) with a determination coefficient of 0.981.

$$B = -4337.4\Lambda^2 + 15298.9\Lambda - 13473.2, \tag{9}$$

$$\ln A = -0.7057B - 8.3678. \tag{10}$$

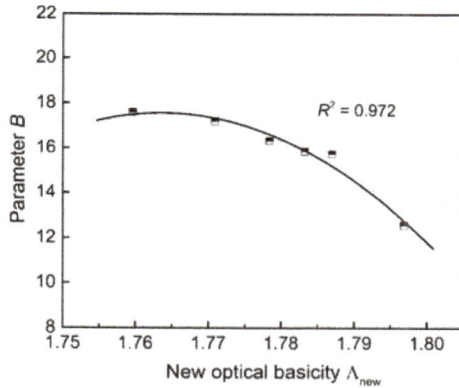

Figure 10. The relationship between the new optical basicity (Λ_{new}) and parameter B.

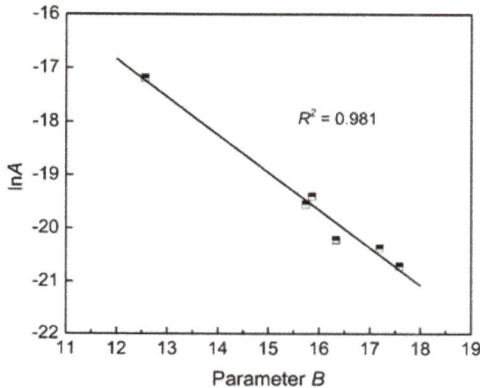

Figure 11. The relationship between parameters A and B.

Viscosity of the slags now can be calculated by Equations (7)–(10). The estimated viscosity compared with the measured viscosity is shown in Figure 12. It can be seen that the predicted viscosity by the modified NPL model is in a reasonable agreement with the experimental data with the deviation below 20%.

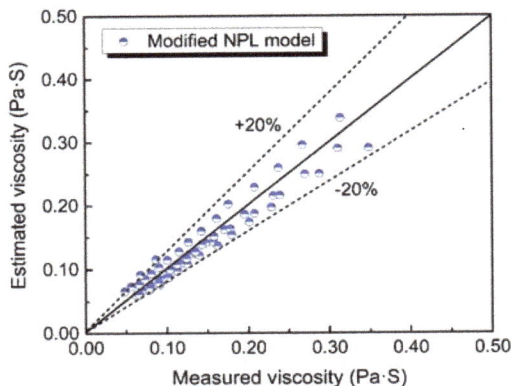

Figure 12. The estimated viscosity against the measured viscosity using modified NPL model.

To sum up, the modified parameters A and B, as well as a new optical basicity, were employed in the modified NPL model to reveal the dependence of viscosity on composition of $CaO–Al_2O_3$-based slag system in this study. As there is no measured viscosity data for slags with similar compositions in the other literature, an additional mold flux (S-vali) was measured with the composition listed in Table 1. Using the modified NPL model in this paper, it is found that the predicted and measured values also fit very well as shown in Figure 13.

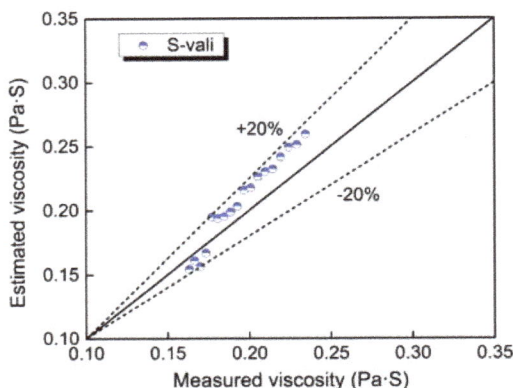

Figure 13. The estimated viscosity against the measured viscosity of S-vali using the modified NPL model.

5. Conclusions

In this paper, the effect of substituting CaO with BaO and CaO/Al_2O_3 ratio on the viscosity of $CaO–BaO–Al_2O_3–CaF_2–Li_2O$ mold flux system was investigated. It was found that the viscosity of the slags increased with increasing BaO as a substitute for CaO, while the viscosity decreased with the increase of CaO/Al_2O_3 ratio due to the combination effect of depolymerization by free oxygen and polymerization by charge compensation on Al-O network structure. The deviation of the viscous activation energy calculated by Arrhenius and Weymann–Frenkel equations is less than 10%

and the values of the viscous activation energy according to Arrhenius equation are in the range of 92.1–133.4 kJ·mol^{-1}. The modified NPL model and the new optical basicity regarding Al$_2$O$_3$ as an acid oxide were employed to predict the viscosity. The prediction results show that the estimated viscosity is in a reasonable agreement with the measured value, with the deviation less than 20%.

Author Contributions: Z.L., S.H., Q.W. (Qian Wang), and Q.W. (Qiangqiang Wang) conceived and designed the experiment; Z.L., X.Y., and M.L. performed the experiment and analyzed the experimental data; Z.L. and Q.W. (Qiangqiang Wang) wrote the paper.

Funding: This research was funded by the Key Program of the National Natural Science Foundation of China (grant number U1660204) and the National Natural Science Foundation of China (grant numbers 51874057, 51804057) and the Fundamental Research Funds for the Central Universities in China (grant number 2018CDXYCL0018).

Conflicts of Interest: The authors declare no conflict of interest.

References

1. Neu, R. Performance and characterization of TWIP steels for automotive applications. *Mater. Perform. Charact.* **2013**, *2*, 244–284. [CrossRef]
2. Grässel, O.; Krüger, L.; Frommeyer, G.; Meyer, L. High strength Fe–Mn–(Al,Si) TRIP/TWIP steels development–properties–application. *Int. J. Plast.* **2000**, *16*, 1391–1409. [CrossRef]
3. He, S.; Li, Z.; Chen, Z.; Wu, T.; Wang, Q. Review of mold fluxes for continuous casting of high-alloy (Al,Mn,Ti) steels. *Steel Res. Int.* **2019**, *90*, 1800424. [CrossRef]
4. Fu, X.; Wen, G.; Liu, Q.; Tang, P.; Li, J.; Li, W. Development and evaluation of CaO–SiO$_2$ based mould fluxes for casting high aluminum TRIP steel. *Steel Res. Int.* **2015**, *86*, 110–120. [CrossRef]
5. Becker, J.J.; Madden, M.A.; Natarajan, T.T.; Piccone, T.; Serrano, E.; Story, S.; Ecklund-Baker, S.; Nickerson, I.; Schlichting, W. Liquid/Solid Interactions during Continuous Casting of High-Al Advanced High Strength Steels. In Proceedings of the AISTech-Conference, Charlotte, NC, USA, 9–12 May 2005.
6. Blazek, K.; Yin, H.; Skoczylas, G.; McClymonds, M.; Frazee, M. Evaluation of Lime-silica and Lime-alumina Mold Powders Developed for Casting High Aluminum TRIP Steel Grades. In Proceedings of the METEC InSteelCon 2011, Düsseldorf, Germany, 27 June–1 July 2011.
7. Street, S.; James, K.; Minor, N.; Roelant, A.; Tremp, J. Production of high-aluminum steel slabs. *Iron Steel Technol.* **2008**, *5*, 38–49.
8. Xiong, Y.; Wen, G.-H.; Ping, T.; Huan, W. Behavior of mold slag used for 20Mn23Al nonmagnetic steel during casting. *J. Iron Steel Res. Int.* **2011**, *18*, 20–25.
9. Wu, T.; He, S.; Zhu, L.; Wang, Q. Study on reaction performances and applications of mold flux for high-aluminum steel. *Mater. Trans.* **2016**, *57*, 58–63. [CrossRef]
10. Ji, C.; Yang, C.; Zhi, Z.; Tian, Z.; Zhao, C.; Zhu, G. Continuous casting of high-Al steel in Shougang Jingtang steel works. *J. Iron Steel Res. Int.* **2015**, *22*, 53–56. [CrossRef]
11. He, S.; Wang, Q.; Zeng, J.; Zhang, M.; Xie, B. Properties control of mold fluxes for high aluminum steel. *J. Iron Steel Res.* **2009**, *12*, 59–62.
12. Cho, J.; Blazek, K.; Frazee, M.; Yin, H.; Park, J.H.; Moon, S.W. Assessment of CaO–Al$_2$O$_3$ based mold flux system for high aluminum TRIP casting. *ISIJ Int.* **2013**, *53*, 62–70. [CrossRef]
13. Liu, Q.; Wen, G.; Li, J.; Fu, X.; Tang, P.; Li, W. Development of mould fluxes based on lime–alumina slag system for casting high aluminium TRIP steel. *Ironmak. Steelmak.* **2014**, *41*, 292–297. [CrossRef]
14. Yu, X.; Wen, G.; Tang, P.; Wang, H. Investigation on viscosity of mould fluxes during continuous casting of aluminium containing TRIP steels. *Ironmak. Steelmak.* **2009**, *36*, 623–630. [CrossRef]
15. Wu, T.; He, S.; Guo, Y.; Wang, Q. Study on reactivity between mould fluxes and high-Al molten steel. *Charact. Miner. Met. Mater.* **2014**, 265–270.
16. Nakano, T.; Kishi, T.; Koyama, K.; Komai, T.; Naitoh, S. Mold powder technology for continuous casting of aluminum-killed steel. *Trans. Iron Steel Inst. Jpn.* **1984**, *24*, 950–956. [CrossRef]
17. Zhang, Z.; Wen, G.; Tang, P.; Sridhar, S. The influence of Al$_2$O$_3$/SiO$_2$ ratio on the viscosity of mold fluxes. *ISIJ Int.* **2008**, *48*, 739–746. [CrossRef]
18. Xu, J.F.; Zeng, T.; Sheng, M.Q.; Jie, C.; Wan, K.; Zhang, J.Y. Viscosity of low silica CaO–5MgO–Al$_2$O$_3$–SiO$_2$ slags. *Ironmak. Steelmak.* **2014**, *41*, 486–492. [CrossRef]

19. Zhang, X.; Jiang, T.; Xue, X.; Hu, B. Influence of MgO/Al$_2$O$_3$ ratio on viscosity of blast furnace slag with high Al$_2$O$_3$ content. *Steel Res. Int.* **2016**, *87*, 87–94. [CrossRef]
20. Kim, G.H.; Sohn, I. Effect of Al$_2$O$_3$ on the viscosity and structure of calcium silicate-based melts containing Na$_2$O and CaF$_2$. *J. Non-Cryst. Solids* **2012**, *358*, 1530–1537. [CrossRef]
21. Li, J.; Shu, Q.; Chou, K. Effect of Al$_2$O$_3$/SiO$_2$ mass ratio on viscosity of CaO–Al$_2$O$_3$–SiO$_2$–CaF$_2$ slag. *Ironmak. Steelmak.* **2015**, *42*, 154–160. [CrossRef]
22. Behera, R.; Mohanty, U. Viscosity of Molten Al$_2$O$_3$–Cr$_2$O$_3$–CaO–CaF$_2$ Slags at Various Al$_2$O$_3$/CaO Ratios. *ISIJ Int.* **2001**, *41*, 834–843. [CrossRef]
23. Wu, T.; Wang, Q.; Yao, T.; He, S. Molecular dynamics simulations of the structural properties of Al$_2$O$_3$-based binary systems. *J. Non-Cryst. Solids* **2016**, *435*, 17–26. [CrossRef]
24. Kang, Y.; Morita, K. Thermal conductivity of the CaO–Al$_2$O$_3$–SiO$_2$ system. *ISIJ Int.* **2006**, *46*, 420–426. [CrossRef]
25. Wu, T.; Wang, Q.; He, S.; Xu, J.; Long, X.; Lu, Y. Study on properties of alumina-based mould fluxes for high–Al steel slab casting. *Steel Res. Int.* **2012**, *83*, 1194–1202. [CrossRef]
26. Urbain, G. Viscosity estimation of slags. *Steel Res.* **1987**, *58*, 111–116. [CrossRef]
27. Riboud, P.; Roux, Y.; Lucas, L.; Gaye, H. Improvement of continuous casting powders. *Fachberichte Huttenpraxis Metallweiterverarbeitung* **1981**, *19*, 859–869.
28. Iida, T.; Sakai, H.; Kita, Y.; Shigeno, K. An equation for accurate prediction of the viscosities of blast furnace type slags from chemical composition. *ISIJ Int.* **2000**, *40*, S110–S114. [CrossRef]
29. Iida, T.; Sakai, H.; Kita, Y.; Murakami, K. Equation for estimating viscosities of industrial mold fluxes. *High. Temp. Mater. Processes* **2000**, *19*, 153–164. [CrossRef]
30. Mills, K.; Sridhar, S. Viscosities of ironmaking and steelmaking slags. *Ironmak. Steelmak.* **1999**, *26*, 262–268. [CrossRef]
31. Wang, Z.; Sohn, I. Effect of substituting CaO with BaO on the viscosity and structure of CaO–BaO–SiO$_2$–MgO–Al$_2$O$_3$ slags. *J. Am. Ceram. Soc.* **2018**, *101*, 4285–4296. [CrossRef]
32. Sukenaga, S.; Saito, N.; Kawakami, K.; Nakashima, K. Viscosities of CaO–SiO$_2$–Al$_2$O$_3$–(R$_2$O or RO) melts. *ISIJ Int.* **2006**, *46*, 352–358. [CrossRef]
33. Whittaker, E.; Muntus, R. Ionic radii for use in geochemistry. *Geochim. Cosmochim. Acta* **1970**, *34*, 945–956. [CrossRef]
34. Kou, T.; Mizoguchi, K.; Suginohara, Y. The Effect of Al sub 2 O sub 3 on the Viscosity of Silicate Melts. *J. Jpn. Inst. Met.* **1978**, *42*, 775–781. [CrossRef]
35. Licheron, M.; Montouillout, V.; Millot, F.; Neuville, D.R. Raman and 27Al NMR structure investigations of aluminate glasses: $(1 − x)$Al$_2$O$_3$ − xMO, with M = Ca, Sr, Ba and $0.5 < x < 0.75$. *J. Non-Cryst. Solids* **2011**, *357*, 2796–2801.
36. Gao, E.; Wang, W.; Zhang, L. Effect of alkaline earth metal oxides on the viscosity and structure of the CaO–Al$_2$O$_3$ based mold flux for casting high-Al steels. *J. Non-Cryst. Solids* **2017**, *473*, 79–86. [CrossRef]
37. Xu, J.; Zhang, J.; Jie, C.; Ruan, F.; Chou, K. Experimental measurements and modelling of viscosity in CaO–Al$_2$O$_3$–MgO slag system. *Ironmak. Steelmak.* **2011**, *38*, 329–337. [CrossRef]
38. Neuville, D.R.; Henderson, G.S.; Cormier, L.; Massiot, D. The structure of crystals, glasses, and melts along the CaO–Al$_2$O$_3$ join: Results from Raman, Al L-and K-edge X-ray absorption, and 27Al NMR spectroscopy. *Am. Mineral.* **2010**, *95*, 1580–1589. [CrossRef]
39. Mills, K.C. The influence of structure on the physico-chemical properties of slags. *ISIJ Int.* **1993**, *33*, 148–155. [CrossRef]
40. Brandaleze, E.; Di Gresia, G.; Santini, L.; Martín, A.; Benavidez, E. Mould Fluxes in the Steel Continuous Casting. In *Science and Technology of Casting Processes*; Srinivasan, M., Ed.; IntechOpen: London, UK, 2012; pp. 212–215.
41. Sridhar, S.; Mills, K.; Afrange, O.; Lörz, H.; Carli, R. Break temperatures of mould fluxes and their relevance to continuous casting. *Ironmak. Steelmak.* **2000**, *27*, 238–242. [CrossRef]
42. Kim, J.W.; Choi, J.; Kwon, O.H.; Lee, I.R.; Shin, Y.K.; Park, J.S. Viscous Characteristics of Synthetic Mold Powder for High Speed Continuous Casting. In Proceedings of the 4th International Conference on Molten Slags and Fluxes, Sendai, Japan, 8–11 June 1992.
43. Mills, K.; Yuan, L.; Jones, R. Estimating the physical properties of slags. *J. South. Afr. Inst. Min. Metall.* **2011**, *111*, 649–658.

44. Urbain, G.; Cambier, F.; Deletter, M.; Anseau, M.R. Viscosity of silicate melts. *Trans. J. Br. Ceram. Soc.* **1981**, *80*, 139–141.

45. Ji, F.Z.; Sichen, D.; Seetharaman, S. Experimental studies of the viscosities in the $CaO–Fe_nO–SiO_2$ slags. *Metall. Mater. Trans. B* **1997**, *28*, 827–834. [CrossRef]

46. Ray, H.; Pal, S. Simple method for theoretical estimation of viscosity of oxide melts using optical basicity. *Ironmak. Steelmak.* **2004**, *31*, 125–130. [CrossRef]

47. Shankar, A.; Görnerup, M.; Lahiri, A.; Seetharaman, S. Estimation of viscosity for blast furnace type slags. *Ironmak. Steelmak.* **2007**, *34*, 477–481. [CrossRef]

metals

MDPI

Article

Structure of Solidified Films of CaO-SiO₂-Na₂O Based Low-Fluorine Mold Flux

Jianhua Zeng [1], Xiao Long [2,3], Xinchen You [1], Min Li [1], Qiangqiang Wang [1] and Shengping He [1,*]

[1] College of Materials Science and Engineering, Chongqing University, Chongqing 400044, China; yjyzjh@126.com (J.Z.); vanmeuen@cqu.edu.cn (X.Y.); sjzlimin@cqu.edu.cn (M.L.); wangqiangq@cqu.edu.cn (Q.W.)

[2] School of Materials and Metallurgical Engineering, Guizhou Institute of Technology, No.1 Caiguan Road, Guiyang 550003, China; xiaolong@git.edu.cn

[3] Key Laboratory of Light Metal Materials Processing Technology of Guizhou Province, Guizhou Institute of Technology, Guiyang 550003, China

* Correspondence: heshp@cqu.edu.cn; Tel.: + 86-13883296352

Received: 4 December 2018; Accepted: 11 January 2019; Published: 16 January 2019

Abstract: As an essential synthetic material used in the continuous casting of steels, mold fluxes improve the surface quality of steel slabs. In this study, a CaO-SiO₂-Na₂O-based low-fluorine mold flux was solidified by an improved water-cooled copper probe with different temperatures of molten flux and different probe immersion times. The heat flux through solid films and the film structures were calculated and inspected, respectively. Internal cracks (formed in the glassy layer of films during solidification) were observed. The formation and evolution of those cracks contributed to the unstable heat flux density. The roughness of the surface in contact with the water-cooled copper probe formed as films were still glassy and the roughness had no causal relationship with crystallization or devitrification. Combeite with columnar and faceted dendritic shapes were the main crystal in the film.

Keywords: mold flux; low fluorine; internal crack; surface roughness; slag film

1. Introduction

In the continuous casting of steels, mold fluxes control the heat flux from steel shells to the mold [1–3] by forming a solid slag film. For medium-carbon steels, large volume shrinkages caused by peritectic transition (delta ferrite to austenite) at meniscus make some steels quite crack-sensitive [4,5]. Thus, slow and uniform cooling are required to prevent the formation of longitudinal cracks on slab surfaces. The usual method to solve this problem is the use of mold fluxes to lower the heat flux from initial steel shells to the mold near meniscus. Conventional mold fluxes for peritectic steels usually have high fluorine contents and high basicities (binary basicity from 1.2 to 1.5, CaO-SiO₂-CaF₂-based slags). As an important element in fusion agents (e.g., NaF, CaF₂, etc.), fluorine simplifies the microstructure of molten fluxes, decreases its high-temperature viscosities, and promotes the formation of cuspidine (3CaO·2SiO₂·CaF₂) in solid slag films [6–8]. Based on previous approaches [9,10], the precipitation of cuspidine in slag films is widely considered as a major contribution of high-basicity mold fluxes to decrease the heat flux. Therefore, fluorine is considered as a significant component in mold fluxes for peritectic steels, and is usually present in levels up to 10 mass percentage or more. However, fluorine-containing gases (mostly reported as NaF and SiF₄) evaporate from molten fluxes at high temperatures. These gases potentially pollute the environment and damage the health of workers. High-fluorine mold fluxes also corrode casters as fluorine dissolves into the second-cooling water to form hydrofluoric acid [11,12].

Based on these disadvantages, efforts towards the development of low-fluorine and fluorine-free mold fluxes have been reported frequently, mostly in CaO-SiO$_2$-Na$_2$O- and CaO-SiO$_2$-TiO$_2$-based systems [13–16]. However, industrial trials gave unstable performances of these fluorine-free or low-fluorine fluxes for peritectic steels—even their physical properties (i.e., melting temperature, viscosity, break temperature, crystallization capacity, etc.) are similar with conventional high-fluorine ones. Related research gives some possible explanations [17]: TiO$_2$-containing mold fluxes are likely to deteriorate the lubrication capacity of liquid slag films and lead to an unstable casting (sticking or breakouts) because of the formation of titanium carbide and/or titanium carbonitride particles in molten fluxes (TiO$_2$ reacts with the carbon from sintering-control components and dissolved nitrogen in molten fluxes). Besides, the structure and evolution of the fluorine-decreased films during solidification (the roughness of surfaces in contact with the mold, porosity, growth rate, crystallization characteristics, etc.) could be very different from conventional ones with a high-fluorine content, which also affect the performance of mold fluxes directly and are still unclear.

This study selected a typical low-fluorine CaO-SiO$_2$-Na$_2$O-based mold flux (which gives unstable performances in commercial practices) and investigated the structural evolution of its films upon solidification. The structure differences were then compared with conventional high-fluorine ones.

Previous approaches [18] used a water-cooled copper probe to solidify slag films from molten fluxes and obtain heat flux densities. However, some details of the previous probe likely affected the results. First, the intensity cooling of the large probe surpasses the capacity of resistance furnaces to maintain a constant temperature of molten fluxes, resulting in a poor uniformity of film structures (especially the thickness and feature of the surface in contact with the probe). Second, the probe is in a cubic shape with small width–thickness ratio, and the obtained solid films are more likely to be affected by two-dimensional cooling, which could affect the structure of films. To solve those problems, a water-cooled copper probe with a much smaller volume and large width–thickness ratio was developed to obtain consistent film thicknesses, structures, and reliable heat flux densities [19].

2. Experimental Method

2.1. Mold Flux Selection and Slag Film Acquisition

A typical CaO-SiO$_2$-Na$_2$O-based low-fluorine (LF) mold flux was used as the basis of this study (see the composition in Table 1). Compared with conventional high-fluorine fluxes, this mold flux gave unstable performance in the casting of peritectic grade steels. Experimental samples were prepared using analytical-grade reagents (CaCO$_3$, SiO$_2$, Na$_2$CO$_3$, CaF$_2$, and Al$_2$O$_3$).

Table 1. Composition of mold flux and mass percentages.

CaO%/SiO$_2$%	Al$_2$O$_3$	Na$_2$O	F
0.7	1	23	3

The physical properties were measured (see Table 2) [20–22]. The melting temperature was measured by the hemisphere point method; the high temperature viscosity and viscosity–temperature curve were measured by a high-temperature rotational viscometer (using graphite bobs with 15 mm diameter and graphite crucibles with 55 mm inner diameter). The break temperature was defined as the temperature below which the flux viscosity increases sharply upon cooling at speed 6 K/min. For mold fluxes with a strong crystallization tendency, the break temperature was approximately equal to the liquidus temperature.

Table 2. Physical property of the mold flux.

Viscosity$_{1300°C}$, Pa·s	Melting Temperature, K	Break Temperature, K
0.183	1431	1480

For each experiment, 300 g of pre-melted mold flux was loaded into a graphite crucible (60 mm inner diameter) and melted in a resistance furnace. The water-cooled probe was immersed into the melted flux to solidify slag films (probe size: 6 mm in thickness, 20 mm in length, 15 mm in height; immersion depth: 13 mm; cooling water of the probe: 1.75 dm³/ min). The heat flux densities through solid slag films were calculated using Equation (1), where Q is heat flux density, W is flow rate of cooling water, $(T_{out} - T_{in})$ is the temperature increase of water after it passes through the probe, A is the surface area that the probe contacted liquid fluxes, and C_p is the specific heat capacity of water. The solidified films were recovered after different probe immersion times (i.e., 60, 90, and 120 s). Three slag bulk temperatures (1573, 1623, and 1673 K) were used to reveal the effect of bulk temperature on the structure and thermal property of solid films.

$$Q = \frac{WC_p(T_{out} - T_{in})}{1000A} \tag{1}$$

2.2. Measurements on Solidified Films

The thickness of solid films was measured with a point micrometer (six measurements for each wide-face film). The roughness of film surfaces in contact with the copper probe was measured with a contact profilometer. The roughness (R_a) measurements were performed on the probe-side surface of each wide-face film (measuring six times). The illustrated positions, length, and direction of the roughness measurements are given in Figure 1a. The illustrated positions of thickness measurements are shown in Figure 1b.

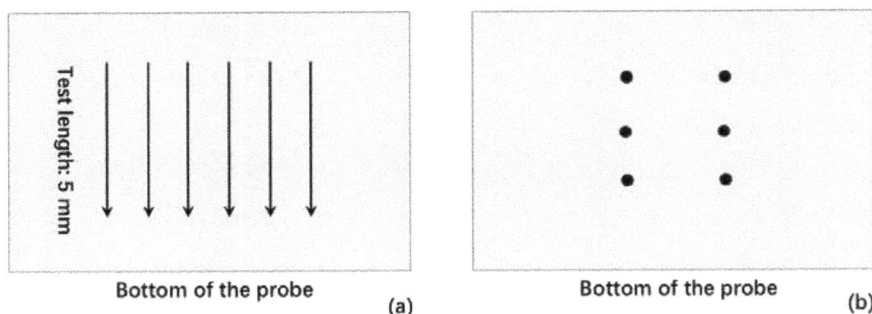

Figure 1. (**a**) Illustrated positions of surface roughness measurements and (**b**) thickness measurements. Films for measurement were recovered from the wide face of the probe.

The overall closed porosity of films was calculated. The apparent densities (d_a) of solid films and true densities (d_t, films pulverized into powders with size smaller than 45 µm) were measured by a gas pycnometer. The closed porosity of films can be calculated as $(d_t - d_a)/d_t$.

Surface and internal features of films were inspected with scanning electron and optical microscopies. The cross section samples were prepared after mounting films in resin and polishing with Al₂O₃ suspension. Samples for SEM were sputter-coated with 2–3 nm of Pt. X-ray diffraction (Cu-K$_\alpha$ radiation) was used to identify the major crystal of the pulverized film (the film sample recovered from molten flux with 1623 K bulk temperature after 120 s immersion of the probe).

3. Results and Discussions

3.1. Thickness and Heat Flux Density of Solid Films

The results of the thickness and heat flux density of films are shown in Figure 2. As expected, the thickness of films increased gradually with decreased slag bulk temperatures and increased probe immersion times. Unstable heat flux densities were detected, especially at higher slag bulk temperature (i.e., 1623 or 1673 K) and short immersion time (within 20 s, as marked with a rectangle in Figure 2). Besides intensive cooling, non-uniform cooling conditions at meniscuses also contributed to the formation of longitudinal cracks on initial steel shells. The unstable heat flux through low-fluorine films may contribute to their unstable performances. Other recent works have shown a similar fluctuation of heat flux through solidified fluorine-free films [19], but for conventional and ultrahigh-basicity mold fluxes (with high F content), no apparent fluctuation was detected [23,24]. The fluctuation of heat flux densities was partially contributed by the formation and evolution of internal cracks of glassy films upon solidification, which is discussed below.

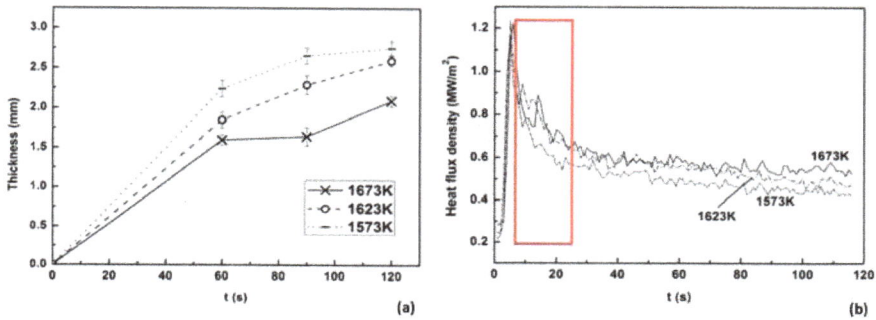

Figure 2. (**a**) Thickness of solid films after different immersion times in molten fluxes with different bulk temperatures (error bars show standard deviations); and (**b**) Measured heat flux density.

3.2. Cracks Formed in the Glassy Layer

Apparent internal cracks were observed (see Figure 3) in the glassy layer of solid films. The obvious fusion tendency of the crack boundaries indicated that those cracks formed and fused during the immersion of probe in molten fluxes (not formed upon cooling after taking solidified films out of the molten flux). Intensive cooling of the water-cooled copper probe and the large temperature gradients of initial solidified films could cause the formation of cracks. The mechanism of the formation and evolution of those cracks is shown in Figure 4, which indicates that the small pores with irregular shapes in the glass layer (see Figure 3a) are likely formed by the evolution (fusion) of cracks. As the formation of large cracks in the initial glassy film increased its thermal contact resistance (as Figure 4a and 4b show), the fusion of crack boundaries in contrast decreased the thermal contact resistance (Figure 4c). The continual formation and fusion of cracks naturally contributes to the heat flux fluctuation (especially in the early stage of solidification, as shown in Figure 2). No similar cracks and large heat flux fluctuations were observed in solid films with high ratio of crystals [19,23,24].

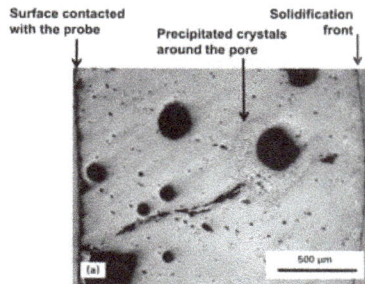

Figure 3. *Cont.*

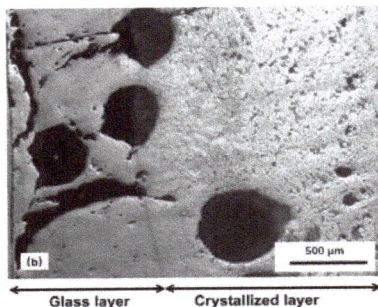

Figure 3. Cross sections of films recovered from the molten flux with 1623 K bulk temperature, after (a) 60 s and (b) 90 s of probe immersion. The left side of micrographs is the probe side; crystals started to precipitate around the edge of pores in (a). (Optical micrographs; black regions and spots are pores and cracks).

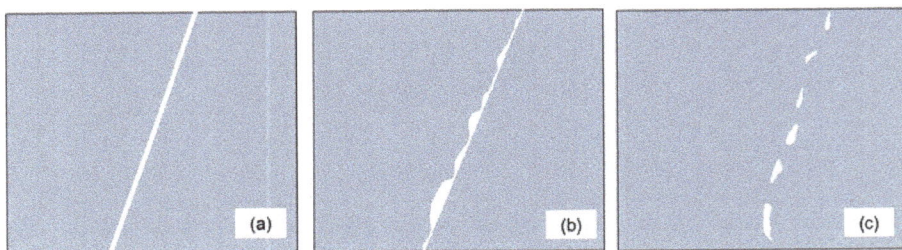

Figure 4. Schematic of the formation and evolution of cracks in the glassy layer (left side of each graph refers to the probe side): (a) formation of initial cracks; (b) the probe-side surface of the crack had a lower temperature (contributed by the interfacial thermal resistance of the crack) and larger shrinkage ratio upon cooling; (c)some parts of the crack interface melted (molten flux side), and isolated pores with irregular shapes formed (the white areas in (c) are pores).

Several recent studies have suggested that films with higher basicity and cuspidine ratio tend to have a higher conductivity [25,26]. The roughness of film surfaces in contact with the probe has no causal relationship with crystallization [23,24]. Although those approaches indicated that crystals contribute less to heat flux control directly, a low glass-ratio film is still expected to stabilize the heat flux within a micro zone. The worse performance of the low-fluorine mold flux in this work was likely partially caused by unstable heat fluxes through the slag films, because of the formation and fusion of internal cracks in the initial glassy films.

Thus, besides the larger thermal contact resistance provided by the larger surface roughness of solid slag film, which was already proved in other works [23,24], another explanation of the observation that mold fluxes with high basicity and high fluorine content can provide more stable performances than the mold flux applied in this study is that cracks forming in a fully crystallized film do not have the evolution process shown in Figure 4b and 4c. Besides, a thicker film with a smaller ratio of glassy layer also decreases the effect of internal cracks on the heat flux fluctuations.

3.3. Surface Roughness of Solid Films

The roughness of film surfaces in contact with the copper wall increases the thermal contact resistance between steel shells and the mold, which is considered as one of the major contributions to heat flux control in continuous casting. The typical feature of film surfaces in contact with the probe is shown in Figure 6 and the measured roughness is given in Figure 5. For films obtained at higher bulk temperatures (i.e., 1623 and 1673 K), the difference of surface roughness was not statistically obvious when the immersion time was less than 90 s. A longer probe immersion time gave lower roughness of films obtained at a slag bulk temperature 1673 K. The lower slag bulk temperature (1573 K) tended to result in lower roughness. Some results demonstrate that the formation of surface roughness for those films was not caused by crystallization or devitrification. The roughness did not increase with increased probe immersion times (even decreased continuously at a slag bulk temperature of 1673 K). Figure 3a shows that the majority of the film cross section was glass, the surface in contact with the probe was amorphous (the small particle with size around several micrometers detailed in Figure 6 is not crystal, which was confirmed by the microscopic analysis), but this film already had a very high roughness (as shown in Figure 5). Compared with the high-basicity and high-fluorine films [23,24], the effect of bulk slag temperatures on roughness fluctuation of low-fluorine films was more obvious in this case.

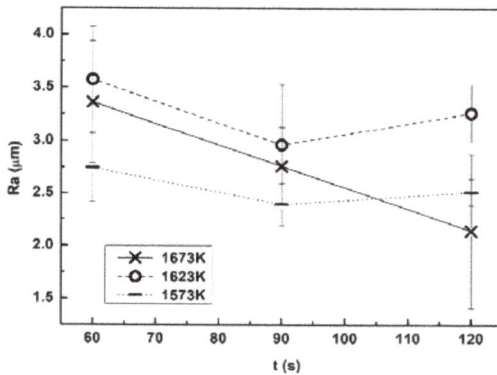

Figure 5. Roughness of film surfaces in contact with the water-cooled copper probe, for films formed in molten flux with different bulk temperatures after different probe immersion times. Error bars give the standard deviations.

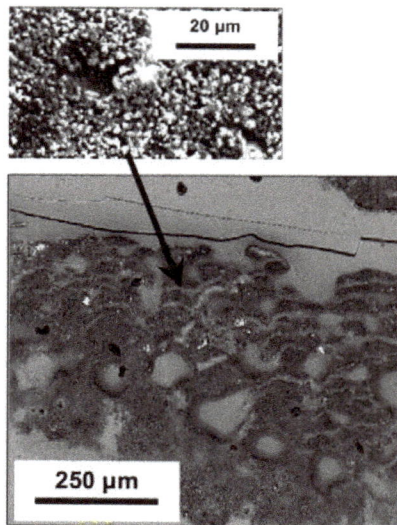

Figure 6. Appearance of the surface of a slag film in contact with the probe. Film recovered after 60 s of immersion of the probe in 1623 K mold flux (the film was almost glassy; secondary electron image).

3.4. Closed Porosity and Crystallization of Films

Closed pores in solid slag films decrease the effective thermal conductivity of films. The measured closed porosity of films is shown in Figure 7. The porosity tended to increase upon solidification. This result indicates that pores also formed in the outer layer of solid films, which increased the overall porosity of films (especially for films obtained at 1623 K, their porosity increased from about 2.17 to 8.40 volume percentage from 60 to 120 s immersion of the probe). Regarding the relationship between the formation of pores and crystallization by solidification and devitrification, Figure 3 demonstrates that for this study—similar to high-fluorine mold fluxes [23,24]—before crystallization or devitrification, large pores already existed in the glass films. This indicates that the formation of those large round-shaped pores at the probe side of films had no causal relationship with crystallization or devitrification (although the devitrification may have increased the porosity by forming micro-pores). The X-ray diffraction pattern in Figure 8 shows that the major crystal in the film was combeite $(Ca_{1.543}Na_{2.914}Si_3O_9)$ and those crystals had columnar and faceted dendritic shapes (similar shape with conventional high-fluorine mold fluxes for peritectic steels, see Figure 9). Compared with conventional mold fluxes, this low-fluorine flux had a much lower crystallization and devitrification rate [23,24].

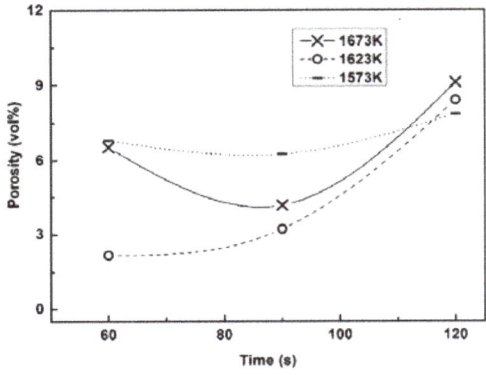

Figure 7. Closed porosity of films obtained at different slag bulk temperatures and probe immersion times.

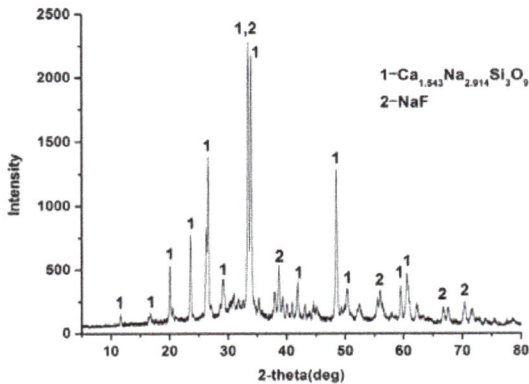

Figure 8. X-ray diffraction pattern of pulverized solid film (probe immersion time: 120 s; slag temperature: 1623 K; Cu-K$_\alpha$ radiation).

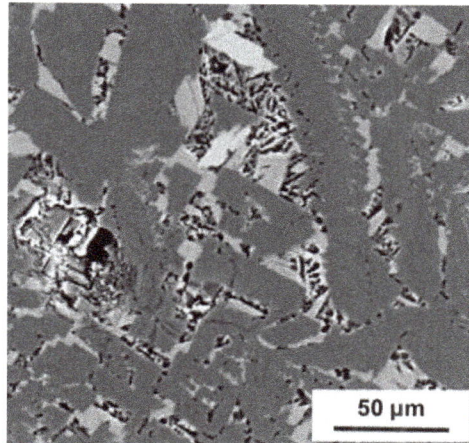

Figure 9. Typical backscattered electron micrograph of crystals in the solid film. The film was recovered after immersing the probe for 120 s in molten flux with a bulk slag temperature of 1623 K.

3.5. Density Evolution of Solidified Films

To calculate the closed porosity of films, true densities and apparent densities of slag films in solidification were measured accurately. The density evolution of films was discussed, as it is an important feature of films. The apparent density and true density of films are shown in Figure 10. Results showed that the temperature of liquid slag affected the density of solidified films directly. For the apparent density, films recovered from 1623 K flux gave higher density values than ones recovered from the flux with other temperatures at short probe immersion times (60 and 90 s). This was partially caused by the difference of their closed porosity (see Figure 7). The temperature of molten flux and probe immersion time clearly affected the true density of films. As the initial films were almost glass (as proved by the cross section morphology of films—see Figure 3a), which had a relative low density, films recovered after 60 s of the probe immersion with different bulk temperatures gave similar true densities around 2.59 g/cm^3. When the immersion time reached 90 s, higher temperatures of molten fluxes resulted in lower true densities of solidified films. As crystals of mold fluxes tend to have higher densities than the glass matrix, the true density change of solid film directly reflects the crystal ratio of films in solidification. The results indicated that for this mold flux, crystals began to precipitate after 60 s of the probe immersion. For the probe immersion period 60 to 90 s, a higher bulk temperature resulted in a relatively lower crystallization rate and ratio. This was related to the larger temperature gradient of films solidified in high-temperature molten fluxes, because crystallization requires sufficient supercooling at the solidification front. However, in the probe immersion period 90 s to 120 s, the crystallization rate of films solidified in the higher-temperature flux increased obviously. This was caused by a faster devitrification rate in the glassy film at the probe side (i.e., higher bulk temperature gives a larger temperature gradient and also increases the temperature of the probe-side part of films for devitrification).

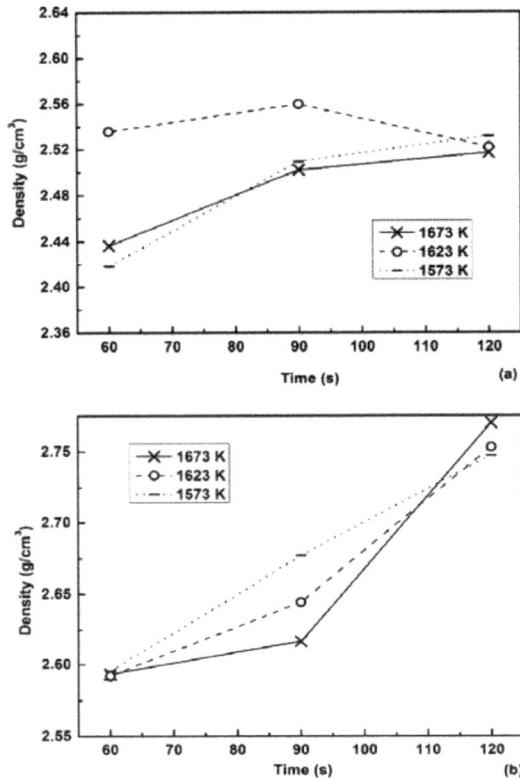

Figure 10. (**a**) Apparent densities and (**b**) true densities of slag films obtained at different slag bulk temperatures and probe immersion times.

4. Conclusions

In this study, a low-fluorine mold flux was solidified by an improved water-cooled copper probe. The heat flux through solid films was calculated, and film samples were recovered and inspected. Based on the results, the following conclusions can be drawn:

(1) Internal cracks in initial solidified glassy films were observed. The formation and fusion of those cracks during solidification likely contributed to the fluctuation of heat flux density, especially at the early stage of film solidification.

(2) The roughness of surfaces in contact with the water-cooled probe formed early before crystallization or devitrification, which was obviously affected by the temperature of molten fluxes and probe immersion time. The formation of closed pores in the probe-side glassy films had no causal relationship with crystallization.

(3) Combeite ($Ca_{1.543}Na_{2.914}Si_3O_9$) with columnar and faceted dendritic shapes was the major crystal in the solid films.

Author Contributions: conceptualization, X.L. and S.H.; methodology, X.L.; formal analysis, X.L.; investigation, J.Z. and X.L.; resources, X.Y.; data curation, M.L. and Q.W.; writing—original draft preparation, X.L.; writing—review and editing, J.Z.; supervision, S.H.; funding acquisition, S.H.

Funding: This research was funded by Natural Science Foundation of China, grant number 51874057 and U1660204.

Acknowledgments: The authors would like to appreciate the support from Natural Science Foundation of China (project No. 51874057 and U1660204). Professor P. C. Pistorius of Carnegie Mellon University is deeply appreciated for the valuable discussion and guidance on the improved water-cooled copper probe device.

Conflicts of Interest: The authors declare no conflict of interest.

References

1. Nakada, H.; Susa, M.; Seko, Y.; Hayashi, M.; Nagata, K. Mechanism of Heat Transfer Reduction by Crystallization of Mold Flux for Continuous Casting. *ISIJ Int.* **2008**, *48*, 446–453. [CrossRef]
2. Yamauchi, A.; Sorimachi, K.; Sakuraya, T.; Fujii, T. Heat Transfer between Mold and Strand through Mold Flux Film in Continuous Casting of Steel. *ISIJ Int.* **1993**, *33*, 140–147. [CrossRef]
3. Mills, K.C.; Fox, A.B. The Role of Mould Fluxes in Continuous Casting-So Simple Yet So Complex. *ISIJ Int.* **2003**, *43*, 1479–1486. [CrossRef]
4. Mills, K.C.; Fox, A.B.; Li, Z.; Thackray, R.P. *VII International Conference on Molten Slags, Fluxes & Salts: 25–28 January 2004, Cape Town, South Africa]*; The South African Institute of Mining and Metallurgy: Johannesburg, South Africa, 2004.
5. Brimacombe, J.; Sorimachi, K. Crack formation in the continuous casting of steel. *Metall. Trans. B* **1977**, *8*, 489–505. [CrossRef]
6. Fox, A.; Mills, K.; Lever, D.; Bezerra, C.; Valadares, C.; Unamuno, I.; Laraudogoitia, J.; Gisby, J. Development of Fluoride-Free Fluxes for Billet Casting. *ISIJ Int.* **2005**, *45*, 1051–1058. [CrossRef]
7. Nakada, H.; Fukuyama, H.; Nagata, K. Effect of NaF Addition to Mold Flux on Cuspidine Primary Field. *ISIJ Int.* **2006**, *46*, 1660–1667. [CrossRef]
8. Nakada, H.; Nagata, K. Crystallization of $CaO–SiO_2–TiO_2$ slag as a candidate for fluorine free mold flux. *ISIJ Int.* **2006**, *46*, 441–449. [CrossRef]
9. Cho, J.W.; Emi, T.; Shibata, H.; Suzuki, M. Heat transfer across mold flux film in mold during initial solidification in continuous casting of steel. *ISIJ Int.* **1998**, *38*, 834–842. [CrossRef]
10. Seo, M.D.; Shi, C.B.; Cho, J.W.; Kim, S.H. Crystallization behaviors of $CaO–SiO_2–Al_2O_3–Na_2O–CaF_2$ -$(Li_2O–B_2O_3)$ mold fluxes. *Metall. Mater. Trans. B* **2014**, *45*, 1874–1886. [CrossRef]
11. Persson, M.; Sridhar, S.; Seetharaman, S. Kinetic Studies of Fluoride Evaporation from Slags. *ISIJ Int.* **2007**, *47*, 1711–1717. [CrossRef]
12. Wang, Z.; Shu, Q.; Chou, K. Viscosity of Fluoride-Free Mold Fluxes Containing B_2O_3 and TiO_2. *Steel Res. Int.* **2013**, *84*, 766–776. [CrossRef]
13. Fan, G.; He, S.; Wu, T.; Wang, Q. Effect of Fluorine on the Structure of High Al_2O_3-Bearing System by Molecular Dynamics Simulation. *Metall. Mater. Trans. B* **2015**, *46*, 2005–2013. [CrossRef]
14. Lu, B.; Wang, W. Effects of Fluorine and BaO on the Crystallization Behavior of Lime–Alumina-Based Mold Flux for Casting High-Al Steels. *Metall. Mater. Trans. B* **2015**, *46*, 852–862. [CrossRef]
15. He, S.; Wang, Q.; Xie, D.; Xu, C.; Li, Z.S.; Mills, K.C. Solidification and crystallization properties of $CaO-SiO_2-Na_2O$ based mold fluxes. *Int. J. Min. Metall. Mater.* **2009**, *16*, 261–264. [CrossRef]
16. Wen, G.; Sridhar, S.; Tang, P.; Qi, X.; Liu, Y. Development of fluoride-free mold powders for peritectic steel slab casting. *ISIJ Int.* **2007**, *47*, 1117–1125. [CrossRef]
17. Wang, Q.; Lu, Y.; He, S.; Mills, K.; Li, Z.S. Formation of TiN and Ti (C, N) in TiO_2 containing, fluoride free, mould fluxes at high temperature. *Ironmak. Steelmak.* **2011**, *38*, 297–301. [CrossRef]
18. Ryu, H.; Zhang, Z.; Wen, J.W.C.G.; Sridhar, S. Crystallization behaviors of slags through a heat flux simulator. *ISIJ Int.* **2010**, *50*, 1142–1150. [CrossRef]
19. Lara Santos Assis, K. Heat Transfer through Mold Fluxes: A New Approach to Measure Thermal Properties of Slags. Ph.D. Thesis, Carnegie Mellon University, Pittsburgh, PA, USA, February 2016.
20. Long, X.; He, S.; Xu, J.; Huo, X.; Wang, Q. Properties of high basicity mold fluxes for peritectic steel slab casting. *J. Iron Steel Res. Int.* **2012**, *19*, 39–45. [CrossRef]
21. He, S.; Long, X.; Xu, J.; Wu, T.; Wang, Q. Effects of crystallisation behaviour of mould fluxes on properties of liquid slag film. *Ironmak. Steelmak.* **2012**, *39*, 593–598. [CrossRef]
22. Wu, T.; Wang, Q.; He, S.; Xu, J.; Long, X.; Lu, Y. Study on properties of alumina-based mould fluxes for high-Al steel slab casting. *Steel Res. Int.* **2012**, *83*, 1194–1202. [CrossRef]

23. Long, X.; Wang, Q.; He, S.; Pistorius, P.C. Structure evolution of slag films of ultrahigh-basicity mold flux during solidification. *Metall. Mater. Trans. B* **2017**, *48*, 1938–1942. [CrossRef]
24. Long, X.; He, S.; Wang, Q.; Pistorius, P.C. Structure of solidified films of mold flux for peritectic steel. *Metall. Mater. Trans. B* **2017**, *48*, 1652–1658. [CrossRef]
25. Kromhout, J.A.; Dekker, E.R.; Kawamoto, M.; Boom, R. Challenge to control mould heat transfer during thin slab casting. *Ironmak. Steelmak.* **2013**, *40*, 206–215. [CrossRef]
26. Andersson, S.P.; Eggertson, C. Thermal conductivity of powders used in continuous casting of steel, part 1—Glassy and crystalline slags. *Ironmak. Steelmak.* **2015**, *42*, 456–464. [CrossRef]

Article

Influence of Al Content on the Inclusion-Microstructure Relationship in the Heat-Affected Zone of a Steel Plate with Mg Deoxidation after High-Heat-Input Welding

Longyun Xu [1] , **Jian Yang** [1,*] **and Ruizhi Wang** [2]

[1] State Key Laboratory of Advanced Special Steel, School of Materials Science and Engineering, Shanghai University, Shanghai 200444, China; xulongyun@shu.edu.cn

[2] Steelmaking Research Department, Research Institute, Baosteel Group Corporation, Shanghai 201900, China; wangruizhi@baosteel.com

* Correspondence: yang_jian@t.shu.edu.cn; Tel.: +86-021-6613-6580

Received: 21 November 2018; Accepted: 3 December 2018; Published: 6 December 2018

Abstract: The effects of Al content on inclusions, microstructures, and heat-affected zone (HAZ) toughness in a steel plate with Mg deoxidation have been investigated by using simulated high-heat-input welding and an automated feature system. The studies indicated that the main kind of oxysulfide complex inclusions in two steels without and with Al addition were both MgO-MnS. The number densities and mean sizes of inclusions were 96.65 mm^{-2} and 3.47 μm, 95.03 mm^{-2} and 2.03 μm, respectively. The morphologies of MgO-MnS complex inclusions in steel were changed obviously with the addition of Al. When containing 0.001 wt.% Al, they consisted of a central single MgO particle and outside, the MnS phase. When containing 0.020 wt.% Al, they comprised several small MgO particles entrapped by the MnS phase. Because the former could nucleate intragranular acicular ferrites (IAFs) and the latter was non-nucleant, the main intragranular microstructures in HAZs were ductile IAFs and brittle ferrite side plates (FSPs), respectively. Therefore, HAZ toughness of the steel plate without Al addition after high-heat-input welding of 400 kJ/cm was significantly better than that of the steel plate with Al addition.

Keywords: Mg deoxidation; inclusions; Al addition; high-heat-input welding; heat-affected zone; toughness

1. Introduction

In steelmaking, extensive efforts have been made to remove massive inclusions, which deteriorate the final qualities of products. In this regard, Takamura et al. [1] introduced the concept termed "oxide metallurgy" for the first time at the sixth international iron and steel congress in 1990. Since then, increasing attention has been drawn to the functions of fine inclusions dispersed in steels. Nowadays, oxide metallurgy technology has been deemed as the most effective method to improve the heat-affected zone (HAZ) toughness of steel plates after high-heat-input welding [2–4]. Inclusions can act as intragranular nucleation sites for acicular ferrite, resulting in interlocked fine intragranular acicular ferrites (IAFs) and consequently improved HAZ toughness of the steel plate [5]. The characteristics including chemistry, size, and morphology of inclusions play important roles in the nucleation of IAF.

Currently, strong deoxidizers, such as Mg, have been utilized to develop advanced oxide metallurgy technology. Mg deoxidation is an effective method to control inclusions. Kim et al. [6] reported that Mg addition in Mn-Si-Ti deoxidized steels benefited the refining and modification of inclusions. With the increase of Mg content in steels, the mean size of inclusions decreased, and the number density increased. The oxide phase of inclusions changed from Ti_2O_3 to Ti-Mg-O, to $MgTiO_3$, and then to MgO.

Chai et al. [7] also found that the addition of Mg in steel was able to refine and modify Ti-based inclusions. Besides, Park et al. [8] comprehensively investigated the evolution of inclusions in Mn-Si-Ti-Al-Mg deoxidized steel containing Mg of about 10 ppm (in mass fraction) and Al in the range of 6 and 147 ppm. They found that the major oxide inclusions changed from the Mg-Ti-O to $MgAl_2O_4$ with increasing content of Al in steels. They also observed large $MgO-Al_2O_3$ aggregates entrapped in MnS in steels containing 87 and 147 ppm Al. Li et al. [9] studied the effect of Mg addition on the nucleation of IAF in as-cast Al-killed low carbon steel. They suggested that the Mn-depleted zone (MDZ) around the $MgO-Al_2O_3$-MnS inclusion induced by MnS precipitation on the inclusion promoted the IAF formation. However, the influence of Al content on the relationship between inclusions and HAZ microstructures has not been comprehensively investigated in steel plates for high-heat-input welding.

In our previous study, it has been reported that HAZ toughness of Al-killed steel was deteriorated after high-heat-input welding, resulting from coarse prior-austenite grains (PAG) and brittle microstructures [10]. It was observed that the major kind of inclusions in Al-killed steel was the Al_2O_3-MnS inclusion, which was unable to nucleate an acicular ferrite during the high-heat-input welding process [10,11]. According to reports in recent literature, in Mg deoxidized steels containing a low level of Al content, Mg-containing inclusions, such as MgO-MnS [12], (Mg-Ti-O)-MnS [7], (Mg-Al-Ti-O)-MnS [11], and (Ti-Ca-Mg-O)-MnS [13], were favorable to promote the formation of IAF, resulting in enhanced HAZ toughness. However, the effect of inclusions formed in Mg deoxidized steel containing a high level of Al content on the HAZ microstructure and toughness has not yet been understood comprehensively.

In the present study, the influence of Al content on inclusions and microstructures in HAZ of steel plates with Mg deoxidation after high-heat-input welding is studied. In order to evaluate the characteristics of inclusions, two experimental steels with Mg deoxidation were prepared by the addition of Al or not. The morphology, size, composition, and number density of inclusions were investigated. Inclusions were correlated with the HAZ microstructure of each specimen in order to reveal the relationship between them. The main purpose of this paper is to clarify the role of Al content on HAZ toughness in steel plates with Mg deoxidation after high-heat-input welding of 400 kJ/cm.

2. Materials and Methods

2.1. Experimental Steel Preparation

Two kinds of experimental steels with Mg deoxidation, namely 3Mg1Al and 3Mg20Al, were melted in a 50-kg vacuum induction furnace with sintered magnesia lining, respectively. Table 1 shows the major chemical compositions of experimental steel samples. As shown in Table 1, Al contents in 3Mg1Al and 3Mg20Al were 10 and 200 ppm, respectively, and other elements were nearly kept at the same levels. Deoxidation experiments were carried out in this furnace under Ar atmosphere. Firstly, about 40 kg of pure iron were melted, and lime (CaO) was added as the top slag to assure considerably low oxygen potential in the slag. Then, the proper amount of deoxidants, Mn, Si, Al, Ti, and Mg, were added into molten steel to obtain the target composition. These melts were then cast into ingots with the size of $120 \times 180 \times 240$ mm^3. Each ingot was hot rolled into a steel plate with a thickness of 50 mm. Roughing rolling was conducted at a temperature above 930 °C with a reduction ratio greater than 30%, and finishing rolling was carried out at a temperature of about 800 °C with a reduction ratio greater than 30%. Then, the steel plate was cooled down from 760–400 °C at a cooling rate of about 10 °C/s.

Table 1. Measured chemical compositions of steel samples (wt.%).

Steels	C	Si	Mn	P	S	Ti	Mg	Al	Als *	O	N
3Mg1Al	0.082	0.22	1.56	0.006	0.005	0.011	0.0027	0.001	0.001	0.0011	0.0032
3Mg20Al	0.082	0.22	1.56	0.006	0.004	0.011	0.0027	0.020	0.019	0.0007	0.0032

* Als means the content of acidic soluble Al in steel.

2.2. Simulated High-Heat-Input Welding Experiments

In order to evaluate the HAZ toughness of experimental steel plates, simulated high-heat-input welding experiments were carried out by using of Gleeble 3800 thermal simulator (Dynamic Systems Inc., New York, NY, USA). It was designed to simulate electrogas arc welding with a heat input of 400 kJ/cm for a steel plate with a thickness of 50 mm. The steel specimen for simulated welding experiments with a size of $11 \times 11 \times 71$ mm^3 (Figure 1a) was cut down from a position about 1/4 width and 1/4 length from the edge of the steel plate. As shown in Figure 1a, two thermocouples were welded in the center position of the steel specimen for the simulated high-heat-input welding experiment. During the experiment, the peak welding temperature was 1400 °C with a holding time of 3 s. The cooling rates from 1400–800 °C, from 800–500 °C, and from 500–300 °C were controlled to be 3.41, 0.78, and 0.17 °C/s, respectively. Figure 1b shows the steel specimen after the simulated welding thermal cycle process. A relatively large area near the welding spot was obtained, namely HAZ. It was then machined into a standard Charpy specimen with the size of $10 \times 10 \times 55$ mm^3 (Figure 1c). A V-notch was opened at the location of welding spots. Then, Charpy impact tests of HAZ specimens were conducted at −20 °C. Afterward, the fracture surfaces were examined by a scanning electron microscope (SEM, EVO 18, Carl Zeiss, Oberkochen, Germany).

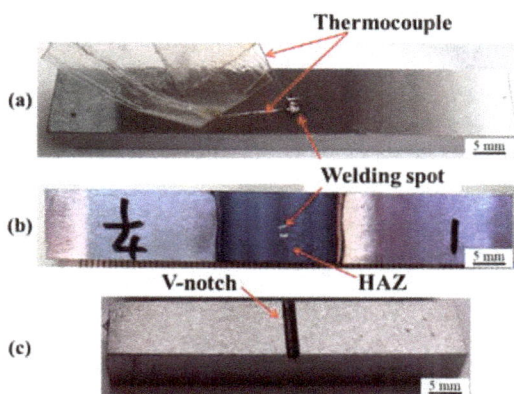

Figure 1. Steel specimens for the simulated high-heat-input welding experiment: (**a**) before welding, (**b**) after welding, and (**c**) the Charpy V-notch impact test. HAZ, heat-affected zone.

2.3. Characterization of Inclusions and Microstructures

These HAZ specimens were taken from the surfaces in parallel with the fracture cross-sections after Charpy impact tests. Firstly, these specimens were polished and then analyzed by an automated feature system operating on SEM. This technique combines the advantages of energy dispersive X-ray spectrometry (EDS, Oxford Instruments, Abingdon, UK) with digital image analysis of backscattered electron (BSE) micrographs. It provides fast measurements of composition, size, and morphology simultaneously for thousands of inclusions embedded in a steel matrix [14]. Then, these specimens were slightly etched by 4-pct nital solution to observe HAZ microstructures with optical microscopy (OM, DM 2700 M, Leica, Germany) and SEM-EDS. The present study mainly focused on these inclusions with the equal-circle diameter (ECD) larger than 1 μm. The ECD was defined as the diameter of a circle whose area is equal to the area occupied by the inclusion in the BSE micrograph. Mapping scanning of typical inclusions in each sample was performed to examine the elemental distribution inside inclusions.

3. Experimental Results

3.1. Morphology and Composition of Typical Inclusions

Figure 2 shows the typical distributions of inclusions in 3Mg1Al and 3Mg20Al. Inclusions embedded in the steel matrix are observed as small black dots in the BSE micrographs. It is seen that inclusions in both steel samples are evenly distributed in steel matrixes. It is also seen that the amount of inclusions with relatively large sizes in 3Mg1Al (Figure 2a) is more than that in 3Mg20Al (Figure 2b).

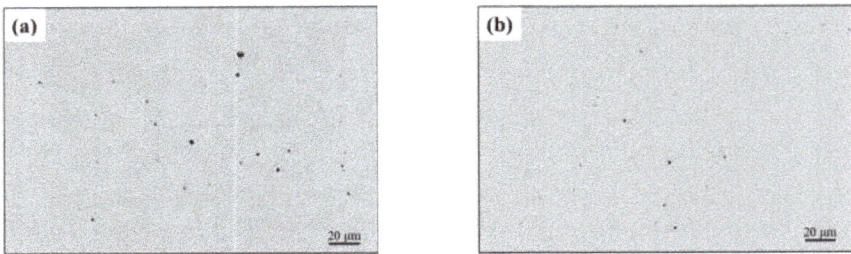

Figure 2. Distributions of inclusions in typical field of views with 500 magnifications in experimental steels. (**a**) 3Mg1Al; (**b**) 3Mg20Al.

Figure 3 shows the morphologies and compositions of typical MgO-MnS inclusions in 3Mg1Al and 3Mg20Al. The mapping analysis result reveals that the MgO-MnS inclusion in 3Mg1Al was composed of a central single MgO particle with a diameter about 2 μm and a peripheral MnS phase. On the contrary, the morphology of the MgO-MnS inclusion in 3Mg20Al was quite different. It was composed of four small MgO particles with a diameter smaller than 0.5 μm and an irregular MnS phase as the "matrix".

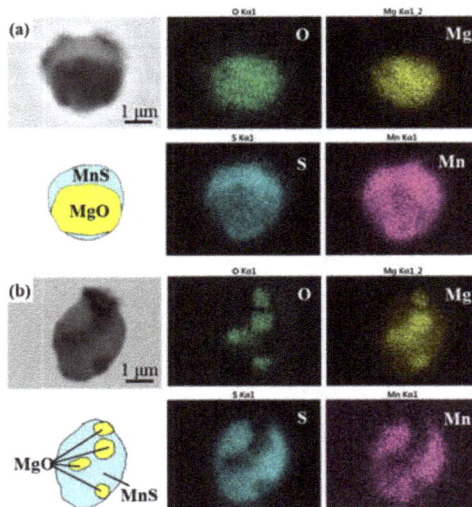

Figure 3. Morphologies and compositions of typical MgO-MnS inclusions. (**a**) 3Mg1Al; (**b**) 3Mg20Al.

Figure 4 shows the morphologies and compositions of typical complex inclusions comprising oxide, sulfide, and nitride in 3Mg1Al and 3Mg20Al. According to the mapping analysis results, it was confirmed that the nitride phase was TiN, which was located at the edge of the complex inclusion. As shown in Figure 4a, a single MgO particle of the complex inclusion in 3Mg1Al was covered by a small amount TiN and MnS. It is seen in Figure 4b that in the complex inclusion of 3Mg20Al, two small MgO-Al$_2$O$_3$ particles were embedded in the MnS phase, and TiN was precipitated at the edge of the MnS phase. According to the thermodynamic calculation results reported by Park et al. [8], the precipitate temperature of TiN was under the solidus of steel in Ti-Al-Mg deoxidized steels, in which the contents of Ti and N were similar to the experimental steel samples in the present study. Thus, it is reasonable that a pure TiN precipitate was hardly observed in both steels. In the present study, TiN phases were generally observed on the surfaces of MgO (Figure 4a) or MnS (Figure 4b), probably due to the lower interfacial energy [8,15].

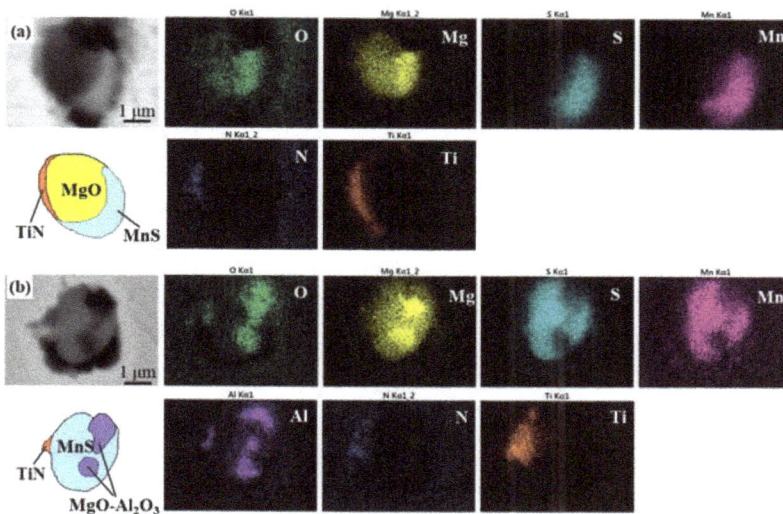

Figure 4. Morphology and composition of typical inclusions. (a) MgO-MnS-TiN in 3Mg1Al; (b) MgO-Al$_2$O$_3$-MnS-TiN in 3Mg20Al.

3.2. Number Density and Size Distribution of Inclusions

Table 2 shows the statistical results of inclusions for steel samples 3Mg1Al and 3Mg20Al. They were based on the examination of inclusions with the number of 3750 and 3003 in the area of 38.8 and 31.6 mm^2, respectively. The number densities of inclusions in 3Mg1Al and 3Mg20Al were 96.65 and 95.03 mm^{-2}, respectively. The mean sizes of inclusions in both steel samples were 3.47 and 2.03 μm, respectively. Figure 5 shows the size distribution of inclusions in both experimental steels. The cumulative frequencies of inclusions with the size less than 5 μm in 3Mg1Al and 3Mg20Al were 86.61% and 96.97%, respectively. The frequencies of inclusions with the sizes between 2 and 5 μm in 3Mg1Al and 3Mg20Al were 72.29% and 29.77%, respectively. In comparison, the frequency of inclusions with the sizes between 1 and 2 μm in 3Mg20Al were as high as 63.73%. It was concluded that the inclusion size of 3Mg20Al was much smaller than that of 3Mg1Al.

Table 2. Statistical results of inclusions in experimental steels.

Steels	Area (mm^2)	Number of Inclusions	Number Density (mm^{-2})	Mean Size (μm)
3Mg1Al	38.8	3750	96.65	3.47
3Mg20Al	31.6	3003	95.03	2.03

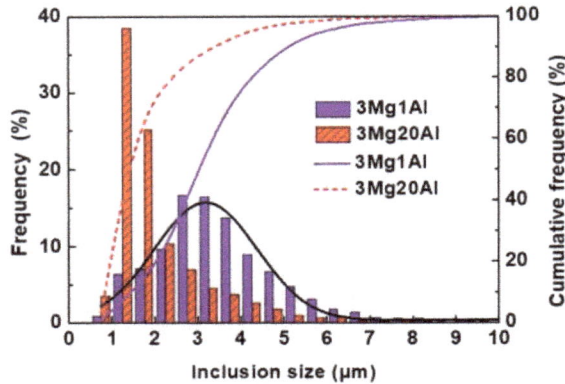

Figure 5. Size distribution of inclusions in experimental steels.

3.3. Composition of Inclusions

Figure 6 shows the average compositions of inclusions in 3Mg1Al and 3Mg20Al according to the statistical results of 3750 and 3003 inclusions, respectively. As shown in Figure 6a, the sums of Mg, O, Mn, and S contents of inclusions in 3Mg1Al and 3Mg20Al were 90.77 and 86.46 wt.%, respectively. It was inferred that inclusions in both steel samples were mainly composed of MgO and MnS. The average Al content of inclusions in 3Mg1Al was as low as 0.11 wt.%. Comparatively, that in 3Mg20Al was 3.51 wt.%, which was 30-times more than that in 3Mg1Al. Other major elements' contents were nearly at the same level. According to the analysis results in Section 3.1, the elemental contents of inclusions in both steels can be converted into the components of inclusions, as shown in Figure 6b. Main components of inclusions in 3Mg1Al and 3Mg20Al were both MgO and MnS. In addition, the content of Al$_2$O$_3$ in inclusions of 3Mg1Al was much less than that of 3Mg20Al.

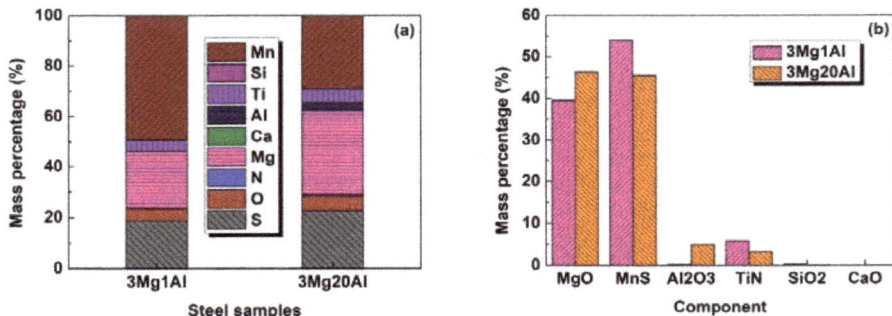

Figure 6. Comparison of the average composition of inclusions in 3Mg1Al and 3Mg20Al.

3.4. Types of Inclusions

According to the analysis results of inclusions' morphology and composition, these inclusions could be classified into 12 types by the major compositional elements (Mg, Al, Ti, Mn, O, N, and S), namely (1) MgO, (2) Al_2O_3, (3) $MgO-Al_2O_3$, (4) MgO-MnS, (5) Al_2O_3-MnS, (6) $MgO-Al_2O_3$-MnS, (7) MgO-TiN, (8) MgO-MnS-TiN, (9) $MgO-Al_2O_3$-MnS-TiN, (10) MnS-TiN, (11) TiN, and (12) MnS. Figure 7 shows the number density and average size of different types of inclusions in 3Mg1Al and 3Mg20Al. The main inclusion types in 3Mg1Al were MnS, MgO-MnS, and MgO-MnS-TiN. Their number densities were 30.95, 22.73, and 22.45 mm^{-2} with average sizes of 3.08, 3.43, and 3.73 µm, respectively, as shown in Figure 7a,c. In 3Mg20Al, the main inclusion types were MgO-MnS, MgO-MnS-TiN, and $MgO-Al_2O_3$-MnS with number densities and average sizes of 41.93, 12.15, 12.12 mm^{-2} and 2.24, 2.29, 1.80 µm, respectively, as shown in Figure 7b,d. In comparison, inclusion types containing Al_2O_3 were hardly found in 3Mg1Al, while they were relatively easy to observe in 3Mg20Al. In addition, the number density of MnS in 3Mg1Al, 30.95 mm^{-2} was five-times larger than that in 3Mg20Al, 5.13 mm^{-2}. This may because the S content in 3Mg1Al was higher than that in 3Mg20Al, as shown in Table 1.

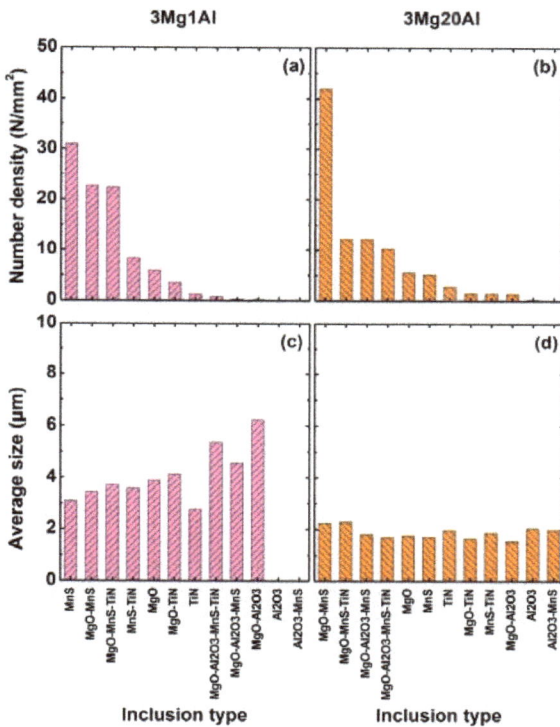

Figure 7. Number density and average size of different types of inclusions in experimental steels.

3.5. HAZ Microstructure and Toughness

Figure 8 shows the HAZ microstructures of experimental steels after high-heat-input welding of 400 kJ/cm. It is obvious that the grain boundary ferrites (GBFs) were network-like distributed in HAZ microstructures of 3Mg1Al and 3Mg20Al, as shown in Figure 8a,d. GBF was nucleated from the austenite grain boundary (AGB) and distributed along with AGB. Thus, prior-austenite grain (PAG) can be characterized by these network-like GBFs. It was observed that the size of PAGs in 3Mg1Al was similar to that in 3Mg20Al, though GBF in the former was finer than that in the latter. However, the microstructure inside the austenite grain in 3Mg1Al was quite different from that in 3Mg20Al. In 3Mg1Al, well-developed IAFs almost occupied the whole inner part of this austenite grain, as shown in Figure 8b. In contrast, intragranular microstructures in 3Mg20Al were mainly composed of ferrite side plates (FSPs), which went through the grain, as shown in Figure 8e. It is seen in Figure 8c that IAFs observed at 1000 magnifications with OM showed an interlocked morphology. It is also observed that a micrometer-sized inclusion was surrounded by several acicular ferrite lathes. In Figure 8f, it is clear that lathes of FSPs showed a parallel morphology.

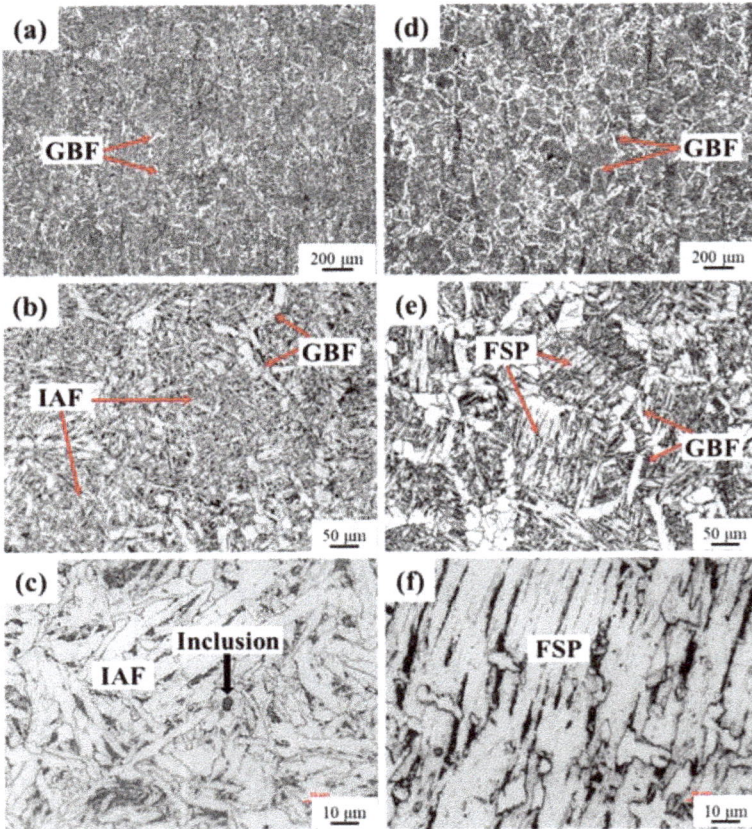

Figure 8. HAZ microstructures of experimental steels after high-heat-input welding with 400 kJ/cm. (**a–c**) 3Mg1Al; (**d–f**) 3Mg20Al. GBF, grain boundary ferrite; IAF, intragranular acicular ferrite; FSP, ferrite side plate.

Figure 9 shows morphologies and SEM mapping images of inclusions in HAZ microstructures for 3Mg1Al and 3Mg20Al. It was observed in a field of view of 500-times magnification in 3Mg1Al that the inclusion with a size of about 3.31 μm was located in the inner part of austenite grain, which was occupied by IAFs, marked with a square in Figure 9a. At a magnification of 5000-times, shown in Figure 9b, this inclusion in Figure 9a was located in the center of four ferrite emanations. It was inferred that these four acicular ferrite laths were nucleated directly from the same inclusion. This kind of inclusion was reasonably considered as an effective nucleant of IAF [16]. According to SEM mapping images of this nucleant, shown in Figure 9c, it was composed of a spherical MgO particle with a diameter of 2.38 μm and a peripheral MnS phase.

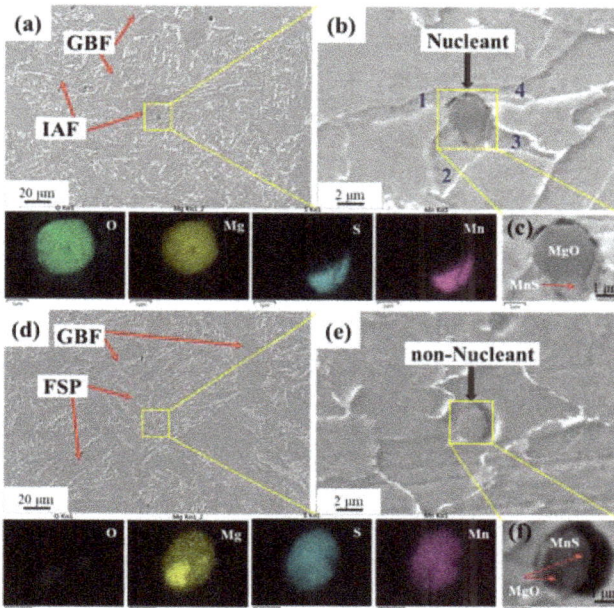

Figure 9. Morphologies and SEM mapping images of inclusions in HAZ microstructures for experimental steels. (**a–c**) 3Mg1Al; (**d–f**) 3Mg20Al.

The HAZ microstructure observed by SEM in 3Mg20Al is shown in Figure 9d. It is seen that parallel FSPs were surrounded by coarse GBFs. IAFs were hardly found in the HAZ microstructure of 3Mg20Al. An intragranular inclusion with a size of about 2.46 μm was also observed. At a magnification of 5000-times, shown in Figure 9e, this inclusion was located within the ferrite matrix and without ferrite emanations. This kind of inclusion is regarded as the non-nuclent of IAF because of the absence of ferrite emanation [16]. The element mapping analysis result, shown in Figure 9f, indicates that this inclusion consisted of two small MgO particles with diameters under 1 μm distributed in the MnS phase as the "matrix".

Figure 10 shows the macro fractographs and SEM images of the fracture surfaces of HAZs for 3Mg1Al and 3Mg20Al. The fractograph of the specimen of 3Mg1Al after the standard Charpy impact test shows obvious transverse deformation, as indicated by apparently lateral expansion, as shown in Figure 10a. In comparison, the steel sample of 3Mg20Al had less transverse deformation, as shown in Figure 10d. The values of the lateral expansion of 3Mg1Al and 3Mg20Al were 2.26 and 0.98 mm, respectively. Table 3 shows the results of quantitative statistical analysis of fracture surfaces in 3Mg1Al and 3Mg20Al. The area fractions of the fibrous zone and the shear lip zone associated with ductile fracture in 3Mg1Al were 56.03% and 21.95%, respectively. They were 20.63% and 2.51% in 3Mg20Al. The area faction of radical zone relating to brittle fracture in 3Mg1Al was 22.01%. That in 3Mg20Al was as high as 78.86%. The sizes of fractures in the radical zone in 3Mg1Al were smaller than those in 3Mg20Al, as shown in Figure 10b,e, respectively. Dimples were apparently observed in fibrous zones of both steel samples, as shown in Figure 10c,f, respectively. The average Charpy absorbed energy values at −20 °C of 3Mg1Al and 3Mg20Al were 201 and 75 J, respectively. It was indicated that the HAZ toughness of 3Mg1Al was much better than that of 3Mg20Al.

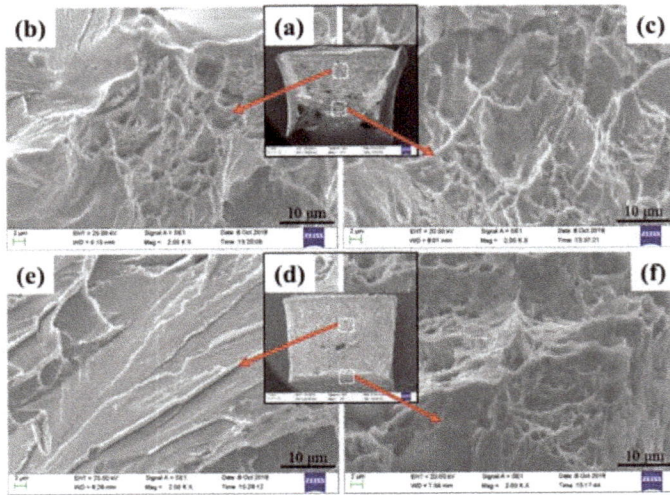

Figure 10. Fracture morphology features of HAZs. (**a**–**c**) 3Mg10Al; (**d**–**f**) 3Mg20Al.

Table 3. Results of the quantitative statistical analysis of fracture surfaces and Charpy absorbed energy of HAZ at −20 °C.

Steels	Impact Fracture			Impact Toughness (J)	
	Fibrous Zone (%)	Shear Lip Zone (%)	Radical Zone (%)	Individual Value	Mean
3Mg1Al	56.03	21.95	22.01	195, 186, 222	201
3Mg20Al	20.63	2.51	76.86	59, 91, 75	75

4. Discussion

4.1. Effect of Al on Inclusions

Al has a relative strong affinity with oxygen in molten steel. It is a kind of strong deoxidizer widely used during the deoxidation process in steel plants. Al_2O_3 is the unique direct product of Al deoxidation. Thus, Al_2O_3 inclusion is the major oxide inclusion appearing in normal Al-killed steel. The strong attractive force between Al_2O_3 inclusions causes the coagulation and formation of Al_2O_3 clusters. However, the attractive force between a pair of MgO-Al_2O_3 inclusions in Mg-added Al-killed steel and that of MgO inclusions in Mg-killed steel is only one-tenth of that between a pair of Al_2O_3 inclusions, resulting in a much weaker tendency to form MgO-Al_2O_3 or MgO clusters [17]. This is an effective method to refine and disperse inclusions by the use of Mg addition in steel. Thus, inclusions observed in both experimental steels 3Mg1Al and 3Mg20Al were dispersed uniformly, as shown in Figure 2. In the present study, the 3Mg1Al steel was a Mg-killed steel. It was prepared without intentional Al addition. As shown in Figure 3a, the oxide phase of inclusion is a pure MgO particle with a diameter of about 2 μm. In our previous study, it was observed that oxide inclusions in steel containing the same level of Mg content were Mg-Ti-O or Mg-Al-Ti-O [18] when the oxygen level was up to 40 ppm. In comparison, the oxygen content in steel sample 3Mg1Al, 11 ppm, was much lower, as shown in Table 1. In addition, the affinity with oxygen of Mg was much stronger than that of Ti [19]. Therefore, MgO was preferentially formed in 3Mg1Al, when it contained relatively low oxygen content.

Al_2O_3 inclusions easily formed clusters and separated from molten steel by flotation [20]. Generally, the total oxygen content in steel was decreased with increasing acidic soluble Al content [21]. In addition, the coarsening rate of particles in molten steel was largely dependent on the dissolved oxygen content [22].

In the present study, Mg was added into molten steel as a final deoxidizer. Due to the relatively low level of oxygen in molten steel, the diameter of MgO particles formed in 3Mg20Al was much smaller than that in 3Mg1Al, as shown in Figure 3. Consequently, size distributions of inclusions in 3Mg1Al and 3Mg20Al were concentrated in the range from 2.5–4.5 µm and from 1.5–2.5 µm, respectively, as shown in Figure 5.

Oxysulfide complex inclusions are intentionally formed in oxide metallurgy technology [5,23]. As can be seen in Figure 7, the major oxysulfide complex inclusions in 3Mg1Al and 3Mg20Al were both MgO-MnS. However, the morphologies between them were quite different, as shown in Figure 3. Firstly, the oxide phase in the MgO-MnS complex inclusion in 3Mg1Al was a single MgO particle with a diameter larger than 1 µm, as shown in Figure 3a. In 3Mg20Al, the oxide phases were several separate MgO particles with a diameter smaller than 0.5 µm, as shown in Figure 3b. Secondly, the MnS phase in 3Mg1Al was precipitated on the surface of the MgO particle. In 3Mg20Al, several small MgO particles were entrapped by MnS. Park et al. [8] also found oxysulfide complex inclusions with a similar morphology. They observed small MgO-Al_2O_3 granules entrapping a large MnS in the steel containing Al of more than 40 ppm. It is not certain how this kind of inclusion is formed in the present state. However, the following mechanism may be postulated. Since the dissolved oxygen content in molten steel is relatively low resulting from a relatively high level of acidic soluble Al content, a large quantity of MgO particles with small sizes was formed after final deoxidation by Mg addition. These small MgO particles were easily encapsulated by MnS during solidification. Besides, there was a very weak interaction between MgO particles, so that no MgO aggregates formed in the steel [17]. Therefore, most of the inclusions in 3Mg20Al were fine and evenly dispersed.

4.2. Effect of Al on HAZ Microstructures

As shown in Figure 8, the intragranular microstructure in HAZ of 3Mg1Al was quite different from that of 3Mg20Al. In the previous studies, several mechanisms of inclusion nucleating IAF have been proposed, including (1) reduction in interfacial energy [24], (2) decrease in lattice mismatch [25], (3) lessening thermal strains [5], and (4) the formation of a solute depletion zone [26,27]. Although the exact mechanism has not been elucidated yet, it is widely accepted that the chemistry and size of inclusions play an important role in the nucleation of acicular ferrite.

According to the inert interface mechanism, an inclusion larger than 1.0 µm is advantageous to nucleate IAF [16]. Zhang et al. [24] suggested that the optimal size for inclusion to become the nucleation site of IAF was about 3 µm. However, TiO_x-MnS with sizes from 1.0–3.0 µm [28] and ZrO_2-MnS with a size of 0.8 µm [29] were also able to nucleate IAF. In the present study, as shown in Figure 9a–c, MgO-MnS complex inclusion with a size of 3.31 µm can act as the nucleation center of IAFs. However, the MgO-MnS complex inclusion with a size of 2.46 µm in 3Mg20Al was just located in the ferrite matrix and unable to nucleate acicular ferrite, as shown in Figure 9d–f. Although the difference of the size between inclusions in 3Mg1Al and 3Mg20Al was relatively small, the former was nucleant and the latter non-nucleant. Thus, it should be concluded that inclusion size was not the definitive factor for inclusions acting as nucleants in the present study.

Instead, the chemistry and morphology of inclusions may play key roles in inclusion nucleating IAF. Generally, oxysulfide complex inclusions comprising central oxides and, outside, the MnS phase are considered as effective nucleants for IAF [5,23]. As shown in Figure 9c, in 3Mg1Al, the MgO-MnS complex inclusion acting as the nucleant comprises a central MgO particle and outside, the MnS phase. However, the morphology of MgO-MnS complex inclusion in 3Mg20Al was several small MgO particles entrapped by MnS, as shown in Figure 9f. This kind of complex inclusion in 3Mg20Al was non-nucleant, as shown in Figure 9e. It was confirmed that pure MnS phase inclusion in steel matrix was unable to nucleate IAF [23]. It was inferred that the MgO-MnS complex inclusion with such a morphology shown in Figures 3 and 9 was similar to the pure MnS phase inclusion. It should be noted that in addition to chemistry and size, the morphology of inclusions also played an important role in the nucleation of acicular ferrite. In 3Mg1Al, due to well-developed IAFs induced by MgO-MnS complex inclusions in the HAZ microstructure, excellent HAZ toughness was obtained after high-heat-input welding of 400 kJ/cm. However, in 3Mg20Al, the formation of FSP cannot be inhibited, resulting in the lack of a nucleant for IAF. As a result, the HAZ toughness of 3Mg20Al deteriorated.

5. Conclusions

The effects of Al content on the inclusion-microstructure relationship in the heat-affected zone (HAZ) of steel plates with Mg deoxidation after high-heat-input welding were investigated based on experimental studies. HAZ toughness was also measured after the simulated welding process with a heat input of 400 kJ/cm. The following conclusions were obtained:

1. The main inclusion types in 3Mg1Al without Al addition were MnS, MgO-MnS, MgO-MnS-TiN, and those in 3Mg20Al with Al addition were MgO-MnS, MgO-MnS-TiN, MgO-Al_2O_3-MnS. The number density of inclusions in 3Mg1Al, 96.65 mm^{-2}, was similar to that in 3Mg20Al, 95.03 mm^{-2}. However, the mean size of inclusions in the former, 3.47 μm, was larger than that in the latter, 2.03 μm.

2. Although the chemistries of the main kind of oxysulfide complex inclusions in 3Mg1Al and 3Mg20Al were both MgO-MnS, the morphologies were quite different. The former consisted of a central single MgO particle and outside, the MnS phase. The latter comprised several small MgO particles entrapped by the MnS phase.

3. Because the MgO-MnS complex inclusions in 3Mg1Al could nucleate intragranular acicular ferrites (IAFs) and these in 3Mg20Al were non-nucleant, the main intragranular microstructure in HAZs for 3Mg1Al was ductile IAFs, while that for 3Mg20Al was brittle ferrite side plates (FSPs). Therefore, the HAZ toughness of Mg deoxidized the steel plate without Al addition was much better than that with Al addition.

Author Contributions: L.X. performed the experiments, analyzed the experimental results, and wrote this manuscript. J.Y. contributed to the guidance, conceived of, and designed this research. R.W. performed the simulations.

Funding: This research received no external funding.

Acknowledgments: The financial support from Baosteel Group Corporation is greatly appreciated.

Conflicts of Interest: The authors declare no conflict of interest.

References

1. Takamura, J.; Mizoguchi, S. Roles of Oxides in Steels Performance. In Proceedings of the Sixth International Iron and Steel Congress, Nagaya, Japan, 21–26 October 1990.
2. Kojima, A.; Kiyose, A.; Uemori, R.; Minagawa, M.; Hoshino, M.; Nakashima, T.; Ishida, K.; Yasui, H. Super high HAZ toughness technology with fine microstructure imparted by fine particles. *Shinnittetsu Giho* **2004**, 2–5.
3. Suzuki, S.; Ichimiya, K.; Akita, T. High tensile strength steel plates with excellent HAZ toughness for shipbuilding. *JFE Tech. Rep.* **2005**, *5*, 24–29.
4. Yang, J.; Zhu, K.; Wang, R.Z.; Shen, J.G. Improving the toughness of heat affected zone of steel plate by use of fine inclusion particles. *Steel Res. Int.* **2011**, *82*, 552–556. [CrossRef]
5. Sarma, D.S.; Karasev, A.V.; Jönsson, P.G. On the role of non-metallic inclusions in the nucleation of acicular ferrite in steels. *ISIJ Int.* **2009**, *49*, 1063–1074. [CrossRef]
6. Kim, H.S.; Chang, C.H.; Lee, H.G. Evolution of inclusions and resultant microstructural change with Mg addition in Mn/Si/Ti deoxidized steels. *Scripta Mater.* **2005**, *53*, 1253–1258. [CrossRef]
7. Chai, F.; Yang, C.F.; Su, H.; Zhang, Y.Q.; Xu, Z. Effect of magnesium on inclusion formation in Ti-killed steels and microstructural evolution in welding induced coarse-grained heat affected zone. *J. Iron Steel Res. Int* **2009**, *16*, 69–74. [CrossRef]
8. Park, S.C.; Jung, I.H.; Oh, K.S.; Lee, H.G. Effect of Al on the evolution of non-metallic inclusions in the Mn-Si-Ti-Mg deoxidized steel during solidification: Experiments and thermodynamic calculations. *ISIJ Int.* **2004**, *44*, 1016–1023. [CrossRef]

9. Li, X.B.; Min, Y.; Yu, Z.; Liu, C.J.; Jiang, M.F. Effect of Mg addition on nucleation of intra-granular acicular ferrite in Al-killed low carbon steel. *J. Iron Steel Res. Int.* **2016**, *23*, 415–421. [CrossRef]
10. Xu, L.Y.; Yang, J.; Wang, R.Z.; Wang, Y.N.; Wang, W.L. Effect of Mg content on the microstructure and toughness of heat-affected zone of steel plate after high heat input welding. *Metall. Mater. Trans. A* **2016**, *47*, 3354–3364. [CrossRef]
11. Xu, L.Y.; Yang, J.; Wang, R.Z.; Wang, W.L.; Wang, Y.N. Effect of Mg addition on formation of intragranular acicular ferrite in heat-affected zone of steel plate after high-heat-input welding. *J. Iron Steel Res. Int.* **2018**, *25*, 433–441. [CrossRef]
12. Xu, L.Y.; Yang, J.; Wang, R.Z.; Wang, W.L.; Ren, Z.M. Effect of welding heat input on microstructure and toughness of heated-affected zone in steel plate with Mg deoxidation. *Steel Res. Int.* **2017**, *88*, 1700157. [CrossRef]
13. Lou, H.; Wang, C.; Wang, B.; Wang, Z.; Li, Y.; Chen, Z. Inclusion Evolution behavior of Ti-Mg oxide metallurgy steel and its effect on a high heat input welding HAZ. *Metals* **2018**, *8*, 534. [CrossRef]
14. Van, E.M.; Guo, M.; Zinngrebe, E.; Blanpain, B.; Jung, I. Evolution of non-metallic inclusions in secondary steelmaking: Learning from inclusion size distributions. *ISIJ Int.* **2013**, *53*, 1974–1982.
15. Pervushin, G.V.; Suito, H. Effect of primary deoxidation products of Al_2O_3, ZrO_2, Ce_2O_3 and MgO on TiN precipitation in Fe-10 mass% Ni alloy. *ISIJ Int.* **2001**, *41*, 748–756. [CrossRef]
16. Lee, T.K.; Kim, H.J.; Kang, B.Y.; Hwang, S.K. Effect of inclusion size on the nucleation of acicular ferrite in welds. *ISIJ Int.* **2000**, *40*, 1260–1268. [CrossRef]
17. Kimura, S.; Nakajima, K.; Mizoguchi, S. Behavior of alumina-magnesia complex inclusions and magnesia inclusions on the surface of molten low-carbon steels. *Metall. Mater. Trans. B* **2001**, *32*, 79–85. [CrossRef]
18. Yang, J.; Xu, L.Y.; Zhu, K.; Wang, R.Z.; Zhou, L.J.; Wang, W.L. Improvement of HAZ toughness of steel plate for high heat input welding by inclusion control with Mg deoxidation. *Steel Res. Int.* **2015**, *86*, 619–625. [CrossRef]
19. Karasev, A.V.; Suito, H. Characteristics of fine oxide particles produced by Ti/M (M = Mg and Zr) complex deoxidation in Fe-10 mass%Ni alloy. *ISIJ Int.* **2008**, *48*, 1507–1516. [CrossRef]
20. Ohta, H.; Suito, H. Characteristics of particle size distribution of deoxidation products with Mg, Zr, Al, Ca, Si/Mn and Mg/Al in Fe-10 mass% Ni alloy. *ISIJ Int.* **2006**, *46*, 14–21. [CrossRef]
21. Wang, Y.; Tang, H.; Wu, T.; Wu, G.; Li, J. Effect of acid-Soluble aluminum on the evolution of non-metallic inclusions in spring steel. *Metall. Mater. Trans. B* **2017**, *48*, 943. [CrossRef]
22. Ohta, H.; Suito, H. Effects of dissolved oxygen and size distribution on particle coarsening of deoxidation product. *ISIJ Int.* **2006**, *46*, 42–49. [CrossRef]
23. Lee, J.L.; Pan, Y.T. Effect of sulfur content on the microstructure and toughness of simulated heat-affected zone in Ti-killed steels. *Metall. Trans. A* **1993**, *24*, 1399–1408. [CrossRef]
24. Zhang, C.; Gao, L.; Zhu, L. Effect of inclusion size and type on the nucleation of acicular ferrite in high strength ship plate steel. *ISIJ Int.* **2018**, *58*, 965–969. [CrossRef]
25. Zhu, L.; Wang, Y.; Wang, S.; Zhang, Q.; Zhang, C. Research of microalloy elements to induce intragranular acicular ferrite in shipbuilding steel. *Ironmaking Steelmaking* **2017**, 1–9. [CrossRef]
26. Kang, Y.B.; Lee, H.G. Thermodynamic analysis of Mn-depleted near Ti oxide inclusions for intragranular nucleation of ferrite in steel. *ISIJ Int.* **2010**, *50*, 501–508. [CrossRef]
27. Mabuchi, H.; Uemori, R.; Fujioka, M. The role of Mn depletion in intra-granular ferrite transformation in the heat affected zone of welded joints with large heat input in structural steels. *ISIJ Int.* **1996**, *36*, 1406–1412. [CrossRef]
28. Zheng, C.C.; Wang, X.M.; Li, S.R.; Shang, C.J.; He, X.L. Effects of inclusions on microstructure and properties of heat-affected-zone for low-carbon steels. *Sci. Chin. Technol. Sci.* **2012**, *55*, 1556–1565. [CrossRef]
29. Shi, M.H.; Zhang, P.Y.; Zhu, F.X. Toughness and microstructure of coarse grain heat affected zone with high heat input welding in Zr-bearing low carbon steel. *ISIJ Int.* **2014**, *54*, 188–192. [CrossRef]

Article

Improvement of Heat-Affected Zone Toughness of Steel Plates for High Heat Input Welding by Inclusion Control with Ca Deoxidation

Ruizhi Wang [1,2], Jian Yang [1,3,*] and Longyun Xu [3]

1   School of Metallurgy, Northeastern University, Shenyang 110083, China; wangruizhi@baosteel.com
2   Steelmaking Research Department, Research Institute, Baosteel Group Corporation, Shanghai 201900, China
3   State Key Lab of Advanced Special Steel, School of Materials Science and Engineering, Shanghai University, Shanghai 200444, China; xulongyun@shu.edu.cn
*   Correspondence: yang_jian@t.shu.edu.cn; Tel.: +86-021-3604-7721

Received: 22 October 2018; Accepted: 13 November 2018; Published: 14 November 2018

Abstract: The characteristics of inclusions and microstructure in heat-affected zones (HAZs) of steel plates with Ca deoxidation after high heat input welding of 400 kJ·cm^{-1} were investigated through simulated welding experiments and inclusions automatic analyzer systems. Typical inclusions in HAZs of steels containing 11 ppm and 27 ppm Ca were recognized as complex inclusions with the size in the range of 1~3 μm. They consisted of central Al_2O_3 and peripheral $CaS + MnS$ with TiN distributing at the edge ($Al_2O_3 + CaS + MnS + TiN$). With increasing Ca content in steel, the average size of inclusions decreased from 2.23 to 1.46 μm, and the number density increased steadily from 33.7 to 45.0 mm^{-2}. $Al_2O_3 + CaS + MnS + TiN$ complex inclusions were potent to induce the formation of intragranular acicular ferrite (IAF). Therefore, the HAZ toughness of steel plates after high heat input welding was improved significantly by utilizing oxide metallurgy technology with Ca deoxidation.

Keywords: heat-affected zone; high heat input welding; Ca deoxidation; inclusion control; intragranular acicular ferrite

---

1. Introduction

In recent years, the high heat input welding technology with the heat input larger than 400 kJ·cm^{-1} has been widely used for welding heavy steel plates in the areas of shipbuilding, architectural construction, etc. For example, the heat input of tandem electrogas welding could reach 500 kJ·cm^{-1}. Also, that this kind of welding process is applicable to a very limited portion of material, since the welding spot is very limited. During the high heat input welding process, the heat-affected zone (HAZ) of steel plate is exposure at temperatures as high as 1400 °C for a long time, with a relatively low cooling rate. As a consequence, the HAZ microstructure is coarsened, leading to a significant degradation of the HAZ toughness [1,2].

It is widely accepted that the oxide metallurgy technology is an effective method for the improvement of HAZ toughness of steel plates after high heat input welding [3]. Oxide metallurgy technology refers to oxide inclusions and other kinds of inclusions and precipitates, serving as nucleation sites of intragranular acicular ferrite (IAF) or pinning the growth of austenite grain during the welding process. Nowadays, it has been extensively studied to develop advanced oxide metallurgy technology utilizing strong deoxidizers, such as Mg and Zr. Yang et al. [1,4–6] have developed third-generation oxide metallurgy technology to obtain excellent HAZ toughness by inclusion control with Mg deoxidation. They found that the formation of micro-meter inclusions and the nano-meter precipitates in steel plates can effectively promote the formation of IAF and restrain the growth of γ grains, respectively. Shi et al. [7] reported the excellent toughness (more than 100 J) of HAZ with the heat input of 400 kJ·cm^{-1} in Zr bearing low carbon steel, attributing to the proper austenite grain size that favored the development of IAF. Kojima et al. [8] reported that the HAZ toughness was improved significantly by using fine particles with sizes of 10–100 nm containing Mg or Ca.

Ca treatment is a well-established method to transform solid alumina clusters to liquid calcium aluminates in order to avoid nozzle clogging problems in continuous casting of Al-killed steels [9]. Although Ca is also a kind of strong deoxidizing element and is widely used in the area of inclusion modification, termed as Ca treatment, limited research on oxide metallurgy technology with Ca deoxidation has been reported [10–13]. In view of the advantages of steel industrial application of Ca, it has scientific and industrial values to investigate the oxide metallurgy technology by inclusion control with Ca deoxidation to improve the HAZ toughness.

In this paper, the characteristics of inclusions and microstructure in HAZs of steel plates with Ca deoxidation are studied after high heat input welding of 400 kJ·cm^{-1}. In addition, the mechanism of improvement of HAZ toughness after high-heat input welding by inclusion control with Ca deoxidation is illustrated.

2. Experimental Procedures

2.1. Experimental Steel Preparation

Steel samples were melted in a 50-kg vacuum induction furnace with sintered magnesia lining. Table 1 shows the major chemical compositions of experimental steel samples. The method for steel chemical analysis is Method for Photoelectric Emission Spectroscopic Analysis by JIS G 1253-2002. The measurement of C, O and N are Combustion and Melting Technology by ASTM E1019-11. As shown in Table 1, steel samples C11 and C27 were both prepared by oxide metallurgy technology with Ca deoxidation, which Ca contents were measured to be 11 and 27 ppm, respectively. For comparison, a conventional Al-killed steel without Ca deoxidation denoted as A was used in the present study. Deoxidation process was carried out in the furnace under the Ar atmosphere. About 40 kg of pure iron was first melted, and lime (CaO) was added as top slag to assure considerably low oxygen potential in slag. Proper amount of deoxidants of Mn, Si, Al, Ti and Ca was added in order to obtain the target steel compositions. The melt was then cast into an ingot with the size of 120 × 180 × 240 (mm^3), and each ingot was hot rolled into a steel plate with the thickness of 50 mm. Roughing rolling was conducted at the temperature above 930 °C with the reduction ratio larger than 30%, and finishing rolling was carried out at the temperature of about 800 °C with the reduction ratio larger than 30%. Subsequently, the steel plate was cooled down from 760 °C to 400 °C at the cooling rate of about 10 °C·s^{-1}.

Table 1. Chemical compositions of the experimental steels (wt%).

Steel	C	Mn	P	S	Ca	Al	Ti	O	N
A	0.078	1.5	0.01	0.005	0.0005	0.026	0.012	0.0022	0.0014
C11	0.078	1.5	0.01	0.005	0.0011	0.011	0.014	0.0012	0.0029
C27	0.078	1.5	0.01	0.005	0.0027	0.015	0.013	0.0013	0.0030

2.2. Welding Thermal Simulation Experiments

In order to evaluate the HAZ toughness of the experimental steel plates, simulated welding experiments were carried out by use of Gleeble 3800 (Dynamic Systems Inc., New York, NY, USA). It was designed to simulate an electrogas arc welding with the heat input of 400 kJ·cm^{-1} for the steel plate with a thickness of 50 mm. The steel sample for simulated welding experiments with the size of 11 × 11 × 55 (mm^3) was cut down from the position about 1/4 width and 1/4 length from the edge of steel plate. As shown in Figure 1, the peak welding temperature was 1400 °C with a holding time of 3 s. The cooling rates from 1400 to 800 °C, from 800 to 500 °C and from 500 to 300 °C were controlled to be 3.41, 0.78, 0.17 °C·s^{-1}, respectively. Then, the standard V notch Charpy impact tests of HAZs were conducted at −20 °C for the samples after simulated welding experiments.

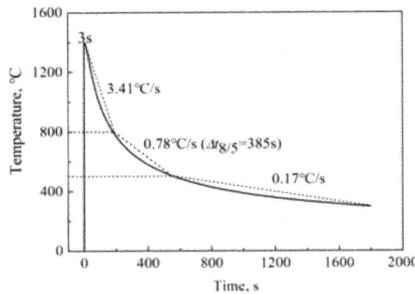

Figure 1. Thermal cycle of simulated heat-affected zone (HAZ) of high heat input welding with 400 kJ·cm^{-1}.

2.3. Characterization of Inclusions and Microstructures

These HAZ specimens were taken from the surfaces in parallel with the fracture cross sections after Charpy impact tests. Firstly, these specimens were polished and then analyzed by the inclusion automatic analyzer system (IAAS) made in Germany. IAAS is consisted of scanning electron microscope (SEM), energy-dispersive spectroscopy (EDS) and software which can analyze the inclusions automatically. Then HAZ specimens were etched by 4% nital to observe the HAZ microstructures with optical microscopy and SEM. The present study mainly focuses on the inclusions with the equal-circle size larger than 1 μm. The equal-circle size is given by the diameter of a circle whose area is equal to the area occupied by the inclusion in the SEM picture. Mapping scanning of typical inclusions in each sample was performed in order to determine the elemental distribution inside the inclusions. A schematic drawing with the sampling procedure is shown in Figure 2.

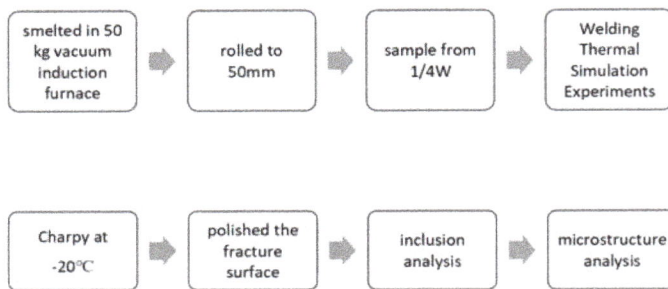

Figure 2. Schematic draw with the sampling procedure.

3. Results

3.1. HAZ Toughness

The Charpy impact test results of HAZs at −20 °C of steel samples A, C11, and C27 are shown in Table 2. All three Charpy-absorbed energy values of steel sample A are very low with the average value of only 27 J. Obviously, the HAZ toughness of the conventional Al-killed steel of sample A without Ca deoxidation was greatly deteriorated after high heat input welding. The average Charpy-absorbed energy values of steel samples C11 and C27, both with Ca deoxidation, are 187 J and 123 J, respectively. By comparison, the HAZ toughness values of these two steel samples were increased by more than four times. It is clear that excellent HAZ toughness of steel plate after high heat input welding with 400 kJ·cm^{-1} has been obtained by utilizing oxide metallurgy technology with Ca deoxidation.

Table 2. Charpy-absorbed energy of HAZ at -20 °C.

Specimen	Individual Value (J)	Mean Value (J)	FA (%)	Standard Deviation
A	14, 33, 34	27	0, 5, 5	127
C11	180, 189, 192	187	70, 70, 70	39
C27	88, 136, 144	123	35, 45, 50	917

Figure 3 shows the macro fractographs and SEM images of the fracture surfaces of the HAZs in steel samples A, C11 and C27 after Charpy impact tests. The fractographs of steel samples C11 and C27 show the obvious transverse deformation, indicated by the lateral expansion or notch contraction as shown in Figure 3d,g. In contrast, steel sample A does not show any transverse deformation (Figure 3a). Figure 3b,c shows that the fracture surface of steel sample A is covered with radial zones and river-like patterns, indicating a brittle fracture and inferior HAZ toughness. The ratios of radial zones in fractures for steel samples C11 and C27 are less than that of steel sample A, as shown in Figure 3d,g. Both are full of fibrous zones containing large quantity of dimples, as shown in Figure 3e,f,h,i. It manifests that the HAZ fracture mode of steel samples C11 and C27 is ductile fracture. Thus, steel samples C11 and C27 show excellent HAZ toughness.

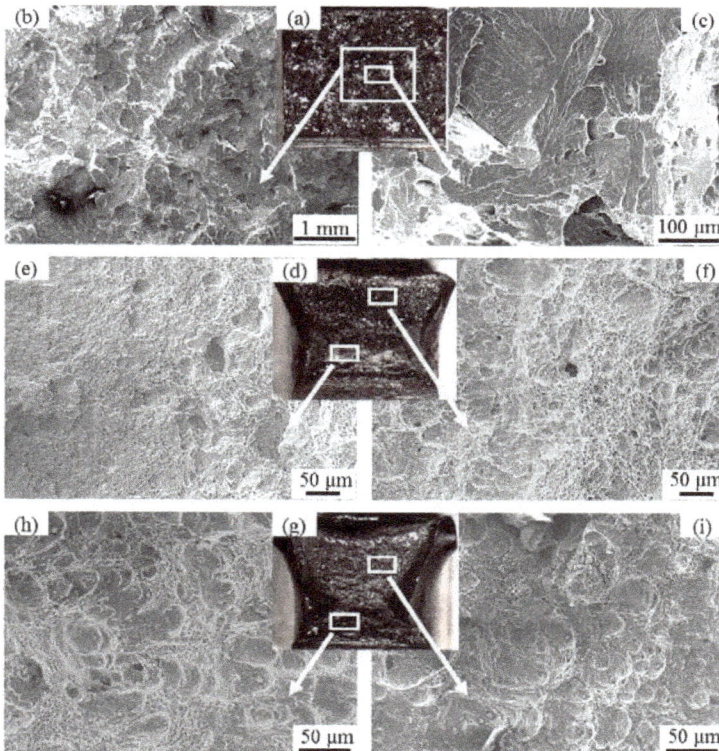

Figure 3. Macro fractographs and SEM images of fracture surfaces of HAZs: (**a–c**) A; (**d–f**) C11; (**g–i**) C27.

3.2. Typical Inclusions

The morphologies and mapping analysis results analyzed by SEM-EDS for typical inclusions in steel samples A, C11 and C27 are presented in Figures 4–6, respectively. Generally, complex inclusions comprising central oxides and peripheral sulfides are in the shape of globular or nearly globular.

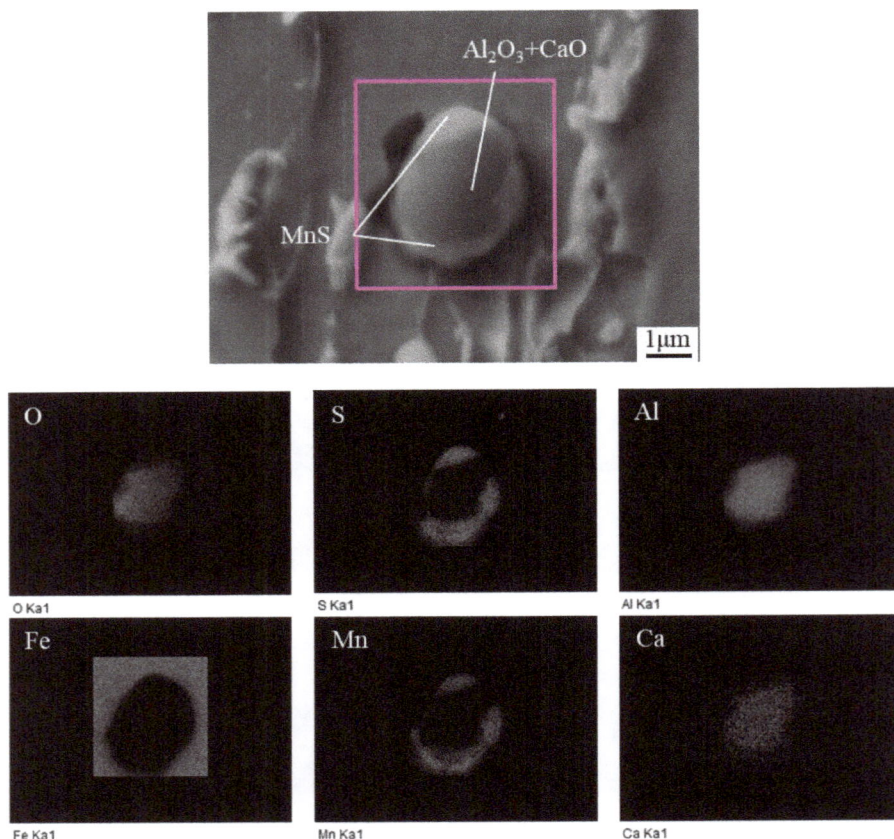

Figure 4. Elemental mapping pattern of typical inclusion in steel sample A.

The mapping analysis results of the typical inclusion in the conventional Al-killed steel sample A is shown in Figure 4. It is observed that the distributions of Ca and Al elements are both overlapped with O element. Meanwhile, the distribution of S element coincides with that of Mn element. It is indicated that the central part of the typical inclusion in steel sample A is composed of Al_2O_3 with a little CaO, denoted as Al_2O_3 + CaO. Besides, the sulfide covering the typical inclusion is MnS phase.

Figure 5 shows the distribution of Ca is not overlapped with O element but S and Mn elements. It is indicated the sulfide should be the solution of MnS phase and CaS phase, termed as (Ca, Mn)S. Meanwhile, in the central part of this inclusion, the distribution of Al element coincides with that of O element. Thus, the typical inclusion in steel sample C11 is mainly composed of irregular Al_2O_3 in the center with (Ca, Mn)S outside, as shown in Figure 5. It is observed that there are small amount of TiN locating at the edge of the typical inclusion in steel sample C11.

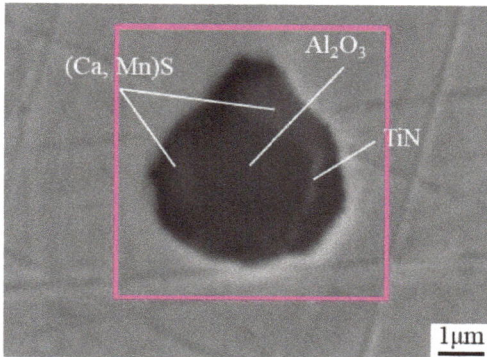

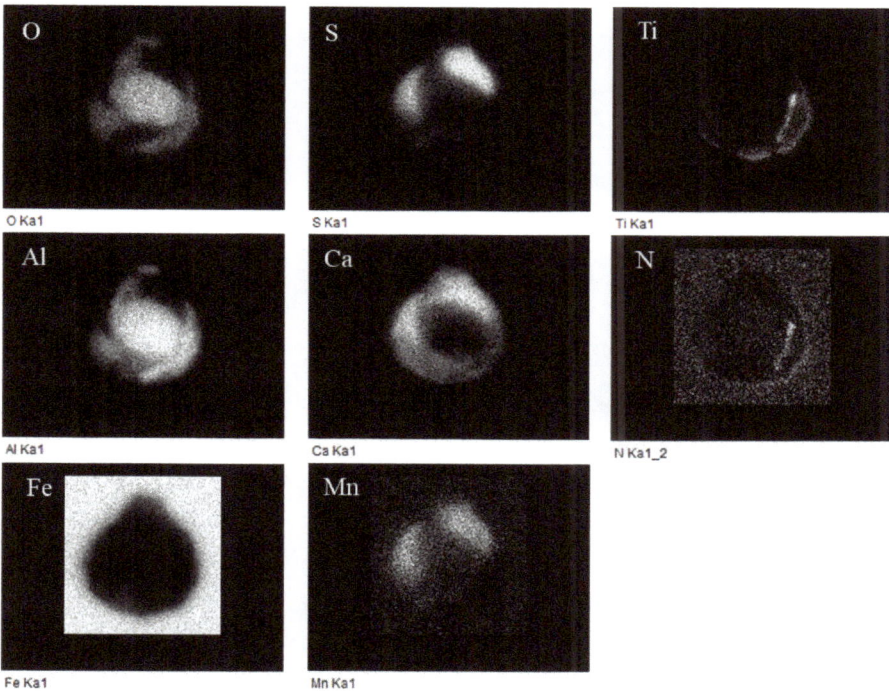

Figure 5. Elemental mapping pattern of typical inclusion in steel sample C11.

As shown in Figure 6, the typical inclusion in steel sample C27 containing 27 ppm Ca reveals a nearly globular shape. The distribution of Ca element is also not overlapped with O element but S element. Meanwhile, Mn element is detected in this inclusion. It is indicated that the sulfide is mainly composed of CaS phase with MnS phase. Therefore, the typical inclusion in steel sample C27 is mainly composed of central Al_2O_3 and peripheral (Ca, Mn)S, along with a small amount of TiN observed at the edge.

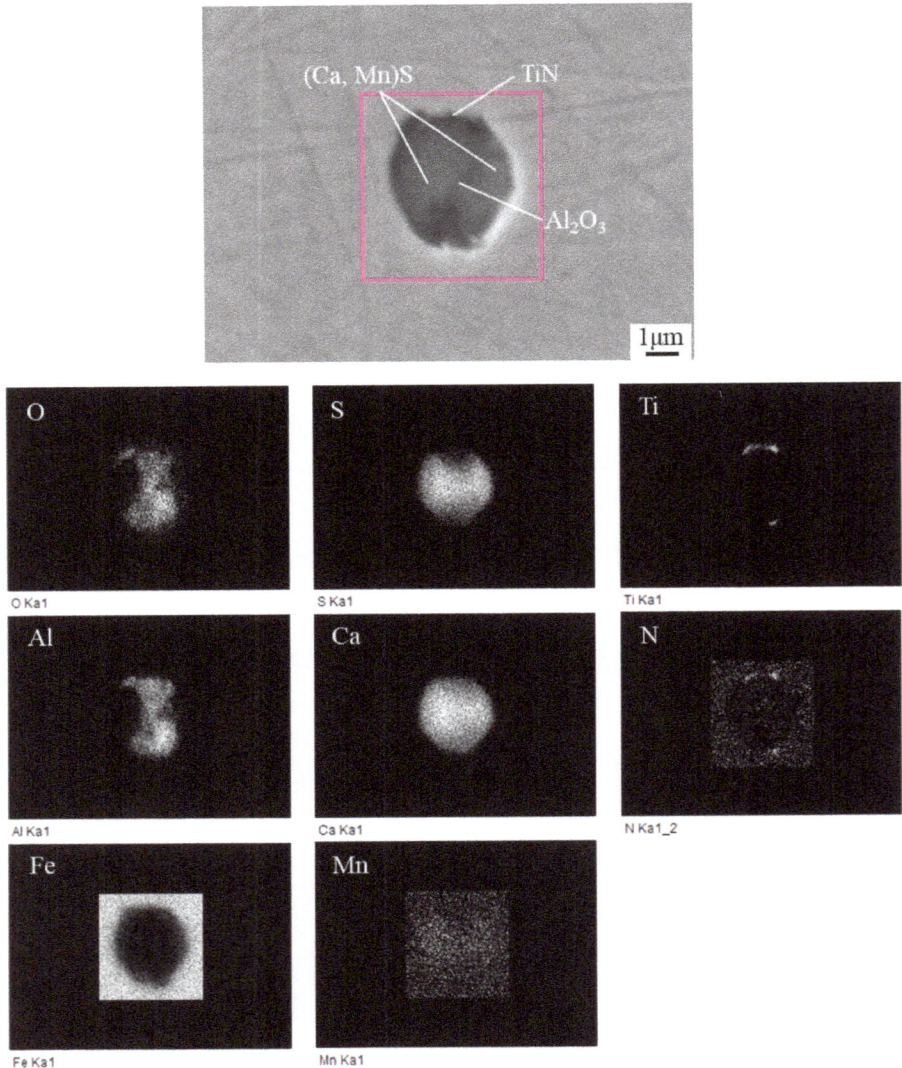

Figure 6. Elemental mapping pattern of typical inclusion in steel sample C27.

3.3. Number Density and Size of Inclusions

The analysis results of inclusions for steel samples A, C11 and C27 as shown from Figures 7–10 are based on the examination of inclusions in HAZs through IAAS. The numbers of inclusions examined in steel samples A, C11 and C27 are 2340, 1892 and 2762 in the observed areas of 69.5, 45.2 and 61.4 mm^2, respectively.

Figure 7 shows the relationship between Ca content in steel and the average size and number density of inclusions with the size larger than 1 μm in HAZs. With increasing the Ca content in steel from 5 to 11 and then to 27 ppm, the number density of inclusions is increased steadily from 33.7 to 41.9 and then to 45.0 mm^{-2} and the standard deviation is 34.09. Meanwhile, the average size of inclusions is decreased from 2.23 to 2.20 and then to 1.46 μm and the standard deviation is 0.19.

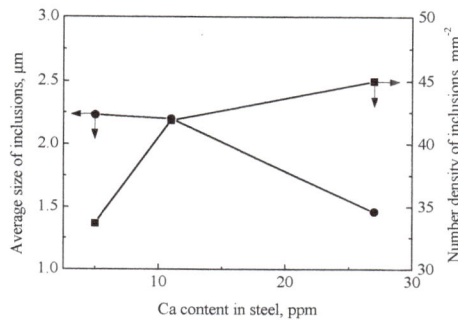

Figure 7. Changes in number density and average size of inclusions in different steels.

Figure 8 gives the size distributions of inclusions in steels with different Ca contents. It is seen that there are few inclusions larger than 10 μm in all three steel samples, and most of inclusions are smaller than 3 μm after Ca deoxidation. The frequencies of inclusions with size between 1 and 2 μm in A, C11 and C27 are 57.2%, 52.2% and 91.4%, respectively, suggesting that the average size of inclusions in steel sample C27 is much smaller than that in steel samples A and C11. The size distribution of steel sample A is similar with that of steel sample C11. Therefore, the average sizes of inclusions in steel samples A and C11 are at the same level, 2.23 and 2.20 μm, respectively, as shown in Figure 7.

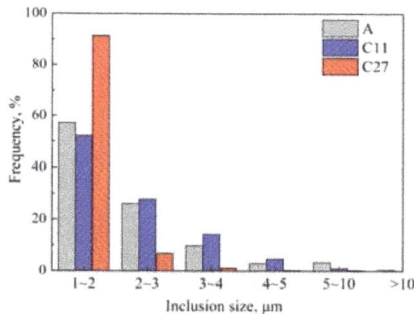

Figure 8. Size distributions of inclusions in the experimental steels.

3.4. Composition of Inclusions

Figure 9 shows the changes of average compositions of inclusions in steel samples A, C11 and C27, respectively. The Ca content in the average composition of inclusions gradually increases from 1.01% to 11.61% and then to 22.09% (in mass percentage) with increasing Ca content from 5 to 11 and then to 27 ppm. There is a similar tendency in the change of Al content in the average composition of inclusions, which increases from 16.54% to 21.95% and then to 25.22%. It is also noted that the Ti contents in the average composition of inclusions in both steel samples with Ca deoxidation, C11 and C27, 3.43% and 5.62%, are much higher than that in the conventional Al-killed steel sample A, 0.94%. As shown in Figures 5 and 6, the Ti element in inclusions existed in the form of TiN precipitate. It is inferred that the precipitation of TiN is promoted by Ca deoxidation. Figure 9 also shows that the content of S element in inclusions is relatively high, nearly more than 20%, but without significant change of S content in inclusions with increasing the Ca content. Besides, the Mn content in inclusions decreases sharply from 39.73% to 20.27% and then to 7.12% with addition of Ca content in steel from 5 to 11 and then to 27 ppm.

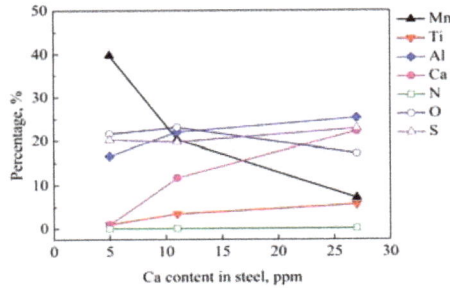

Figure 9. Changes of average composition of inclusions with Ca content in steel.

It is seen from Figures 4–6 that these typical inclusions in all three steel samples are complex inclusions composed of central oxides, outside sulfide and TiN at the edge. The complex inclusions could be classified into nine types [14] by the major compositional elements (Al, Ca, Ti, Mn, O, N and S) as shown in Figure 10. These nine types of inclusions are (1) Al_2O_3, (2) Al_2O_3 + CaO, (3) Al_2O_3 + MnS, (4) Al_2O_3 + CaO + MnS, (5) Al_2O_3 + CaS + MnS + TiN, (6) CaS + MnS + TiN, (7) MnS, (8) TiN, and (9) MnS + TiN. As shown in Figure 10, the frequencies of Al_2O_3 + CaS + MnS + TiN complex inclusions in steel samples C11 and C27 are as high as 88.74% and 68.28%, respectively. Therefore, Al_2O_3 + CaS + MnS + TiN complex inclusion is the main inclusion type in both steel samples with Ca deoxidation. In comparison, this type of inclusion is hardly found in steel sample A, but instead the main inclusions in steel sample A are MnS and Al_2O_3 + MnS with the frequencies of 35.84% and 27.96%, respectively. There are also small amounts of Al_2O_3 + CaO and Al_2O_3 + CaO + MnS complex inclusions in steel sample A with the frequencies of 5.79% and 12.49% respectively. It is concluded that the Ca element in the inclusions of steel sample A exist in the form of CaO in the Al_2O_3 + CaO or Al_2O_3 + CaO + MnS complex inclusions, as shown in Figure 4. However, the existent forms of Ca element in the inclusions of steel samples C11 and C27 are both CaS, as shown in Figures 5 and 6. It is also found, in Figure 10, that the frequencies of isolate MnS decreases from 35.84% to 8.4% and then to 1.92% with increasing Ca content in steel from 5 ppm to 11 ppm and then to 27 ppm. Meanwhile, the change of MnS + TiN inclusions frequency has the similar tendency, decreasing from 2.79% to 1.74% and then to 0.18%. The CaS + MnS + TiN type inclusion is hardly found in steels with Ca contents of 5 and 11 ppm, but the frequency of this type of inclusion increases up to 21.4% when the Ca content increases to 27 ppm, as shown in Figure 10.

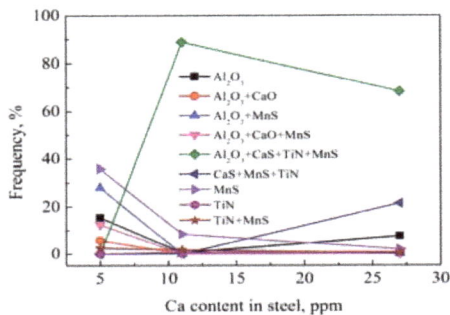

Figure 10. Variations of different types of inclusions with the increase of Ca content.

3.5. Characteristics of HAZ Microstructures

Figure 11 shows the typical HAZ microstructures of steel samples A, C11 and C27 after simulated welding with the heat input of 400 kJ·cm^{-1}. Grain boundary ferrites (GBFs) with network-like distribution are clearly observed in the HAZ microstructures of all three steel samples. Because GBFs form from the prior-austenite grain (PAG) boundaries and grow along them, the size of the cycle formed by GBF can indicate the size of PAG. The PAG sizes were measured by the straight intercept method of ASTM E112 standard. The intercept points per millimeter of three samples are 2.53, 3.65, 2.74, respectively, and the PAG sizes are deprived by the corresponding relation between the intercept point and the PAG sizes. The average PAG sizes of steel samples A, C11 and C27 are approximately 420, 260 and 390 μm, respectively. According to the result reported by Zhang et al. [15], austenite grain with the size larger than 200 μm was appropriate for the formation of IAF. As shown in Figure 11b, the intragranular part in the HAZ microstructure of steel sample A mainly consists of coarse ferrite side plates (FSPs), but it is quite different in both steel samples with Ca deoxidation, C11 and C27. In the steel samples C11 and C27, the intragranular part in the HAZ microstructure is mainly comprised of well-developed IAFs, as shown in Figure 11d,f.

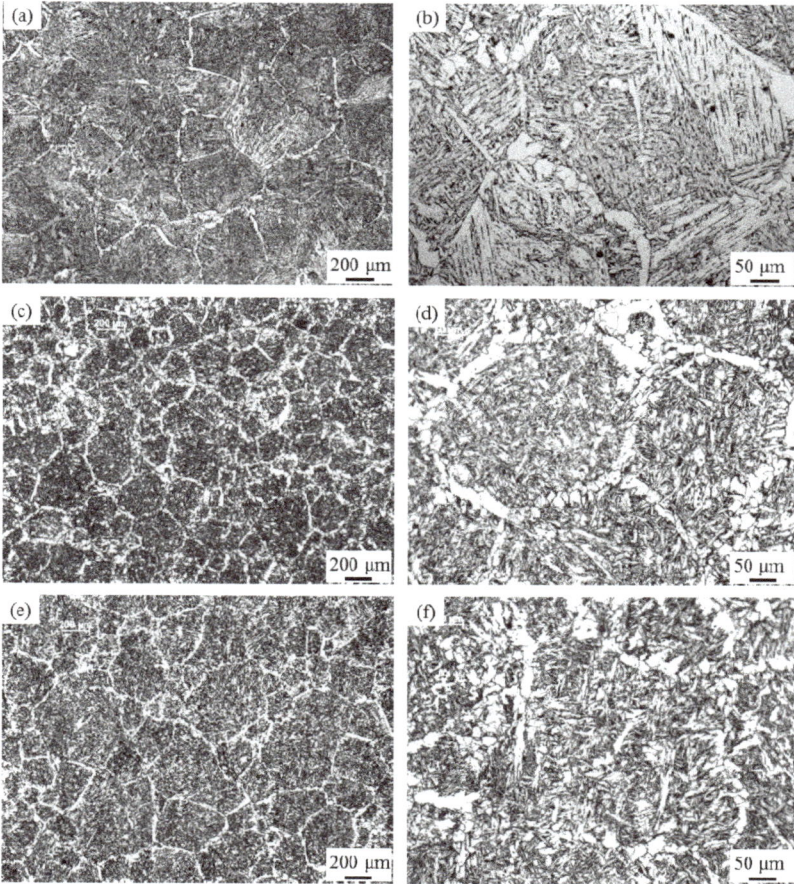

Figure 11. Typical HAZ microstructures for steel plates after simulated welding with heat input of 400 kJ·cm^{-1}: (**a,b**) A; (**c,d**) C11; (**e,f**) C27.

Figure 12 shows SEM micrographs of typical ferrite grains associated with inclusions in steel samples A, C11 and C27. As can be seen in Figure 12a, in steel sample A, the inclusion located in the coarse FSP lath is comprised of mainly Al_2O_3 + CaO and small amount of MnS. SEM-EDS analysis results shown in Figure 12d,f indicate that these inclusions in steel samples C11 and C27, inducing the formation of IAFs, are composed of Al_2O_3 + CaS + MnS + TiN.

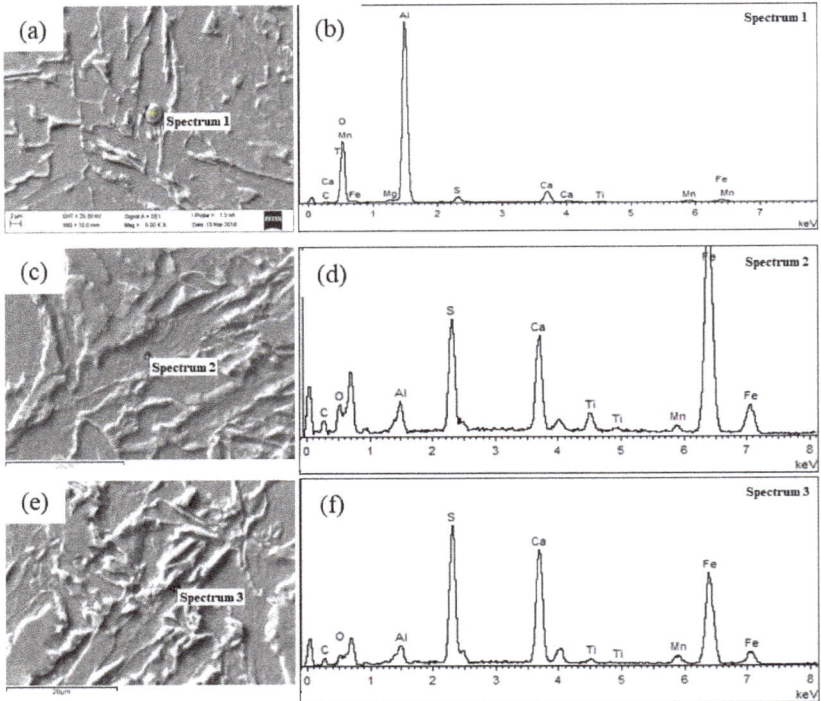

Figure 12. EDS analysis results of inclusions in HAZs: (**a,b**) A, (**c,d**) C11, and (**e,f**) C27.

4. Discussion

4.1. Behavior of Ca Element in Oxide Metallurgy Technology

During the melting process of three samples A, C11 and C27, Ca was added into the melt after Al-deoxidation. The listed reactions would exist.

$$2[Al] + 3[O] = Al_2O_3; \Delta G^\ominus = -864.370 + 222.5T \tag{1}$$

$$Al_2O_3 + 3[Ca] = 3CaO_{(inclusions)} + 2[Al]; \Delta G^\ominus = -1{,}068{,}893 + 215.4T \text{ [16]} \tag{2}$$

$$\lg f_i = \sum_{j=2}^{n} e_i^j[\%j] \tag{3}$$

$$\Delta G_r = \Delta G^\ominus + RT\ln Q_r \tag{4}$$

Q_r is the reaction quotient which can be expressed as

$$Q_r = \frac{\alpha_{CaO \cdot Al_2O_3}}{\alpha_{Ca} \cdot \alpha_{Al} \cdot \alpha_O} \tag{5}$$

The ΔG_r have been calculated by Equation (2) [16]. The content of C, Si, Mn, Al, Ca are shown in Table 2. The coefficient activity has been calculated by Equation (3). If ΔG_r is negative, the reaction proceeds. If ΔG_r is positive, the reaction proceeds in the reverse direction.

The calculated results ΔG_r of reaction Equation (2) of three steels are $-153{,}275$ J·mol^{-1}, $-216{,}899$ J·mol^{-1}, $-249{,}188$ J·mol^{-1}, respectively. So, the reaction Equation (2) exist in A, C11 and C27 samples. During the process of reaction Equation (2), the products of Al deoxidation, Al_2O_3, are modified into complex oxide inclusions.

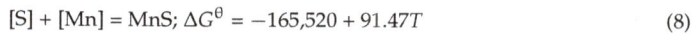

$$CaO_{inclusions} + [S] = CaS + [O]; \Delta G^\ominus = 114{,}521 - 29.8T \text{ [17]} \tag{6}$$

$$[Ca] + [S] = CaS; \Delta G^\ominus = -530{,}900 + 116.2T \text{ [17]} \tag{7}$$

$$[S] + [Mn] = MnS; \Delta G^\ominus = -165{,}520 + 91.47T \tag{8}$$

During the reaction process, the formation of CaS and MnS are discussed. The reaction of S, Ca and Mn are shown as Equations (6)–(8). Figure 13a shows the change of Gibbs free energy of reaction with activity of CaO in CaO-Al_2O_3. As it can be seen, the Gibbs free energy is negative when α_{CaO} is greater than 0.03 and the CaO in CaO-Al_2O_3 change into CaS. The Gibbs free energy is positive when α_{CaO} is less than 0.03 and the CaO in CaO-Al_2O_3 does not change into CaS. The content of CaO in CaO-Al_2O_3 of three samples can be calculated approximately in Figure 9. They are 13%, 28.8%, 37.5% respectively. The activity of CaO in CaO-Al_2O_3 can be obtained by the calculated results of Yang [17] and the content of CaO in CaO-Al_2O_3 of three samples. The α_{CaO} is greater than 0.03 when the content of CaO is greater than 20%, therefore, the reaction of Equation (6) only exists in C11 and C27 samples.

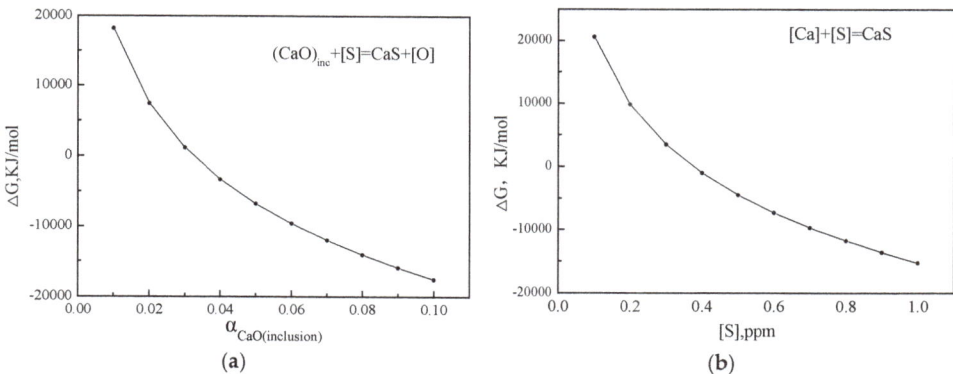

Figure 13. Change of Gibbs free energy of reaction with (**a**) activity of CaO in CaO-Al_2O_3 and (**b**) dissolved S.

There are two sources of CaS [17]. One is the reaction product of CaO in CaO·Al_2O_3 and S in steel expressed by Equation (6). The other is the product of dissolved Ca and S expressed by Equation (8). It is easier to form sulfide during the solidification and cooling processes. The solubility of sulfide is decreased greatly during the solidification and cooling process. So, the MnS formed on the Al_2O_3 and Al_2O_3 + CaO in Sample A by Equation (8).

Figure 13b shows the change of Gibbs free energy of reaction with content of dissolved S. The Gibbs free energy is positive when the content of dissolved S is less than 0.3 ppm and the reaction of Equation (6) proceeds in the reverse. In C11 and C27 samples, the content of dissolved S decrease because of the formation of MnS by Equation (7). So, the reaction of Equation (6) proceeds in reverse direction and decomposes to dissolved S when the content of dissolved S is less than 0.3 ppm. MnS forms continuously by the dissolved S and Mn. So, the major final inclusions are the type (4) in sample

A and type (5) in samples C11 and C27, as shown in Figure 10. In summary, the order of reactions in Sample A are Equations (1), (2) and (7). Al_2O_3 nucleated firstly in liquid steel, and then, MnS precipitated on Al_2O_3 after adding Ca. The order of reactions in samples C11 and C27 are Equations (1), (2), (4), (6) and (7); the main inclusion is type (5).

As shown in Figures 4–6, the Ti contents in both steel samples with Ca deoxidation, in C11 and C27, existed in the form of TiN, which precipitated on the surface of the complex inclusions. By comparison with Al and Ca in steel samples C11 and C27, the deoxidation ability of Ti is weaker. Meanwhile, the contents of oxygen in both steel samples are in a low level (less than 13 ppm). Therefore, TiN is promoted to precipitate on the surface of inclusions. as shown in Figure 10.

4.2. Mechanism of HAZ Toughness Improved by Ca Deoxidation

Oh et al. [17] reported that the austenite grain with the size larger than 100 μm was appropriate for the stable growth of GBF. As shown in Figure 11, sizes of PAGs in all three steel samples are larger than 200 μm, and thus, GBF is observed clearly in the HAZ microstructure. It has been reported that the austenite grain with size around 200 μm benefits the formation of IAF [6,16]. Moreover, Al_2O_3 + MnS inclusions in conventional LCAK steel plates are impotent to induce the nucleation of IAF during the process of high heat input welding with 400 kJ·cm^{-1} [6]. Accordingly, the intragranular part of coarse austenite grain was occupied by brittle HAZ microstructure, namely FSP, as shown in Figure 11b. As a consequence, the HAZ toughness of conventional Al-killed steel sample A was greatly deteriorated.

It is widely accepted that non-metallic inclusions in steel may act as effective nuclei for IAF. Many mechanisms of IAF nucleation on inclusions have been proposed, including (1) reduction of interfacial energy [18], (2) decrease in lattice mismatch [19,20], (3) lessening thermal strains around the inclusions [21,22], and (4) formation of solute depletion zone around the inclusions [23]. In present study, the majority of inclusions are Al_2O_3 + CaS + MnS + TiN complex inclusions, as shown in Figure 10. TiN is distributed at the edge of this type of complex inclusions, as shown in Figures 5 and 6. Kanazawa et al. [24] have reported that TiN particles may promote the formation of ferrite. Mu et al. [25] also confirmed that the TiN phase was an effective nucleation site for the formation of intragranular ferrite. The role of TiN in heterogeneous nucleation of IAF was evaluated by the lattice matching theory [21,26]. According to the literature, the lattice misfit between TiN substrate and ferrite was 3.8% [27]. It is supposed that the narrow TiN layer on the inclusion surface promotes acicular ferrite nucleation due to the low interface energy [21].

As discussed in Section 4.1, in samples C11 and C27, the CaS decomposes to dissolved S and the MnS forms continuously, so the type of inclusions contain (Ca, Mn)S, covering the central Al_2O_3, as shown in Figures 5 and 6 may arouse the formation of Mn-depleted zone (MDZ) around the interface between inclusion and steel matrix. MDZ is an important factor for inclusions to accelerate the nucleation of the IAF [28]. On the contrary, the Al_2O_3 + MnS inclusions in A steel does not deliver S from inside to outside, so the formation of MnS is not continuous. Additionally, most of inclusions in both steel samples C11 and C27 are of sizes between 1 and 3 μm, as shown in Figure 8. It has been reported that the inclusions with sizes around 2 μm could induce the formation of IAF [6,29–31]. In conclusion, it is inferred that the combined action of low lattice misfit and MDZ may be the mechanisms for the formation of well-developed IAF in HAZ during the high heat input welding and the main mechanism is that Mn-depleted zone (MDZ). As a result, excellent HAZ toughness of steel plates after high heat input welding is obtained by utilizing oxide metallurgy technology with Ca deoxidation.

5. Conclusions

The effect of Ca content on the number, size and composition of inclusions in the heat-affected zones (HAZs) of steel plates with Ca deoxidation was investigated based on the experimental study. The HAZ toughness was also measured after simulating welding process with the high heat input of 400 kJ·cm^{-1}. The following conclusions were drawn:

Metals **2018**, *8*, 946

1. The typical inclusions found in the HAZs of steel samples with Ca deoxidation containing 11 and 27 ppm Ca were Al_2O_3 + CaS + MnS + TiN complex inclusions with the size in the range of 1~3 μm, together with TiN formed at the edge of this type of inclusions.

2. In conventional Al-killed steel sample containing 5 ppm Ca, the existence form of Ca element was CaO in the Al_2O_3 + CaO or Al_2O_3 + CaO + MnS complex inclusions. On the other hand, the Ca element in the inclusions of the steels with Ca deoxidation mainly existed in the form of (Ca, Mn)S covering the central Al_2O_3.

3. With increasing Ca content in the steels from 5 to 11 and 27 ppm, the size of inclusions increased from 2.23 to 2.20 and then to 1.46 μm, and the number density of inclusions increased steadily from 33.7 to 41.9, and then to 45.0 mm^{-2}.

4. The average size of prior-austenite grains in HAZs of conventional Al-killed deoxidation and developed steel with Ca deoxidation were all larger than 200 μm. In the steel sample with Ca deoxidation, Al_2O_3 + CaS + MnS + TiN complex inclusions were potent to induce the formation of intragranular acicular ferrite (IAF) so that well-developed IAF formed in the HAZ microstructures. Therefore, excellent HAZ toughness of steel plates after welding with heat input of 400 $kJ \cdot cm^{-1}$ was obtained by utilizing oxide metallurgy technology with Ca deoxidation.

Author Contributions: J.Y. contributed the guidance, conceived and designed this research. R.W. analyzed the experimental results and wrote this manuscript. L.X. performed the experiments.

Funding: This research was funded by the National Key Research and Development Program of China, grant number 2016YFB0300602.

Acknowledgments: The financial support from Baosteel Group Corporation is great appreciated.

Conflicts of Interest: The authors declare no conflict of interest.

References

1. Yang, J.; Zhu, K.; Wang, R.; Shen, J.G. Improving the toughness of heat affected zone of steel plate by use of fine inclusion particles. *Steel Res. Int.* **2011**, *82*, 552–556. [CrossRef]

2. Yang, J.; Zhu, K.; Wang, G.D. Progress in the technological development of oxide metallurgy for manufacturing steel plates with excellent HAZ toughness. *Baosteel Tech. Res.* **2008**, *2*, 43–50.

3. Ogibayashi, S. Advances in technology of oxide metallurgy. *Nippon Steel Tech. Rep.* **1994**, 70–76.

4. Xu, L.Y.; Yang, J.; Wang, R.Z.; Wang, Y.N.; Wang, W.L. Effect of Mg content on the microstructure and toughness of heat-affected zone of steel plate after high heat input welding. *Metall. Mater. Trans. A* **2016**, *47*, 3354–3364. [CrossRef]

5. Zhu, K.; Yang, Z.G. Effect of magnesium on the austenite grain growth of the heat-affected zone in low-carbon high-strength steels. *Metall. Mater. Trans. A* **2011**, *42*, 2207–2213. [CrossRef]

6. Xu, L.Y.; Yang, J.; Wang, R.Z.; Wang, W.L.; Wang, Y.N. Effect of Mg addition on formation of intragranular acicular ferrite in heat-affected zone of steel plate after high-heat-input welding. *J. Iron Steel Res. Int.* **2018**, *25*, 433–441. [CrossRef]

7. Shi, M.H.; Zhang, P.Y.; Wang, C.; Zhu, F.X. Effect of high heat input on toughness and microstructure of coarse grain heat affected zone in Zr bearing low varbon steel. *ISIJ Int.* **2014**, *54*, 932–937. [CrossRef]

8. Kojima, A.; Kiyose, A.; Uemori, R.; Minagawa, M.; Hoshino, M.; Nakashima, T.; Ishida, K.; Yasui, H. Super high HAZ toughness technology with fine microstructure imparted by fine particles. *Nippon Steel Tech. Rep.* **2004**, *90*, 2–6.

9. Abraham, S.; Bodnar, R.; Raines, J.; Wang, Y.F. Inclusion engineering and metallurgy of calcium treatment. *J. Iron Steel Res. Int.* **2018**, *25*, 133–145. [CrossRef]

10. Yang, S.F.; Li, J.S.; Wang, Z.F.; Li, J.; Lin, L. Modification of $MgO \cdot Al_2O_3$ spinel inclusions in Al-killed steel by Ca-treatment. *Int. J. Min. Met. Mater.* **2011**, *18*, 18–23. [CrossRef]

11. Yang, D.; Wang, X.H.; Yang, G.W.; Wei, P.Y.; He, J.P. Inclusion evolution and estimation during secondary refining in calcium treated aluminum killed steels. *Steel Res. Int.* **2014**, *85*, 1517–1524. [CrossRef]

12. Ren, Y.; Zhang, Y.; Zhang, L. A kinetic model for Ca treatment of Al-killed steels using FactSage macro processing. *Ironmak. Steelmak.* **2017**, *44*, 497–504. [CrossRef]

13. Liu, Y.; Zhang, L.F.; Zhang, Y.; Duan, H.J.; Ren, Y.; Yang, W. Effect of sulfur in steel on transient evolution of inclusions during calcium treatment. *Metall. Mater. Trans. B* **2018**, *49*, 610–626. [CrossRef]
14. Wang, Y.N.; Yang, J.; Xin, X.L.; Wang, R.Z.; Xu, L.Y. The effect of cooling conditions on the evolution of non-metallic inclusions in high manganese TWIP steels. *Metall. Mater. Trans. B* **2016**, *47*, 1378–1389. [CrossRef]
15. Zhang, D.; Terasaki, H.; Komizo, Y. In situ observation of the formation of intragranular acicular ferrite at non-metallic inclusions in C-Mn steel. *Acta Mater.* **2010**, *58*, 1369–1378. [CrossRef]
16. Yang, G.W.; Wang, X.H. Inclusion evolution after calcium addition in low carbon Al-killed steel with ultra low sulfur content. *ISIJ Int.* **2015**, *55*, 126–133. [CrossRef]
17. Oh, Y.J.; Lee, S.Y.; Byun, J.S.; Shim, J.H.; Cho, Y.W. Non-metallic inclusions and acicular ferrite in low carbon steel. *Mater. Trans.* **2000**, *41*, 1663–1669. [CrossRef]
18. Ricks, R.A.; Howell, P.R.; Barritte, G.S. The nature of acicular ferrite in HSLA steel weld metals. *J. Mater. Sci.* **1982**, *17*, 732–740. [CrossRef]
19. Gregg, J.M.; Bhadeshia, H.K.D.H. Titanium-rich mineral phases and the nucleation of bainite. *Metall. Mater. Trans. A* **1994**, *25*, 1603–1611. [CrossRef]
20. Enomoto, M. Nucleation of phase transformations at intragranular inclusions in steel. *Met. Mater.* **1998**, *4*, 115–123. [CrossRef]
21. Yamada, T.; Terasaki, H.; Komizo, Y. Relation between inclusion surface and acicular ferrite in low carbon low alloy steel weld. *ISIJ Int.* **2009**, *49*, 1059–1062. [CrossRef]
22. Nako, H.; Okazaki, Y.; Speer, J.G. Acicular ferrite formation on Ti-rare earth metal-Zr complex oxides. *ISIJ Int.* **2015**, *55*, 250–256. [CrossRef]
23. Shim, J.H.; Cho, Y.W.; Chung, S.H.; Shim, J.D.; Lee, D.N. Nucleation of intragranular ferrite at Ti_2O_3 particle in low carbon steel. *Acta Mater.* **1999**, *47*, 2751–2760. [CrossRef]
24. Kanazawa, S.; Nakashima, A.; Okamoto, K.; Kanaya, K. Improved toughness of weld fussion zone by fine TiN particles and development of a steel for large heat input welding. *Tetsu-to-Hagané* **1975**, *11*, 2589–2603. [CrossRef]
25. Mu, W.Z.; Jönsson, P.G.; Shibata, H.; Nakajima, K. Inclusion and microstructure characteristics in steels with TiN additions. *Steel Res. Int.* **2016**, *87*, 339–348. [CrossRef]
26. Bramfitt, B.I. The effect of carbide and nitride additions on the heterogeneous nucleation behavior of liquid iron. *Metall. Trans.* **1970**, *1*, 1987–1995. [CrossRef]
27. Ohkita, S.; Horii, Y. Recent development in controlling the microstructure and properties of low alloy steel weld metals. *ISIJ Int.* **1995**, *35*, 1170–1182. [CrossRef]
28. Mabuchi, H.; Uemori, R.; Fujioka, M. The role of Mn depletion in intra-granular ferrite transformation in the heat affected zone of welded joints with large heat input in structural steels. *ISIJ Int.* **1996**, *36*, 1406–1412. [CrossRef]
29. Chai, F.; Yang, C.F.; Su, H.; Zhang, Y.Q.; Xu, Z. Effect of magnesium on inclusion formation in Ti-killed steels and microstructural evolution in welding induced coarse-grained heat affected zone. *J. Iron Steel Res. Int.* **2009**, *16*, 69–74. [CrossRef]
30. Wen, B.; Song, B.; Pan, N.; Hu, Q.Y.; Mao, J.H. Effect of SiMg alloy on inclusions and microstructhres of 16Mn steel. *Ironmak. Steelmak.* **2011**, *38*, 577–583. [CrossRef]
31. Zheng, C.C.; Wang, X.M.; Li, S.R.; Shang, C.J.; He, X.L. Effects of inclusions on microstructure and properties of heat-affected-zone for low-carbon steels. *Sci. China Technol. Sci.* **2012**, *55*, 1556–1565. [CrossRef]

metals

MDPI

Article

Non-Destructive Evaluation of Steel Surfaces after Severe Plastic Deformation via the Barkhausen Noise Technique

Miroslav Neslušan [1,*], Libor Trško [1] , Peter Minárik [2] , Jiří Čapek [3], Jozef Bronček [1], Filip Pastorek [1], Jakub Čížek [4] and Ján Moravec [1]

[1]   University of Žilina, Univerzitná 8215/1, 010 26 Žilina, Slovakia; libor.trsko@rc.uniza.sk (L.T.); jozef.broncek@fstroj.uniza.sk (J.B.); filip.pastorek@rc.uniza.sk (F.P.); jan.moravec@fstroj.uniza.sk (J.M.)
[2]   Faculty of Mathematics and Physics, Charles University, Ke Karlovu 5, 121 16 Praha 2, Czech Republic; peter.minarik@mff.cuni.cz
[3]   Faculty of Nuclear Sciences and Physical Engineering, ČVUT Praha, Trojanova 13, 120 00 Praha 2, Czech Republic; jiri.Capek@fjfi.cvut.cz
[4]   Faculty of Mathematics and Physics, Charles University, V Holešovičkach 2, 180 00 Praha 8, Czech Republic; jcizek@mbox.troja.mff.cuni.cz
*   Correspondence: miroslav.neslusan@fstroj.uniza.sk, Tel.: +421-908-811-973

Received: 18 November 2018; Accepted: 4 December 2018; Published: 6 December 2018

Abstract: This paper reports about the non-destructive evaluation of surfaces after severe shot peening via the Barkhausen noise technique. Residuals stresses and the corresponding Almen intensity, as well as microstructure alterations, are correlated with the Barkhausen noise signal and its extracted features. It was found that residual stresses as well as the Barkhausen noise exhibit a valuable anisotropy. For this reason, the relationship between the Barkhausen noise and stress state is more complicated. On the other hand, the near-the-surface layer exhibits a remarkable deformation induced softening, expressed in terms of the microhardness and the corresponding crystalline size. Such an effect explains the progressive increase of the Barkhausen noise emission along with the shot-peening time. Therefore, the Barkhausen noise can be considered as a promising technique capable of distinguishing between the variable regimes of severe shoot peening.

Keywords: shot peening; Barkhausen noise; crystallite size

1. Introduction

Conventional shot peening (CSP) is a widely employed technique applied for the final surface processing of components, which improves the mechanical properties and the corresponding fatigue or/and corrosion behavior under applied stress [1–4]. Shot peening (SP) usually alters the surface morphology; near surface microstructure expressed in terms of microhardness, grain size, and/or dislocation density; and the stress state [1–4]. The CSP surfaces contain compressive residual stresses of a variable magnitude and penetration depth. These stresses strongly correspond with the Almen intensity as a parameter that can be easily measured on the commonly used Almen strips. The Almen intensity refers to the arc height of the Almen strip after shot peening, and the required Almen intensity can be obtained by adjusting the SP conditions, such air pressure, shot size, peening time, and others. Unal [5] reports that the Almen intensity is directly considered as a major causal agent of all of the mechanical and metallurgical changes in the surface.

Severe shot peening (SSP) is usually employed when a higher magnitude of compressive stresses and higher degree of microstructure alterations are required (especially grain refinement). For these reasons, higher Almen intensities and degrees of coverage are required for the SSP than those for the CSP. Based on the value of the exerted kinetic energy (except CSP and SSP), the over and re-shot peening processes can be employed as well. The shot peening process is widely reported from many points of view, and published studies discuss a variety of shot peened materials under variable conditions. Kleber and Barroso [1] reported about strain-induced martensite transformations of austenitic steel AISI 304L of a different extent and degree under the free surface, using the variable surface coverage during the CSP. Unal and Varol [2] also refer to the mechanical twins in AISI 304 after the CSP, SSP, and re-peening. Fargas et al. [6] and Chen et al. [7] clearly showed that extensive plastic deformation initiates phase transformations in duplex steel when the fraction of martensite increases at the expense of austenite. Moreover, the CSP process also remarkably increases the dislocation density and decreases the domain size in the near surface region. Fu et al. [3] analyzed the surface hardness and stress state after the CSP and the consecutive annealing. Unal [5] carried out an optimization of the Almen intensity, surface roughness, and hardness, varying the input processing parameters. Segurado et al. [8] investigated the effect of different types of shots on surface morphology, residual stresses, and fatigue behaviour. Maleki et al. [9] studied the influence of various CSP regimes on grain refinement, stress, and microhardness profiles, as well as on S-N curves. Maleki and Unal [10] studied the influence of surface coverage and the re-peening process on the properties of AISI 1045 steel after CSP as well as SSP. Trško et al. studied the CSP and SSP process in relation to the fatigue properties of steel [11] and aluminium alloys [12], together with influence on the surface texture and fracture surface character [13]. His results showed that the SSP process can be more beneficial to the fatigue life of a material, however, exceeding a certain point leads to significant surface damage and a rapid drop of fatigue properties.

The CSP and SSP alter the surface integrity in the complexity of this term. The surface state after the SP is affected by many input variables. Therefore, the SP cycle performed on the real components requires the preliminary phase when the SP parameters are adjusted using the Almen strips. On the other hand, the surface state of the real components after SP can vary, especially when the components of a complicated geometry are shot peened, even when the Almen strips are mounted directly on the component in various places. For this reason, the non-destructive technique would be beneficial for monitoring such components in the real production. Magnetic Barkhausen noise (MBN) is a magnetic technique based on an irreversible and discontinuous domain wall motion. The domain wall motion and the corresponding MBN signal (as well as the extracted MBN parameters) are sensitive to the stress state and to the microstructure alterations. The stresses affect mainly the alignment of the domain walls, whereas the microstructure features affect the pinning strength of the matrix and the free path of the motion of the domain walls. It is well known that a higher MBN can be obtained under tensile stresses rather than compressive ones [14–19]. Furthermore, the domain wall motion can be strongly pinned by dislocation cells [20,21], precipitates [22], and/or non-ferromagnetic phases [23]. The concept in which the SP surface could be accessed via the MBN technique has been already published. Kleber and Barroso [1] found that the degree of strain-induced martensite transformation in austenitic steel AISI 304L strongly correlates with the MBN and extracted MBN envelopes. Theiner and Hauk [24] reported that macro and micro stresses, the work hardening depth, and the homogeneity of the CSP surface can be determined using the MBN parameters. Sorsa et al. [25], Gur and Savas [26], and Tiitto and Francino [27] correlated the CSP parameters to the residual stress profiles and MBN. Marconi et al. [28] employed the CSP for the hydraulic Pelton wheels in order to increase their fatigue resistance. The authors found that the CSP drastically influences the magnetic properties of the surface and correlates MBN with the stress state and volume of the retained austenite.

As it has been reported, the potential of MBN technique for the SP surface has been investigated for many years. Compared to CSP, SSP represents the production of excessively peened surfaces containing a more developed matrix alteration, especially in the near surface region and in the deeper extent of the compressive stresses. There is no exact boundary between the CSP and SSP treatments, because it significantly depends on the type of the treated material. For example, the SSP parameters for steels are much more severe than for aluminum alloys. However, the CSP process is mainly connected with accumulation of residual stresses in the subsurface volume, while the SSP treatment is considered when, besides the compressive residual stresses, significant changes in the microstructure (mainly grain refinement) occur. For these reasons, the magnetic properties and the corresponding MBN emission (due to the superimposing effects of the microstructure and residual stresses alterations) after SSP should also be affected. This is a case study focused on the SSP surfaces obtained via different regimes (the corresponding Almen intensities), in which the influence of the residual stresses and microstructure on the MBN emission is investigated.

2. Materials and Methods

Experiments were carried out on Almen A strips made of AISI 1070 spring steel (cold rolled, tensile strength 640 MPa, yield strength 495 MPa, strips of thickness 1.295 ± 0.025 mm, width 19.05 ± 0.127 mm, and length 76.2 ± 0.381 mm, shot peening center of Polytechnic University of Milan, Italy). Heat treatment: oil quenched from a heating temperature of $810\,°C \pm 20\,°C$, and subsequently high tempered at $440\,°C$ for hardness $44 \div 45$ HRC. The chemical composition is indicated in Table 1. Strips length is referred in the rolling direction (RD), whereas the width of the strips is referred to as the transversal direction (TD). The SSP parameters are specified in Table 2. Figure 1 illustrates the evolution of the Almen intensity for the different shot peening regimes. The measurements were carried out on the deformed Almen strips. The shots (S170) hardness occurs in the range of 40 to 51 HRC, whereas the strips hardness is 468 ± 22 Vickers microhardness (HV), which corresponds 46 ± 2 HRC.

The surface roughness was measured using the Hommel Tester T 2000 in RD (three repetitive measurements on each strip, length 12.5 mm in length, Hommel Werke, Jena, Germany). To reveal the microstructural or other transformations induced by the SSP, 10 mm long pieces were routinely prepared for the SEM observations (hot molded, ground, polished, and etched by 1% Nital for 10 s). The microstructure was observed in RD.

Vickers microhardness (HV) testing was conducted using a Zwick Roel ZHm microhardness tester (Zwick Roell, Ulm, Germany) by applying a force of 50 g for 10 s. The microhardness was determined by averaging three repeated measurements (three microhardness profiles spaced at 0.15 mm). The bulk microhardness was measured at a depth of approximately 0.5 mm below the free surface.

The MBN was measured by using a RollScan 350 (Stresstech, Jyväskylä, Finland), and was analyzed using MicroScan 500 software (magnetizing voltage 7.5 V, magnetizing frequency 125 Hz, sensor type S1-18-12-01, frequency range of MBN pulses in the range from 10 to 1000 kHz). The MBN values were obtained by averaging ten MBN bursts (five magnetizing cycles). The MBN refers to the root mean square (RMS) value of the signal. Taking into consideration the hardness of the samples, the estimated sensing depth of the MBN signal is ~70 μm [29]. The magnetization of the strips was carried out in the RD and TD directions. In addition to the conventional MBN parameter (RMS value of the signal), the peak position (PP) of the MBN envelope was also analyzed.

Table 1. Chemical composition of the AISI 1070 steel in wt%.

Fe	C	Mn	P	S
balance	0.70	0.75	max. 0.04	max. 0.05

Table 2. Shot peening parameters.

Shot Peening Time (s)	Shot Size (mm-S170)	Almen Intensity (mm A)	Ra (μm)	Rz (μm)
10	0.4318	14.1	4.41 ± 0.20	31.53 ± 2.50
20	0.4318	15.6	4.70 ± 0.18	34.00 ± 1.73
40	0.4318	15.8	4.59 ± 0.14	33.02 ± 4.01
80	0.4318	16.6	5.48 ± 0.41	37.67 ± 3.06

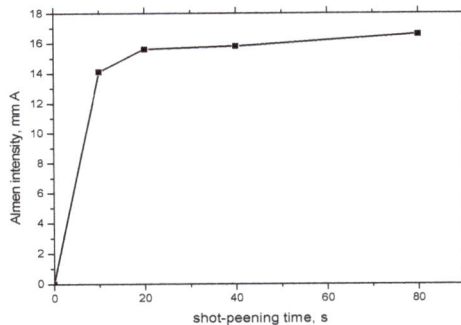

Figure 1. Evolution of Almen intensities with the shot peening time.

The determination of the residual stresses was performed using the XRD technique carried out on a Proto iXRD Combo diffractometer (Proto manufacturing Ltd., Ontario, Canada) using CrK$_\alpha$ radiation. The average effective penetration depth of the XRD measurements was ~5 µm, the scanning angle was ±39°, and the Bragg angle was 156.4°. The residual stress was calculated from the shifts of the 211 reflection. The Winholtz and Cohen method and X-ray elastic constants $\frac{1}{2}s_2 = 5.75$ TPa^{-1} and $s_1 = -1.25$ TPa^{-1} were used for the residual stress determination. In order to analyze the stress gradients beneath the surface of the samples, layers of material were gradually removed using electro-chemical polishing in the center of the sample. The crystallite size (the size of coherently diffracted domains) and dislocation density ρ_D were determined using the XRD patterns obtained on a X'Pert PRO diffractometer (Panalytical Ltd., Eindhoven, Netherlands), using the CoK$_\alpha$ radiation and subsequent Rietveld refinement performed on Mstruct software (version 0.1, Charles University, Czech Republic) [30–34] (used for the fitting of the measured XRD patterns). The effective penetration depth of the XRD measurements was, in this case, $1 \div 5$ µm. Using the XRD data, the crystallite size (*t*) was calculated via the Scherrer formula (Equation (1)), as follows:

$$t = \frac{K \cdot \lambda}{B \cdot \cos \theta} \tag{1}$$

where K, λ, B, and θ are the shape factor, X-ray wavelength, peak broadening and half of the diffraction angle, and the Bragg angle, respectively.

A brief sketch of the analyses' positioning on the strips (as well as their order) is illustrated in Figure 2. The first phase represents non-destructive measurements, followed by the XRD stress profile investigations. Consequently, the shot peened strips were cut and hot molded for SEM observations and microhardness measurements.

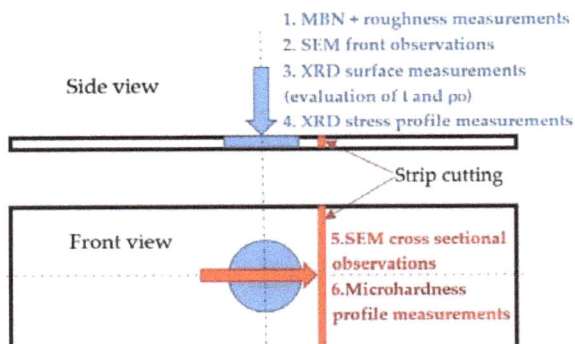

Figure 2. Brief sketch of experimental setup. MBN—Magnetic Barkhausen noise.

3. Results

3.1. SEM Observations

The cross sectional images after SSP (Figure 3) exhibited no distinctive difference in the microstructure for all of the shot peening times. The deeper regions remain untouched by the SSP process, and the near surface layer does not contain indications of remarkable structure transformations and/or severe plastic deformation.

On the other hand, the front view of the SEM images of the shot peened surfaces (Figure 4) illustrate that the original surface was remarkably notched by cold rolling (see Figure 4a). This is particularly visible in the areas not covered by the shot peening process after the 10 s treatment, and the notches are preferentially oriented in the rolling direction. A further increase in the shot peening time, and consequently, an increase in the surface coverage produces a surface fully covered by shot peening dimples. However, the notches that originated from cold rolling are still preserved on the surface (except SSP time 80 s), but tend to diminish with the increasing shot peening time. Such behavior is associated with the increasing intensity of plastic deformation as a result of the repetitive effect of shot impacts and the accumulation of their energy in the surface. Finally, the surface shot peened for 80 s contains no notches; however, it seems to be over peened and contained multiple scurfs.

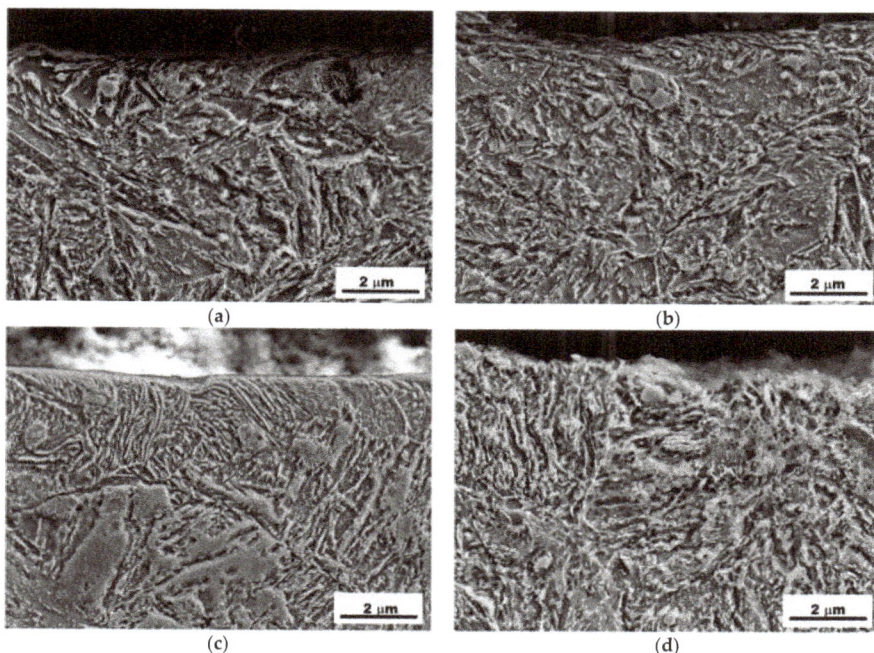

Figure 3. Cross sectional images of shot peened surfaces for the different shot peening times. (**a**) 10 s, (**b**) 20 s, (**c**) 40 s, (**d**) 80 s.

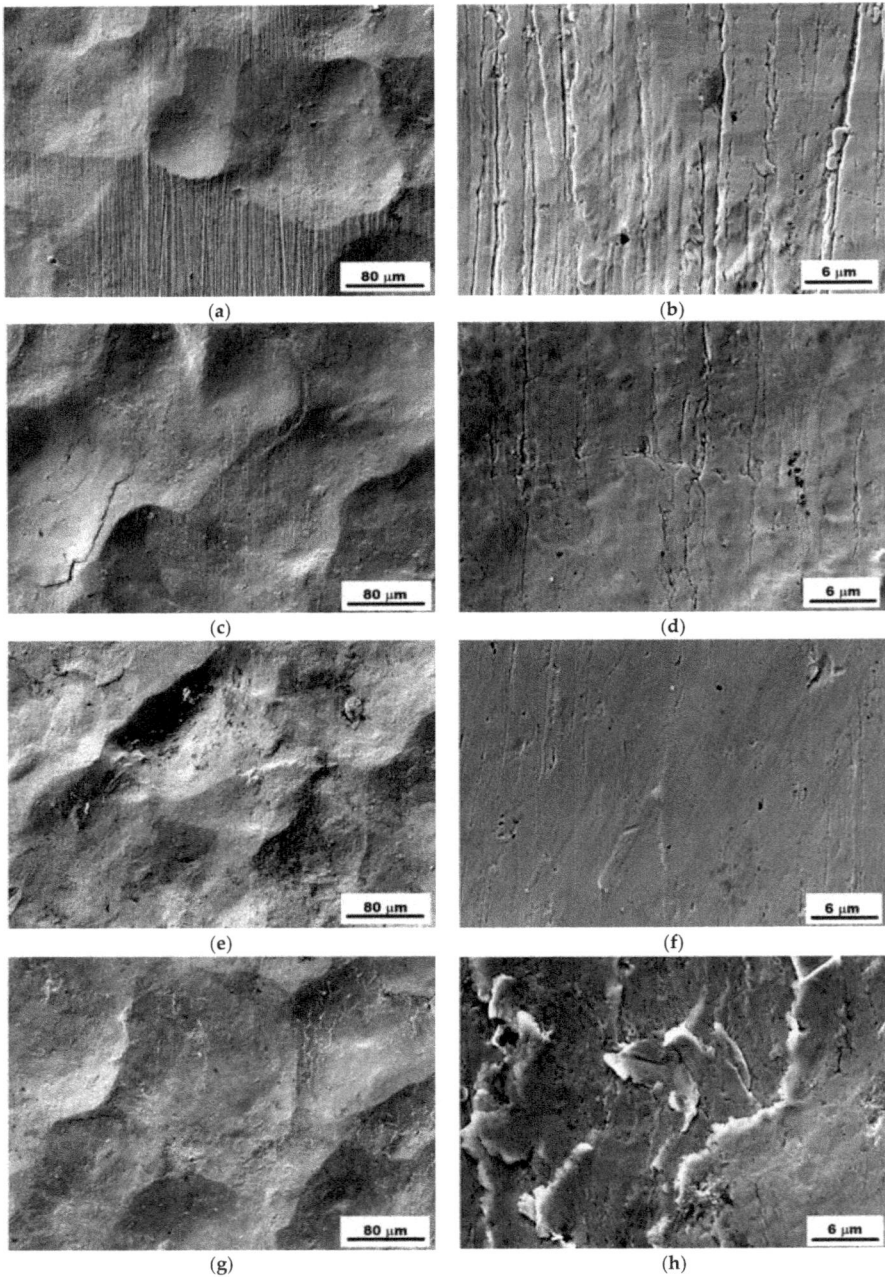

Figure 4. Front images of shot peened surfaces for the different shot peening times. (**a**) 10 s, (**b**) 10 s - detail, (**c**) 20 s, (**d**) 20 s - detail, (**e**) 40 s, (**f**) 40 s - detail, (**g**) 80 s, (**h**) 80 s - detail.

Figure 4 also depicts that the produced surface is typical for a shot peening process containing peaks and valleys. The height of those peaks and valleys increases along with the shot peening time and the corresponding degree of energy accumulated in the shot peened surfaces (see Figure 4, as well as information about surface roughness in Table 2).

3.2. XRD Measurements

Compared to CSP [9,10,12], SSP produces stress profiles containing a high magnitude of compressive stresses penetrating quite deep beneath the free surface (see Figure 5). The maximum of these stresses can be found at a depth of 0.1 to 0.2 mm, followed by a steep decrease and change into tensile stresses in the deeper regions. The compressive stresses for the RD and TD directions are quite similar, but TD exhibits about 200 MPa higher magnitudes of compressive stresses. The surface stress for the TD direction is about −600 MPa for all of the SSP regimes. The surface stress for the RD direction is about −390 MPa for all of the shot peening regimes, except for the 40 s regime, which exhibits about 70 MPa higher compressive stresses. The stress profiles for 10 and 20 s (for both directions) are very similar, whereas the 40 s regime produces stress profiles that are gently shifted to the higher stress magnitude. Moreover, the depth in which the maximum of the compressive stresses can be found is higher than those for the 10 and 20 s regimes. The 80 s regime exhibits lower compressive stresses in the descending part of the stress profiles; however, the compressive stresses penetrate deeper than those for the 10 and 20 s regimes (such a statement is also valid for the 40 s regime). The residual stress anisotropy after SP is influenced by the original microstructure of the material after rolling, as well as its significant anisotropy. In addition, the presence of the retained austenite was verified using the XRD technique, however no austenite peak was measured in the irradiated surface, and thus, the presence of retained austenite was excluded.

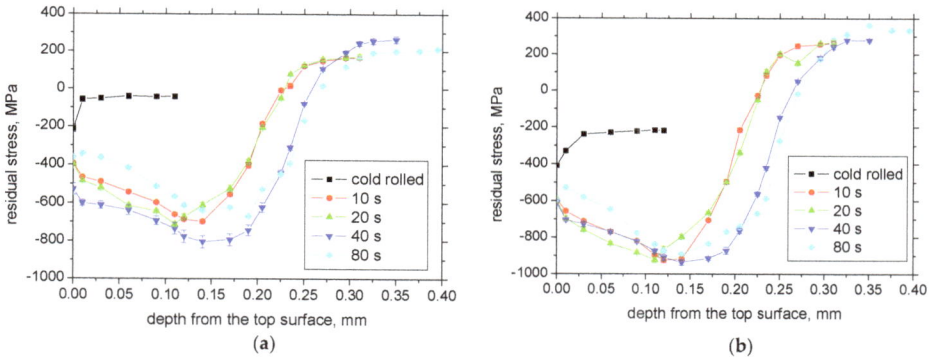

Figure 5. Stress profiles for the rolling direction (RD) and transversal direction (TD) directions. (**a**) RD direction, (**b**) TD direction.

The full width at half maximum (FWHM) of the XRD diffraction peaks' profiles are very similar for all of the shot peening regimes, as well as for the RD and TD directions (see Figure 6). The FWHM decreases along with the increasing depth from the free surface. However, the local maximum occurs for all of the FWHM profiles. This maximum can be found at a depth of about 0.20 mm for the 10 and 20 s regimes, whereas it is shifted slightly deeper for the 40 and 80 s regimes. It can be found that this maximum occurs at the depth where the steepest stress gradient can be found (the steepest decrease of compressive stresses).

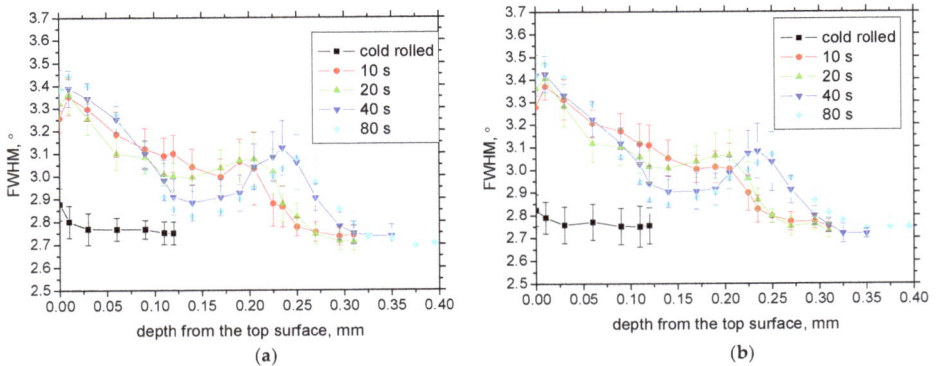

Figure 6. The full width at half maximum (FWHM) of XRD for the RD and TD directions. **a**) RD direction, (**b**) TD direction.

3.3. MBN Measurements

Figure 7 shows that the MBN increases with the shot peening time. The RD direction exhibits a higher MBN emission than that for the TD direction. As opposed to the evolution of the Almen intensity (exhibits early saturation along with the shot peening time), the MBN increases nearly linearly with the shot-peening time.

The MBN envelopes are shown in Figure 8. An increasing MBN emission is usually associated with an alteration of the stress state and/or the decreasing pinning strength of the magnetizing matrix; therefore, the shift of the MBN envelope to the lower magnetizing field and the corresponding lower magnetic field in which an MBN envelope reaches the maximum (usually referred as PP). However, the MBN envelopes, as well as Figure 9 (PP of the MBN envelopes), clearly demonstrate that the MBN envelopes are shifted to the higher magnetizing fields (only 80 s regime and RD direction exhibits certain decrease).

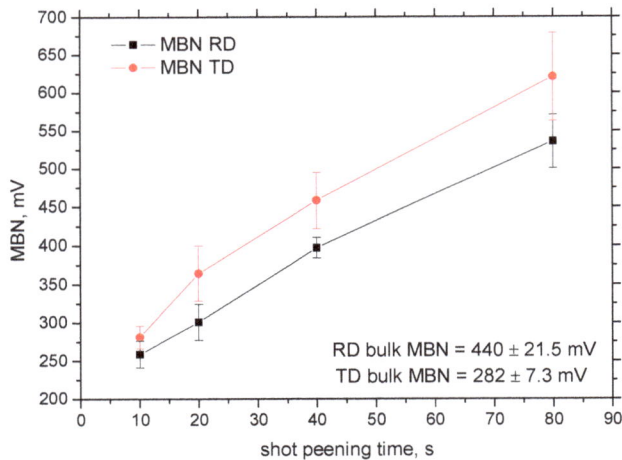

Figure 7. MBN (root mean square (RMS) value) after shot peening.

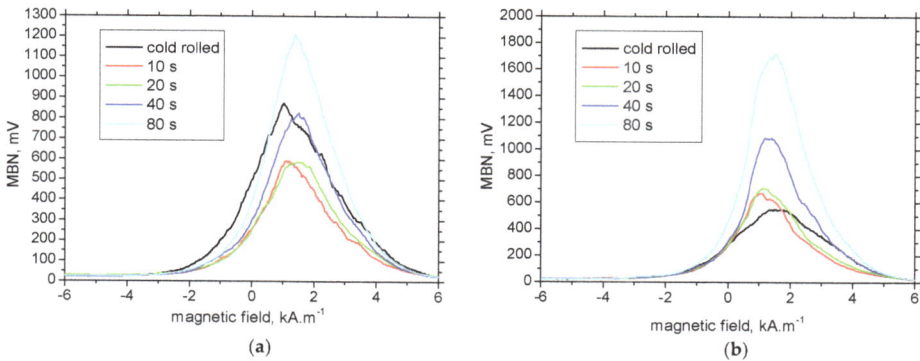

Figure 8. MBN envelopes for the RD and TD directions. (**a**) RD direction, (**b**) TD direction.

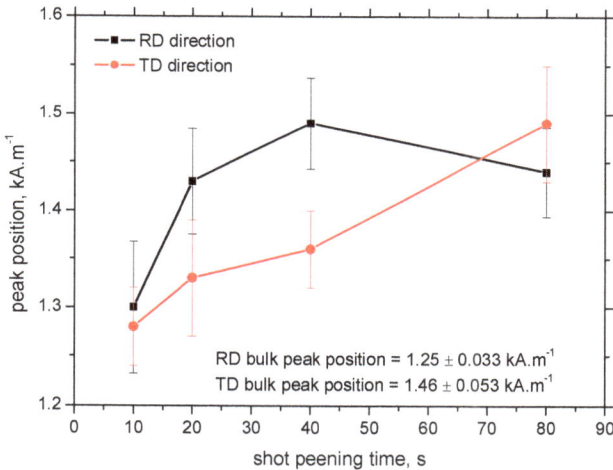

Figure 9. The peak position (PP) of the MBN envelopes for the RD and TD directions.

4. Discussion of Obtained Results

The evolution of the MBN (as that shown in Figure 7) is controversial with respect to the stress state as well as the accumulation of energy (plastic deformation) with the increasing shot peening time (as shown in Figure 4). Therefore, one might expect that the MBN would decrease with the shot peening time. Taking into consideration the estimated sensing depth of MBN (~70 μm [29]), in this particular case, increasing the magnitude of the compressive stresses with the shot peening time, should contribute to lower MBN emissions [25,26]. The 80 s regime exhibits a certain decrease of residual stresses within the MBN sensing depth (as compared to the other regimes), however such stress differences could not satisfactory and fully explain the much higher MBN for this shot peening time. Furthermore, the TD direction exhibits a higher magnitude of compressive residual stresses, but the MBN for this direction exceeds that for the RD (see Figures 5 and 7). A strong correlation between the residual stresses and the MBN is usually reported for uniaxial elastic stresses. The tensile stresses tend to align the domain walls in the direction of the exerted stress, which, in turn, increases the magnitude of the MBN pulses, whereas decreasing the MBN under compressive stresses is due to the alignment of the domain walls in the direction perpendicular to the direction of the stress [35]. However, under the biaxial or multiaxial stress state, the alignment initiated by the stress in a certain direction has a compensation effect in the other direction [36]. For these reasons, the effect of the residual stresses on the MBN, in this case, is inferior to the microstructure effects. As Kleber and Vincent [37] reported, the increase of MBN versus plastic strains can only be attributed to the microstructure, especially to the dislocation tangles-domain walls interaction.

The RMS of the MBN signals is defined as follows [38]:

$$\text{RMS} = \sqrt{\frac{1}{n}\sum_{i=1}^{n} X_i^2} \tag{2}$$

where n is the total number of MBN pulses (events) captured at the specific frequency range, and X_i is the amplitude of the individual pulses. The increasing MBN emission, expressed in term of its RMS value, is therefore the result of (i) increasing the number of MBN pulses, (ii) increasing the magnitude of the MBN pulses, and (iii) increasing the number and amplitude of the MBN pulses. Figure 10 depicts that the number of MBN pulses for the RD and TD direction slightly decreases with the shot peening time; thus, the increasing MBN has to be associated with their increasing magnitude. The magnitude of the MBN pulses is mainly driven by the free path of the domain wall motion (domain wall thickness could also contribute) [39]. The accumulation of the energy (deformation) in the shot peened surface with the shot peening time (as Figure 4 clearly illustrates) would increase the dislocation density. Bayramoglu et al. [21] and Ng et al. [40] report that the MBN emission in the case of plastic the deformation depends on two major factors that counteract each other. The first factor is the elongation of the grains, which leads to the reorientation of the domains and the corresponding domain walls, and thus the increasing MBN. The second factor is the presence of the dislocation cells, which hinder the domain wall motion. The higher deformation ratios induce a large number of smaller cell structures, so the boundary area increases, which results in the low MBN emission [32]. The effect of the domains and the corresponding reorientation of the domain walls results in an increase of MBN in the direction of the plastic deformation at the expense of the perpendicular direction. For this reason, this factor in this case does not take place, as the MBN increases in the RD and TD direction.

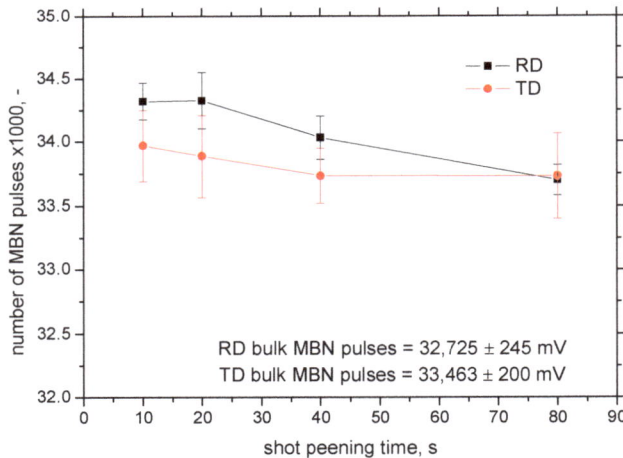

Figure 10. Number of detected MBN pulses.

On the other hand, Figure 11 shows the microhardness profiles of the shot peened surfaces. The deeper regions remain untouched by the SSP, whereas the near-the-surface layer exhibits a certain drop in microhardness. It is worth mentioning that the hardness of the strips before (46 ± 2 HRC) and also after shot peening (41 HRC) occurs in the range of the shots hardness ($40 \div 51$ HRC). Such a relationship could explain why the microhardness is not affected more, even though the FWHM of

XRD patterns exhibits quite a high response. In this case, the microhardness strongly correlates with the dislocation density. It is assumed that the shot impacts initiate the dislocation slip in the thin near surface layer.

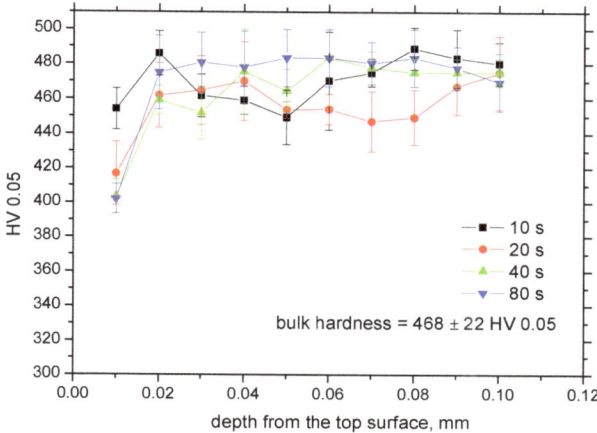

Figure 11. Microhardness profiles.

Figure 12 clearly depicts that the average dislocation density, ρ_D, gradually increases along with the shot penning time, and exhibits a certain decrease in the shot peening time of 80 s. Such behavior explains the evolution of the PP of the MBN envelope (especially for the RD direction) when the MBN envelopes are shifted to higher magnetic fields along with increasing shot peening time. On the other hand, the increasing MBN is a result of the remarkable redistribution of these dislocations in the near surface region, expressed in terms of the M_W parameter. This parameter presents information about the distribution of dislocations in the matrix, and can be defined as follows [30–34]:

$$M_W = R_c \cdot \sqrt{\rho_D} \tag{3}$$

where R_c is defined as the cut of radius (size of dislocation stress field). Low M_W values indicate a strong non-homogeneity of dislocations density and vice versa.

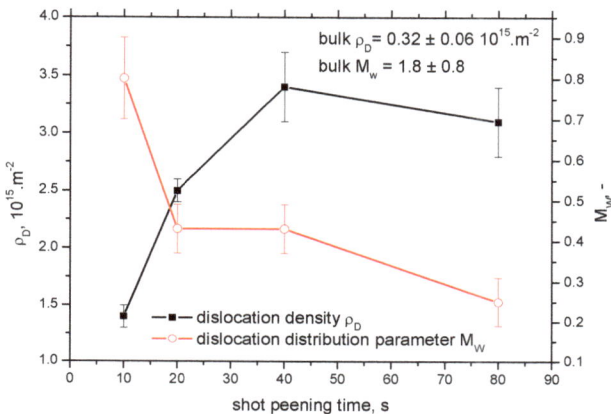

Figure 12. Dislocations density and their distribution versus shot peening time.

The size of the coherently diffracting domains (in other words, the crystallite size obtained from the XRD patterns) represents a measure of the average size of the structural units of the material, with the lattice not being distorted by the defects. In the present case, it is estimated that the crystallite size corresponds approximately to the mean size of the dislocation cells, that is, to the regions with a low dislocation density separated by dislocation walls (tangles). In Figure 13, one can see that the size of the coherently scattering domains is higher for the lower microhardness, and vice versa.

The decreasing microhardness (in near-the-surface region), increasing crystallite size, and decreasing M_W with the shot peening time indicate that these aspects play major roles when considering the evolution of MBN. The effective value (RMS) of MBN is mainly driven by the increasing size of the regions containing a low dislocation density, whereas the influence of the dislocation tangles (of high dislocation density) is only minor, see Figure 14. The low M_W indicates that the dislocations are clustered in the dislocation cells and in the matrix containing low dislocation density neighbors with dislocation tangles of very high dislocation density. Figure 12 clearly illustrates that the increasing average dislocation density for the shot peening times of 10, 20 and 40 s is compensated by the increasing non-homogeneity of their distribution (decreasing M_W with the shot peening time). Therefore, the accumulation of the energy of shot impacts (with shot peening time) gradually increases the dislocations density in the dislocation tangles, at the expense of the neighboring cells of a low dislocation density. Figure 13 depicts that the increasing dislocation density in the dislocation tangles is compensated by decreasing the dislocation density of the neighboring cells containing a much lower dislocation density. However, such an evolution tends to be saturated, and the further enlarging of the SP time (80 s regime) results in the gentle decrease of the average dislocation density (see Figure 12). Increasing the dislocation density in the dislocation tangles counteracts against decreasing the dislocation density in the neighboring regions for the 10, 20, and 40 s regimes. MBN values increase with the SP time as a result of the predominating effect of the regions containing the low dislocation density. On the other hand, it is considered that the further increase of MBN for the 80 s regime is due to the synergistic effect of decreasing the dislocation density over the entire grain. Such behavior also confirms the findings of Kleber and Vincent [37], in which a higher MBN is also attributed to the higher mobility of the domain walls in the regions of the low dislocation density, while the high-density regions are much less extended. The crystallite size increases and the surface microhardness decreases versus the shot peening time, as a result of the predominating influence of the cells of a low dislocation density. However, the surface microhardness tends to saturate for longer shot peening times, whereas the crystallite size increases gradually.

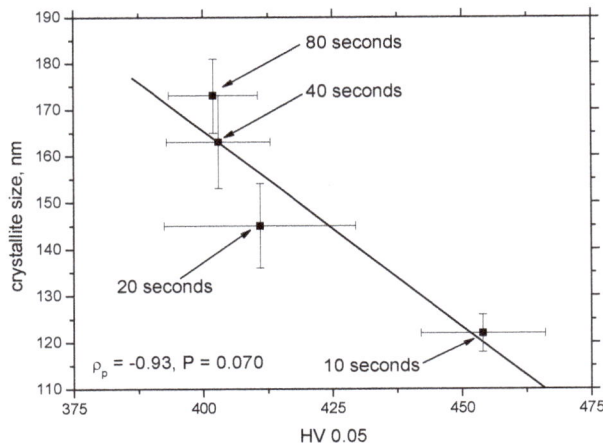

Figure 13. Surface microhardness versus crystallite size.

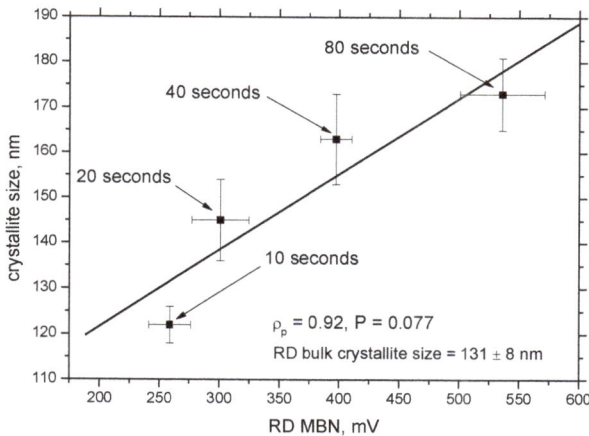

Figure 14. MBN (RMS value) versus crystallite size.

5. Conclusions

It is worth mentioning that shot peening is very often referred as a process decreasing the MBN, because of the compressive residual stresses and/or surface hardening [25,26]. However, such studies usually report about the CSP or shot peening of samples that behave in a malleable manner. On the other hand, AISI 1070 spring steel is usually employed in the elastic regime of loading, and exhibits poor plastic properties. Thus, the response of the surface to the accumulated shots impacts (in regime of the SSP) and the evolution of MBN against the accumulated energy differ. This study demonstrates a very good correlation between the MBN and the shot peening time, whereas the Almen intensities saturate early. Therefore, the MBN technique could be employed for the non-destructive evaluation of the surface after the SSP.

Author Contributions: Conceptualization, M.N., L.T., and J.B.; methodology, M.N. and L.T.; software, M.N., J.Č. (Jakub Čížek), J.Č. (Jiří Čapek), and J.M.; validation, L.T., F.P., and J.M.; formal analysis, J.B. and J.M.; investigation, J.Č. (Jiří Čapek), P.M., M.N., and F.P.; resources, J.B. and J.M.; data curation, M.N. and L.T; writing (original draft preparation), M.N. and L.T.; writing (review and editing), visualization, J.B. and J.M.; supervision, M.N. and L.T.; project administration, M.N.; funding acquisition, M.N., L.T., and P.M.

Funding: This study was supported by APVV project no. 16-0276 and 14-0284, VEGA project no. 1/0121/17, and KEGA project no. 008ŽU-4/2018. P.M. acknowledges the financial support of the Czech Science Foundation under project 14-36566G, and J.Čí acknowledges the financial support of the Czech Science Agency project P108/12/G043.

Conflicts of Interest: The authors declare no conflict of interest.

References

1. Kleber, X.; Barroso, S.P. Investigation of shot-peened austenitic stainless steel 304L by means of magnetic Barkhausen noise. *Mater. Sci. Eng. A* **2010**, *527*, 6046–6052. [CrossRef]
2. Unal, O.; Varol, R. Surface severe plastic deformation of AISI 304 via conventional shot peening, severe shot peening and re-peening. *Appl. Surf. Sci.* **2015**, *351*, 289–295. [CrossRef]
3. Fu, P.; Chu, R.; Xu, Z.; Ding, G.; Jiang, C. Relation of hardness with FWHM and residual stress of GCr15 steel after shot peening. *Appl. Surf. Sci.* **2018**, *431*, 165–169. [CrossRef]
4. Sekine, Y.; Soyama, H. Evaluation of the surface of alloy tool steel treated by cavitation shot less peening using an eddy current method. *Surf. Coat. Technol.* **2009**, *203*, 2254–2259. [CrossRef]
5. Unal, O. Optimization of shot peening parameters by response surface methodology. *Surf. Coat. Technol.* **2016**, *305*, 99–109. [CrossRef]
6. Fargas, G.; Roa, J.J.; Mateo, A. Effect of shot peening on metastable austenitic stainless steels. *Mater. Sci. Eng. A* **2015**, *641*, 290–296. [CrossRef]

7. Chen, M.; Liu, H.; Wang, L.; Wang, C.; Zhu, K.; Xu, Z.; Jiang, C.; Ji, V. Evaluation of the residual stress and microstructure character in SAF 2507 duplex stainless steel after multiple shot peening process. *Surf. Coat. Technol.* **2018**, *344*, 132–140. [CrossRef]
8. Segurado, E.; Belzunce, F.J.; Pariente, I.F. Effect of low intensity shot peening treatments applied with different types of shots on the fatigue performance of a high-strength steel. *Surf. Coat. Technol.* **2018**, *340*, 25–35. [CrossRef]
9. Maleki, E.; Unal, O.; Kashyzadeh, K.R. Effect of conventional, severe, over and re-shot peening process on the fatigue behaviour of mild carbon steel. *Surf. Coat. Technol.* **2018**, *344*, 62–74. [CrossRef]
10. Maleki, E.; Unal, O. Roles of surface coverage increase and re-peening on properties of AISI 1045 carbon steel in conventional and severe shot peening processes. *Surf. Interfaces* **2018**, *11*, 82–90. [CrossRef]
11. Trško, L.; Bokůvka, O.; Nový, F.; Guagliano, M. Effect of severe shot peening on ultra-high-cycle fatigue of a low-alloy steel. *Mat. Des.* **2014**, *57*, 103–113. [CrossRef]
12. Trško, L.; Guagliano, M.; Bokůvka, O.; Nový, F.; Jambor, M.; Florková, Z. Influence of Severe Shot Peening on the Surface State and Ultra-High-Cycle Fatigue Behavior of an AW 7075 Aluminum Alloy. *J. Mater. Eng. Perform.* **2017**, *26*, 2784–2797. [CrossRef]
13. Trško, L.; Fintová, S.; Nový, F.; Bokůvka, O.; Jambor, M.; Pastorek, F.; Florková, Z.; Oravcová, M. Study of Relation between Shot Peening Parameters and Fatigue Fracture Surface Character of an AW 7075 Aluminium Alloy. *Metals* **2018**, *8*, 111. [CrossRef]
14. Moorthy, V.; Shaw, B.A.; Mountford, P.; Hopkins, P. Magnetic Barkhausen noise emission technique for evaluation of residual stress alteration by grinding in case—Carburized En36 steel. *Acta Mater.* **2005**, *53*, 4997–5006. [CrossRef]
15. Dhar, A.; Clapham, L.; Atherton, D.L. Influence of uniaxial plastic deformation on magnetic Barkhausen noise in steel. *NDTE Int.* **2001**, *34*, 507–514. [CrossRef]
16. Fiorillo, F.; Küpferling, M.; Appino, C. Magnetic hysteresis and Barkhausen noise in plastically deformed steel sheets. *Metals* **2018**, *8*, 15. [CrossRef]
17. Baak, N.; Schaldach, F.; Nickel, J.; Biermann, D.; Walther, F. Barkhausen noise assessment of the surface conditions due to deep hole drilling and their influence on the fatigue behavior of AISI 4140. *Metals* **2018**, *8*, 720. [CrossRef]
18. Santa-aho, S.; Vippola, M.; Sorsa, A.; Leiviskä, K.; Lindgren, M.; Lepistö, T. Utilization of Barkhausen noise magnetizing sweeps for case-depth detection from hardened steel. *NDTE Int.* **2012**, *52*, 95–102. [CrossRef]
19. Baak, N.; Garlich, M.; Schmiedt, A.; Bambach, M.; Walther, F. Characterization of residual stresses in austenitic disc springs induced by martensite formation during incremental forming using micromagnetic methods. *Mater. Test.* **2017**, *59*, 309–314. [CrossRef]
20. Čížek, J.; Neslušan, M.; Čilliková, M.; Mičietová, A.; Melikhova, O. Modification of steel surfaces induced by turning: Non-destructive characterization using Barkhausen noise and positron annihilation. *J. Phys. D Appl. Phys.* **2014**, *47*, 1–17. [CrossRef]
21. Bayramoglu, S.; Gür, C.H.; Alexandrov, I.V.; Abramova, M.M. Characterization of ultra-fine grained steel samples produced by high pressure torsion via magnetic Barkhausen noise analysis. *Mater. Sci. Eng. A* **2010**, *527*, 927–933. [CrossRef]
22. Neslušan, M.; Čížek, J.; Kolařík, K.; Minárik, P.; Čilliková, M.; Melikhová, O. Monitoring of grinding burn via Barkhausen noise emission in case-hardened steel in large-bearing production. *J. Mater. Process. Technol.* **2017**, *240*, 104–117. [CrossRef]
23. Neslušan, M.; Čížek, J.; Zgútová, K.; Kejzlar, P.; Šrámek, J.; Čapek, J.; Hruška, P.; Melikhova, O. Microstructural transformation of a rail surface induced by severe thermoplastic deformation and its non-destructive monitoring via Barkhausen noise. *Wear* **2018**, *402*, 38–48. [CrossRef]
24. Theiner, W.A.; Hauk, V. Non-Destructive Characterization of Shot Peened Surface States by the Magnetic Barkhausen Noise Method. In Proceedings of the 12th World Conference on Non-Destructive Testing, Amsterdam, The Netherland, 23–28 April 1989; pp. 583–587.
25. Sorsa, A.; Santa-Aho, S.; Wartiainen, J.; Souminen, L.; Vippola, M.; Leviskä, K. Effect of shot peening parameters to residual stress profiles and Barkhausen noise. *J. Non-Destruct. Eval.* **2018**, *37*, 1–11. [CrossRef]
26. Gur, C.H.; Savas, S. Measuring of surface residual stresses in shot peened steel components by magnetic Barkhausen noise method. In Proceedings of the 18th World Conference on Non-Destructive Testing, Durban, South Africa, 16–20 April 2012.

27. Tiitto, K.; Francino, P. Testing shot peening stresses in the field. *Int. Newslett. Shot Peening Surf. Finish. Ind.* **1991**, *4*, 1–2.

28. Marconi, P.; Lauro, M.; Bozzolo, W. Shot peening on Pelton wheels: Methods of control and results. In Proceedings of the 4th International Conference on Barkhausen Noise and Micromagnetic Testing, Brescia, Italy, 3–4 July 2003; pp. 151–158.

29. Moorthy, V.; Shaw, B.A.; Evans, J.T. Evaluation of tempering induced changes in the hardness profile of case-carburised EN36 steel using magnetic Barkhausen noise analysis. *NDT&E Int.* **2003**, *36*, 43–49. [CrossRef]

30. Scardi, P.; Leoni, M. Line profile analysis: Pattern modeling versus profile fitting. *J. Appl. Crystallogr.* **2006**, *39*, 24–31. [CrossRef]

31. Ribárik, G.; Ungár, T.; Gubicza, J. MWP-fit: A program for multiple whole-profile fitting of diffraction peak profiles by ab initio theoretical functions. *J. Appl. Crystallogr.* **2001**, *34*, 669–676. [CrossRef]

32. Ribárik, G.; Ungár, T. Characterization of the microstructure in random and textured polycrystals and single crystals by diffraction line profile analysis. *Mater. Sci. Eng. A* **2010**, *528*, 112–121. [CrossRef]

33. Scardi, P.; Leoni, M. Whole powder pattern modeling. *Acta Crystallogr. A* **2002**, *58*, 190–200. [CrossRef]

34. Matěj, Z.; Kužel, R.; Nichtová, L. XRD total pattern fitting applied to study of microstructure of TiO_2 films. *Powder Diffr.* **2010**, *25*, 125–131. [CrossRef]

35. Karpuschewski, B.; Bleicher, O.; Beutner, M. Surface integrity inspection on gears using Barkhausen noise analysis. *Proc. Eng.* **2011**, *19*, 162–171. [CrossRef]

36. Krause, T.W.; Mandala, K.; Atherton, D.L. Modeling of magnetic Barkhausen noise in single and dual easy axis systems in steel. *J. Magn. Magn. Mater.* **1999**, *195*, 193–205. [CrossRef]

37. Kleber, X.; Vincent, A. On the role of residual internal stresses and dislocations on Barkhausen noise in plastically deformed steel. *NDTE Int.* **2004**, *37*, 439–445. [CrossRef]

38. He, Y.; Mehdi, M.; Hilinski, E.J.; Edrisy, A. The coarse-process characterization of local anisotropy on non-oriented electrical steel using magnetic Barkhausen noise. *J. Magn. Magn. Mater.* **2018**, *453*, 149–162. [CrossRef]

39. Varga, R. *Domain Walls and Their Dynamics*, 1st ed.; Pavol Jozef Šafárik University: Košice, Slovakia, 2014; pp. 32–41.

40. Ng, D.H.L.; Cho, K.S.; Wong, M.L.; Chan, S.L.I.; Ma, X.Y.; Lo, C.C.H. Study of microstructure, mechanical properties, and magnetization process in low carbon steel bars by Barkhausen emission. *Mater. Sci. Eng. A* **2003**, *358*, 186–198. [CrossRef]

metals

MDPI

Article

Research on the Bonding Interface of High Speed Steel/Ductile Cast Iron Composite Roll Manufactured by an Improved Electroslag Cladding Method

Yulong Cao [1], Zhouhua Jiang [1,*], Yanwu Dong [1,*], Xin Deng [2], Lev Medovar [3,4] and Ganna Stovpchenko [3,4]

[1]   School of Metallurgy, Northeastern University, Shenyang 110819, China; caoyl_neu@163.com
[2]   School of Materials and Metallurgy, University of Science and Technology Liaoning, Anshan 114051, China; dengxin_neu@163.com
[3]   E.O. Paton Welding Institute of Ukraine, 03150 Kyiv, Ukraine; lmedovar@gmail.com (L.M.); anna_stovpchenko@ukr.net (G.S.)
[4]   Elmet-Roll, 03150 Kyiv, Ukraine
*   Correspondence: jiangzh@smm.neu.edu.cn (Z.J.); dongyw@smm.neu.edu.cn (Y.D.); Tel.: +86-24-8368-6453 (Z.J.); +86-24-8369-1689 (Y.D.)

Received: 17 April 2018; Accepted: 25 May 2018; Published: 28 May 2018

Abstract: In the present study, a new electroslag cladding method by using of the advanced current supplying mold technology was used for manufacturing the high speed steel (HSS)/ductile cast iron (DCI) composite roll. The graphite morphology, matrix microstructure, elements distribution, carbides morphology, and carbides composition have been investigated by means of optical microscope (OM), scanning electron microscope (SEM), and energy dispersive spectroscopy (EDS). With increasing distance from the HSS side, a transition of graphite morphology from naught to existence and from small and dispersed to large and nonuniform was obtained at the interface. It was closely related to the fact that graphite in DCI participated in the phase change and the roll core surface and its nearby positions was heated to a high temperature by the liquid slag during the whole electroslag cladding process. Due to the combined effects of melting and elements diffusion, a significant migration of the alloying elements have occurred through the line scan analysis. Based on this, different types of carbides with the morphology and composition were found at the bonding interface. In addition, no obvious slag inclusions, porosity, shrinkage and other defects at the bonding interface were found. Results of the tensile test also illustrated that the bonding interface had a good quality and it could fully meet the requirements of the roll.

Keywords: electroslag cladding; high speed steel; ductile cast iron; composite roll; bonding interface

1. Introduction

In the field of hot rolling plants, rolling conditions have gradually become complicated due to the increasing demands for not only high quality of sheet product shape and sheet surface condition but also high productivity and energy saving in rolling. High speed steel (HSS) rolls are generally characterized of much higher hardness and more excellent resistance to wear, oxidization and roughness by adding strong carbide forming elements like V, W, Mo, and Cr to form very hard carbides of MC, M_2C, and M_7C_3 than the conventional work rolls such as indefinite chilled roll, Ni-grain cast iron roll and high chromium cast iron roll. In addition, the high speed steel also has a good capacity to retain a high level of hardness at high temperatures. It is widely used to manufacture hot rolling rolls to produce strips of good shape and small crown with extended roll service life [1–3].

Ductile cast irons are a family of alloys consisting of graphite spheroids dispersed in a matrix similar to that of steel, which combine the principle advantages of gray iron (low melting point, good fluidity and castability, excellent machinability, and good wear resistance) with the engineering advantages of steel (high strength, toughness, ductility, hot workability, and hardenability).

To date, centrifugal casting method has been widely used for manufacturing the composite roll especially in the production field of high speed steel (HSS)/ductile cast iron (DCI) composite roll [4–6]. However, it has some inevitable weakness points as described below. Firstly, as the high speed steel contains many carbide forming elements like V, W, Mo and Cr, the density (g/cm^3) difference between V (5.8), W (19.1), Mo (10.2), and Fe (7.86) is very large, it is not suitable to be manufactured by the conventional centrifugal casting method because of the composition segregation and the nonuniformity of microstructure and mechanical properties in the roll outer shell [1]. Secondly, when the liquidus temperature of the roll core material is higher than that of the composite layer, the final poured liquid metal of the roll core is fused with the inner surface of the solidified composite layer. This molten layer has a lower liquidus temperature than that of the roll core and becomes the final site of the solidification, which can easily lead to casting defects and reduce the strength of the bonding interface [7]. Thirdly, the centrifugal casting method can't use forged steel as the roll core material, which limits its application in the production field of high strength and toughness composite roll. In comparison with the centrifugal casting method, the improved electroslag cladding method [8] does not have the above problems, it not only can improve the purity of the molten steel but also can effectively improve the composition segregation during the solidification process of composite layer molten steel. However, it is worth mentioning that the improved electroslag cladding method has a larger production cost and more complex operation control. Based on the comprehensive consideration of the advantages and disadvantages of this two methods, the improved electroslag cladding method is a good choice for the production of high quality composite roll.

Performance of the composite roll is closely related to its bonding interface properties which can be said to be the key to the success of producing composite roll. However, the bonding interface is always the weakest region of a composite roll [9] as the composite layer easily cracks and flakes when the metallurgical bonding is incomplete, so, it is necessary to check carefully and improve the performance at the bonding interface effectively. This study attempts to reveal the bonding characteristic the bimetallic interface between HSS and DCI of the composite roll.

2. Experimental

In the present study, an improved electroslag cladding method with the electrode—current supplying mold conductive circuit [8] was used to manufacture the HSS/DCI composite roll. Schematic diagram of the new improved electroslag cladding method is shown in Figure 1. It mainly includes a T-type mold, consumable electrodes, slag bath, composite layer, roll core, current supplying mold and a pair of metal level sensors. The upper part of the T-type mold is a Φ420 mm current supplying mold, while solidification of the molten steel used as the composite layer occurs in the lower Φ350 mm water-cooled mold.

According to the numerical simulation results [8], by adoption of the current supplying mold in the improved electroslag cladding process, the highest temperature value occurs in the slag pool that between the consumable electrode and mold which is far away from the roll core surface, it reduces the heat transfer from slag pool to roll core effectively which results in an flexible control of the roll core surface temperature. It is worth mentioning that the melting temperature difference between the two materials is very important for the operation and technological parameter match during the bimetallic composite process.

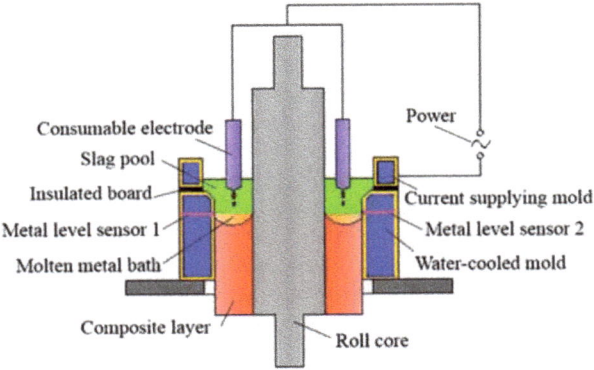

Figure 1. Schematic diagram of the new improved electroslag cladding method.

For the experimental research, the ductile cast iron is used as the roll core with a diameter of Φ240 mm and the high speed steel is used as the composite layer material. Twenty consumable electrodes of AISI M2 HSS with the size of Φ30 × 1500 mm were welded evenly on a disc and then they were inserted in the space between roll core and the molds. The chemical composition of the DCI and M2 HSS are given in Table 1.

Table 1. Chemical composition of the two materials (wt. %).

Composition in wt.%	C	Si	Mn	P	S	Cr	Mo	V	W	Ni	Mg	Fe
DCI	3.30	2.10	0.63	0.030	0.014	0.64	0.67	-	-	3.20	0.04	Bal.
M2 HSS	0.89	0.37	0.32	0.025	0.004	4.03	4.76	1.85	6.19	0.19	-	Bal.

In the present study, a HSS/DCI composite roll with the diameter of 350 mm and composite height of 264 mm was produced by the above method and it was mainly used to investigate the bonding characteristics of the bimetallic interface between HSS and DCI during the electroslag cladding process. It was cut after a stress relief annealing heat treatment at 1023 K for 6 h. Figure 2 shows the sampling schematic of roll core, composite layer and bimetallic bonding interface as described below.

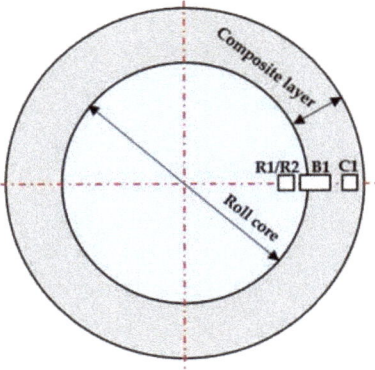

Figure 2. Sampling schematic of roll core, composite layer, and bimetallic bonding interface.

Metallographic samples of roll core before and after the electroslag cladding process were taken from the position marked by the letters "R1/R2" in Figure 2, respectively, to analyze its influence on the DCI microstructure. A sample of composite layer was taken from the position marked by the letter "C1" and it was used to observe the HSS microstructure. In order to research the graphite morphology and microstructure of the bimetallic bonding interface, a sample was taken from the position marked by the letter "B1" in Figure 2. It was polished and then etched with 4% nitric acid alcohol to observe the carbides morphology and measure the carbides composition as well as the matrix by using the UltraPlus scanning electron microscope (SEM, Carl Zeiss AG, Oberkochen, Germany) and energy dispersive spectroscopy (EDS, Carl Zeiss AG, Oberkochen, Germany). At last, a tensile test and hardness test were carried out to evaluate the mechanical properties of the bimetallic bonding interface. The tensile specimen was taken from the bimetallic interface and it was a plane plate specimen with the thickness of 4 mm and the total length of 108 mm. The tensile tests process were carried out by the AG-Xplus100kN electronic universal testing machine (Shimadzu, Tokyo, Japan) with the strain rate of $0.278\ s^{-1}$ in the present study. During the hardness test process, an HRS-150D digital Rockwell hardness tester (Shanghai JvJing Precision Instrument manufacturing Co., Ltd., Shanghai, China) was used and the testing load was 1471 N. The measuring distance between two adjacent points in the radial direction was 5 mm, the hardness value of a point was obtained after averaging three measurements.

3. Results and Discussion

3.1. Graphite Morphology and Microstructure of Roll Core before and after the Cladding Process

The mechanical properties of ductile cast iron are seriously influenced by the morphology of the graphite nodule, including count, diameter, sphericity, and spatial distribution. For instance, the sphericity and spatial distribution of nodules affect strength and toughness. Fatigue strength is influenced by count and diameter [10,11].

Graphite spheroidization rate is a representation of the degree for the graphite morphology nearly spherical and the graphite rating is based on the percentage of graphite spheroidization rate. Figure 3 shows the graphite morphology and microstructure of DCI before the cladding process. Figure 3a illustrates that the graphite morphology is basically spheroidal and agglomerated (isolated distribution, irregular shape) without chunky graphite. The graphite rating is second grade through a comparison with the graphite rating maps. For the cast iron roll, requirement of the graphite rating is not less than third grade. So, it can well meet the requirements of the cast iron roll. As Ni element can stabilize austenite and promote the formation of pearlite, a small amount of Ni element can inhibit the ferrite formation. Therefore, before the cladding process, the matrix of the DCI is a pearlitic matrix as shown in Figure 3c. In addition, iron atoms in Fe_3C can be replaced by other metal atoms to form solid solutions which are generally referred to as "alloy cementite" in the present study. During this process, the Cr element is finitely soluble in Fe_3C, while the W and Mo elements are only slightly soluble in Fe_3C, V is almost insoluble in Fe_3C. As the DCI contains some carbide forming elements like Cr and Mo, some alloy cementite has been generated during its solidification process as shown in Figure 3b.

Figure 4 shows the graphite morphology and microstructure of DCI after the cladding process of bimetallic metals. Compared with that in Figure 3, the graphite spheroidization rate, number of graphite particles per unit area and area percentage of the graphite after the cladding process are changed from second grade, $21.56/mm^2$ and 8.37% to third grade, $8.79/mm^2$ and 11.41%, respectively. Through the comparison, a less sphericity, smaller count and larger diameter of the graphite are obtained after the cladding process. During the electroslag cladding process, the roll core of DCI was preheated continuously by the heat conduction from the high temperature liquid slag bath, and then it participated in the cladding process with the molten electrodes metal which used to form the composite layer. In the present study, the roll core surface and its nearby positions have been heated to a very high temperature during the whole cladding process. As the austenitizing temperature of DCI is only 1080 K through the calculation of thermo-calc software, a high temperature austenitizing process have been performed in the heat affected zone (the position near the roll core surface) of roll core during the electroslag cladding process. When the heated temperature of DCI exceeds the austenitizing temperature by a certain value, the graphite can be disintegrated with different degrees and it will be described in detail below. The disintegration methods include peripheral exfoliation type, center melting type, and cracking type. Different disintegration methods will lead to different graphite morphology [12]. The disintegrated graphite is shown in Figure 4a.

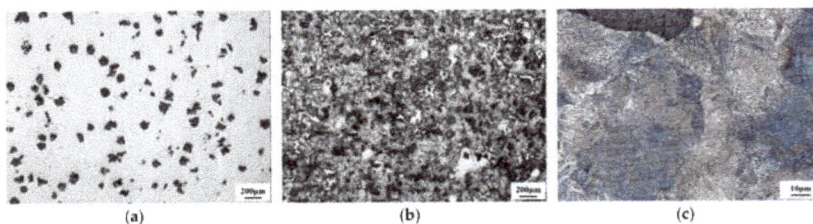

(a)   (b)   (c)

Figure 3. Graphite morphology (**a**) and microstructure of ductile cast iron (DCI) before the cladding process at low (**b**) and high (**c**) magnification.

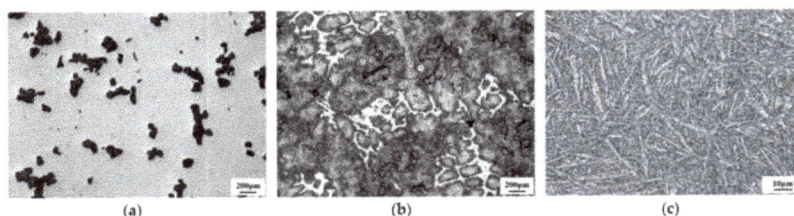

(a)   (b)   (c)

Figure 4. Graphite morphology (**a**) and microstructure of DCI after the cladding process at low (**b**) and high (**c**) magnification.

Due to the preheating process and the subsequent bimetallic cladding and cooling process of the roll core, the content of alloy cementite has an obvious increase and it always precipitates in network between eutectic cells as shown in Figure 4b. The matrix microstructure of the DCI after the cladding process mainly consists of needle-like bainite and partial granular bainite as shown in Figure 4c. During the high temperature austenitizing process, the graphite is also involved in the phase change process as a phase in DCI. The presence of graphite is equivalent to a "carbon storage pool" and when the DCI is heated to a certain temperature, the graphite begins to disintegrate and dissolve to some extent which leads to an increase of carbon content in the austenite. Because of the rapid cooling during the electroslag cladding process, the microstructure transformation of austenitized DCI will be based on the Fe-Fe$_3$C metastable equilibrium phase diagram. As this time, graphite is no longer precipitated from the supersaturated austenite but secondary cementite. This secondary cementite precipitated from the austenite is usually distributed in network along the austenite grain boundary. The presence of network cementite will greatly reduce the mechanical properties of the DCI, especially the plasticity and toughness.

3.2. Microstructure of the Composite Layer

Microstructure of the composite layer (HSS) etched with 4% nitric acid alcohol are shown in Figure 5 and the chemical composition of the carbides as indicated by arrows in Figure 5b are shown in Table 2. It shows that the eutectic carbides are mostly located in the intercellular regions, forming continuous networks of densely populated carbides. Through the carbides composition, two types of carbides such as spherical or blocky MC carbides (point C) and plate-like M$_2$C carbides (point A and B) are found in these micrographs. As the low content of V element in the steel, MC carbides are concentrated on cell boundaries rather than inside cells. MC carbides are rich in V and W whereas plate-like M_2C carbides are rich in W and Mo but poor in Fe as shown in Table 2. It is known that M_2C carbides are created by the eutectic reaction, $L \rightarrow \gamma + M_2C$ [13]. In austenite-carbide eutectics, carbide with higher fusion entropy is generally considered to be a faceted phase, which grows anisotropically during solidification. The morphology and microstructure of plate-like M_2C exhibit representative characteristics of a faceted phase [14]. From the Figure 5, it can be seen that the M2 HSS ingots consist of numerous M_2C carbides with a small amount of MC carbides which is consistent with the report by Zhou et al. [14] through the X-ray diffraction analysis.

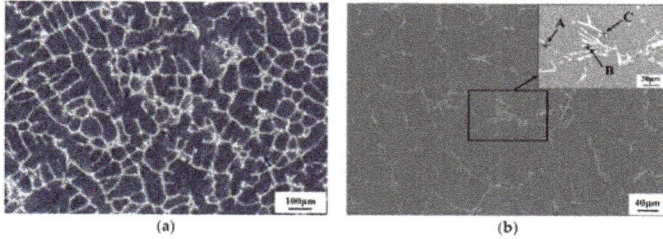

(a)   (b)

Figure 5. Microstructure of the composite layer at low (a) and high (b) magnification.

Table 2. Chemical composition of the carbides (wt. %).

Point	W	Mo	Cr	V	Fe
A	42.51	36.70	7.00	6.32	7.47
B	43.84	34.00	7.01	8.41	6.74
C	31.35	14.76	2.27	48.01	3.61

3.3. Graphite Morphology Analysis in the Bimetallic Bonding Interface

Figure 6 gives a clear display of the graphite morphology changes in the bimetallic bonding interface of the HSS/DCI composite roll through the optical microscope.

Though the Figure 6, it can be seen that there are no defects such as slag inclusion and porosity in the bonding interface and a good metallurgical bonding status is achieved. There also exists an obvious transition layer between HSS and DCI and in where the graphite morphology shows obvious changes. At the DCI side, a nonuniform distribution with a poor sphericity, big size, and flowering morphology of the graphite is obtained as shown in Figure 6e. Near the DCI side, the flowering graphite gradually disappeared and turned into spherical graphite with a relatively good sphericity and uniform distribution as shown in Figure 6d. During the transition from DCI to HSS side, the graphite morphology changes again and transforms into fine and dispersed graphite particles as shown in Figure 6b,c. Near the HSS side, the graphite particles gradually disappears.

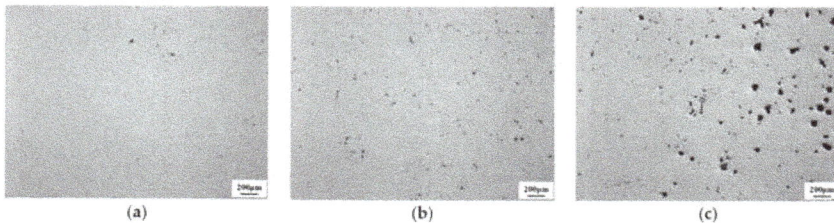

(a)   (b)   (c)

Figure 6. *Cont.*

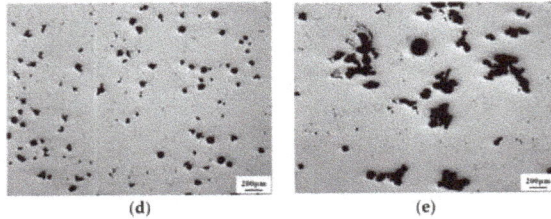

Figure 6. Graphite morphology in the bonding interface. (a–e) → high speed steel (HSS) side to DCI side.

During the electroslag cladding process of the bimetallic metals, the roll core (DCI) undergoes a continuous heating process. As a phase of the DCI, the graphite must have some changes during this heating process. For this objective, the morphological changes of spheroidal graphite under the continuous heating were observed by HM-350 high temperature metallographic microscope [11]. It illustrates that the transformation of pearlite into austenite after 1023 K has been very obvious. From about 1173 K, the austenitization of pearlite has been completed. As the temperature continues to rise to 1358 K, the matrix around the graphite particles melts firstly due to the interfacial reaction between austenite and graphite which results in a low melting point liquid phase. Presence of the liquid phase accelerates the dissolution of graphite particles in turn. As the inhomogeneous structure of the graphite, the melting of carbon atoms shows a nonuniform development. With the disintegration of the original spheroidal and agglomerated graphite, the deformity graphite as flowering graphite are gradually formed. As it has been described above, different disintegration methods will lead to different graphite morphology. As the heating temperature of DCI continues to increase, the disintegrated graphite (flowering graphite) dissolves into the matrix much further, with a different amount of dissolution, it gradually forms the morphological changes of graphite as shown in Figure 6a–e.

3.4. Microstructure Analysis in the Bimetallic Bonding Interface

As a partial melting phenomenon of the roll core surface has occurred during the electroslag cladding process, the combined effects of the melting and element diffusion occurs in the bimetallic bonding interface. With the combined effects, different microstructures are formed as shown in Figure 7. However, the measurement area moves 3 mm towards the HSS side when compared with that in the Figure 6.

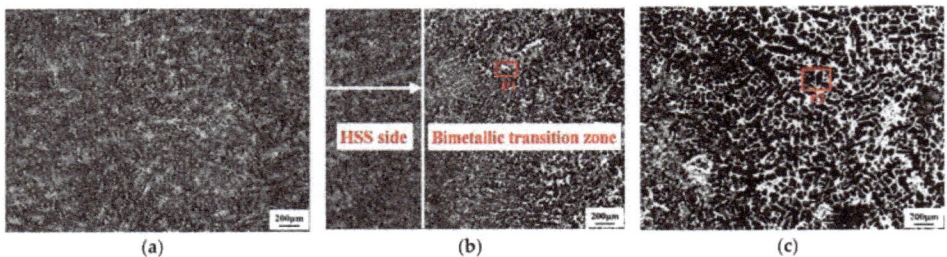

Figure 7. *Cont.*

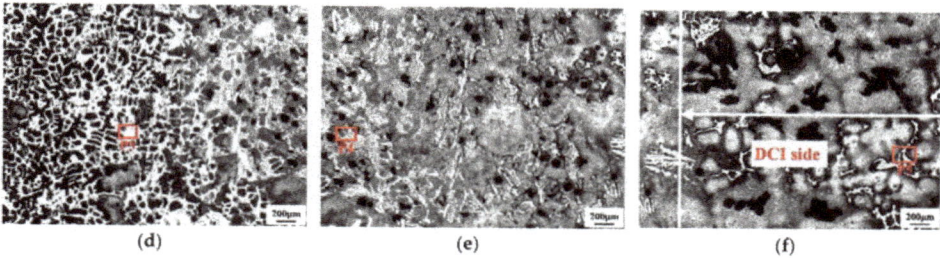

Figure 7. Microstructure in the bonding interface. (a–f) → HSS side to DCI side.

From Figure 7, three areas of the HSS side (Figure 7a), bimetallic transition zone (Figure 7b–e) and DCI side (Figure 7f) are defined according to the different microstructures morphology in the bimetallic bonding interface. Under this definition, width of the bimetallic transition zone is about 11.44 mm and in where four categories of carbides morphology are formed due to the combined effects of the melting and elements diffusion of C, W, Mo, Cr, and V between the HSS side and DCI side. At the HSS side, the fine and dispersed carbides are formed with a uniform distribution. In the bimetallic transition zone, with an increasing distance from the HSS side, the small network carbides gradually become coarse and its content also increases dramatically. In the spheroidal and agglomerated graphite position (Figure 7e), significant transformed ledeburite are formed and the eutectic cementite are very coarse with a nonuniform distribution. Finally, at the DCI side, the small amount and dispersed network carbides are formed which are located in the matrix of the DCI.

In order to give a better understanding of the carbides morphology and composition in the bimetallic bonding interface, the SEM-EDS analyses have been made for the carbides of different positions and the results are shown in Figure 8 and Table 3, respectively. Figure 8a shows the carbides morphology at some area of the HSS side as it is difficult to mark the actual analysis position and Figure 8b–f shows that of the positions marked by the red frames and letters "P1" to "P5" in Figure 7b–f respectively.

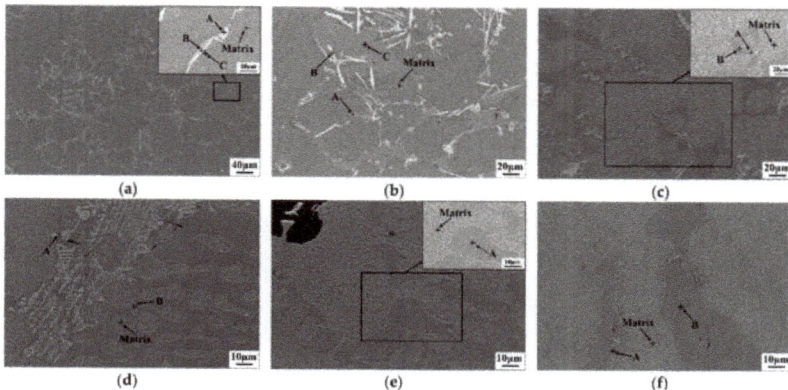

Figure 8. Carbides morphology in different positions of bimetallic transition zone. (a–f) → HSS side to DCI side.

At the HSS side (Figure 8a), since the area marked by the letter A mainly contains V and W elements and its small spherical or blocky morphology, it is identified to be the MC carbide. The white-colored area marked by the letter B indicates a plate-like M_2C carbide containing mostly

W and Mo. The gray-colored area C mainly consists of Fe and Cr as well as few amount of W, Mo, and V, it is identified to be coarse network (or blocky) alloy cementite in where the iron atoms can be replaced by other metal atoms like Cr, W, Mo, and V. It is worth noting that the M_2C carbide and MC carbide mainly occurs on the blocky alloy cementite. Compared with Figure 8a, the biggest difference is that the volume fraction of alloy cementite increases obviously as shown in Figure 8b. In Figure 8c, a large amount of coarse blocky alloy cementite are formed, which are distributed in the network shape structure to a certain extent. In this position, no MC carbides are found. Figure 8d shows that the fishbone-like M_6C carbide is also found on the blocky alloy cementite and there is a significantly increase of the carbides content and size when compared to that in the Figure 8a–c. Near the DCI side as shown in Figure 8e, it only contains the coarse eutectic cementite in which a certain amount of Cr as well as few W, Mo, and V alloying elements are also found as shown in Table 3. At the DCI side, a very small amount of phase precipitation was found at the edge of the cementite as shown in Figure 8f.

Table 3. Chemical composition of the different carbides (wt. %).

Position	Point	W	Mo	Cr	V	Fe
HSS Side	A	35.38	20.26	4.11	36.39	3.86
	B	38.49	34.25	7.23	11.93	8.10
	C	4.61	5.53	11.26	1.85	76.75
	Matrix	2.94	1.89	2.77	0.63	91.77
P1	A	35.02	18.09	3.82	40.40	2.67
	B	43.28	35.25	4.55	9.86	7.06
	C	4.92	5.08	10.51	2.33	77.16
	Matrix	3.01	1.74	3.05	0.52	91.68
P2	A	41.84	37.85	1.90	5.55	12.86
	B	3.25	2.62	6.27	2.55	85.31
	Matrix	1.92	1.38	1.61	0.25	94.84
P3	A	22.00	38.52	0.58	0.55	38.35
	B	2.38	2.99	3.00	1.05	90.58
	Matrix	3.30	3.99	0.51	0.00	92.20
P4	A	0.52	1.05	2.69	0.74	95.00
	Matrix	0.00	0.38	0.66	0.00	98.96
DCI Side(P5)	A	0.01	75.23	1.97	0.22	22.57
	B	0.00	2.76	5.96	0.00	91.28
	Matrix	0.00	1.55	2.35	0.10	96.00

Composition of the carbides and matrix in different positions of bimetallic bonding interface corresponding to Figure 8 are shown in Table 3.

As shown in Table 3, from the HSS side to DCI side, the contents of W, Mo, Cr, and V elements in the alloy cementite (M_3C) show a gradual decrease trend which has a close contact with the diffusion and migration of the alloying elements between roll core and composite layer. From the position P1 to P3 in the bimetallic transition zone, the contents of W, Mo, Cr, V alloying elements in the matrix do not change obviously, but, near the DCI side (position P4), the contents of W and V elements decrease to zero which indicate that the diffusion of W and V elements in this zone is very slight. A certain contents of Cr and Mo elements detected in the matrix in this zone are mainly due to the original composition of DCI which contains the Cr and Mo elements as shown in Table 1. In the DCI side (position P5), the distribution characteristics of alloying elements in the carbides and matrix are similar to that in the position P4. From the Table 3, it also can be seen that MC carbides are V-rich carbides containing mostly V with small amounts of W, Mo, Cr, and Fe while M_2C carbides are mainly containing Mo and W.

3.5. Elements Distribution in the Bimetallic Bonding Interface

Sample of the bimetallic bonding interface taken from the B1 position in Figure 2 was used to measure the elements distribution of C, Fe, W, Mo, Cr, and V in the bimetallic bonding interface through a line scan analysis. In order to make it be corresponding to the microstructure in Figure 7, width of the tested line for line scan analysis was same as that from Figure 7a–f (a region with continuous change) and the analysis results were shown in Figure 9. The vertical lines for the position of the bimetallic transition zone are consistent with that in Figure 7b,f.

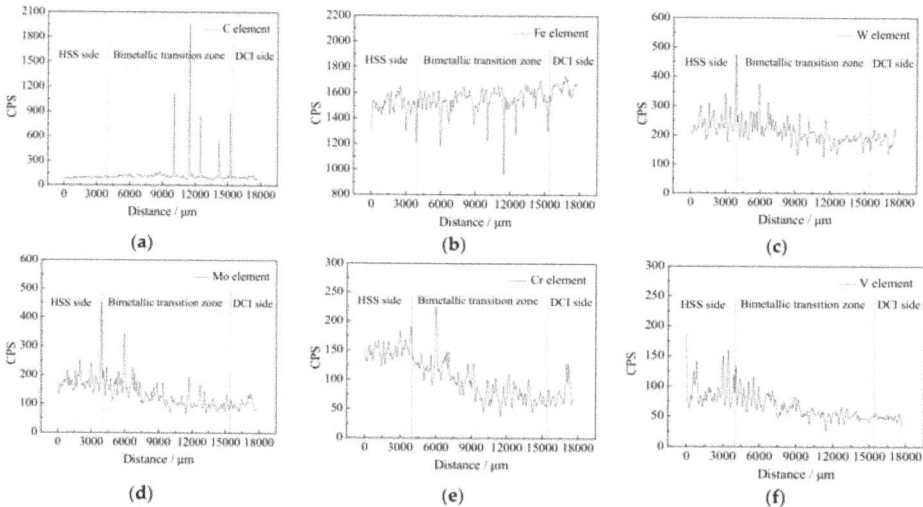

Figure 9. Elements distribution in the bimetallic bonding interface: (**a**) C element; (**b**) Fe element; (**c**) W element; (**d**) Mo element; (**e**) Cr element; and (**f**) V element.

Through the Figure 9, from the HSS side to DCI side, a gradual decrease trend of W, Mo, Cr, and V alloying elements content and a certain width of bimetallic transition zone are obtained. For a detailed description, from 0 to 3000 µm, the contents of the above elements are basically maintained at a constant level, while from 3000 to 12,000 µm, the content of each alloying element changes very obviously due to the diffusion and migration of the alloying elements between high speed steel and ductile cast iron. After 12,000 µm, the content of each alloying element tends to be flat, this is due to the fact that the atomic number of W, Mo, Cr, and V alloying elements are relatively large and their diffusion coefficients are small, meanwhile, the time spent on the bimetallic cladding process is only 42 min (for the total composite height of 264 mm), therefore, the diffusion distance of each alloying element is limited. For C and Fe elements, severe fluctuations in their contents are caused by the sweep of high carbon graphite during the line scan analysis process. At this moment, the peak in the C element spectrum corresponds to the trough in the Fe element spectrum. In addition, it will also lead to a severe decrease of the Fe element content and increase of alloying elements contents when the scanning line passes through the alloy carbides formed by W, Mo, Cr, and V elements. The elements distribution results as shown in Figure 9 is helpful for us to have a good understanding of the formation of different carbides and its containing alloying elements as shown in Table 3 in different positions in the bimetallic bonding interface.

3.6. Hardness Analysis

As we all know, the inhomogeneous microstructure will inevitably affect the mechanical properties, in order to give a further analysis of the mechanical properties changes of the bimetallic transition zone, samples included the bonding interface were taken from the HSS/DCI composite roll in the radial direction and the properties of macro Rockwell hardness and tensile strength were measured.

In the course of our investigation, a rectangular sample with the size of $140 \times 20 \times 8$ mm^3 was used for the hardness test and it started at the edge of the composite layer. The measuring distance between two adjacent points in the radial direction was 5 mm, the hardness value of a point was obtained after averaging three measurements as shown in Figure 10.

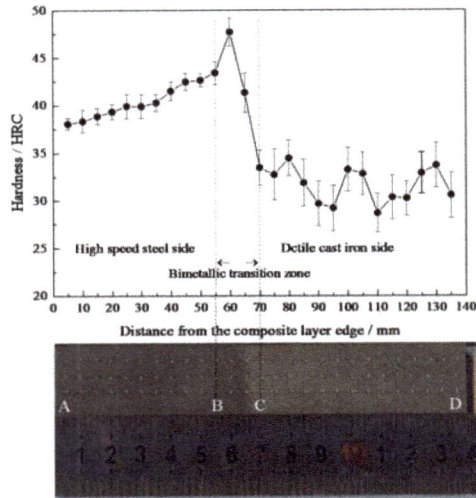

Figure 10. Hardness test of the bimetallic bonding interface in the radial direction.

From Figure 10, it can be seen that, the hardness value increases from 38.1 HRC to 43.4 HRC at the HSS side, this is due to the fact that as the distance from the bimetallic transition zone decreases, the C element diffusion from the DCI side becomes more conspicuously (as the carbon atom is a interstitial atom and the atomic radius is very small, the diffusion speed is fast). With the increase of C content, it is easy to form high hardness MC type carbide with the other alloying elements such as V, W, Mo, and Cr in high speed steel. So, the hardness value shows a trend of gradual increase when it moves from point A to point B.

When coming to the bimetallic transition zone (from point B to point C), a conspicuous change of the hardness value has taken place and it shows a trend of first increasing and then decreasing (43.4 HRC → 47.7 HRC → 41.3 HRC → 33.5 HRC), the maximum hardness value can reach 47.7 HRC and it appears at the bimetallic transition zone near the HSS side. From the previous analysis, a large number of continuous coarse network carbides are formed in the bimetallic transition zone near the HSS side. In addition, these formed carbides are mostly alloy cementite and compound carbides rich for W and Mo elements. While, near the DCI side, only some cementite (eutectic cementite) are formed containing few Mo and Cr elements and a dispersed distribution is obtained. As we all know, the cementite (M_3C carbide) has the lowest hardness among the so many types of carbides [15,16], so, the hardness value increases first and then decreases in the bimetallic transition zone.

At the DCI side (from point C to point D), since a large amount of carbon is present in the form of graphite, the formation of cementite is reduced and the hardness of ductile cast iron is reduced too, so the hardness value is maintained around 31.6 HRC.

3.7. Tensile Strength Analysis

According to the tensile tests, the tensile strength of the HSS/DCI composite sample is 452 MPa. As there is no corresponding standard for HSS/DCI composite roll, we can't give a proper evaluation of the tensile strength. As the requirement of tensile strength for cast iron roll is not less than 350 MPa, it is well illustrated that the tensile strength of the HSS/DCI bimetallic composite specimens in the present study is very good.

Figure 11 shows the tensile sample state before and after the tensile test. It shows that the tensile fracture location of the composite sample occurs in the bimetallic transition zone, which indicates that this is a weak zone of the bimetallic composite roll. In addition, it also can be seen that no significant necking occurs before the tensile fracture of the composite sample, which indicates that it has a large brittleness and small plasticity. As a poor plasticity, no significant plastic deformation occurs during the tensile test process.

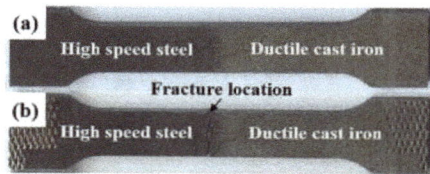

Figure 11. The tensile sample state before (**a**) and after (**b**) the tensile test.

In order to have a further understanding for the cause of the tensile fracture, microstructure of the fracture location was analyzed by using the DSX510 OLYMPUS metallographic microscope (OLYMPUS Corp., Tokyo, Japan). The results are shown in Figure 12.

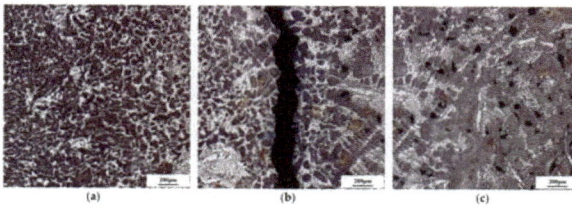

Figure 12. Microstructure of the tensile fracture of the composite specimen. (**a**) Towards the HSS side; (**b**) Fracture position; (**c**) Towards the DCI side.

Figure 12 shows that the coarse network carbides region in the bimetallic transition zone is more fragile than the fine network carbide and eutectic cementite region, where the tensile fracture occurs. Though the above analysis from the Figure 10, the coarse network carbides region has the maximum hardness value which can help us understand the tensile fracture occurring at this region.

Figure 13 is the tensile fracture at high magnification, under the tensile load, crack initiated and propagated along the coarse network carbides, and finally the fracture occurred as shown in Figure 13.

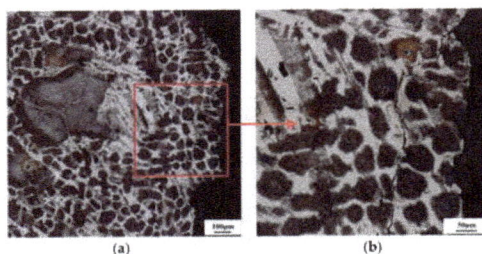

Figure 13. The crack initiated and propagated along the coarse network carbides at low (**a**) and high (**b**) magnification.

As the carbides are harder than the matrix, it is more difficult to deform under the tensile load, so, a stress concentration will occur between carbides and matrix which eventually leads to the fracture. Investigations on the hardness analysis of bimetallic bonding interface reveal that it is influenced by microstructural factors as kind, size, volume fraction, distribution of carbides located in the intercellular regions. As most carbides are located along cell boundaries and are much harder than the matrix, microcracks initiate primarily along cell boundaries.

4. Conclusions

The following becomes clear after the investigation on the microstructure and mechanical properties of the bimetallic bonding interface of HSS/DCI composite roll.

(1)   During the electroslag cladding process, the roll core of ductile cast iron was preheated continuously by the heat conduction from the high temperature liquid slag bath as well as the liquid metal of composite layer, therefore, a high temperature austenitizing process have been performed in the heat affected zone (the position near the roll core surface associated with obvious changes of graphite morphology, carbides, and matrix).

(2)   Due to the combined effects of the melting and elements diffusion between high speed steel and ductile cast iron, a large amount of carbides with different morphology, size, and composition generated in the bimetallic transition zone which leads to a trend of first increasing and then decreasing of the hardness value.

(3)   The tensile strength of the HSS/DCI bimetallic composite sample is 452 MPa and it can well meet the requirement of tensile strength for the cast iron roll. However, no significant plastic deformation occurs during the tensile test process due to the large amount of carbides and the tensile fracture occurs in the coarse network carbides region in the bimetallic transition zone.

According to the present study, basic characteristics and microstructure of the bimetallic bonding interface of HSS/DCI composite roll manufactured by the improved electroslag cladding method have been analyzed and a good understanding has been obtained.

Author Contributions: Y.C., Z.J., and Y.D. conceived and designed the experiments; Y.C., Y.D., and X.D. performed the experiments; L.M. and G.S. contributed the metal level sensor and technical consultation about the current supplying mold technology; Y.C. analyzed the data. Y.C. wrote the paper with the support of Z.J. and Y.D.

Funding: This work was supported by National Natural Science Foundation of China (N51274266 and N51674140), Joint Research Fund of National Natural Science Foundation of China and Baosteel Group Corporation (No. U1360103), Fundamental Research Funds for the Central Universities of China (N150202003 and N172507002).

Conflicts of Interest: The authors declare no conflict of interest.

References

1. Sano, Y.; Hattori, T.; Haga, M. Characteristics of high-carbon high speed steel rolls for hot strip mill. *ISIJ Int.* **1992**, *32*, 1194–1201. [CrossRef]
2. Hashimoto, M.; Oda, T.; Hokimoto, K.; Kawakami, T.; Kurahashi, R. Development and application of high-speed tool steel rolls in hot strip rolling. *Nippon Steel Tech. Rep.* **1995**, *66*, 82–90.
3. Okabayashi, A.; Morikawa, H.; Tsujimoto, Y. Development and characteristics of high speed steel roll by centrifugal casting. *SEAISI Q.* **1997**, *26*, 30–40.
4. Fu, H.G.; Zhao, A.M.; Xing, J.D.; Fu, D.M. Centrifugal casting of high speed steel/nodular cast iron compound roll collar. *J. Iron Steel Res. Int.* **2002**, *9*, 32–35.
5. Wang, Z.C.; Fu, H.M.; Li, J.P.; Feng, C.H. Production process of centrifugally cast high speed steel-nodular iron composite roller. *Mod. Cast Iron* **2009**, *29*, 44–48.
6. Wu, R.H.; Wu, C.J.; Zhang, X.P.; Gan, Z.P.; Zhao, W.Z.; Chuan, X.Z.; Ma, Y. Effect of heat treatment on property of core of high speed steel-ductile cast iron compound roll. *Foundry Technol.* **2007**, *28*, 190–194.
7. Shi, J.W.; Yang, D.X.; Ni, F.; Long, R. Development of high speed steel compound roll. *Res. Stud. Foundry Equip.* **2005**, *27*, 28–31.
8. Cao, Y.L.; Jiang, Z.H.; Dong, Y.W.; Deng, X.; Medovar, L.; Stovpchenko, G. Research on the bimetallic composite roll produced by an improved electroslag cladding method: Mathematical simulation of the power supply circuits. *ISIJ Int.* **2018**, *58*, 1052–1060. [CrossRef]
9. Lee, D.J.; Ahn, D.H.; Yoon, E.Y.; Hong, S.I.; Lee, S.; Kim, H.S. Estimating interface bonding strength in clad metals using digital image correlation. *Scr. Mater.* **2013**, *68*, 893–896. [CrossRef]
10. Cocco, V.D.; Iacoviello, F.; Rossi, A.; Cavallini, M.; Natali, S. Graphite nodules and fatigue crack propagation micromechanisms in a ferritic ductile cast iron. *Fatigue Fract. Eng. Mater. Struct.* **2013**, *36*, 893–902. [CrossRef]
11. Shiraki, N.; Usui, Y.; Kanno, T. Effects of number of graphite nodules on fatigue limit and fracture origins in heavy section spheroidal graphite cast iron. *Mater. Trans.* **2016**, *57*, 379–384. [CrossRef]
12. Qian, H.C.; Qi, M.D.; Zhao, K.J.; Chen, Y.L.; Shi, C.H.; Lin, X.R.; Lou, J.C. Degradation process of spheroidal graphite in vacuum heated ductile cast iron—Study on spheroidal decline mechanism. *J. Chongqing Univ.* **1979**, *2*, 1–15.
13. Fischmeister, H.F.; Riedl, R.; Karagöz, S. Solidification of high-speed tool steel. *Metall. Trans. A* **1989**, *20*, 2133–2148. [CrossRef]
14. Zhou, X.F.; Liu, D.; Zhu, W.L.; Fang, F.; Tu, Y.Y.; Jiang, J.Q. Morphology, microstructure and decomposition behaviour of M_2C carbides in high speed steel. *J. Iron Steel Res. Int.* **2017**, *24*, 43–49. [CrossRef]
15. Gong, K.L.; Dong, Y.J.; Gao, C.L. Research and manufacture of compound high speed steel rolls. *Iron Steel* **1998**, *33*, 1–7.
16. Han, J.W. Research on the Microstructural Characteristics and Properties of Cladding Layer and Composite Interface of W6Mo5Cr4V2 High Speed Steel/35CrMo Low Alloy Steel Composite Roll. Master's Thesis, Jiangsu University, Zhenjiang, China, 2016.

metals

MDPI

Review

CO$_2$ Utilization in the Ironmaking and Steelmaking Process

Kai Dong [1,*] and Xueliang Wang [1,2]

[1]   School of Metallurgical and Ecological Engineering, University of Science and Technology Beijing, Beijing 100083, China

[2]   ENFI Research Institute, China ENFI Engineering Corporation, Beijing 100038, China; xueliang1019@126.com

*   Correspondence: dongkai@ustb.edu.cn; Tel.: +86-106-233-2122

Received: 11 December 2018; Accepted: 18 February 2019; Published: 28 February 2019

Abstract: Study on the resource utilization of CO$_2$ is important for the reduction of CO$_2$ emissions to cope with global warming and bring a beneficial metallurgical effect. In this paper, research on CO$_2$ utilization in the sintering, blast furnace, converter, secondary refining, continuous casting, and smelting processes of stainless steel in recent years in China is carried out. Based on the foreign and domestic research and application status, the feasibility and metallurgical effects of CO$_2$ utilization in the ferrous metallurgy process are analyzed. New techniques are shown, such as (1) flue gas circulating sintering, (2) blowing CO$_2$ through a blast furnace tuyere and using CO$_2$ as a pulverized coal carrier gas, (3) top and bottom blowing of CO$_2$ in the converter, (4) ladle furnace and electric arc furnace bottom blowing of CO$_2$, (5) CO$_2$ as a continuous casting shielding gas, (6) CO$_2$ for stainless steel smelting, and (7) CO$_2$ circulation combustion. The prospects of CO$_2$ application in the ferrous metallurgy process are widespread, and the quantity of CO$_2$ utilization is expected to be more than 100 kg per ton of steel, although the large-scale industrial utilization of CO$_2$ emissions is just beginning. It will facilitate the progress of metallurgical technology effectively and promote the energy conservation of the metallurgical industry strongly.

Keywords: carbon dioxide; injection; blast furnace; converter; combustion

1. Introduction

Almost two tons of carbon dioxide (CO$_2$) is exhausted per ton of steel in the ferrous industry because of its energy-intensive feature [1]. Carbon dioxide utilization is beginning to attract worldwide attention, because it transmits CO$_2$ waste emissions into valuable products [2–6]. Study on the resource utilization of CO$_2$ in the ferrous metallurgy process is important for the reduction of CO$_2$ emissions to cope with global warming and bring a beneficial metallurgical effect. There are three main methods of emission reduction or utilization of CO$_2$. The first is the use of new technology or energy to reduce the use of fossil energy, the second is CO$_2$ storage technology, and the last is using CO$_2$ as a recycling resource. Currently, however, CO$_2$ emission reduction in metallurgical processes mainly relies on energy saving and waste heat utilization.

Carbon dioxide is a linear three-atom molecule, which is a weak acid, colorless and tasteless at room temperature. Its isobaric heat capacity is about 1.6 times that of nitrogen at steelmaking temperature and its infrared radiation ability is strong. CO$_2$ with CO generated from CO$_2$ can play a role in stirring in the ferrous metallurgy process. It can also play a role in controlling the temperature of the molten bath because of the carbon endothermic reaction, which protects molten steel from being oxidized and dilutes oxidants in the combustion. In the realization of CO$_2$ emission reduction, it is possible to save energy and reduce costs, and at the same time, dephosphorize and remove inclusions in steel.

CO_2 is a weaker oxidizing agent compared with O_2. The reactions of CO_2 oxidizing carbon, iron, silicon, or manganese in a molten bath may occur at steelmaking temperature, and the reactions are endothermic or slightly exothermic reactions, respectively. Therefore, it is possible to control the temperature and atmosphere by adopting CO_2 in the steelmaking process, hence realizing the aims of (1) reducing dust generation by reducing the temperature in the fire zone, (2) purifying the liquid steel by promoting dephosphorization, (3) minimizing the loss of valuable metals by low oxidation, (4) saving energy by increasing gas recovery, and (5) reducing the total consumption. The current CO_2 applications in the ferrous metallurgy process are shown in Figure 1.

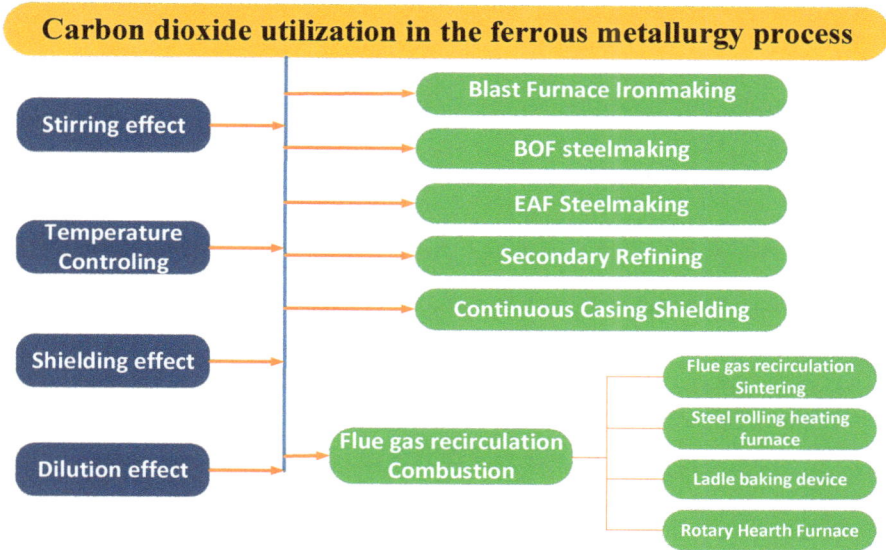

Figure 1. Carbon dioxide utilization in the ferrous metallurgy process.

2. Application in the Steelmaking Process.

2.1. Top-Blowing Carbon Dioxide in Converter

As Table 1 shows, compared with pure oxygen, the chemical heat release decreases when CO_2 is used in the steelmaking process as an oxidizing agent due to the endothermic or micro exothermic reaction of CO_2 in the molten bath, which may mean a reduced capacity for scrap melting. Therefore, when a certain proportion of CO_2 is blown into the converter in the dephosphorization process, the temperature is controlled, and suitable thermodynamic conditions for the dephosphorization reaction are created. At the same time, the reaction produces more stirring gas, which is conducive to strengthen bath stirring and creates favorable dynamic conditions for the dephosphorization reaction too.

Since 2004, Zhu Rong et al. [7–14] have carried out basic exploration research and industrial tests involving CO_2 utilization in steelmaking. After researching this area for a decade, they applied CO_2 in the top-blowing of the Basic Oxygen Furnace (BOF) and researched the CO_2 injection technology involved in top-blowing in the converter. When blowing CO_2 in the converter, the dust is reduced by 19.13%, the total Fe in the dust decreases by 12.98%, and the slag iron loss is reduced by 3.10%. As an agitation effect achieved from the temperature controlling improvement, the dephosphorization rate increases by 6.12%, and the nitrogen content of the molten steel reduces too, which leads to a higher quality of the molten steel. According to some test results, about 80~90% of CO_2 blown into the BOF participates in the chemical reaction at 1600 °C [15].

Table 1. Chemical reaction thermodynamic data of CO_2 with elements in the molten iron.

Elements	Chemical Reactions	$\Delta G^{\ominus}$ (J/mol)	$\Delta G^{\ominus}$ (kJ/mol) 1773 K	ΔH (kJ/mol) 298 K
C	$1/2O_2 + [C] = CO(g)$	$-140,580 - 42.09\ T$	-215.21	-139.70
	$O_2 + [C] = CO_2(g)$	$-419,050 + 42.34\ T$	-343.98	-393.52
	$CO_2(g) + [C] = 2CO(g)$	$137,890 - 126.52\ T$	-86.43	172.52
Fe	$1/2O_2(g) + Fe(l) = (FeO)$	$-229,490 + 43.81\ T$	-151.81	-272.04
	$CO_2(g) + Fe(l) = (FeO) + CO(g)$	$48,980 - 40.62\ T$	-23.04	40.37
Si	$O_2 + [Si] = (SiO_2)$	$-804,880 + 210.04\ T$	-432.48	-910.36
	$2CO_2(g) + [Si] = (SiO_2) + 2CO(g)$	$-247,940 + 41.18\ T$	-174.93	-344.36
Mn	$1/2O_2 + [Mn] = (MnO)$	$-412,230 + 126.94\ T$	-187.17	-384.93
	$CO_2(g) + [Mn] = (MnO) + CO(g)$	$-133,760 + 42.51\ T$	-58.39	-101.91

Top-blowing CO_2 in the converter is a major innovation in China and has been applied in Shougang Jingtang Company, who have achieved good results. CO_2 sources are wide in the iron and steel enterprises, providing a convenient condition for the converter blowing CO_2.

2.2. Bottom-Blowing Carbon Dioxide in the BOF

In the 1970s, the scholars began to study bottom-blowing CO_2 in the converter steelmaking process and found [9] that CO_2 can participate in the bath reaction, and its bottom-blowing agitation ability is stronger than that of Ar and N_2, when compared with blowing N_2/Ar from the bottom, which is an easy way to make [N] increase, and blowing O_2/C_xH_y from the bottom, which is an easy way to make [H] increase. CO_2 is an effective alternative to high-cost Ar and potentially harmful N_2 [16–18].

In the 1990s, researchers from AnGang Company [19] studied the bottom blowing of CO_2 in the top and bottom blowing converter and found that CO_2 can be used for bottom blowing of the top-bottom blowing converter. To prevent the strong cooling effect of the bottom blowing CO_2 gas from the nozzle being clogged, some oxygen was mixed in the bottom blowing gas. Unfortunately, the use of this technique was stopped due to issues with the blowing brick's life.

Recent studies have found that by bottom blowing CO_2, the slag iron loss can be reduced, the bath stirring can be strengthened, and the dephosphorization rate can be improved. In 2009, industrial experiment of bottom-blowing CO_2 on the 30 t converter was conducted in Fujian Sanming Steel company (Sanming, Fujian, China). The results showed that the converter with CO_2 bottom blowing is feasible, with no obvious erosion of the hearth [20].

2.3. Bottom-Blowing Stirring in the Ladle Furnace

Previously, to avoid re-oxidation and hydrogen and nitrogen absorption, the bottom blowing of CO_2 instead of Ar for stirring in a ladle was studied. The stirring mechanism of CO_2 bottom blowing during the ladle furnace (LF) refining process was studied [21]. As an exploratory industrial experiment, different proportions of CO_2 and Ar gas mixtures were bottom blown in a 75 t LF. The results showed that the stirring was reinforced when blowing CO_2, and the desulfurization rate was increased from 49.7% to 65.1%, and the average slag (FeO) content was less than 0.5%, which meets the oxidizing slag requirements. Though the type, the morphology, and the composition of the inclusions in molten steel changed little, the average number of inclusions per analyzed area decreased, which means the cleanliness of the liquid steel improved. The tests showed that the LF furnace can use CO_2 gas for refining.

2.4. Bottom-Blowing Stirring in the Electric Arc Furnace

Since the emergence of Electric Arc Furnace (EAF) bottom blowing technology in the 1980s, scholars have studied and found that the EAF bottom blowing technology can improve the bath agitation ability, promote the inter-slag reaction, uniform bath temperature, and composition, and improve the alloy yield. It is of great significance to improve the furnace dynamics.

Industrial experiments [21] of bottom-blowing CO_2 in a 65 t Consteel EAF verified that bottom-blowing CO_2 instead of Ar is feasible. The studies showed that compared with the conventional bottom-blown Ar process, bottom-blowing CO_2 increases the end [C] content and oxidizes a small amount of [Cr], but the contents of [Mn], [Mo], [O] and [N] are not influenced. This method can also enhance the bath stirring, raise the basicity, and reduce the slag (FeO) content. It provides a suitable kinetic and thermodynamic condition for EAF desulfurization and dephosphorization, with the desulfurization degree increasing by 7%.

2.5. Shielding Gas in Continuous Casting Process

In 1989, reference [22] stated that a US company applied CO_2 instead of Ar to protect the injection flow when casting special steel rod. Benefiting from a higher density, CO_2 will fall in parallel with the injection flow through the upper portion of the upper casing or spiral holes, maintaining the positive pressure of the stream around to prevent air suction. It shows a great effect in cutting off the liquid steel from the air and preventing the oxidation of the molten steel.

To solve problems of using CO_2 with specific process issues, our team conducted experiments and found that when using CO_2 instead of Ar in submerged nozzle seal protection, the [N] content increased. When bottom blowing Ar for 40Cr steel and 45 steel, the [N] content increased by 10.4% and 53.6%, respectively. While bottom blowing CO_2, the [N] content increased by 17.6% and 54.4%, respectively. When observing CO_2-protected steel, the [O] content was shown to be decreased. CO_2 can play a protective role in casting, which can reduce secondary oxidation.

The amount of CO_2 emission utilization in the BOF steelmaking process will be greater than 40 kg/ton of steel. Unfortunately, reports of sustainable utilization of CO_2 emissions in the steelmaking industry are almost non-existent, because the industrial supply of CO_2 gas has not yet been established yet.

3. Application in Sintering and Blast Furnace

3.1. Flue Gas Recirculation (FGR) Sintering

The main combustion produce gas in sintering is CO_2. Typical Flue Gas Recirculation (FGR, Figure 2) sintering [23] includes the waste gas regional recirculation process developed by Nippon Steel [24], and similar technologies have been developed now, such as the Exhaust Gas Recirculation (EGR) sintering technology developed by Hata [25], the Emission Optimized Sintering (EOS) process developed by Corus Ijmuiden in Netherlands [26], the LEEP (low emission and energy optimized sintering process) developed by HKM [24] and the EPOSINT (environmental process optimized sintering) process developed by Siemens VAI [27–29].

The existing recycling technology still has the following disadvantages [23,24]:(1) In the Nippon Steel recycling process, only high oxygen flue gas is circulated. The flue gas emission reduction rate is relatively low, about 28%. The circulation process is complex, and it is difficult for modifying a sintering machine. (2) EOS technology has not considered the characteristics of sintering flue gas emissions, and the effect of dealing with different components in the flue gas is not the best. (3) For the LEEP technology, the heat of the high temperature flue gas is not fully utilized. The rear part of the flue gas has a high content of SO_2, which results in the sintering ore [S] content increasing. (4) The EPOSINT process only makes a high sulfur gas cycle, and the reduction in emission rates is small, only 28~25%, and the sintering ore [S] content increases too. In addition, the high temperature flue gas is not circulating, and the energy saving rate and the dioxin emission reduction rate are both low.

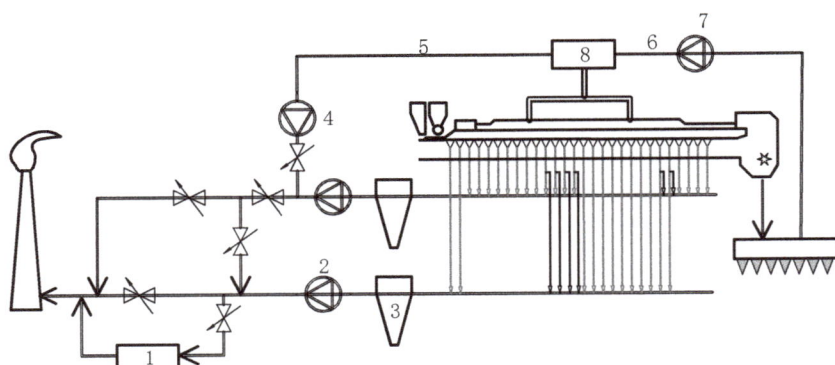

1-Desulfurization system, 2-Main exhaust fan, 3-Dust catcher, 4-Circulating air blower, 5-Large flue circulating flue gas, 6-Air cooling hot exhaust gas, 7-Air cooling heat recovery fan, 8-Flue gas mixing chamber.

Figure 2. Flow diagram of flue gas circulation sintering process.

3.2. CO$_2$ Injection Through the Blast Furnace Tuyere

In 2010, Fu Zhengxue [30] et al. developed a method of injecting carbon dioxide into a Blast Furnace (BF). Blowing CO$_2$ or the exhaust gas containing CO$_2$ into BF can effectively solve some problems, such as resource saving, CO$_2$ emissions reduction, and environmental pollution elimination. The technical scheme is as follows: Firstly, CO$_2$ or waste gas containing CO$_2$ is blown into BF cold air pipes. After heating in the hot blast stove, CO$_2$ is blown or sprayed into the BF tuyere zone by the hot air pipeline. Blown oxygen and CO$_2$ react with the burning carbon in the tuyere zone, and then CO is generated. Because of CO$_2$ absorbing part of the exhausted heat of oxygen, a higher oxygen enrichment can be achieved, and the CO generated is the smelting reduction agent, so the reductant rate of the Blast Furnace is increased, and the production efficiency of BF will be improved. However, the technology has not been applied industrially.

3.3. Carrier Gas of Pulverized Coal Injection

In 2011, Chinese researchers [31] invented a method to use CO$_2$ as the transmission medium of pulverized coal injected into BF (Figure 3). Mixed coal powder was used in BF tuyere to replace coke, provide heat, and act as a reducing agent.

Pulverized coal in the tuyere zone reacts not only with the enriched CO$_2$, but also with the oxygen from hot air. It is necessary to adjust the amount of pulverized coal injection and the oxygen enrichment level to achieve the best ratio and promote complete combustion of the pulverized coal in front of the tuyere. Due to the use of CO$_2$ instead of compressed air or nitrogen as the transmission medium, the total amount of exhaust gas through BF should be reduced, which will significantly reduce the nitrogen content and increase the amount of CO$_2$ + CO. Therefore, the purity and calorific value of BF gas increase.

So far, there has been no industrial application reported for the above two patents. CO$_2$ injection through a Blast Furnace tuyere creates a new route of CO$_2$ utilization. The amount of CO$_2$ utilization in the BF process will be greater than 50 kg/ton of iron, but the utilization of CO$_2$ emissions in the BF process is still being researched.

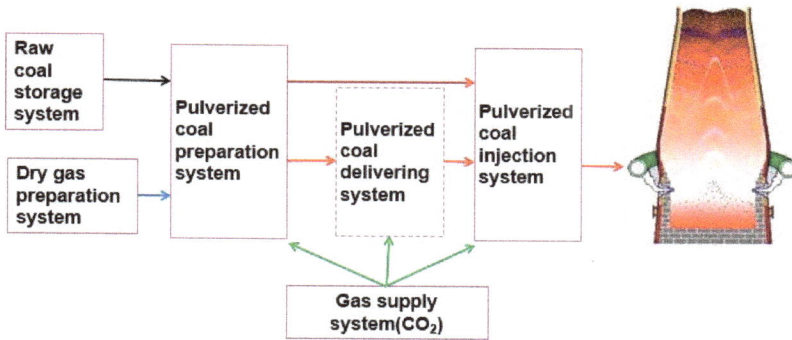

Figure 3. Flow diagram of pulverized coal injection of BF.

4. Applications in Other Ferrous Metallurgy Processes

4.1. Application in Smelting Stainless Steel

In 2011, Anshan Iron and Steel Co., Ltd. [32] invented a smelting method of AOD by blowing CO_2 to produce stainless steel, which is mainly about injecting CO_2 into molten steel to enhance the decarburization. This suits smelting steel whose carbon content ranges from 0.001% to 0.3%. CO_2 injected into the molten steel not only can decarburize, but can also enhance the bath stirring, and it can promote the oxygen reaction and cool the oxygen lance, whose life is improved by 20%.

Our team studied the mechanism of Cr retention and decarburization when injecting CO_2 to AOD from the aspects of thermodynamics and kinetics. In the laboratory, tube furnace experiments found that the Cr retention and decarburization effects of CO_2 are very good, and the carbon content can reach 0.5% with little chromium oxidation occurring [33]. When the O_2 proportion of smelting gas was increased, large amounts of chromium oxidation occurred, and the decarburization was relatively reduced. An analysis of the changes in materials and energy balance found that it could meet the requirements for refining the temperature when the CO_2 injection ratio was less than 9.13%. As the proportion of CO_2 increased within this range, the AOD furnace surplus heat reduced, and the CO proportion of furnace gas increased.

4.2. CO_2 Circulation Combustion

CO_2 circulation combustion technology uses recycled flue gas to replace the nitrogen in the air (Figure 4), which can reduce nitric oxide emissions and improve the thermal efficiency of combustion [34]. Its application in steel rolling heating furnaces and pit furnaces can reduce fuel consumption, shorten the heating time, and reduce emissions.

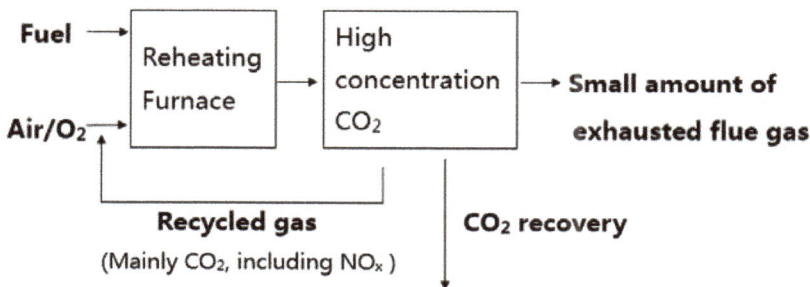

Figure 4. Flow diagram of flue gas circulation combustion technology.

Metals **2019**, *9*, 273

In the 1920s, the Linde Group [35] tried applying flameless combustion for heating in the ferrous industry, reducing fuel consumption and emissions of NO_x and CO_2 through the development of a special burner to achieve a special method of combustion. In addition, the Swedish Hofors works Ovako also used the technology in the heating furnace, finding that the production increased by 30–50% and the fuel reduced by 30–45% with more uniform heating and reductions in CO_2 and NO_x emissions.

5. Conclusions

There are many methods of CO_2 utilization in the ferrous metallurgy process, such as directly being blown in the BF and Converter, serving as a carrier gas for coal injection in BF, use as a shielding gas and mixing gas in refinement, the continuous casting process and stainless steelmaking, and use in circulation in the exhaust gas of the steel rolling heating furnace. All of these uses reflect the special role that CO_2 plays in the ferrous metallurgy process.

In China, policies of CO_2 utilization and emission reduction are actively carried out. The historical carbon emissions are being investigated to build a unified nationwide carbon emissions trading system. Meanwhile, some Chinese steel plants have carried out procedures involvingCO_2 utilization. The technology of top and bottom blowing CO_2 in the 300 t converter has been applied, which achieved good results.

With the continuous improvement and further expansion of CO_2 application in the ferrous metallurgy process, carbon dioxide usage in the ferrous metallurgy process is expected to be more than 100 kg per ton of steel. At present, the annual steel output of China is about 800 million tons, and the annual amount of recycled CO_2 utilization is around 80 million tons in metallurgical processes, which could effectively facilitate the progress of metallurgical technology, strongly promoting energy conservation in the metallurgical industry. It meets the need for sustainable development in China.

Author Contributions: Writing-Original Draft Preparation, K.D. and X.W; Writing-Review & Editing, K.D.; Visualization, K.D. and X.W; Supervision, X.W.; Project Administration, K.D.

Conflicts of Interest: The authors declare no conflict of interest.

References

1. Shangguan, F.Q.; Zhang, C.X.; Hu, C.Q.; Li, X.P.; Zhou, J.C. Estimation of CO_2 Emission in Chinese Steel Industry. *China Metall.* **2010**, *19*, 37–42.
2. Aresta, M.; Dibenedetto, A.; Angelini, A. The changing paradigm in CO_2 utilization. *J. CO$_2$ Util.* **2013**, *1*, 65–73. [CrossRef]
3. Cuéllar-Franca, R.M.; Azapagic, A. Carbon capture, storage and utilisation technologies: A critical analysis and comparison of their life cycle environmental impacts. *J. CO$_2$ Util.* **2015**, *3*, 82–102. [CrossRef]
4. Olajire, A.A. Valorization of greenhouse carbon dioxide emissions into value-added products by catalytic processes. *J. CO$_2$ Util.* **2013**, *1*, 74–92. [CrossRef]
5. Hu, B.; Guild, C.; Suib, S.L. Thermal, electrochemical, and photochemical conversion of CO_2 to fuels and value-added products. *J. CO$_2$ Util.* **2013**, *1*, 18–27. [CrossRef]
6. Saeidi, S.; Amin, N.A.S.; Rahimpour, M.R. Hydrogenation of CO_2 to value-added products—A review and potential future developments. *J. CO$_2$ Util.* **2014**, *2*, 66–81. [CrossRef]
7. Yin, Z.J.; Zhu, R.; Yi, C.; Chen, B.Y.; Wang, C.R. Fundamental Research on Controlling BOF Dust by COMI Steel Making Process. *Iron Steel* **2009**, *53*, 92–94.
8. Jin, R.J.; Zhu, R.; Feng, L.X.; Yin, Z.J. Experimental study of steelmaking with CO_2 and O_2 mixed blowing. *J. Univ. Sci. Technol. Beijing* **2007**, *29*, 77–80.
9. Lv, M.; Zhu, R.; Bi, X.R.; Lin, T.C. Application research of carbon dioxide in BOF steelmaking process. *J. Univ. Sci. Technol. Beijing* **2011**, *33*, 126–130.
10. Yi, C.; Zhu, R.; Yin, Z.J.; Hou, N.N.; Chen, B.Y. Experimental Research of COMI Steelmaking Process Based on 30t Converter. *Chin. J. Process Eng.* **2009**, *9*, 222–225.
11. Ning, X.J.; Yin, Z.J.; Yi, C.; Zhu, R.; Dong, K. Experimental research on dust reduction in steelmaking by CO_2. *Steelmaking* **2009**, *35*, 32–34.

12. Zhu, R.; Yi, C.; Chen, B.Y.; Wang, C.R.; Ke, J.X. Inner circulation research of steelmaking dust by COMI steelmaking process. *Energy Metall. Ind.* **2010**, *29*, 48–51.

13. Lv, M.; Zhu, R.; Bi, X.R.; Wei, N.; Wang, C.R. Fundamental research on dephosphorization of BOF by COMI steelmaking process. *Iron Steel* **2011**, *55*, 31–35.

14. Bi, X.R.; Liu, R.Z.; Zhu, R.; Lv, M. Research on Mechanism of Dust Generation in Converter. *Ind. Heat.* **2010**, *39*, 13–16.

15. Li, Z.Z. Investigations on Fundamental Theory of CO_2 Applied in Steelmaking Processes. Ph.D. Thesis, University of Science and Technology Beijing, Beijing, China, 2017.

16. O'Hara, R.D.; Spence, A.G.R.; Eisenwasser, J.D. Carbon Dioxide Shrouding and Purging at Ipsco's Melt Shop. *Iron Steelmak.* **1986**, *13*, 24–27.

17. Guo, X.; Chen, W. Action mechanism of bottom-blown CO_2 in the bath of combined-blown converter. *J. Iron Steel Res.* **1993**, *5*, 9–14.

18. Gu, Y.L.; Zhu, R.; Lv, M.; Chen, L.; Liu, R.Z. Exploratory research on bottom blowing CO_2 during the LF process. *Iron Steel* **2013**, *57*, 34–39.

19. Li, C.B.; Han, Y. The application of the combined blowing technique with CO_2 bottom blowing to the converters of Anshan Iron and Steel Company. *Steelmaking* **1996**, *12*, 19–25.

20. Zhu, R.; Bi, X.R.; Lv, M. Application and development of CO_2 in the steelmaking process. *Iron Steel* **2012**, *56*, 1–5.

21. Wang, H.; Zhu, R.; Liu, R.Z. Application Research of Carbon Dioxide in EAF Bottom Blowing. *Ind. Heat.* **2014**, *43*, 12–14.

22. Anderson, S.A.H.; Foulard, J.; Lutgen, V. Inert Gas Technology for the Production of Low Nitrogen Steels. In Proceedings of the 47th Electric Furnace Conference, Orlando, FL, USA, 29 October–1 November 1989; pp. 365–375.

23. Yu, H.; Wang, H.; Zhang, C. Analysis on advantages and disadvantages of sintering waste gas recirculation process. *Sinter. Pelletizing* **2014**, *6*, 51–55.

24. Eisen, H.P.; Hüsig, K.R.; Köfler, A. Construction of the exhaust recycling facilities at a sintering plant. *Stahl Eisen* **2004**, *124*, 37–40.

25. Hu, B.; He, X.H.; Wang, Z.C.; Ye, H.D.; Chen, Y.Y. Wind and Oxygen Balance based on Flue Gas Circulation Sintering Process. *Iron Steel* **2014**, *58*, 15–20.

26. Menad, N.; Tayibi, H.; Carcedo, F.G.; Hernández, A. Minimization methods for emissions generated from sinter strands. *J. Clean. Prod.* **2006**, *14*, 740–747. [CrossRef]

27. Fleischanderl, A. MEROS-Improved dry sintering waste gas treatment process. *Rev. Métall.* **2006**, *29*, 481–484. [CrossRef]

28. Brunnbauer, G.; Ehler, W.; Zwittag, E.; Schmid, H.; Reidetschlaeger, J.; Kainz, K. New waste-gas recycling system for the sinter plant at Voestalpine Stah. *MPT Metall. Plant Technol. Int.* **2006**, *29*, 38–42.

29. Alexander, F.; Christoph, A.; Erwin, Z. New Developments for Achieving Environmentally Friendly Sinter Production—Eposint and MEROS. *China Metall.* **2008**, *17*, 41–46.

30. Fu, Z.X.; Liu, S.J.; Fu, X.; Liu, L.M.; Liu, D.P.; Cui, F.H. A Method of Injecting Carbon Dioxide into Blast Furnace. CN Patent No. 201010199232.6, 12 June 2010.

31. Guo, X.Q.; Yuan, L.; Zhang, J.; Wu, S.Q.; Ma, X.Y. The Method of Using Carbon Dioxide as Transport Medium for Pulverized Coal Injection in Blast Furnace. CN Patent No. 201010268748.1, 30 August 2010.

32. Tang, F.P.; Zhao, G.; Li, D.G.; Cao, D.; Zhao, C.L.; Xue, J. A Method of AOD Injecting CO_2 for Producing Stainless Steel. CN Patent No. 201010108211.9, 5 February 2010.

33. Bi, X.R.; Zhu, R.; Liu, R.Z.; Lv, M.; Yi, C. Fundamental research on CO_2 and O_2 mixed injection stainless steelmaking process. *Steelmaking* **2012**, *28*, 67–70.

34. Okazaki, K.; Anda, T. NOx reduction mechanism in coal combustion with recycled CO_2. *Energy* **1997**, *22*, 207–215. [CrossRef]

35. Scheele, J.V.; Gartz, M.; Paul, R.; Lantz, M.T.; Riegert, J.P.; Söderlund, S. Flameless oxyfuel combustion for increased production and reduced CO_2 and NOx emissions. *Stahl Eisen* **2008**, *128*, 35–41.

metals

MDPI

Article

Comparison of Energy Consumption and CO$_2$ Emission for Three Steel Production Routes—Integrated Steel Plant Equipped with Blast Furnace, Oxygen Blast Furnace or COREX

Jiayuan Song [1], Zeyi Jiang [1,*], Cheng Bao [1] and Anjun Xu [2]

[1] School of Energy and Environmental Engineering, University of Science and Technology Beijing, Beijing 100083, China; songjiayuan2015@outlook.com (J.S.); baocheng@mail.tsinghua.edu.cn (C.B.)

[2] School of Metallurgical and Ecological Engineering, University of Science and Technology Beijing, Beijing 100083, China; anjunxu@metall.ustb.edu.cn

* Correspondence: zyjiang@ustb.edu.cn; Tel.: +86-10-6233-2741; Fax: +86-10-6233-2741

Received: 31 December 2018; Accepted: 20 March 2019; Published: 21 March 2019

Abstract: High CO$_2$ emissions and energy consumption have greatly restricted the development of China's iron and steel industry. Two alternative ironmaking processes, top gas recycling-oxygen blast furnace (TGR-OBF) and COREX®, can reduce CO$_2$ emissions and coking coal consumption in the steel industry when compared with a conventional blast furnace (BF). To obtain parameters on the material flow of these processes, two static process models for TGR-OBF and COREX were established. Combining the operating data from the Jingtang steel plant with established static process models, this research presents a detailed analysis of the material flows, metallurgical gas generation and consumption, electricity consumption and generation, comprehensive energy consumption, and CO$_2$ emissions of three integrated steel plants (ISP) equipped with the BF, TGR-OBF, and COREX, respectively. The results indicated that the energy consumption of an ISP with the TGR-OBF was 16% and 16.5% lower than that of a conventional ISP and an ISP with the COREX. Compared with a conventional ISP, the coking coal consumption in an ISP with the TGR-OBF and an ISP with the COREX were reduced by 39.7% and 100% respectively. With the International Energy Agency factor, the ISP with the TGR-OBF had the lowest net CO$_2$ emissions, which were 10.8% and 35.0% lower than that of a conventional ISP and an ISP with the COREX. With the China Grid factor, the conventional ISP had the lowest net CO$_2$ emissions—2.8% and 24.1% lower than that of an ISP with the TGR-OBF and an ISP with the COREX, respectively.

Keywords: oxygen blast furnace; COREX; static process model; integrated steel plant; material flow; energy consumption; CO$_2$ emissions

1. Introduction

Steel is the world's most popular construction material due to its durability, processability, and cost. However, producing steel creates high energy consumption and CO$_2$ emissions. According to the World Steel Association, the production of crude steel reached 1691 million tons in 2017 and the large amount of crude steel production resulted in about 3.1×10^{11} GJ of energy consumption and 3043.8 million tons of CO$_2$ emissions [1]. The blast furnace (BF)—basic oxygen furnace (BOF) route is the dominant steel production route in the world, and its ironmaking process contributes approximately 70% of the above energy consumption and CO$_2$ emissions [2]. Many studies have shown that adopting commercially available energy-saving technologies for BF has significantly reduced the energy consumption of the ironmaking process [3] and almost reached the thermodynamic limits of BF [4]. Thus, in order to minimize the energy consumption and CO$_2$ emissions of the ironmaking process, alternative liquid iron production technologies to BF such as top gas recycling-oxygen blast furnace (TGR-OBF) and COREX have been proposed and developed.

The oxygen blast furnace (OBF) is a type of metallurgical furnace used for producing liquid iron, which was proposed by Wenzel and Gudenau in 1970 [5]. However, the first generation of OBF had the problem of overheating in the lower part of the furnace and thermal shortage in the upper part of the furnace. In order to solve these problems, some TGR-OBF processes such as Fink [6], Lu [7], Nippon Kokan Steel (NKK) [8], Tula [9], full oxygen blast furnace (FOBF) [10], and Ultra-Low CO_2 Steelmaking initiative (ULCOS) [11] have been proposed. In the TGR-OBF process, iron ore, coke, and flux are supplied through the top of OBF in succession, while normal temperature oxygen and pulverized coal are injected into the OBF by tuyeres at the bottom. The preheated top gas is recycled into the OBF at the lower stack, after the removal of CO_2 and H_2O. Compared with the BF process, the two key characteristics of the TGR-OBF process are the injection of pure oxygen as an oxidant instead of a hot blast and the recycling of top gas as a reductant after the removal of CO_2 and H_2O and preheating. These characteristics increase the amount of pulverized coal in the OBF to more than 300 kg/t-hot metal [12] and reduced the coke ratio to less than 200 kg/t-hot metal [13]. As a result of the study of a pilot plant, an 8 m^3 experimental TGR-OBF built during the Ultra-low CO_2 steelmaking initiative phase I (ULCOS I) showed that the TGR-OBF process was feasible and could significantly reduce the consumption of coke and CO_2 emissions in the ironmaking process [14]. Moreover, many researchers have conducted experiments and numerical simulations of the TGR-OBF process, which showed that the TGR-OBF process is a promising alternative liquid iron production technology given its low coke consumption and CO_2 emissions [15–18].

COREX is an industrially and commercially proven smelting reduction ironmaking process and was created by Siemens Voest-Alpine Industrieanlagenbau Gmbh & Co. (VAI) in the 1970s [19]. There are eight COREXs in the world that have been successfully commercialized [20]; two of which are the latest generation of COREX with a capacity of 1.5 million tons of liquid iron per year and were built in China at the Baosteel Luojing steel plant [21]. The characteristic of COREX is the separation of the iron reduction and smelting operations into two separate reactors, namely the upper reduction shaft and the lower melter-gasifier [22]. In the upper reduction shaft, iron ore and flux are continuously charged from the top of the shaft. They descend by gravity and are reduced into approximately 95% direct reduced iron by the reduction gas generated from the melter-gasifier. In the melter-gasifier, the reduction shaft products—direct reduced iron—is fed from the top of the melter-gasifier for further reduction and melting. Non-coking coal and room temperature oxygen, as the main reductant and oxidant, are injected via the lock hopper system at the top of the melter-gasifier and tuyeres at the bottom of the melter-gasifier, respectively. After being purified by hot gas cyclones, a major part of the top gas from the melter-gasifier is cooled and subsequently added to the reduction shaft. The excess gas and the reduction shaft gas are mixed before the take-over point, called the COREX export gas. When compared with the BF or TGR-OBF process, the COREX process can directly produce liquid iron without using coking coal, sinter, or pellets. The COREX process combines the coking plant, sinter plant, and blast furnace or oxygen blast furnace into a single ironmaking process. Therefore, as an alternative liquid iron production technology to BF, further research into the TGR-OBF and COREX steel production processes are of great significance.

Most of the studies on TGR-OBF and COREX have focused on a reaction unit rather than the ISP. Wenlong et al. estimated the impact of different metallization rates and fuel structures on the energy consumption of the COREX process by using a modified Rist operating diagram [23]. Hu et al. compared the CO_2 emissions between the COREX and blast furnace ironmaking system based on carbon element flow analysis [24]. Lianzhi et al. used a two-dimensional numerical simulation model to analyze the effect of top gas recycling on the blast furnace status, productivity, and energy consumption. The results showed that TGR-OBF productivity increased by 5.3–35.3% and the energy saving was 27.3–35.9% when compared to blast furnaces [17]. Wei et al. conducted exergy analyses of the TGR-OBF ironmaking process [18]. Xuefeng et al. investigated the carbon saving potential of the TGR-OBF ironmaking process from the perspective of the relationship between the degree of direct reduction and carbon consumption [25]. These studies demonstrated well the effect of operating parameters on energy consumption and CO_2 emissions in the COREX or TGR-OBF process. However, focusing only on a specific unit like the ironmaking process will make it difficult for policymakers to fully understand the alternative ironmaking process from a mill-wide perspective. Up to now, many studies conducted from a plant-wide perspective are based on only the one steel production route. Hooey et al. made a techno-economic comparison of an ISP equipped with the OBF and CO_2 capture by using process and economic models and found that the OBF with CO_2 capture offered a significant potential to reduce the overall CO_2 emissions from an ISP, achieving 47% CO_2 avoidance at a cost of ~\$56/t CO_2 for the given assumptions [26]. Huachun et al. analyzed the energy consumption and carbon emissions of a conventional ISP by using an integrated material flow analysis model [27]. Jin et al. analyzed the energy consumption and carbon emissions of an ISP with the TGR-OBF route, according to an established heat transfer and reaction kinetics mathematical model for the TGR-OBF ironmaking process. The results indicate that the energy consumption of an ISP with the TGR-OBF route was reduced to 14.4 GJ per ton of crude steel; moreover, the direct CO_2 emissions were reduced by 26.2% per ton of crude steel when compared with a conventional ISP [28]. Arasto et al. analyzed the technical assessment of the application of OBF with CCS to an ISP. The analysis showed that the CO_2 emissions from an ISP could be significantly reduced by the application of an oxygen blast furnace and CCS [29]. Based on the technical analysis, the economic profitability was further evaluated in Reference [30], which found that the investment on OBF or CCS depended on CO_2, the fuels, and electricity prices. However, the comprehensive energy consumption and CO_2 emissions analysis on three different routes based on the static process model such as an ISP with BF, an ISP with the TGR-OBF, and an ISP with the COREX were not analyzed in detail.

Therefore, in this study, TGR-OBF and COREX static process models were developed based on the mass and heat balance to obtain the material and energy flow parameters of the two ironmaking processes. Combining the operating data from the Jingtang steel plant with established static process models, this research presents a detailed analysis on the material flow, metallurgical gas generation and consumption, electricity consumption and generation, comprehensive energy consumption, and CO_2 emissions of the BF, TGR-OBF, and COREX in an ISP.

2. Models and Methods

2.1. Static Process Models

The objective of this study is to compare different steel production routes under the same composition and temperature of the hot metal, as well as the same chemical composition of raw materials (For the TGR-OBF, the raw materials with the same composition as the BF are flux, coke, coal, sinter, and pellets. For the COREX, the raw materials with the same composition as the BF are flux, and coal). Thus, on the basis of the mass and heat balance principles, a TGR-OBF static process model based on our previous research—References [17,28,31] and a COREX static process model based on the Baosteel COREX3000 were established by using the gPROMS Modelbuilder platform. According to the calculation of the two static process models, the consumption of raw material and fluxes, the production and composition of slag and metallurgical gas per ton of hot metal can be obtained. Combining the two static process models with the ISP of the Jingtang, this research presents a detailed comparison and analysis of the three ISPs equipped with the BF, TGR-OBF, and COREX, respectively.

2.1.1. The TGR-OBF Process Model

The flow diagram of the TGR-OBF process is shown in Figure 1. In the TGR-OBF process, most of the top gas is recycled into the furnace as the recycled gas after CO_2 removal and preheating. Part of the recycled gas is injected at the hearth tuyeres, another part of the recycled gas is injected at the lower shaft. This kind of gas recycling in the TGR-OBF process could effectively supplement the heat and reducing atmosphere in the shaft area, and also reduced the combustion temperature in the tuyere area [17]. In this study, the total amount of the recycled gas assumed was 600 m^3/t-HM, the upper and lower recycled gas assumed were 300 m^3/t-HM, respectively (Supplementary Materials Table S7). This condition could ensure the successful operation of the TGR-OBF and provide high calorific value gas for other processes, as well as ensure the balance of the metallurgical gas in the ISP [31]. Moreover, based on our previous research [31], the temperature of the recycled gas at the hearth tuyeres had little effect on the burn-out rate of the pulverized coal (when the temperature of the recycled gas raised from 298 K to 1500 K, the burnout rate of the pulverized coal only changed by 1.5%). Therefore, considering the technical difficulty and the cost, the recycled gas was assumed to be preheated to 1173 K (Table S7). The main reactions in the furnace include the reduction of various iron oxides by CO/H$_2$, the decomposition of carbonate, coke, and coal, and the combustion of carbon and reducing gas. In order to make the TGR-OBF process operate successfully under the certain conditions of the raw materials and fuel, it is very important to select reasonable slag-forming practices to make the slag have a better performance. The performance of the slag is closely related to its chemical composition. The basicity is an important parameter of measuring the viscosity of the slag, at the same time, the basicity has great influence on the melting and stability of the slag. Therefore, the binary basicity R_2 (Equation (3)) of the slag were included as the constrained equation. Based on the following balance Equations (1), (3), and (4) in the TGR-OBF process, the consumption of raw materials (sinter, pellets, limestone, and dolomite), the amount and chemical composition of slag can be obtained by giving the value of chemical composition of raw materials and hot metal (Tables S1, S2, S4, and S6).

- Mass balance:

$$\sum_{in} m_i \times w(x)_i = \sum_{out} m_j \times w(x)_j. \tag{1}$$

When the consumption of raw materials was calculated, the mass balance equations included the Fe balance equation, the S balance equation, the CaO balance equation, the SiO$_2$ balance equation, the Al$_2$O$_3$ balance equation, the MgO balance equation, the MnO balance equation, and the P$_2$O$_5$ balance equation. Taking the Fe balance equation as an example,

$$\sum_i m_i \times w(Fe)_i = m_{HM} \times w(Fe)_{HM} + m_{slag} \times w(FeO)_{slag} \times 56/72 \tag{2}$$

- Binary basicity:

$$R_2 = m_{CaO\text{-slag}} / m_{SiO_2\text{-slag}} \tag{3}$$

- Quantity of slag:

$$m_{slag} = m_{FeO\text{-slag}} + m_{CaS\text{-slag}} + m_{CaO\text{-slag}} + m_{SiO_2\text{-slag}} + m_{Al_2O_3\text{-slag}} + m_{MgO\text{-slag}} + m_{MnO\text{-slag}} + m_{P_2O_5\text{-slag}} \tag{4}$$

where m_i is the mass of sinter, pellets, coke, coal, limestone, and dolomite for producing 1 ton of hot metal, kg; $w(Fe)_i$ is the mass fraction of Fe in material i; $w(FeO)_{slag}$ is the mass fraction of FeO in slag; m_{HM} is the mass of hot metal, which in this study was 1 ton; m_{slag} is the mass of slag, kg; and $w(x)_i$ is the composition of x in material i, %.

The utilization of hydrogen in the top gas and the oxidation degree of the top gas are important parameters that can affect the chemical composition of the top gas. Therefore, the utilization of hydrogen in the top gas and the oxidation degree of the top gas were included as the constrained equations. Based on the operating conditions and parameters (Tables S3 and S7) of the TGR-OBF process, the volume and chemical composition of top gas were obtained by calculating the following balance Equations (5)–(9), and (11).

- C balance:

$$\sum_i m_i \times \omega(C)_i + V_{rg} \times [v(CO)_{rg} + v(CO_2)_{rg}] \times 12/22.4 = (V_{CO} + V_{CO_2} + V_{CH_4}) \times 12/22.4 + m_{HM} \times \omega(C)_{HM} + m_{dust} \times \omega(C)_{dust} \quad (5)$$

- H balance:

$$\sum_i m_i \times \omega(H)_i + V_{rg} \times [v(H_2)_{rg} + v(H_2O)_{rg}] \times 2/22.4 = (4 \times V_{CH_4} + 2 \times V_{H_2} + 2 \times V_{H_2O})/22.4 \quad (6)$$

- N balance:

$$\sum_i m_i \times \omega(N)_i + V_{rg} \times v(N_2)_{rg} \times 28/22.4 = V_{N_2} \times 28/22.4 \quad (7)$$

- The utilization of hydrogen in the top gas:

$$\eta_{H_2} = V_{H_2O_r}/V_{H_2O_r} + V_{H_2} \quad (8)$$

- The oxidation degree of the top gas:

$$OD = (V_{H_2O} + V_{CO_2})/(V_{H_2O} + V_{CO_2} + V_{H_2} + V_{CO}) \quad (9)$$

here, V_{rg} is the volume of recycled gas; V_{CO}, V_{H2O}, V_{CO2}, V_{H2}, V_{CH4}, and V_{N2} are the volume of CO, H$_2$O, CO$_2$, H$_2$, CH$_4$, and N$_2$ in top gas for producing 1 ton of hot metal, m^3, respectively; V_{H2Or} is the volume of water in the top gas, which was produced by the reduction, m^3; $v(CO)_{rg}$, $v(CO_2)_{rg}$, $v(H_2)_{rg}$, $v(H_2O)_{rg}$, $v(N_2)_{rg}$ are the volume fraction of CO, CO$_2$, H$_2$, H$_2$O, and N$_2$ in recycled gas, respectively; $\omega(C)_i$, $\omega(H)_i$, $\omega(N)_i$ and $\omega(C)_{HM}$ are the mass fraction of C, H, and N in material i and hot metal, respectively; and T_{gas} is the temperature of the metallurgical gas, K.

- Heat balance:

$$\sum_{in} m_i w_i(x) \cdot h_i + \sum_{in} m_i \cdot q_i = \sum_{out} m_j w_j(x) \cdot h_j + \sum_{out} m_j \cdot q_j + H_l \quad (10)$$

where h_i, h_j are the enthalpy of the chemical reaction, kJ/kg; q_i, q_j are the sensible heat, kJ/kg; and H_l is the enthalpy of loss, kJ.

Based on the above calculation results and the oxygen balance equation of the furnace, the volume of oxygen blast can be obtained. Based on the mass balance of the gas in top, bosh, and hearth of the furnace, and the heat balance (Equation (10)) of the whole furnace, the volume of recycled gas can be obtained. The volume of gas required for preheating the recycled gas is obtained by the heat balance (Equation (10)) of the heater unit. The parameters calculated by the TGR-OBF static process model are shown in Tables S1–S7 (Supplementary Materials).

In the VPSA unit, the recovery rate of CO$_2$ was set at 90% [32], and the energy consumption was set at 645.7 kJ/kg-CO$_2$ [33]. In the heater unit, the thermal efficiency was set at 86% [11].

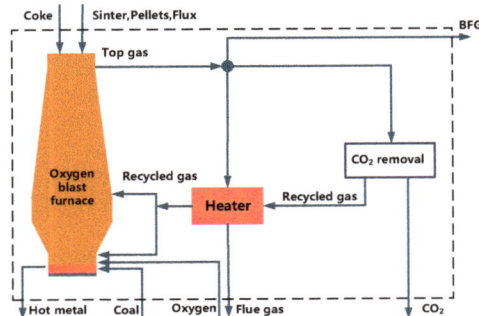

Figure 1. The flow diagram of top gas recycling-oxygen blast furnace (TGR-OBF) ironmaking process.

2.1.2. The COREX Process Model

The COREX process flow chart is shown in Figure 2. The COREX process is comprised of the shaft furnace and melter-gasifier. In the shaft furnace, the main reactions include the reduction of various iron oxides and carbonate decomposition, while the main reactions inside the melter-gasifier include coal decomposition, carbon and reducing gas combustion, reduction of residual iron oxides, and decomposition of residual carbonate. In the COREX process, the raw gas produced by the melter-gasifier is first purified by hot gas cyclones. The dust after cyclone removal is recycled back to the melter-gasifier by dust burners. The major part of the purified raw gas is made available as reduction gas for the reduction shaft. Additionally, the rest of the purified top gas is scrubbed and mixed with the cleaned reduction shaft top gas to form COREX gas. The static process model and its operating parameters are based on Baosteel COREX 3000.

Based on the Tables S1, S2, S4, and S6, the consumption of raw materials (iron ore, limestone, and dolomite), the amount and chemical composition of slag can be calculated by using balance Equations (1), (3), and (4) in the COREX process. Based on the operating conditions and parameters (supplementary materials Tables S3 and S10) and the equation 11 of the COREX process, the volume and chemical composition of raw gas can be obtained by calculating balance Equations (9), and (11)–(14).

- H balance:

$$\sum_i m_i \times w(H)_i = (4 \times V_{CH_4} + 2 \times V_{H_2} + 2 \times V_{H_2O})/22.4 \tag{11}$$

- C balance:

$$\sum_i m_i \times w(C)_i = (V_{CO} + V_{CO_2} + V_{CH_4}) \times 12/22.4 + m_{HM} \times w(C)_{HM} \tag{12}$$

- N balance:

$$\sum_i m_i \times w(N)_i = V_{N_2} \times 28/22.4 \tag{13}$$

- The equilibrium constant of the water gas shift reaction:

$$K = (V_{CO_2} + V_{H_2})/(V_{H_2O} + V_{CO}) = \exp[(29,490 - 26.8 \times T_{gas})/(8.314 \times T_{gas})] \tag{14}$$

where, V_{CO}, V_{H2O}, V_{CO2}, V_{H2}, V_{CH4}, and V_{N2} are the volume of CO, H_2O, CO_2, H_2, CH_4, and N_2 in top gas for producing 1 ton hot metal, m³; $w(C)_i$, $w(H)_i$, $w(N)_i$ and $w(C)_{HM}$ are the mass fraction of C, H, N in material i and hot metal, respectively; T_{gas} is the temperature of the raw gas, K.

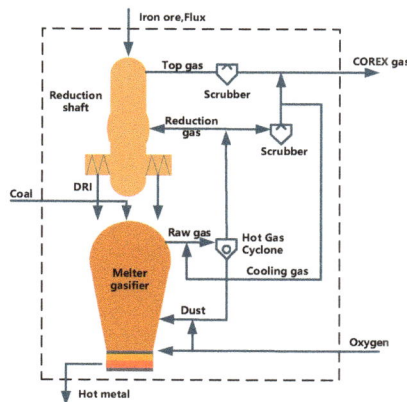

Figure 2. The flow diagram of the COREX® ironmaking process.

Based on the volume, temperature, and composition of the raw gas, the volume of the cooling gas can be calculated by the heat balance (Equation (10)) of the hot gas cyclone. The consumption of coal is obtained by the heat balance (Equation (10)) of the melter-gasifier. The oxygen consumption can be obtained by the oxygen balance of the melter-gasifier. The volume of top gas can be obtained by the mass balance of the shaft furnace and gas oxidation degree. Finally, based on the mass balance of the metallurgical gas, the volume of COREX gas can be obtained. The parameters of the COREX process calculated by the COREX static process model are shown in Tables S1–S4, S6, and S9 (Supplementary Materials).

2.2. Integrated Steel Plant

As shown in Figure 3, the ISP consists of the steel manufacturing system and the energy system. The steel manufacturing system is comprised of the raw material handling process, ironmaking process, steelmaking process, and steel rolling process. In the steel manufacturing system, the iron ore is made into steel production with the help of energy inputs. At the same time, the generated metallurgical gases (including coke oven gas, blast furnace gas, and Linz-Donawitz gas), generated electricity, generated oxygen, and steam are controlled by the energy system. As a buffer, the power plant and oxygen plant can convert the surplus metallurgical gas from the steelmaking system into electricity and oxygen. For the BF and TGR-OBF steelmaking routes, the raw material handling processes include coking, sintering, and pelletizing, while the COREX steel production route does not. The analysis scale of an ISP was based on 1 t crude steel. In the analysis of a conventional ISP, the conventional ISP of Jingtang (which is located in the north of China) was taken as a reference case. By calculating the static process models of the TGR-OBF process and COREX process, the input and output parameters of per ton of hot metal in the two ironmaking processes can be obtained. On the basis of the hot metal consumption per ton of crude steel in the jingtang steel plant, combining the two static process models with the operation data of Jingtang steel plant (without BF process), an ISP with the TGR-OBF and an ISP with the COREX can be analyzed. In the analysis of the three ISPs, the following assumptions were made in this study:

1. In the calculation of the TGR-OBF and COREX process models, the temperature and composition of hot metal, the chemical composition of raw materials were considered to be the same as that of the BF process in the Jingtang steel plant, and it was feasible for the two ironmaking processes. For the TGR-OBF process, the raw materials with the same composition as the BF were flux, coke, coal, sinter, and pellets. For the COREX process, the raw materials with the same composition as the BF were flux and coal.

2. For a better comparison of the three ironmaking technology in the ISP, the static process model of the TGR-OBF established in this paper was based on the current feasible technology, and the CCS technology was not considered in this study.

3. In the analysis of an ISP with the TGR-OBF and an ISP with the COREX, the operation data for the raw materials handling process, steelmaking process, rolling process, lime making process, oxygen plant, and power plant were all from the Jingtang steel plant. The input and output values of the raw materials handling process were proportional to the raw material consumption of the TGR-OBF process and COREX process, respectively.

4. In an ISP, the metallurgical gas was preferentially supplied to the steel manufacturing system for use. The surplus gas was transported to the power plant, and the generating efficiency was 33%.

5. The blast furnace blower was driven by electricity and the power consumption was 0.103 kWh/m³.

6. Since the calculation of TRT power generation is very complex. In this study, the generating capacity was simplified to be only related to the flow of top gas, it was 0.028 kWh/m³.

7. The electricity consumption of cryogenic oxygen process is 0.86 kWh/m³-oxygen.

Therefore, based on the same composition and temperature of the hot metal, without changing the raw materials composition, the comparison and analysis of the three steel production routes can be obtained.

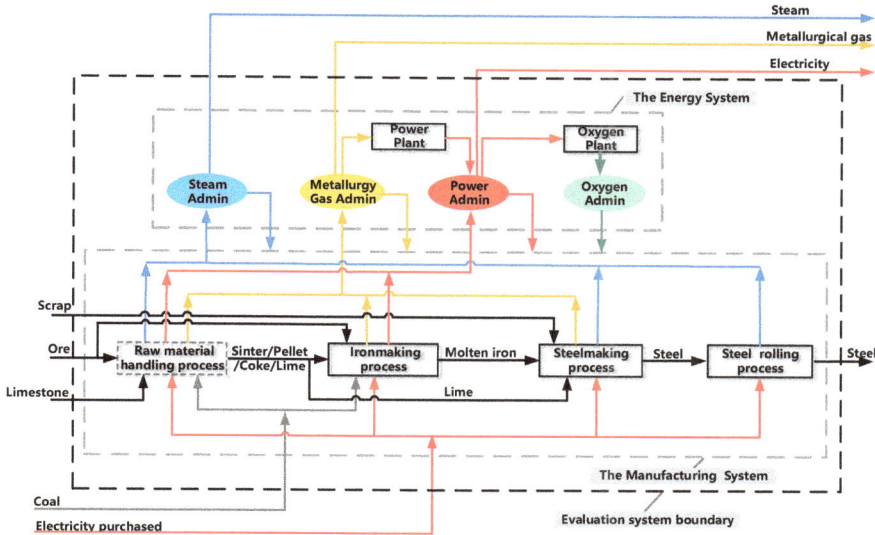

Figure 3. The integrated steel plant (ISP) in this study.

2.3. Methods to Calculate Comprehensive Energy Consumption

The calculation of comprehensive energy consumption of per ton steel is based on the energy balance of an ISP by using the energy equivalent value of each energy substance [34]. In a conventional ISP or an ISP with TGR-OBF, the energy substances input into the system mainly include injection coal, coking coal, and electricity, and the outputs mainly include electricity, chemical by-products, and steam. In an ISP with COREX, the energy substance input into the system mainly includes coal and electricity, and the output mainly includes electricity and steam. The equation of comprehensive energy consumption of per ton steel is as follows:

$$E = \sum_{i=1}^{m} \left(e_i^{in} \times p_i \right) - \sum_{i=1}^{n} \left(e_i^{out} \times p_i \right) \tag{15}$$

where e_i^{in}, e_i^{out} are the input and output energy substances i (kg/t crude steel); p_i is the energy conversion coefficient of substance i (GJ/kg); and m, n are the number of energy substances.

2.4. Methods to Calculate CO$_2$ Emissions

As proposed in the WRI & WBCSD (World Resources Institute & World Business Council for Sustainable Development) guidelines, the net CO$_2$ emissions are calculated by the direct CO$_2$ emissions, the indirect CO$_2$ emissions and the CO$_2$ credit [35]. In an ISP, the combustion of fossil fuel, metallurgical gas, and the decomposition of carbonate contribute to the direct CO$_2$ emissions, the purchased electricity and steam contribute to the indirect CO$_2$ emissions, and the excess steam, electricity, and chemical by-products can be considered as CO$_2$ credit. The net CO$_2$ emissions can be calculated by the following equations,

$$CE = CE_d + CE_i - CE_c \tag{16}$$

$$CE_d = (C_{in} - C_p - C_{byp})/12 \times 44 \qquad (17)$$

where CE, CE_d, CE_i, CE_c represent the net, direct, indirect, and credit of the CO_2 emissions, t/t crude steel, respectively; and C_{in}, C_p, C_{byp} represent the amount of total carbon input to the system, the amount of carbon fixed in the product, and the amount of carbon fixed in the byproduct, t/t crude steel.

3. Results and Discussion

3.1. Material Flows Analysis

Based on the operating data of the Jingtang steel plant, the material flows of a conventional ISP were investigated and presented in Figure 4. For a better comparison with a conventional ISP, the material flows of an ISP with the TGR-OBF and an ISP with the COREX were studied by combining the operating data in the Jingtang steel plant with established static process models, as shown in Figures 5 and 6. The parameters of the TGR-OBF process and the COREX process analyzed in the ISP are shown in Figures S1 and S2 (Supplementary Materials). As shown in Figure 4, a large amount of coking coal was consumed by a conventional ISP with 487.4 kg/t-crude steel, and the consumptions of coal and oxygen were 134.3 kg/t-crude steel and 96.9 m³/t-crude steel, respectively. In a conventional ISP, the total production of blast furnace gas (BFG) and coke oven gas (COG) were 1336.3 m³/t-crude steel and 166.8 m³/t-crude steel. Most were used to maintain production for each main process, and the surplus BFG and COG were transferred to the power plant for power supply and network pressure maintenance. As Figures 5 and 6 demonstrate, the material flows of an ISP with the TGR-OBF and an ISP with the COREX changed significantly in comparison with a conventional ISP. Compared with a conventional ISP, the coking coal consumption in the ISP with the TGR-OBF reduced by 39.7%. Since coking coal is not used in the production process of the COREX, the amount of coking coal was zero. However, the coal consumption in the ISP with the TGR-OBF and the ISP with the COREX increased by 23.6% and 535.2%, respectively. This is because the top gas recycling in the TGR-OBF reduced the coke consumption and oxygen blast increased the coal consumption. In the ISP with the COREX, there was no coking process and the consumption of coal and oxygen greatly increased in the melter-gasifier. At the same time, oxygen consumption increased by 174.7% and 521.5% in the TGR-OBF and COREX, respectively. In addition, due to the reduction in coke consumption and top gas recycling, the metallurgical gas produced in the ISP with the TGR-OBF was significantly reduced in comparison with the conventional ISP, i.e., the BFG and COG were reduced by 84% and 39.7%, respectively. In this steel production route, almost no surplus metallurgical gas was transported to the power plant for power generation. However, in the ISP with the COREX, a large amount of coal was used to make iron, so the generated COREX gas was very large, which was 23.9% more than the BFG process.

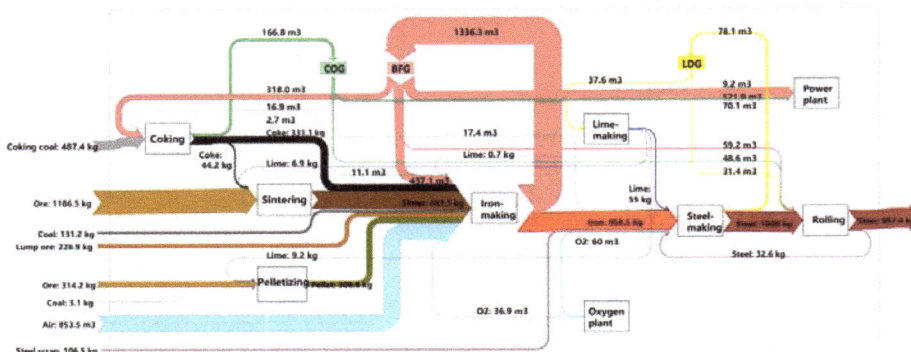

Figure 4. Chart of the material flows of a conventional ISP.

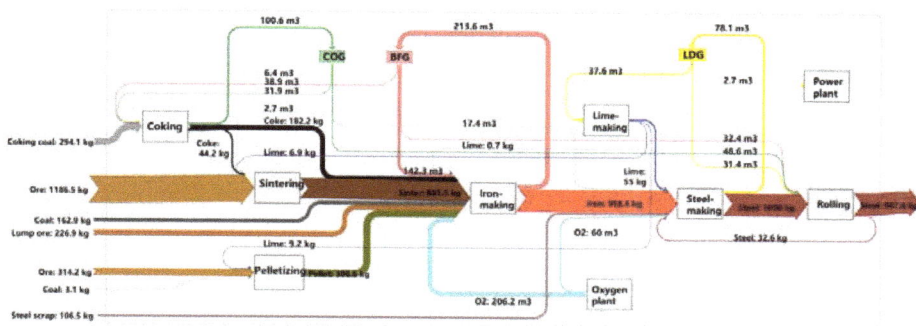

Figure 5. Chart of the material flows of an ISP with the TGR-OBF.

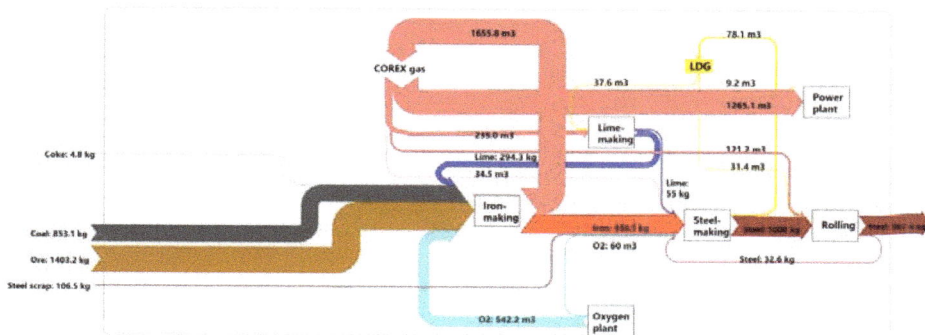

Figure 6. Chart of the material flows of an ISP with the COREX.

3.2. Metallurgical Gas Analysis

Figure 7 illustrates the generation and consumption of metallurgical gas by each process in a conventional ISP, an ISP with the TGR-OBF, and an ISP with the COREX. The order of metallurgical gas generation in the three ISPs is: ISP with the COREX (12.3 GJ/t-steel) > conventional ISP (7.8 GJ/t-steel) > ISP with the TGR-OBF (3.6 GJ/t-steel). This is because the oxygen consumption and coal consumption increased to 602.2 m³/t-crude steel and 853.1 kg/t-crude steel in an ISP with the COREX. The ISP with the TGR-OBF generated the least amount of metallurgical gas because of the use of recycled top gas. In addition, the ironmaking process of the ISP with the COREX contributed most of the metallurgical gas with 11.6 GJ/t-steel. This is because the ISP with COREX does not have a coking process, and a large amount of oxygen and coal react in the melter-gasifier to produce a large amount of high heat value COREX gas. In the coking process, the amount of COG in the conventional ISP and the ISP with the OBF were different. Compared with a conventional ISP (2.6 GJ/t-steel), the ISP with the TGR-OBF generated less COG with 1.6 GJ/t-steel. This is because the coke usage dropped to 294.1 kg/t-steel in the ironmaking process. The processes of metallurgical gas consumption included the power plant, ironmaking, coking, rolling, lime making, and steelmaking. In the ISP with COREX, 9.0 GJ/t-steel metallurgical gas was applied to the power plant, while only 2.9 GJ/t-steel and 0.027 GJ/t-steel were applied to power plants in a conventional ISP and an ISP with TGR-OBF.

Since an ISP with COREX does not include the coking process, sintering process, and pelletizing process, the amount of gas used in the steel manufacturing system is low and the amount of gas generated is large. Most of the surplus gas has to be sent to power plants for power generation.

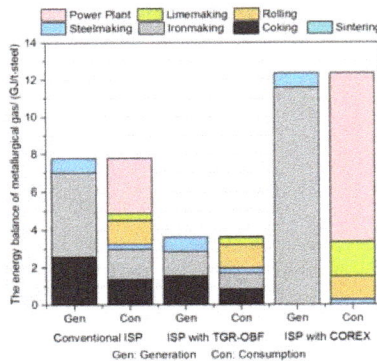

Figure 7. The metallurgical gas balance in the conventional ISP, ISP with the TGR-OBF, and ISP with the COREX.

3.3. Electricity Analysis

Figure 8 presents the generation and consumption of electricity by each process in a conventional ISP, an ISP with the TGR-OBF, and an ISP with the COREX. The electricity generated by the power plant was from the surplus metallurgical gas. Among the three steel production routes, the generation of electricity in an ISP with the COREX was the highest, followed by the conventional ISP, and the ISP with the TGR-OBF, which were 823.9 kWh/t-crude steel, 305.2 kWh/t-crude steel, and 35.4 kWh/t-crude steel, respectively. This is mainly because there is a large amount of surplus high heating value of metallurgical gas in an ISP with the COREX. The top gas recycling in an ISP with the TGR-OBF reduced the metallurgical gas production of the ironmaking and coking process, so almost no surplus metallurgical gas was transported to the power plant. Furthermore, the consumption of electricity in the ISP with the COREX was also the highest, followed by an ISP with the TGR-OBF, and the conventional ISP was the lowest. The increased demand for oxygen from the COREX and TGR-OBF was responsible for the significant increase in electricity consumption in the oxygen plant. As a result, the electricity consumed by the oxygen plant increased by 521.6% and 174.7%, respectively in comparison with the conventional ISP. Nevertheless, the ironmaking process of the ISP with the TGR-OBF consumed the most electricity in comparison with that of the other two steel production routes because the CO_2 removal unit in the TGR-OBF process raises the demand for electricity. When compared with a conventional ISP, the electricity consumption in the ironmaking process of the ISP with the TGR-OBF increased by 89%. Therefore, as presented in Figure 8, the ISP with TGR-OBF purchased the most electricity, followed by the conventional ISP, and the ISP with COREX purchased the least, which were 710.1 kWh/t-crude, 212.4 kWh/t-crude, and 77.3 kWh/t-crude, respectively.

3.4. Comprehensive Energy Consumption Analysis

The comprehensive energy consumptions of the conventional ISP, the ISP with the TGR-OBF, and the ISP with the COREX are shown in Figure 9. Among these three steel production routes, the ISP with the TGR-BOF consumed the least energy to produce 1 t of steel. The energy consumption was 14.16 GJ/t-steel, 16% and 16.5% lower than the conventional ISP and the ISP with the COREX, respectively. This is because the top gas recycling in the TGR-OBF ironmaking process can significantly reduce carbon consumption; despite requiring the most electricity to be purchased, the energy conversion coefficient of electricity is smaller than that of coal. Therefore, the comprehensive energy consumption in the ISP with the TGR-OBF is the lowest. Moreover, the energy consumed by the conventional ISP and the ISP with the COREX were 16.85 GJ/t-steel and 16.96 GJ/t-steel, respectively. However, the coking coal consumption accounted for 80.2% of the total energy input in the conventional ISP, which was the most out of the three steel production routes.

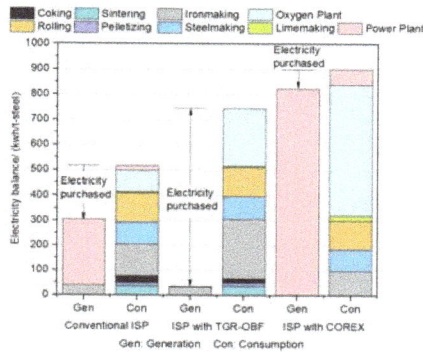

Figure 8. The electricity balance in the conventional ISP, the ISP with the TGR-OBF, and the ISP with the COREX.

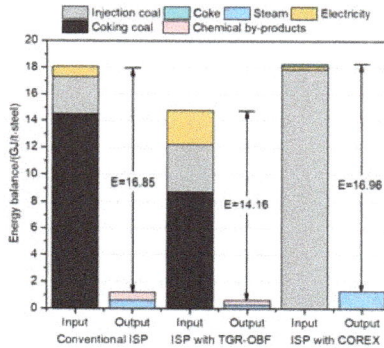

Figure 9. The energy balance in the conventional ISP, the ISP with the TGR-OBF, and the ISP with the COREX.

3.5. CO_2 Emissions Analysis

Figure 10 shows the net, direct, indirect, and the credit of CO_2 emissions in the three steel production routes with different CO_2 emissions factors. CCS technology was not considered in this research. When considering the impact of the CO_2 emissions factor of electricity, the International Energy Agency factor (0.504) and the China Grid factor (1.0302) were used to analyze the CO_2 emissions. In the conventional ISP, the ISP with the TGR-OBF, and the ISP with the COREX, the direct CO_2 emissions accounted for the largest proportion of the total CO_2 emissions. With the International Energy Agency factor, the direct CO_2 emissions accounted for 93.2%, 79.8%, and 97.9% of the total CO_2 emissions in the three routes, respectively. With the China Grid factor, the direct CO_2 emissions accounted for 88.5%, 66.0% and 96.5% of the total CO_2 emissions of the three routes. The net CO_2 emissions in the conventional ISP, the ISP with the TGR-OBF, and the ISP with the COREX increased from 1.94 t/t-steel, 1.73 t/t-steel, and 2.66 t/t-steel (International Energy Agency factor) to 2.05 t/t-steel, 2.11 t/t-steel, and 2.70 t/t-steel (China Grid factor), respectively. The increases were 5.7%, 22%, and 3.8%, respectively. This is because the indirect CO_2 emissions can be influenced by the CO_2 emissions factor. Furthermore, the more the electricity purchased by an ISP, the greater the impact of the CO_2 emissions factor. Therefore, if the International Energy Agency factor was used, the net CO_2 emissions of the ISP with the TGR-OBF was the lowest, followed by the conventional ISP, and the ISP with the COREX. However, if the China Grid factor was used, the net CO_2 emissions of the ISP with the TGR-OBF was not the lowest. That is to say, the net CO_2 emissions of the ISP with the TGR-OBF rely heavily on the electricity CO_2 emissions factor. When compared with the conventional ISP and the

ISP with the COREX, the ISP with the TGR-OBF had the lowest direct CO_2 emissions and the highest indirect CO_2 emissions. This is due to its lowest coal consumption and highest electricity consumption when compared with the other routes. The ISP with the COREX had the highest net CO_2 emissions for consuming a large amount of coal, which was 37.1% and 53.8% higher (International Energy Agency factor) as well as 31.7% and 28.0% higher (China Grid factor) than that of the conventional ISP and the ISP with the TGR-OBF, respectively.

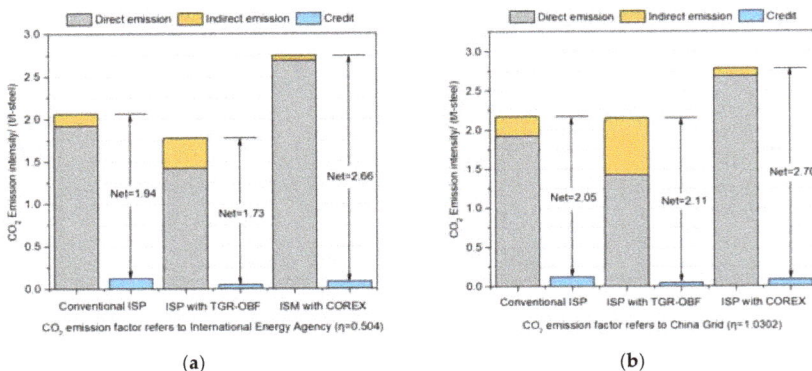

Figure 10. The CO_2 emissions intensity in the conventional ISP, the ISP with the TGR-OBF, and the ISP with the COREX with the CO_2 emissions factor referring to the world average (a) or North China Grid (b).

4. Conclusions

Based on two static process models, the material flows of the TGR-OBF and COREX ironmaking processes were investigated. Combining the operating data of the Jingtang steel plant, three steel production routes of the ISP with the BF, TGR-OBF, and COREX were analyzed comparatively on the material flows, metallurgical gas generation and consumption, electricity consumption and generation, comprehensive energy consumption, and CO_2 emissions. Compared with the conventional ISP, coking coal consumption in the ISP with the TGR-OBF and the ISP with the COREX were reduced by 39.7% and 100%. The coal consumption in the ISP with the TGR-OBF and the ISP with the COREX increased by 23.6% and 535.2%. Among the three routes, the electricity purchased for the ISP with the TGR-OBF was the highest with 710.1 kWh/t-steel and the ISP with the COREX was the lowest with 77.3 kWh/t-steel. The energy consumption of the ISP with the TGR-OBF was 16% and 16.5% lower than that of the conventional ISP and the ISP with the COREX, respectively. The energy consumption of the ISP with the COREX was the highest with 16.96 GJ/t-steel. The ISP with the COREX had the highest net CO_2 emissions, which was 37.1% and 53.8% higher (International Energy Agency factor) as well as 31.7% and 28.0% higher (China Grid factor) than that of the conventional ISP and the ISP with TGR-OBF. Although the ISP with COREX had no advantages in energy consumption and carbon emissions, it had the least purchased electricity and coke consumption. In addition, the ISP with TGR-OBF had the lowest energy consumption. Therefore, the ISP with the COREX is suitable for the areas with high electricity prices and where the coke is not available. Moreover, the ISP with the TGR-OBF should be carefully considered when the CCS technology is not applied.

Supplementary Materials: The following are available online at http://www.mdpi.com/2075-4701/9/3/364/s1, Table S1: The chemical composition of ore in the TGR-OBF process and the COREX process. Table S2: The chemical composition of flux in the TGR-OBF process and the COREX process. Table S3: The chemical composition of coal and coke in the TGR-OBF process and the COREX process. Table S4: The chemical composition of hot metal in the TGR-OBF process and the COREX process. Table S5: The chemical composition of oxygen blast furnace gas in the TGR-OBF process. Table S6: The distribution ratio of elements in the slag and hot metal (the TGR-OBF process & the COREX process) Table S7: The operating parameters of the TGR-OBF process. Table S8: The chemical composition of COREX gas in the COREX process. Table S9: The operating parameters of the COREX process. Figure S1: The parameters of the TGR-OBF process. Figure S2: The parameters of the COREX process.

Author Contributions: Data curation, J.S.; funding acquisition, Z.J. and A.X.; software, C.B. and J.S.; methodology, J.S. and Z.J.; writing—original draft preparation, J.S.; writing—review and editing, Z.J. and J.S. All authors read and approved the manuscript.

Funding: This research was supported by the National Key Research and Development Program of China (2016YFB0601301).

Acknowledgments: Funding from the National Key Research and Development Program of China (2016YFB0601301) is gratefully acknowledged.

Conflicts of Interest: The authors declare no conflict of interest.

Nomenclature

m	mass of substance (kg)
e_i^{in}, e_i^{out}	the input and output energy substances (kg/t crude steel)
CE, CE_d, CE_i, CE_c	the net, direct, indirect, and credit of the CO_2 emissions (t/t crude steel)
C_{in}, C_p, C_{byp}	the amount of total carbon input to the system, the amount of carbon fixed in the product, the amount of carbon fixed in the byproduct (t/t crude steel)
h	Specific enthalpy of chemical reaction (kJ/kg)
H	enthalpy value (kJ)
K	equilibrium constant
OD	oxidation degree
p_i	energy conversion coefficient for substance i (GJ/kg)
q	sensible heat (kJ/kg)
R	basicity
T	temperature (K)
v	gas volume fraction
V	gas volume (m^3)
ω	mass fraction

References

1. World Steel Association. *World Steel in Figures 2018 Now Available*; World Steel Association: Brussels, Belgium, 2018.
2. Chen, W.; Yin, X.; Ma, D. A bottom-up analysis of China's iron and steel industrial energy consumption and CO_2 emissions. *Appl. Energy* **2014**, *136*, 1174–1183. [CrossRef]
3. Shen, X.; Chen, L.; Xia, S.; Xie, Z.; Qin, X. Burdening proportion and new energy-saving technologies analysis and optimization for iron and steel production system. *J. Cleaner Prod.* **2018**, *172*, 2153–2166. [CrossRef]
4. Smith, M.P. Blast furnace ironmaking—A view on future developments. *Procedia Eng.* **2017**, *174*, 19–28. [CrossRef]
5. Wenzel, W. Hochofenbetrieb mit gasformigen Hilfsreduktionsmitteln. Germany Patent 2030468, 1970.
6. Fink, F. Suspension smelting reduction—A new method of hot iron production. *Steel Times (UK)* **1996**, *224*, 398–399.
7. Lu, W.; Kumar, R.V. The feasibility of nitrogen free blast furnace operation. *ISS Trans.* **1984**, *5*, 25.
8. Ohno, Y.; Hotta, H.; Matsuura, M.; Mitsufuji, H.; Saito, H. Development of oxygen blast furnace process with preheating gas injection into upper shaft. *Tetsu-to-Hagané* **1989**, *75*, 1278–1285. [CrossRef]
9. Pukhov, A. Introduction of blast furnace technology involving injection of hot reducing gases. *Publ. Steel USSR* **1991**, *21*, 333–338.
10. Qin, M.; Gao, K.; Wang, G. Study on operation of blast furnace full oxygen blast. *Iron Steel* **1987**, *22*, 1–7. (In Chinese)
11. Danloy, G.; Berthelemot, A.; Grant, M.; Borlée, J.; Sert, D.; Van der Stel, J.; Jak, H.; Dimastromatteo, V.; Hallin, M.; Eklund, N.; et al. ULCOS-Pilot testing of the low-CO_2 blast furnace process at the experimental BF in Luleå. *Rev. Mét. Inter. J. Metall.* **2009**, *106*, 1–8. [CrossRef]
12. Yamaoka, H.; Kamei, Y. Theoretical study on an oxygen blast furnace using mathematical simulation model. *ISIJ inter.* **1992**, *32*, 701–708. [CrossRef]

13. Yamaoka, H.; Kamei, Y. Experimental study on an oxygen blast furnace process using a small test plant. *ISIJ Inter.* **1992**, *32*, 709–715. [CrossRef]

14. Babich, A.I.; Gudenau, H.W.; Mavrommatis, K.T.; Froehling, C.; Formoso, A.; Cores, A.; Garcia, L. Choice of technological regimes of a blast furnace operation with injection of hot reducing gases. *Revista De Metalurgia* **2002**, *38*, 288–305. [CrossRef]

15. Jin, P.; Jiang, Z.; Bao, C.; Lu, Y.; Zhang, J.; Zhang, X. Mathematical modeling of the energy consumption and carbon emission for the oxygen blast furnace with top. *Gas Steel Res. Int.* **2016**, *87*, 320–329. [CrossRef]

16. Mitra, T.; Helle, M.; Pettersson, F.; Saxén, H.; Chakraborti, N. Multiobjective optimization of top gas recycling conditions in the blast furnace by genetic algorithms. *Mater. Manuf. Processes* **2011**, *26*, 475–480. [CrossRef]

17. Liu, L.; Jiang, Z.; Zhang, X.; Lu, Y.; He, J.; Wang, J.; Zhang, X. Effects of top gas recycling on in-furnace status, productivity, and energy consumption of oxygen blast furnace. *Energy* **2018**, *163*, 144–150. [CrossRef]

18. Zhang, W.; Zhang, J.; Xue, Z. Exergy analyses of the oxygen blast furnace with top gas recycling process. *Energy* **2017**, *121*, 135–146. [CrossRef]

19. Zhu, K. Analysis of COREX process as a new ironmaking technology. *Fuel Energy Abstr.* **1995**, *4*, 284.

20. Guo, Z.C.; Fu, Z.X. Current situation of energy consumption and measures taken for energy saving in the iron and steel industry in China. *Energy* **2010**, *35*, 4356–4360. [CrossRef]

21. Eberle, A.; Siuka, D.; Bohm, C. New Corex C-3000 plant for Baosteel and status of the Corex technology. *Stahl Und Eisen* **2006**, *126*, 31–+.

22. Hasanbeigi, A.; Arens, M.; Price, L. Alternative emerging ironmaking technologies for energy-efficiency and carbon dioxide emissions reduction: A technical review. *Renew. Sustain. Energy Rev.* **2014**, *33*, 645–658. [CrossRef]

23. Zhan, W.; Wu, K.; He, Z.; Liu, Q.; Wu, X. Estimation of energy consumption in COREX process using a modified rist operating diagram. *J. Iron. Steel Res. Int.* **2015**, *22*, 1078–1084. [CrossRef]

24. Hu, C.; Han, X.; Li, Z.; Zhang, C. Comparison of CO_2 emission between COREX and blast furnace iron-making system. *J. Environ. Sci.* **2009**, *21*, S116–S120. [CrossRef]

25. She, X.; An, X.; Wang, J.; Xue, Q.; Kong, L. Numerical analysis of carbon saving potential in a top gas recycling oxygen blast furnace. *J. Iron. Steel Res. Int.* **2017**, *24*, 608–616. [CrossRef]

26. Hooey, L.; Tobiesen, A.; Johns, J.; Santos, S. Techno-economic study of an integrated steelworks equipped with oxygen blast furnace and CO_2 capture. *Energy Procedia* **2013**, *37*, 7139–7151. [CrossRef]

27. He, H.; Guan, H.; Zhu, X.; Lee, H. Assessment on the energy flow and carbon emissions of integrated steelmaking plants. *Energy Rep.* **2017**, *3*, 29–36. [CrossRef]

28. Jin, P.; Jiang, Z.; Bao, C.; Hao, S.; Zhang, X. The energy consumption and carbon emission of the integrated steel mill with oxygen blast furnace. *Resour. Conserv. Recycl.* **2017**, *117*, 58–65. [CrossRef]

29. Arasto, A.; Tsupari, E.; Kärki, J.; Lilja, J.; Sihvonen, M. Oxygen blast furnace with CO_2 capture and storage at an integrated steel mill—Part I: Technical concept analysis. *Int. J. Greenh. Gas Control* **2014**, *30*, 140–147. [CrossRef]

30. Tsupari, E.; Kärki, J.; Arasto, A.; Lilja, J.; Kinnunen, K.; Sihvonen, M. Oxygen blast furnace with CO2 capture and storage at an integrated steel mill—Part II: Economic feasibility in comparison with conventional blast furnace highlighting sensitivities. *Int. J. Greenhouse Gas Control* **2015**, *32*, 189–196. [CrossRef]

31. Jin, P. Feasibility Investigation on Oxygen Blast Furnace with Top Gas Recycling Based on Multi-427 Level Models. PhD Thesis, University of Science and Technology Beijing, Beijing, China, 2 November 2015.

32. Arasto, A.; Tsupari, E.; Kärki, J.; Pisilä, E.; Sorsamäki, L. Post-combustion capture of CO_2 at an integrated steel mill—Part I: Technical concept analysis. *Int. J. Greenhouse Gas Control* **2013**, *16*, 271–277. [CrossRef]

33. Liu, Z.; Grande, C.; Li, P.; Yu, J.; Eodrigues, A. Multi-bed vacuum pressureswing adsorption for carbon dioxide capture from flue gas. *Sep. Purif. Technol.* **2011**, *81*, 307–317. [CrossRef]

34. Standardization Administration of China. *General Principles for Calculation of the Comprehensive Energy Consumption*; Standardization Administration of China: Beijing, China, 2008.

35. World Resources Institute & World Business Council for Sustainable Development. *The Greenhouse Gas Protocol*; WRI: Washington, DC, USA, 2013.

metals

MDPI

Article

Occupational Exposure to Fine Particles and Ultrafine Particles in a Steelmaking Foundry

Gabriele Marcias [1,2,*], Jacopo Fostinelli [3], Andrea Maurizio Sanna [1], Michele Uras [1], Simona Catalani [3], Sergio Pili [1], Daniele Fabbri [1], Ilaria Pilia [1], Federico Meloni [1], Luigi Isaia Lecca [1], Egidio Madeo [3], Giorgio Massacci [2], Luca Stabile [4], Ernesto D'Aloja [1], Giorgio Buonanno [4,5,6], Giuseppe De Palma [3] and Marcello Campagna [1]

[1] Department of Medical Sciences and Public Health, University of Cagliari, 09042 Monserrato, Italy; andrea.sanna18@gmail.com (A.M.S.); michele_uras@hotmail.com (M.U.); serginho.pili@gmail.com (S.P.); daniele.fabbri@hotmail.it (D.F.); drssa.pilia@gmail.com (I.P.); federicomeloni@hotmail.it (F.M.); isaialecca@gmail.com (L.I.L.); ernestodaloja@gmail.com (E.D.); mam.campagna@gmail.com (M.C.)
[2] Department of Civil and Environmental Engineering and Architecture, University of Cagliari, 09123 Cagliari, Italy; massacci@unica.it
[3] Department of Medical and Surgical Specialties, Radiological Sciences, and Public Health, University of Brescia, 25123 Brescia, Italy; j.fostinelli@unibs.it (J.F.); simona.catalani@unibs.it (S.C.); madeoegidio@gmail.com (E.M.); giuseppe.depalma@unibs.it (G.D.P.)
[4] Department of Civil and Mechanical Engineering, University of Cassino and Southern Lazio, I-03043 Cassino, Italy; l.stabile@unicas.it (L.S.); giorgio.buonanno@uniparthenope.it (G.B.)
[5] International Laboratory for Air Quality and Health, Queensland University of Technology (QUT), 4001 Brisbane, Australia
[6] Department of Engineering, University of Naples "Parthenope", 80133 Naples, Italy
* Correspondence: gabriele.marcias@libero.it; Tel.: +39-070-6754-435

Received: 30 December 2018; Accepted: 28 January 2019; Published: 1 February 2019

Abstract: Several studies have shown an increased mortality rate for different types of tumors, respiratory disease and cardiovascular morbidity associated with foundry work. Airborne particles were investigated in a steelmaking foundry using an electric low-pressure impactor (ELPI+™), a Philips Aerasense Nanotracer and traditional sampling equipment. Determination of metallic elements in the collected particles was carried out by inductively coupled plasma mass spectrometry. The median of ultrafine particle (UFP) concentration was between 4.91×10^3 and 2.33×10^5 part/cm^3 (max. 9.48×10^6 part/cm^3). Background levels ranged from 1.97×10^4 to 3.83×10^4 part/cm^3. Alveolar and deposited tracheobronchial surface area doses ranged from 1.3×10^2 to 8.7×10^3 mm^2, and 2.6×10^1 to 1.3×10^3 mm^2, respectively. Resulting inhalable and respirable fraction and metallic elements were below limit values set by Italian legislation. A variable concentration of metallic elements was detected in the different fractions of UFPs in relation to the sampling site, the emission source and the size range. This data could be useful in order to increase the knowledge about occupational exposure to fine and ultrafine particles and to design studies aimed to investigate early biological effects associated with the exposure to particulate matter in the foundry industries.

Keywords: ultrafine particles exposure; steelmaking factory; chemical composition

1. Introduction

The exposure to contaminants generated by iron and steel melting processes has been included in the monograph of the International Agency for Research on Cancer (IARC) as a Group 1 human carcinogen [1]. Several studies have shown an increased mortality rate for different types of tumors, respiratory disease and cardiovascular morbidity associated with foundry work [2–6]. Foundry workers, during the processing stages, could be exposed to a multitude of breathable dust types and aerosols, such as metal fumes, polycyclic aromatic hydrocarbons (PAH), mineral powders, resins and isocyanates [7]. Among the several toxic and carcinogenic substances contained in foundry dust, heavy and transition metal fumes represent a major health concern, as they can induce local inflammation in the lung tissue, lipid peroxidation of cell membranes and oxidative damage to the genome [8,9].

Several studies have shown that different hot processes in the metallurgical industry have the capacity to generate high concentrations of sub-micrometric particles. In particular, important number concentrations of ultrafine particles (UFPs, <100 nm in diameter) were generated as combustion products or in saturated vapors [10–18]. UFPs may have more pronounced toxic effects than larger particles, due to their larger surface area to unit mass ratio, which determines their peculiar physicochemical properties and increased biological activity [19–23]. Recently, some studies have shown an association between ultrafine particulate exposure and health effects on the cardiovascular and respiratory tract [24–26], however, epidemiological evidence on UFP-related adverse health effects is still limited and subject to disagreement [27–31].

Some studies have focused on surface-related effects [24,32–34], particle-related effects [25,35–37], mass-related effects [38] or effects related to metallic elements contained in the particulate matter [39–41]; however, the role that the different (size- or non-size-related) components in particulate matter play in determining the adverse health effects observed, and the most appropriate metric (or metrics) for exposure assessment and control, remain unclear [42–44].

Although in recent decades research has increased into UFP exposure in living and working environments [45,46], there is limited evidence of the epidemiological studies about UFP-related adverse health effects, probably attributable to the lack of available data for UFP exposure assessment. Therefore, more knowledge is needed on the different metrics that may be associated with health effects, which may provide data for the realization of job-exposure matrices. The latter are indispensable for designing epidemiological studies aimed at investigating the health effects of the airborne dispersed particulate matter and of the various components that make it up. The main objective of this study was to assess the occupational exposure to fine and ultrafine particles in a steelmaking factory, with a multi-metric and multi-instrumental approach, in order to increase knowledge about sources of fine and ultrafine particles and possible health implications.

2. Materials and Methods

2.1. Sampling Site and Study Design

Sampling was performed in a foundry that uses the "Mini Mills" electric arc furnace technology (EAF) for the treatment of molten steel in the ladle and subsequent continuous casting line for the production of steel billets intended for feeding the rolling plant. Iron scrap is used as raw material for feeding the furnace. The factory produces steel of different qualities and diameter intended for concrete reinforcing in the construction industry.

The exposure assessment strategy was mainly based on a previous study conducted in the same working environment for testing assessment of fine and ultrafine particle emissions [47]. Furthermore, the deposited particle surface area per unit volume of inhaled air in some regions of the respiratory tract (particularly in the tracheobronchial and alveolar regions) was assessed.

The basic strategy combined with additional monitoring equipment to obtain additional information is described below. The monitoring strategy (for six days in the summer season) consisted of stationary, quasi-personal and personal samples in 16 different work environments during standard working conditions. The sampling time varied according to work activities. For logistical reasons, it was not possible to use all the sampling equipment at all sampling sites at the same time. The sampling sites were identified as the areas where worker exposure could be more relevant. The quasi-personal samplings were carried out in the welding laboratory at approximately 30 cm from the worker's breathing zone. In addition, where stationary sampling was not feasible, personal samplings were carried out close to the worker's breathing zone. Table 1 summarizes all the sampling methods, sampling sites, sampling equipment, and sampling times.

Table 1. Summary of sampling methods carried out in steelmaking foundry.

Sampling Site	Equipment	Sampling Methods	Sampling Time	Sampling Site	Equipment	Sampling Methods	Sampling Time
BG	IF and RF	Stationary	6 h 26 min	W2	IF and RF	ND	ND
	ELPI+	Stationary	6 h 4 min		ELPI+	Quasi-personal	1 h 18 min
	NT	Stationary	6 h 26 min		NT	Quasi-personal	1 h 18 min
P-EAF	IF and RF	Stationary	5 h 34 min	W3	IF and RF	Quasi-personal	1 h 33 min
	ELPI+	Stationary	5 h 34 min		ELPI+	Quasi-personal	1 h 33 min
	NT	Stationary	1 h 9 min		NT	Quasi-personal	1 h 33 min
LF	IF and RF	Stationary	4 h 14 min	EAF	NT	Stationary	1 h 27 min
	ELPI+	Stationary	4 h 14 min	P-LF	NT	Personal	2 min
	NT	Stationary	4 h 57 min	P-CC	NT	Personal	51 min
CC	IF and RF	Stationary	3 h 51 min	AG	NT	Personal	1 h 41 min
	ELPI+	Stationary	3 h 51 min	BT	NT	Personal	41 min
	NT	Stationary	2 h 21 min	OC1	NT	Personal	25 min
W1	IF and RF	Quasi-personal	1 h 24 min	OC2	NT	Personal	35 min
	ELPI+	Quasi-personal	1 h 24 min	QDW	NT	Personal	25 min
	NT	Quasi-personal	1 h 24 min	SC	NT	Personal	1 h

Abbreviations: IF = inhalable fraction; RF = respirable fraction; ELPI+ = electric low pressure impactor; NT = Philips Aerasense Nanotracer; ND = not detected. Sampling sites are described below in the text.

Figure 1 shows sampling sites inside and outside the plant. Monitoring was carried out in the following areas or workstations:

- outside the plant, to measure general environmental background levels (BG) not influenced by the factory emissions;
- at a distance of 50 m from the electric arc furnace (EAF), 2 m from the ladle furnace (LF), 2 m from the continuous casting (CC), and within the control consoles (respectively P-EAF, P-LF and P-CC);
- in three welding stations (W1, W2, W3), respectively, with CASTOLIN 5006 electric welding on steel (55 electrodes), Nicro HLS on cast iron and electrode welding on knife (special iron) (21 electrodes);
- inside the mechanical workshop (BT), in which various activities were carried out, including the use of an oxide flame and a grinder;
- inside the rolling mill department (AG), welding station with use of angle grinder;
- inside the overhead crane cabin in the finished product department (OC1);
- in the scrap yard and in the overhead crane in the scrap yard (OC2);
- inside the quality department workshop (QDW);
- inside the company canteen during the lunch break (SC).

2.2. Sampling Equipment

The UFP distribution and number concentration were measured using an electric low pressure impactor and a portable particle counter. The electric low pressure impactor, model ELPI+™ (electric low pressure impactor—Dekati Ltd., Kangasala, Finland) allows the measurement of particulate matter at a stationary location. This instrument, through the dimensional selection of airborne particulates, detects in real time the particle diameter (sizes between 6 nm and 10 μm), the concentration and, based on the data collected, provides an estimate of the concentration in surface area/mass/volume of sampled particulates [48]. The ELPI+ was connected to an air intake pump with 0.6 m^3/h flow rate and a pressure of 40 mbar at the final stage of the impactor (absolute filter). The number of ultrafine particles was calculated as the sum of the particles having a central geometric mean diameter (Di) between 10 nm and 314 nm (D50% range 6 nm–257 nm), assuming a density of 1 g/cm^3. Data provided by ELPI+ were processed with the ELPI+ VI 2.0 software (Dekati Ltd., Kangasala, Finland). It was not possible to carry out measurements with the ELPI+ in all workstations investigated due to logistical reasons. From the second to the fifth stage of the ELPI+, the polycarbonate foils not greased were mounted for subsequent chemical analysis of the collected particulate matter, in order to determine the concentration of metals contained in it, by inductively coupled plasma inductivity mass spectrometry (ICP-MS). In this study the substrates were not greased to avoid any potential interference with the chemical analyses [49].

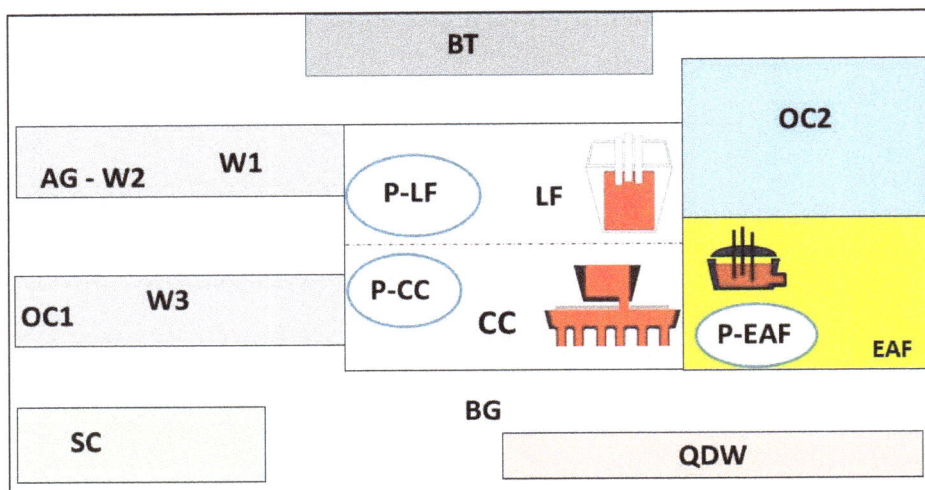

Figure 1. Sampling sites inside and outside the factory.

Personal samplings were carried out using a Philips Aerasense Nanotracer (NT—Koninklijke Philips Electronics N.V., Eindhoven, Netherlands) portable particle counter, which allows the real time measurement of particles number concentration with a diameter between 10 nm and 300 nm. The NT is a portable sampler that measures particle concentration up to 1×10^6 cm^3 in the 10 nm to 300 nm size range for an airflow 0.3–0.4 L/min. The NT design and operation characteristics, as well as sensitivity and limitations, were discussed in detail in a previous study [50]; charging time and battery life is seven hours. The NT was operated in advanced mode, measuring particle concentration and average particle diameter at a fixed sampling interval of 10 s.

The lung-deposited surface area concentration was calculated using data recorded by NT. The NT monitor provides real time information about their concentration, average size, and surface area per unit volume of inhaled air that deposits in the various compartments of the respiratory tract [50]. Marra et al. [50] report that the data are obtained from the International Commission on Radiological Protection (ICRP) Publication 66 [51] with an air volume assumed for normal flow (light exercise) of workers of 1.5 m^3/h. The dose (in terms of deposited alveolar or tracheobronchial surface area particles per mm^2) received by workers in different areas was determined with the means of the particle surface area concentration in the alveolar or tracheobronchial tract (μm^2/cm^3), weighted for a time of six hours exposure. In addition, samples of powders, inhalable and respirable fraction (respectively, IF and RF), were performed, according to the Italian UNI EN 481 standard method [52], by means of samplers with 2 L/min constant flow for the inhalable fraction and 1.7 L/min for the respirable fractions. The airborne inhalable fraction was collected by filtration, using the Institute of Occupational Medicine (IOM) selector (IOM Sampler, SKC Inc., Eighty Four, PA, USA), while a Dorr-Oliver selector was used for the respirable fraction. Both fractions were collected on cellulose ester membranes with a diameter of 25 mm and porosity of 0.8 μm, according to the Unichim 1998:13 and 2010:11 methods [53,54]. The dust analysis was conducted with the microgravimetric method on the conditioned membranes, before and after collection, in the Activa Climatic box (at constant temperature and humidity for 24 h) and weighed with a fifth decimal place electronic analytical balance. The difference in weight, related to the volume of air intake, allowed the calculation of dustiness in mg/m^3. The limit of detection of the method is 0.03 mg and the coefficient of variation is 0.2%.

2.3. Chemical Characterization

Particulate collected through sampling performed by ELPI+ and through the traditional methods (inhalable fraction) was analyzed by ICP-MS for the determination of the metallic elements. The analysis was aimed at determining the following metallic elements: Al, Mn, Co, Ni, Cu, Zn, As, Sr, Mo, Cd, Sn, Sb, Ba, Hg, Pb, Be, Fe and Cr. These particle samples were analyzed by ICP-MS analysis on a Perkin Elmer ELAN DRC II instrument (Perkin Elmer Sciex, Woodbridge, ON, Canada) equipped with dynamic cell reaction (DRC) to analyze chromium and iron. The analytical method and specific technical details have been reported in previous studies [47,55]. The mixed cellulose ester membrane filters and the polycarbonate foil substrates were extracted overnight in a nitric acid (HNO$_3$) American Chemical Society (ACS)Reagent (Purity 90.0%; Sigma, Milan, Italy) 70% (v/v), and the extracted samples were diluted into Ultrapure deionized water (Tracepure$^®$ water for inorganic analysis, Merck, Rome, Italy). The reagent blank was made from blank membranes, acid, and deionized water used for the sampled membranes The limits of detection (LOD) were determined on the basis of three standard deviations (SDs) of the background signal; LOD ranged from 0.0001 µg to 0.0006 µg and the coefficient of variation ranged from 6.5% to 9%. The accuracy of the method was determined on the basis of the mean values obtained on certified reference materials submitted to the same treatment as the samples (trace elements in water National Institute of Standards and Technology (NIST) 1640). Our method of determination of metallic elements in environmental and biological samples is validated and the laboratory participates in the inter-comparison program for toxicological analysis in biological materials (G-EQUAS of the German Society of Occupational and Environmental Medicine). The limit of detection of the laboratories was accredited (ISO 9001:2000 No. 9122 SP 16).

3. Results

3.1. Particle Size Distribution

Figure 2 shows the distribution of particle number concentration measured by ELPI+ in the BG, P-EAF, LF, CC, W1, W2 and W3 samples.

The BG and P-EAF distributions show a modal value at 10 nm. Measurements carried out near the ladle furnace (LF) show a bimodal distribution, with the highest peak centered at 10 nm and a second peak at 41 nm. The distribution measured near the continuous casting shows a modal value at about 71 nm, with an additional peak at 10 nm. The distributions of W2 and W3 show a mode at about 22 nm, while W1 shows a bimodal distribution with two peaks at 41 nm and 314 nm.

3.2. Particle Number Concentration

Figure 3 shows median, interquartile range, minimum and maximum UFP number concentration, measured by ELPI+ and NT.

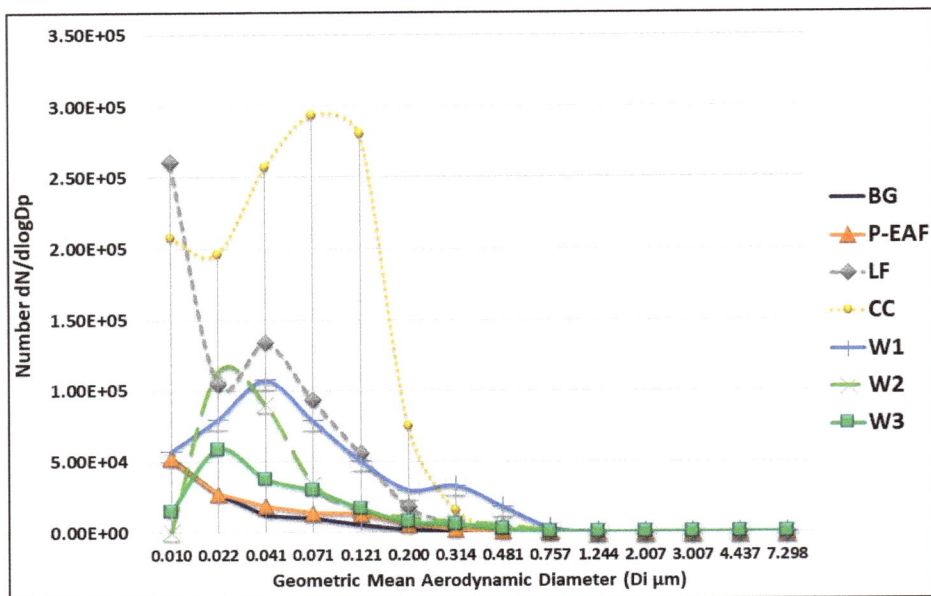

Figure 2. Number distribution measured by ELPI+ in sampling sites: BG, P-EAF, LF, CC, W1, W2 and W3.

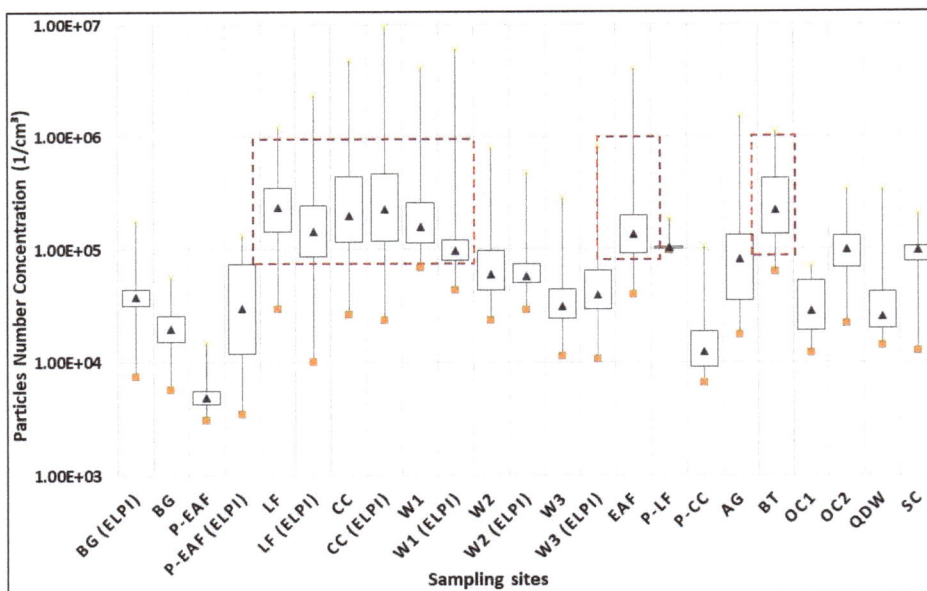

Figure 3. Median, interquartile range, minimum and maximum UFP number concentration measured through stationary, quasi-personal and personal samplings. Red boxes show the highest UFPs median number concentrations measured.

The UFP median of background levels was found to range from 1.97×10^4 to 3.83×10^4 part/cm^3. The median of UFP ranged from 4.91×10^3 to 2.33×10^5 part/cm^3, respectively, inside the EAF control pulpit and next to the ladle furnace sampling site. The maximum concentration was measured in close proximity to the continuous casting line (9.48×10^6 part/cm^3), while in the welding positions the median of UFP ranged between 3.15×10^4 and 1.57×10^5. Finally, the UFP median measured during the lunch break was 9.64×10^4 part/cm^3.

3.3. Particle Surface Area Concentration

Table 2 shows average particle size range, median and mean of particle surface area concentration ($\mu m^2/cm^3$) deposited in the alveolar and tracheobronchial tract.

Table 2. Particle size range (average, nm), particle surface area concentration deposited in alveolar and tracheobronchial tract ($\mu m^2/cm^3$) measured by NT in each sampling site.

Sampling Site	Particle Average Size Range (nm)	Particle Surface Area Concentration ($\mu m^2/cm^3$)			
		Alveolar Tract		Tracheobronchial Tract	
		Mean	Median	Mean	Median
BG	41.3	3.93×10^1	3.91×10^1	7.95×10^0	7.90×10^0
P-EAF	56.5	1.45×10^1	1.30×10^1	2.93×10^0	2.63×10^0
LF	29.87	3.79×10^2	3.01×10^2	7.65×10^1	6.09×10^1
CC	32.68	4.66×10^2	3.08×10^2	9.42×10^1	6.23×10^1
W1	52.39	7.16×10^2	4.28×10^2	1.45×10^2	8.64×10^1
W2	38.17	1.73×10^2	1.06×10^2	3.50×10^1	2.14×10^1
W3	59.19	1.05×10^2	9.10×10^1	2.12×10^1	1.84×10^1
EAF	46.71	4.22×10^2	3.12×10^2	8.53×10^1	6.30×10^1
P-LF	42.83	2.24×10^2	2.08×10^2	4.52×10^1	4.20×10^1
P-CC	47.58	3.52×10^1	2.91×10^1	7.11×10^0	5.88×10^0
AG	34.7	1.49×10^2	1.19×10^2	3.01×10^1	2.41×10^1
BT	61.4	9.69×10^2	4.52×10^2	9.33×10^1	9.13×10^1
OC1	33.04	5.20×10^1	4.79×10^1	1.05×10^1	9.68×10^0
OC2	43	2.13×10^2	2.01×10^2	4.31×10^1	4.07×10^1
QDW	61.4	1.62×10^2	7.53×10^1	3.28×10^1	1.52×10^1
SC	40.69	1.68×10^2	1.73×10^2	3.40×10^1	3.49×10^1

The maximum UFP surface area concentration ($\mu m^2/cm^3$) in the alveolar tract was found in BT, and the maximum UFP surface area concentration ($\mu m^2/cm^3$) in the tracheobronchial tract was found in W1 (mean) and BT (median). The minimum value of UFP surface area concentration was found in P-EAF. Figure 4 shows the estimated doses of UFP surface areas in the alveolar and tracheobronchial tracts for each measuring point. The highest average values of surface area, in terms of dose deposited in the alveolar tract, were measured at BT, W1, CC, EAF and LF, whereas, the highest average surface area values, in terms of dose deposited in the tracheobronchial tract, were measured at W1, CC, BT, EAF and LF.

3.4. Particle Mass Concentration

Table 3 shows the mass concentration (mg/m^3) of inhalable and respirable fraction measured by gravimetric method. The concentrations measured for both fractions collected in the external environment (BG) were found to be below the analytical detection limit. The highest concentrations were measured at welding station 1 (W1), both for the inhalable and the respirable fractions.

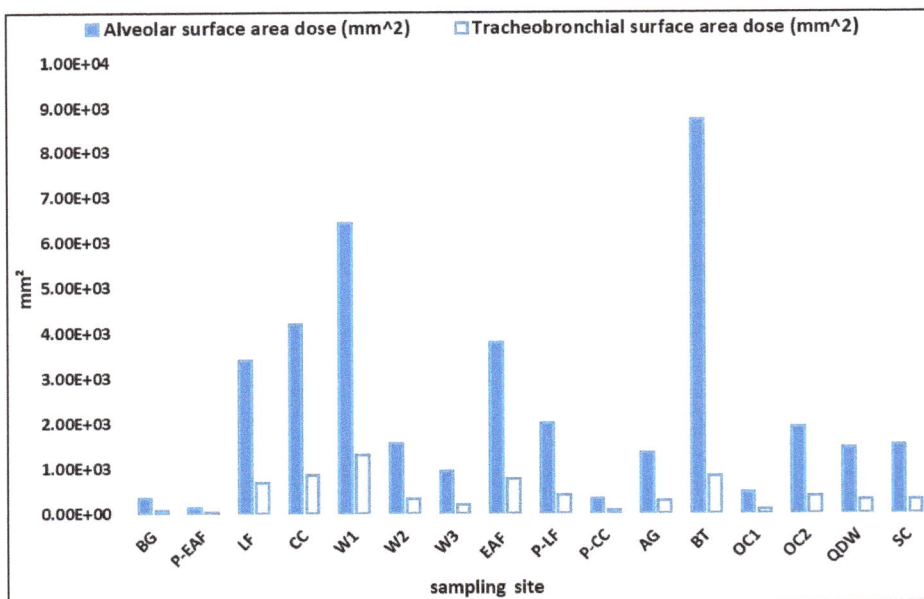

Figure 4. Alveolar deposited surface area dose (mm^2) and the tracheobronchial deposited surface area dose (mm^2). Both values were weighted for a six-hours exposure.

Table 3. Mass concentration (mg/m^3) of the inhalable and respirable fractions measured in BG, P-EAF, LF, CC, W1 and W3 sampling locations.

Sampling Site	BG	P-EAF	LF	CC	W1	W3
Inhalable Fraction	<LOD	0.11	0.7	0.77	1.63	0.5
Respirable Fraction	<LOD	0.11	0.46	0.08	0.92	0.59

3.5. Chemical Composition

Table 4 shows metallic element concentration (µg/m^3) determined in the inhalable infraction sampled at the sampling sites corresponding to BG, P-EAF, LF, CC, W1 and W3. Different concentrations of the analyzed metal elements were observed in relation to the different sampling sites investigated. Overall, the concentrations of the determined metallic elements, for which occupational exposure limits are available, were below the limits set by Italian legislation [56]. The highest levels of Al, As, Ba, Cu, Mo, Pb, Sb and Sn were found in the particles collected near the continuous casting line. The highest levels of Cd and Zn were determined in the particle collected inside P-EAF. The highest levels of Co, Mn and Sr were measured close to LF. The highest levels of Cr and Fe were found in W1 and the highest levels of Ni in W3. Overall, the lowest levels were measured in the background (BG).

Figure 5 shows the concentration of the metal elements analysed in the ultrafine particulate collected by ELPI+. Variable concentrations of the metallic elements were observed in relation to the different sampling sites and to the different granulometric fractions analysed. In particular, Al 38%, Fe 33%, Zn 9%, Ni 5% and Cu 4% were the metallic elements most represented in BG. Fe 38%, Zn 26%, Cu 10%, Mn 8% and Pb 8% were the metallic elements most represented in P-EAF. Fe 44%, Cu 17%, Mn 11% and Zn 10% are the most represented metallic elements in LF. Fe 70%, Cu 14%, Zn 6%, Mn 3%, Pb 3% were the most represented metallic elements in CC. Fe 67%, Cr 20% and Fe 61%, Mo 11%, Cr 10% were the major metallic elements represented in W1 and W3, respectively. Figure 5 shows the concentrations of the metallic elements (in percentage) analysed in the ultrafine particulate collected in BG, P-EAF, LF, CC and in two welding stations (W1 and W3).

Table 4. Concentration in $\mu g/m^3$ of the metallic elements determined in the inhalable fraction.

Metallic Element	Sampling Site					
	BG	P-EAF	LF	CC	W1	W3
Al	0.43	0.46	3.58	6.4	2.59	4.82
As	<LOD	<LOD	0.078	0.136	0.087	<LOD
Ba	0.05	0.034	0.24	0.248	0.07	<LOD
Be	<LOD	<LOD	<LOD	<LOD	<LOD	<LOD
Cd	0.0006	0.007	0.001	0.002	0.001	0.001
Co	0.001	0.002	0.042	0.024	0.031	<LOD
Cr	0.07	0.009	0.21	0.078	38.2	3.63
Cu	0.07	0.13	1.1	2.7	0.97	0.21
Fe	0.01	1.16	15.7	14.11	129.1	2.64
Hg	<LOD	<LOD	<LOD	<LOD	<LOD	<LOD
Mn	0.18	0.63	44.64	16.21	3.94	8.05
Mo	0.009	0.013	0.08	0.157	0.02	<LOD
Ni	0.05	0.053	0.4	0.346	1.5	6.47
Pb	0.09	1.82	0.64	1.83	0.42	0.07
Sb	0.003	0.004	0.02	0.048	0.01	<LOD
Sn	0.01	0.066	0.22	0.288	0.22	0.03
Sr	0.01	0.005	0.08	0.066	0.07	0.02
Zn	0.68	7.89	1.74	6.19	1.04	0.73

Table 5 shows mass concentration (ng/m^3) of the metallic elements determined in the UFPs collected by ELPI+ in BG, P-EAF, LF, CC, W1 and W3.

Overall, the metallic elements determined in LF, CC and P-EAF showed a trend comparable with a greater concentration of metallic elements in the fractions of 71 nm and 121 nm. The metallic elements determined in the particles collected outside the plant did not show a clear trend. Some of these elements are more present in particulates of 22 nm, others in the fraction of 71 nm or 121 nm. The metallic elements determined in the UFPs collected in W1 show a trend similar to that observed in the other sampling sites inside the plant, however an important concentration is present in the size range of 41 nm. In W3, the metallic elements show a less clear trend compared to the elements determined in W1, however, most of the metallic elements are present in the size range between 71 and 121 nm. Figure 6 shows the chemical composition in percent (left) and in ng/m^3 (right) of the metallic elements in the different particle size ranges, for each area and working station.

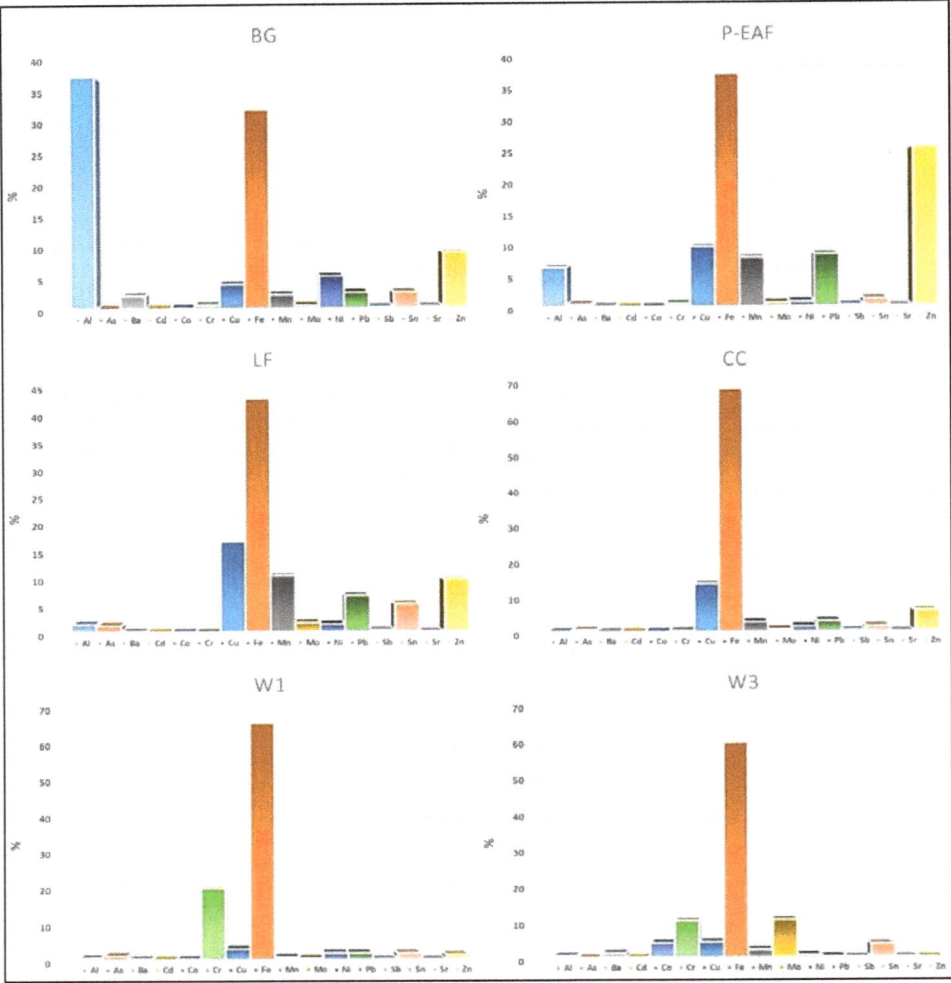

Figure 5. Concentration of metallic elements (percentage values) analysed in the ultrafine particulate collected by ELPI+ at the several sampling sites.

Table 5. Concentration (ng/m³) of the metallic elements determined in the UFPs collected by ELPI+ for each size range in BG, P-EAF, LF, CC, W1 and W3.

Sampling Site	Di nm	Metallic Elements															
		Al	As	Ba	Cd	Co	Cr	Cu	Fe	Mn	Mo	Ni	Pb	Sb	Sn	Sr	Zn
BG	22	<LOD	<LOD	<LOD	<LOD	0.11	1.18	0.93	40.22	0.33	0.14	10.03	0.38	0.05	0.49	<LOD	3.71
P-EAF		2.39	<LOD	<LOD	<LOD	<LOD	<LOD	0.42	3.71	0.24	0.24	<LOD	0.21	<LOD	<LOD	<LOD	<LOD
LF		2.09	<LOD	<LOD	<LOD	<LOD	<LOD	8.03	<LOD	<LOD	1.02	<LOD	3.54	0.16	2.72	<LOD	7.95
CC		<LOD	1.73	0.69	<LOD	<LOD	<LOD	17.1	26.49	1.13	2.64	0.3	10.39	0.3	0.95	<LOD	7.57
W1		<LOD	<LOD	<LOD	<LOD	0.47	7.71	10.08	10.44	0.47	<LOD	5.34	0.47	<LOD	1.3	<LOD	<LOD
W3		<LOD	<LOD	2.64	<LOD	0.82	31.32	2.14	36.27	0.49	1.98	<LOD	<LOD	<LOD	23.9	<LOD	<LOD
BG	41	6.59	<LOD	0.69	<LOD	<LOD	<LOD	1.37	<LOD	0.19	0.08	0.16	0.3	0.05	0.55	<LOD	2.61
P-EAF		0.54	<LOD	<LOD	<LOD	<LOD	<LOD	2.63	21.55	0.75	0.48	<LOD	1.5	0.09	<LOD	<LOD	1.92
LF		5.24	1.69	1.69	<LOD	<LOD	<LOD	25.79	3.74	2.09	3.27	<LOD	11.97	0.51	10.47	0.12	14.33
CC		4.33	7.23	1.21	<LOD	<LOD	1.17	116.87	372.24	10.39	7.79	8.44	32.46	1.73	14.72	<LOD	39.39
W1		4.86	3.56	1.66	<LOD	0.36	7.94	18.97	175.5	2.85	2.25	6.52	3.79	<LOD	2.25	<LOD	6.88
W3		<LOD	<LOD	2.97	<LOD	1.81	<LOD	5.93	65.94	4.95	7.75	<LOD	<LOD	<LOD	<LOD	<LOD	<LOD
BG	71	<LOD	<LOD	0.22	<LOD	0.08	<LOD	2.14	4.95	1.13	0.41	0.71	1.37	0.08	3.21	0.49	2.47
P-EAF		20.51	0.51	<LOD	<LOD	<LOD	<LOD	12.3	28.74	6.5	0.57	1.41	8.38	0.21	2.72	<LOD	22.9
LF		4.8	5.98	<LOD	<LOD	<LOD	<LOD	72.64	125.39	1.97	7.76	8.58	26.18	1.14	20.67	0.16	35.83
CC		3.46	28.57	1.95	<LOD	1.34	7.36	716.35	3185.71	107.34	30.08	70.55	116.87	12.55	76.83	0.13	227.24
W1		<LOD	9.49	1.66	<LOD	0.95	28.82	20.75	583.42	2.85	2.37	16.01	20.16	0.95	20.16	<LOD	13.04
W3		<LOD	<LOD	3.3	<LOD	8.57	15	14.01	105.51	4.78	32.15	0.99	<LOD	0.49	4.45	<LOD	<LOD
BG	121	71.84	<LOD	3.43	<LOD	<LOD	<LOD	3.71	21.98	3.02	0.3	<LOD	3.3	0.16	1.15	0.05	10.08
P-EAF		8.23	1.53	0.36	<LOD	<LOD	2.39	32.48	131.72	31.58	1.5	2.39	31.43	0.54	2.99	0.15	101.93
LF		5.71	7.09	0.2	<LOD	0.28	<LOD	89.96	376.18	118.9	8.07	8.66	37.99	1.73	27.17	0.24	55.91
CC		9.52	24.46	1.95	<LOD	1.56	17.53	668.74	4051.39	186.12	25.32	78.34	170.97	10.17	67.74	0.26	380.9
W1		7.11	12.45	<LOD	<LOD	0.95	415.04	20.16	758.92	8.89	2.49	15.42	18.38	0.71	17.79	<LOD	10.08
W3		4.95	<LOD	2.8	<LOD	25.72	46.16	17.31	329.71	9.89	52.75	5.44	2.14	<LOD	5.77	0.66	<LOD

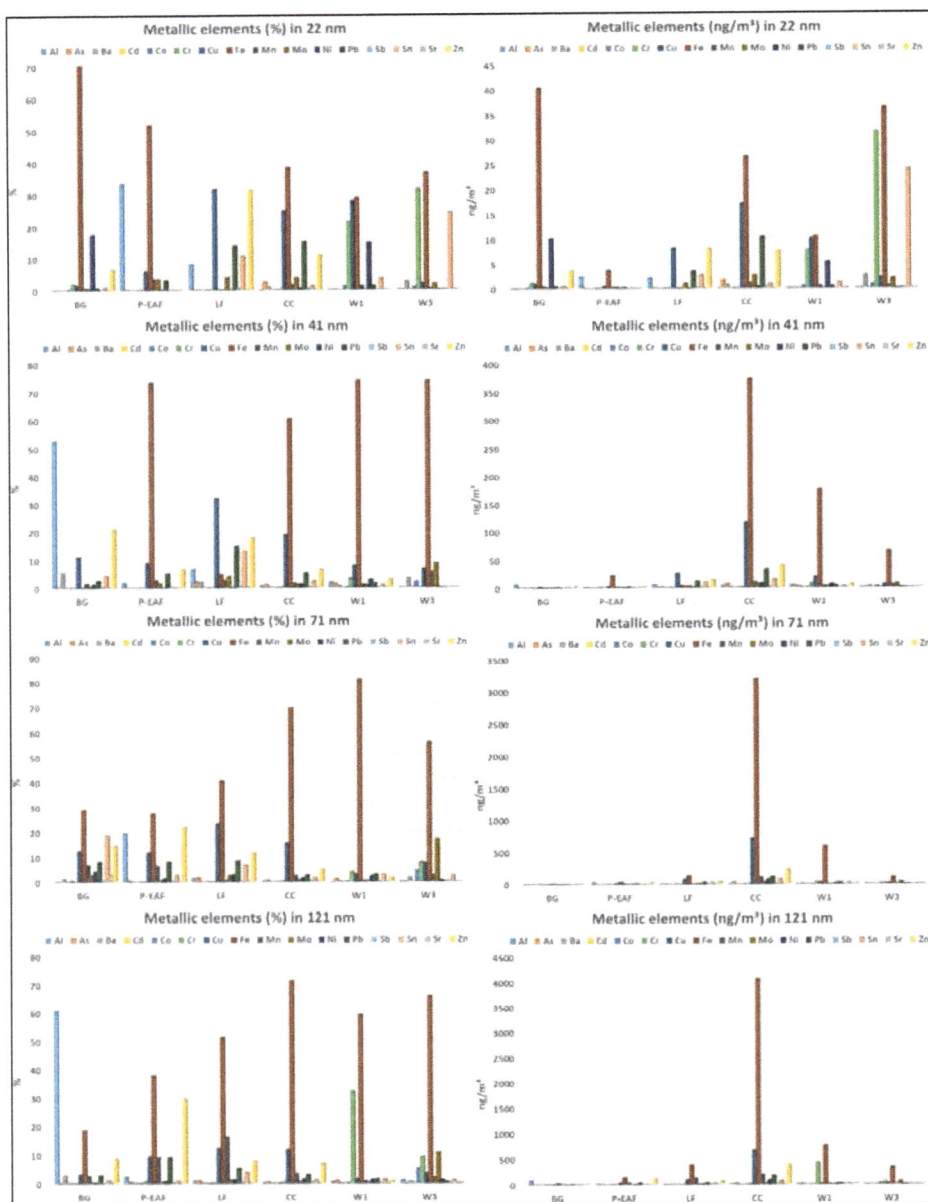

Figure 6. Distribution of the metallic elements (percentage value (**left**) and ng/m^3 (**right**) for each fraction) in the different UFP size ranges collected by ELPI+, for each area and working station. The percentages shown in the figure represent the fraction of each metallic element on the total of the metallic elements determined in each size range for each sampling site.

Overall, considering the total of the metallic elements determined in all the fractions of UFPs, in the particles collected near the LF, Mn and Co were present at more than 95% in the size range of 121 nm. Ba was present at more than 89% in the size range of 41 nm. Al showed a substantially homogeneous distribution in all size fractions. The others metallic elements were present for more than 75% in the size range between 71 nm and 121 nm.

In the particles collected near the CC, Al and Ba were present at more than 65% in the size range between 71 nm and 121 nm. As, Mo and Pb were present at more than 80% in UFPs of 71 nm and 121 nm. Co and Sr were present at 100% in the size range of 71 nm and 121 nm. The other metallic elements were present at more than 90% in the size range between 71 nm and 121 nm.

In the UFPs collected in the P-EAF, Cr, Ba and Sr were present at 100% in the size range of 121 nm. Ni, As and Sn were found at 100% in the size range between 71 nm and 121 nm. Fe, Sb, Mo, Al, presented a homogeneous distribution in the nanometric fractions, although Al was present at about 64% in the size range of 71 nm. The other metallic elements were present at more than 90% in particles between 71 nm and 121 nm.

Overall, the metallic element concentrations determined in the UFPs collected outside the plant were lower than the concentrations determined in the samples collected inside the plant. The determination of the different nanometric fractions allowed the observation that Cr, Fe, Co, and Ni were present at higher concentrations in the size range of 22 nm (100%, 59%, 57%, 91%, respectively). Al, Mn, Pb were present at about 90% in particles between 71 nm and 121 nm. Sr was present at 90% in the particles of 71 nm, whereas Ba was present at 79% in the size range of 121 nm. As was below the limit of detection. The other metallic elements showed a substantially homogeneous distribution in the UFPs.

In the UFPs collected in W1, Mn, Co, Ni and Cu showed a substantially homogeneous concentration distribution. Zn, As and Mo showed a homogeneous distribution in the particles between 41 nm and 121 nm. Sn, Fe and Pb were present at 90% in the size range of 71 nm and 121 nm. Al was present at 100% in the size range of 41 and 121 nm, Sb was present at 100% in the fractions of 71 and 121 nm, and Ba was present at 100% in particulate matter between 41 nm and 71 nm. Cr was present at 90% in the 121 nm fraction. Sr was below the limit of detection.

In the nanometric particles collected in W3, As and Zn were below the limit of detection. Al, Sr and Pb were present at 100% in 121 nm particles. Sb was present at 100% in the 71 nm fraction, while Sn was present at 70% in the 22 nm fraction. Co, Ni and Mo were present at more than 85% in the fractions of 71 nm and 121 nm. Mn, Cu, Ba, Fe and Cr showed a homogeneous concentration distribution.

Cd was lower than the limit of detection in all UFPs collected in all sampling sites.

4. Discussion

The assessment of occupational fine and ultrafine particles carried out in the steelmaking factory allowed the detection of variations in particle number distribution, number, mass and surface area concentration and chemical composition, in different areas and work stations in the factory.

The particle number distribution measured next to continuous casting (CC) shows a main mode around 71 nm, with additional peak at 10 nm (Figure 2). Measurements performed near the ladle furnace (LF) show a bimodal distribution, with the highest peak centered at 10 nm and a second peak at 41 nm. A previous study has observed a higher presence of small particles in the size range between 72 and 316 nm and an additional peak in the 22 nm size range next to the casting process [47], whereas, next to the ladle furnace, particle number distribution measured showed a peak at 10 nm, accounting for 63% of the total particle number. As suggested by previous studies, the largest particle size of the emission fumes of the casting process could depend on a vapor species available for condensation and coagulation. In contrast, the smallest particles measured next to the ladle furnace (diameter 10 nm) fumes were likely composed of freshly nucleated particles [18]. The distributions of W2 and W3 show a mode about 22 nm, while W1 shows a bimodal distribution with two peaks at 41 nm and 314 nm. A previous study has showed that the particle number size distributions resulting from gas metal arc welding activity was multi-modal and may change with respect to time. The authors have highlighted that welding particles are initially formed from the nucleation of vapors emanating from the superheated metal droplets located within the arc, and from spatter particles ejected from the welding process, and they suggest that coagulation can be responsible for scavenging of smaller particles by larger particles [57].

Background UFP number concentrations have proved to be similar to UFP levels measured by three previous studies of similar areas outside plants [12,47,58], which measured levels ranging from 3.30 to 3.69×10^4 and 1.26 to 1.89×10^4, and of 4.00×10^4, respectively.

Some studies have shown that particle bounce could lead to an increase in particle number at the lower working range of the ELPI with greased foil and not greased foil [59,60]. Although the concentration and number distribution measured are in agreement with previous studies, further studies are needed in order to estimate if and how the particle bounce that could occur in the different stages of the ELPI influences the particle number concentration measured in a steelmaking foundry. The melting, casting, and welding operations and the activity inside the mechanical workshop resulted in the primary sources of UFPs, compared to all the investigated activities, with UFPs' number concentration higher than for the outdoor background. These findings are in agreement, in terms of concentration and number distribution, with previous studies, which were conducted in similar working environments, such as iron and steel foundries, engine machining and assembly facilities [12–14,47,58]. In particular, Evans et al. [12], in an automotive grey iron foundry, and Cheng et al. [58] in the casting area in an iron foundry, have measured particle number concentration between 7.0×10^4 and 2.39×10^5 particles/cm^3 and between 2.07×10^4 and 2.82×10^5 particles/cm^3, respectively. In a previous study carried out in a steelmaking foundry it was observed that next to the ladle furnace and continuous casting the median UFP number concentrations were 1.64×10^5 particles/cm^3 and 2.92×10^5 particles/cm^3, respectively [47]. Heitbrink et al. [13] have observed in an engine machining and assembly facility a very fine particle concentration which ranged from 3.0×10^5 to 7.5×10^5 (geometric mean); Peters et al. [14] have measured the maximum particle number concentrations (>1,000,000 particles/cm^3) from the operation of direct-fire natural gas burners. In the present study, the highest UFP number concentration was measured next to continuous casting (9.48×10^6 particles/cm^3). However, UFP number concentrations above 1,000,000 particles/cm^3 were measured in, LF, W1, EAF, AG and BT. Several studies have observed that welding activity can determine a high emission of UFP. Zimmer et al. [57], during a characterization of the aerosols generated by arc welding processes, measured an average number concentration range from 1.6×10^7 particles/cm^3 near the arc (0 cm horizontal, 4.8 cm vertical) to 3.2×10^6 particles/cm^3 at the point corresponding to the farthest point measured (15 cm horizontal, 19.2 cm vertical). A previous study, carried out in automotive plants at a distance of 3 m from welding activities, showed an average UFP concentration of 1×10^5 particles/cm^3, with a peak of concentrations, particularly for surface area (3×10^3 mm^2/cm^3, max. 3×10^4 mm^2/cm^3) observed in the area characterized by high density of manual resistance welding activities or close to oxyacetylene welding activities [61]. Others authors [62] have reported high concentrations of fine particles in welding and grinding activities at a distance of 1.5 m from the job activities (total particles between 9.9×10^4 and 1.0×10^5 particles/cm^3), highlighting that the welding and/or grinding activities can produce a greater number of UFPs compared with brazing operations. In the present study, welding activities showed UFP number concentrations and surface area concentrations in the alveolar tract from 3.15×10^4 (W3) part/cm^3 to 1.57×10^5 part/cm^3 (W1) (median), and from 9.10×10^1 μm^2/cm^3 to 4.28×10^2 μm^2/cm^3 (median), respectively. While the grinding activity resulted in a lower average concentration compared to welding activity, in agreement with previous studies, a concentration of 3.51×10^4 particles/ cm^3 during abrasive blasting/grinding operations has been reported [12].

Furthermore, it has been possible to observe that in some work stations in the steelworks, in particular in the control consoles (P-EAF, P-CC), in the cabin of the overhead crane used for the finished products (OC1), and in the quality department workshop (QDW), the median UFP number concentrations were comparable with outdoor background levels.

The UFP number concentrations measured in QDW may depend on the restricted use of particulate sources. The main activities carried out in QDW mainly involve the assessment of the quality of the finished products. The UFP number concentration measured in control consoles (P-EAF, P-CC) and in the cabin of the overhead crane used for handling finished products (OC1), may be influenced by the efficiency of the ventilation system installed inside such work environments. The UFP number concentrations measured inside the pulpit of the ladle furnace were found to be higher than the outdoor background, however, the median concentration measured was low compared to the concentration detected near the ladle furnace (LF) and substantially overlapping the levels measured during the lunch break in the company canteen (far from industrial emission sources).

The dose estimated, in terms of deposited alveolar or tracheobronchial surface area in particles per mm^2 received by workers in the different working stations (weighted for a 6-h exposure), ranged from a minimum of 1.3×10^2 mm^2 for the alveolar tract and a minimum of 2.6×10^1 mm^2 for the tracheobronchial tract to a maximum of 8.7×10^3 mm^2 for the alveolar tract and a maximum of 1.3×10^3 for the tracheobronchial tract. The highest deposition levels for the alveolar tract and the tracheobronchial tract were recorded in the mechanical workshop and in the first welding station (Figure 4). Several studies have suggested that a large surface area or number may play an important role to causing adverse health effects [24,26,32,37,63–65]. The respiratory dose could be a key factor for assessing potential health effects of inhaled particles. Lung dose assessment can help to verify the effective dose relating to possible subclinical and clinical adverse health effects [34]. In this study, the dose of UFPs in terms of surface area deposition in the alveolar tract is greater by one order of magnitude compared to the dose values measured in Italian children living in urban or rural areas [66] and below the total daily deposited dose for typical Italian smokers [67].

Indoor levels of mass concentration of the inhalable and respirable fraction and airborne concentrations of metallic elements in the inhalable fraction were higher than those measured outside the plant, even if they were below the limits established by the Italian legislation [56]. Dust concentration and metallic element concentration were in line with other studies carried out in the iron and steel industry in Italy [68], but they differ from the findings of Nurul et al. [69], who reported for a steelmaking plant, a mean concentration and a range of total particulate matter of 2.76 mg/m^3, and 0.13–11.18 mg/m^3 respectively, with Co, Cr (VI) and Ni concentration at 2.36 mg/m^3, 8.36 mg/m^3 and 1.10 mg/m^3, respectively. In our study, the highest dust concentration, in terms of inhalable and respirable fractions, was measured in W1 (IF 1.63 mg/m^3, RF 0.99 mg/m^3), while in the steelmaking section (next to LF and CC), dust concentration did not exceed 0.77 mg/m^3 (Table 3). The highest metal concentration measured in the inhalable fraction was found in W1 (Fe 129 µg/m^3) and in P-EAF the highest metal concentration measured was Zn (7.89 µg/m^3), which was the highest measured concentration of Zn among all samples. The highest metal concentration measured in the inhalable fraction collected in LF, CC and W3 was Mn with a concentration of 44.64 µg/m^3 16.21 µg/m^3 and 8.05 µg/m^3, respectively. In the inhalable fraction collected outside the plant (BG) the highest metal concentration was Zn (0.68 µg/m^3) (Table 4).

Overall, the chemical characterization of UFPs shows that the highest total metallic element (of the all metallic elements determined) mass concentration was found in the UFPs collected in CC, followed by UFPs collected in W1 and LF, while the lowest was measured in the UFPs collected in BG and P-EAF (Table 5 and Figure 5). However, the distribution of metallic elements in the different fractions of UFP collected in P-EAF shows a pattern more similar to LF and CC compared to the trend observed in the UFPs collected outside the plant. This could show a greater contribution of the melting and casting operations in the issuance of UFPs within the control consoles compared to the UPFs measured outside the plant, which means the particles may have different sources. In the UFPs collected during welding activity, an important presence of chrome was detected compared to the UFPs collected in the other sampling sites. Moreover, except for the UFPs collected in BG, which showed a greater presence of Al than the other elements, in UFPs collected in LF, CC, P-EAF, W1 and W3, iron (Fe) was found to be relatively higher compared to metals across all size ranges (Figure 6), which was consistent with results from previous studies [47,62,70,71].

Although further studies are needed in order to investigate more and other workstations and also to include chemical characterization (not only of metallic elements) in different fractions of particulates through personal sampling, this study provided useful information on the possible exposure to particulate-dispersed and metallic elements of workers within the steel factory. Indeed, to the best of our knowledge, this is the first study to measure airborne particle exposure in a steel factory by simultaneously assessing size distribution, number, mass, surface area concentration, dose deposited in the respiratory system, and the composition of the airborne metals, and also with reference to different nanometric fractions.

The exposure assessment carried out allowed observation of a wide spatial distribution of the airborne particulate levels and the contained metallic elements, thus allowing identification of the main sources of exposure in term of mass, number and lung-deposited surface area of particles (in terms of deposited alveolar or tracheobronchial surface area, mm^2). Furthermore, it was possible to detect the concentration of low doses of metallic elements in the different fractions of UFPs. Chemical composition in terms of metallic elements determined in the UFPs varied depending on the sampling site, the emission source and the size range. In particular, Fe, Cu, Mn, Zn, Pb and Al were the most represented elements in that context and this result is in agreement with other studies conducted in foundries [47,71,72].

An in-depth assessment that takes into account the different chemical–physical characteristics of the airborne particulate may provide useful information for increasing the knowledge about occupational exposure to fine and ultrafine particles. In addition, although further research is needed to confirm the observations, the results achieved could also prove useful for designing studies aimed to investigate early biological effects associated with exposure to particulate matter and to several components within metal industries. Future studies based on job-exposure matrices could clarify the role of the different components (both size and non-size related) which could determine adverse health effects on respiratory and cardiovascular systems, in particular.

5. Conclusions

This study measured and assessed the occupational exposure concentrations of fine particles in a steel factory. Stationary and personal samples were carried out in different workstations and during different work phases in standard working conditions. UFP number, surface area concentration and metallic element composition were measured. Results confirmed the findings of previous studies conducted in similar industrial contexts, and improved the knowledge about ultrafine particle exposure and the fractions of metallic elements in nanometric particles.

These results may be useful for identifying preventive measures aimed at limiting workers' exposure and could lead to a better knowledge of the characterization of occupational exposure to UFPs. Furthermore, our results provided relevant information for the development of work-based exposure matrices, for epidemiological studies design, and for the planning of studies on early biological effects, in order to improve knowledge on health effects related to exposure to UFPs in the workplaces.

Author Contributions: Conceptualization, G.M. (Gabriele Marcias), J.F., A.M.S., M.U., S.C., S.P., D.F., I.P., F.M., L.I.L., E.M., G.D.P. and M.C.; data curation, G.M. (Gabriele Marcias), J.F., A.M.S., M.U., S.C., S.P., D.F., I.P., F.M., L.I.L., E.M.; formal analysis, G.M. (Gabriele Marcias), J.F., A.M.S., M.U., S.C., S.P., D.F., I.P., F.M., L.I.L., E.M., L.S.; investigation, G.M. (Gabriele Marcias), S.C., M.U., A.M.S., D.F., E.M. and L.I.L.; project administration, G.D.P., E.D., M.C.; validation, J.F., S.C., G.M. (Giorgio Massacci), L.S., E.D., G.B., G.D.P., M.C.; visualization, G.M. (Gabriele Marcias), J.F., A.M.S., M.U., S.C., S.P., I.P.; writing—original draft, G.M. (Gabriele Marcias) and J.F.; writing—review and editing, G.M. (Gabriele Marcias), J.F., S.C., S.P., G.M. (Giorgio Massacci), L.S., E.D., G.B., G.D.P., M.C.

Funding: This research received no external funding.

Acknowledgments: The authors are grateful to Denise Festa and Roberta Ghitti of the University of Brescia, for the support in carrying out the environmental sampling.

Conflicts of Interest: The authors declare no conflict of interest.

References

1. IARC Working Group on the Evaluation of Carcinogenic Risk to Humans. *Chemical Agents and Related Occupations Volume 100 F—A Review of Human Carcinogens*; International Agency for Research on Cancer (IARC) Monographs on the Evaluation of Carcinogenic Risks to Humans, No. 100F; OCCU: Lyon, France, 2012.

2. Hobbesland, A.; Kjuus, H.; Thelle, D.S. Study of cancer incidence among 8530 male workers in eight Norwegian plants producing ferrosilicon and silicon metal. *Occup. Environ. Med.* **1999**, *56*, 625–631. [CrossRef] [PubMed]

3. Kjuus, H.; Andersen, A.; Langård, S.; Knudsen, K.E. Cancer incidence among workers in the Norwegian ferroalloy industry. *Br. J. Ind. Med.* **1986**, *43*, 227–236. [CrossRef] [PubMed]

4. Tossavainen, A. Estimated risk of lung cancer attributable to occupational exposures in iron and steel foundries. *IARC Sci. Publ.* **1990**, *104*, 363–367.

5. Koskela, R.-S.; Mutanen, P.; Sorsa, J.-A.; Klockars, M. Respiratory disease and cardiovascular morbidity. *Occup. Environ. Med.* **2005**, *62*, 650–655. [CrossRef] [PubMed]

6. Hałatek, T.; Trzcinka-Ochocka, M.; Matczak, W.; Gruchała, J. Serum Clara cell protein as an indicator of pulmonary impairment in occupational exposure at aluminum foundry. *Int. J. Occup. Med. Environ. Health* **2006**, *19*, 211–223. [CrossRef] [PubMed]

7. Liu, X.; Lee, S.; Pisaniello, D. Measurement of fine and ultrafine dust exposure in an iron foundry in South Australia. *J. Heal.* **2010**, *26*, 5–9.

8. Leonard, S.S.; Chen, B.T.; Stone, S.G.; Schwegler-Berry, D.; Kenyon, A.J.; Frazer, D.; Antonini, J.M. Comparison of stainless and mild steel welding fumes in generation of reactive oxygen species. *Part. Fibre Toxicol.* **2010**, *7*, 32. [CrossRef]

9. Antonini, J.M.; Leonard, S.S.; Roberts, J.R.; Solano-Lopez, C.; Young, S.-H.; Shi, X.; Taylor, M.D. Effect of stainless steel manual metal arc welding fume on free radical production, DNA damage, and apoptosis induction. *Mol. Cell. Biochem.* **2005**, *279*, 17–23. [CrossRef]

10. Vincent, J.H.; Clement, C.F. Ultrafine particles in workplace atmospheres. *Philos. Trans. R. Soc. A Math. Phys. Eng. Sci.* **2000**, *358*, 2673–2682. [CrossRef]

11. Wake, D.; Mark, D.; Northage, C. Ultrafine Aerosols in the Workplace. *Ann. Occup. Hyg.* **2002**, *46*, 235–238. [CrossRef]

12. Evans, D.E.; Heitbrink, W.A.; Slavin, T.J.; Peters, T.M. Ultrafine and Respirable Particles in an Automotive Grey Iron Foundry. *Ann. Occup. Hyg.* **2007**, *52*, 9–21. [CrossRef] [PubMed]

13. Heitbrink, W.A.; Evans, D.E.; Peters, T.M.; Slavin, T.J. Characterization and Mapping of Very Fine Particles in an Engine Machining and Assembly Facility. *J. Occup. Environ. Hyg.* **2007**, *4*, 341–351. [CrossRef] [PubMed]

14. Peters, T.M.; Heitbrink, W.A.; Evans, D.E.; Slavin, T.J.; Maynard, A.D. The Mapping of Fine and Ultrafine Particle Concentrations in an Engine Machining and Assembly Facility. *Ann. Occup. Hyg.* **2005**, *50*, 249–257. [CrossRef] [PubMed]

15. Kero, I.; Naess, M.K.; Tranell, G. Particle size distributions of particulate emissions from the ferroalloy industry evaluated by electrical low pressure impactor (ELPI). *J. Occup. Environ. Hyg.* **2015**, *12*, 37–44. [CrossRef] [PubMed]

16. Kero, I.T.; Jørgensen, R.B. Comparison of Three Real-Time Measurement Methods for Airborne Ultrafine Particles in the Silicon Alloy Industry. *Int. J. Environ. Res. Public Health* **2016**, *13*, 871. [CrossRef] [PubMed]

17. Debia, M.; Weichenthal, S.; Tardif, R.; Dufresne, A. Ultrafine Particle (UFP) Exposures in an Aluminium Smelter: Soderberg vs. Prebake Potrooms. *Environ. Pollut.* **2011**, *1*, 2. [CrossRef]

18. Chang, M.-C.O.; Chow, J.C.; Watson, J.G.; Glowacki, C.; Sheya, S.A.; Prabhu, A. Characterization of Fine Particulate Emissions from Casting Processes. *Aerosol Sci. Technol.* **2005**, *39*, 947–959. [CrossRef]

19. Oberdörster, G. Pulmonary effects of inhaled ultrafine particles. *Int. Arch. Occup. Environ. Health* **2001**, *74*, 1–8. [CrossRef] [PubMed]

20. Oberdörster, G.; Oberdörster, E.; Oberdörster, J. Nanotoxicology: An emerging discipline evolving from studies of ultrafine particles. *Environ. Health Perspect.* **2005**, *113*, 823–839. [CrossRef] [PubMed]

21. Donaldson, K.; Brown, D.; Clouter, A.; Duffin, R.; MacNee, W.; Renwick, L.; Tran, L.; Stone, V. The Pulmonary Toxicology of Ultrafine Particles. *J. Aerosol Med.* **2002**, *15*, 213–220. [CrossRef] [PubMed]

22. Manke, A.; Wang, L.; Rojanasakul, Y. Mechanisms of nanoparticle-induced oxidative stress and toxicity. *Biomed Res. Int.* **2013**, *2013*, 942916. [CrossRef] [PubMed]

23. Cho, W.-S.; Duffin, R.; Poland, C.A.; Howie, S.E.M.; MacNee, W.; Bradley, M.; Megson, I.L.; Donaldson, K. Metal Oxide Nanoparticles Induce Unique Inflammatory Footprints in the Lung: Important Implications for Nanoparticle Testing. *Environ. Health Perspect.* **2010**, *118*, 1699–1706. [CrossRef] [PubMed]

24. Hennig, F.; Quass, U.; Hellack, B.; Küpper, M.; Kuhlbusch, T.A.J.; Stafoggia, M.; Hoffmann, B. Ultrafine and Fine Particle Number and Surface Area Concentrations and Daily Cause-Specific Mortality in the Ruhr Area, Germany, 2009–2014. *Environ. Health Perspect.* **2018**, *126*. [CrossRef] [PubMed]

25. Lanzinger, S.; Schneider, A.; Breitner, S.; Stafoggia, M.; Erzen, I.; Dostal, M.; Pastorkova, A.; Bastian, S.; Cyrys, J.; Zscheppang, A.; et al. Associations between ultrafine and fine particles and mortality in five central European cities—Results from the UFIREG study. *Environ. Int.* **2016**, *88*, 44–52. [CrossRef] [PubMed]

26. Clark, J.; Gregory, C.C.; Matthews, I.P.; Hoogendoorn, B. The biological effects upon the cardiovascular system consequent to exposure to particulates of less than 500 nm in size. *Biomarkers* **2016**, *21*, 1–47. [CrossRef] [PubMed]

27. HEI Review Panel on Ultrafine Particles. *Understanding the Health Effects of Ambient Ultrafine Particles*; HEI Perspectives 3; Health Effects Institute: Boston, MA, USA, 2013.

28. Stone, V.; Miller, M.R.; Clift, M.J.D.; Elder, A.; Mills, N.L.; Møller, P.; Schins, R.P.F.; Vogel, U.; Kreyling, W.G.; Alstrup Jensen, K.; et al. Nanomaterials Versus Ambient Ultrafine Particles: An Opportunity to Exchange Toxicology Knowledge. *Environ. Health Perspect.* **2017**, *125*, 106002. [CrossRef] [PubMed]

29. Magalhaes, S.; Baumgartner, J.; Weichenthal, S. Impacts of exposure to black carbon, elemental carbon, and ultrafine particles from indoor and outdoor sources on blood pressure in adults: A review of epidemiological evidence. *Environ. Res.* **2018**, *161*, 345–353. [CrossRef] [PubMed]

30. Downward, G.S.; van Nunen, E.J.H.M.; Kerckhoffs, J.; Vineis, P.; Brunekreef, B.; Boer, J.M.A.; Messier, K.P.; Roy, A.; Verschuren, W.M.M.; van der Schouw, Y.T.; et al. Long-Term Exposure to Ultrafine Particles and Incidence of Cardiovascular and Cerebrovascular Disease in a Prospective Study of a Dutch Cohort. *Environ. Health Perspect.* **2018**, *126*, 127007. [CrossRef] [PubMed]

31. Regional Office for Europe, World Health Organization. *Review of Evidence on Health Aspects of Air Pollution—REVIHAAP Project Technical Report*; Regional Office for Europe: Copenhagen, Denmark, 2013.

32. Maynard, A.D.; Maynard, R.L. A derived association between ambient aerosol surface area and excess mortality using historic time series data. *Atmos. Environ.* **2002**, *36*, 5561–5567. [CrossRef]

33. Buonanno, G.; Marks, G.B.; Morawska, L. Health effects of daily airborne particle dose in children: Direct association between personal dose and respiratory health effects. *Environ. Pollut.* **2013**, *180*, 246–250. [CrossRef]

34. Kim, C.S.; Jaques, P.A. Analysis of Total Respiratory Deposition of Inhaled Ultrafine Particles in Adult Subjects at Various Breathing Patterns. *Aerosol Sci. Technol.* **2004**, *38*, 525–540. [CrossRef]

35. Franck, U.; Odeh, S.; Wiedensohler, A.; Wehner, B.; Herbarth, O. The effect of particle size on cardiovascular disorders—The smaller the worse. *Sci. Total Environ.* **2011**, *409*, 4217–4221. [CrossRef] [PubMed]

36. Peters, A.; Wichmann, H.E.; Tuch, T.; Heinrich, J.; Heyder, J. Respiratory effects are associated with the number of ultrafine particles. *Am. J. Respir. Crit. Care Med.* **1997**, *155*, 1376–1383. [CrossRef] [PubMed]

37. Stafoggia, M.; Schneider, A.; Cyrys, J.; Samoli, E.; Andersen, Z.J.; Bedada, G.B.; Bellander, T.; Cattani, G.; Eleftheriadis, K.; Faustini, A.; et al. Association Between Short-term Exposure to Ultrafine Particles and Mortality in Eight European Urban Areas. *Epidemiology* **2017**, *28*, 172–180. [CrossRef] [PubMed]

38. Dockery, D.W.; Pope, C.A.; Xu, X.; Spengler, J.D.; Ware, J.H.; Fay, M.E.; Ferris, B.G.; Speizer, F.E. An Association between Air Pollution and Mortality in Six U.S. Cities. *N. Engl. J. Med.* **1993**, *329*, 1753–1759. [CrossRef] [PubMed]

39. Chen, L.C.; Lippmann, M. Effects of Metals within Ambient Air Particulate Matter (PM) on Human Health. *Inhal. Toxicol.* **2009**, *21*, 1–31. [CrossRef] [PubMed]

40. Gray, D.L.; Wallace, L.A.; Brinkman, M.C.; Buehler, S.S.; La Londe, C. Respiratory and Cardiovascular Effects of Metals in Ambient Particulate Matter: A Critical Review. *Rev. Environ. Contam. Toxicol.* **2015**, *234*, 135–203. [PubMed]

41. Cakmak, S.; Dales, R.; Kauri, L.M.; Mahmud, M.; Van Ryswyk, K.; Vanos, J.; Liu, L.; Kumarathasan, P.; Thomson, E.; Vincent, R.; et al. Metal composition of fine particulate air pollution and acute changes in cardiorespiratory physiology. *Environ. Pollut.* **2014**, *189*, 208–214. [CrossRef] [PubMed]

42. Cassee, F.R.; Héroux, M.E.; Gerlofs-Nijland, M.E.; Kelly, F.J. Particulate matter beyond mass: Recent health evidence on the role of fractions, chemical constituents and sources of emission. *Inhal. Toxicol.* **2013**, *25*, 802–812. [CrossRef] [PubMed]

43. Atkinson, R.W.; Mills, I.C.; Walton, H.A.; Anderson, H.R. Fine particle components and health—A systematic review and meta-analysis of epidemiological time series studies of daily mortality and hospital admissions. *J. Expo. Sci. Environ. Epidemiol.* **2015**, *25*, 208–214. [CrossRef] [PubMed]

44. Wyzga, R.E.; Rohr, A.C. Long-term particulate matter exposure: Attributing health effects to individual PM components. *J. Air Waste Manag. Assoc.* **2015**, *65*, 523–543. [CrossRef] [PubMed]

45. Kumar, P.; Morawska, L.; Birmili, W.; Paasonen, P.; Hu, M.; Kulmala, M.; Harrison, R.M.; Norford, L.; Britter, R. Ultrafine particles in cities. *Environ. Int.* **2014**, *66*, 1–10. [CrossRef]

46. Viitanen, A.-K.; Uuksulainen, S.; Koivisto, A.J.; Hämeri, K.; Kauppinen, T. Workplace Measurements of Ultrafine Particles—A Literature Review. *Ann. Work Expo. Heal.* **2017**, *61*, 749–758. [CrossRef] [PubMed]

47. Marcias, G.; Fostinelli, J.; Catalani, S.; Uras, M.; Sanna, A.; Avataneo, G.; De Palma, G.; Fabbri, D.; Paganelli, M.; Lecca, L.; et al. Composition of Metallic Elements and Size Distribution of Fine and Ultrafine Particles in a Steelmaking Factory. *Int. J. Environ. Res. Public Health* **2018**, *15*, 1192. [CrossRef] [PubMed]

48. Dekati Ltd. *ELPI VI Software Manual*, version 4.1 0; Dekati Ltd.: Kangasala, Finland, 2008.

49. Cernuschi, S.; Giugliano, M.; Ozgen, S.; Consonni, S. Number concentration and chemical composition of ultrafine and nanoparticles from WTE (waste to energy) plants. *Sci. Total Environ.* **2012**, *420*, 319–326. [CrossRef] [PubMed]

50. Marra, J.; Voetz, M.; Kiesling, H.-J. Monitor for detecting and assessing exposure to airborne nanoparticles. *J. Nanopart. Res.* **2010**, *12*, 21–37. [CrossRef]

51. ICRP. *Human Respiratory Tract Model for Radiological Protection*; ICRP Publication 66. Ann; ICRP: Ottawa, ON, Canada, 1994; Volume 24.

52. *UNI EN 481:1994—Atmosfera Nell'ambiente di Lavoro. Definizione Delle Frazioni Granulometriche per la Misurazione Delle Particelle Aerodisperse.* Available online: http://store.uni.com/catalogo/index.php/uni-en-481-1994.html (accessed on 30 December 2018).

53. *Ambienti di Lavoro—Determinazione Della Frazione Inalabile Delle Particelle Aerodisperse—Metodo Gravimetrico—UNICHIM—Associazione Per l'Unificazione Nel Settore Dell'industria Chimica, Federato All'uni (Ente Nazionale di Unificazione)*; Italy. Available online: https://www.unichim.it/metodi/ (accessed on 30 December 2018).

54. *Ambienti di Lavoro—Determinazione Della Frazione Respirabile Delle Particelle Aerodisperse—Metodo Gravimetrico—UNICHIM—Associazione Per l'Unificazione Nel Settore Dell'industria Chimica—Federato All'uni (Ente Nazionale di Unificazione)*; Italy. Available online: https://www.unichim.it/metodi/ (accessed on 30 December 2018).

55. Apostoli, P.; De Palma, G.; Catalani, S.; Bortolotti, F.; Tagliaro, F. Multielemental analysis of tissues from Cangrande della Scala, Prince of Verona, in the 14th century. *J. Anal. Toxicol.* **2009**, *33*, 322–327. [CrossRef]

56. *D.lgs. 9 Aprile 2008, n. 81 Testo Coordinato con il D.Lgs. 3 Agosto 2009, n. 106*; Italy, 2008. Available online: http://www.gazzettaufficiale.it/eli/id/2008/04/30/008G0104/sg (accessed on 30 December 2018).

57. Zimmer, A.T.; Biswas, P. Characterization of the aerosols resulting from arc welding processes. *J. Aerosol Sci.* **2001**, *32*, 993–1008. [CrossRef]

58. Cheng, Y.-H.; Chao, Y.-C.; Wu, C.-H.; Tsai, C.-J.; Uang, S.-N.; Shih, T.-S. Measurements of ultrafine particle concentrations and size distribution in an iron foundry. *J. Hazard. Mater.* **2008**, *158*, 124–130. [CrossRef]

59. Virtanen, A.; Joutsensaari, J.; Koop, T.; Kannosto, J.; Yli-Pirilä, P.; Leskinen, J.; Mäkelä, J.M.; Holopainen, J.K.; Pöschl, U.; Kulmala, M.; et al. An amorphous solid state of biogenic secondary organic aerosol particles. *Nature* **2010**, *467*, 824–827. [CrossRef]

60. Leskinen, J.; Joutsensaari, J.; Lyyränen, J.; Koivisto, J.; Ruusunen, J.; Järvelä, M.; Tuomi, T.; Hämeri, K.; Auvinen, A.; Jokiniemi, J. Comparison of nanoparticle measurement instruments for occupational health applications. *J. Nanopart. Res.* **2012**, *14*. [CrossRef]

61. Buonanno, G.; Morawska, L.; Stabile, L. Exposure to welding particles in automotive plants. *J. Aerosol Sci.* **2011**, *42*, 295–304. [CrossRef]

62. Iavicoli, I.; Leso, V.; Fontana, L.; Cottica, D.; Bergamaschi, A. Characterization of inhalable, thoracic, and respirable fractions and ultrafine particle exposure during grinding, brazing, and welding activities in a mechanical engineering factory. *J. Occup. Environ. Med.* **2013**, *55*, 430–445. [CrossRef] [PubMed]

63. Weichenthal, S. Selected physiological effects of ultrafine particles in acute cardiovascular morbidity. *Environ. Res.* **2012**, *115*, 26–36. [CrossRef] [PubMed]

64. Oberdörster, G.; Ferin, J.; Gelein, R.; Soderholm, S.C.; Finkelstein, J. Role of the alveolar macrophage in lung injury: Studies with ultrafine particles. *Environ. Health Perspect.* **1992**, *97*, 193–199. [CrossRef] [PubMed]

65. Donaldson, K.; Li, X.; MacNee, W. Ultrafine (nanometre) particle mediated lung injury. *J. Aerosol Sci.* **1998**, *29*, 553–560. [CrossRef]

66. Buonanno, G.; Marini, S.; Morawska, L.; Fuoco, F.C. Individual dose and exposure of Italian children to ultrafine particles. *Sci. Total Environ.* **2012**, *438*, 271–277. [CrossRef] [PubMed]

67. Fuoco, F.; Stabile, L.; Buonanno, G.; Scungio, M.; Manigrasso, M.; Frattolillo, A.; Fuoco, F.C.; Stabile, L.; Buonanno, G.; Scungio, M.; et al. Tracheobronchial and Alveolar Particle Surface Area Doses in Smokers. *Atmosphere* **2017**, *8*, 19. [CrossRef]

68. Soleo, L.; Lovreglio, P.; Panuzzo, L.; Nicolà D'errico, M.; Basso, A.; Gilberti, M.E.; Drago, I.; Tomasi, C.; Apostoli, P. Valutazione del rischio per la salute da esposizione a elementi metallici nei lavoratori del siderurgico e nella popolazione generale di Taranto (Italia). *G. Ital. Med. Lav. Erg.* **2012**, *34*, 381–391.

69. Nurul, A.H.; Shamsul, B.M.T.; Noor Hassim, I. Assessment of dust exposure in a steel plant in the eastern coast of peninsular Malaysia. *Work* **2016**, *55*, 655–662. [CrossRef]

70. Chang, C.; Demokritou, P.; Shafer, M.; Christiani, D. Physicochemical and toxicological characteristics of welding fume derived particles generated from real time welding processes. *Environ. Sci. Process. Impacts* **2013**, *15*, 214–224. [CrossRef] [PubMed]

71. Mohiuddin, K.; Strezov, V.; Nelson, P.F.; Stelcer, E.; Evans, T. Mass and elemental distributions of atmospheric particles nearby blast furnace and electric arc furnace operated industrial areas in Australia. *Sci. Total Environ.* **2014**, *487*, 323–334. [CrossRef] [PubMed]

72. Marris, H.; Deboudt, K.; Augustin, P.; Flament, P.; Blond, F.; Fiani, E.; Fourmentin, M.; Delbarre, H. Fast changes in chemical composition and size distribution of fine particles during the near-field transport of industrial plumes. *Sci. Total Environ.* **2012**, *427–428*, 126–138. [CrossRef] [PubMed]

MDPI

St. Alban-Anlage 66

4052 Basel

Switzerland

Tel. +41 61 683 77 34

Fax +41 61 302 89 18

www.mdpi.com

Metals Editorial Office

E-mail: metals@mdpi.com

www.mdpi.com/journal/metals